Liquid Crystalline Polymers

Vijay Kumar Thakur • Michael R. Kessler
Editors

# Liquid Crystalline Polymers

## Volume 2—Processing and Applications

Springer

*Editors*
Vijay Kumar Thakur
Mechanical and Materials Engineering
Washington State University
Pullman, WA, USA

Michael R. Kessler
Mechanical and Materials Engineering
Washington State University
Pullman, WA, USA

ISBN 978-3-319-20269-3 ISBN 978-3-319-20270-9 (eBook)
DOI 10.1007/978-3-319-20270-9

Library of Congress Control Number: 2015946615

Springer Cham Heidelberg New York Dordrecht London


Printed on acid-free paper

Springer International Publishing AG Switzerland is part of Springer Science+Business Media (www.springer.com)

# Preface

Liquid-crystalline polymers (LCPs) constitute an important class of materials that represents new advanced organic functional materials combining both anisotropic order of liquid crystals and excellent mechanical properties of polymers. These are one of the most important macromolecular systems having the capability of self-organization because of their unique molecular architectures with high anisotropy. The most distinguished feature of LCPs lies in that they exhibit intermediate morphological states between crystalline (or glassy) and liquid states, where macromolecular chains can be self-organized into a wide spectrum of ordered structures such as nematic, smectic, and cholesteric mesophases. Since last few years, study on LCPs has been a flourishing research field due to their broad promising applications in various high-technology areas such as various organic optoelectronic applications, nonlinear optical materials, tunable diffraction gratings, thermal insulated materials, and the high-performance Kevlar fiber. Over the past decades, numerous endeavors have been made to rationally design and synthesize LCPs with controllable supramolecular structures. When mesogenic groups, such as calamitic, discotic, and bent-cored ones, are introduced into polymers, liquid crystalline (LC) phases may form, resulting in LC polymers (LCPs). LCPs can be classified into several general categories based on their architecture and position of mesogens in the polymer chain. Most frequently, LCPs are categorized into three types, namely, main-chain (MCLCPs), side-chain (SCLCPs), or main-chain/side-chain combined (MCSCLCPs) LCP, depending on where mesogenic units are incorporated.

The advancement in the knowledge of structure, chemistry, processing, and applications of LCPs coincided with a period of rapid expansion in terms of the physical techniques available to the materials. This, together with great advances in the theoretical study and the rapid progress in computer simulation techniques, has made liquid crystals an exciting area for scientific research with many fundamental challenges. Indeed, the study of LCPs is an attractive research field that has engrossed significant curiosity, and their synthesis from different sources has awakened quite an enormous deal of scientific awareness. The exploration of these innovative materials is still a challenge since the rapid development of different technologies demands new advanced materials, which possess as wide a range of properties as possible.

Scientists in collaborations with industries are extensively developing new classes of "Liquid Crystalline Polymers." Different kinds of advanced materials can be obtained by the exploration of LCPs. This book is solely focused on the "Liquid Crystalline Polymers" and deals with the "Processing and Application" aspects of these materials. Several critical issues and suggestions for future work are comprehensively discussed in this book with the hope that the book will provide a deep insight into the state-of-the-art of "Liquid Crystalline Polymers." We would like to thank Brian Halm from Springer for the editorial assistance and patience as well as the "Springer Publisher" for the invaluable help in the organization of the editing process.

Pullman, WA, USA Vijay Kumar Thakur
Pullman, WA, USA Michael R. Kessler

# Contents

# About the Editors

**Michael R. Kessler** is an expert in the mechanics, processing, and characterization of polymer matrix composites and nanocomposites. His research thrusts include the development of multifunctional materials (including the development of self-healing structural composites), polymer matrix composites for extreme environments, bio-renewable polymers and composites, and the evaluation of these materials using experimental mechanics and thermal analysis. These broad-based topics span the fields of organic chemistry, applied mechanics, and processing science. He has extensive experience in processing and characterizing thermosets, including those created through ring-opening metathesis polymerization (ROMP), such as poly dicyclopentadiene, and the cyclotrimerization of cyanate ester resins. In addition to his responsibilities as professor of Mechanical and Materials Engineering at Washington State University, he serves as the Director of the school. He has developed an active research group with external funding of over 10 million dollars—including funding from the National Science Foundation, ACS Petroleum Research Fund, Strategic Environmental Research and Development Program (SERDP), Department of Defense, Department of Agriculture, and NASA. His honors include the Army Research Office Young Investigator Award, the Air Force Office of Scientific Research Young Investigator Award, the NSF CAREER Award, and the Elsevier Young Composites Researcher Award from the American Society for Composites. He has >110 journal papers and 3700 citations, holds six patents, edited a book on the characterization of composite materials, presented >200 talks at national and international meetings, and serves as a frequent reviewer and referee in his field.

**Vijay Kumar Thakur** has been working as a Research Faculty (Staff Scientist) in the School of Mechanical and Materials Engineering at Washington State University, United States, since September 2013. His former appointments include being a Research Scientist in Temasek Laboratories at Nanyang Technological University, Singapore, and a Visiting Research Fellow in the Department of Chemical and Materials Engineering at LHU, Taiwan. His research interests include the synthesis and processing of biobased polymers, nanomaterials, polymer micro/nanocomposites, nanoelectronic materials, novel high dielectric constant materials, electrochromic materials for energy storage, green synthesis of nanomaterials, and surface functionalization of polymers/nanomaterials. He did his Post Doctorate in Materials Science at Iowa State University, USA, and his Ph.D. in polymer science (2009) at the National Institute of Technology. In his academic carrier, he has published more than 80 SCI journal research articles in the field of polymers/materials science and holds one United States patent. He has also published 20 books and 35 book chapters on the advanced state-of-the-art of polymers/materials science with numerous publishers. He is an editorial board member of several international journals and also is member of scientific bodies around the world. In addition to being on the editorial board of journals, he also serves as the *Regional Editor* for the journal *Recent Patents on Materials Science*, *Academic Editor* for the journal *British Journal of Applied Science & Technology*, and Guest Editor for the *Journal of Nanomaterials*, *International Journal of Polymer Science*, *Journal of Chemistry*, and *American Journal of Applied Chemistry*.

# Chapter 1
# Liquid Crystals Order in Polymeric Microchannels

**Giovanna Palermo, Luciano De Sio, Roberto Caputo, Cesare Umeton, and Roberto Bartolino**

## 1.1 Introduction

In the last few decades, Liquid crystals (LCs) have shown a deep impact on versatile applications ranging from electronics (Armitage 1980) and photonics (Beeckman et al. 2011), to plasmonics (Kossyrev et al. 2005) and metamaterials (Pratibha et al. 2009). This is because LCs simultaneously exhibit some characteristics that are common to both ordinary isotropic liquids and solid crystals. The capability of LCs to change their properties as functions of external perturbations, such as electric fields, magnetic fields, optical fields and temperature variations (de Gennes and Prost 1993), makes these materials the best candidates for the development of a new class of optical devices (Tanggaard Larsen et al. 2003),

G. Palermo • R. Caputo
Department of Physics, Centre of Excellence for the Study of Innovative Functional Materials, University of Calabria, Arcavacata di Rende (CS) 87036, Italy

L. De Sio (✉)
Beam Engineering for Advanced Measurements Company, Winter Park, FL 32789, USA
e-mail: luciano@beamco.com; luciano.desio@fis.unical.it

C. Umeton
Department of Physics, Centre of Excellence for the Study of Innovative Functional Materials, University of Calabria, Arcavacata di Rende (CS) 87036, Italy

CNR-Lab. Licryl, Institute NANOTEC-87036 Arcavacata di Rende (Cs)-Italy

R. Bartolino
Department of Physics, Centre of Excellence for the Study of Innovative Functional Materials, University of Calabria, Arcavacata di Rende (CS) 87036, Italy

CNR-Lab. Licryl, Institute NANOTEC-87036 Arcavacata di Rende (Cs)-Italy

Interdisciplinary institute B. Segre of the National Academy dei Lincei, 00165 Rome, Italy



V.K. Thakur, M.R. Kessler (eds.), *Liquid Crystalline Polymers*,
DOI 10.1007/978-3-319-20270-9_1

sensors (Shiyanovskii et al. 2005) and many other among them. The most popular application of LCs for photonics products is related to the design of liquid crystal displays (LCDs) on rigid (Kirsch and Bremer 2000) and flexible substrates (De Sio et al. 2012) for flat screen televisions, laptop screen, digital clock, smartphone (Slikkerveer et al. 2004; Varghese et al. 2004; Osterman et al. 2005; Konovalov et al. 2004). The ambivalence of both solid and liquid response of LCs is also found in the effects of surfaces on these systems, which lead to a great diversity of phenomena. In the absence of geometrical constraints and external perturbations, the LCs molecular director **n**, a versor that represents the average orientation of elongated molecules, is free to point in any direction. It is possible, however, to force the director to follow specific direction by conveniently functionalizing the surface of LC substrates by means of specific chemical compounds, or mechanically, in order to obtain a uniform alignment direction; in this way an organization of the liquid crystalline structure close the surface is obtained. Beyond this transition region, the bulk liquid crystalline structure is determined by the orientation fixed by the surface. This phenomenon of orientation of LCs induced by surfaces is the so-called *anchoring* (Jerome 1991); the anchoring of a LC is crucial for the performance of LC based devices. A typical LC device consists of a pair of indium tin oxide (ITO) treated conducting glass substrates, filled with LC. The substrates are also covered with a substance that induces an orientational monodomain within the LC. The cell can be used in a display device only if it has high transmissivity in the visible range, wide viewing angle, good optical contrast and fast response time (Kelly and O'Neill 2001). All these characteristics depend on the quality of the LC alignment that is obtained by interaction between the aligning layer and the LC. For this reason, the quality of the director alignment of LCs plays a fundamental role and lot of efforts are addressed to obtain a good alignment of the LC, free of any chemical treatment of substrates. In this way, the impact of this technology on the environment is also minimized, the technique becomes less expensive and ensures a better stability in time of the device, by avoiding hysteresis and image sticking phenomena (Kelly and O'Neill 2001). In this chapter we report on the major techniques currently available to align LCs, by investigating the benefits and drawbacks of each one. A particular attention is devoted to a noninvasive, all-optical, holographic technique for obtaining a permanent alignment of LCs in a microperiodic template realized in soft-composite (polymeric) materials. All the knowledge related to the alignment of the used LC can be exploited with other kinds of LCs, such as Polymeric Liquid Crystals (PLCs). Results open new doors for further research involving these topics.

## 1.2 Surface Anchoring

In general, the LC alignment originates from a symmetry breaking at the surface of the functionalized substrate. Every kind of surface causes a particular orientation of the molecular director ***n*** close to it: the alignment present at the surface propagates in the material over macroscopic distances, due to the elastic properties of LCs.

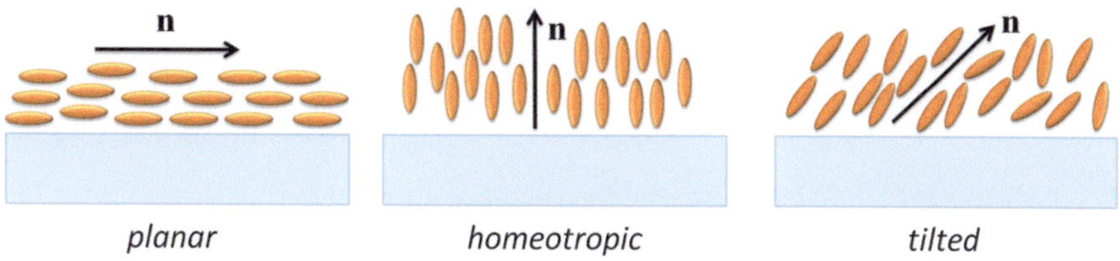


**Fig. 1.1** Types of liquid crystal alignment near the surface: planar, homeotropic and tilted

**Fig. 1.2** Orientation of **n** nearby the surface

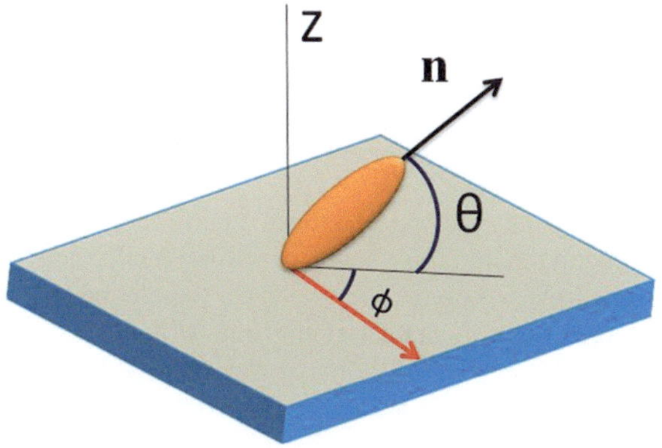


There are three main configurations for ***n*** nearby the surface of the substrate: planar, homeotropic and tilted (Fig. 1.1).

The surface in contact with the LC is usually considered flat at the microscale, and the orientation of ***n*** close to it is determined by polar ($\theta$) and azimuthal ($\varphi$) angles (Fig. 1.2), where $\theta$, also called *pre-tilt angle*, is the angle between ***n*** and the plane of the surface, while $\varphi$ is the angle between the projection of ***n*** on the plane of the surface and the "easy axis" (red arrow in Fig. 1.2), which represents the direction induced by the alignment treatment of the surface.

The pre-tilt angle is a very important parameter in the realization of LC based devices and can allow a defect-free alignment and an improvement of electro-optical characteristics such as driving voltage, response time, color performance and viewing angle. It also represents a control parameter for estimating the alignment quality.

In the case of *homeotropic alignment* the polar angle is equal to 90° and ***n*** is orthogonal to the surface. When the ***n*** lies in the plane of the surface ($\theta = 0°$, *planar alignment*), two possible orientation cases exist:

- *homogeneous planar* orientation, in which ***n*** is oriented uniformly all over the surface and $\varphi$ remains fixed;
- *heterogeneous planar* orientation, in which the orientation of ***n*** is not uniform over the surface and $\varphi$ assumes different values in different points of the surface.

In the case of tilted orientation $\theta$ is fixed and $\varphi$ is arbitrary. To date, the two main groups of techniques used to align LCs are: mechanical treatment and surface chemical functionalization.

## 1.3 Rubbing

A well known method to promote a given alignment of LC molecules in contact with a surface is the mechanical rubbing (Fig. 1.3) (Ito et al. 2000). The surface of the glass substrate is covered with a thin organic film, such as polymide (PI), nylon, polyvinyl alcohol, etc. by means of spin coating or spray coating techniques. The most widely used polymer for the rubbing process is the PI due to its high curing temperature (~350 °C), good chemical resistance and excellent mechanical properties (Zheng 2010).

Spin coating allows to deposit a uniform layer on a flat substrate: a small amount of coating material is usually deposited on the center of the substrate which is then rotated at a certain speed (rounds per minute, rpm) in order to spread out the coating material. The thickness of the film will depend on the number of rpm and on viscosity and concentration of the coating material. After the deposition of the coating film, a cloth (velvet) with short fibers is moved over the surface in order to create microgrooves (Fig. 1.4). The rubbing process induces a certain orientation of the polymer molecules, which is transmitted to the LC molecules anchored on the substrates; moreover, the same kind of orientational order propagates in the bulk of the LC molecules via intermolecular forces (Stohr et al. 1998).

By means of the rubbing technique it is possible to obtain both planar and tilted alignments, with a strong anchoring energy (defined as the energy needed to fix the LC molecules to the easy axis). Although this technique is quite simple and can be easily scaled up (to large area devices), it intrinsically exhibits several drawbacks, such as contaminations, static charge accumulation and mechanical damages. These drawbacks have important consequences in the realization of LCD systems, where the presence of even very small amounts of dust particles or static surface charges can lead to a deterioration of display performances and higher costs (Kelly and O'Neill 2001). Furthermore, this technique is not applicable for achieving patterned alignment layers, necessary for different kinds of LC based devices.

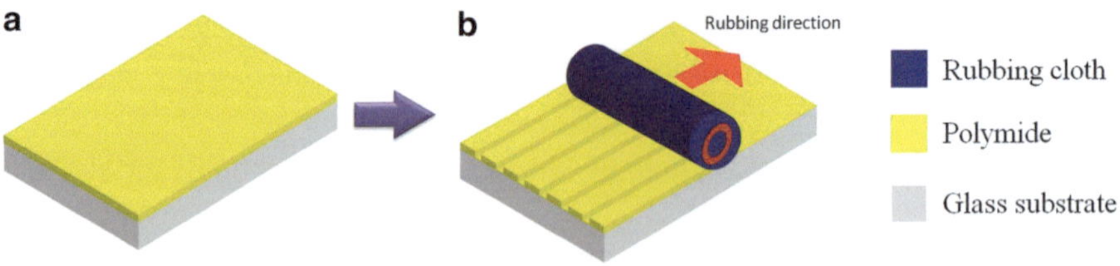


**Fig. 1.3** Rubbing method: glass substrate covered with a polymide layer (**a**); process of rubbing by means of cloth rotation (**b**)

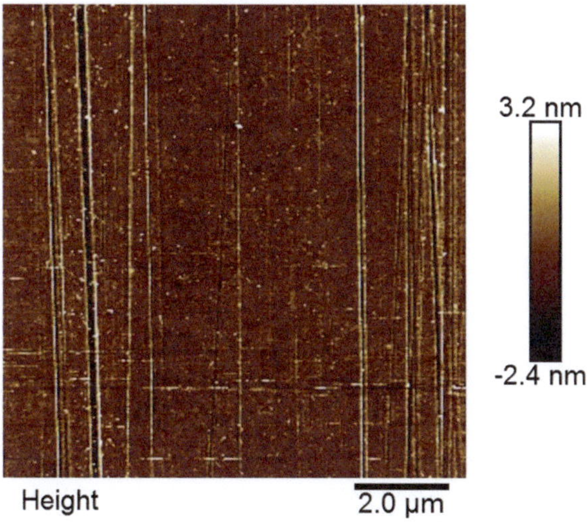


**Fig. 1.4** Atomic force microscopic image of the microgrooves morphology on a rubbed polymide surface

## 1.4 Evaporation of SiOx

An alternative, non contact, method to align LCs consists in functionalizing the surface of the glass substrate with an amorphous film of SiOx. This method exploits a vacuum deposition technique that deposits thin films onto various surfaces by condensation of a vaporized form of the selected material, at a specific incident angle (Monkade et al. 1988). Characteristics of the deposited film strongly depend on the value of the incident evaporation angle and on other evaporation parameters such as pressure, temperature, etc. For silicon oxide films deposited at an incidence angle of 60°, the LC director $\boldsymbol{n}$ lies parallel to the surface and is perpendicular to the plane of incidence. With an angle of about 30° with the substrate normal, a uniform planar alignment is achieved with a 0° tilt. Finally, evaporating at an angle of 80° results in a tilted alignment (Goodman et al. 1977). However, this method turns out to be quite expensive, due to the cost of the used equipment (vacuum systems, evaporator, etc.).

## 1.5 Photoalignment

Another important, non contact, method used to align LCs is the photoalignment. By means of spin coating or spray coating, a photosensitive film of polymer is deposited on the surface of the glass substrate. Then, the substrate is exposed to linearly polarized UV light (Fig. 1.5). Due to the dependence of the absorption of light by molecules of the photosensitive film on the polarization direction, an anisotropic distribution of the orientation of molecules of the alignment material occurs; in this

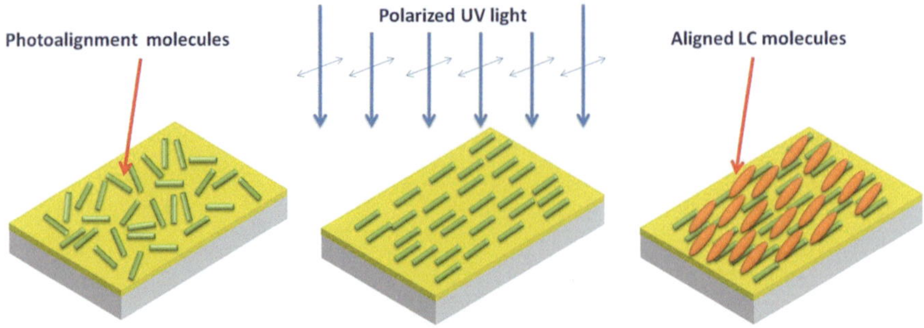


**Fig. 1.5** Photoalignment method

way the induced degree of surface ordering on the layer is further transferred to the LC molecules, that become oriented in a given direction. Although the creation of orientational anisotropy by means of polarized light absorption has been known since 1920 (Weigert effect) (Wardosanidze 2011), the photoalignment mechanism in some materials is not yet fully understood, because photochemical reactions are different in different photoalignment materials (O'Neill and Kelly 2000).

In fact, the most studied type of photochemical reactions for photoalignment of LCs are: photoisomerization, photodimerization, photodecomposition, photo-linking, and a combination of decomposition and linking (Chigrinov et al. 2003, 2008). The strongest advantage of this technique is related to the possibility of easily establish multiple domains of LC alignment without mechanical and electrostatic damage.

## 1.6 POLICRYPS Composite Structures

In recent years, many efforts have been made to find new, inexpensive, alignment techniques that do not involve the use of chemical compounds, in order to make device components recyclable materials, thus minimizing the environmental impact. The new alignment mechanisms must therefore ensure a minor charge storing, more stability over time, and guarantee immunity from the effects of hysteresis and image sticking. For this purpose, we have developed a noninvasive, all-optical, holographic technique for permanently aligning liquid crystalline materials in a microperiodic template realized in soft-composite (polymeric) materials (Caputo et al. 2004). POLICRYPS is the acronym of Polymer Liquid Crystal Polymer Slices and represents a nano/micro composite structure made of films of well-aligned Nematic Liquid Crystal (NLC) alternated to slices of almost pure polymer. The structure is obtained by irradiating a homogeneous mixture of NLC (BL-001 by Merck) and curing agent molecules (NOA-61 by Norland) with an interference pattern of UV/visible light, under suitable experimental and geometrical conditions. The structure can be utilized either in a transmission or a reflection configuration, with negligible scattering losses (Caputo et al. 1999, 2000, 2005, 2009). In fact, a POLICRYPS structure can be utilized with excellent performances as switchable diffraction grating (Caputo et al. 2006), switchable optical phase modulator (De Sio et al. 2008), array of mirror-less optical micro-resonators for lasing effects and as a tuneable Bragg filter (d'Alessandro et al. 2008). A fascinating new application of POLICRYPS is due to the possibility of aligning LCs by exploiting a selective microfluidic etching process (De Sio et al. 2011).

## 1.7 POLICRYPS Realization

The necessary protocol to fabricate a good POLICRYPS template is the following: a mixture of NLC, photoinitiator and monomer is heated above the nematic-isotropic transition temperature of the NLC component (65 °C) in order to exploit the high diffusivity of the NLC molecules in the isotropic state. In this condition, the mixture is cured with a UV interference pattern radiation. When the curing process is over, the sample is slowly cooled down to room temperature, below the isotropic-nematic transition temperature. The optical holographic setup for UV curing is reported in Fig. 1.6.

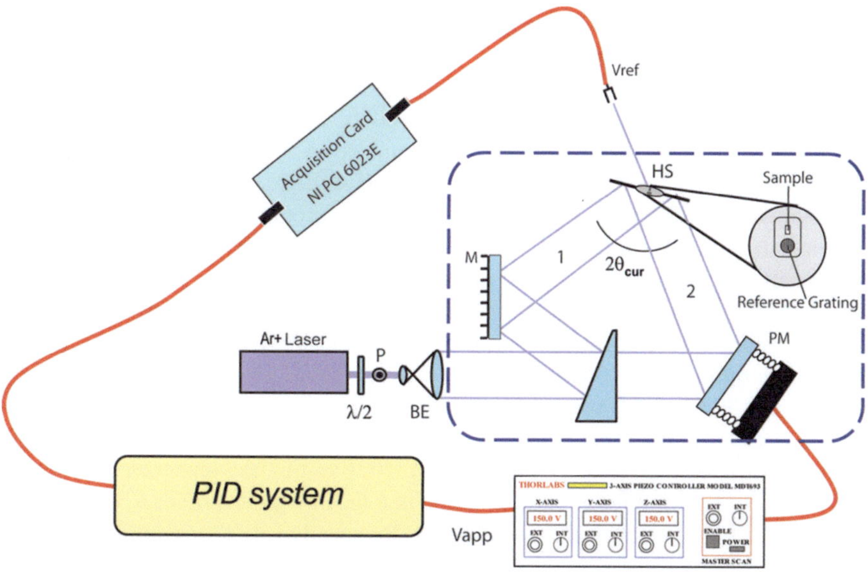


**Fig. 1.6** Optical holographic set-up for UV curing gratings with stability check: *P* polarizer, *λ/2* half-wave plate, *BE* beam expander, *BS* beam splitter, *2θcur* total curing angle, *M* mirrors, *S* sample, *PD1* first beam photodetector, *PD2* second beam photodetector, *PD3* diffracted and reflected beam photodetector, *Vapp* ramp generator. *Inset*: reference grating (positioned immediately below the sample area) that enables the stability check. Reproduced with permission from De Sio et al. (2006); copyright 2006, Optical Society of America

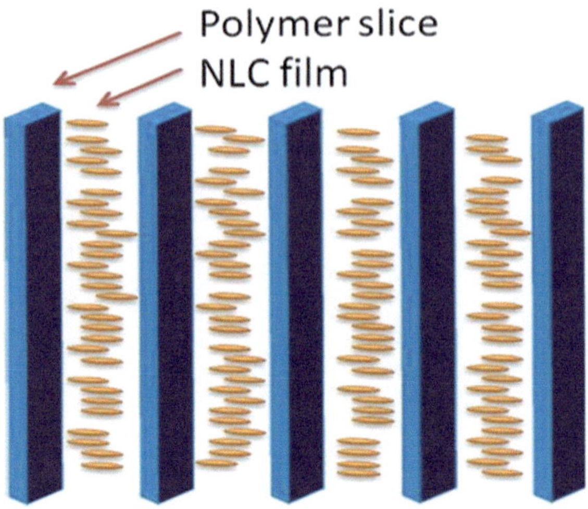


**Fig. 1.7** Sketch of a POLICRYPS structure

The UV curing radiation is provided by a Ar-Ion laser ($\lambda = 305$ nm), whose spot is enlarged to a diameter of about 25 mm by means of a beam expander (BE); the beam is then divided into two parts of same intensity by means of a beam splitter (BS). The two beams overlap on the sample cell S, which is, therefore, acted on by the interference pattern of the two beams. In this way, the homogeneous mixture becomes a grating made of slices of pure polymer alternated to films of pure NLC, whose director is uniformly aligned perpendicularly to the polymeric slices (Fig. 1.7). Most importantly, the NLC alignment is spontaneously obtained without any kind of surface treatments. The diffraction efficiency of this structure turns out to be quite high (98 %) (Caputo et al. 1999, 2000, 2005, 2009).

A hot stage is used to control the temperature of the sample and the cooling process. The spatial periodicity of the structure is settled in the range 0.2–15 μm by adjusting the interference angle ($2\theta_{cur}$). The whole process is realized in condition of high mechanical stability by exploiting a stabilization system (PID system). By means of a real-time control of vibrations of the sample and a signal feedback, the system drives a mirror-holder whose position can be controlled by a piezoelectric mechanism with a sensitivity of 6–7 nm; in this way, a continuous compensation of the optical path changes can be provided (De Sio et al. 2006).

Figure 1.8a is a side view of the POLICRYPS structure acquired by means of a scanning electron microscope after removing the top cover glass. The analysis has been performed at the edge of the structure, after cutting the glass substrate along the middle of the sample. It is worth noting that the structure exhibits a morphology with a sinusoidal-like profile, a topology that is due to the sinusoidal spatial modulation of the intensity of the curing interference pattern. Figure 1.8b is a polarized optical microscope (POM) view of the sample and shows the almost complete phase separation that is usually achieved in this kind of systems, evidencing the absence of LC droplets; in fact, their possible presence could yield some scattering of the impinging light and some depolarization of the

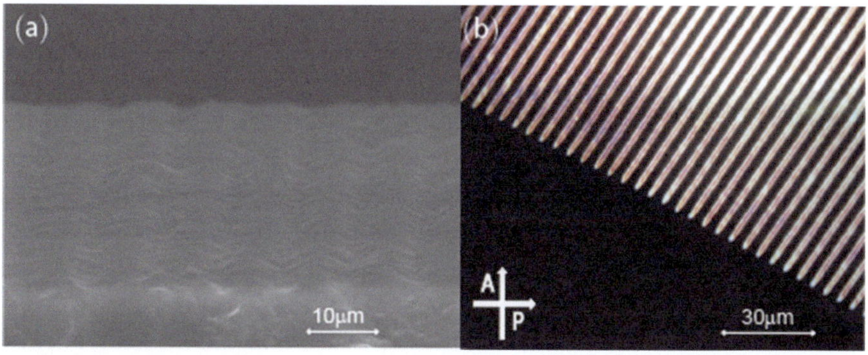


**Fig. 1.8** SEM side view (**a**) and POM micrograph (**b**) of a typical POLICRYS structure. Copyright 2013, Material view

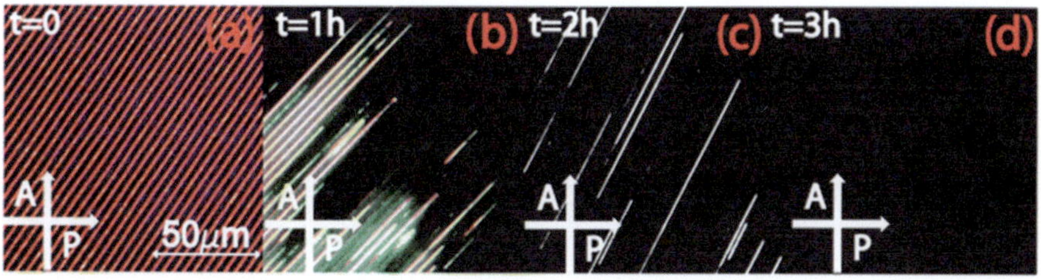


**Fig. 1.9** POM view of the POLICRYPS structure during the micro-fluidic etching process at different timescales (**a–d**). Reproduced with permission from De Sio et al (2011); copyright 2011, Royal Society of Chemistry

transmitted/diffracted one. Furthermore, the absence of LC droplets inside the sample represents a key point in order to remove the LC molecules from the sample and obtain an empty polymeric template. The wet etching process of the POLICRYPS is carried out by dipping the sample (without opening the cell) in a solution of water and tetrahydrofuran (THF); then, by capillarity flow, the solvent washes out the LC from the polymeric structure. Indeed, for short time intervals (3–4 h), the THF acts as a selective agent, removing the LC and the unpolymerized component, without affecting the regularity of the polymer slices (Fig. 1.9). The process is carried out above the nematic–isotropic transition temperature of the LC (65 °C), to ensure a low viscosity of the LC component; vibrating the system at ultrasonic frequencies accelerates the removal of the LC from the sample.

The complete removal of the LC component is checked by monitoring the sample under a polarizing optical microscope (POM) before and during the washing procedure. In Fig. 1.9, different optical microscope pictures show the sample during the emptying process: The micrograph in Fig. 1.9a was acquired between crossed polarizers before immersing the sample in the solvent; Fig. 1.9b shows a sample removed from the solvent after 1 h and Fig. 1.9c after 2 h. The completely black area in Fig. 1.9d confirms that the LC has been completely washed out if removal of the sample from the solvent is made after 3 h (De Sio et al. 2011).

## 1.8 Universal Soft-Matter Template

A fascinating application of the polymeric template derived from a POLICRYPS structure is related to the possibility to induce the so called "uniform lying helix" (ULH) geometry in cholesteric liquid crystals (CLCs, De Sio et al. 2011). In fact, when the CLC is aligned in this particular configuration, an in-plane rotation of the optical axis under the action of an electric field can be obtained. The effect can be exploited for temperature independent applications of the CLCs, characterized by a response time in the microsecond range. For this purpose, we have used an ITO coated POLICRYPS cell, L = 10 μm in thickness and Λ = 5 μm in spatial periodicity; the realized empty template has been filled by capillarity with CLC (BL088 by Merck, helix pitch ~ 400 nm), according to the previously described method (De Sio et al. 2011).

Figure 1.10a is a POM micrograph of the CLC infiltrated sample at the edge of the template, which shows, on the left, the existence of a focal conic texture, induced by a random distribution of the helical axes (high magnification in Fig. 1.10c). On the right, the ULH geometry induced by the POLICRYPS structure is well evident (high magnification in Fig. 1.10d).

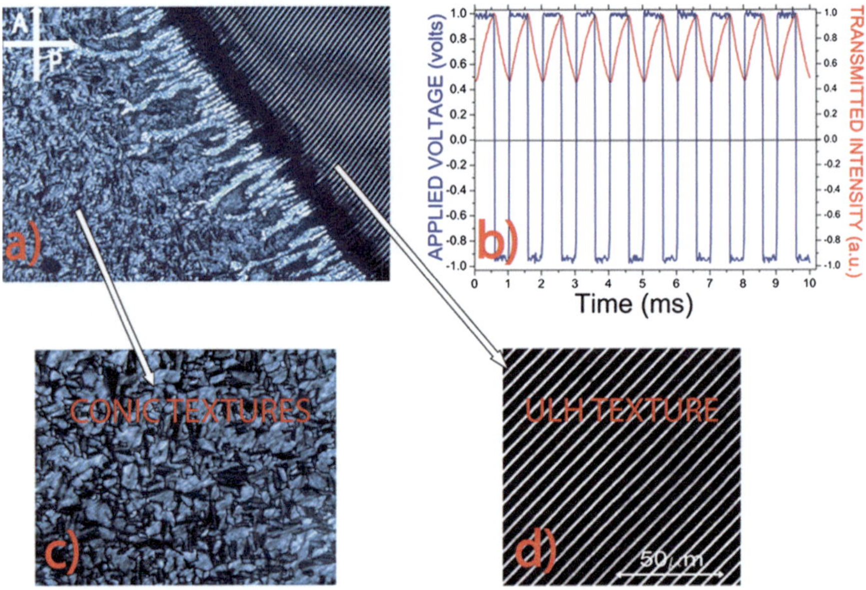


**Fig. 1.10** POM view of the empty POLICRYPS template filled with short pitch CLC at the edge of the grating area (**a**). The high magnification of the random oriented CLC area is reported in (**c**) while the CLC area aligned in ULH geometry is shown in (**d**); its electro-optical response is reported in (**b**). Reproduced with permission from De Sio et al. (2011); copyright 2011, Royal Society of Chemistry

An electro-optical characterization of the area depicted in Fig. 1.10d has been carried out: An external voltage (1 KHz square wave, Fig. 1.10b), applied across the CLC cell, perpendicularly to the helix axis, induces an in-plane tilt of the optical axis of the CLC, aligned in ULH texture; the tilt is inverted if the polarity of the electric field is reversed. The magnitude of the transmitted intensity (red curve in Fig. 1.10b), which is related to the tilt of the optical axis, turns out to be proportional to the applied voltage that induces the in-plane rotation of the sample optical axis. It is important to note that, in this configuration, LC devices exhibit response times that are one order of magnitude lower than typical response times of LC cells; The similar results can be obtained by exploiting Ferroelectric LCs (FLCs), characterized by fast response, a wide view angle and a bistable memory capability, but at the expense of much greater difficulties in the alignment of the material. Thus, we decided to explore also the capability of our polymeric template to align FLCs, by infiltrating the structure with FLC (CS-1024, by Chisso), always according to the previously described procedure (De Sio et al. 2011). We have analysed several samples, all with a thickness of $L = 10$ μm, but different periodicity $\Lambda$, which has been varied from 3 to 6 μm; the best results in terms of uniformity of the induced FLC geometry (verified by POM analysis) were obtained in the sample with $\Lambda = 3$ μm and a corresponding channel width of about 1.0 μm.

Figure 1.11a is a POM micrograph of the FLC infiltrated sample at the edge of the grating area, which shows the typical focal conic texture of the non aligned FLC molecules (right view, Fig. 1.11a, b). On the other hand, it is well evident that, once realized, the cell of FLC behaves as a uniaxial plate (left view, Fig. 1.11a, b). By rotating the sample of 45° under the microscope, the "Surface Stabilized Ferroelectric Liquid Crystal" SSFLC area (left) becomes dark while the simple FLC region (right) appears almost unchanged. This is the demonstration that a director alignment occurred in the SSFLC region: The helices underwent a restraint by the polymeric template channels, whose width is of the same order of magnitude of the helical pitch; due to boundary conditions, the molecular director has been constrained at a given angle (22°) with the normal to the polymeric slices. The electro-optical effect is obtained by applying a voltage that changes the director orientation (±22); consequently, the polarization axis (which is perpendicular to the molecular director) is switched between two stable states ("Clark–Lagerwall effect") (Clark and Lagerwall 1980). Figure 1.11d shows the bistability of the SSFLC area (depicted in Fig. 1.11c), measured by using a bipolar electric pulse with a period of 1 ms (blue curve, Fig. 1.11d). The red curve of Fig. 1.11d shows the two stable states of the transmitted intensity; jumps between them are driven by the applied bipolar voltage. A characterization of the dynamics shows a 4–5 ms duration of the stability states (for single pulses), a time interval which is strongly related to the geometry of the sample.

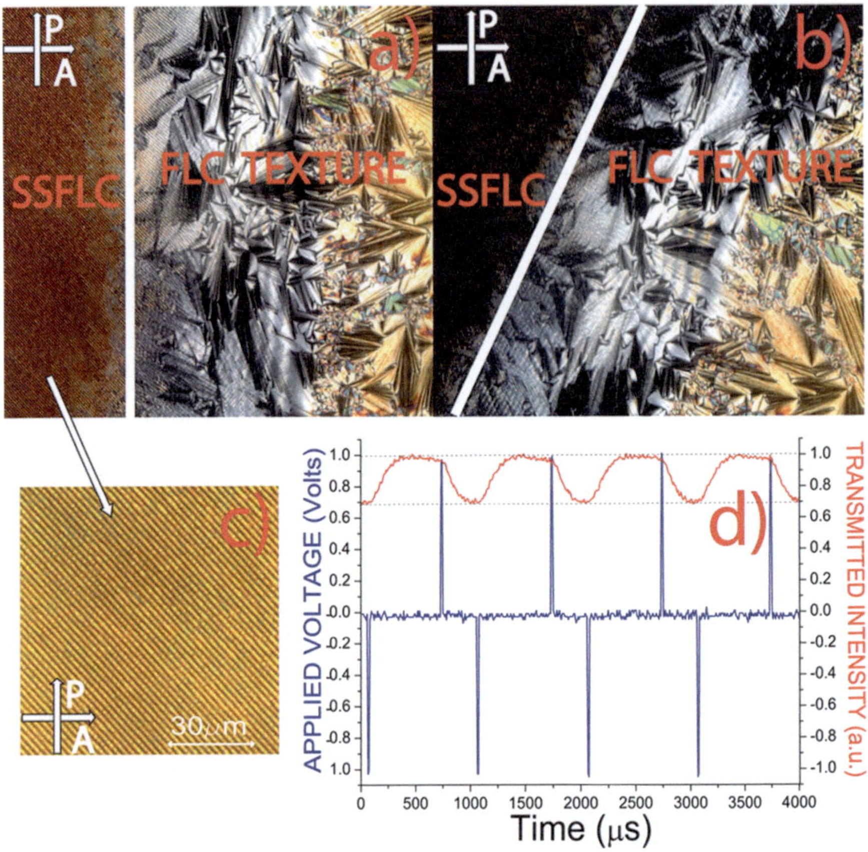


**Fig. 1.11** POM view of the template (grating vector aligned at 45° with the polarization direction) filled with short pitch FLC at the edge of the grating area (**a**) and after rotating the grating vector of 45° (**b**). The bistable electro-optical response of the depicted area (**c**) is reported in (**d**). Reproduced with permission from De Sio et al. (2011); copyright 2011, Royal Society of Chemistry

## 1.9 Conclusion

In this chapter we have reported a review of the currently available techniques to align LCs by focusing the attention on a new fascinating surfactant-free method based on the exploitation of a polymeric template. The use of an empty polymeric microstructure is particularly useful for eliminating the future need of any mechanical alignment and rubbing process for realizing novel LC based photonic devices. The empty polymeric template, fabricated by combining a holographic optical setup and a chemical etching process, proved suitable to be employed for inducing long range order in several kinds of Liquid Crystalline materials. In particular, it has been possible to exploit Cholesteric and Ferroelectric LCs to realize ultra-fast electro-optical switches. The universality and simplicity of exploitation of the new structure are demonstrated by the easy overcoming of some standard and well known drawbacks of LC devices, related to chemical and/or mechanical treatments of cell surfaces.

**Acknowledgment** The research was partially supported by the Air Force Office of Scientific Research (AFOSR), Air Force Research Laboratory (AFRL), U.S. Air Force, under grant FA8655-12-1-003 (P.I. L. De Sio, EOARD 2014/2015) and the Materials and Manufacturing Directorate, AFRL. The authors would also like to acknowledge the contribution of the COST Action IC1208. www.ic1208.eu.

## References

Armitage D (1980) Liquid crystal voltage controlled retardation display. Appl Opt 19:2235–2239

Beeckman J, Neyts K, Vanbrabant JM (2011) Liquid-crystal photonic applications. Opt Eng 50:081202–081209

Caputo R, Sukhov AV, Tabiryan N, Umeton C (1999) Efficiency dynamics of diffraction gratings recorded in liquid crystalline composite materials by a UV interference pattern. Chem Phys 245:463–471

Caputo R, Sukhov AV, Umeton C, Ushakov RF (2000) Formation of a Grating of Submicron Nematic Layers by Photopolymerization of Nematic-Containing Mixtures. J Exp Theor Phys 91:1190–1197

Caputo R, De Sio L, Sukhov AV, Veltri A, Umeton C (2004) Development of a new kind of switchable holographic grating made of liquid-crystal films separated by slices of polymeric material. Opt Lett 29:1261–1263

Caputo R, Umeton C, Veltri A, Sukhov AV, Tabiryan N (2005–2007) Holographic diffraction grating, process for its preparation and opto-electronic devices incorporating it. European Patent Request 1,649,318; US Patent Request 2007/0019152A1

Caputo R, De Sio L, Veltri A, Umeton C, Sukhov AV (2006) POLICRYPS switchable holographic grating: a promising grating electro-optical pixel for high resolution display application. J Display Technol 2(1):38–51

Caputo R, De Luca A, De Sio L, Pezzi L, Strangi G, Umeton C, Veltri A, Asquini R, D'Alessandro A, Donisi D, Beccherelli R, Sukhov AV, Tabiryan N (2009) POLICRYPS: a liquid crystal composed nano/microstructure with a wide range of optical and electro-optical applications. J Opt A: Pure Appl Opt 11:024017 (13 pp)

Chigrinov V, Kozenkov VM, Kwok HS (2003) Optical applications of liquid crystals. In: Vicari L (ed) Inst. Physics, Bristol, p 201

Chigrinov V, Kozenkov V, Kwok HS (2008) Photoalignment of liquid crystalline materials: physics and applications. Wiley, Chichester

Clark NA, Lagerwall ST (1980) Submicrosecond bistable electro-optic switching in liquid crystals. Appl Phys Lett 36:899–902

d'Alessandro A, Donisi D, De Sio L, Beccherelli R, Asquini R, Caputo R, Umeton C (2008) Tunable integrated optical filter made of a glass ion-exchanged waveguide and an electro-optic composite holographic grating. Opt Express 16(13):9254–9260

de Gennes PG, Prost J (1993) The physics of liquid crystals. Oxford Science, Oxford

De Sio L, Caputo R, De Luca A, Veltri A, Umeton C, Sukhov AV (2006) In situ optical control and stabilization of the curing process of holographic gratings with a nematic film-polymer-slice sequence structure. Appl Opt 45:3721–3727

De Sio L, Tabiryan N, Caputo R, Veltri A, Umeton C (2008) POLICRYPS structures as switchable optical phase modulators. Opt Express 16(11):7619–7624

De Sio L, Ferjani S, Strangi G, Umeton C, Bartolino R (2011) Universal soft matter template for photonic applications. Soft Matter 7(8):3739–3743

De Sio L, Romito M, Giocondo M, Vasdekis AE, De Luca A, Umeton C (2012) Electro-switchable polydimethylsiloxane based optofluidics. Lab Chip 12(19):3760–3765

Goodman LA, Mcginn JT, Anderson CH, Digeronimo F (1977) Topography of obliquely evaporated silicon oxide films and its effect on liquid-crystal orientation. IEEE Trans Electron Devices Ed-24(7):795–804

Ito N, Sakamoto K, Arafune R, Ushioda S (2000) Relation between the molecular orientations of a very thin liquid crystal layer and an underlying rubbed polyimide film. J Appl Phys 88:3235–3238

Jerome B (1991) Surface effects and anchoring in liquid crystals. Rep Prog Phys 54:391–452

Kelly SM, O'Neill M (2001) Liquid crystals for electro-optic applications. In: Nalwa HS (ed) Handbook of advanced electronic and photonic materials and devices, vol 7, Liquid crystals, display and laser materials. Academic, New York, pp 2–66

Kirsch P, Bremer M (2000) Nematic liquid crystals for active matrix displays: molecular design and synthesis. Angew Chem Int Ed 39(23):4216–4235

Konovalov V, Chigrinov V, Kwok HS, Takada H, Takatsu H (2004) Photoaligned vertical aligned nematic mode in liquid crystals. Jpn J Appl Phys 43:261–266

Kossyrev PA, Yin A, Cloutier SG, Cardimona DA, Huang D, Alsing PM, Xu JM (2005) Electric field tuning of plasmonic response of nanodot array in liquid crystal matrix. Nano Lett 5:1978–1982

Monkade M, Boix M, Durand G (1988) Order electricity and oblique nematic orientation on rough solid surfaces. Europhys Lett 5(8):697–702

O'Neill M, Kelly SM (2000) Photoinduced surface alignment for liquid crystal displays. J Phys D Appl Phys 33(10):R67

Osterman J, Adås C, Madsen L, Skarp K (2005) Properties of azo-dye alignment layer on plastic substrates. In: SID symposium digest of technical papers, vol 36, pp 772–775

Pratibha R, Park K, Smalyukh II, Park W (2009) Tunable optical metamaterial based on liquid crystal-gold nanosphere composite. Opt Express 17:19459–19469

Shiyanovskii SV, Lavrentovich OD, Schneider T, Ishikawa T, Smalyukh II, Woolverton CJ, Niehaus GD, Doane KJ (2005) Lyotropic chromonic liquid crystals for biological sensing applications. Mol Cryst Liq Cryst 434:259–270

Slikkerveer P, Bouten P, Cirkel P, de Goede J, Jagt H, Kooyman N, Nisato G, van Rijswijk R, Duineveld P (2004) A fully flexible colour display. In: SID symposium digest of technical papers, vol 35, pp 770–773

Stohr J, Samant MG, Cossy-Favre A, Dìaz J, Momoi Y, Odahara S, Nagata T (1998) Microscopic origin of liquid crystal alignment on rubbed polymer surfaces. Macromolecules 31:1942–1946

Tanggaard Larsen T, Bjarklev A, Sparre Hermann D, Broeng J (2003) Optical devices based on liquid crystal photonic bandgap fibres. Opt Express 11(20):2589–2596

Varghese S, Narayanankutty S, Bastiaansen CWM, Crawford GP, Broer DJ (2004) Patterned alignment of liquid crystals by μ-rubbing. Adv Mater 16:1600–1605

Wardosanidze ZV (2011) Holography based on the Weigert's effect. In: Naydenova I (ed) Holograms—recording materials and applications. ISBN: 978-953-307-981-3

Zheng W (2010) Surface wetting characteristics of rubbed polyimide thin films. In: Hashim AA (ed) Polymer thin films. ISBN: 978-953-307-059-9, InTech. doi:10.5772/8403

# Chapter 2
# Novel Liquid Crystal Polymers with Tailored Chemical Structure for High Barrier, Mechanical and Tribological Performance

**Shintaro Komatsu, Bernd Wetzel, and Klaus Friedrich**

## 2.1 High Performance Liquid Crystal Polymer (HP-LCP)

### *2.1.1 Introduction*

Thermotropic liquid crystal polymers (TLCPs) are dominant as matrix materials of precision products such as connectors and relays that are integrated into electric and electronic devices. This is due to their extraordinary properties, e.g. their high heat resistance, high melt flowability, excellent thin-wall strength as well as low out-gassing and moisture absorption. At the same time, TLCPs also possess excellent chemical resistance. Aiming towards their applications as structural parts, TLCPs are attractive because of other key properties such as high gas barrier, mechanical damping, dimension stability, flame retardancy and low smoke, and particularly one of the lowest coefficients of thermal expansion of all polymers. Because of these benefits, TLCPs have a high potential to be utilized in primary and secondary structural components especially for transportation applications.

One of the unique properties of TLCPs is their exceptional gas barrier. Natural gas provides less $CO_2$ emission than petroleum and coal when burned, and it is the major energy resource in the world. For the near future a hydrogen society, which is the sustainable energy society exploiting hydrogen gas as one of the clean energy resources, sounds promising. Keeping up with the rapid pace of times, advanced materials and technologies for storage and transport of gases have been developed

S. Komatsu (✉)
IT-related Chemicals Research Laboratory, Sumitomo Chemical Co., Ltd., Kitahara 6, Tsukuba, Ibaraki 300-3294, Japan
e-mail: komatsus2@sc.sumitomo-chem.co.jp

B. Wetzel • K. Friedrich
Department of Materials Science, Institut für Verbundwerkstoffe GmbH, Erwin-Schroedinger-Strasse 58, Kaiserslautern 67663, Germany
e-mail: bernd.wetzel@ivw.uni-kl.de; klaus.friedrich@ivw.uni-kl.de



V.K. Thakur, M.R. Kessler (eds.), *Liquid Crystalline Polymers*,
DOI 10.1007/978-3-319-20270-9_2

(Mori and Hirose 2009). Other polymers, e.g. ethylene vinyl alcohol polymer (EVOH), have an equivalent or even better gas barrier property than TLCPs under dry condition. However, the water absorption rate of EVOH is high due to its hydrophilic nature, which results in a sharp drop of barrier properties with increasing water absorption rate. Moreover, the barrier property of EVOH diminishes at temperatures above the glass transition temperature $T_g$ (about 60 °C); meanwhile, the $T_g$ of EVOH diminishes with the increasing water absorption rate. These effects accelerate the decrease in the barrier property (Massey 2003).

On the contrary, TLCPs have great advantages because they combine high heat resistance with excellent gas barrier properties and low water absorption rates. Their ductile nature, even under cryogenic temperatures, is also quite attractive for gas storing applications. This behavior should be easily confirmed by the observation of the fracture surface of the specimen immersed in liquid nitrogen for a certain period.

In the automotive sector, the damping property of TLCPs has attracted much attention. The weight reduction of automobiles has been rapidly progressing in order to reduce $CO_2$ emissions. It is widely known that replacing metals, such as aluminum alloy with polymer composites, allows lowering the vehicle weight. However, when parts around the engine are replaced with lighter materials, vibrations and noises emerging from the engine compartment become more intense due to the proportional relation between the material weight and the damping property, making both the ride and the handling of the vehicle less comfortable. In order to solve the vibration and noise problem, it is inevitable to make the walls of the parts thicker, resulting in a contradiction against lightweight design which is the essential purpose of the replacement. This substantial issue led to hesitations concerning the replacement of metal materials with lightweight materials for parts around the car engine. Although high strength and high modulus materials have poor damping properties in general, TLCPs have both high strength/modulus and superior damping property, promising the reduction of vibration and noise generated from the replacement.

It is, however, difficult to handle TLCPs because they have some drawbacks, which are poor strength at the weld line in the injection molding, anisotropic nature and weak hydrolysis resistance (Cox 1987). Therefore, the use of TLCPs as structural components has not been progressing. Solutions for the poor weld strength of TLCPs have been widely reported (Cox 1987; Agarwal and Mashelkar 1992; Fellahi et al. 1995; Nguyen-Chung 2004). The most effective way is to optimize the product design. An appropriate mold design should minimize the effects of the weld line by locating them in non-critical positions. Another solution is to use molding techniques. Wang et al. (1995) propose a molding technique called shear controlled orientation technology, which can improve the strength of weld lines up to 16 times compared to conventional molding methods.

TLCPs have an oriented molecular structure and therefore possess anisotropic material properties. It is well known that the anisotropic nature of TLCPs can be reduced by the addition of fillers. This effect was reported by Cox (1987) who incorporated fibrous and particulate fillers into molded components made of

TLCPs. Plummer et al (1993a) show that the addition of glass fibers could be the most effective way to reduce the anisotropy of TLCPs.

Hydrolysis resistance is regarded as one of the key parameters for using polymer materials as a matrix for structural components. The improvement of hydrolysis resistance of TLCPs is, however, substantially difficult because the weakness of hydrolysis resistance is attributed to the nature of the ester linkages in the chemical structure of TLCPs. Under higher temperature and with higher moisture the ester linkages of TLCPs are easily cut and then TLCPs gradually deteriorate. Improving the hydrolysis resistance of TLCPs should therefore strongly advance their applicability and open new markets. One way to improve hydrolysis resistance is to prevent the penetration of water molecules into the polymer matrix. In case of common poly(ethylene terephthalate) (PET) and poly(ethylene naphthalate) (PEN), for example, their barrier properties against water vapor and gas have been thoroughly evaluated (Light and Seymour 1982; Rueda and Varkalis 1995; Brolly et al. 1996; McGonigle et al. 2001; Pavel and Shanks 2003). The gas permeability of PEN is much lower than that of PET. This is due to the effect that molecular motions are more restrained in PEN than in PET, which strongly affects the diffusion coefficient and accordingly influences the permeability. The apparent diffusion coefficient of water is reported to be larger in the case of PET than in that of PEN. The barrier properties of water vapor and gases have also been reported for TLCPs (Chiou and Paul 1987; De Candia et al. 1990; Weinkauf and Paul 1992a, b; Hu et al. 2003; Kanehashi et al. 2010). Weinkauf and Paul (1992b) examined the gas permeability by changing the composition ratio of hydroxybenzoic acid (HBA) and 2-hydroxy-6-naphthoic acid (HNA) of TLCPs. They showed that the highest HNA content exhibits the best barrier properties because the non-linear nature of the naphthalene moiety leads to a more hindered rotation around the chain axis and the more hindered motions of the naphthalene moiety restrict penetrant mobility. It has been generally said that naphthalene moiety has a kink structure, so that copolyesters with a large number of HNA units disturb chain linearity and tend to lose liquid crystallinity of TLCPs. Well known TLCPs with naphthalene moiety are Vectra A950 and B950. According to reports, the composition of A950 is 27/73 molar ratio of HNA/HBA (Tjong 2003; Plummer et al. 1993a; Rath et al. 2007; Brooks 1992; Calundann 1979) and B950 is made of 60/20/20 of HNA/p-amino phenol/terephthalic acid (TA) (Tjong 2003; East et al. 1982). Corresponding to the heat resistance of these TLCPs, the melting temperatures of A950 and B950 are both 280 °C and the deflection temperature under load (DTUL) of 1.8 MPa for A950 is 187 °C.

In order to use TLCPs as matrix materials for structural components, a high heat resistance is obligatory. It has been revealed that TLCP with swivel molecular units such as HBA, terephthalic acis (TA) or hydroquinone (HQ) creates a dramatic increase in the melting temperature (Tjong 2003). The molecular design of TLCPs with high heat resistance traditionally starts from HBA homopolymer, which is a semi-crystalline polymer with melting temperature higher than the decomposition temperature. The crystallinity of HBA is disturbed through introducing copolymer components, such as 4.4-biphenol (BP), HQ, TA and isophthalic acid (IA), and then the liquid crystal phase is developed as an intermediate phase.

The development of the novel LCP reported in this work, focused on TLCP with high mechanical performance, high heat resistance, superior hydrolysis resistance, and excellent gas barrier, which are the most important properties for automobile and aircraft applications. The development strategy followed a two-step approach:

1. The first step was to incorporate more naphthalene moiety into the chemical structure of LCP in order to improve the hydrolysis resistance and to generate the liquid crystalline phase without a HBA unit, which is commonly adopted to commercial TLCPs.
2. The second step focused to optimize the monomer composition in order to keep a high heat resistance of type I, which is the highest level in the classification of heat resistance of TLCPs.

The resulting novel high-performance thermotropic liquid crystal polymer, designated as HP-LCP, excels by its high heat resistance and hydrolysis resistance, subsequently providing extraordinary gas barrier, high strength and high modulus as well as unique electric properties.

### *2.1.2 Synthesis of High Performance Liquid Crystal Polymer*

Improving the heat resistance of TLCPs by molecular design and increasing the productivity is generally a trade-off relationship, since a heat resistant polymer has to be handled under higher process temperature, which requires more energy and more expensive equipment. There are several measurement techniques to classify the heat resistance of polymers. Considering a wide applicability, the soldering heat resistance was selected as criteria for durability in a reflow process. The reflow process is generally applied in the manufacturing process of electronic devices, and it allows lead-free solder melting to connect copper wires on a printed circuit board with surface mounted parts. The maximum temperature of the reflow process is around 260 °C. The soldering heat resistance is easily evaluated by the degree of deflection of a molded part after soaked directly into a lead-free solder bath for a certain period.

The high performance LCP (HP-LCP) was made of copolyester of 2-hydroxy-6-naphthoic acid (HNA), hydroquinone (HQ), 2,6-naphthalene dicarboxylic acid (NDCA) and terephthalic acid (TA). HNA and NDCA were selected as naphthalene moiety because of imparting hydrolysis resistance to copolyester. Isophthalic acid (IA) was not used as one of the components because it was observed that the IA moiety, a bent-shaped monomer, deteriorates hydrolysis resistance. The ratio of HQ was specifically selected for adjusting the target melt viscosity of the copolymer. The level of heat resistance depends mainly on the ratio of both TA and HQ. Although many TLCPs have been developed and each commercial TLCP has a different combination of monomers, the composition of the HP-LCP developed and presented here, has not been known. The polymerization procedure is described in the work of Okamoto and Hosoda (2006). The amount of HNA and NDCA obtained by the polymer was more than 50 mol% and for stoichiometric reasons the relative amount between HQ and the sum of NDCA and TA stayed the same.

The resulting thermotropic liquid crystal polymer was designated as a high performance LCP (HP-LCP). HP-LCP enables large-scale manufacturing by an established commercial processing method and technology which can suppress the increase of costs. However, because of the unique viscoelastic nature of HP-LCP, an inflation film production process can be applied to obtain a biaxially oriented film. By controlling the blow-up ratio, i.e. the ratio between the tube diameter after blowing and the tube diameter before blowing for molding, it is realized that the HP-LCP film has isotropic properties in both machine direction (MD) and transversal direction (TD).

In order to evaluate its performance, the HP-LCP was compared with "Sumikasuper® LCP E6000" (LCP-E), which is one of the highest heat resistant commercial LCPs. LCP-E was supplied by Sumitomo Chemical Co., Ltd. It is composed of HBA, BP, IA and TA (Kalika et al. 1991; Wang et al. 2003; Okamoto et al. 2004).

## 2.1.3 *Properties of HP-LCP*

### 2.1.3.1 Thermal Analysis of HP-LCP

Differential scanning calorimetry (DSC) analysis was carried out under nitrogen atmosphere. The HP-LCP powder was heated from 25 to 380 °C at a heating rate of 10 K/min. Figure 2.1 depicts the first heating curve. The endothermal peak was

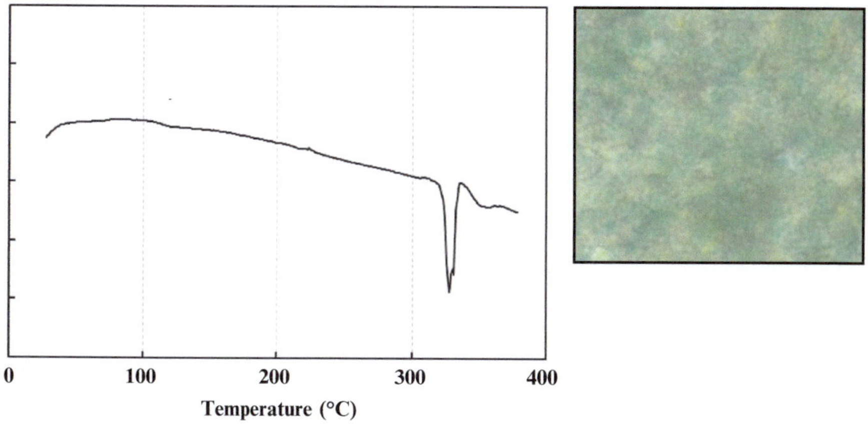


**Fig. 2.1** DSC curve and polarized optical microscope image of HP-LCP

clearly observed at 328 °C. Furthermore, the material was examined with a polarization microscope in order to clarify the phase transition of HP-LCP. Under crossed polarized light, the sample stage of the microscope was continuously heated from 25 to 360 °C. The image in Fig. 2.1 shows the typical morphology of the molten HP-LCP at 335 °C. A clear optical anisotropic phase was observed, revealing that HP-LCP has a liquid crystal phase, which was stable up to 360 °C. The onset temperature of solid-liquid crystal transition temperature ($T_{S\text{-}L}$) extrapolated from the DSC pattern was 321 °C. The DSC analysis and the observation by polarization microscope proved that HP-LCP with naphthalene-rich structure has a liquid crystalline nature without HBA moiety which is generally applied in commercial LCPs.

#### 2.1.3.2 Melt Viscosity of HP-LCP

The temperature dependence of the apparent melt viscosity of pure HP-LCP significantly differs from that of pure LCP-E. The melt viscosity of pure HP-LCP and pure LCP-E were measured by a Capilograph-1B capillary rheometer (Toyo Seiki Seisaku-sho, Ltd.) which can detect the shear rate and shear stress of molten polymer flowing out of the capillary. The geometry of the capillary die used in this measurement was 0.5 mm in diameter and 10 mm in length.

Figure 2.2 (left) compares the melt viscosity data of pure HP-LCP with that of pure LCP-E for a shear rate of 1000 $s^{-1}$. With increasing temperature the melt viscosity of pure LCP-E decreased sharply by more than one decade after reaching the solid-liquid crystal temperature $T_{S\text{-}L}$ of LCP-E. This is the typical behavior of TLCPs. On the other hand, the drop in melt viscosity of pure HP-LCP was rather small at the $T_{S\text{-}L}$ and a wide plateau at a value of approximately 100 Pa s was observed after the $T_{S\text{-}L}$ revealing that the temperature dependence of the melt viscosity of pure HP-LCP is minimal.

The low temperature dependence of melt viscosity correlates well with the slow solidification rate from liquid crystal phase to solid phase. A fast solidification of

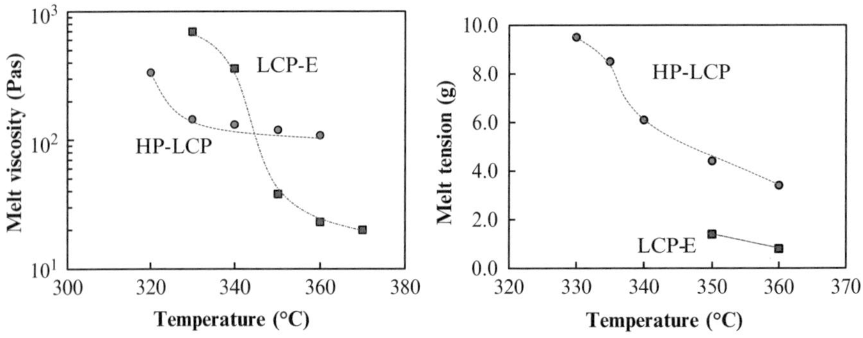


**Fig. 2.2** Temperature dependence of apparent melt viscosity and melt tension of HP-LCP

TLCPs causes weak weld strength. Rigid TLCP molecules align parallel to the weld line and rapidly solidify, and this effect allows only little inter-diffusion of TLCP chains at the weld line. Therefore, the weld strength of TLCP with a rapid solidification rate, such as LCP-E, is poor. On the other hand, the solidification rate of HP-LCP is so slow that the weld strength tends to be high. Furthermore, the slow solidification rate of HP-LCP is quite favorable in order to produce film and fiber. This nature allows the realization of minimum viscosity differences and results in constant quality during the processing of film and fiber. These rheological characteristics of HP-LCP play an important role in film and fiber production. Although the melt viscosity of HP-LCP is high compared to that of LCP-E, the level is considerably lower than that of other heat resistant polymers such as PPS or PEEK. Therefore, components with complex geometry can be easily realized by using HP-LCP as a matrix polymer.

#### 2.1.3.3 Melt Tension of HP-LCP

One of the most interesting features of HP-LCP is its high melt tension. High melt tension is an advantage in the manufacturing process because stable resin flow from an extrusion opening is guaranteed without breakage of film and fiber during the processing. Melt tension was measured by a capillary rheometer with a capillary die of 1 mm in diameter and 10 mm in length. The piston speed was fixed at 5 mm/min and the drawing speed was changed gradually. Melt tension data was then collected at steady drawing conditions. The results show that the melt tension of pure HP-LCP substantially differed from that of pure LCP-E, and Fig. 2.2 (right) depicts the influence of the temperature on the melt tension. The melt tension of pure LCP-E was low at both 350 and 360 °C. The reason for such a low melt tension is that the molecular chains are easily oriented along the flow direction under even low shear stress. On the other hand, pure HP-LCP exhibited a higher melt tension over the complete temperature range from 330 to 360 °C. At 350 and 360 °C, the melt tension of pure HP-LCP was more than three times higher as those of pure LCP-E. Such high melt tension of HP-LCP incidentally enhances mechanical properties and toughness of a molded part.

#### 2.1.3.4 Physical Properties of HP-LCP

The following section demonstrates some key aspects of physical properties for HP-LCP. Table 2.1 shows the physical properties of pure HP-LCP and pure LCP-E. A deflection temperature under load (DTUL) of 1.8 MPa was measured in order to identify the short-term heat resistance of pure HP-LCP and pure LCP-E. The DTUL of pure HP-LCP was 240 °C and thus lower than that of pure LCP-E, because the chemical structure of HP-LCP has no HBA moiety which would increase the melting temperature of LCP. However, the DTUL of HP-LCP composite reinforced with 30 wt% short glass fibers (GF) was 285 °C, and it remained on the same level

**Table 2.1** Physical properties of HP-LCP, LCP-E, composites, HP-LCP and sLCP films

| A | | HP-LCP | LCP-E |
|---|---|---|---|
| Specific gravity | g/cm$^3$ | 1.39 | 1.37 |
| DTUL | °C | 240 | 265 |
| Tensile strength | MPa | 180 | 156 |
| Tensile strain | % | 11.0 | 7.7 |
| Flexural strength | MPa | 115 | 79 |
| Flexural modulus | GPa | 4.5 | 3.7 |
| Water absorption | % | 0.02 | 0.10 |
| Dielectric constant@1 GHz | | 3.1 | 3.2 |
| Dielectric tangent@1 GHz | | 0.001 | 0.003 |

| B | | HP-LCP composite | | | LCP-E composite |
|---|---|---|---|---|---|
| | | Carbon fiber | Glass fiber | Mineral | Glass fiber |
| Reinforcement | | 30 wt% | 30 wt% | 25 wt% | 30 wt% |
| Specific gravity | g/cm$^3$ | 1.47 | 1.60 | 1.58 | 1.61 |
| DTUL | °C | 280 | 285 | 260 | 285 |
| Tensile strength | MPa | 270 | 210 | 230 | 150 |
| Tensile elongation | % | 2.9 | 2.7 | 8.0 | 1.9 |
| Flexural strength | MPa | 400 | 330 | 235 | 230 |
| Flexural modulus | GPa | 38.3 | 25.7 | 18.8 | 25.4 |
| Specific volume resistance | Ω m | $10^3$ | $10^{13}$ | $10^{13}$ | $10^{13}$ |

| C | | HP-LCP film | sLCP film |
|---|---|---|---|
| Processing method | | Inflation process | Solvent casting process |
| Specific gravity | g/cm$^3$ | 1.39 | 1.39 |
| $T_m$ | °C | 328 | 339 |
| Hydrogen gas permeability | cm$^3$ 25 μm/(m$^2$ 24 h bar) | 114 | 218 |
| Oxygen gas permeability | | 4.1 | – |
| Water vapor permeability | | 0.01 | – |
| Water absorption | % | 0.02 | 0.46 |
| Tensile strength (MD/TD) | MPa | 330/300 | 145/142 |
| Tensile modulus (MD/TD) | GPa | 13/12 | 4/4 |
| Tensile elongation (MD/TD) | % | 13/17 | 24/21 |
| CTE (MD/TD) | ppm/°C | −2/0 | 32/32 |
| Thermal conductivity | W/m K | 0.38 | – |
| Breakdown voltage | kV/mm | 210 | – |
| Dielectric constant@1 GHz | | 3.0 | 3.1 |
| Dielectric tangent@1 GHz | | 0.001 | 0.003 |

as LCP-E composite containing 30 wt% GF. HP-LCP shows high heat resistance; further data on the physical properties of HP-LCP composites can be seen in Sect. 3.3.

The soldering heat resistance of HP-LCP and LCP-E was evaluated by detecting the degree of a sample's deflection after soaking in a lead-free solder bath at 280 °C for 120 s. The specimens used for this test were reinforced with 40 wt% short glass fibers. The specimen geometry was a dumbbell shape according to JIS K7113 1(1/2) standard, i.e. the overall length was 75 mm, the length of the narrow parallel-sided portion was 30 mm, the parallel part was 5 mm, and the thickness was 1.2 mm. No deflection at all was observed for both HP-LCP and LCP-E composite after the test, which demonstrated a high heat resistance of both materials. It is therefore obvious that the heat resistance is high enough to allow the application of HP-LCP not only in structural components but also for electronic device applications.

The tensile and flexural properties of pure HP-LCP were superior to those of pure LCP-E. The tensile strength of pure HP-LCP is nearly 15 % and the flexural strength is 45 % higher than those of pure LCP-E. Although the improvement of the mechanical performance was not the main focus of this development, these properties could be substantially enhanced due to the high melt tension of HP-LCP.

The water absorption of pure HP-LCP was measured for the following conditions: temperature 85 °C, relative humidity 85 % RH, time period 168 h. With a value of 0.02 % the water absorption of pure HP-LCP was nearly five times lower than that of pure LCP-E (Table 2.1(A)). In fact, the chemical structure of HP-LCP is composed of a high ratio of naphthalene moiety, and it is believed that the naphthalene moiety restricted the diffusion of water molecules into the polymer, since the rotation of the chain axis of naphthalene moiety is more hindered. This result is one of the reasons for the superior hydrolysis resistance of HP-LCP.

The excellent dielectric properties of pure HP-LCP are also of great interest. It was specifically observed that the dielectric tangent was significantly low in a wide frequency range from 1 MHz to 1 GHz. These properties allow to use HP-LCP even as an insulator for high frequency applications.

#### 2.1.3.5 Hydrolysis Resistance of HP-LCP

In order to determine the hydrolysis resistance of HP-LCP, the highly accelerated temperature and humidity stress test (HAST) based on IEC 68-2-66 standard was carried out. HAST, which is known as the 'pressure cooker test' (PCT), is a highly accelerated method of electronic component reliability testing using temperature and humidity as the environmental parameters. This test temporarily accelerates the infiltration of moisture into the sample. The specimen used in the PCT had the same dumbbell shape that was used to measure the soldering heat resistance (Sect. 1.3.4). The saturated PCT method was done at a temperature of 121 °C, a relative humidity of 100 % and a water vapor pressure of 0.2 MPa (2 bar). After exposing the materials to these conditions, the alteration of mechanical properties due to hydrolysis was determined by tensile tests, regarding tensile strength as an indicator.

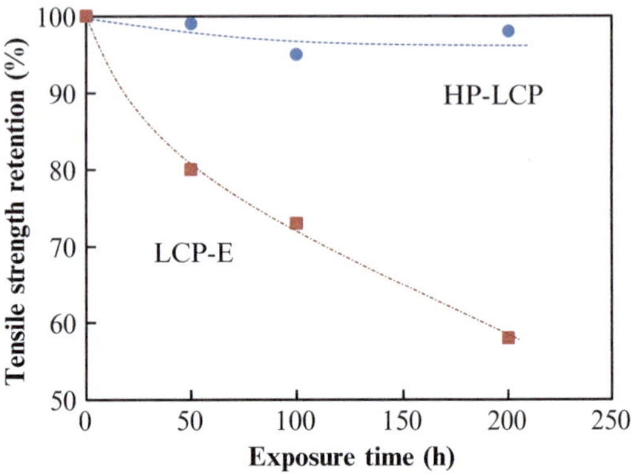


**Fig. 2.3** Hydrolysis resistance of HP-LCP and LCP-E (E6000)

The tensile strengths of HP-LCP and LCP-E were measured after the PCT for 50, 100 and 200 h, respectively, and the retention percentage of tensile strength of each sample after the PCT compared to those before the PCT was calculated. The deformation speed of the tensile tests was 10 mm/min. Figure 2.3 shows the results of PCT, and it is observed that in the case of LCP-E, the retention of tensile strength after the PCT for 50 h dropped to 80 % of the initial strength and then continuously decreased to 60 % of the initial strength after the PCT for 200 h. On the other hand, the strength retention of HP-LCP maintained 98 % of the initial strength even after 200 h. These results demonstrate the excellent hydrolysis resistance of HP-LCP. It is believed that the diffusion of water vapor was restricted by the hindered motions of the naphthalene moiety (Weinkauf and Paul 1992b).

## 2.2 Soluble LCP (sLCP)

### *2.2.1 Introduction*

Commercial TLCPs are known to be unable to dissolve in any common organic solvents except for expensive fluorinated reagents. From the commercial point of view, the manufacturing of TLCPs film is limited to molten processes such as extrusion and inflation technologies. It is difficult to control the orientation of TLCP molecules by the molten processes in order to obtain a high quality film because TLCP molecules spontaneously align along the shear direction.

For developing a TLCP which is able to dissolve in common organic solvents, a solvent casting method widely used to make films of polycarbonate and polyimide can be employed, anticipating a TLCP film with isotropic nature and with competitive manufacturing costs.

The industrialization of soluble TLCPs has been considered to be fundamentally difficult because TLCPs can be dissolved only in costly fluorine phenol solvents which are difficult to handle (Makhija et al. 1997). In addition, even if it is dissolved in a solvent for a certain period, the length of the molecules with rigidity (Kuhn length) could be shorter, and the concentration of rigid parts in the molecules becomes too low in the solution. It turns into a normal liquid and a solution without liquid crystal, and therefore, it has been believed that the unique properties of TLCP never generate.

In this work, however, it is demonstrated that soluble TLCPs can be successfully developed. It is possible to dissolve the TLCP in n-methylpyrrolidone by optimizing the molecular design (Okamoto et al. 2006). Here, we have designated this soluble TLCP as sLCP.

A solvent casting film obtained from sLCP exhibits isotropic characteristics as well as a high temperature tolerance, an excellent gas barrier and a low water absorption. Furthermore, the sLCP film has a low dielectric loss at high frequencies and a high thermal conductivity, and it possesses high tear strength because the material is less anisotropic in comparison to extruded LCP films. With many advantages of sLCP film generated via the casting method, the film can be applied in electronic devices such as flexible printed circuit boards, film condensers, film dampers and gas barrier films. In addition, the solution viscosity of sLCP is low enough to allow a compounding of the polymer with a high amount of functional fillers and fibers in order to obtain multifunctional composite properties.

### *2.2.2 Film Preparation and Molecular Orientation of Soluble Liquid Crystal Polymer*

Soluble LCP (sLCP) was supplied by Sumitomo Chemical Co., Ltd. (Tokyo, Japan) (Okamoto et al. 2005, 2006). The sLCP was dissolved in n-methylpyrrolidone. A LCP film was obtained from sLCP by a solvent casting method described as follows: The sLCP was applied to a substrate such as copper foil by bar coating and subjected to heat treatment at 100 °C for 1 h. The sLCP film was in an amorphous (non-crystalline) state and therefore transparent. The film was heated to 300 °C for 3 h under nitrogen atmosphere, and then the liquid crystal phase in the film was reached. After that, the film turned semitransparent or opaque.

The molecular orientation of both sLCP films made by an extrusion method and the solvent casting method were measured by Microwave Molecular Orientation Analyzer at 25 °C and a frequency of 12.58 GHz (MOA-5012A, Oji Scientific Instrument, Japan). Each film was inserted into a narrow gap between a pair of waveguides constituting the cavity resonator system. The polarized microwaves were irradiated perpendicular to the plane of the film that rotates for 6.0 s around the regular axis of the film plane, and the transmitted microwave intensity was measured at every degree of rotation angle (Osaki 1990, 1997). The angular dependence of transmitted microwave intensity, called the orientation pattern, was measured at a fixed frequency. The orientation pattern provides the molecular orientation angle (MOA) and the molecular orientation ratio (MOR). The direction, in which the transmitted microwave intensities are at a minimum, is designated as the orientation

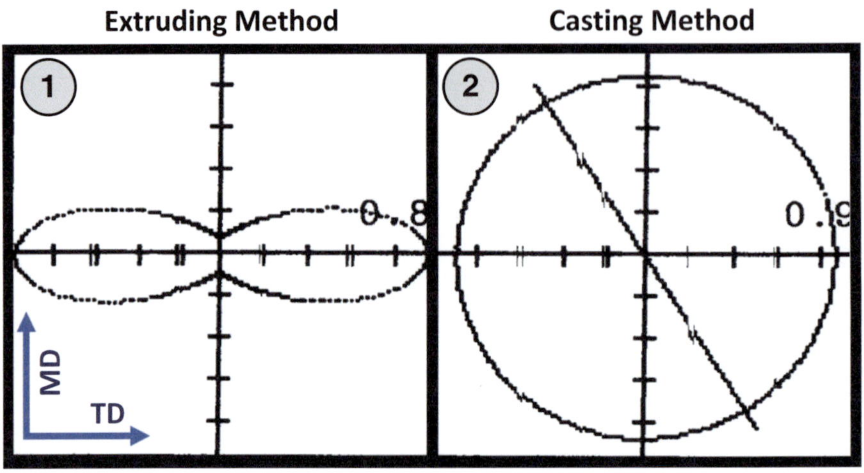


**Fig. 2.4** Molecular orientation of sLCP film with extruding method (1) and casting method (2)

angle corresponding to the angle between the main axis of the molecular chain and the flow direction. Both parameters of MOA and MOR describe the extent of alignment of molecular chains with reference to the flow direction. The maximum-to-minimum ratio of transmitted microwave intensity is defined as MOR reflecting the mechanical anisotropy. For example, MOR of 1.0 reflects random orientation. This technique allows the easy determination of the molecular orientation in the film.

Figure 2.4 shows the typical orientation patterns of sLCP films made via extrusion and the solvent casting method by means of the Microwave Molecular Orientation Analyzer. The vertical axis is corresponding to the MD (machine direction) and the horizontal axis is corresponding to the TD (transverse direction). The orientation pattern of sLCP by the solvent casting method was round in shape and MOR was 1.0, representing a random orientation of sLCP in MD and TD. Because the solvent casting film of sLCP was obtained from the amorphous state in a solution by heat treatment, the anisotropy was not observed in the film. On the other hand, the lower transmitted microwave intensity was detected in the sLCP film made via extrusion, and MOR was 9.9 showing a higher quantity of sLCP chains aligned close to the MD.

## 2.3 Applications and Potential Applications

### *2.3.1 Film Applications of HP-LCP and sLCP*

Multifunctional thin films of both HP-LCP and sLCP have been successfully developed in response to the needs of an expanding market of heat resistant films. Different film production methods were adapted to HP-LCP and sLCP, respectively. In the case of HP-LCP, the inflation film production process was applied to obtain a biaxially oriented film. On the other hand, sLCP film was made via solvent casting method and exhibited intrinsic isotropic nature. In principle, the sLCP has the potential to allow the manufacturing of super thin films with thicknesses of less than 20 μm.

Table 2.1(C) shows the physical properties of HP-LCP inflation film and sLCP solvent casting film. In this case, the thickness of HP-LCP film and sLCP film were both 25 μm. In particular, an unique property of the HP-LCP film is its excellent gas barrier. Permeability tests were performed under hydrogen gas, oxygen gas and water vapor with the following conditions: 23 °C, 50 %RH and 0.1 MPa (1 bar), 23 °C, 60 % RH and 0.1 MPa (1 bar) and 40 °C, 90 % RH and 0.1 MPa (1 bar), respectively. All gas permeabilities measured were extremely low and belonged to the lowest values known for thermoplastic films. The reason for this effect should be the hindered motion of the naphthalene moiety in HP-LCP that restricts penetration mobility of gas molecules. The combination of the excellent gas barrier properties and the low water absorption rate are key properties and great advantages of HP-LCP film.

The biaxial orientation of molecular chains gained through the inflation process results in a high tensile strength of HP-LCP film with values of 330 MPa in MD and 300 MPa in TD, which showed minimal anisotropy. As it can be expected from the natural characteristic of TLCPs, the coefficient of mean linear thermal expansion (CTE) between 50 and 150 °C was very low (−2 ppm/°C) and the values in MD and TD were almost the same. It is interesting to note that the thermal conductivity of HP-LCP film was 0.38 W/m K which belongs to the highest level in pure polymers so far. Besides, a high breakdown voltage of 210 kV/mm of HP-LCP was observed. This means that HP-LCP might be suitable to serve as an insulation film for automotive motors and electric devices. HP-LCP film also demonstrated a low dielectric loss (tanδ) which is a key property for being applied as a substrate material of flexible printed circuit boards for high frequency signal transmission.

The sLCP solvent casting film shows that the tensile properties of MD and TD were almost the same, resulting from the isotropic nature of sLCP. With a value of 145 MPa, the tensile strength was high and the tensile elongation at maximum stress (24 %) showed ductile character. Additionally, with a tensile modulus value of 4 GPa, the sLCP film exhibits high stiffness so that it is possible to design even complex three-dimensional parts. No anisotropy of the CTE (32 ppm/°C) was observed between MD and TD. Furthermore, the hydrogen gas permeability of the sLCP film was significantly low as well as that of HP-LCP film because sLCP also has naphthalene moiety. The vibration damping property of sLCP is excellent in a wide frequency range, and due to the combination with other superior properties such as high heat resistance and high stiffness, the sLCP already has been commercially adopted as a vibration damper film of headphone diaphragm.

Another quite attractive feature of sLCP is that multifunctional properties can be imparted into the sLCP film by incorporating functional fillers and fibers with high loading since the solution viscosity of sLCP is significantly low.

In summary, both the HP-LCP film and the sLCP film have excellent multipurpose properties such as high heat resistance, high gas barrier, low water absorption, unique dielectric properties and high dimension stability, which provide promising options for applications in high potential growth markets, e.g. energy (solar cells), high frequency (flexible printed circuit boards), display and electric/electronic applications such as electric motor insulators.

### 2.3.2 Microcellular Foam Injection Molding of HP-LCP

TLCPs mainly consist of two layers, a skin layer and a core layer. The skin layer has a dominant effect on the mechanical properties because of the high orientation of molecular chains (Cox 1987; Plummer et al. 1993b). The core layer of LCP does not have positive effects on mechanical properties and therefore it is possible to add another functionality to the core layer. One special option is to introduce a microcellular foam which was already successfully realized by using the Mucell® technology for the HP-LCP with high melt tension. The main idea was to reduce the total weight of the molded parts by incorporating air cells into the core layer. The specific flexural modulus, i.e. the ratio of flexural modulus and specific gravity, was measured for a microcellular foam article and a value of 5.0 MNm/kg was achieved. This value represents a more than 50 % increase compared to a standard injection molding part with only 3.2 MNm/kg. Therefore, a structural component made of HP-LCP and the Mucell® technology could be much lighter than those manufactured by the standard injection method.

### 2.3.3 Filler Reinforced HP-LCP Composite for Injection Molding

HP-LCP exhibits a quite attractive mechanical performance when incorporating fillers or fibers. Table 2.1(B) shows the physical properties of HP-LCP composite and LCP-E composite. For measuring mechanical properties Dumbbell-shaped specimens according to the JIS K7113 1(1/2) standard were used. Their overall length was 75 mm, the length of the narrow parallel-sided portion was 30 mm, the width of narrow portion was 5 mm and the thickness was 0.5 mm.

First, the physical properties of HP-LCP composite and LCP-E composite were compared in combination with short glass fiber (GF, 30 wt%). The deflection temperature under load (DTUL) of HP-LCP/GF30 composite was 285 °C and thus the same as that of LCP-E/GF30 composite, revealing that both composites possess high heat resistance. The tensile strength of HP-LCP/GF30 composite and LCP-E/GF30 composite were 210 and 150 MPa, respectively, showing that HP-LPC/GF30 composite has excellent mechanical performance. Similarly, the flexural strength of HP-LCP/GF30 composite was significantly increased compared to the LCP-E/GF30, while the flexural modulus was only marginally affected.

In particular, the incorporation of short carbon fiber (CF) to HP-LCP generated a markedly reinforcing effect on the mechanical properties. The tensile strength increased to 270 MPa and the flexural strength to 400 MPa for HP-LCP/CF30 composite. Indeed, the flexural modulus of 38 GPa for HP-LCP/CF30 composite is close to that for a magnesium alloy (45 GPa). Thus, HP-LCP/CF30 composite can be a promising candidate for replacing metals in structural components. The creep strain of HP-LCP/CF30 composite under a constant load of 6.2 MPa (62 bar) and at a temperature of 150 °C was only 1 % after 1000 h, and under the same testing conditions this value revealed better creep resistance than that of PPS/GF40 composite which showed a creep strain of 2 %. PPS/GF40 composite, however, is one of the most popular thermoplastic composites for structural components.

Because HP-LCP/CF composites possess high heat resistance and high strength and modulus as well as excellent dimension stability and thin-wall flowability, even complicated mechanical parts around an automobile engine have been considered for adaption.

It is interesting to note that after adding a reinforcing mica mineral filler (MC) with plate-like shape and high aspect ratio, the tensile strength of the HP-LCP/MC25 composite was measured to be 230 MPa, which is significantly higher than that of LCP-E/GF30 composite. In particular, the elongation at break dramatically increased to 8 %, which also indicates high fracture toughness of HP-LCP/MC25 composite. Furthermore, the data for the specific volume resistance of filler reinforced composites depended on the incorporated filler which is in agreement with the expectation.

### *2.3.4 Tribological Application of HP-LCP Composites*

One of the important applications requiring materials with high heat resistance and high strength/modulus are tribological applications. In automotive applications, for example, bearings and seal rings have been used under severe sliding conditions, such as high temperature environment, high pressure and/or high speed. High performance polymer composites based on polyimide (PI) and polyether-etherketone (PEEK) are successfully used in many tribological applications and have already replaced metals. PI composites can impart excellent dimension stability to final parts because of being machined from plate-like or rod-like raw materials. The disadvantages of PI composites are their higher manufacturing costs and longer cycle times, which are typical challenges for thermoset materials. PEEK composites have the great advantage of lower manufacturing costs compared to thermosetting materials because an injection molding can be performed. However, in the case of thin-walled structures such as seal rings with a thickness of 1 mm, more than two injection gates need to be designed in the mold because the high melt viscosity for PEEK composites makes it difficult for the molten polymer to flow into a thin-walled mold. This means that at least one weld line exists in the final parts. The strength at the weld lines tends to be weak and the dimensions of final parts could change at the weld line.

Advanced tribology materials have been continuously required in industries. For example, a seal ring is a product for sealing engine oil in the gear box of automatic transmissions. Downsizing a car engine has been rapidly progressing for reducing $CO_2$ emission. In order to provide a shorter overall length of the engine, the diameter of the engine shaft must become larger, and as a result, the size of the seal ring increases. Therefore, a thermoplastic composite with low melt viscosity should be beneficial. Additionally, since fine shapes are designed in each seal ring to improve the sealing performance, thermoplastic composites with low melt viscosity are also favorable in this aspect.

There are several advantages for the design of a single gate in a mold of injection molding. First, there is no concern about weld lines. Second, when the gate is on the edge of a molded article, the flow direction of molten polymer is parallel to the major axis of the product, expecting high strength attributed to the orientation of molecular chains along the major axis. Furthermore, a single gate makes the polymer flow simpler, anticipating the uniform dimensional stability of the molded article. Finally, multi-cavity molds can be designed for materials with low melt viscosity, resulting in a reduction of manufacturing costs by improving productivity.

HP-LCP is a promising matrix polymer for tribological applications because of its low melt viscosity, which is the typical nature of TLCPs, as well as high heat resistance and high strength and modulus. However, TLCPs have an intrinsic nature that results from the orientation of molecular chains in the skin layer of a molded article along MD, which generates fibrils easily from the surface.

It has been reported that tribological properties, such as coefficient of friction and wear rate, are significantly enhanced by incorporating nanoparticles into polymer composites (Zhang et al. 2004; Chang et al. 2005; Chang and Friedrich 2010). It has been proposed that the friction force is reduced by the rolling effect of hard nanoparticles at the contact surfaces between the polymer composite and the counterpart. In case of LCP composites, nanoparticles are expected to favorably affect the tribology behavior by modifying the surface skin layer as well as by inducing the rolling effect.

An HP-LCP composite with excellent tribological performance has been successfully developed (Komatsu and Maeda 2013). Short carbon fibers, PTFE and $Al_2O_3$ nanoparticles were incorporated into this material. The compounding of these fillers with HP-LCP was carried out by a twin screw extruder followed by injection molding to prepare test specimens. The injection molded specimens were cut into pins with a contact surface of $4 \times 3$ mm$^2$ and then examined under dry sliding conditions without lubricant. Sliding wear tests were performed in a pin-on-disc configuration under different conditions. The disc counterpart was made of 100Cr6 ball bearing steel as specified in EN 10027-1. The roughness was $R_a = 0.2$ μm and the hardness HRC 60. All tests were conducted for a total running time of 20 h under a systematically varying combination of pressure and sliding speed. The friction force and the height change, i.e. wear of the specimen were measured continuously during this period. The characteristic wear properties are the coefficient of friction (CoF) and the specific wear rate ($w_s$). The CoF relates the applied normal force $F_N$ with the resulting friction force $F_R$, CoF = $F_R/F_N$, while the specific wear rate provides information about the material volume ablated under the specific running conditions, load, speed and time. These were calculated as average values between the running time of 15 and 20 h after reaching steady state wear conditions. The specific wear rate ($w_s$) was then calculated as follows:

$$w_S = \frac{\Delta h \times A}{F_N L} \left(\frac{mm^3}{Nm}\right)$$

where $\Delta h$ is the height change of the specimen during steady state wear process, $A$ is the contact area, $F_N$ is the normal load and $L$ is the total sliding distance.

In order to compare the tribological performance of HP-LCP composite with two reference materials, i.e. a PI composite containing PTFE and graphite, and a PEEK composite with PTFE and short carbon fibers, all composites were tested under the same sliding conditions. The PEEK composite was conditioned at 240 °C for 3 h before the sliding test. Table 2.2 shows the friction and wear properties of HP-LCP, PEEK and PI composite under a pressure of 1 MPa and a sliding speed of 1 m/s. One test was conducted at a room temperature of 23 °C, the other at controlled counterpart surface temperature of 190 °C.

At a temperature of 23 °C all composites showed excellent friction and wear properties and the CoF of the HP-LCP composite was the lowest among them. Although the $w_s$ of the HP-LCP composite was higher than that of the PI composite, the value was still much lower than that of the PEEK composite. The excellent tribological properties of the HP-LCP composite were attributed to a synergistic effect between superior mechanical properties of HP-LCP and probably a rolling effect induced by nanoparticles. Figure 2.5 shows the worn surface of HP-LCP composite under a pressure of 1 MPa, a sliding speed of 1 m/s and at 23 °C. The worn surface of HP-LCP composite was smooth and without fibrillation of molecules after the sliding test. A stable transfer film was formed on the counterpart, which is a key mechanism to obtain low friction and wear (Zhang 2010).

**Table 2.2** Comparison of friction and wear properties of HP-LCP with other high performance polymer composites, polyetheretherketone (PEEK) and polyimide (PI) at a pressure of 1 MPa and velocity of 1 m/s at low and high counterpart temperatures

| | 1 MPa, 1 m/s, 23 °C | | 1 MPa, 1 m/s, 190 °C | |
|---|---|---|---|---|
| | CoF | $w_s$ $10^{-7}$ mm³/Nm | CoF | $w_s$ $10^{-7}$ mm³/Nm |
| HP-LCP composite | 0.38 | 4.9 | 0.06 | 48.5 |
| PEEK composite | 0.47 | 7.3 | 0.79 | 102.2 |
| PI composite | 0.43 | 0.4 | 0.10 | 31.5 |

CoF represents the coefficient of friction, and $w_s$ is the specific wear rate

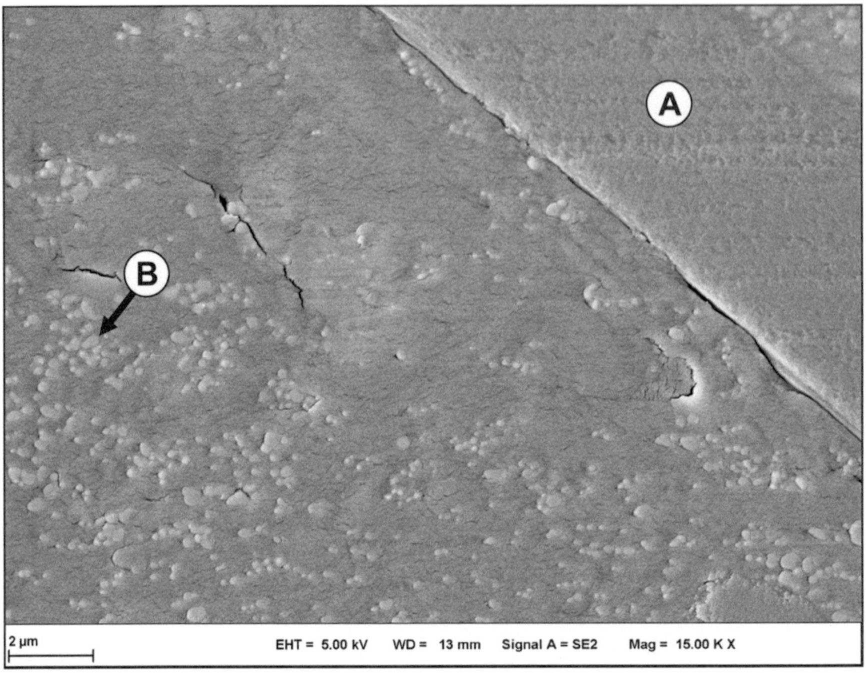


**Fig. 2.5** Micrograph of the surface of HP-LCP composite after sliding wear test at a pressure of 1 MPa, sliding speed 1 m/s, 23 °C. Item A shows SCF, and item B represent $Al_2O_3$ nanoparticles

Under more severe sliding conditions, i.e. a pressure of 16 MPa and a velocity of 1 m/s at 23 °C, the CoF of HP-LCP composite was measured to be 0.16 and the $w_s$ was $8.4 \times 10^{-7}$ $mm^3/Nm$. No failure of the HP-LCP composite could be observed, which demonstrates the suitability of HP-LCP composite for a wide range of *pv* (pressure × velocity) conditions.

During the steady state of each test, the surface temperature of the counterpart against HP-LCP, PEEK and PI composite reached 33 °C, 35 °C, and 33 °C, respectively. Because the CoF of the PEEK composite was higher than the CoF of other composites, more friction heat was generated, resulting in a higher surface temperature of the counterpart.

A significant difference in the tribological performance among the three composites was observed at a temperature of 190 °C. Thermal destruction was not observed for the HP-LCP composite which is attributed to the high heat resistance of HP-LCP. Again, the CoF of the HP-LCP composite was the lowest among the three composites, and the $w_s$ was almost at the same level with that of the PI composite. The PEEK composite, on the other hand, showed a stick-slip phenomenon that resulted in a high CoF, high wear rate and unstable running conditions.

The processability of both the HP-LCP and the PEEK composite should be superior to that of the PI composite because these thermoplastics can be easily injection molded, anticipating significant lower manufacturing costs and cycle time.

Compared to the PEEK composite, the melt viscosity of the HP-LCP composite was significantly lower, so that thin-walled parts or complex shape parts can be manufactured more easily.

### 2.3.5 Substrates of sLCP for Electronic Devices

The sLCP has excellent dielectric characteristics at high frequencies and low moisture absorption as shown in Table 2.1(C). The dielectric properties of the sLCP film were found to be constant up to a frequency of 25 GHz. Trials of sLCP as a substrate material of flexible printed circuit boards for higher frequency applications have been performed. Interestingly, the sLCP itself has a relatively high thermal conductivity. Its low viscosity allows the incorporation of a larger quantity of fillers to even further improve thermal conductivity. Trials have been continued in order to make use of sLCP as a substrate for metal based copper clad laminates (MCCL).

### 2.3.6 Coatings of sLCP for Tribological Application

Heat resistant and chemically resistant coating materials are attractive for applications in the automotive, aircraft and other industries. Guerriero et al. (2011) reported that coatings made of liquid crystal thermosetting polymers (LCT) showed markedly increased interface adhesion between LCT and an aluminum alloy substrate. Compared to a commercial LCP the LCT possessed higher hardness and stiffness, and this was favorable for their wear resistance. They concluded that LCTs have potential to serve as protective coatings for aerospace applications.

The solution viscosity of sLCP is significantly low, so that multifunctional properties can be induced by incorporating large amounts of functional fillers and fibers into the solution. Furthermore, a low solution viscosity makes it possible to utilize not only a solvent casting method but also a spray coating method which is a great advantage in respect of saving manufacturing costs and cycle time.

Using sLCP as a matrix new coating materials with low friction and wear were developed for tribological applications. The fillers were chosen to be graphite (Gr, Type RGC A39, Superior Graphite Europe Ltd.) with particle size of 10 μm as solid lubricant, alumina nanoparticles ($Al_2O_3$, Type Aeroxide Alu C, Evonik) with primary particle size of 13 nm, and short carbon fibers (SCF, Type A385, Tenax) with a diameter of 7 μm and a length of 40 μm. The characteristic specific surface area of powdery alumina is 100 $m^2/g$.

The individual nanoparticles adhered to each other due to interactive forces and formed micro-agglomerates. Since the properties of nanocomposites may be influenced by the interface area, it is necessary to disperse and distribute individual nanoparticles homogeneously within the sLCP. This was done by applying mechanical shear forces using a dissolver and a three-roll calender dispersing device to generate a highly filled masterbatch. The masterbatch was subsequently thinned down and mixed with graphite and/or short carbon fibers.

The Brookfield-viscosity of the neat sLCP at 22 °C was only 300 mPa s, and it was only 900 mPa s in case of the highly filled composite containing 23 vol% of $Al_2O_3$ nanoparticles, Gr and SCF in total. At a temperature of 60 °C the viscosity of the composite decreased to 260 mPa s. Therefore, reproducible thin films with constant thickness could be generated according to ASTM D 823 by an automatic film applicator coater (Zehntner ZAA 2300) on various substrates, e.g. copper, aluminum and steel. After the coating process, the mixtures were dried in an oven and then polymerized at elevated temperature of 300 °C for 5 h under vacuum. The neat sLCP coatings and the sLCP composite coatings showed a good adhesion to both copper and aluminum substrate, but not to steel.

Micrographs of cross sections of fractured coatings of sLCP/SCF/Gr/$Al_2O_3$ nanocomposites confirmed that Gr (Fig. 2.6(2), item C) and SCF (Fig. 2.6(2), item B) were homogeneously distributed in the polymer. $Al_2O_3$ nanoparticles were, however, difficult to be observed because of their spherical shape and small dimension, and they could not be distinguished from the fibrillar structures of the ductile fracture of sLCP (Fig. 2.6(1)). Furthermore, it was observed that sLCP adhered well to both SCF and Gr, which was demonstrated by plastically deformed and cavitated areas in the interface regions between SCF, Gr and sLCP (Fig. 2.7, items D, E). Tiny spherical structures might represent $Al_2O_3$ nanoparticles (Fig. 2.7, item F) and/or the ends of fractured sLCP fibrils.

Tribological properties were measured by using a plate-on-ring test setup. The counterpart was made from polished 100Cr6 bearing steel with a roughness Ra of 0.2 μm. The constant normal force was set to 10 N and the velocity to 1 m/s under dry (unlubricated) running conditions. The coefficient of friction (CoF) was measured continuously over the whole running time of 4 h for each test. Six specimens were tested for each material.

The tribological behavior of sLCP composites is shown in Fig. 2.8. Data for the friction coefficient (CoF) and the specific wear rate ($w_s$) are presented. This data demonstrates the improvement of tribological properties of sLCP due to solid lubricant Gr, $Al_2O_3$ nanoparticles, and SCF. It is observed that the $Al_2O_3$ nanoparticles did not exert a strong friction or wear reducing effect. Instead, it is Gr that slightly reduced CoF and also $w_s$. The highest friction and wear reduction was indeed gained by the combination of Gr with SCF. In case of the composite materials containing Gr/SCF or Gr/SCF/$AL_2O_3$ nanoparticles (5 %) a decrease of the CoF by nearly 40 % from 0.38 to 0.23 was observed, and the $w_s$ was reduced by more than one decade from $3.95 \times 10^{-6}$ to $0.27 \times 10^{-6}$ $mm^3/Nm$. The latter could be verified for both substrate types, copper and steel. The tribological performance of sLCP/Gr/SCF/$Al_2O_3$ nanoparticles is comparable to that of high performance composite coatings based on polyamidimide (PAI), e.g. PAI/Gr/SCF (Rasheva 2014). Such PAI composites may serve e.g. as coatings on piston skirts and slide bearings used in the automotive industry. The special advantage of sLCP over PAI

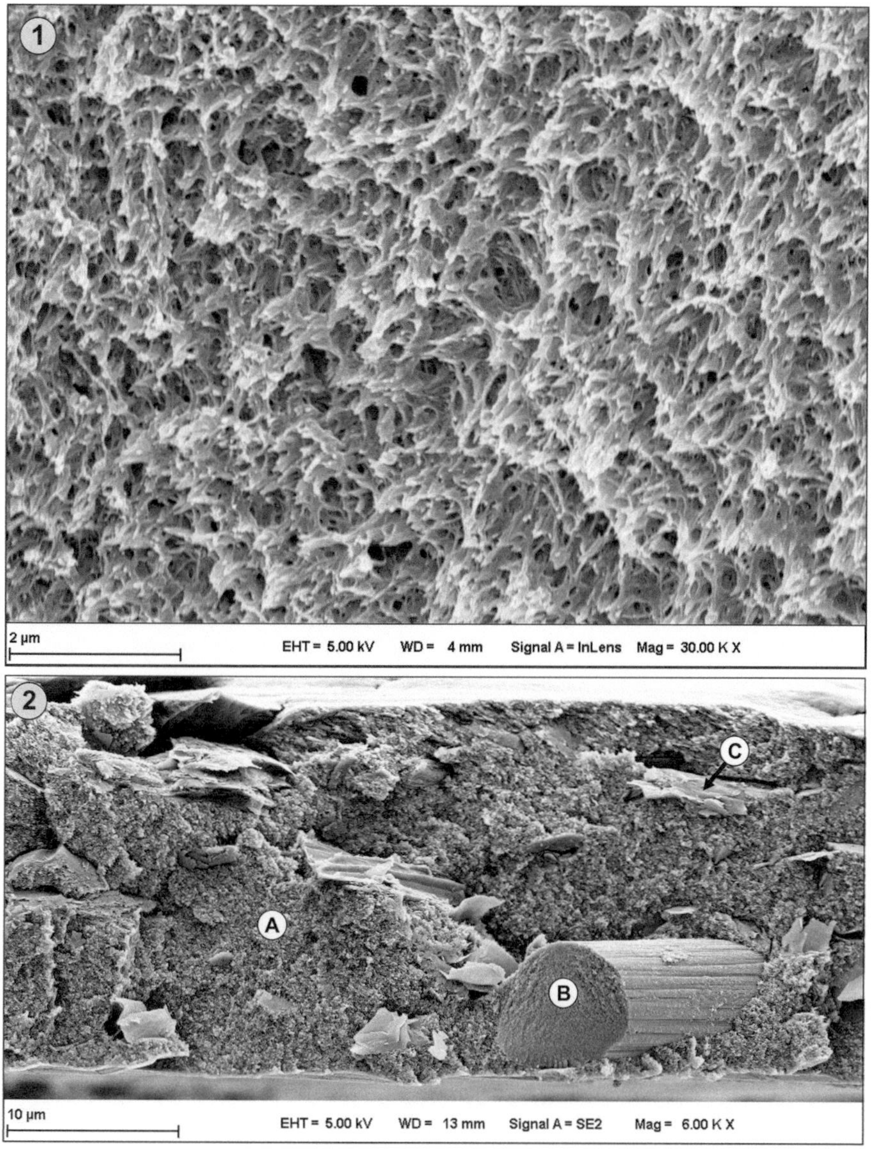


**Fig. 2.6** Micrographs of the cryo-fractured coatings' cross sections of (1) neat sLCP and (2) sLCP/Gr/SCF/$Al_2O_3$ nanocomposite with areas containing the $Al_2O_3$ nanoparticles (2, item A), the graphite (Gr) (2, item C) and the short carbon fibers SCF (2, item B)

is its low viscosity, which allows reinforcing the polymer with a large amount of fillers and especially fibers to further improve mechanical and wear performance, and even extended functionality for cutting edge performance and broad applicability in the future.

**Fig. 2.7** SCF (item B) shows good adhesion to sLCP in interface regions (item D) and also to Gr (item E). The tiny spherical structures might represent $Al_2O_3$ nanoparticles (item F) but could also be the fractured loose ends of sLCP fibrils

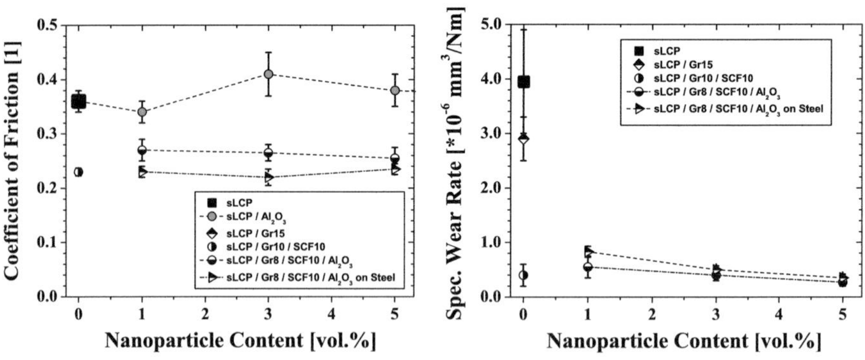


**Fig. 2.8** Coefficient of friction (CoF) and specific wear rate ($w_s$) of sLCP composites

## 2.4 Conclusions and Future Perspectives

Two kinds of LCP with naphthalene-rich structure and without the commonly used HBA monomer were reported. High performance LCP (HP-LCP) has demonstrated to possess excellent hydrolysis resistance, high strength and modulus, low gas permeability and unique dielectric properties. HP-LCP composites for injection molding are a promising candidate for structural components of transportation industries. The flexural modulus of carbon-fiber reinforced HP-LCP composite was significantly high and close to that for magnesium alloy, which might promote the replacement of metals for primary and secondary structural components. The tribological performance of HP-LCP was excellent even under high temperatures because of the synergistic effect of HP-LCP with high heat resistance and incorporated nanoparticles. Furthermore, the low melt viscosity and the relatively high melt tension qualify HP-LCP as quite suitable for film and fiber production as well as for injection molding applications. The HP-LCP film made via inflation method has biaxial orientation, resulting in the high mechanical performance in both machine direction (MD) and transversal direction (TD). Additionally, TLCP fiber was successfully developed from HP-LCP, exhibiting the tensile strength of 4200 MPa and the tensile modulus of 77 GPa.

Soluble LCP (sLCP) can be dissolved in NMP which is widely used in industries. The solvent casting film of sLCP shows isotropic nature in the in-plane direction, low gas permeability, excellent dielectric and tribological performance. The sLCP allows the coating and production of complex parts based on low solution viscosity. Additionally, high filler loading into sLCP is possible, and therefore wide opportunities exist to impart multifunctional properties to the material. Ongoing efforts focus on developing new applications for HP-LCP and sLCP. It is expected that this work strongly contributes to opening the door towards advanced applications for TLCPs.

**Acknowledgements** The authors are grateful for the support of the technical staff at IVW, and especially the fruitful discussions with Dr. Ga Zhang and the contribution of Mr. Andreas Gutmacher.

## References

Agarwal US, Mashelkar RA (1992) Diffusion of rigid rodlike molecules across interfaces: implications in welding of liquid-crystalline polymers. Macromolecules 25:6703–6704

Brolly JB, Bower DI, Ward IM (1996) Diffusion and sorption of $CO_2$ in poly(ethylene terephthalate) and poly(ethylene naphthalate). J Polym Sci: Part B Polym Phys 34:769–780

Brooks GT (1992) Crystalline polyphthalamide composition having improved properties. US Patent 5,098,940, 24 Mar 1992

Calundann GW (1979) Polyester of 6-hydroxy-2-naphthoic acid and para-hydroxy benzoic acid capable of readily undergoing melt processing. US Patent 4,161,470, 17 July 1979

Chang L, Friedrich K (2010) Enhancement effect of nanoparticles on the sliding wear of short fiber-reinforced polymer composites: a critical discussion of wear mechanisms. Tribol Int 43:2355–2364

Chang L, Zhang Z, Breidt C, Friedrich K (2005) Tribological properties of epoxy nanocomposites: I. Enhancement of the wear resistance by nano-$TiO_2$ particles. Wear 258:141–148

Chiou JP, Paul DR (1987) Gas transport in a thermotropic liquid-crystalline polyester. J Polym Sci: Part B Polym Phys 25:1699–1707

Cox MK (1987) The application of liquid crystal polymer properties. Mol Cryst Liq Cryst 153:415–422

De Candia F, Renzulli A, Vittoria V, Roviello A, Sirigu A (1990) Transport properties of a thermotropic liquid-crystalline polyester. J Polym Sci: Part B Polym Phys 28:203–211

East AJ, Charbonneau LF, Calundann GW (1982) Poly(ester-amide) capable of forming an anisotropic melt phase derived from 6-hydroxy-2-naphthoic acid, dicarboxylic acid, and aromatic monomer capable of forming an amide linkage. US Patent 4,330,457, 18 May 1982

Fellahi S, Meddad A, Fisa B, Favis BD (1995) Weldlines in injection-molded parts: a review. Adv Polym Technol 14:169–195

Guerriero G, Alderliesten R, Dingemans T, Benedictus R (2011) Thermotropic liquid crystalline polymers as protective coatings for aerospace. Prog Org Coat 70:245–251

Hu YS, Schiraldi DA, Hiltner A, Baer E (2003) Structural model for oxygen permeability of a liquid crystalline polymer. Macromolecules 36:3606–3615

Kalika DS, Yoon DY, Iannelli P, Parrish W (1991) Structural, dielectric, and rheological characterization of thermotropic liquid crystalline polyesters based on 4-hydroxybenzoic acid, 4,4′dihydroxybiphenyl, terephthalic acid, and isophtalic acid. Macromolecules 24:3413–3422

Kanehashi S, Kusakabe A, Sato S, Nagai K (2010) Analysis of permeability; solubility and diffusivity of carbon dioxide; oxygen; and nitrogen in crystalline and liquid crystalline polymers. J Membr Sci 365:40–51

Komatsu S, Maeda M (2013) Liquid crystal polyester composition. JP Patent pending

Light RR, Seymour RW (1982) Effect of sub-Tg relaxations on the gas transport properties of polyesters. Polym Eng Sci 22:857–864

Makhija S, Haider MI, Chenevey EC, Jaffe M (1997) High barrier transparent films. US Patent 5,672,426, 30 Sep 1997

Massey LK (2003) Plastics design library: Permeability properties of plastics and elastomers, a guide to packaging and barrier materials, 2nd edn. William Andrew, New York, pp 259–279

McGonigle EA, Liggat JJ, Pethrick RA, Jenkins SD, Daly JH, Hayward D (2001) Permeability of $N_2$, Ar, He, $O_2$ and $CO_2$ through biaxially oriented polyester films—dependence on free volume. Polymer 42:2413–2426

Mori D, Hirose K (2009) Recent challenges of hydrogen storage technologies for fuel cell vehicles. Int J Hydrogen Energy 34:4569

Nguyen-Chung T (2004) Flow analysis of the weld line formation during injection mold filling of thermoplastics. Rheol Acta 43:240–245

Okamoto S, Hosoda T (2006) Aromatic liquid-crystalline polyester. US Patent 7,014,921 B2, 21 Mar 2006

Okamoto S, Ohtomo S, Katagiri S (2006) Liquid-crystalline polyester solution composition. US Patent 7,022,807 B2, 4 Apr 2006

Okamoto S, Yano K, Fujiwara M, Aitsu K (2004) Method for producing liquid crystalline polyester. JP Patent 2004083778A, 28 Aug 2002

Okamoto S, Yu R, Djourelov N, Suzuki T (2005) Study on the thermal behavior of a solution-cast liquid-crystalline polymer film by positron-annihilation lifetime spectroscopy. Polymer 46:6455–6460

Osaki S (1990) Explanation of orientation patterns determined for sheet materials by means of microwaves. J Appl Phys 67:6513–6519

Osaki S (1997) A new microwave cavity resonator for determining molecular orientation and dielectric anisotropy of sheet materials. Rev Sci Instrum 68:2518–2523

Pavel D, Shanks R (2003) Molecular dynamics simulation of diffusion of $O_2$ and $CO_2$ in amorphous poly(ethylene terephthalate) and related aromatic polyesters. Polymer 44:6713–6724

Plummer CJG, Wu Y, Davies P, Zulle B, Demarmels A, Kausch HH (1993a) The short- and long-term mechanical properties of filled and unfilled thermotropic liquid crystalline polymer injection moldings. J Appl Polym Sci 48:731–740

Plummer CJG, Zulle B, Demarmels A, Kausch HH (1993b) The structure of filled and unfilled thermotropic liquid crystalline polymer injection moldings. J Appl Polym Sci 48:751–766
Rasheva Z (2014) Functional coatings for fast running machine elements with integrated tribological functionality. PhD thesis, Institut für Verbundwerkstoffe GmbH, University of Kaiserslautern, Germany
Rath T, Kumar S, Mahaling RN, Mukherjee M, Das CK, Pandey KN, Saxena AK (2007) Flexible composite of PEEK and liquid crystalline polymer in presence of polyphosphazene. J Appl Polym Sci 104:3758–3765
Rueda RD, Varkalis A (1995) Water sorption/desorption kinetics in poly(ethylene naphthalene-2,6-dicarboxylate) and poly(ethylene terephthalate). J Polym Sci: Part B Polym Phys 33:2263–2268
Tjong SC (2003) Structure, morphology, mechanical and thermal characteristics of the in situ composites based on liquid crystalline polymers and thermoplastics. Mater Sci Eng R 41:1–60
Wang L, Allan PS, Bevis MJ (1995) Enhancement of internal weld line strength in thermotropic liquid crystal polymer moldings. Plast Rubber Compos Process Appl 23:139–150
Wang Y, Xu J, Cheng SX, Pramoda KP, Chung TS, Goh SH (2003) Thin-film polymerization and characterization of Sumitomo's Sumikasuper®-type liquid crystalline polymers. Liq Cryst 30:753–764
Weinkauf DH, Paul DR (1992a) Gas transport properties of thermotropic liquid-crystalline copolyesters. I. The effects of orientation and annealing. J Polym Sci: Part B Polym Phys 30:817–835
Weinkauf DH, Paul DR (1992b) Gas transport properties of thermotropic liquid-crystalline copolyesters. 2. The effect of copolymer composition. J Polym Sci: Part B Polym Phys 30:837–849
Zhang Z, Breidt C, Chang L, Haupert F, Friedrich K (2004) Enhancement of the wear resistance of epoxy: short carbon fibre, graphite, PTFE and nano-$TiO_2$. Compos Part A 35:1385–1392
Zhang G (2010) Structure–tribological property relationship of nanoparticles and short carbon fibers reinforced PEEK hybrid composites. J Polym Sci: Part B Polym Phys 48:801–811

# Chapter 3
# Selected Mechanical Properties of Uniaxial Side Chain Liquid Crystalline Elastomers

**Philippe Martinoty**

## 3.1 Introduction and Historical Overview

Liquid Crystal elastomers have been pioneered by Finkelmann's group (Finkelmann et al. 1981). These are a particular class of materials composed of a network and liquid crystal molecules of low-molecular-weight, called mesogens. Depending on whether the mesogens are incorporated into the polymer backbone or attached to it via a flexible spacer, these elastomers are called main chain elastomers or side chain elastomers, respectively. They combine the elastic properties of usual elastomers with the thermal and orientational properties of liquid crystals. The mesogens can be either macroscopically disoriented (polydomain sample) or macroscopically oriented (monodomain sample) in the liquid crystal state. One of the most fundamentally interesting issues in this field is the question of the physical consequences of the coupling between the director, which defines the macroscopic orientation of the mesogens, and the polymer network (Warner and Terentjev 2003; Brand et al. 2006; Ohm et al. 2010). This chapter deals only on the monodomain (uniaxial) nematic side chain elastomers, or briefly NEs.

Several methods are used for preparing a NE. The usual method, initiated by Finkelmann, is based on a two-step cross-linking process, in which a liquid crystalline polymer is first slightly cross-linked, and then stretched to orient the mesogens and finally cross-linked again to chemically fix the orientation of the mesogens (Küpfer and Finkelmann 1991, 1994). Most of the physical experiments were performed with these systems. However, there are other methods that exist, which consist of a cross-link by UV irradiation a nematic side chain liquid crystal polymer (called thereafter NP) oriented by an electric or a magnetic field (Brehmer et al. 1994; Komp et al. 2005; Brömmel et al. 2013) or by a surface

P. Martinoty (✉)
Institut Charles Sadron, UPR 22, CNRS/UDS, 23 rue du Loess, BP 84047, Strasbourg 67034, France
e-mail: p.martinoty@unistra.fr; philippe.martinoty@ics-cnrs.unistra.fr



V.K. Thakur, M.R. Kessler (eds.), *Liquid Crystalline Polymers*,
DOI 10.1007/978-3-319-20270-9_3

treatment of the sample-bearing glass slides (Urayama et al. 2007). As we will see later, the elastic properties of the resulting NEs strongly depend on the preparation process.

The obtaining of NEs is particularly interesting because of the coupling between the director, which defines the macroscopic orientation of the mesogens, and the polymer network leads to unique mechanical properties, such as an elastic anisotropy, a spontaneous contraction or elongation of the NE induced by a temperature change, or the presence of an elastic plateau in the stress-strain curve when the NE is stretched perpendicularly to the initial direction of the director. The elasticity of these materials has been the subject of numerous theories, which will be discussed below.

This chapter focuses on the elastic properties of the NEs deduced from shear mechanical measurements. As the comparison between the theoretical descriptions and the experimental results have led to considerable controversies (Martinoty et al. 2004a, b, c; Terentjev and Warner 2004; Stenull and Lubensky 2004) it is useful to understand the evolution of the topic in order to make a brief historical presentation of the various static and dynamic theories—conventional linear elastic theory, soft elasticity (original version), soft elasticity (version 2), bifurcation-type theory—which were progressively introduced for describing the elastic properties of these materials.

The first approach is based on conventional linear elastic theory. It predicts that one of the five elastic coefficients of an uniaxial elastic medium, $C_5$, is renormalized by the coupling between the strain and the director. The renormalized coefficient, called $\widetilde{C}_5$ (in the literature this coefficient is sometimes called $\widetilde{C}_{44}$), is smaller than $C_5$, but nevertheless exhibits a value similar to those of the other elastic coefficients (de Gennes 1980).

The second approach is based on the soft elasticity concept that is characterized by a zero value of $\widetilde{C}_5$. This concept assumes that a NE is formed via a spontaneous symmetry breaking from an isotropic elastomer (Golubovic and Lubensky 1989). Soft elasticity was then transformed into semi-soft elasticity for taking into account that the NE is not isotropic at high temperatures, but weakly anisotropic. Semi-soft elasticity is characterized by a value of $\widetilde{C}_5$ that is non-zero but small (Verwey and Warner 1995; Warner 1999). Soft or semi-soft elasticity has crucial consequences on (a) the mechanical anisotropy that must be much larger than the conventional mechanical anisotropy, (b) the slope of the elastic plateau in the stress-strain curve associated with the NEs stretched perpendicularly to the initial direction of the director, which must be small since it is proportional to $\widetilde{C}_5$ (Warner 1999) and (c) the dynamic mechanical response for which a second plateau is expected at low frequency in addition to the usual rubbery plateau, in the case where the shear is applied in a plane containing the director (Terentjev and Warner 2001).

The third approach that we will call version 2 of soft elasticity or soft elasticity induced by stretching, differs from soft elasticity by the fact that the elastic modulus $\widetilde{C}_5$ exhibits now a classical value for the un-stretched state of the NE, but vanishes for two singular values $\lambda_1$ and $\lambda_2$ of the strain when the NE is stretched in a direction perpendicular to the initial orientation of the director (i.e. at the beginning and at the end of the director rotation) (Biggins et al. 2008). Such behavior of $\widetilde{C}_5$ can only be observed in the particular situation where the applied sinusoidal shear is parallel to the stretching direction and perpendicular to the initial orientation of the director. This behavior can be viewed as a soft mode associated with a second-order phase transition induced by an external stress. Although this behavior has nothing to do with semi-soft elasticity, it has unfortunately been also named semi-soft (Ye et al. 2007; Biggins et al. 2008). In order to avoid any confusion, we have called this approach, version 2 of soft (semi-soft) elasticity.

The last approach is a non-linear extension of conventional theory, which is called bifurcation-type model. It leads to an effective shear modulus which we will call also $\widetilde{C}_5$ in the following for convenience, and which follows the same qualitative behavior as the one predicted by version 2 of soft elasticity (Menzel et al. 2009b).

From this short presentation, it turns out that the elastic coefficient $\widetilde{C}_5$ is the crucial parameter that governs the elasticity of NEs. This coefficient can be directly measured for an un-stretched or stretched NE by applying a sinusoidal shear in a plane containing the director.

Another key issue that will be discussed in this chapter, concerns the influence of the preparation process on the nature—Gaussian or non-Gaussian—of the elasticity. This issue arises because, as mentioned above, the orientation of the NEs can be obtained either by a mechanical stretching of the network formed during the two-step cross-linking process, or by orienting a NP before the photo cross-linking. This question can be answered by doing shear measurements on both types of NEs swollen by a nematic solvent.

Most of the studies on the shear mechanical properties of the NEs have been performed by Martinoty and collaborators, using piezorheology (Gallani et al. 1996; Weilepp et al. 1999a, b; Stein et al. 2001; Zanna et al. 2002; Martinoty et al. 2004a, b; Rogez et al. 2006a, b, 2011; Rogez and Martinoty 2011). The review and the analysis of these properties constitute the major part of this chapter. Section 2 describes the piezorheometer used to do the shear experiments, and in Sect. 3 two typical examples of the NEs are studied, with their preparation and characterization. Section 4 gives a review of the various theories developed for describing the elasticity of these systems. Section 5 reports the experimental results and their detailed analysis. This section includes two paragraphs respectively dealing with the shear mechanical experiments performed on dry NEs and on NEs swollen by a nematic solvent, a paragraph dealing with the consequence of the Gaussian or non-Gaussian character of the elasticity of the network, and a paragraph dealing with the experiments performed on mechanically stretched NEs.

## 3.2 Piezoelectric Rheometer

$G$ was measured in the linear response regime with the piezo-rheometer developed over the last 10 years. This device is a plate-plate rheometer that uses piezoelectric ceramics vibrating in the shear mode to apply a very small strain $\varepsilon$ (typically $10^{-4}$) to the sample and to measure the stress $\sigma$ transmitted through the sample. The complex shear modulus $G$ of the sample is given by the stress-strain relationship $G = \sigma/\varepsilon$. One of the interesting aspects of this device is the wide frequency range available (typically from 0.1 to a few $10^3$ Hz). In practice, the sample is placed between two glass slides stuck to the emitting and the receiving ceramics, respectively. The measuring cell is placed in an oven allowing taking measurements between −60 and 150 °C with a temperature controlled to within a tenth of a degree Celsius. The NEs studied were films with a thickness varying from about 50 to about 300 μm. Their surface areas A were in-between 1 and 2 $cm^2$. To ensure thermal equilibrium during the measurements, the cooling and heating rates between two temperatures where the data were taken were 1 h (3 K $h^{-1}$), and the time required for taking a frequency spectrum at each temperature of measurement was 1 h.

The three main geometries characterizing the shear mechanical properties of the NEs have been studied. Figure 3.1a shows that there are two in-plane geometries (director n in the plane of the film) for which the shear is either parallel to the director (thereafter called parallel geometry) or perpendicular to the director (thereafter called perpendicular geometry), and a geometry for which n is normal to the plane of the film and to the shear (thereafter called homeotropic geometry or normal geometry). The parallel and homeotropic geometries correspond to the situation where the director and the shear are coupled. This coupling does not exist for the perpendicular geometry.

For the study of the reorientation transition induced by stretching the NE in a direction perpendicular to the initial orientation of the director, a specific shear cell has been built for measuring $G$ as a function of the elastomer elongation (Rogez and Martinoty 2011).

## 3.3 Materials: Preparation and Characterization

To illustrate the elastic properties of NEs, we have chosen two compounds that are representative of the NEs which have been studied. They were prepared either by the two-step cross-linking process or by UV cross-linking of an oriented NP. In this latter case, the orientation was obtained by an electric field or a magnetic field in order to obtain a macroscopic orientation of the mesogens, respectively normal (homeotropic geometry) or parallel (planar geometry) to the plane of the elastomeric film. Only the planar geometry can be obtained with the two-step cross-linking process.

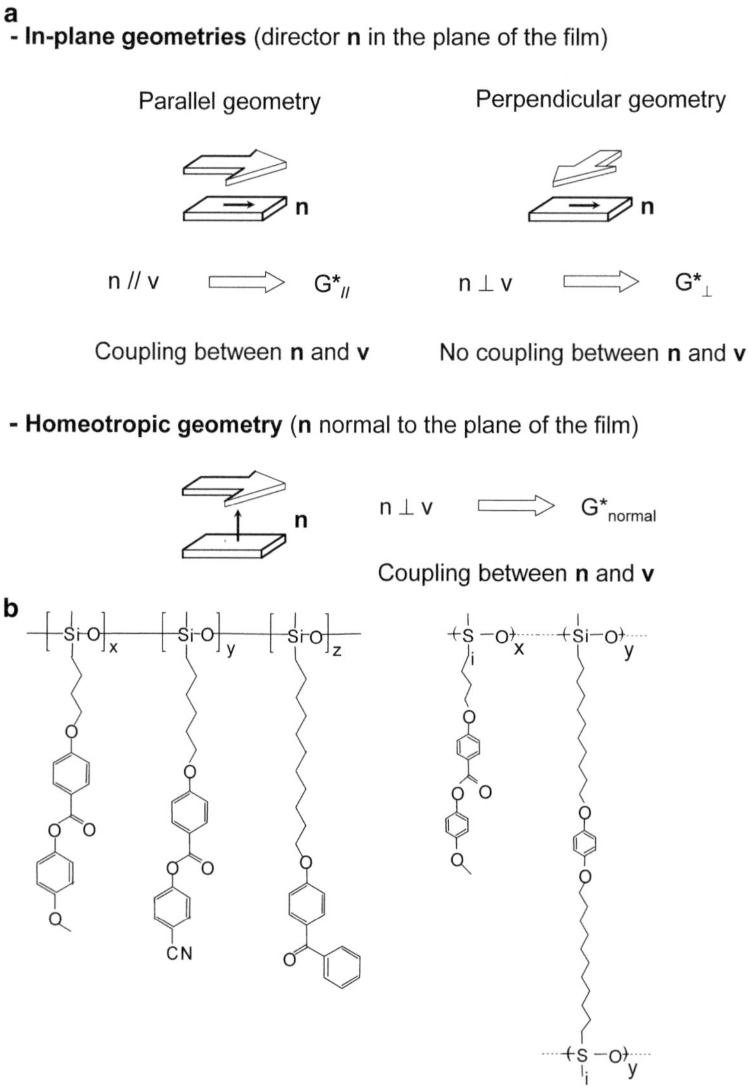


**Fig. 3.1** Panel (**a**): The three experimental geometries used for characterizing the shear mechanical properties of the NEs. The *small* and *large arrows* show the orientation of the director **n** and of the applied shear, respectively. **v** is the velocity of the shear. Panel (**b**), *left*: Chemical composition of the NE obtained by photo cross-linking a NP oriented with an E- or H-field. $x = 78.5$, $y = 18.5$, $z = 3$ are the concentrations in mol%. Panel (**b**), *right*: Chemical composition of a NE prepared with the two-step cross-linking process and oriented by a mechanical stretching of the network formed after the first cross-linking step. $x = 90$, $y = 10$ are the concentrations in mol%. Reprinted and adapted with kind permission of Springer Science + Business Media (Rogez and Martinoty 2011)

The chemical composition of the NEs prepared by UV cross-linking is shown in Figure 3.1b (left). It shows that the elastomer exhibits two types of mesogens and a photo-cross-linker. The role of the mesogen with the cyano end group is to make the orientation easier with the E-field. The concentration of the mesogens with four carbons in the spacer is 78.5 mol%, and that of the mesogens with the cyano group is 18.5 mol%. The concentration of the photo-cross-linkers is 3 mol%. The chemical composition of the NE prepared by the two-step cross-linking process with the second cross-linking step performed in the nematic phase is shown in Fig. 3.1b (right). It can be seen that the polymer backbone is also made of siloxane chains, to which the mesogens (90 mol%) and the bifunctional cross-linkers (10 mol%) are attached. This NE is supercritical (Disch et al. 1994; Martinoty et al. 2004a; Lebar et al. 2005) and does not show the typical features of the nematic-to-isotropic phase transition. Therefore, we will not use the term "phase transition" since there is only a single phase. In the same way, we will replace the transition temperature $T_{NI}$ by the temperature $T_{DSC}$ of the DSC peak that is associated with the existence of an underlying nematic-to-isotropic phase transition. In the following, the temperature range located around, above and below $\mathrm{T_{DSC}}$ will be respectively called "transition region", weakly anisotropic state and nematic phase. We will also use the symbol $T_{\mathrm{DSC}}$ instead of $T_{\mathrm{NI}}$ for the samples oriented with E- or H-fields and cross-linked by UV irradiation. The values of the glass transition temperature $T_G$ and of $T_{\mathrm{DSC}}$ are ~11 and ~96 °C for the film prepared by UV cross-linking, and ~5 and ~74 °C for the film prepared by the two-step cross-linking process. The thicknesses of the films prepared by UV cross-linking ranged from 50 to 100 μm, and those of the films prepared by the two-step cross-linking process were of the order of 300 μm. A more detailed account of the preparation and the characterization of the two films is given by Brömmel et al. (2013) for the film prepared by UV cross-linking, and by Küpfer and Finkelmann (1994) for the film prepared by the two-step cross-linking process.

Homogeneous swollen and partially swollen elastomers were prepared by immersing each dry elastomer in a solution composed of a nematic solvent and of cyclohexane. A small amount of toluene was subsequently added to the solution to progressively and non-destructively swell the sample. After waiting for a few hours, cyclohexane and toluene were evaporated. Most of the experiments were performed with the 4-n-pentyl-4′-cyanobiphenyl (5CB) as nematic solvent. Some additional experiments were performed with p-methoxybenzylidene p-butylaniline (MBBA), a non-polar nematic solvent. The weight concentration $X$ of the solvent, for example 5CB, in the swollen sample is given by

$$X = \mathrm{m(5CB)}/[\mathrm{m(5CB)} + \mathrm{m(NE)}] \tag{3.1}$$

where m(5CB) and m(NE) are the weights of 5CB and of the dry NE, respectively.

## 3.4 Theoretical Overview

### 3.4.1 *Mechanical Properties Predicted by the Conventional Linear Theory*

The existence of anisotropy in the elastic modulus $G$ has been first predicted by de Gennes who introduced in the free energy, terms coupling the director to both the network and the strain, (de Gennes 1980). When the shear is applied in a plane containing the director, the shear modulus is given by (de Gennes 1980; Terentjev and Warner 2001):

$$G'_{//} = G'_{normal} = 4\widetilde{C}_5 \tag{3.2}$$

with

$$\widetilde{C}_5 = C_5 - \frac{D_2^2}{4D_1} \tag{3.3}$$

The symbols // and normal indicate that the shear is applied in a direction parallel to the director (for the in-plane geometry) and normal to the director (for the homeotropic geometry), respectively (see Fig. 3.1a). $C_5$ is one of the five conventional elastic moduli of an uniaxial elastic medium. $D_1$ and $D_2$ are the coupling constants introduced by de Gennes (1980), which are associated with the relative rotations of the director with respect to the network and to the applied shear, respectively. Equation (3.3) shows that $C_5$ is renormalized by these couplings. The renormalized modulus $\widetilde{C}_5$ is smaller than $C_5$ because $D_1$ is always positive for energy considerations. As a result, $G'_{//}$ and $G'_{normal}$ are reduced by these couplings. When the shear is applied in a direction perpendicular to the director (for the in-plane geometry), there is no coupling between the shear and the director, and $G'_{\perp}$ therefore exhibits the conventional behavior of a rubber. As a consequence, the shear modulus exhibits an anisotropy that is governed by the value of the elastic modulus $\widetilde{C}_5$. A small anisotropy corresponds to a value of $\widetilde{C}_5$ comparable to that of the other elastic constants, and a large anisotropy to a small value of $\widetilde{C}_5$.

### 3.4.2 *Soft and Semi-soft Elasticity Concept*

Soft elasticity originates from the hypothetical spontaneous formation of an elastic solid with an internal degree of freedom from an isotropic phase. This phase transition breaks the rotational symmetry of the isotropic phase and induces a Goldstone mode characterized by the vanishing of one of the elastic modulus (Golubovic and Lubensky 1989; Warner et al 1994; Olmsted 1994). Thus, this "ideal elastic solid" has a "soft elasticity" compared to the conventional elasticity characterizing elastic solids. In the case of NEs, the vanishing elastic modulus is $\widetilde{C}_5$ (Olmsted 1994; Warner et al. 1994). However, the NEs prepared in the laboratory do not form spontaneously from an isotropic phase, because, as seen above, their formation requires some aligning field, even at high temperature in the quasi-isotropic phase. The resulting NEs are therfore weakly anisotropic at high temperature above the transition region and more anisotropic in the low temperature nematic range. Since there is no breaking of the rotational symmetry, the soft elasticity concept is not applicable. It was later replaced by the semi-soft elasticity concept for which $\widetilde{C}_5$ is small but non-zero (Verwey and Warner 1995; Warner 1999). Thus, the NEs are said "soft" or "semi-soft" depending on whether $\widetilde{C}_5$ is respectively zero or very small. Semi-soft elasticity leads to a large shear mechanical anisotropy due to the small value of $\widetilde{C}_5$.

The notion of soft (semi-soft) elasticity was also introduced to describe the flat or nearly flat plateau with a practically zero modulus observed in the stress-strain curves when the deformation is applied to the sample perpendicularly to the initial orientation of the director (Verwey et al 1996; Warner and Terentjev 2003). The nematic extension of the Gaussian rubber elasticity (the so-called neo-classical model), which will be reviewed in the next section, allows one to make the link with the previous definition of soft elasticity by demonstrating that this modulus is $\widetilde{C}_5$. This model has also been used for numerical simulations of the behavior of the elastic plateau (Conti et al. 2002).

Later on, semi-soft elasticity concept has been extended to the dynamical case. The theory, which is based on the separation of time scales between the director and the network, describes the mechanical properties of monodomain NEs in the linear response regime, when the sample is subjected to a sinusoidal shear of small amplitude (Terentjev and Warner 2001). When the shear is applied in a plane containing the director, the theory predicts the existence of a low frequency semi-soft elastic plateau, in addition to the usual rubbery plateau observed at higher frequencies.

### 3.4.3 Stress Associated with the Reorientation Transition Induced by a Strain λ Applied in a Direction Perpendicular to the Initial Orientation of the Director

For a NE stretched perpendicularly to the initial orientation of the director, the neo-classical model (Bladon et al. 1993; Verwey and Warner 1995, 1997a, b; Verwey et al. 1996; Warner and Terentjev 2003) based on Gaussian rubber elasticity predicts that the stress-strain curve exhibits three distinct regions associated with low strains, intermediate strains and high strains. The low strains region (I) is characterized by an increase of the stress with the strain $\lambda$, until $\lambda = \lambda_1$. Above $\lambda_1$, the stress increases more slowly or remains constant up to a threshold $\lambda_2$. This region of the intermediate strains (II) corresponds to the reorientation of the director. The region above $\lambda_2$ is the region of high strains (III) for which the stress increases again with the strain, as for a conventional rubber. The threshold strains $\lambda_1$ and $\lambda_2$ that mark the beginning and the end of the elastic plateau are respectively given by (Warner 1999; Warner and Terentjev 2003):

$$\lambda_1 = \frac{(r-1)^{1/3}}{(r-1-\alpha r)^{1/3}} \tag{3.4}$$

$$\lambda_2 = \sqrt{r}\lambda_1 \tag{3.5}$$

$r$ is the chain anisotropy given by $r = l_{//}/l_{\perp}$ where $l_{//}$ and $l_{\perp}$ are the step lengths parallel and perpendicular to the director, respectively, and $\alpha$ is a coefficient given by $\alpha = (\langle 1/r \rangle - 1/\langle r \rangle)$. $\alpha$ measures the degree of compositional fluctuations of the strands between cross-links (i.e. how $\langle 1/r \rangle$ differs from $1/\langle r \rangle$), and is considered as a semi-soft parameter that takes the system away from ideal softness for which $\alpha = 0$ (Warner and Terentjev 2003). Equation (3.4) shows that $\alpha$ can be deduced from the measured values of the threshold $\lambda_1$ and of the chain anisotropy $r$.

The nominal stresses associated with the low (I), intermediate (II) and high (III) strain regions are respectively given by (Warner 1999; Warner and Terentjev 2003):

$$\sigma_I = \mu\lambda\left(1 - \frac{1}{\lambda^3}\right) \tag{3.6}$$

$$\sigma_{II} = \mu\lambda\left(1 - \frac{1}{\lambda_1^3}\right) \tag{3.7}$$

$$\sigma_{III} = \mu\lambda\left(1 - \frac{1}{\lambda_1^3} + \frac{1}{\lambda_2^2\lambda_1} - \frac{\lambda_2}{\lambda_1\lambda^3}\right) \tag{3.8}$$

where $\lambda_1$ and $\lambda_2$ are the threshold strains defined above and $\mu$ is the shear elastic modulus that sets the energy scale for the NE ($\mu \sim nk_BT$, with $n$ the number of cross-links between chains, and $k_B$ the Boltzmann constant). $\mu$ is given by the slope of the stress-strain curve in the initial low-strain regime. There are therefore no adjustable parameters when comparing the theoretical stress-strain curve to the measured one in the intermediate and high-strain regimes.

The slope $p$ of the elastic plateau can be deduced from (3.7), and is given by

$$p = \mu\left(1 - \frac{1}{\lambda_1^3}\right) \tag{3.9}$$

$\mu$ is related to the shear modulus $\widetilde{C}_5$ (Warner 1999; Warner and Terentjev 2003). If one uses (3.11) (see next section) with $\lambda = 1$, $p$ can be rewritten as:

$$p = \frac{\widetilde{C}_5}{\lambda_1^3}\left(\frac{r\lambda_1^3 - 1}{r}\right) \tag{3.10}$$

which shows that $p$ is proportional to $\widetilde{C}_5$. It is therefore possible to check whether the value of $\widetilde{C}_5$ deduced from the stress-strain experiments is consistent with the value directly measured by the shear measurements.

### *3.4.4 Models Describing the Behavior of the Shear Modulus $\widetilde{C}_5$ for the Reorientation Transition Induced by a Strain $\lambda$ Applied in a Direction Perpendicular to the Initial Orientation of the Director*

Recently, several models have been developed to describe the mechanical response of NEs when they are stretched in a direction perpendicular to the initial orientation of the director. These models include an extension of the neoclassical model (Biggins et al. 2008), and the biaxial model that is equivalent to the one describing a nematic liquid-crystal in orthogonal electric and magnetic fields (Ye et al. 2007). Both focus on the behavior of the shear modulus $\widetilde{C}_5$ and both predict that $\widetilde{C}_5$, which is different from zero for the un-stretched state of the NE, decreases when the elongation $\lambda$ increases up to $\lambda_1$ and becomes zero for $\lambda = \lambda_1$ and for $\lambda = \lambda_2$ (i.e. at the beginning and at the end of the elastic plateau). Such behavior of $\widetilde{C}_5$ can only be observed in the particular situation where the applied sinusoidal shear is parallel to the stretching direction and perpendicular to the initial orientation of the director.

The elasticity involved in Ye et al. (2007) and in Biggins et al. (2008) differs from semi-soft elasticity by the fact that the elastic modulus $\widetilde{C}_5$ exhibits now a classical value for the un-stretched state of the NE, and vanishes for only the two singular values $\lambda_1$ and $\lambda_2$ of the strain when the NE is stretched in a direction perpendicular to the initial orientation of the director. This behavior that can be viewed as a soft mode associated with a second-order phase transition induced by external stress has been named semi-soft (Ye et al. 2007; Biggins et al. 2008). However, this term is misleading, because it suggests that the physical underlying phenomenon is the same as the one predicted by semi-soft elasticity theory, which is clearly not the case.

According to the extension of the neo-classical model (Biggins et al. 2008), the shear modulus $\widetilde{C}_5$, which is different from zero when $\lambda = 1$ (un-stretched sample), decreases when $\lambda$ increases and becomes zero when $\lambda = \lambda_1$. For $\lambda < \lambda_1$, the variation of $\widetilde{C}_5$ is given by:

$$\widetilde{C}_5(\lambda) = \frac{\mu r}{\lambda}\left(\frac{\lambda^3 - \lambda_1^3}{\lambda^3 - r\lambda_1^3}\right) \tag{3.11}$$

The parameters $r$, $\lambda_1$ and $\mu$ were previously defined. $\lambda_1$ and $\mu$ can be determined from the stress-strain measurements, and $r$ from the thermo-elastic measurements.

The behavior of $\widetilde{C}_5$ has also been calculated for the intermediate-strains region ($\lambda_1 \leq \lambda \leq \lambda_2$) and for the high-strains region ($\lambda > \lambda_2$), and is given for each region by the following equations:

- for $\lambda_1 \leq \lambda \leq \lambda_2$

$$\widetilde{C}_5(\lambda) = \frac{4\mu r\left[\beta^2(1+r) - \beta^4 - r\right]}{\lambda\beta^3(r-1)^2} \tag{3.12}$$

where $\beta = \lambda/\lambda_1$

- for $\lambda > \lambda_2$

$$\widetilde{C}_5(\lambda) = \frac{\mu}{\sqrt{r}\lambda}\left(\frac{\lambda^3 - r^{3/2}\lambda_1^3}{\lambda^3 - \sqrt{r}\lambda_1^3}\right) \tag{3.13}$$

The experimental behavior of $\widetilde{C}_5$ as a function of $\lambda$, can be compared to that theoretically predicted, since all the parameters $\mu$, $r$, $\lambda_1$ in (3.11)–(3.13) are experimentally determined. A similar behavior of $\widetilde{C}_5(\lambda)$ is predicted by the biaxial model (Ye et al. 2007).

### 3.4.5 Behavior of the Stress-Strain Relationship and of the Effective Shear Modulus Predicted by the Bifurcation-Type Model for the Reorientation Transition Induced by a Strain λ Applied in a Direction Perpendicular to the Initial Orientation of the Director

The bifurcation-type model (Menzel et al. 2009b) is based on a non-linear extension of the conventional elasticity theory, and does not involve any notion of soft or semi-soft elasticity. It also predicts that the director starts to reorient above the critical strain $\lambda_1$ occurring in the stress-strain curve when the sample is stretched in a direction perpendicular to the initial orientation of the director. At $\lambda_1$, the reorientability of the director diverges allowing critical fluctuations to develop, as for a second-order phase transition, leading to a soft mode. The stress-strain curve exhibits a plateau between the critical strains $\lambda_1$ and $\lambda_2$ associated with the reorientation of the director (Menzel et al. 2009a). Above $\lambda_2$, there is no reorientation of the director anymore, and only the network contributes to the stress-strain curve. The curve giving the variation of the effective shear modulus, which we also call $\widetilde{C}_5$ for convenience, as a function of $\lambda$ exhibits a behavior with a vanishing at $\lambda_1$ and $\lambda_2$ that is similar to that predicted by the extension of the neoclassical model or by the biaxial model. The bifurcation-type model also studies the influence of an oblique stretching, and shows that a slightly oblique stretching dramatically changes the $\widetilde{C}_5$ behavior, which becomes non-zero at $\lambda_1$.

## 3.5 Behavior of the Shear Moduli of the Dry and Swollen NEs for the Planar and the Homeotropic Geometries

### 3.5.1 *Shear Mechanical Experiments Performed on Dry NEs*

In this section we present and discuss the results of shear mechanical experiments performed on the dry NEs associated with the two preparation methods.

Figure 3.2a gives an example of the behavior of $G'_{//}$ and $G'_{\perp}$ observed at 1 Hz as a function of temperature, for a NE prepared by the usual two-step cross-linking process and oriented by stretching the network formed after the first cross-linking step. The data have been normalized in the weakly anisotropic state. It can be seen that the response of the NE, which is isotropic at high temperature, becomes anisotropic in the transition region around $T_{DSC}$, when the temperature is decreasing. A similar behavior was observed by Clarke et al. (2001)). The mechanical anisotropy comes from the decrease in $G'_{//}$, which is associated with the coupling between the shear and the director, and from the regular behavior of $G'_{\perp}$, which is due to the absence of this coupling (see (3.2) and (3.3)). The strong increase of $G'_{//}$ and $G'_{\perp}$, which occurs for the lowest temperatures, reflects the influence of the dynamics of the glass transition, which is not the same for the two geometries.

Figure 3.2b shows the results obtained for a NE prepared by photo cross-linking a NP oriented by a magnetic ($G'_{//}$, $G'_{\perp}$) or an electric field ($G'_{normal}$). Contrary to the case of the NE prepared by the two-step cross-linking process, this preparation method also allows one to obtain a NE with a director aligned perpendicular to the plane of the film, and therefore to measure the shear modulus $G'_{normal}$. Like for the previous NE, the data have been taken at 1 Hz and normalized at high temperature. It can be seen that the temperature dependence of $G'_{normal}$ shows a slight decrease around the transition region when entering the nematic phase. Like for $G'_{//}$, this effect is due to the coupling between the shear and the director. The comparison between the behaviors of $G'_{normal}$ and $G'_{//}$ shows that the relationship $G'_{//} = G'_{normal}$ predicted by the theory (Terentjev and Warner 2001) is verified in the temperature range studied. The data also show that $G'_{\perp}$ becomes smaller than $G'_{//}$ below 70 °C. This inversion of the anisotropy comes from the dynamics of the glass transition, which is different for the two geometries.

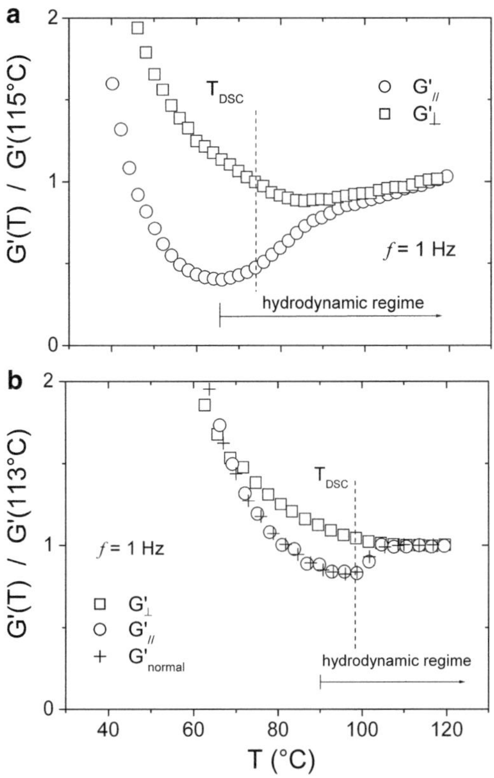


**Fig. 3.2** Panel (**a**): Temperature variation of the shear moduli $G'_{//}$ and $G'_{\perp}$ measured at 1 Hz for the NE prepared with the two-step cross-linking process and oriented by a mechanical stretching of the network formed after the first cross-linking step. The data have been normalized in the high temperature phase at $T = 115$ °C. $T_{\mathrm{DSC}}$ is the temperature determined by DSC. Panel (**b**): Temperature variation of the three main shear moduli $G'_{//}$, $G'_{\perp}$ and $G'_{normal}$ measured at 1 Hz for the NEs obtained by photo cross-linking a NP oriented with a H-field (*open circle, open square*) or an E-field (*plus*). The data have been normalized in the high temperature phase at $T = 113$ °C. $T_{\mathrm{DSC}}$ is the temperature determined by DSC. For the meaning of the symbols //, ⊥ and normal, see Fig. 3.1a. Reprinted and adapted with kind permission of Springer Science + Business Media (Rogez and Martinoty 2011)

The comparison between the results of Fig. 3.2a, b shows that the shear anisotropy is much smaller for the NE prepared by photo cross-linking a NP oriented by a H-field or an E-field, than for the NE prepared with the two-step cross-linking process. The mechanical anisotropy of the network is therefore strongly influenced by the preparation process used. The results also show that the temperature range where the shear anisotropy persists above $T_{DSC}$ is much smaller for the NE prepared by photo cross-linking than for the NE prepared with the mechanical stretching. Given that the latter is supercritical (Disch et al. 1994; Martinoty et al. 2004a, Lebar et al. 2005) with strongly elongated polymer chains, this observation suggests that the NEs oriented by E- or H-fields are subcritical with polymer chains that are only slightly elongated. The rigidity resulting from the elongation of the polymer chains can explain the larger anisotropy of the NE prepared by stretching.

Finally, for the two types of NEs, the value of the elastic modulus $\widetilde{C}_5$ deduced from the measurements of $G'_{//}$ or $G'_{normal}$ in the temperature range located slightly below $T_{DSC}$, where the mechanical response corresponds to the hydrodynamic regime, remains comparable to the value of the elastic modulus measured above the transition region. This shows that these NEs are not semi-soft (see Sect. 4.2).

The next question is to clarify whether the difference between the mechanical anisotropies of the two NEs reflects the anisotropy of the polymer chains. This latter anisotropy is experimentally obtained by measuring the spontaneous change of the film length $L$ in the direction parallel to the director. It is given by $r = (L/L_0)^3$ where the length $L$ is normalized with respect to the length $L_0$ measured at high temperature (Warner and Terentjev 1996). Figure 3.3 shows the variation of $r$ as a function of temperature for the films corresponding to an in-plane orientation of the director. The data shows that $r$ is much smaller for the film prepared by photo cross-linking than for the film prepared by the two-step cross-linking process, as in the case of the shear mechanical anisotropy. These similar behaviors show that the elasticity of the polymer chains contributes to the elasticity of the network. The more stretched the polymer chains are, the larger the chain anisotropy and the shear mechanical anisotropy. It should be noted that the relationship between $r$ and the spontaneous change in the film length $L$ refers to Gaussian elasticity. If the elasticity is non-Gaussian, as it is the case for the film prepared with the two-step cross-linking process, the relation $r = (L/L_0)^3$ probably does not rigorously apply, but is expected to qualitatively describe the thermal variation of $r$.

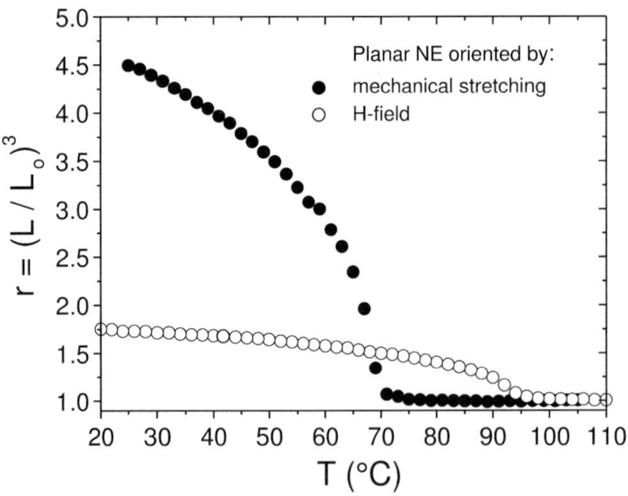


**Fig. 3.3** Temperature variation of the chain anisotropy $r$ deduced from the thermoelastic measurements. $L$ is the length of the films in the direction of the nematic director, normalized with respect to the length $L_0$ measured at high temperature in the weakly anisotropic state. The data for the NE obtained by photo cross-linking a NP oriented by an H-field were taken by F. Brömmel (unpublished result)

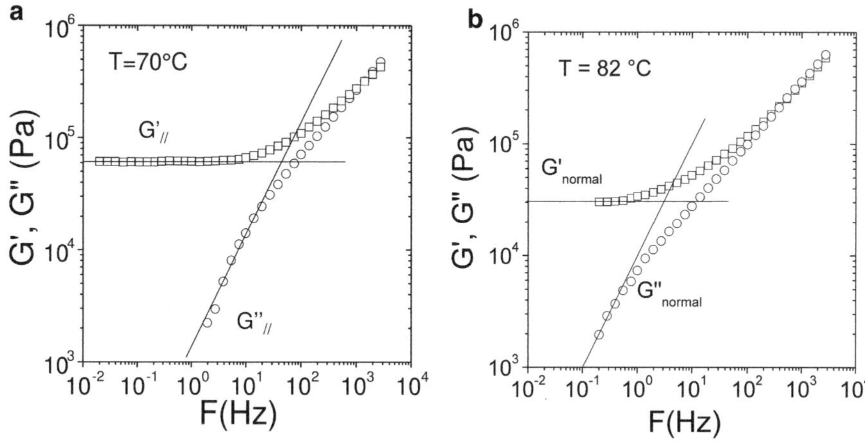


**Fig. 3.4** Typical behavior of $G'$ and $G''$ measured as a function of frequency in the nematic phase of a NE prepared by the two-step cross-linking process (Panel (**a**)) and by UV cross-linking process (Panel (**b**)). The *straight lines* indicate the hydrodynamic behaviors of $G'$ and $G''$. The subscripts // and normal are defined in Fig. 3.1a. Reprinted and adapted with kind permission of Springer Science + Business Media (Rogez et al. 2006a; Rogez and Martinoty 2011)

We close this section by briefly discussing the response of the NEs as a function of frequency. The experiments performed on both types of NEs between 0.2 Hz and a few $10^3$ Hz show that the frequency response is composed of two parts: a low frequency part, which corresponds to the hydrodynamic behavior for which $G'$ is frequency independent and $G'' \sim f$, and a high frequency viscoelastic part characterized by a power law $G' \sim G'' \sim f^n$. The value 0.5 of the exponent $n$ indicates that the high frequency response is Rouse-type (Rogez and Martinoty 2011; Rogez et al. 2006a; Zanna et al. 2002). A typical example of the response is illustrated by Fig. 3.4a for a NE prepared by the two-step cross-linking process and by Fig. 3.4b for a NE prepared by UV cross-linking. Increasing the temperature increases the frequency range where the response is hydrodynamic. Decreasing the temperature decreases the frequency range where the response is viscoelastic. A modified Rouse model taking explicitly into account the chain structure of the network has been developed and is in good agreement with the experimental data (Toshchevikov and Gotlib 2009). Finally, it should be noted that the data show that the semi-soft elastic plateau expected at low frequency is not observed, and that there is no separation between the time scales of the director and the network (Martinoty et al. 2004a).

### *3.5.2 Shear Mechanical Experiments Performed on NEs Swollen by a Nematic Solvent*

The large difference in the shear mechanical anisotropy and in the chain anisotropy between the two types of NE suggests that the nature of their elasticity is not the same. Since Gaussian elasticity is characterized by a decrease of $G'$ when the sample is swollen (Treloar 1949), the answer to this question can be directly obtained by measuring the complex shear modulus as a function of the swelling degree of the NEs.

#### 3.5.2.1 Influence of Swelling on the Mechanical Properties of NEs Prepared by the Two-Step Cross-linking Process

As previously noted, these dry NEs are characterized by an in-plane orientation of the director. Birefringence measurements and observations between crossed polarizers show that the dry and the swollen elastomers are nematic in the whole temperature range investigated (i.e. between 25 and 120 °C). The swelling of the dry elastomer with the nematic solvent has several effects such as:

- a change in the response of the dry NE from viscoelastic to hydrodynamic, resulting from a shift of the glass transition to the low temperature range
- a shift of the internal modes (Rouse-type modes) of the dry NE to higher frequencies due to a higher mobility of the mesogens
- a strong reduction of the coupling between the director and the shear, as shown by the disappearance of the dip observed in the case of the parallel geometry of the dry NE, which is precisely associated with this coupling.

However, the most important point concerns the influence of the weight concentration $X$ of the solvent on the mechanical response, which is shown on Fig. 3.5a.

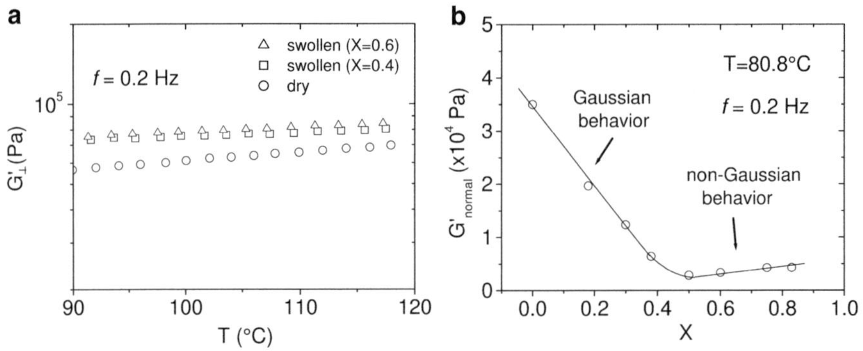


**Fig. 3.5** Panel (**a**): Variation of $G'_{\perp}$ as a function of temperature for different values of the weight concentration $X$ of 5CB for the NE prepared by the two-step cross-linking process. The increase of $G'_{\perp}$ with $X$ indicates that the elasticity of the network is non-Gaussian. Panel (**b**): Variation of $G'_{normal}$ as a function of the weight concentration $X$ of 5CB for the homeotropic NE prepared by a UV cross-linking of a NP oriented by an E-field. The decrease in $G'_{normal}$ observed for $X < 0.5$ is the behavior expected for Gaussian elasticity. The increase of $G'_{normal}$ observed for the higher values of $X$ reflects the additional rigidity stemming from the elongation of the polymer chains. The line is a guide for the eyes. $G'_{\perp}$ and $G'_{normal}$ are defined in Fig. 3.1a. Reprinted and adapted with kind permission of Wiley-VCH (Rogez et al. 2011)

The data were taken for the perpendicular geometry for which there is no coupling between the director and the shear. The solvent is 5CB. The key result is that $G'$ increases when $X$ increases. This behavior is opposite to the one predicted by the Gaussian rubber elasticity, which indicates that the elasticity of the NE is non-Gaussian. It can be seen that the increase of $G'$ is small, which suggests that the polymer chains are already considerably elongated by the mechanical stretching used to orient the mesogens. The data obtained for the parallel geometry also shows that $G'$ increases when $X$ increases, so confirming that the elasticity of the network is non-Gaussian in the temperature range studied.

#### 3.5.2.2 Influence of Swelling on the Mechanical Properties of NEs Prepared by Photo Cross-linking a NP Oriented Either by an Electric or a Magnetic Field

The swelling of these dry NEs has several effects, which are similar to those observed for the NEs prepared with the two-step cross-linking process, and which have been recalled above: a change in the frequency response from viscoelastic to hydrodynamic in the low temperature range, a shift of the Rouse modes towards higher frequencies and a reduction of the coupling between the director and the shear. In contrast, the behavior of $G'$ as a function of the weight concentration $X$ of the solvent is completely different. This is illustrated by Fig. 3.5b which shows the data taken at $T = 80.8$ °C for the homeotropic geometry with 5CB as solvent. It can be seen that $G'$ strongly decreases when $X$ increases, and then increases when $X$ becomes higher than ~0.5. The decrease in $G'$ when $X$ increases is the behavior expected for Gaussian elasticity (Treloar 1949). It can be considered as a dilution effect leading to a reduction of the cross-linking density per unit volume. The slight increase in $G'$ observed for $X > 0.5$, reflects the additional rigidity of the NE resulting in the elongation of the polymer chains. Similar behaviors were also observed for the NE with an in-plane orientation of the director, or when the NEs have been swollen with MBBA.

The comparison between the shapes of the swollen and the dry NE shows that the thickness ratio, the length ratio and the width ratio are roughly the same. This shows that the NE swells in the three dimensions. This 3D swelling is in contrast to the 2D swelling observed with the NEs prepared with the two-step cross-linking process, which do not swell in the direction parallel to the director because of the presence of a smectic-like ordering induced during to the stretching of the network (Yusuf et al. 2004). The difference in the swelling behavior between the two types of NEs shows again that the conformation of the polymer chains strongly depends on the preparation process.

One of the key results of this study is that the elasticity of the network is non-Gaussian for the NEs prepared by the two-step cross-linking process, which are the most studied NEs to date, and Gaussian for the NEs prepared by photo cross-linking an oriented NP. The Gaussian or non-Gaussian character of the elasticity is associated with the stretching of the polymer chains. For the NEs oriented before cross-linking, the elongation of the chains is small because it is obtained by orienting the mesogens in the nematic phase at thermodynamic equilibrium. For the NEs oriented by stretching the network, the elongation of the chains is large because the orientation of the mesogens has been obtained by stretching the network beyond its equilibrium state, which also explains why the resulting NE is supercritical.

A major consequence of the Gaussian versus non-Gaussian elasticity will be illustrated in the next section by analyzing some typical stress-strain curves with the neo-classical model.

### 3.5.3 Consequence of the Gaussian or Non-Gaussian Character of the Elasticity on the Stress-Strain Curve Associated with the Reorientation of the Director Induced by Stretching a NE in a Direction Perpendicular to the Initial Orientation of the Director

First, we recall that the various stress-strain experiments, which have been performed for the two types of NEs, revealed that the stress-strain curve is composed of three regimes separated by two threshold strains $\lambda_1$ and $\lambda_2$. The initial regime is characterized by an increase of the stress with the imposed deformation $\lambda$ until $\lambda = \lambda_1$. This is the classical rubber response. Above $\lambda_1$, the stress increases more slowly or remains constant up $\lambda_2$; this is the region, often called the elastic plateau region, associated with the reorientation of the director. Above $\lambda_{2,}$ the stress increases again as $\lambda$ increases further. This final region corresponds again to the characteristic response of usual rubbers. In this region the director rotation is completed; if the director was initially perpendicular to the stretching direction it is now parallel to the stretching direction.

In the following, typical examples of stress-strain curve corresponding to the two types of NEs will be analysed with the neo-classical model. Equations (3.6)–(3.8) of this model show that the three different regions characterizing the stress-strain curve are only governed by three parameters: the shear modulus $\mu$, the threshold strain $\lambda_1$ and the chain anisotropy $r$. For all the analyses presented below, the $\mu$-value was determined by fitting the data taken in the low strain regime (region I) with (3.6), and the $\lambda_1$-value determined from the intercept of this curve with that of the elastic plateau. The stress-strain behavior for regions II and III was not fitted, but calculated with these values. The value of $\lambda_2$ was calculated with the value of $r$ deduced from the thermo-elastic measurements.

#### 3.5.3.1 Analysis of the Stress-Strain Curves Associated with NEs Prepared by Photo Cross-Linking a NP Oriented by an E-field or a Surface Treatment of the Sample-Bearing Glass Slides

Figure 3.6a shows the stress-strain curve of a NE prepared by photo cross-linking a NP oriented in the homeotropic configuration by a surface treatment of the sample-bearing glass slides (Urayama et al. 2007). The stress is the nominal stress and the elongation $\lambda$ is defined by $\lambda = L/L_0$, where $L$ and $L_0$ are the lengths of the sample in the stretched and the un-stretched states, respectively. Infrared spectroscopy measurements performed during elongation have shown that the elastic plateau is associated with the reorientation of the director and that no further reorientation occurs above $\lambda_2$ (Urayama et al. 2007). The solid lines are the theoretical curves calculated using (3.6)–(3.8). $\lambda_1$ is determined by the intercept of the straight lines characterizing the small and the intermediate strains regions, $\mu$ from the fit of the data belonging to the small strains region ($\mu = 1.2\ 10^5$ Pa) and $\lambda_2$ from the relationship $\lambda_2 = \sqrt{r}\lambda_1$ with $r = 1.46$ (Urayama et al. 2007). We underline again that the theoretical curves of regions 2 and 3 are not fits, but curves calculated with the known parameters $r$, and $\lambda_1$. It can also be seen that the calculated value of $\lambda_2$, which marks the end of the elastic plateau, is in perfect agreement with its experimental determination. The excellent agreement between the data and the theoretical curves derived from the neo-classical model confirms that the elasticity of the network is Gaussian. The figure also reveals a deviation of the data from the theoretical curve for $\lambda > 1.5$, which suggests that the elasticity becomes no longer Gaussian above this value of $\lambda$.

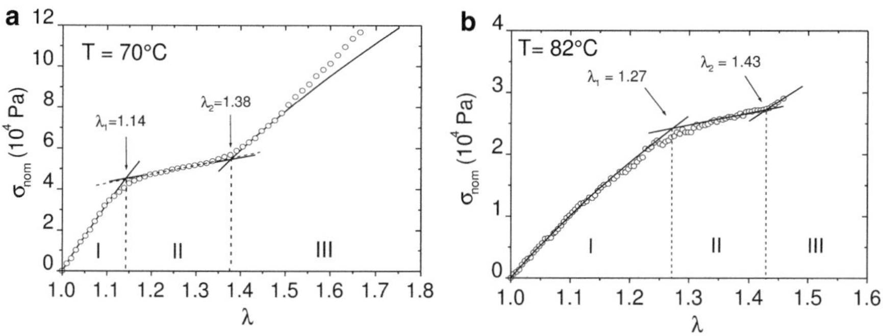


**Fig. 3.6** Panel (**a**): Stress-strain curve of a homeotropic NE oriented by a surface treatment of the sample-bearing glass slides (Urayama et al. 2007). The data have been extracted from this reference. The stress is the nominal stress and the extension ratio $\lambda$ is defined by $\lambda = L/L_0$, where $L$ and $L_0$ are the lengths of the NE for the stretched and un-stretched states. Panel (**b**): Stress-strain curve of the homeotropic NE oriented by an E-field (Rogez et al. 2011). For both samples, the *solid lines* for regimes II and III are not fitted curves but theoretical curves calculated from (3.6) to (3.8) derived from the neo-classical model, using the known values of parameters $\lambda_1$ and $\mu$ (see text). The excellent agreement between the experimental and the calculated curves shows that the elasticity of the network is Gaussian. Reprinted and adapted with kind permission of Springer Science + Business Media (Rogez and Martinoty 2011)

Figure 3.6b shows that a similar agreement between the experimental and the theoretical stress-strain curves is obtained for the NE prepared by photo cross-linking a NP oriented in the homeotropic configuration by an E-field (Rogez and Martinoty 2011). Furthermore, the value of $\widetilde{C}_5$, which is deduced from the slope of the elastic plateau (3.10), is consistent with the value deduced from the $G'$ measurement performed in the hydrodynamic regime of the un-stretched sample, as predicted by the neo-classical model.

Thus, these analyses with the neo-classical model of the stress-strain curves of the NEs homeotropically oriented by an E-field or by a surface treatment of the sample-bearing glass slides confirm that the elasticity of these NEs is Gaussian.

#### 3.5.3.2 Analysis of the Stress Strain Curves Associated with NEs prepared by the Two-Step Cross-Linking Process and Oriented by Stretching the Network Obtained After the First Cross-linking Step

Figure 3.7 shows two typical stress-strain curves of NEs with an in-plane orientation of the director, prepared with the two-step cross-linking process and cross-linked with bi-functional cross-linkers. Both NEs have the same chemical composition and the same cross-linking density, but the second cross-linking was performed either well below $T_{DSC}$ in the nematic phase (Fig. 3.7a), or above $T_{DSC}$ in the weakly anisotropic state (Fig. 3.7b). The solid lines are the theoretical curves calculated with (3.6) and (3.7) derived from the neo-classical model. As for Fig. 3.6, $\lambda_1$ is given by the interception point between the solid line characterizing the small strains region (region I) and the dashed line characterizing the intermediate strains region (region II), and $\mu$ is deduced from the fit of the data belonging to the small strains region ($\mu = 7 \times 10^4$ Pa for curve a, $\mu = 1.4 \times 10^4$ Pa for curve b). For both samples, it can be seen that there is no agreement between the experimental and the theoretical stress-strain curves in region II, which confirms the non-Gaussian character of the elasticity of these NEs. The data also show that the slope of the measured elastic plateau is smaller for the NE cross-linked above the transition region than for the NE cross-linked below the transition region. We will discuss this point at the end of this section.

Another type of NEs has been recently studied (Petelin and Copic 2009, 2010). They were also prepared with the two-step cross-linking process with the second cross-linking step performed either above the transition region in the weakly anisotropic state, or below the transition region in the nematic phase, but they differ from the previous ones by the fact that they were prepared with tri-functional cross-linkers instead of bi-functional cross-linkers. For the NEs prepared above the transition region, the neo-classical model does not accurately describe part II of the stress-curve curve and does not give the value of $\lambda_2$, as for the NEs with the bi-functional cross-linkers prepared with the two-step cross-linking process, suggesting that the elasticity of this NE is non-Gaussian (Rogez and Martinoty

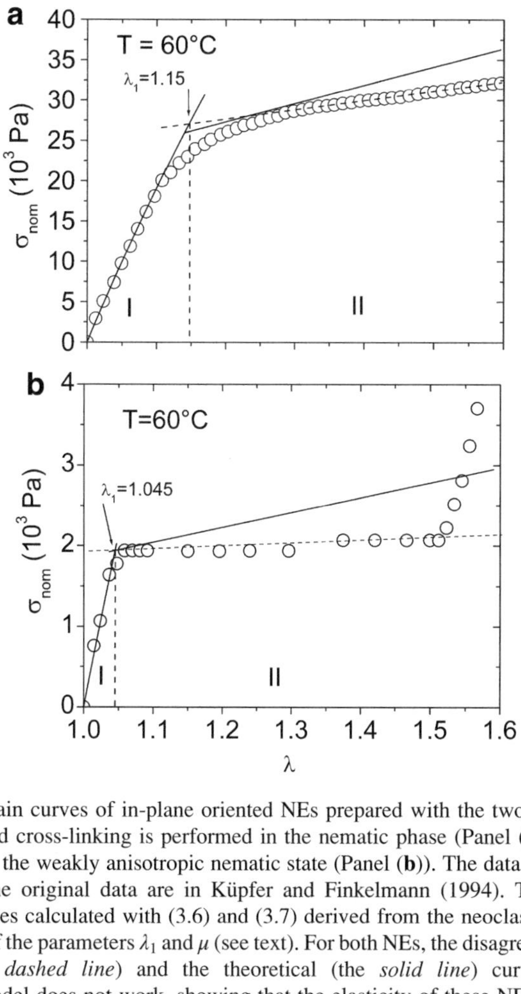


**Fig. 3.7** Stress-strain curves of in-plane oriented NEs prepared with the two-step cross-linking process. The second cross-linking is performed in the nematic phase (Panel (**a**)), and above the transition region in the weakly anisotropic nematic state (Panel (**b**)). The data are extracted from Warner (1999). The original data are in Küpfer and Finkelmann (1994). The *solid lines* are the theoretical curves calculated with (3.6) and (3.7) derived from the neoclassical model, using the known values of the parameters $\lambda_1$ and $\mu$ (see text). For both NEs, the disagreement between the experimental (the *dashed line*) and the theoretical (the *solid line*) curves indicates that the neoclassical model does not work, showing that the elasticity of these NEs is non-Gaussian. Reprinted with kind permission of Springer Science + Business Media (Rogez and Martinoty 2011)

2011). The situation is more complicated for the NEs prepared in the nematic phase, because the analysis with the neo-classical model suggests the existence of a cross-over between Gaussian and non-Gaussian elasticity occurring by decreasing the temperature for a value of r slightly higher than 2 (Rogez and Martinoty 2011).

We close this section by discussing the origin of the elastic plateau. The experimental data of Fig.3.7 show that the slope of the elastic plateau is smaller for the NEs twice cross-linked above the transition region in the weakly anisotropic state, than the slope of the NEs with the second cross-linkage performed below the transition region in the nematic phase. Since, according to the argument of Golubovic and Lubensky (1989), the cross-link in the weakly anisotropic state is closer of the conditions for ideal softness, the reduction of the slope value has been interpreted as evidence for soft or semi-softness elasticity (Verwey et al. 1996; Warner and Terentjev 2003). However, this interpretation is questioned by the experiments of Petelin and Copic discussed above, which show that in the nematic phase the value of the slope of the elastic plateau of the NE twice cross-linked in the weakly anisotropic state is higher or similar to that of the NE cross-linked in the nematic phase (for example, at $T = 70$ °C, $p \sim 5.8\ 10^3$ Pa for the NE prepared in the weakly anisotropic state and $p \sim 3.6\ 10^3$ Pa for the NE prepared in the nematic phase). Furthermore, the value of the parameter $\alpha$ which, according to the theory of soft or semi-soft elasticity, measures the degree of softness must be much smaller for the NE prepared in the weakly anisotropic state than for the NE prepared in the nematic phase, but the results show that the values of $\alpha$ are close (for example, at $T = 70$ °C, $\alpha \sim 6.10^{-2}$ for the NE prepared in the weakly anisotropic state and $\alpha \sim 8.10^{-2}$ for the NE prepared in the nematic phase). Finally, the behaviors of $p$ and $\alpha$ between the NEs prepared in the weakly anisotropic state and in the nematic phase, which should be similar in the framework of soft elasticity, is not observed, since, as shown above for $T = 70$ °C, $p$ is smaller for the NE prepared in the nematic phase than for the NE prepared in the weakly anisotropic phase, whereas $\alpha$ shows the opposite behavior. The slope of the elastic plateau is therefore not governed by the formation history, and $\alpha$ is not a parameter measuring the deviation from ideal softness but rather a parameter measuring the internal aligning field.

### 3.5.4 Shear Mechanical Experiments Performed on a NE Mechanically Stretched in a Direction Perpendicular to the Initial Orientation of the Director

These experiments were performed to test version 2 of the soft elasticity theory, which predicts that $G'$ should be zero at the thresholds $\lambda_1$ and $\lambda_2$ marking the beginning and the end of the elastic plateau of the stress-strain curve. This behavior is expected only when the applied shear is parallel to the stretching direction and perpendicular to the initial orientation of the director (Biggins et al. 2008). The studied NEs were films prepared by photo cross-linking a NP oriented in the homeotropic configuration. The experiments were performed in the following way. The NE was first stretched with a small micrometric device embedded in the piezorheometer cell, without any contact with the plates of the cell. Then the plates of the cell were gently close together, and the $G$ measurements taken when the plates were in full contact with the stretched NE (Rogez and Martinoty 2011). This process was followed for each elongation of the NE. Testing the theory requires that the measurements belong to the hydrodynamic regime. This regime occurs only at high temperature for this material, even for the low frequency ($f = 0.2$ Hz) used in the experiment, as illustrated by panel b of Fig. 3.4.

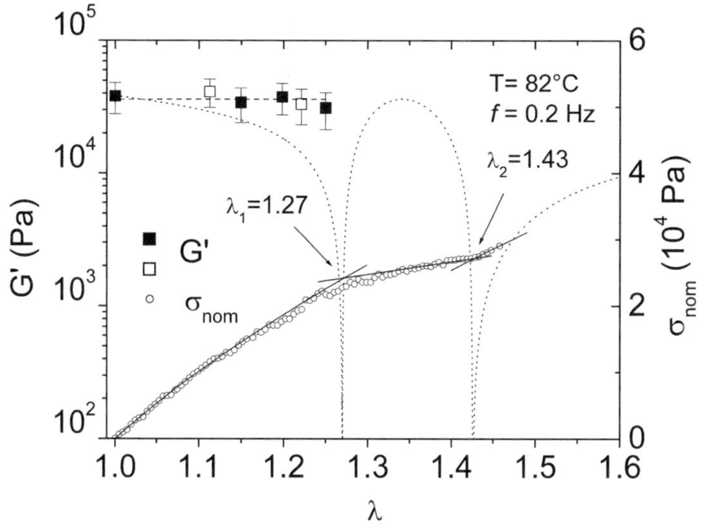


**Fig. 3.8** Variation of the real part $G'$ of the shear modulus of the homeotropic NE cross-linked by UV irradiation and oriented with an E-field, as a function of the elongation $\lambda$, showing that $G'$ does not decrease when $\lambda$ approaches $\lambda_1$. For each $G'$-value, the bars represent the dispersion of the data taken on several samples. The *dashed line* is a guide for the eyes, and the *dotted line* shows the behavior calculated with (3.11)–(3.13) and the known values of $\lambda_1$ and $r$ (see text). The *solid* and the *open squares* correspond to a shear applied in a direction respectively parallel (soft-response geometry according to the theory), and perpendicular (non-soft-response geometry according to the theory) to the stretching direction. The stress-strain curve of Fig. 3.6 (panel b) is also reported on the figure to show the position of the thresholds $\lambda_1$ and $\lambda_2$. Reprinted with kind permission of Springer Science + Business Media (Rogez and Martinoty 2011)

Figure 3.8 shows the behavior of $G'$ at $T = 82$ °C, determined as a function of $\lambda$ for the homeotropic NE, when the shear is parallel to the stretching direction and perpendicular to the initial orientation of the director. The stress-strain curve measured at the same temperature is also plotted in the figure in order to identify the position of the critical strains $\lambda_1$ and $\lambda_2$. The lack of shear data above $\lambda = 1.25$ comes from the fact that the NEs systematically broke above this value. In the stress-strain measurements, it was possible to stretch slightly more the NEs up to $\lambda \sim 1.45$, probably because the stretching step was much smaller than for the shear experiments. The theoretical curve calculated from (3.11) to (3.13) using the $\lambda_1$, $\mu$ and $r$ values, which were deduced respectively from the stress-strain measurements and from the thermo-elastic measurements of Fig. 3.3, has also been plotted in the figure for comparison with the experimental data. It can be seen that the stretching of the sample does not modify significantly the $G'$ value measured without stretching, and that the decrease in $G'$ predicted for $\lambda < \lambda_1$ is not observed. It can also be pointed out that the $G'$ values are similar to those measured for the geometry of the non-soft response (the open squares), for which the shear is perpendicular to both the stretching direction and the initial orientation of the director. Thus, when $\lambda$ tends towards $\lambda_1$, the decrease in $G'$ predicted by the different models (extension of the neo-classical model, biaxial model and bifurcation type model) is not experimentally observed. Before accepting this result, it is necessary to examine if there are mechanisms that could prevent the observation of the dip in $G'$ in the vicinity of $\lambda_1$. The presence of a spatial non-uniformity in the bulk induced by the clamping conditions at both ends of the film and/or by the stretching could be one possibility. Another possibility could be due to the fact that the direction of stretching was not perfectly perpendicular to the initial orientation of the director. However, these two possibilities seem to be excluded, since the use of a new sample for each measurement gave a similar result. On the other hand, it should be noted that the bifurcation-type model predicts that a deviation from the perfectly perpendicular condition by an angle smaller than $1°$ leads to a very strong reduction of the anomalous decrease in $G'$ at $\lambda_1$ (Menzel et al. 2009b). This prediction would make the observation of this effect, if it exists, very difficult.

On the other hand, a dynamic light scattering study performed on NEs as a function of a deformation applied perpendicular to the director, revealed the existence of a dynamic soft optical mode (Petelin and Copic 2009, 2010) when $\lambda$ increases and tends to $\lambda_1$. This experiment was done on the NEs prepared with the tri-functional cross-linkers mentioned above, by using the two-step cross-linking process. If this soft optical mode is a proof of version 2 of soft elasticity as it is claimed (Petelin and Copic 2009, 2010), there is clearly a contradiction with the mechanical measurements that do not show a decrease of $\widetilde{C}_5$ when $\lambda$ increases and tends to $\lambda_1$. However, it should be noted the $\widetilde{C}_5$ measurements were performed on a NE prepared by cross-linking an oriented NP for which the elasticity is Gaussian, whereas the optical measurements were performed on NEs prepared by the two-step cross-linking process, for which the elasticity is either non-Gaussian or at the crossover between Gaussian and non-Gaussian behaviors. This observation suggests that the nature of the elasticity, Gaussian or non-Gaussian, could govern version 2 of soft elasticity. In other words, it is possible that the decrease of $\widetilde{C}_5$ when $\lambda \rightarrow \lambda_1$ could be observed for the NEs prepared by the two-step cross-linking process. This experiment is in preparation.

## 3.6 Conclusions and Future Perspective

In this short review we have summarized our current knowledge of two key issues concerning the elasticity of uniaxial nematic side chain elastomers (NEs). The measurements taken on the dry and swollen NEs, show that there are two types of NEs.

The first type, the most studied to date, is prepared by the two-step cross-linking process and oriented by stretching the network formed after the first cross-linking step. These NEs are characterized by strongly stretched chains leading to a large chain anisotropy, a large mechanical anisotropy, non-Gaussian elasticity and to a supercritical state (i.e. the elastomer remains slightly anisotropic at high temperature).

The second type is prepared from a nematic polymer oriented by an E-field, an H-field or a surface treatment of the sample-bearing glass slides, and cross-linked by UV irradiation. These NEs are characterized by slightly stretched chains leading to a small chain anisotropy, a small mechanical anisotropy, Gaussian elasticity and to a subcritical state.

The non-Gaussian elasticity of the NEs prepared with the two-step cross-linking process leads to the failure of the neo-classical model, which is based on Gaussian rubber elasticity, to describe some physical properties of these NEs. Typical examples are given by the stress-strain curves, in particular those with the nearly flat plateau associated with the sample twice cross-linked at high temperature. In contrast, the neo-classical model describes well the NEs prepared by a UV cross-linking of an oriented nematic polymer. For both types of NEs, the value of $\widetilde{C}_5$ is not small and comparable to that of the other elastic coefficients. This clearly shows that the NEs are not semi-soft materials. On the other hand, the elastic plateau of the stress-strain curve is usually called semi-soft because it refers to the notion of ideal softness. This term has become incorrect since the elastic plateau is not due to the vicinity of a hypothetical ideal state and should be modified.

The measurements as a function of frequency performed on the dry NEs show that $G$ has two components whatever the temperature range studied, above or below the transition region. The component occurring at low frequency corresponds to the hydrodynamic regime for which $G'$ is frequency independent and $G''$ increases with frequency; the other component that occurs at higher frequencies is viscoelastic in nature. This latter component is characterized by a scaling law showing that the internal modes of the NEs are Rouse-like. These behaviors are observed for all the studied geometries—parallel, perpendicular, and homeotropic—and for both types of NEs. The swollen NEs exhibit the same type of frequency behavior, but with a shift towards the higher frequencies of the viscoelastic behavior when the swelling increases, which is due to an increase in the mobility of the mesogens. The experiments on the dry NEs also show that there is no separation of time scales between the director and the network, and that the second elastic plateau is not observed, which demonstrates again that soft or semi-soft elasticity does not apply. The NEs can therefore be described as uniaxial rubber networks with an internal variable, the director, which are characterized by a small mechanical anisotropy in the hydrodynamic regime and a Rouse-like frequency behavior in the viscoelastic region.

The measurements taken on the NE as a function of its stretching, indicate that version 2 of soft elasticity (soft elasticity induced by stretching) does not apply. However, it should be noted that these measurements were performed on a NE prepared by UV cross-linking, for which the elasticity is Gaussian. Before to claim that version 2 of soft elasticity does not represent a typical feature of NEs, as is the case of the original version of the soft (semi-soft) elasticity, it would be necessary to do a similar experiment with a NE prepared with the two-step cross-linking process, for which the elasticity is non-Gaussian.

For the future, there are several outstanding challenges associated with the mechanical properties of the NEs. The first one concerns the problem of the soft response predicted for $\lambda = \lambda_1$ when the network is stretched in a direction perpendicular to the initial orientation of the director (version 2 of soft elasticity). As we have seen above, the expected decrease of $G'$ when $\lambda$ goes to $\lambda_1$ is not observed for the NE prepared by UV cross-linking for which the polymer chains are flexible (Gaussian elasticity). The question is to know whether this absence of decrease in $G'$ is a general effect or an effect that is associated with the nature of the network elasticity. It can be answered by studying NEs prepared by the two step cross-linking process for which the polymer chains are elongated (non-Gaussian elasticity) and therefore exhibit a better orientation than that of the flexible chains. The second one concerns the field of main-chain NEs. Some of their mechanical properties should be different from those of the side-chain elastomers, as a result of their different architecture based on the fact that the mesogens are incorporated into the network, whereas they are more or less coupled to the network via a flexible spacer for the side-chain elastomers. Their shear mechanical properties in the linear regime have already been studied (Rogez et al. 2006b) and shown a behavior different of that observed for the side chain elastomers, as a result of the presence of static smectic clusters in the nematic and isotropic phases resulting of their preparation. Another issue of interest concerns the mechanical properties of the gels, which have to be studied in much greater detail. One interesting topic is the study of the sol-gel transition of a gel directly formed in the measuring cell of the piezorheometer, for example by photo-polymerization of monomers and flexible cross-linkers in a non-reactive liquid crystal solvent whose chemical structure is close to that of the monomers. It would therefore be possible to characterize the formation process in the isotropic phase, and in the nematic phase of the solution, and to test the concept of directed percolation when the gel is formed in the nematic phase. This study would also determine whether the mechanical behavior of the gel in the nematic phase depends on the phase—nematic or isotropic—of the solution where the gel was formed. Another question also related to the formation of a liquid crystalline gel is whether the mechanical properties of a gel formed by invading a NE by a nematic solvent (like those studied here) are the same as those of a gel directly formed in the same nematic solvent.

**Acknowledgment** First, I would like to thank my colleague D. Rogez for the fruitful and enjoyable collaboration during this research. I also wish to thank H. Pleiner, H. Brand, H. Finkelmann and their Ph.D. students for the numerous stimulating discussions we had over the long period covering the whole of the work on the topic. I also thank F. Brömmel for having working with us on the nematic elastomers prepared by UV irradiation, A. Sanchez-Ferrer for interesting discussions and I. Kulic for a critical reading of the manuscript.

## References

Biggins JS, Terentjev EM, Warner M (2008) Semisoft elastic response of nematic elastomers to complex deformations. Phys Rev E 78:041704 1-4

Bladon P, Terentjev EM, Warner M (1993) Transitions and instabilities in liquid-crystal elastomers. Phys Rev E 47:R3838–R3840

Brand HR, Pleiner H, Martinoty P (2006) Selected macroscopic properties of liquid crystalline elastomers. Soft Matter 2:182–189

Brehmer M, Zentel R, Wagenblast G, Siemensmeyer K (1994) Ferroelectric liquid-crystalline elastomers. Macromol Chem Phys 195:1891–1904

Brömmel F, Benzie P, Osterwinter GJ, Kramer D, Finkelmann H, Elston S, Hoffmann A (2013) Orientation behaviour of the minor director of homeotropically oriented nematic elastomers in mechanical fields. Soft Matter 9:2646–2651

Clarke SM, Tajbakhsh AR, Terentjev EM, Warner M (2001) Anomalous viscoelastic response of nematic elastomers. Phys Rev Lett 86:4044–4047

Conti S, deSimone A, Dolzmann G (2002) Semisoft elasticity and director reorientation in stretched sheets of nematic elastomers. Phys Rev E 66:061710 1-8

de Gennes PG (1980). In: Helfrich W, Heppke G (eds) Liquid crystals of one- and two-dimensional order. Springer, Berlin, p 231–237

Disch S, Schmidt C, Finkelmann H (1994) Nematic elastomers beyond the critical point. Macromol Rapid Commun 15:303–310

Finkelmann H, Kock HJ, Rehage H (1981) Liquid crystalline elastomers: a new type of liquid crystalline material. Makromol Chem Rapid Commun 2:317–322

Gallani JL, Hilliou L, Martinoty P, Doublet F, Mauzac M (1996) Mechanical behavior of side-chain liquid crystalline networks. J Phys II France 6:443–452

Golubovic L, Lubensky TC (1989) Non linear elasticity of amorphous solids. Phys Rev Lett 63:1082–1085

Komp A, Rühe J, Finkelmann H (2005) A versatile preparation route for thin free-standing liquid single crystal elastomers. Macromol Rapid Commun 26:813–818

Küpfer J, Finkelmann H (1991) Nematic liquid single elastomers. Makromol Chem Rapid Commun 12:717–726

Küpfer J, Finkelmann H (1994) Liquid crystal elastomers: influence of the orientational distribution of the crosslinks on the phase behaviour and reorientation processes. Macromol Chem Phys 195:1353–1367

Lebar A, Kutnjak Z, Zumer S, Finkelmann H, Sanchez-Ferrer A, Zalar B (2005) Evidence of supercritical behavior in liquid single crystal elastomers Phys Rev Lett 94:197801 1-4

Martinoty P, Stein P, Finkelmann H, Pleiner H, Brand HR (2004a) Mechanical properties of monodomain side-chain nematic elastomers. Eur Phys J E Soft Matter 14:311–321

Martinoty P, Stein P, Finkelmann H, Pleiner H, Brand HR (2004b) Reply to the commentary by Terentjev EM, Warner M on "Mechanical properties of monodomain side chain nematic elastomers". Eur Phys J E 14:329–332

Martinoty P, Stein P, Finkelmann H, Pleiner H, Brand HR (2004c) Reply to the commentary by Stenull O, and Lubensky TC on "Mechanical properties of monodomain side chain nematic elastomers". Eur Phys J E 14:339–340

Menzel AM, Pleiner H, Brand HR (2009a) On the nonlinear stress-strain behaviour of nematic elastomers – materials of two coupled preferred directions. J Appl Phys 105:013503 1-13

Menzel AM, Pleiner H, Brand HR (2009b) Response of prestretched nematic elastomers to external fields. Eur Phys J E 30:371–377

Ohm C, Brehmer M, Zentel R (2010) Liquid crystalline elastomers as actuators and sensors. Adv Mater 22:3366–3387

Olmsted P (1994) Rotational invariance and goldstone modes in nematic elastomers and gels. J Phys II 4:2215–2230

Petelin A, Copic M (2009) Observation of a soft mode of elastic instability in liquid crystal elastomers. Phys Rev Lett 103:077801 1-4

Petelin A, Copic M (2010) Strain dependence of the nematic fluctuation relaxation in liquid-crystal elastomers. Phys Rev E 82:011703 1-8

Rogez D, Francius G, Finkelmann H, Martinoty P (2006a) Shear mechanical anisotropy of side chain liquid crystal elastomers: influence of sample preparation. Eur Phys J E 20:369–378

Rogez D, Brandt H, Finkelmann H, Martinoty P (2006b) Shear mechanical properties of main chain liquid crystalline elastomers. Macromol Chem Phys 207:735–745

Rogez D, Martinoty P (2011) Mechanical properties of monodomain nematic side-chain liquid-crystalline elastomers with homeotropic and in-plane orientation of the director. Eur Phys J E 34(69):1–13
Rogez D, Brömmel F, Finkelmann H, Martinoty P (2011) Influence of swelling on the shear mechanical properties of monodomain side-chain liquid-crystal elastomers: Gaussian versus non-Gaussian elasticity. Macromol Chem Phys 212:2667–2673
Stein P, Aßfalg N, Finkelmann H, Martinoty P (2001) Shear modulus of polydomain, monodomain and non-mesomorphic side-chain elastomers: influence of the nematic order. Eur Phys J E 4:255
Stenull O, Lubensky TC (2004) Commentary on "Mechanical properties of monodomain side chain nematic elastomers" by Martinoty P, et al. Eur Phys J E 14:333–337
Terentjev EM, Warner M (2001) Linear hydrodynamics and viscoelasticity of nematic elastomers. Eur Phys J E 4:343–353
Terentjev EM, Warner M (2004) Commentary on "Mechanical properties of monodomain side chain nematic elastomers" by Martinoty P, Stein P, Finkelmann H, Pleiner H, Brand HR. Eur Phys J E 14:323–327
Toshchevikov VP, Gotlib YY (2009) Shear dynamic modulus of nematic elastomers: modified Rouse model. Macromolecules 42:3417–3429
Treloar LRG (1949) The physics of rubber elasticity. Clarendon, Oxford
Urayama K, Mashita R, Kobayashi I, Takigawa T (2007) Stretching-induced director rotation in thin films of liquid crystal elatomers with homeotropic alignment. Macromolecules 40:7765–7670
Verwey G, Warner M (1995) Soft rubber elasticity. Macromolecules 28:4303–4306
Verwey G, Warner M, Terentjev EM (1996) Elastic instability and stripe domains in liquid crystalline elastomers. J Phys II France 6:1273–1290
Verwey G, Warner M (1997a) Compositional fluctuations and semisoftness in nematic elastomers. Macromolecules 30:4189–4195
Verwey G, Warner M (1997b) Nematic elastomers cross-linked by rigid rod linkers. Macromolecules 30:4196–4204
Warner M, Bladon P, Terentjev EM (1994) "Soft elasticity"—deformation without resistance in liquid crystal elastomers. J Phys II 4:93–102
Warner M, Terentjev EM (1996) Nematic elastomers—a new state of matter? Prog Polym Sci 21:853–891
Warner M (1999) New elastic behaviour arising from the unusual constitutive relation of nematic solids. J Mech Phys Solids 47:1355–1377
Warner M, Terentjev EM (2003) Liquid crystal elastomers. Clarendon, Oxford and references therein
Weilepp J, Stein P, Assfalg N, Finkelmann H, Martinoty P, Brand HR (1999a) Rheological properties of mono- and polydomain liquid crystalline elastomers exhibiting a broad smectic A phase. Eur Phys Lett 47:508–514
Weilepp J, Zanna J, Assfalg N, Stein P, Hilliou L, Mauzac M, Finkelmann H, Brand HR, Martinoty P (1999b) Rheology of liquid crystalline elastomers in their isotropic and smectic A state. Macromolecules 32:4566–4574
Ye F, Mukhopadhyay R, Stenull O, Lubensky TC (2007) Semisoft nematic elastomers and nematics in crossed electric and magnetic fields. Phys Rev Lett 98:147801 1-4
Yusuf Y, Ono Y, Sumisaki Y, Cladis PE, Brand HR, Finkelmann H, Kai S (2004) Phys Rev E 69:021710 1-9
Zanna J, Stein P, Marty D, Mauzac M, Martinoty P (2002) Influence of molecular parameters on the elastic and viscoelastic properties of side-chain liquid crystalline elastomers. Macromolecules 35:5459–5465

# Chapter 4
# Recent Advances in the Rheology of Thermotropic Liquid Crystal Polymers

**Ahmed O. Rahman, Rahul K. Gupta, and Sati N. Bhattacharya**

## 4.1 Introduction

### *4.1.1 Liquid Crystal Polymer*

Liquid crystals (LCs) have material properties between those of a conventional liquid and a solid crystal (Chandrasekhar 1994). It has long-range molecular order somewhere between the crystalline state, which exhibits three-dimensional order, and the disordered isotropic fluid state (Marrucci 1991). The molecules in a crystal are ordered whereas in a liquid they are not. Liquid crystalline phase is also known as mesomorphic phase or simply mesophase. Mesomorphic phase can be defined as the intermediate form between liquid and crystal (de Gennes and Prost 1993). These materials are not crystalline and may not even be liquid, but exist in a stable intermediate phase between the liquid and crystalline states. Considerable research work has been carried out to understand properties of LCPs and its relationship with molecular structure however only a limited understanding exists of how changes in molecular structure affect material properties. Liquid crystal polymers (LCPs) consist of rod like molecules and often called "self-reinforcing thermoplastic" (Ophir and Ide 1983). The rheological behaviour as well as development of orientation is similar to that of the short fiber-filled composite. Numerous studies have established that liquid crystal polymers (LCPs) exhibit many anomalous rheological properties. These anomalous behaviours attract much attention from the researcher.

A.O. Rahman • R.K. Gupta (✉) • S.N. Bhattacharya
Rheology and Materials Processing Group, School of Civil, Environmental and Chemical Engineering, RMIT University, Melbourne, VIC 3001, Australia
e-mail: rahul.gupta@rmit.edu.au



V.K. Thakur, M.R. Kessler (eds.), *Liquid Crystalline Polymers*,
DOI 10.1007/978-3-319-20270-9_4

The commercial introduction of liquid crystal polymers did not occur until 1984 since liquid crystal polymers could not be injection moulded till then (Tai-Shung 2001). Nowadays, liquid crystal polymers can be melt processed on conventional equipment at fast speeds with excellent replication of mould. Some of these LCPs, which constitute as simple common displays, are found in calculators or mobile phones. However, the numerous and increasingly sophisticated applications, relying upon the use of liquid crystalline materials, require a complexity of superior properties to achieve improved devices performance. Currently, LCPs play a dominant role in a large part of the display technology.

### *4.1.2 Types of LCPs and Fillers*

In general, LCPs can be classified as Thermotropic Liquid Crystal Polymer (TLCP) and Lytropic Liquid Crystal Polymer. Liquid Crystal Polymers (LCPs) whose phase transition to the liquid crystalline phase occurs due to change in temperature, are called thermotropic liquid crystal while a phase transition by changing the polymer concentration in a solvent (as well as temperature) is called lyotropic liquid crystal (Brostow 1992).

To increase the strength and stiffness of TLCPs, they have been combined with fibers to create composites. Typically, this reinforcement has been generated by compounding fiber into the matrix or impregnating a fiber weave with polymer. The fiber reinforcement has ordinarily been glass, graphite, carbon, or aromatic polyamide. Glass fiber has been the most common form of reinforcement because of its low cost and the experience in using glass as a reinforcing agent (Shibley 1982).

Due to the recent development in nanotechnology, TLCPs are usually compounded with nanoparticles to form particulate composites to enhance their properties, such as barrier properties, electrical properties, mechanical properties and thermal properties. Carbon-based nanofillers such as carbon nanotube (CNT), graphene and graphene oxide are the most common fillers used for the TLCP matrices (Cheng et al. 2012).

### *4.1.3 Phases and Order Parameter in LCP*

Friedel proposed the first classification scheme of LCPs (Felter and Dourson 1997) dividing them into three different types of mesogens (materials able to sustain mesophases), based upon the level of order the molecules possessed in the bulk material. Three major mesophases have been identified to describe the liquid crystalline phase: nematic, cholestoric and smectic (Singh 2000). In nematic mesophase the molecules have no positional order but have a long range orientational order. But in cholestoric phase, molecules possess different positional order and different positions exhibit different orientational order. In smectic phase molecules rotate with positional and orientational order. Molecular orientation is of the central significance in the field of liquid crystal polymers (Donald et al. 2006).

### 4.1.4 Morphology and Orientation in LCPs

The molecular shape and the arrangement in a solid or liquid greatly influence the properties of polymers. The rigid rod-like molecules of unfilled LCPs act like filler in polymers (Ophir and Ide 1983). These rod-like molecules easily tumble with shear. Hence, the viscosity reduces at low shear rate. In many commercial polymers, LCPs are being used as filler to reduce the initial viscosity (Hashimi and Takeshi 2007).

Morphology of LCP molecules with their dimension of glass fibers and their orientation have been examined using an FEI Quanta 200 Environmental Scanning Electron Microscope (ESEM). Figure 4.1(1)a, b represent the ESEM images of molten unfilled LCPs. These figures demonstrate the linear orientation of molecules of unfilled LCPs. The cross sectional view of filled LCP is also shown in Fig. 4.1(1) c and orientation of filler in surface is shown in Fig. 4.1(1)d. LCP-A and LCP-B were filled with glass fiber of same dimension. It is observed from Fig. 4.1(1)e that the average diameter of glass fiber is 10 μm and from Fig. 4.1(1)f, the length of the glass fiber is 1 mm. Comparative analysis shows that both the LCP-A and LCP-B have fillers of same aspect ratio of 100.

### 4.1.5 Phase Transition in LCPs

In LCPs there is a characteristic increase in viscosity with temperature (Wissbrun 1981; Kim and Han 1994; Fan et al. 2003) and in literature the onset of temperature causing phase change is known as phase transition or nematic-isotropic ($T_{N\text{-}I}$) transition temperature. This anomaly of increasing viscosity with temperature is attributed to changes in the morphological state with the length of the biphasic window increasing with impurities (mixing of nematic and isotropic phase) and molecular weight distribution (Scribben 2004). The low molecular weight portion of the polymer begins to transform into an isotropic region at temperatures below the nematic-isotropic transition due to the poly-disperse nature of rigid rod like molecules whereas in monodisperse system, a discontinuous increase between the liquid crystalline and isotropic state would be expected. Since the viscosities of both the isotropic and nematic phases decrease with temperature, this anomaly is possible due to increase of impurity (i.e. mixing of nematic and isotropic phases) (Kiss 1986). The gradual increase of viscosities at higher temperatures can be described by the biphasic nature of these LCPs, that is, due to thermal activation,

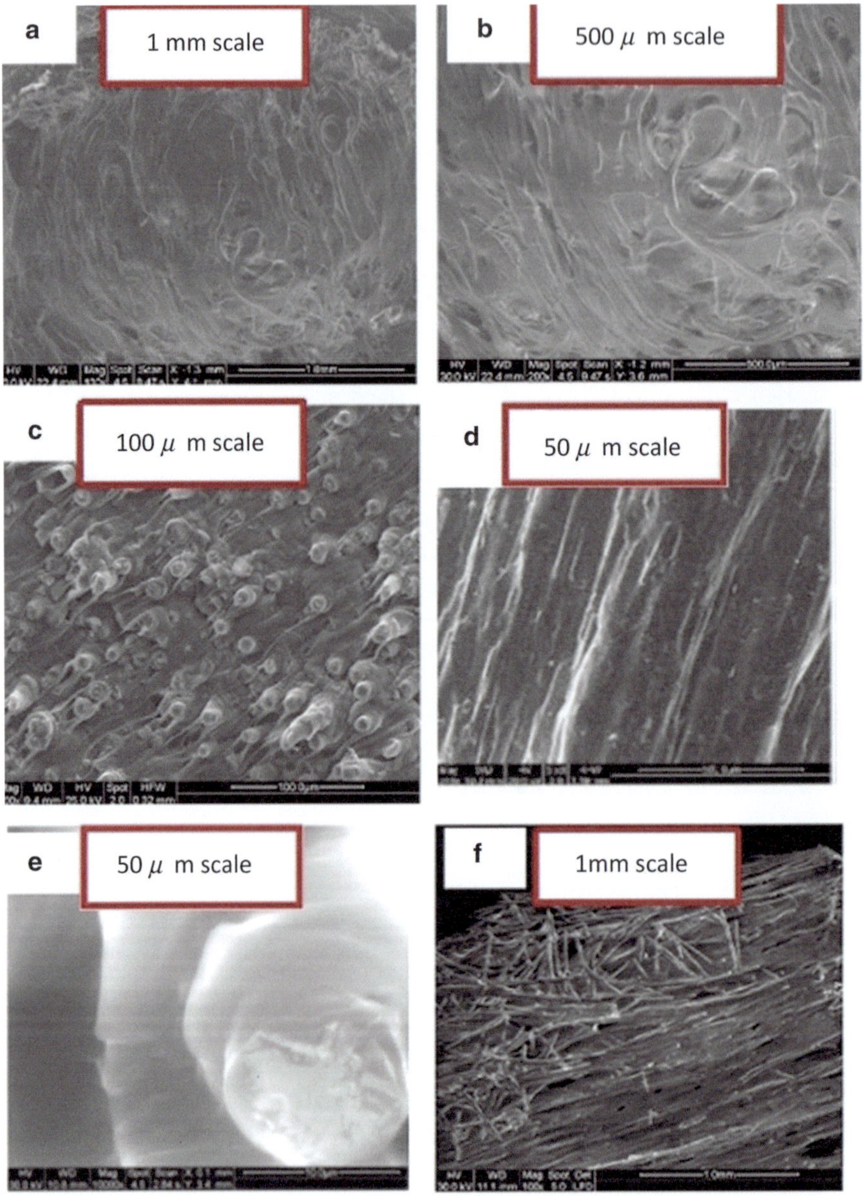


**Fig. 4.1** (1) ESEM images of rod like orientation of Unfilled LCPs at magnification scale (**a**) 1 mm, (**b**) 500 m, (**c**) cross sectional view of LCP pellets, (**d**) surface view of filler arrangement in pellet (Rahman 2013), (**e**) diameter of filler and (**f**) length of filler (2) ESEM images of glass fiber orientation at different shear rates (**a**) 0 $s^{-1}$ magnification scale 500 m, (**b**) 1 $s^{-1}$ scale 1 mm, (**c**) 10 $s^{-1}$ scale 200 m, (magnification scale 500 ) 100 $s^{-1}$ scale 1 mm (Rahman et al. 2013)

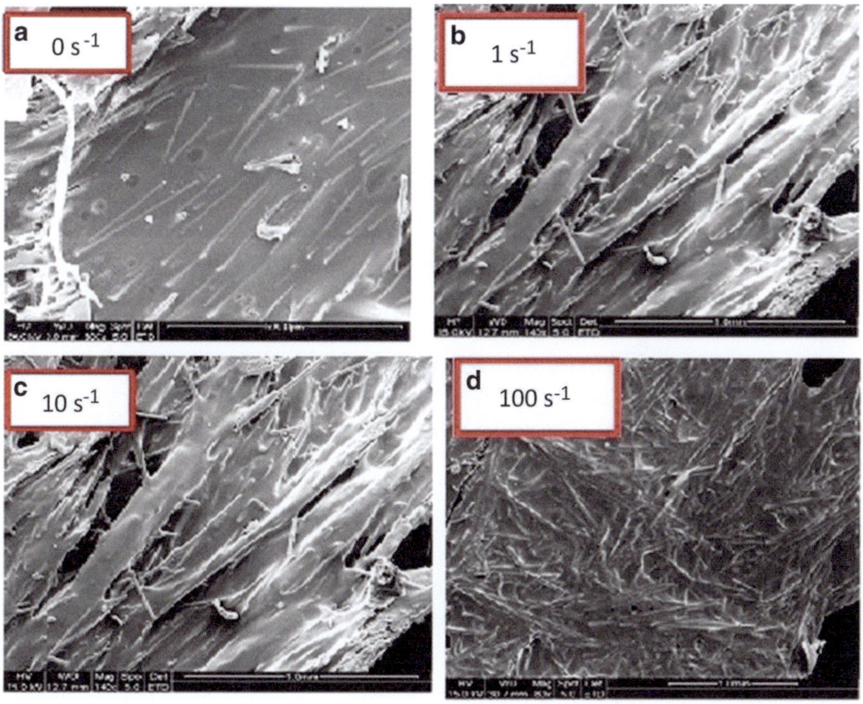


**Fig. 4.1** (continued)

part of the rod-like molecules of nematic phase are transformed to the isotropic phase. The existence of biphasic nature increases more resistance to flow and thus viscosities increase in this region (Gao et al. 1996; Hsies et al. 1999).

## 4.2 Fundamentals of Rheology

Rheology is fundamentally a science studying the flow behaviour of complex materials under different flow conditions. Chemists and physicists use rheology to analyze and characterize the structure of various complex fluids at the macromolecular level. The engineers are more concerned with characterization of the viscosity of these complex fluids (Bhattacharya et al. 2010). The focus of this chapter is to provide the recent advances in the rheology of thermotropic liquid crystal polymer. In general, three basic techniques are utilized to determine the rheological behaviour of a material: (1) shear, (2) dynamic, and (3) extensional rheology. Shear rheological techniques are used to determine the flow behaviour and structure evolution of a material under a steady-state deformation condition. Dynamic rheology is a structural analysis tool because it uses a frequency excitation method to determine the molecular-level structure of a polymeric system. Finally, extensional rheology is used to determine the structural changes in a material undergoing stretching in the complete absence of surface effects. Thus, basically the major difference between these rheological tools is the method of deformation employed to generate the rheological response to investigate the material properties.

### *4.2.1 Steady Shear Measurements*

Steady shear rheology is important to determine the non-Newtonian flow of the melt, performance of the extruder (such as excess pressure to stop the motor), shear thinning characteristics, stable melt flow during the stretching, melt fracture and die swelling (Macosko 1994). The range of shear rates that the melt experiences in an extruder can be determined from the steady shear rheology.

A wide variety of pressure distribution is caused due to the flow of a liquid crystal polymer melt or solution. Graziano and Mackley (1984) were the first to analyse the development of textures during the shearing of a conventional nematic liquid crystal polymer and also thermotropic copolymer. Windle et al. (1985) studied the effect of shear deformation of a thermotropic liquid and observed fine-scale texture (Windle et al. 1985).

The flow mechanism of liquid crystalline systems is more complicated than that of isotropic fluids due to the coupling of molecular orientation and polydomain (composed of multiple mesogenic domains) structure. LCPs exhibit abnormal shear dependency of viscosity in the low shear rate region. One method for categorizing the rheology of LCPs is the three-region flow curve (Onogi and Asada 1980). It proposes that the 3-region curve is characteristic of all LCPs, They also suggested that this behaviour can be recognized as a characteristic property of liquid crystal polymer (Yanase and Asada 1987). However, this behaviour may not be so general for lyotropic LCPs and is very uncommon for TLCPs (Kiss and Porter 1980; Wissbrun 1980).

LCPs exhibit shear thinning behaviour at low shear rate, i.e., the viscosity decreases with increasing shear rate in concentrated as well as thermotropic LCPs (Walker and Wagner 1994; Wagner 1977; Baek et al. 1993). But for the conventional polymers, their viscosity always approaches a constant value as the shear rate tends to zero. The general consensus appears to be that the higher viscosities at lower strain rates are due to some aspect of initial microstructure stability. At extremely low shear rate, the polydomain structure of LCPs hardly deforms. The structural evolution of such a defect 'suspension' would generate the shear thinning behaviour at low shear rates (Marrucci and Maffettone 1993). Both thermotropic and lyotropic LCPs exhibit the shear thinning behaviour at low shear rate region.

When the viscosity is almost constant with shear rate (Newtonian behaviour), this section is referred as region II. With the increase of shear rate, the effect of coalescence also increases and the molecules offer more resistance to flow. This Newtonian behaviour is only observed in some lyotropic LCPs. At region III (very high shear rate zone), the molecules of LCPs start to move in the flow aligning directions, which leads to strong shear thinning behavior (Muir and Porter 1989). At a very high shear rate region all domains of LCPs act as a monodomain and oriented as in isotropic flow (Tao et al. 2006).

In general, the rheological properties in high shear rate region are conducted by capillary rheometers. These rheometers are popular because of their simplicity in design and use. In capillary rheometers, the viscosities are measured by the measurement of the pressure drop of a given flow rate. Capillary rheometers are broadly categorized as constant speed rheometers and constant pressure rheometers.

Very few researchers have studied the rheological properties of LCPs at very high shear rates. The technical report from IUPAC in 2004 presented a comparative experimental study of shear rheological properties of thermotropic LCPs conducted at eight different laboratories (White et al. 2004). Shear thinning behaviour was observed for all the LCPs investigated.

### *4.2.2 Dynamic Shear Measurement*

The oscillatory rheological data are very useful to understand the microstructure of the polymeric material while subjected to deformation in the linear viscoelastic region usually with a small amplitude sinusoidal strain, measuring the resultant sinusoidal stress (Khan et al. 1997). Large amplitude oscillatory strain can also be applied.

Several researches have reported on transient and/or steady shear rheology of TLCPs (Isayev and Nicolais 2011; Han et al. 2001). However, very few studies have been reported on the oscillatory flow of TLCPs. The measurements of oscillatory flow properties have certain advantages over the steady flow properties (Seung and Chang 1993). Viscoelastic parameters are directly related to the structure of the materials and these information are mainly used for characterizing the molecules in their equilibrium state to compare different resins, for quality control by the resin manufacturers (Dealy and Wissburn 1990). Many useful information that can be derived from the dynamic shear rheology is elastic (storage, $G'$) modulus, viscous (loss, $G''$) modulus and complex viscosity ($\eta^*$).

Time evolution of the morphological state in the nematic region of a specimen after being subjected to steady or transient flow, can be monitored in terms of storage modulus ($G'$) or complex viscosity. Moreover, the oscillatory measurement allows applying high angular frequency compared to steady shear measurement. The LCP molecules are very sensitive to thermal history. It has been observed that the rheological properties of LCPs are continuously varied with annealing time at below nematic–isotropic transition temperature. Moreover, LCP molecules exhibit pronounced phase transition behaviour during oscillatory measurements. At low frequency, the molecules of nematic phase start to transform to isotropic phase (Fan et al. 2003). The co-existence of nematic and isotropic phase is also observed at certain critical temperature. With increase of temperature, all the nematic phase is transformed to isotropic phase.

### 4.2.3 Extensional Rheological Measurement

Extensional rheological measurements are complex to initiate and involve more complicated theoretical analysis. The extensional rheological methods are the best method to determine the structure of a surface in the presence of a highly variable flow field (or high shear rates) simply because extensional rheological analysis is normally carried out in the absence of surface and hence the structural changes occur as well as high degree of molecular orientation is produced within the materials (Bhattacharya et al. 2010).

The unusual structure and orientation of thermotropic liquid crystalline polymers (TLCPs) can significantly affect the rheological properties of TLCP. Several techniques have been applied to measure the extensional viscosity of lyotropic LCPs. Whereas, among the various techniques for TLCPs, Wagner et al. (1997) measured the extensional viscosity of commercial TLCP using a Rheotens fiber-spinning instrument, Done and Baird (1990) measured extensional flow by the lubricated squeezing technique, Wilson and Baird (1992) measured the transient extensional viscosity on a rotary clamp extensional rheometer and Ruobing et al. (Yu et al. 2005) used thread disintegration, which is based on the disintegration of TLCPs. In case of extensional flow of TLCPs, the extensional rheological properties of TLCPs, such as extensional viscosity, are crucial to understand the microstructural difference in extensional flow (Yu et al. 2005). The shear rheology of TLCPs has been studied extensively for several decades. However, little consideration was given to the extensional behaviour of TLCPs. In the processing of TLCPs, such as extrusion, blow moulding and fiber spinning, extensional viscosity is very important. Conversely, it is very difficult to measure the extensional viscosity compared to shear viscosity of TLCPs.

### 4.2.4 Orientation of Fillers and LCPs with Shear

Incorporation of fillers greatly influences the rheological properties of LCPs. Moreover, the fiber orientation is also influenced by the rod like molecules of LCPs (Ramazani et al. 2001). Fiber-fiber collisions may also influence suspension viscosity by creating anisotropic structures in the fluid. It is also observed that this anomalous orientation is enhanced due to the presence of fillers. At low shear rates, the interaction between the fibers offsets the force attempting to align the fibers with flow. Consequently, substantial resistances to deformation are observed. With increasing shear rate, polymer matrix overcomes the interactions, caused by the fiber disrupting the structure formed by orienting the fibers with the flow (Kamal and Mutel 1985). In addition, with the increase of shear the aspect ratio of fiber-filled polymers decreases due to breakage of the fibers. Furthermore, the orientations of molecules also change. It is observed that at low shear rate the viscosity and moduli for composite are higher than those of unfilled polymers (Barnes 2003; Eberle et al. 2008).

Incorporation of filler material in polymers and subsequently cross linking them to form a filled polymer network structure leads to a system of great complexity. Numerous theoretical attempts have been made to describe this complicated rheological behaviour of the filled polymers (Leonov 1990). There is either a particle-particle or particle–matrix "network" formed, depending on the strength of inter-particle attractions or the interactions between particles and polymer chains (Wolff and Wang 1992).

When a shear field is applied to the system, the particulate network is quickly destroyed and the successive groups start to rupture. Other parameters such as temperature, activation energies, and the hydrodynamics resistance due to visco-elastic polymer matrix surrounding filler particles govern the kinetics of restructuring the filler particles (Prashant and Leonov 2001). At very high levels of stresses in the composite the particle network groups can be destroyed with almost complete separation of the particles (Leonov et al. 1977). In the high shear rate region, the highly filled systems with a low molecular weight matrix behave like suspensions of rigid particles in Newtonian liquids, whereas with higher matrix viscosity the filled polymers demonstrate usual viscoelastic behaviour, well known for the polymeric liquids (Leonov 1990).

The characteristics of incorporated fillers influence the rheological properties of TLCPs. Moreover, the fibre orientation is also influenced by the rod like molecules of LCPs (Ramazani et al. 2001). In general, the diameter of TLCP fibers is 1 μm (or less) compared to 10–15 μm length typically observed for glass fibers (Laun 1984). For particles of 10 μm or larger, the hydrodynamic interactions can cause a change in the velocity distribution in the vicinity of other fibers (Barnes 2003). Fiber-fiber collisions may also influence suspension viscosity by creating anisotropic structures in the fluid.

Figure 4.1(2) represents the orientation of fibers in LCPs at different shear rates. The ESEM images were taken after freezing the sample at different shear rates. Figure 4.1 (2)a represents the orientation of fiber in molten state (shear rate 0 $s^{-1}$) and the fibers are almost in nematic state. At the shear rate of 1 $s^{-1}$, the fibers start to rotate from the nematic state (Fig. 4.1(2)b). Breakage occurred in fibers at the shear rate of 10 $s^{-1}$ (Fig. 4.1(2)c) and the fibers started to rotate randomly in different directions. Figure 4.1 (2)d shows the isotropic phase of fiber at very high shear rate of 100 $s^{-1}$.

### *4.2.5 Thermo-rheological Behaviour of TLCPs*

The existence of biphasic (mixture of solid and liquid phase even in molten state) nature makes the rheological properties of LCPs extraordinary. For crystalline polymers, the modulus depends on the degree of crystallinity. The degree of crystallinity for LCPs also change with the variation of frequency as well as temperature (Dae Han and Kim 1993). The amorphous portions go through the glass transition, but the crystalline portion remains hard. Thus a composite modulus is found for LCPs.

The other useful parameter i.e. shift factor (for time temperature superposition) is not possible to measure for liquid crystal polymer. The use of time-temperature superposition is meaningful as long as the morphology of the polymer remains the same over the temperature held and thus it has been used primarily for homo polymers. There is no "general" theoretical guideline to calculate the shift factor ($a_T$) for LCPs.

### 4.2.6 *Anomalous Rheological Behaviour of TLCPs*

Rheological behaviour of thermotropic liquid crystal polymers has been the subject of considerable theoretical and experimental analysis (Antoun et al. 1981; Lenz 1985). The molecules of liquid crystals can be oriented by electromagnetic fields and by mechanical stress or shear. Rienacker and Hess (1999) studied theoretically shear induced orientational dynamics of nematic liquid crystals (Rienacker and Hess 1999). In their study, they reported different shear rate regions for different types of orientation. The rigid rods like molecules rotate on their own axis, which is known as tumble. The rod like molecules starts to tumble with shear and rotate from tumble to wagging after certain critical shear rate. With the increase of shear the wagging region transforms to flow aligning region.

The rheological behavior of LCPs is known to be dependent on the nature and also on the texture of its mesophase. Moreover, the relative values of melt viscosities depend on the composition of copolymers which contain both mesogenic and nonmesogenic units (Isayev and Nicolais 2011). These units can be used as a measure of the "degree of liquid crystallinity" in such systems in which both anisotropic LC and isotropic phases are present (Lenz 1985). The rheology of nematic LCPs exhibits some similarities to and difference from isotropic polymer melts and solutions. The physical properties of LCPs are extremely important not only for theoretical interest but also for the optimization of processing condition as well as modeling and simulation. These molecules possess extraordinary characteristics compared to the conventional polymers. TLCPs exhibit anomalous rheological behaviour such as; abnormal thermal dependency of viscosity, shear induced crystallization, negative first normal stress difference for filled system and filled TLCPs show more shear thinning behaviour than unfilled TLCPs at high shear rate.

### 4.2.7 *Rheological Models of LCPs*

The shear rheology of LCPs can only be accounted properly if a microstructural parameter is also included in the conventional rheology models. The Williams-Landel-Ferry (WLF) equation due to its empirical nature is found to be suitable for LCPs. The constants present in the WLF equation can be made to fit any kind of shear flow behavior, hence can be used to describe the shear flow behaviour and the temperature dependency for LCPs as well. However, the complexity arises because the matrix model for LCP should take into account the microstructural evolution within the material since the final properties of a moulded part, for example, will depend on the local microstructure of the material. Also the texture of the part will be a function of the overall macrostructure. Hence the inclusion of the microstructural parameters in the matrix model is justified.

Much research in the last few decades focused on the simulation of LCPs for various processes. Suitable rheological constitutive equations are required for this simulation. Leslie-Ericksen (LE) theory describes the flow behaviour and molecular orientation of many LCPs. LE model is limited to low shear rates and weak molecular distortions. But at high shear rate, the rate of molecular distortions is too fast. Doi and Edwards developed their model to describe the complex dynamics of macromolecules at high shear rate (Doi and Edwards 1978). Doi theory is applicable for lyotropic LCPs of small and moderate concentrations. Due to the complex nature of Doi theory, it is also challenging for simulation. Leonov's continuum theory of weak viscoelastic nematodynamics, developed on the basis of thermodynamics and constitutive relations, consider the nematic viscoelasticity, deformation of molecules as well as evolution of director.

## 4.3 Materials

Most liquid crystalline molecules are highly anisotropic, and to a good approximation can be regarded as rigid rods. In general, the units of liquid crystal compounds consist of two or more aromatic rings (Tai-Shung 1986). Depending upon the architecture of TLCPs, it can be classified as main-chain TLCP or side-chain TLCP. In some cases, the side chain groups control the anisotropic nature of TLCPs (Donald et al. 2006). The aromatic rings can be totally saturated cyclohexane, unsaturated biphenyl, triphenyl, or combinations of them. Isayev (2012) has studied the influence of phenyl group on the melting point of TLCPs (Isayev 2012). Usually, the longer the ring, the higher is the melting temperature of TLCPs. The linking group makes an important contribution to the phase transition and superior physical properties. In general, LCPs have a high mechanical strength at high temperatures, extreme chemical resistance, inherent flame retardancy, and good weatherability. They are also well resistant to stress cracking in the presence of most chemicals at elevated temperatures, including aromatic or halogenated hydrocarbons, strong acids, bases, ketones, and other aggressive industrial substances (Chandrasekhar 1994).

Rahman (2013) investigated the shear rheological properties of filled and unfilled TLCPs at low and high shear rates. The materials used were industrial grade TLCPs. Due to propriety rights, only the major constituents of TLCPs used are mentioned. The following nomenclatures are used for different filled and unfilled LCPs as reported by Rahman et al. (2013).

1. LCP Polymer A Unfilled: (copolymer of p-hydroxy benzoic acid (HBA), 6-hydroxy-2 naphthoic acid (HNA), terephthalic acid (TA), hydroxyl quinone, HQ) and amide group.
2. LCP Polymer A filled: Filled LCP-A with 35 % glass fibers.
3. LCP Polymer B Unfilled: (copolymer of p-hydroxy benzoic acid (HBA), 6-hydroxy-2 naphthoic acid (HNA) and terephthalic acid (TA))
4. LCP Polymer B filled: Filled LCP-B with 30 % glass fibers.

## 4.4 Shear Rheology for Filled and Unfilled LCPs

The structural orientation makes the rheological properties of LCPs often extraordinary and partially understood. It was reported that only complex viscosities increased with temperature at the low shear rate region while no abnormal thermal dependency of steady shear viscosities was observed at the same condition. In addition, very limited rheological data was available for the high shear rate region. Rahman et al. (2013) studied the rheological properties (steady shear viscosity and first normal stress difference) of two filled and unfilled TLCPs from low (0.01–10 $s^{-1}$) to high shear rate (10–4000 $s^{-1}$) region at different temperatures. The contributions of fillers in LCP matrix were also investigated.

The flow mechanism of liquid crystalline systems is more complicated than that of isotropic fluids due to the coupling of molecular orientation and polydomain structure. LCPs exhibit abnormal shear dependency of viscosity in the low shear rate region compared to conventional polymer. LCPs exhibit shear thinning behaviour, while the conventional polymers display zero shear viscosity (viscosity approaches a constant value as the shear rate tends to zero) at low shear rate region (Marrucci and Guido 1995). The rigid rod like molecules of TLCPs slide past each other more readily in the nematic phase. Thus, lower viscosity is observed in the low shear rate region. In addition, in the high shear rate region, the filled LCPs also exhibit unusual behaviour. The low shear rate data were measured by the constant strain rate rheometer, Advanced Rheometrics Expansion system (ARES) whereas the high shear rate data were measured by the capillary rheometer. In the steady shear state, there was no abnormal temperature dependency of viscosity observed throughout the entire region of shear rate, which (viscosity increase with temperature) was observed for complex viscosity (Fan et al. 2003).

LCPs exhibit shear thinning behaviour at low shear rate, i.e., the viscosity decreases with increasing shear rate (Figs. 4.2 and 4.3). Power-law dependence of viscosity was observed throughout the wide range of shear rate. These data showed good agreement with the numerous published studies (Walker and Wagner 1994; Wagner 1977; Baek et al. 1993; Kalika et al. 1989; Wissbrun 1981). In the low shear rate region, the rod like molecules of LCPs tumble with shear. The texture of the nematic phase of LCPs consists of many small domains, in which the local orientations of the director vary from one domain to another. Under this circumstance, each small domain may act as if it is a discrete particle suspended in a very

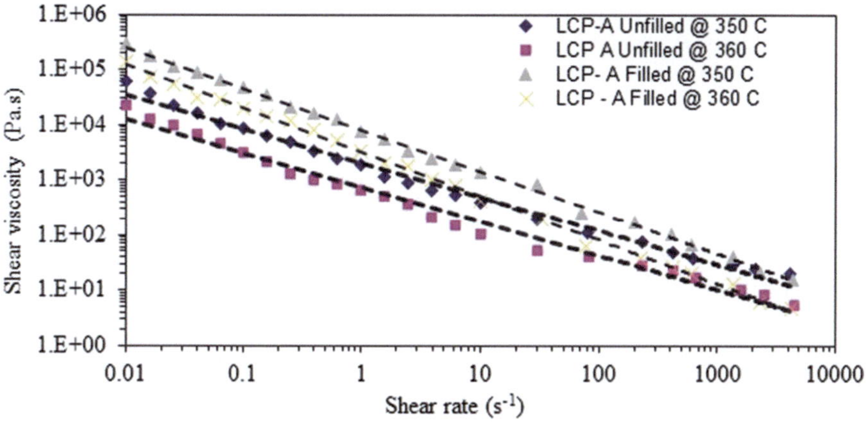


**Fig. 4.2** Shear viscosity data at different temperatures for LCP-A Filled and LCP-A Unfilled (*filled diamond* 350 °C, *filled square* 360 °C, --- trend) (Rahman 2013)

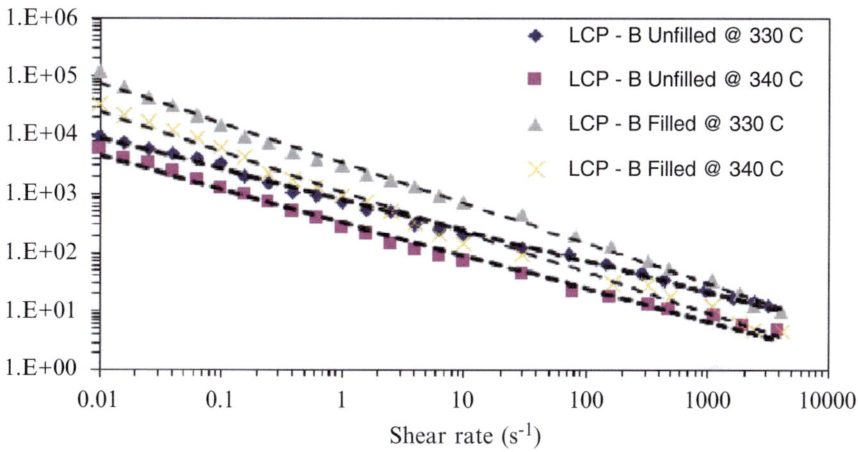


**Fig. 4.3** Shear viscosity data at different temperatures for LCP-B Filled and LCP-B Unfilled (*filled diamond* 330 °C, *filled square* 340 °C, --- trend) (Rahman 2013)

low viscosity fluid (nematic solvent) (Chen 2008). Thus, at extremely low shear rate, the polydomain hardly deforms. The structural evolution of such a defect 'suspension' would generate the shear thinning behaviour at low shear rates (Marrucci and Maffettone 1993).

Rienacker and Hess (1999) studied theoretically shear induced orientational dynamics of nematic liquid crystals (Rienacker and Hess 1999). LCPs exhibit different types of orientation at different shear rate. The rods like molecules start to tumble with shear and rotate from tumble to wagging after certain critical shear rate. With the increase of shear the wagging region transforms to flow aligning region. At the high shear rate region, the molecules of LCPs start to move with flow aligning direction, which leads to strong shear thinning behaviour (Muir and Porter 1989). Moreover, with the increase of shear, the effect of coalescence also increases and the numbers of small domains are also reduced. Hence, in the high shear rate region, all the domains of LCPs act as a monodomain and orient as isotropic flow (Tao et al. 2006).

### 4.4.1 Relaxation Time and Zero Shear Viscosity

The relaxation spectrum of the polymeric materials depends on the molecular properties of the materials as well as the architecture of the molecules. Linear viscoelasticity provides the broad distribution of relaxation modes with which polymeric materials relax. The longer modes originate from the motion of large molecular chain segments. The distribution of relaxation modes represents the material structure. Small changes in the structure can result in large changes in processability and mechanical properties. The relaxation times are sensitive to temperatures since molecular chains relax faster at higher temperatures. Relaxation modulus is generally used for characterising material properties with relaxation behaviour. The existence of solid and liquid phases in the melting process of LCPs, makes the relaxation behaviour of LCPs extraordinary (Gao et al. 1996). Three kinds of relaxation mechanisms are available for LCPs: the relaxation of chain orientation, the relaxation of deformed polydomains, and the coalescence of polydomains (or the relaxation of textures) (Yu et al. 2007).

A small shear strain i.e. 1 % was used for relaxation experiment of the LCPs (Rahman 2013). In all relaxation tests, two distinct steps were observed for all LCPs: initial sharp decay followed by plateau values for long time. The Generalized Maxwell model ((4.1)) has been used to fit the experimental data to obtain the relaxation spectrum of the LCPs (Majumder et al. 2007) (Fig. 4.4).

$$G(t) = \sum G_k e^{-t/\lambda_k} \tag{4.1}$$

$$G'(\omega) = \sum \frac{G_k(\omega\lambda_k)^2}{1 + (\omega\lambda_k)^2} \tag{4.2}$$

$$G''(\omega) = \sum \frac{G_k(\omega\lambda_k)}{1 + (\omega\lambda_k)^2} \tag{4.3}$$

$$\lambda_{average} = \frac{\sum G_k \lambda_k^2}{\sum G_k \lambda_k} \tag{4.4}$$

$$\lambda_{longest} = \underset{\omega \to 0}{Limit} \frac{G'}{G''\omega} \tag{4.5}$$

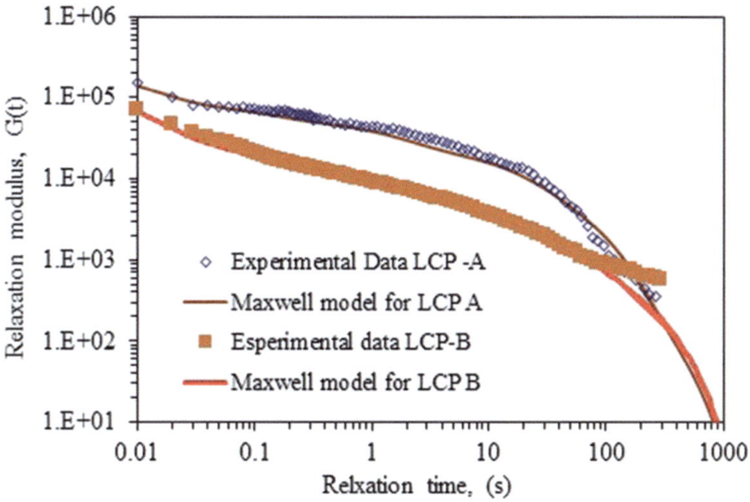


**Fig. 4.4** Relaxation modulus of LCP-A Unfilled at 350 °C and LCP-B Unfilled at 330 °C according to Maxwell model and shear relaxation experiment (Rahman 2013)

Zero Shear Viscosity (ZSV) is the viscosity measured in shear deformation at a shear rate approaching zero. This parameter is an indicator of two polymer characteristics (Phillips and Robertus 1996):

- The stiffness of the polymer material;
- The material resistance to permanent deformation under long term loading.

In the melt flow curve (viscosity versus shear rate), a Newtonian plateau at very low shear rate is usually considered as the ZSV. For polymers of very low molecular weight, this ZSV can be obtained directly from the shear rheology experiment (Dealy and Wissburn 1990). However, this is really difficult to obtain for high molecular weight polymers as well for LCPs. From the experimental data presented in Fig. 4.5, it was impossible to determine the ZSV, since there was no distinct plateau detected at a lower shear rate up to 0.01 $s^{-1}$. Therefore, a modified Cross model (4.6) was used to determine the zero shear viscosity of the LCPs. Data from low and high shear rates were combined to fit into the Cross model predictions (Fig. 4.5). The model prediction showed good agreement with experimental data.

$$\eta = \frac{\eta_0}{1 + K_2 \dot{\gamma}^m} \tag{4.6}$$

Where $\eta_0$ = zero shear viscosity,
$\dot{\gamma}$ = shear rate, and
$K_2$ and $m$ are constants

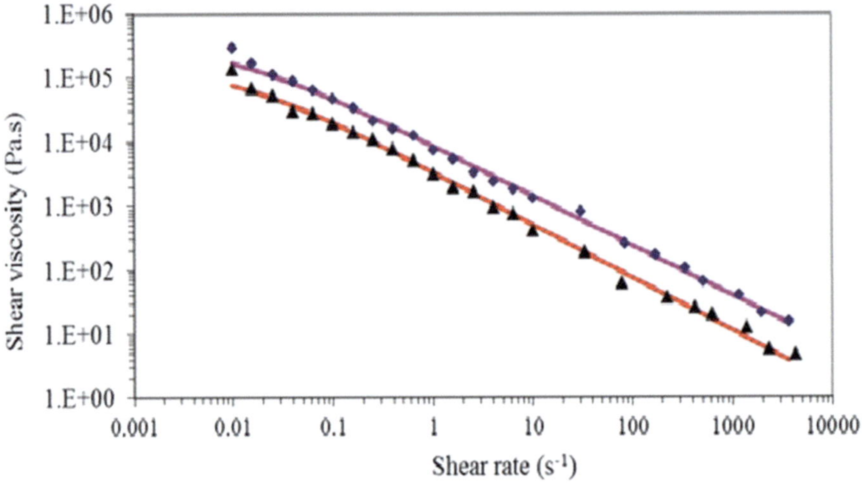


**Fig. 4.5** Steady shear viscosity of LCP-A at (*filled diamond*) 350 °C and (*filled square*) 360 °C using modified Cross model (▬, ▬) and steady shear experiment (Rahman 2013)

## 4.4.2 First Normal Stress Difference

The filled LCPs exhibited negative first normal stress difference ($N_1$) in the low shear rate region (<10 $s^{-1}$). High shear rate results could not be obtained due to the slippage effect. It was observed that the first normal stress difference for filled LCP-A decreased up to the shear rate of 0.09 $s^{-1}$ (Fig. 4.6). After this critical shear rate, although still negative, $N_1$ started to increase. In the case of LCP-B filled, $N_1$ initially decreased up to 0.12 $s^{-1}$ and then started to increase. Filled LCP-A showed a significant decrease in comparison to the filled LCP-B sample. Both the filled LCPs showed anomalous negative first normal stress differences at low shear rate, which is in accordance with previous theoretical studies (Marrucci and Guido 1995). Whereas, in the case of unfilled LCPs the value of $N_1$ remained positive (Fig. 4.7) (Rahman et al. 2013; Narimissa et al. 2013).

The negative normal stress difference is one of the most important rheological characteristics of LCPs. For rod-like molecules of LCPs (Doi 1981) shear induced transitions between tumbling, i.e. rotation of molecules at their own axis, wagging and flow aligned regimes are responsible for the unusual behaviour of the first normal stress differences (Baek et al. 1993; Beak and Magda 1993). LCPs exhibit different types of orientation at different shear rate regions (Rienacker and Hess 1999). The rod-like molecules start to tumble with shear rate and rotate from tumble to wagging beyond certain critical shear rate. The wagging region transforms to flow aligning region with the increase of shear rate.

At a reasonably large shear rate, the ellipsoid shape of domain of LCPs gets deformed and the rotation of molecule tends to become perpendicular to the flow axis. Hence, the value of $N_1$ becomes negative at low shear rates (Fig. 4.6).

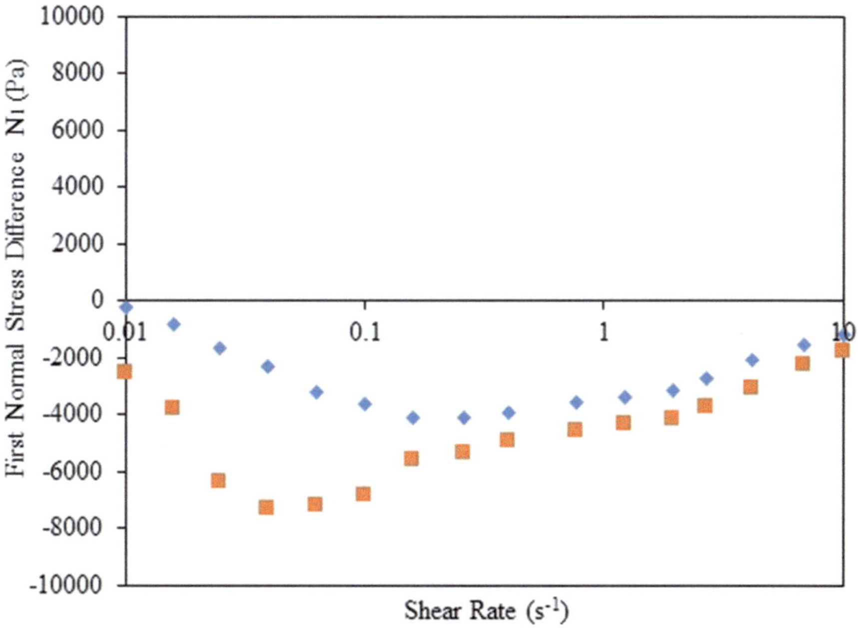


**Fig. 4.6** First normal stress difference for LCP-A (■) Filled at 350 °C and LCP-B Filled (◆) at 330 °C (Narimissa et al. 2013)

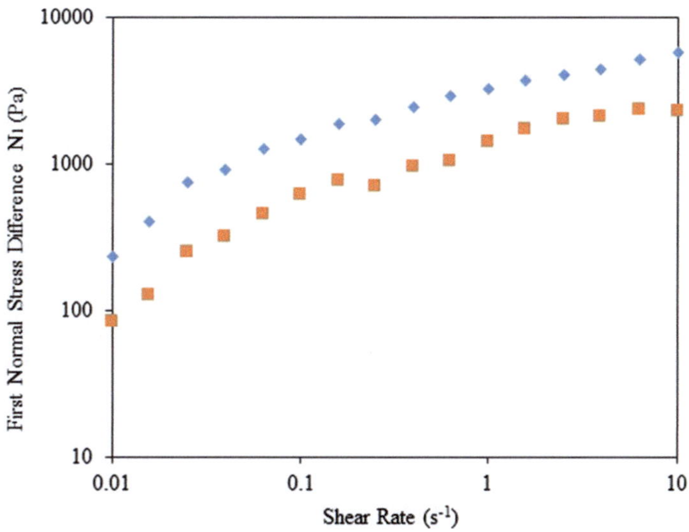


**Fig. 4.7** First normal stress difference for LCP-A (■) and at 350 °C LCP-B (◆) Unfilled at 330 °C (Rahman 2013)

However, when the shear rate is high, the value of $N_1$ becomes positive again. This will occur when the strength of the flow becomes so large that tumbling is suppressed altogether, and the molecules increasingly align in the shear direction (Marrucci and Guido 1995). Besides, the value of $N_1$ also depends on the concentration of fillers within LCP matrix. Beak and Magda (1993)) reported the change of $N_1$ behaviour of lyotropic LCPs from negative to positive as a function of shear rate as well as filler concentration. It has been observed that this anomalous $N_1$ behaviour was enhanced due to the presence of the filler. The interaction between the fiber-like fillers offsets the force attempting to align the fibers with the direction of the flow where substantial resistances to deformation were observed at low shear rates (Litchfield and Baird 2006). The above interactions are overcome due to the orientation of fibers in the direction of flow at higher shear rates; hence, the arrangement of LCPs matrix and fibers is disrupted (Kamal and Mutel 1985).

As mentioned both filled LCPs have glass fibers as the reinforcing material and the aspect ratio of these glass fibers is approximately 100 (Rahman et al. 2013). At low shear rates, the interaction between the fibers compensates the force attempting to align the fibers with flow. Consequently, substantial resistance to deformation occurs at low shear rates. Moreover, fiber-fiber interactions compete with the elasticity of rigid rod-like molecules of LCPs at low shear rate region. Hence, this interaction could suppress the first normal stress difference and leads to negative values. Likewise, such anomalous behaviour of first normal stress difference has also been observed in concentrated suspensions where filler elements with high aspect ratio demonstrated competitiveness with elasticity at the low shear rate region and resulted in the occurrence of negative values of first normal stress difference (Susanne et al. 2002). Litchfield and Baird (2006) demonstrated that the suspension filled with long fiber (aspect ratio $> 100$) showed unusual rheological behaviour and the value of first normal stress difference became negative at low shear rates (Litchfield and Baird 2006).

According to several studies, the magnitude of first normal stress difference may increase as a result of the incorporation of short fibers (Ramazani et al. 2001; Litchfield and Baird 2006; Eberle et al. 2008). However, the abnormal orientation of the existing rigid rod-like molecules in unfilled LCPs is responsible for the increase of the first normal stress differences. During the shear deformation, the rigid rod-like molecules of unfilled LCPs start to tumble (Marrucci 1991) which could enhance the elasticity and, consequently, this could lead to the observed rise in the magnitude of the first normal stress difference samples as depicted for unfilled LCPs (Fig. 4.7).

### *4.4.3 Molecular Weight Distribution (MWD) in TLCPs*

The dynamic rheological measurements, i.e. storage and loss moduli can be used to determine the MWD. Size exclusion chromatography (SEC) has been around for decades for determining MWD. It requires dissolution of the sample for SEC and

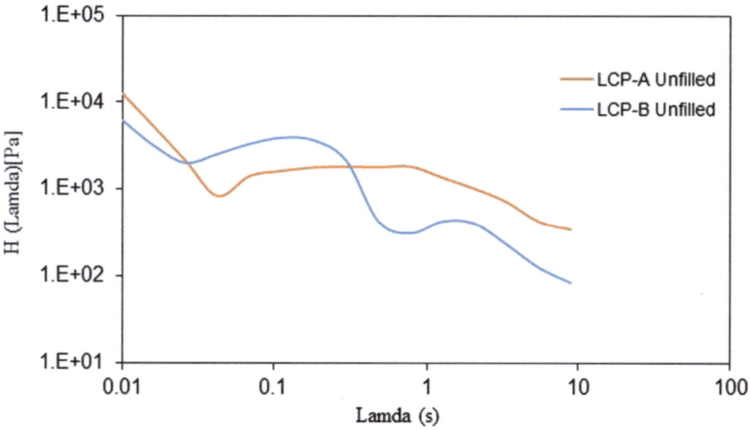


**Fig. 4.8** Relaxation spectrum for LCP-A and LCP-B Unfilled (Rahman 2013)

some polymers are difficult to dissolve, like semi crystalline fluropolymers made from tetrafluroethylene and some liquid crystal polymers (Thomas et al. 1994; Tuminello 1999). Mead and Friedrich et al. (Maier et al. 1998; Mead 1994) developed the algorithm (modulus model) to determine the MWD using rheological data, which carry the same information like SEC. This model formed the basis of the molecular weight distribution module in Rheometric Scientific's Orchestrator software (Costello 2005). In recent study (Rahman et al. 2013), MWD was determined from the experimental rheological data using Orchestrator software. The detail rheological analysis will be discussed in the next section.

To avoid the time dependency, measurements were taken in the region 0.1–100 rad/s (considering 10 points per decade). The relaxation spectrum H ($\lambda$) (at 350 °C for LCP-A and 330 °C for LCP-B) was directly calculated from the dynamic modulus using the Orchestrator software and the spectrum H ($\lambda$) is shown in Fig. 4.8.

The existence of solid and liquid phases in melting process of LCPs, makes the relaxation behaviour of LCPs extraordinary (Gao et al. 1996). In addition, the presence of branches and defects is also dominant in the relaxation process of LCPs. In general, there are three kinds of relaxation mechanisms for nematic LCP: the relaxation of chain orientation, the relaxation of deformed polydomains, and the coalescence of polydomains (or the relaxation of textures) (Yu et al. 2007). The relaxation of chain orientation without constraint is a much faster process (Wiberg et al. 1998). The relaxation of deformed polydomains takes 0.018 s for LCP-A and 0.015 s for LCP-B, which is reflected as the first peak in the spectrum. The relaxation of texture is a slow process, which requires hundreds of seconds. An approximation formula based on the double reptation rule is used to measure the transformation of $H(\lambda)$ into MWD in the rheometric software. The relaxation spectra, in Fig. 4.8 shows good agreement with the published studies (Yu et al. 2007). Equation (4.7) was used for transforming spectrum data into MWD.

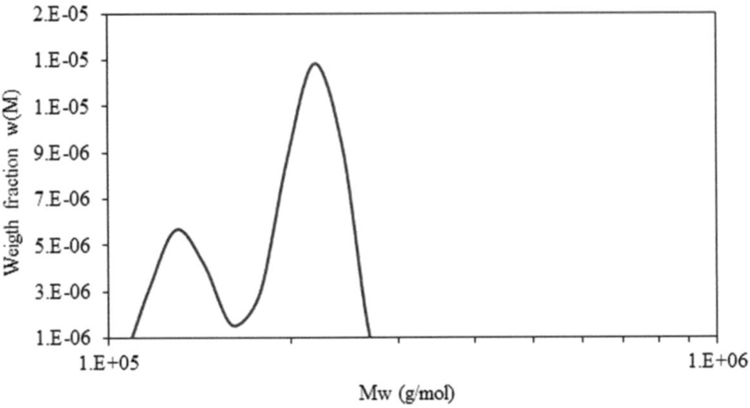


**Fig. 4.9** Molecular weight distribution for LCP-A Unfilled (Rahman 2013)

$$G_r(t) = G_N \left( \int_{Me}^{\infty} F(M,t)^{\frac{1}{\beta}} w(M) \frac{dM}{M} \right)^{\beta} \tag{4.7}$$

In (4.7), $G_r$ is the reptation modulus $G_N$ is the plateau modulus, and $M_e \approx M_c/2$ is the entanglement molecular weight (Mc is the critical molecular weight). F(M, t) denotes the relaxation kernel function, which describes the relaxation behaviour of a molecular weight fraction with a molecular weight of M, and $\beta$ is a parameter which characterizes the mixing behaviour. The required parameters were given to the orchestrator software (ARES machine) from published research data for the transformation of $H(\lambda)$ into MWD i.e. critical molecular weight ($M_c$), activation energy (Tanaka 2007) and $\beta$ parameter (Tuminello 1999). Figures 4.9 and 4.10 represent the MWD for LCP-A and LCP-B.

From this MWD analysis, it has been observed that LCP-A has a higher molecular weight tail and molecular weight distribution (MWD) (Fig. 4.9) in comparison to that of LCP-B (Fig. 4.10). The broad molecular weight distribution (MWD) in LCP-A and a second peak at high molecular weight is likely to indicate the presence of branching with the main chain of LCP (Han and Kwack 1983). From FTIR analysis, it was also observed the existence of branching (side chain) in LCP-A. For commercial grade LCPs , the effects of branching and MWD are significant in melt rheology (Dealy and Wissburn 1990).

### 4.4.4 Dynamic Shear Rheology

Dynamic rheological properties of the melt are very sensitive to the molecular structure of a polymer. It is essential to perform dynamic strain sweep test before

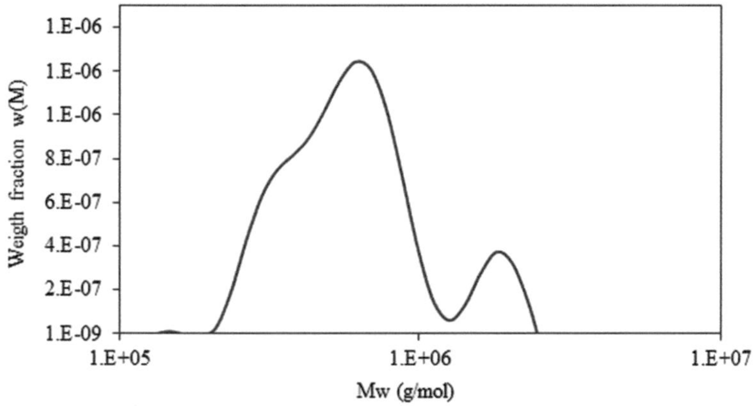


**Fig. 4.10** Molecular weight distribution for LCP-B Unfilled (Rahman 2013)

performing dynamic frequency sweep test to confirm the linear viscoelastic region (LVR). In LVR, the rheological experiments were performed under small deformation. In the LVR, the configuration of the macromolecules would remain unperturbed by the flow history (Carreau et al. 1997). The microstructure of the material would not be affected by shear alignment during experiment. Both the storage and loss moduli are independent of strain applied in the linear viscoelastic region. The molecules of LCPs are rigid rod like and it is very essential to identify the LVR limit of these materials at different temperatures. The complete characterisation of the linear viscoelastic behaviour of the LCP materials can be considered, when the two material functions, storage ($G'(\omega)$) and loss moduli ($G''(\omega)$) are known with sufficient accuracy over the range of frequencies from very low to a frequency higher than the reciprocal of the shortest relaxation of time (Dealy and Larson 2006).

Time evolution of the morphological state in the nematic region of a specimen after being subject to steady or transient flow, can be monitored in terms of storage modulus ($G'$) or complex viscosity. Moreover, the oscillatory measurement allows applying high angular frequency compared to steady shear measurement. The LCP molecules are very sensitive to thermal history. It has been observed that the rheological properties of LCPs are continuously varied with annealing time at below nematic–isotropic transition temperature. Moreover, LCP molecules exhibit pronounced phase transition behaviour during oscillatory measurements. At low frequency, the molecules of nematic phase start to transform to isotropic phase (Fan et al. 2003).

The rheological properties of liquid crystalline polymers (LCPs) have been observed as a function of time at constant temperature and at constant frequency. The time dependent parameter is very important for polymer processing. In general the cycle time of injection moulding is low. However, the cooling process in a given cycle time, which is influenced by time dependent property. Under constant shear rate, the molecules tend to align in the direction of the flow as well as tumble at their own axis. Viscosities, elastic and viscous moduli increased with time at constant frequency and at constant temperature. The abnormal orientation and architecture of LCP molecules lead to continually increasing crystalline phase of LCP molecules at constant shear rate under isothermal condition (Xie et al. 2011). The time dependent behaviour is pronounced at low shear rate. However, the recrystallization processes are opposed by high shear rate.

Rahman et al. (2013) investigated the complex viscosity as a function of temperature and frequency for four LCPs before thermal degradation. The purpose of this study is to identify the nematic– isotropic phase transition temperature as well as the influence of dynamic shear on the phase transition. In LCPs there is a characteristic increase in viscosity with temperature (Wissbrun 1981; Kim and Han 1994; Fan et al. 2003) and in literature the onset of temperature known as phase transition or nematic-isotropic ($T_{N-I}$) transition temperature.

This anomaly of increasing viscosity with temperature is attributed to changes in the morphological state with the length of the biphasic window increasing with impurities (mixing of nematic and isotropic phase) and molecular weight distribution (Scribben 2004). The poly-disperse nature of rigid rod like molecules, the low molecular weight portion of the polymer begins to transform into an isotropic region at temperatures below the nematic-isotropic transition. On the other hand, in monodisperse system, a discontinuous increase between the liquid crystalline and isotropic state would be expected. Since the viscosities of both the isotropic and nematic phases decrease with temperature, this anomaly is possible for the increase of impurity (i.e. mixing of nematic and isotropic) (Kiss 1986). The gradual increase of viscosities at higher temperatures can be described by the biphasic nature of these LCPs, that is, due to thermal activation, part of the rod-like molecules of nematic phase are transformed to the isotropic phase. The existence of biphasic nature increases more repulsion to flow and thus viscosities increase in this region (Gao et al. 1996; Hsies et al. 1999).

Fan et al. (2003) also observed similar types of anomalous thermal dependency for LCPs (Fig. 4.11) (Fan et al. 2003). Many researches have reported nematic to

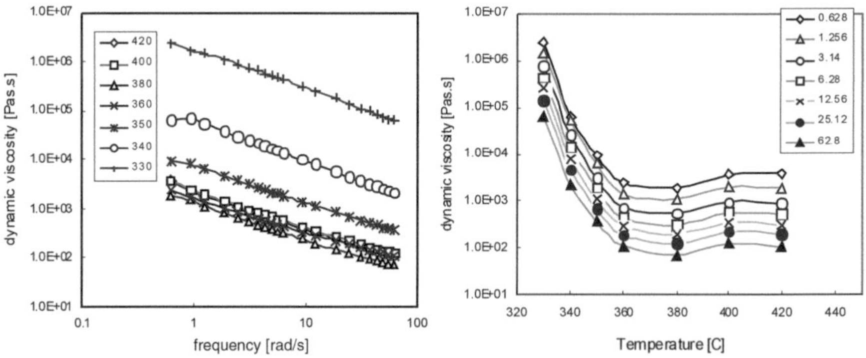


**Fig. 4.11** Measurement of viscosity as function of frequency and temperature (Fan et al. 2003)

isotropic phase transition with temperature only. However, no clear evidence was found about the shear or frequency range where the viscosity increases with temperature. From the recent study (Rahman et al. 2013), it can be concluded that the nematic-isotropic phase transition occurred only at low shear rate region with the increase of temperature.

### 4.4.5 Shear Induced Crystallization

In general, thermotropic liquid crystal polymer (TLCP) melts exhibit a comparable macroscopic rheological properties to that of conventional polymer melts (Mackley 1987) to a certain extent. However, the molecules of TLCPs orient much more spontaneously in shear and extensional flows (Moldenaers et al. 1990; Alderman and Mackley 1985), and exhibit anomalous rheological behaviour in the temperature range immediately above the melting point (Lin and Winter 1988). Nicholson et al. (1992) studied shear-induced crystallization in liquid crystal random copolyester and found a sharp crystalline diffraction peak being sheared above the liquid crystal polymer melting point. They revealed that the crystalline formation in LCPs was the result from similar sequences of monomer units on adjacent chains lining up, which produced non-periodic layer crystals (Nicholson et al. 1992).

The liquid crystal polymers strongly depend on prior thermal history and shear (Berghausen et al. 1997; Chandrasekhar 1994; Guskey and Winter 1991). Rienacker and Hess (1999) studied theoretically shear induced orientational dynamics of nematic liquid crystals (Rienacker and Hess 1999) and reported the existence of different types of orientation in different shear rate regions. Numerous studies have also reported the orientation of rigid rod like molecules and that the transformation of phases from nematic to isotropic are depended on shear. In 1993, Kim and Han studied the oscillatory shear flow behaviour of thermotropic liquid crystal polymer and reported anomalous rheological behaviour in the nematic state. In that study, it was reported that in the nematic phase the rheological properties varied continuously and it was not possible to get meaningful rheological measurements. Zhou et al. (1999) studied dynamic and shear orientation behaviour of thermotropic LCPs and observed no terminal behaviour of moduli (Zhou et al. 1999). It was considered that in descending frequency sweep and at the low frequency region moduli reached plateau. Yu et al. (2007) studied dynamic rheology of thermotropic liquid crystal polymers and also obtained plateau values for the moduli in the low frequency region (Yu et al. 2007). No specific details were mentioned regarding the reason of reaching plateau values.

Rahman (2013) investigated the reason for anomalous behaviour in the low shear rate region. All the tests were conducted in the nitrogen environment and the samples were directly quenched in liquid nitrogen for further *in situ* investigation in (wide angle X-ray diffraction) WAXD analysis. The WAXD analysis reported that the percentage of crystallinity increased after shearing of all four LCPs. Figure 4.12 represents the WAXD analysis for four LCPs. WAXD diffraction is used to

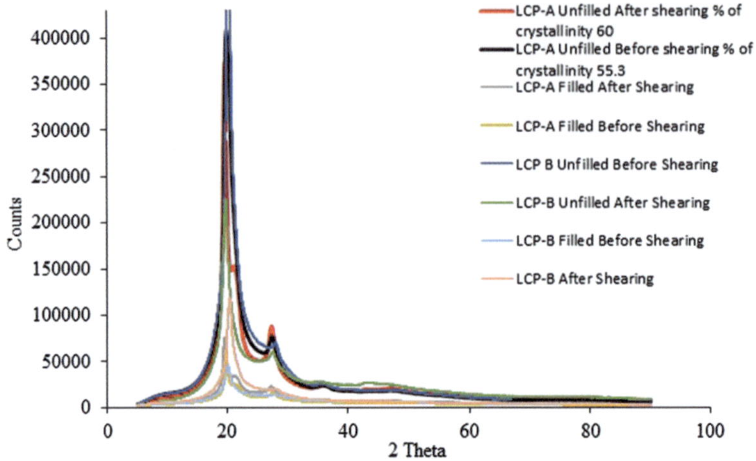


**Fig. 4.12** WAXD analysis for all LCP samples (Rahman 2013)

measure the nature of polymers and the extent of crystallinity present in these LCPs. In these experiments, an X-Ray of a given wavelength impinges on the crystal structure and diffracted into a large number of narrow beams imperceptible by the photographic film or by Geiger counter. The angle of scattering depends on the order of the chains within the polymer. WAXD study of the semicrystalline polymers shows wider peaks on a raised background that indicates the inter atomic distance of the less ordered crystalline planes and sharp peaks indicate the presence of crystalline structure (Halasa et al. 1991).

The TOPAS software of BRUKER machine (WAXD analyzer) was able to calculate the crystalline and amorphous area. For all the four samples, it was observed that the percentage of crystallinity increased after shearing the samples. In molten state, the rigid rod like molecule of LCPs remains in crystalline phase and the molecules are flowing in an ordered way. After applying shear (in the low shear rate region) the rod like molecules starts to form more order. In this region, the nematic phase of LCP molecules are transferred to smectic phase which is more ordered and thus crystallinity increased. With the increase in shear, the polymer chains would move with respect to one another and thus crystals layer formed.

### *4.4.6 Extensional Rheology of TLCPs*

The shear rheological properties of thermotropic liquid crystalline polymers (TLCPs) have been studied extensively whereas, the extensional behaviour of TLCPs has become the subject of very few studies (Gotsis and Odriozola 2000). The unusual orientation and structure of thermotropic liquid crystal polymers (TLCPs) can significantly affect the rheological and mechanical properties of

TLCPs. The extensional flows have been shown to be much more effective compared to shear flow in producing a high degree of molecular orientation (Yu et al. 2005).

Kernick and Wagner (1999) studied the transient extensional viscosity of thermotropic copolyester (HX-8000-210 made by DuPont) (Kernick and Wagner 1999). They reported that the extensional viscosity elongated uniaxially for several extension of the accumulated strain. Gotsis and Odriozola (2000) conducted extensional rheology studies on the thermotropic liquid crystalline polymer Vectra A950. Compression and injection molded samples were used for the extensional rheological experiments. The extensional viscosity of TLCP (Vectra 900) elongated uniaxially for several extension of the accumulated strain. It was demonstrated that all the extensional viscosities for different extension rates collapsed into one master curve and increased strongly with increase in strain. Yu et al. (2005) also investigated the extensional viscosity of thermotropic liquid crystalline polymer; using disintegration of liquid crystalline polymer threads in a quiescent polymer matrix (Yu et al. 2005). Extensional rate thinning behavior and strain-hardening behavior were observed.

### *4.4.7 Leonov's Model for TLCPs*

The unusual characteristic of LCP molecules described earlier makes the rheological properties of TLCPs exceptional compared to conventional polymers. The viscosities of LCPs also exhibit abnormal temperature dependency indicating that LCPs are not thermo-rheologically simple (Brostow et al. 1999). The distinct nematic phenomenon of low-molecular-weight liquid crystals plays a vital role in determining the rheological properties of LCPs. The viscoelastic characteristics of these LCPs are anisotropic and defined by essentially a larger number of constitutive scalar parameters. The experimental determination of these anisotropic properties remains a challenging problem. Recently, Leonov (Leonov 2008a, b; Leonov and Chen 2010) had a significant breakthrough in viscoelastic nematodynamics. In Leonov's nematodynamics some new theoretical techniques (Leonov 2008b) were employed to derive the simplest form of the continuum theory of weakly nonlinear viscoelastic nematodynamics.

Both the continuum and molecular theories dealt with the problem of modelling LCP properties. The continuum theories try to establish a general framework with minimum assumptions about the molecular structures of LCPs. However, many material parameters related to both the basic properties of symmetry and interactions are described by state variables (Leonov and Chen 2010). In contrast, the molecular approaches employ very specific assumptions of structure for these polymers, but unlike the continuum approaches, operate with few molecular parameters. These theories, although well separated, are not contradictory but complementary (Leonov 2008a).

Leonov has modified his model in the last two decades. The term nematodynamics was first introduced by de Gennes and Prost in 1974 (Leonov 2008b). In 2002, Leonov and Volkov, proposed a theory of viscoelastic nematodynamics for nematic liquid crystal polymers (Leonov and Volkov 2002). This theory is applicable for analyzing flows of concentrated polymer suspensions and nano-composites filled with uniaxially symmetric particles. Again, Leonov and Volkov (2005) developed a continuum theory of weak viscoelastic nematodynamics of Maxwell type (Leonov and Volkov 2005). The assumption of small transient (elastic) strains and relative rotations, employed in the theory, seems to be appropriate for LCP's with rigid enough macromolecules, and for slow flows of viscoelastic suspensions with uniaxially shaped particles.

In 2008, Leonov proposed a model of weak viscoelastic nematodynamics. This model was developed based on a continuum theory of weak viscoelastic nematodynamics of Maxwell type, which can describe the molecular elasticity effects in mono-domain flows of liquid crystalline polymers as well as the visco-elastic effects in suspensions of uniaxially symmetric particles in polymer fluids (Leonov 2008b). The viscoelastic nematodynamics combined the viscous and the elastic stresses and expressed them as total stress. Moreover, this model also considers some specific problems in LCP. The nematodynamic studies are important for processing of nematics and prediction of their post-processing properties (Leonov 2008a). In a separate study, Leonov (2008a, b) also developed algebraic theory of viscoelastic nematodynamics, which describes the detail algebra of nematodynamics as well as the use of Leslie-Ericksen-Parodi (LEP) approach for viscous nematics and de Gennes free energy for weakly elastic nematic elastomers (Leonov 2008a).

These assumptions seem to be appropriate for LCPs with rigid enough macro-molecules, and for slow flows of viscoelastic suspensions with uniaxially shaped particles (Leonov 2008b).This theory utilized a specific viscoelastic and nematic kinematics and ignored inertia effects and in cases of LCPs, the effects of Frank elasticity (Leonov 2008b). The rigid rod like molecules are often called as "self-reinforcing thermoplastic" (Ophir and Ide 1983). The rheological behaviour as well as the development of orientation is similar to the short fiber-filled composite. In this study, the continuum theory of viscoelastic nematodynamics is considered to be applicable for filled LCPs.

There are a large number of unknown parameters in the model of continuum or molecular theories. A lot of problem has been encountered during modelling of rheological properties of liquid crystalline polymers using those theories. In a recent study by Rahman et al. (2013), the simulation works for a recently developed weak viscoelastic nematodynamics have been carried out based on the assumption of smallness of elastic (transient) strains and elastic relative rotation. The model parameters were determined by solving the constitutive equations as demonstrated by (Leonov and Chen 2010) and presented in (4.8)–(4.10). Moreover, the model predictions have been validated by comparing with the existing experimental data.

Considering the steady state conditions, ( $\delta\sigma/\delta t = 0$ ) the three differential equations can be rewritten as follow:

$$\theta_0 \dot{\gamma} \frac{\sigma_{11} - \sigma_{12}}{2} + \sigma_{12} + (r_1 - 1)[\sigma_{12} + n_1 n_2(\sigma_{12} + \sigma_{11})] + (\sigma_{11} n_1^2 + \sigma_{22} n_2^2 + 2\sigma_{12} n_1 n_2)$$
$$n_1 n_2 \left[\frac{3}{2}(r_2 - 1) - 2(r_1 - 1)\right] = \frac{\dot{\gamma}\eta_0}{2}\left(1 + \alpha + 2n_1^2 n_2^2\left(\beta - 2\alpha - \frac{3}{2}\right)\right) \tag{4.8}$$

$$-\theta_0 \dot{\gamma} \sigma_{12} + \sigma_{11} + 2(r_1 - 1)(\sigma_{11} n_1^2 + \sigma_{12} n_1 n_2) + \left(\frac{3}{2}(r_2 - 1)(n_1^2 - 1/3) - 2(r_1 - 1)n_1^2\right)$$
$$\times (\sigma_{11} n_1^2 + \sigma_{22} n_2^2 + 2\sigma_{12} n_1 n_2) = \dot{\gamma}\eta_0 n_1 n_2(\alpha(1 - 2n_1^2)) + \left(\beta - \frac{3}{2}\right) \times \left(n_1^2 - \frac{1}{3}\right) \tag{4.9}$$

$$\theta_0 \dot{\gamma} \sigma_{12} + \sigma_{22} + 2(r_1 - 1)(\sigma_{22} n_2^2 + \sigma_{12} n_1 n_2) + \left(\frac{3}{2}(r_2 - 1)(n_2^2 - 1/3) - 2(r_1 - 1)n_2^2\right)$$
$$\times (\sigma_{11} n_1^2 + \sigma_{22} n_2^2 + 2\sigma_{12} n_1 n_2) = \dot{\gamma}\eta_0 n_1 n_2(\alpha(1 - 2n_2^2)) + \left(\beta - \frac{3}{2}\right) \times \left(n_2^2 - \frac{1}{3}\right) \tag{4.10}$$

Equations (4.8)–(4.10) have been solved in simple steady state shear flow using "Mathematica" software (Leonov and Chen 2010). The stress components are expressed as function of shear rate $\dot{\gamma}$ with the values of constitutive parameters $\theta_0, \alpha, \beta, r_1, r_2, \lambda_e, \lambda_v$ and $\eta_0$. Here $\theta_0$ and $\eta_0$ represent relaxation time and zero shear viscosity respectively. The other parameters $\lambda_e \, and \, \lambda_v$ represent the tumbling for elasticity and viscosity. Rest of the characteristic parameters $\alpha, \beta, r_1, r_2$ represent anisotropic properties of liquid crystal polymers. Among the eight parameters, only relaxation time and zero shear viscosity are determined from experimental data. The other six parameters can be obtained from curve fitting data using the Mathematica software.

Rahman et al. (2013) carried out steady shear experiments for four LCP samples at the low and high shear rate region. For low shear experiments, the shear rate considered was between 0.01 and 10 $s^{-1}$, because in the parallel plate geometry, slippage occurred at shear rate above 10 $s^{-1}$. The high shear rate range was approximately from 10 to 6000 $s^{-1}$. Most of the abnormal behaviours of LCPs are observed in the low shear rate region. In this region, tumbling of molecules, overshoots, negative first normal stress difference and abnormal temperature dependencies were observed. The flow in the low shear rate region can be described as anisotropic flow. On the other hand, at the high shear rate region, all the abnormal rotation of rod like molecules are suppressed and molecules are aligned to the flow directions. In this region the flow can be expressed as isotropic flow.

Figures 4.13 and 4.14 represent the experimental viscosity values and the viscosities from model predictions at the high shear rate region. It is observed that in the high shear rate region (10–2000 $s^{-1}$) the model predictions show a good

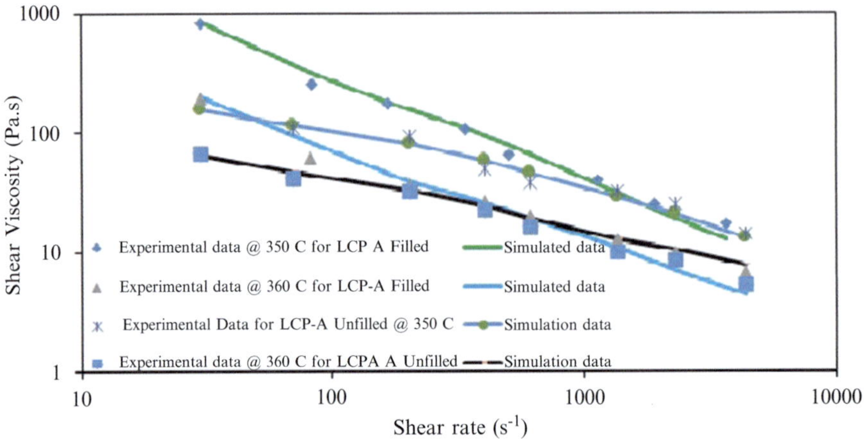


**Fig. 4.13** Experimental and simulated data for LCP-A (Rahman et al. 2013)

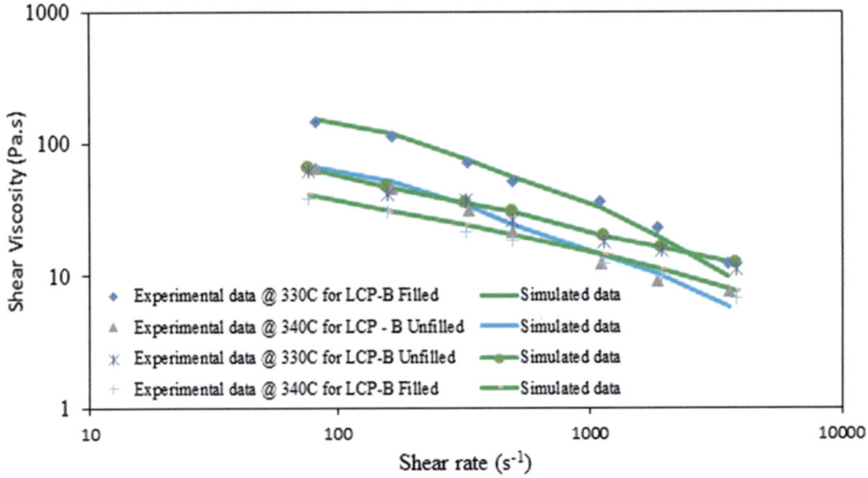


**Fig. 4.14** Experimental and simulated data for LCP-B (Rahman et al. 2013)

fit with the experimental data. In contrast for filled LCPs, at a very high shear rate (above 2000 $s^{-1}$) this model predicted more shear thinning behaviour compared to the experimental data.

Due to enormous measurement difficulties, the elasticity of melt polymeric systems has not been studied to the same extent as shear measurements (Dealy and Wissburn 1990; Shenoy 1999). Some of the common methods of evaluation of the elasticity include normal force measurements, capillary flow entrance pressure drop and extrudate swell. The approach by Rahman et al. was to measure $N_1$, as an indication of the elasticity of melt polymers in steady shear experiment.

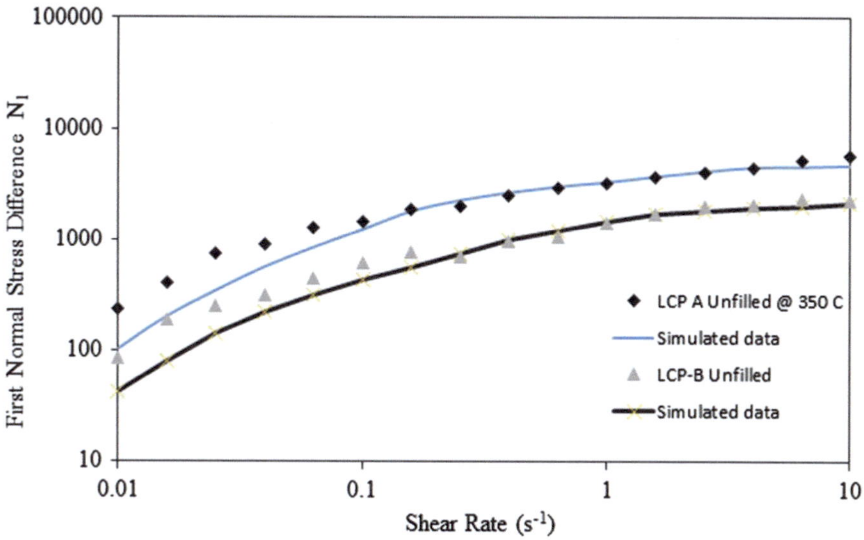


**Fig. 4.15** First normal stress difference data for LCP-A and LCP-B Unfilled (Rahman et al. 2013)

The negative first normal stress difference ($N_1$) could not be predicted by Leonov's nematodynamics model. The values of $N_1$ for filled LCPs are shown as a trend beyond 10 $s^{-1}$..

The experimental trend for both filled LCPs indicated good agreement with the simulation data. For unfilled LCPs, the $N_1$ values were calculated from model predictions. Good agreement is observed between experimental data and model predictions for unfilled LCPs for the entire shear rate region as shown in Fig. 4.15.

## 4.5 Concluding Remarks

The recent advance in rheological behaviour of filled and unfilled thermotropic liquid crystal polymers has been explored. There are many similarities in the rheological behaviour between filled and unfilled TLCPs and the overall rheological response is the same for different types of TLCPs. The TLCPs molecules exhibited different types of orientation at different shear rates i.e. at the low shear rate, anisotropic flow and at high shear rate, isotropic flow. Incorporation of fillers also changed the rheological properties of TLCPs drastically. In the low shear rate region, the filled TLCPs exhibited higher viscosity compared to unfilled TLCPs. However, in the high shear rate region (shear rate > 2000 $s^{-1}$) the viscosity of filled TLCPs showed more shear thinning behaviour compared to the unfilled LCPs due to rheological hybrid effects. Moreover, the filled thermotropic LCPs exhibited negative first normal stress difference in the low shear rate region. The abnormal orientation and architecture of LCP molecules lead to continually increasing crystalline phase of LCP molecules at constant shear rate under isothermal condition. From this study, it was also identified that the rod like molecules of LCPs become more crystalline with low constant agitation due to shearing i.e. its crystallinity increased up to certain critical shear rate and followed by decrease with further shear of the molecules. WAXD analysis also indicated the increase of crystallinity with shear.

Leonov viscoelastic nematodynamics model was chosen for rheological simulation of TLCPs. The simulated viscosities have been compared with experimentally measured viscosity data, indicating a good fit in the shear rate range of 10–2000 $s^{-1}$. In the low shear rate region, the model did not predict negative normal stress difference for filled TLCPs. For unfilled TLCPs, good comparison has been observed between the simulated values of $N_1$ and the experimental data.

## References

Alderman N, Mackley M (1985) Optical textures observed during the shearing of thermotropic liquid-crystal polymers. Faraday Discuss Chem Soc 79:149–160

Antoun S, Lenz RW, Jin I (1981) Liquid crystal polymers. IV. Thermotropic polyesters with flexible spacers in the main chain. J Polym Sci Polym Chem Ed 19:1901–1920

Baek SG, Magda JJ, Cementwala S (1993) Normal stress differences in liquid crystalline hydroxypropyl cellulose solutions. J Rheol 37:935–945

Barnes HA (2003) A review of the rheology of filled viscoelastic systems. In: The British Society of Rheology, pp 1–36

Beak SG, Magda JJ (1993) Rheological differences among liquid-crystalline polymers. 1. The first and second normal stress differences of PBG solutions. J Rheol 37:1201–1224

Berghausen J, Fuchs J, Richtering W (1997) Rheology and shear orientation of a nematic liquid crystalline side-group polymer with laterally attached mesogenic units. Macromolecules 30:7574–7581

Bhattacharya SN, Gupta RK, Bhattacharya SN (2010) The rheology of polymeric nanocomposite. In: Gupta RK, Kennel E, Kim K-J (eds) Polymer nanocomposites handbook. Taylor & Francis, New York

Brostow W (1992) An introduction to liquid crystallinity (Liquid crystalline polymers: from structures to applications). Elsevier Applied Science, New York

Brostow W, D'Souza NA, Kubat J, Maksimov R (1999) Creep and stress relaxation in a longitudinal polymer liquid crystal: prediction of the temperature shift factor. J Chem Phys 110:9706–9712

Carreau PJ, Dekee DCR, Chhabra RP (1997) Rheology of polymeric systems: principles and applications. Hanser, New York

Chandrasekhar S (1994) Liquid crystals. Cambridge University, Cambridge

Chen H (2008) Simulations of shearing rheology of thermotropic liquid crystal polymers. Master of Science, The University of Akron

Cheng HKF, Basu T, Sahoo NG, Li L, Chan SH (2012) Current advances in the carbon nanotube/thermotropic main-chain liquid crystalline polymer nanocomposites and their blends. Polymers 4:889–912

Costello B (2005) Use of rheology to determine the molecular weight distribution of polymers. Annu Trans Nordic Rheol Soc 13:1–5

Dae Han C, Kim JK (1993) On the use of time-temperature superposition in multicomponent/multiphase polymer systems. Polymer 34:2533–2539
De Gennes PG, Prost J (1993) The physics of liquid crystal. Clarendon, Oxford
Dealy J, Wissburn K (1990) Melt rheology and its role in plastics processing. Van Nostrand Reinhold, New York
Dealy JM, Larson RG (2006) Structure and rheology of molten polymers: from polymerization to processability via rheology. Cincinnati Hanser Gardner, Cincinnati
Doi M (1981) Molecular dynamics and rheological properties of concentrated solutions of rodlike polymers in isotropic and liquid crystalline phases. J Phys Sci 19:229–243
Doi M, Edwards SF (1978) Dynamics of rod-like macromolecules in concentrated solution. Part 1. J Chem Soc Mol Chem Phys 74:560–570
Donald AM, Windle AH, Hanna S (2006) Liquid crystalline polymers. Cambridge University, Cambridge
Done D, Baird DG (1990) Transient flow of thermotropic liquid crystalline polymers in step strain experiments. J Rheol 34:62749-7
Eberle PR, Baird DG, Wapperom P (2008) Rheology of non-Newtonian fluids glass fibers: a review of experimental literature. Ind Eng Chem Res 47:3470–3488
Fan Y, Shaocong D, Tanner IR (2003) Rheological properties of some thermotropic liquid crystalline polymers. Korea-Aust Rheol J 15:109–115
Felter SP, Dourson ML (1997) Hexavalent chromium-contaminated soils: options for risk assessment and risk management. Regul Toxicol Pharmacol 25:43–59
Gao P, Lu XH, Chai CK (1996) Rheology of low nematic transition temperature thermotropic liquid crystalline copolyester HBA/HQ/SA. Polym Eng Sci 36:2771–2780
Gotsis AD, Odriozola MA (2000) Extensional viscosity of a thermotropic liquid crystalline polymer. J Rheol 44:1205–1223
Graziano DJ and Mackley MR (1984) Shear induced optical textures and their relaxation behaviour in thermotropic liquid crystalline polymers. Mol Cryst Liq Cryst 106:73–93
Guskey SM, Winter HH (1991) Transient shear behavior of a thermotropic liquid crystalline polymer in the nematic state. J Rheol 35:1191–1207
Halasa AF, Watjen GD, Hsu WL, Martana BA, Massie JM (1991) Relationship between interchain spacing of amorphous polymers and blend miscibility as determined by wide-angle X-ray scattering. J Appl Polym Sci 43:183–190
Han C, Kwack T (1983) Rheology-processing-property relationships in tubular blown film extrusion. J Appl Polym Sci 28:3399–3418
Han CD, Victor MU, Burghardt WR (2001) Shear stress overshoots in flow inception of semi-flexible thermotropic liquid crystal polymers: experimental test of a parameter-free model prediction. Macromolecules 34:3642–3645
Hashimi S, Takeshi K (2007) Shear rate dependence of viscosity and first normal stress difference of LCP/PET blends at solid and molten states of LCP. J Appl Polym Sci 104:2212–2218
Hsies TT, Tiu C, Simon GP, Wu RW (1999) Rheology and miscibility of thermotropic liquid crystalline polymer blends. J Non-Newt Fluid Mech 86:15–35
Isayev AI (2012) Liquid crystalline composite. In: Luigi N, Borzacchiello A (eds) Wiley Encyclopedia of composites, 2nd edn. Wiley, New York
Isayev AI, Nicolais L (2011) Liquid crystalline composite. Wiley Encyclopedia of composites. Wiley, New York
Kalika DS, Giles DW, Denn MM (1989) Shear and time-dependent rheology of a fully nematic thermotropic liquid crystalline copolymer. J Rheol 34:139–154
Kamal MR, Mutel A (1985) Rheological properties of suspensions in Newtonian and non-Newtonian fluids. Polym Eng Sci 5:293–382
Kernick WA, Wagner NJ (1999) Transient viscosity and molecular order of a thermotropic polyester LCP in uniaxial elongational flow. Macromolecules 32:1159–1166
Khan SA, Royer JR, Raghavan SR (1997) Rheology: tools and methods. In: Aviation tools with improved fire safety. National Academic Press, Washington, DC, p 31

Kim SS, Han CD (1994) Effect of shear history on the steady shear-flow behavior of a thermotropic liquid-crystalline polymer. J Polym Sci: Part B Polym Phys 32:371–381
Kiss G (1986) Anomalous temperature dependence of viscosity of thermotropic polyesters. J Rheol 30:585
Kiss G, Porter RS (1980) Rheology of concentrated solution of helical polypeptides. J Polym Sci Polym Phys Ed 18:361–388
Laun HM (1984) Orientation effects and rheology of short glass fiber-reinforced thermoplastics. Colloid Polym Sci 262:257–269
Lenz RW (1985) Characterization of thermotropic liquid crystal polymers. Pure Appl Chem 57:997–984
Leonov AI (1990) On the rheology of filled polymers. J Rheol 34:1039–1045
Leonov AI (2008a) Algebraic theory of linear viscoelastic nematodynamics. Math Phys Anal Geom 11:87–116
Leonov AI (2008b) A model of weak viscoelastic nematodynamics. Z Angew Math Phys 59:333–359
Leonov AI, Bassov NI, Kazankov YV (1977) The basic processing of thermosets and rubbers by injection molding (Russian). Chem Moscow:68–70
Leonov AI, Chen H (2010) Modeling and simulation in polymers. Wiley-VCH, Weinheim
Leonov AI, Volkov VS (2002) A theory of viscoelastic nematodynamics. Rheol Acta
Leonov AI, Volkov VS (2005) Weak viscoelastic nematodynamics Maxwell type. Condensed Matter
Lin YG, Winter HH (1988) Formation of high melting crystal in a thermotropic aromatic copolyester. Macromolecules 21:2439–2443
Litchfield DW, Baird DG (2006) The rheology of high aspect ratio nanocomposite filled liquids. Rheol Rev:1–60
Mackley MR (1987) The rheology and micro structure of flowing thermotropic liquid crystal polymers. Mol Cryst Liq Cryst 153:249–261
Macosko CW (1994) Rheology: principles, measurements and applications. Wiley, New York
Maier D, Eckstein A, Friedrich C, Honerkamp JL (1998) Evaluation of models combining rheological data with molecular weight distribution. J Rheol 42:1153–1173
Majumder KK, Hobbs G, Bhattacharya SN (2007) Molecular, rheological, and crystalline properties of low-density polyethylene in blown film extrusion. Polym Eng Sci 47:1983–1991
Marrucci G (1991) Liquid crystallinity in polymers: principles and fundamental properties. VCH, New York
Marrucci G, Guido S (1995) Shear flow rheology of liquid crystalline polymers. Int J Polym Anal Charact 1:191–199
Marrucci G, Maffettone PL (1993) Liquid crystalline polymers. Pergamon, New York
Mead DW (1994) Determination of molecular weight distributions of linear flexible polymers from linear viscoelastic material functions. J Rheol 38:1797–1827
Moldenaers P, Yanase H, Mewis J (1990) Effect of shear history on the rheological behavior of lyotropic liquid crystals. Liq Cryst Polym 24(1990):370–380
Muir MC, Porter RS (1989) Processing rheology of liquid crystal polymers: a review. Mol Cryst Liq Cryst 169:83–95
Narimissa E, Rahman A, Gupta RK, Kao N, Bhattacharya SN (2013) Anomalous first normal stress difference behavior of polymer nanocomposites and liquid crystalline polymer composites. Polym Eng Sci 54:1300–1312
Nicholson TM, Mackley MR, Windle AH (1992) Shear-induced crystallization in a liquid crystalline random copolyester. Polymer 33:434–435
Onogi S, Asada T (1980) Rheology. In: Astanta G, Marrucci G, Nicolais L (eds) vol 1. Plenum, New York
Ophir Z, Ide Y (1983) Injection molding of thermotropic liquid crystal polymers. Polym Eng Sci 23:792–796

Phillips M, Robertus C (1996) Binder rheology and asphaltic pavement permanent deformation; the zero-shear viscosity. In: Proc. eurasphalt & eurobitume congress
Prashant GJ, Leonov AI (2001) Modelling in steady and time dependent responses in filled, uncured, and crosslinked rubbers. Rheol Acta 40:350–365
Rahman A, Gupta R, Bhattacharya S, Ray S, Costa F (2013) Simulation study of thermotropic LCPs and prediction of normal stress difference at high shear rate. Int Polym Process 28:470–482
Rahman AO (2013) Rheology of thermotropic liquid crystal polymers for injection moulding. Ph.D., RMIT University
Ramazani SAA, Ait-Kadi A, Grmela M (2001) Rheology of fiber suspensions in viscoelastic media: experiments and model predictions. J Rheol 45:945–962
Rienacker G, Hess S (1999) Orientational dynamics of nematic liquid crystals under shear flow. Physica A 267:294–321
Scribben E (2004) Selection of thermotropic liquid crystalline polymers for rotational molding. Ph.D., Virginia Polytechnic Institute and State University, PhD Thesis
Seung SK, Chang DH (1993) Oscillatory shear flow behaviour of a thermotropic liquid-crystalline polymer. Polymer 35:93–103
Shenoy AV (1999) Rheology of filled polymer systems. Kluwer Academic, London
Shibley AM (1982). In: Lubin G (ed) Handbook of composites (Chap 7). Van Nostrand Reinhold, New York
Singh S (2000) Phase transitions in liquid crystals. Phys Rep 324:107–269
Susanne EM, Gleissle W, Mckinley GH, Buggisch H (2002) The normal stress behaviour of suspensions with viscoelastic matrix fluids. Rheol Acta 41:61–67
Tai-Shung C (1986) The recent developments of thermotropic liquid crystalline polymers. Polym Eng Sci 26(13):901–919
Tai-Shung C (2001) Thermotropic liquid crystal polymers: thin-film polymerization. Technomic, Pennsylvania
Tanaka Y (2007) Enthalpy relaxation of liquid crystalline polymer with cyanobiphenyl group in the side chain: activation energy spectrum analysis. Polym J 39:1030–1039
Tao Y-G, Otter WKD, Briels WJ (2006) Shear viscosities and normal stress differences of rigid liquid-crystalline polymers. Macromolecules 39:5939–5945
Thomas V, Giacomin AJ, Wolfenden A (1994) A rheometer to measure the viscoelastic properties of polymer melts at ultrasonic frequencies. Rev Sci Instrum 65(7):2395–2401
Tuminello WH (1999) Determining molecular weight distributions from the rheological properties of polymer melts. In: Proc. 71st Soc. Rheol. Meeting, Madison, Wisconsin
Wagner MH (1977) Prediction of primary normal stress difference from shear viscosity data using a single integral constitutive equation. Rheol Acta 16:43–50
Wagner MH, Ixner T, Geiger K (1997) A note on the melt strength of liquid crystalline polymer. J Rheol 41:1087–1093
Walker L, Wagner N (1994) Rheology of region I flow in a lyotropic liquid-crystal polymer: the effects of defect texture. J Rheol 38:1525–1547
White JL, Dong L, Han P, Laun HM (2004) Rheological properties and associated structural characteristics of some aromatic polycondensates including liquid-crystalline polyesters and cellulose derivatives. Int Union Pure Appl Chem 76(11):2027–2049
Wiberg G, Hillborg H, Gedde UW (1998) Assessment of development and relaxation of orientation in a sheared thermotropic liquid crystalline copolyester. Polym Eng Sci 38:1278–1285
Wilson TS, Baird DG (1992) Transient elongational flow behavior of thermotropic liquid crystalline polymers. J Non-Newt Fluid 44:85–112
Windle AH, Viney C, Golombok R, Donald AM, Mitchel G (1985) Molecular correlation in thermotropic copolyesters. Chem Soc 79:161–173
Wissbrun KF (1980) Observations on the melt rheology of thermotropic aromatic polyesters. Br Polym J 12:163–169
Wissbrun KF (1981) Rheology of rod-like polymers in the liquid crystalline state. J Rheol 25:619

Wolff S, Wang SJ (1992) Filler-elastomer interactions. IV. The effect of surface energies of fillers on elastomer reinforcement. Rubber Chem Technol 65:329–342

Xie H-L, Wang S-J, Zhong G-Q, Liu Y-X, Zhang H-L, Chen E-Q (2011) Combined main-chain/side-chain liquid crystalline polymer with main-chain on the basis of "jacketing" effect and side-chain containing azobenzene groups. Macromolecules 44:7600–7609

Yanase H, Asada T (1987) Rheology of a lyotropic polymer liquid crystal of rodlike polymers and main-chain thermotropic liquid-crystalline polymers. Mol Cryst Liq Cryst Incorporating Nonlinear Opt 153(1):281–290

Yu R, Yu W, Chixing Z, Feng JJ (2007) Rheology and relaxation process in a melting thermotropic liquid crystalline polymer. J Appl Polym Sci 104:3780–3787

Yu R, Yu W, Zhoua C, Feng JJ (2005) Extensional viscosity of a thermotropic liquid crystalline polymer measured by thread disintegration method. Polym Testing 24:513–518

Zhou WJ, Kornfield JA, Ugaz VM, Burghardt WR, Link DR, Clark NA (1999) Dynamics and shear orientation behavior of a main-chain thermotropic liquid crystalline polymer. Macromolecules 32:5581–5593

# Chapter 5
# Liquid Crystalline Polymer and Its Composites: Chemistry and Recent Advances

**Anirban Maitra, Tanya Das, and Chapal Kumar Das**

## 5.1 Introduction

Liquid crystalline polymers (LCPs) are a class of polymer which contains a phase, intermediate between solid crystal state and conventional isotropic liquid phases. This intermediate phase of the liquid crystal polymers are commonly termed as "mesophases" or "mesomorphic phases". (From the Greek word mesos, that means middle or intermediate). Liquid crystalline polymers have got a very typical melting or softening behavior. When we heat a crystalline solid which is not mesomorphic like LCP, it gradually changes from solid phase to isotropic liquid phase directly at its melting temperature region. But in LCP, several different types of mesophases formed before the formation of isotropic melt phases. The mesophases are different from each other which also give rise to various types of disordering in the molecular structure of LCP crystals (Barón and Stepto 2002). For this reason, TLCPs have generated intense interest from both academic and industrial researchers.

Liquid crystalline polymer composites are a special class of advance materials prepared by homogeneous mixing of (1) Either two or more LCP polymer materials having significantly different physical and chemical properties and a filler, or by (2) Mixing LCPs with other type of polymer in presence of a filler. Now if the filler which is distributed or dispersed in one of the LCP phase is in nanometer

A. Maitra
Research scholar in Materials Science Centre, Indian Institute of Technology, Kharagpur 721302, India
e-mail: anirbanm1212@gmail.com

T. Das
Nanyang Technological University, 65 Nanyang Drive, Singapore 637460, Singapore

C.K. Das (✉)
Professor of Materials Science Centre, Indian Institute of Technology, Kharagpur 721302, India
e-mail: chapal12@yahoo.co.in



V.K. Thakur, M.R. Kessler (eds.), *Liquid Crystalline Polymers*,
DOI 10.1007/978-3-319-20270-9_5

dimension, then it is termed as LCP nanocomposites. The advantage of using nanofiller is that it should be added in a very lesser amount as compared to the conventional microfillers. When the ultimate composites are formed, it has been seen that the property of the composites are usually different with respect to that of the starting materials. LCP when mixed with other polymers, it usually decrees the viscosity and hence increases the processability. The fillers and the polymer components may sustain separately or distinguishable within the finished composites. These composites are mainly used for processing applications in various plastic and rubber industries. It can be used for preparing various constructional part designs, building blocks and also in aircraft body part design due to its low weight (Utracki and Favis 1989).

The various types of disordering shown by the liquid crystal polymers are:

1. **Positional disordering**: The extent of disorder in the position of an average molecule or group of molecules shows translational symmetry.
2. **Orientation disordering**: The tendency of the polymer molecules to disorder along the director on a long-range basis.
3. **Conformational disordering**: It appears during the formation of more than one conformation or configurations in the liquid crystal polymer structures. It appears due to rotation about a single bond.

So the liquid crystalline behavior of the polymers depends on the packing of the molecules and the anisotropy present in the molecular structure (Rudko 2002).

There are mainly three types of mesophases identified by Friedel (Brehmer and de Jeu 2012; Dowell 1988), which should be described by their type of ordering.

1. **Smectic mesophases**: Smectic mesophases exhibits highly ordered and well oriented molecular conformations at low temperatures than the nematic phases. In this mesophases, the molecules lying parallel to one another (orientational order) (Fig. 5.1) and long range stratification normal to the chain axis (positional disorder) (Brehmer and de Jeu 2012). This type of alignment of the polymer molecules gives rise to a high viscosity. Smectic mesophases have got a very similar property like that of soap or detergent molecules (National Materials Advisory Board 1990).
2. **Nematic mesophases**: Nematic mesophases exhibits orientational disordering as the polymer molecules lying parallel to one another along one axis (Fig. 5.2). The axis introduces a term named director (A unit vector $\langle n \rangle$ that describes the direction of orientation of long axis of molecules) (National Materials Advisory Board 1990; Brehmer and de Jeu 2012). The rod like polymer molecules has got no positional ordering, but they can self-orient to have a long range directional order with respect to their main axis.

   This alignment giving rises to a low viscosity and makes the LCP phases easier in processing. Nematic mesophases has got fluidity characteristics very similar to that of isotropic liquids but they can be easily oriented by external source like electric or magnetic field. Aligned nematic phases have got very

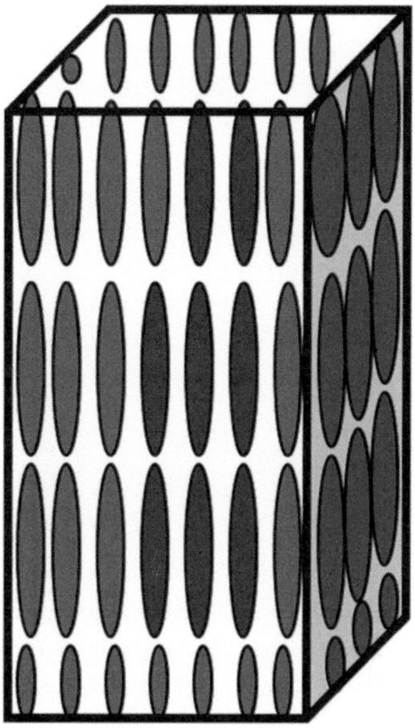

**Fig. 5.1** Schematic diagram of alignment of polymer molecules in smectic mesophases (Rudko 2002)

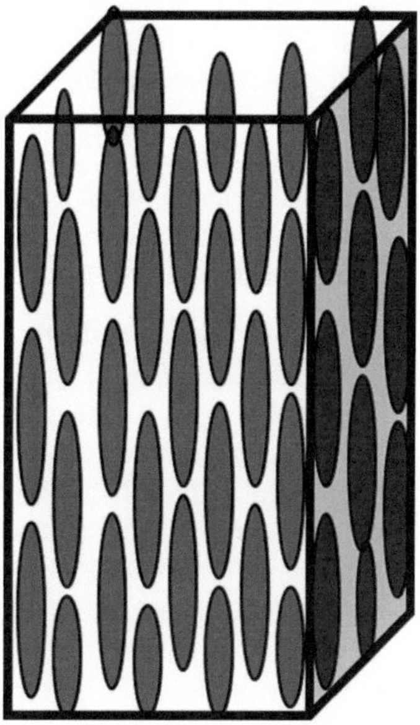

**Fig. 5.2** Schematic diagram of alignment of polymer molecules in nematic mesophases (Rudko 2002)

good optical properties similar to that of uniaxially oriented single crystals. So this type of LCPs can be used in liquid crystal displays.

3. **Cholesteric mesophases**: A sub unit of the nematic mesophases is the Cholesteric mesophases or chiral nematic mesophases. It is nematic because each plain of polymer chains has got a nematic director, but has the additional features that

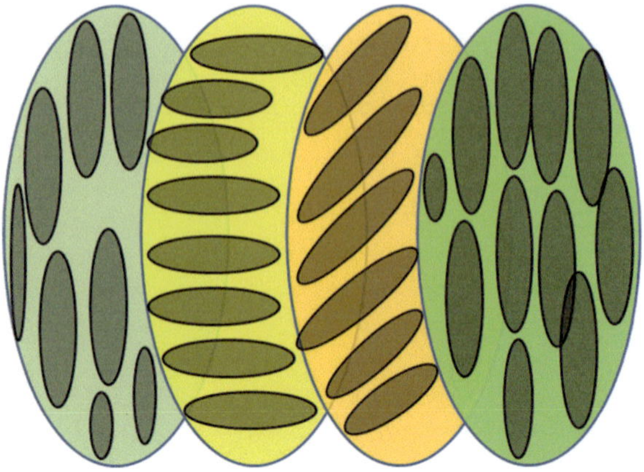

**Fig. 5.3** Schematic diagram of alignment of polymer molecules in *Cholesteric mesophases* (Rudko 2002)

the director shows a cumulative twist (Fig. 5.3). The twist is formed by the chiral centers in the mesogen units showing asymmetrical intermolecular forces in between the nematic layers (National Materials Advisory Board 1990; Brehmer and de Jeu 2012; Shibaev et al. 1984).

These different types of mesophases (mentioned above) can be distinguished by their different optical properties (Utracki and Favis 1989). When viewed under an optical microscope using a polarized light source, different liquid crystal mesophases will appear as a distinct texture. Each region in the texture corresponds to a domain where the liquid crystal mesophases are arranged in a different direction or orientation. Within a domain, the molecules of LCPs are well ordered. Depending on the liquid crystalline behavior in solution or in the melt phase, The LCPs can be divided into two types. (1) **Lytropic LCPs and** (2) **Thermotropic LCPs**.

1. **Lyotropic liquid crystalline polymers (LLCPs)** show its liquid crystallinity in solutions. The liquid crystallinity can be modulated by the type and the nature of the solvent used. The temperature of the solvent at which the polymer dissolves and the concentration of the polymer also plays a significant role in determining the crystallinity of LLCPs in solutions.
2. **Thermotropic liquid crystalline polymers (TLCPs)** show its liquid crystallinity in melt phase (Brehmer and de Jeu 2012; Shibaev et al. 1984; Popa-Nita et al. 2009). Thermotropic phases are those that occur in a certain temperature range. If the temperature is raised too high, thermal motion will destroy the ordering of the LC phase, pushing the material into a conventional isotropic liquid phase. At too low a temperature, most LC phases will form a conventional anisotropic crystal. Many thermotropic LCs shows a variety of phases as temperature is altered (National Materials Advisory Board 1990; Popa-Nita et al. 2009; Davidson 1999). The melt temperature and the thermal history stored within the polymer system plays a vital role in determining the liquid crystallinity of TLCPs.

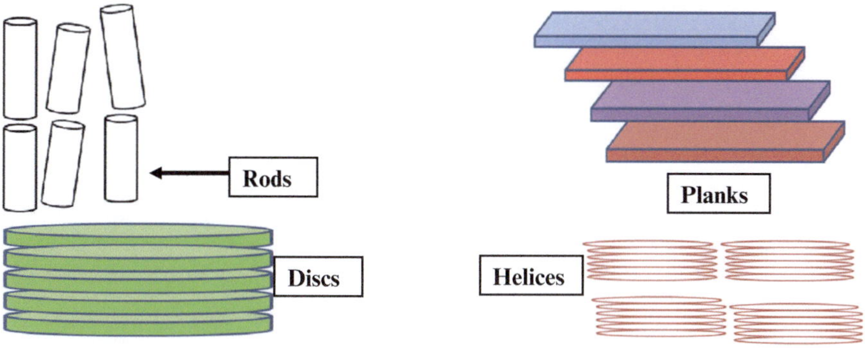


**Fig. 5.4** Different types of anisotropic mesogenic unit's that introduces liquid crystallinity (Rudko 2002)

Typically Lyotropic LCPs cannot able to show its liquid crystallinity in melt phase because they degrade before melting.

The most important part of the liquid crystalline polymers is the rigid anisotropic mesogenic units. These rigid anisotropic mesogenic units can create rod like, plank like, disk or helices like molecules (Fig. 5.4) or branches which can orient to form liquid crystalline order.

### *5.1.1 Structure and Architecture of Liquid Crystal Polymers*

The structure of the mesogen of liquid crystal polymers are usually based on stiff aromatic units with bridging groups linking the individual units. There are various architectures of the stiff aromatic units used in liquid crystal polymers. They may form 1D, 2D, 3D geometry.

The architectures used to form mesogenic units are-

| 1. **Swivel units** | |
|---|---|
| (a) 4-Hydroxybenzoic Acid (HBA) | 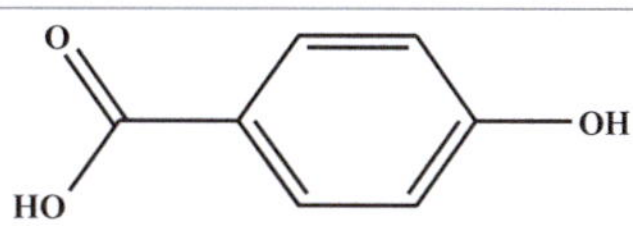  **4-Hydroxybenzoic Acid** |
| (b) Hydroquinone (HQ) | 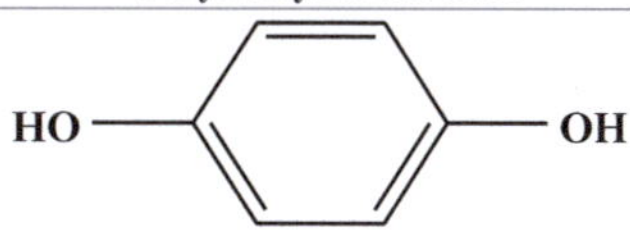  **Hydroquinone** |

(continued)

(continued)

| | |
|---|---|
| (c) Terephthalic acid (TA) | 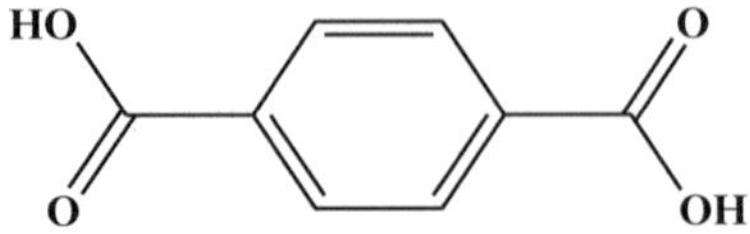 Terephthalic acid |
| (d) 4, 4′ Biphenol (BP) | 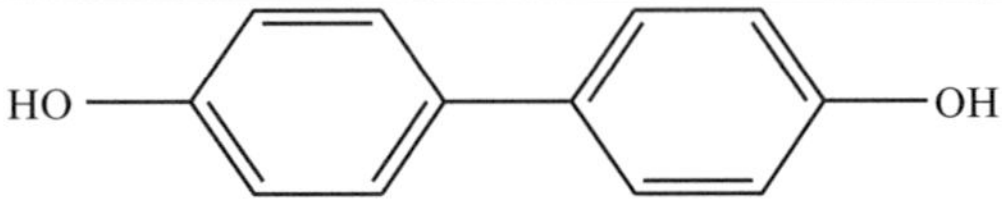 4,4 Biphenol |
| **2. Crank shaft units** | |
| (a) 2, 6-Dihydroxynaphthalene (DHN) | 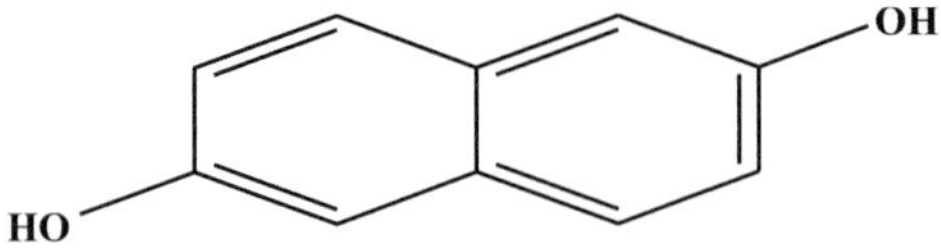 2,6-Dihydroxynaphthalene |
| (b) 6-Hydroxy-2-Naphthoic Acid (HNA) | 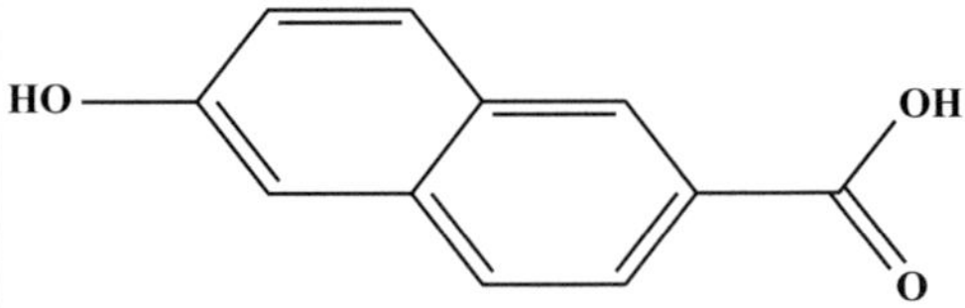 6-Hydroxy-2-Naphthoic Acid |
| (c) 2, 6-Naphthalene Dicarboxylic Acid (NDA) | 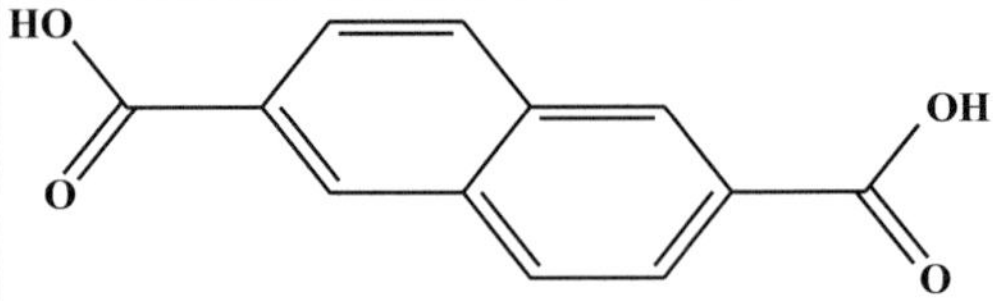 2, 6-Naphthalene Dicarboxylic Acid |
| **3. Bent rigid units** | |
| (a) Isophthalic Acid (IA) | 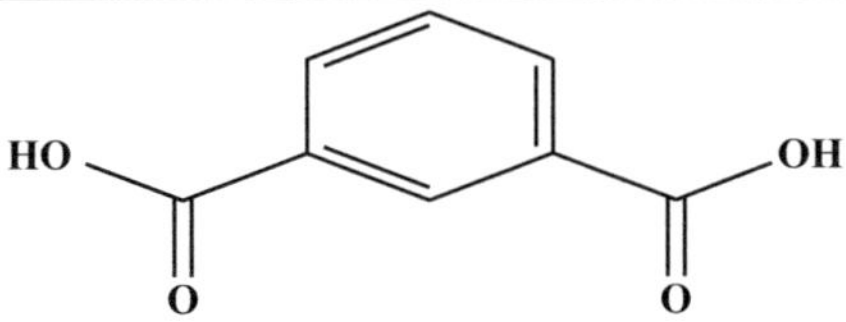 Isophthalic Acid |
| (b) Meta Hydroxy Benzoic Acid (HBA) | 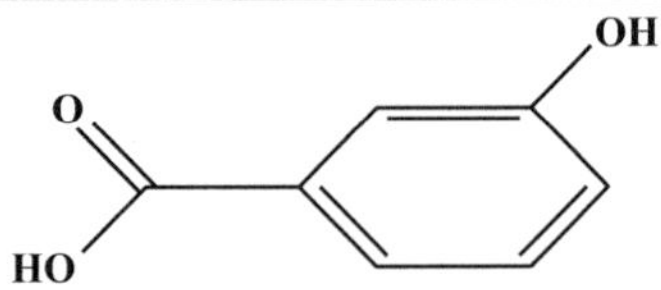 Meta Hydroxy Benzoic Acid |

(continued)

(continued)

| | |
|---|---|
| (c) Resorcinol (R) | 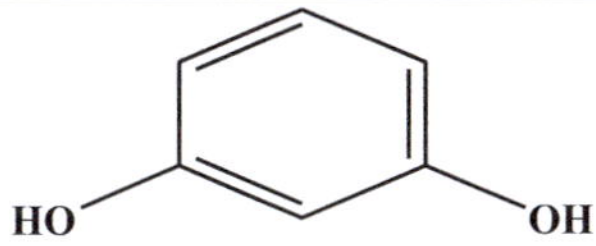  Resorcinol |

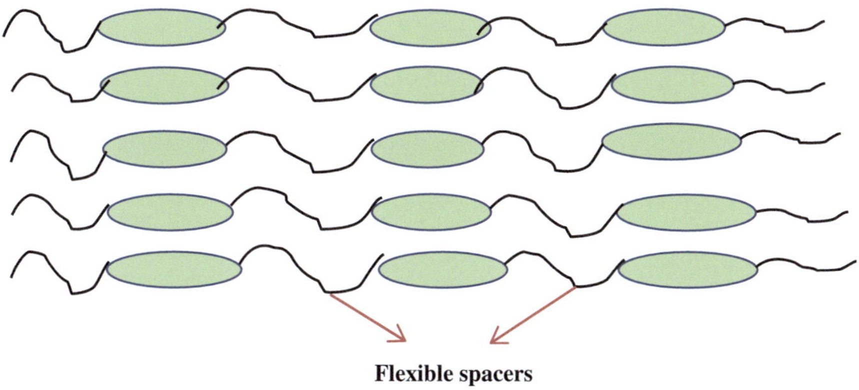


**Fig. 5.5** Schematic pictorial representation of main chain flexible spacers (Utracki and Favis 1989)

Two methods exist to create liquid crystal polymers. (1) Using more than one mesogenic units in the polymer and (2) Incorporating flexible spacers between the mesogenic units (Brehmer and de Jeu 2012; Davidson 1999).

A typical semi rigid liquid crystal polymer which includes more than one mesogenic units are (a) Xydar® which consists a copolyesters of 4, 4′-dihydroxybiphenol (BP), Phthalic acid (TA), and Hydroxy benzoic acid (HBA). (b) Vectra A950, a copolyesters of HBA and 2-Hydroxy-6-napthoic acid (HNA). Vectra B-950 is the only commercially available thermotropic LCPs containing amide linkages on the polymer main-chains.

Vectra B-950 exhibiting the highest glass transition temperature ($T_g$) among Vectra series. This is due to the presence of amide groups which can form strong intermolecular hydrogen bonds that limit chain mobility.

The insertion of flexible segments or spacers to link the mesogenic units along the polymer main chain or backbone is aimed to reduce the transition temperature and to preserve the chemical periodicity of the molecules of LCPs (Barón and Stepto 2002; Brehmer and de Jeu 2012). These are abbreviated as main chain liquid crystalline polymers (MCLCPs). As the result of comprehensive research and development, several LCPs have been successfully commercialized to meet the need for high stiffness, high strength, and lower melting, melts processable molecules for industrial applications. When flexible spacers are introduced between the mesogenic units e.g. PET, polymethylene or poly (oxymethylene), the LCPs become semi flexible (Fig. 5.5). Typical semi flexible TLCPs are copolyesters of

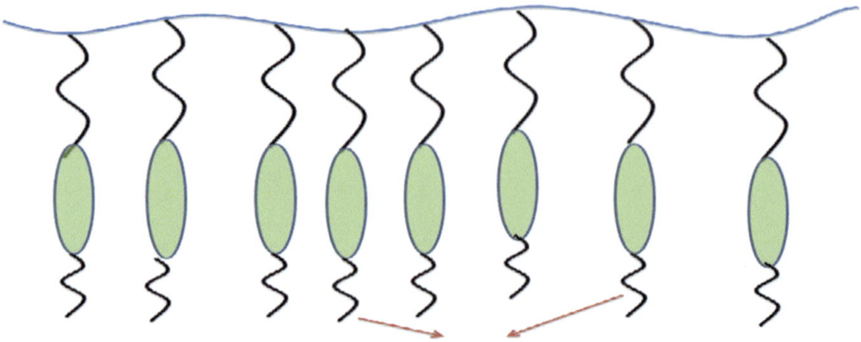


**Fig. 5.6** Schematic pictorial representation of side chain flexible spacers (Utracki and Favis 1989)

Polyethylene terephthalate (PET) and 40–80 mol% of hydroxy benzoic acid (HBA). Here PET acts as flexible spacers.

These flexible units introduced in between the mesogenic units controls the rheology during mixing of liquid crystalline polymers with other type of polymers.

With increase in the number of flexible spacer units in the main chain backbone, more will be the decrease in viscosity of the liquid crystal polymers.

The mesogenic groups can be connected to the main polymer backbone of LCPs with the help of some flexible spacers that can be introduced at the side of LCP backbones (Barón and Stepto 2002). This is termed as SGLCPs. Then also the LCPs become semi flexible (Fig. 5.6).

The incorporation of both rigid and flexible unit in the main chain polymer of LCPs alters the melting temperature, modulus and strength. These materials also possess a higher nematic mesophases. However, LCPs with high melting temperature exhibits poor melt processability. The relation between the melting point $T_m$ and change in entropy is given by

$$\mathbf{T_m = \Delta H/\Delta S}$$

Where, $\mathbf{\Delta H}$ is the difference in enthalpy between the isotropic melts and the anisotropic from of LCPs and $\Delta S$ is the corresponding entropy difference. Thus, $T_m$ could be reduced by decreasing $\Delta H$ or by increasing $\Delta S$, or both. The enthalpy is mainly controlled by interchain separation between adjacent LCP chains. An effective approach of decreasing $\Delta H$ is to insert modifying units randomly along the chain. This leads to the formation of random copolymer in which the crystallinity is disrupted (Utracki and Favis 1989; National Materials Advisory Board 1990; Bates 1991; Rudko 2002; Brehmer and de Jeu 2012; Dowell 1988). Considerable efforts have been devoted to reduce the transition temperature of main chain LCPs in order to reach more favorable conditions for industrial processing.

This includes the following criteria:-

1. Addition of linking units at regular intervals between the mesogenic units.

2. Incorporation of different type cranks shaft or bent rigid units based on essentially rigid bonds.
3. Introducing of bulky side groups or chains, that destroys the regularity of molecular packing.

### *5.1.2 Properties of Liquid Crystalline Polymers*

Thermotropic liquid crystal polymers have several important properties that would make this polymer important in commercial and industrial application.

1. TLCPs have got a high Z axis coefficient of thermal expansion (National Materials Advisory Board 1990; Brehmer and de Jeu 2012; Shibaev et al. 1984; Yoon et al. 1992; Popa-Nita et al. 2009)
2. It has a low melt viscosity and very good melts strength (Roth and Thomas 1989; Brown and Alder 1993).
3. It has got a high chemical resistance because of their negligible solubility in many solvents leading to retention of their mechanical properties (Brehmer and de Jeu 2012; La Mantia et al. 1997; Seo and Seo 2004; Thomas and Roth 1990; Scott and Macosko 1995; Jaarsma et al. 1985; Han and Bhowmik 1997)
4. TLCPs can be used in injection moulding because the change in density of these polymers upon cooling is negligible less. Its mould shrinkage is very less. We can form complicated shapes by using these polymers in injection molding process. The flow behavior of these polymers in the cavity and voids in complex shaped mould is exceptionally good (Brehmer and de Jeu 2012; La Mantia et al. 1997; Seo and Seo 2004; Scott and Macosko 1995; Plummer 1993).
5. TLCPs exhibit a low permeability to many gases relative to that of other thermoplastics (Popa-Nita et al. 2009; Roth and Thomas 1989). Thomas and Roth (Roth and Thomas 1989; Thomas and Roth 1990) checked the oxygen gas permeability of a wide variety of polymers at 23 °C. They found that Vectra A900 film has a permeability value of $1.2 \times 10^{-7}$ ($cm^3$ cm)/($cm^2$ day atmosphere). This is over 100 times lower than the oxygen permeability value of PET.
6. TLCPs exhibit very good mechanical properties. Sometime the mechanical properties of TLCPs exceed the mechanical properties of glass fiber filled thermoplastics. As for example a glass fiber filled PET composite shows a tensile strength of 160 MPa, whereas Vectra B950 shows a much higher tensile strength value of 227 MPa (Brehmer and de Jeu 2012; Thomas and Roth 1990; Kulichikhin and Plate 1991; Roetting and Hinrichsen 1994). Similar phenomenon observed in flexural modulus value. Machine direction flexural modulus value of glass filled PET is 9.7 GPa whereas for Vectra B950 has gat a value of 32.5 GPa which is over three times greater than glass filled PET.
7. LCP is a potential candidate that can impart fiber like reinforcement when melt blended with thermoplastics and elastomers. Their inherent low viscosity and ability to deform into long fibrils are the main attractions for blending TLCP with thermoplastics.

However, the disadvantages of TLCPs is that:

1. They are highly anisotropic which renders them to be weak in the transverse direction to the LCP fiber axis. This has limited their potential in many applications.
2. Presence of highly anisotropic part also limited its uses in industrial application.
3. It has gat a very high cost.

## 5.1.3 Liquid Crystalline Polymer Blends

A liquid crystal is an intermediate structure. The possibility of order existing at equilibrium in a fluid arises when the molecules within the liquid are anisotropic. These anisotropic molecules must also resist entropic forces which would otherwise tend to turn a rod into random coil. Rigid rods readily form liquid crystal phases. Rigidity is, however not as strong a requirement as originally perceived. Several so called rigid rod molecules have in fact been shown to be close to random coils in solution. Almost any form of anisotropy can give rise to a liquid crystal phase. Rods, discs, planks and helices are all well documented examples already showed (Fig. 5.4). There are some improvements in various properties like modulus and tensile strength, barrier properties when LCP is blended to other polymers. Similar results were reported by Zulle (Plummer et al. 1993) using PTFE and Vectra type LCP. The extruded blend again displayed an amorphous skin-core structure that is a recurrent theme in LCP/thermoplastics blends. Improvements in mechanical properties were observed until the LCP concentration exceeded 60 %.

The extent of property improvement is dependent on the method of blending. This effect was demonstrated by Isayev and Swaminathan (1989). They prepared blends of PEI and Vectra A950 using a static mixer which produces blends with a higher degree of fibrillation and corresponding improvement in properties. SEM was used to show that the LCP phase gains a droplets or in elongated morphology at concentrations below 10 % and only forms fibrillar structure when the LCP content was greater than 20 %. It was reported that the LCP fibers formed at the inlet zone of the die where the polymer experienced high elongational stresses. The effect of mixing of polyethersulfone/LCP blends, in static mixer vs. the internal mixer indicates that the better tensile properties can obtained in the static mixer blends. Among four different polymers, the PES matrix was the most efficient in obtaining uniform LCP phase distribution and large aspect ratio in blends.

The effect of miscibility of the host and reinforcing phase formations in in-situ composites have been studied by Zhang and co-workers (Narh et al. 2000a). Seppälä et al. (1992) blended Vectra with various thermoplastics and the blends were injection molded. LCP was found to act as a reinforcing agent in the matrix polymers and improves their dimensional stability. They also found that there is a poor interfacial adhesion between the LCP and the matrix polymers.

In summary, addition of small amount of LCP can enhances the crystallinity of thermoplastics, this shows nucleating capability of LCP for thermoplastics. Most commercial LCPs are incompatible with thermoplastics leading to the properties of LCP/thermoplastic blends to fall short of expectations. Poor interfacial adhesion remains a major obstacle in achieving high performance in situ composites. Thus, suitable compatibilizers additions needed to improve the interfacial adhesion of blend components.

## 5.2 Rheology of Liquid Crystal Polymers and Its Blends

Addition of LCP to the melt processable thermoplastic has an adverse effect on viscosity. It is experimentally proved that addition of a small amount of TLCPs on a thermoplastic polymer matrix reduces the viscosity of the system and thereby enhancing the processability. So, in brief we can say that LCP can be used as processing aid. Weiss et al. (1987) studied the rheology of polystyrene and LCP blends at 220 °C. The die used in the capillary rheometer had an entrance angle of 100° and a length to diameter ratio L:D = 57:1. It was observed that the LCP domains are spherical at low sheer rates and not deformed by sheer field. At high shear rate in the capillary rheometer, the viscosity of the blends was lower than that of the pure PS, decreasing with increasing LCP concentration. For 10 % LCP, a reduction of the viscosity of around 40 % was obtained. This is due to the orientation of the LCP domains caused by tensile stresses at the die entrance and the relatively long relaxation times of LCP which maintains this deformation and orientation, providing slip surface to lubricate the flow.

Rheological properties of the LCP (0.7 HBA/0.3 HNA) with PC, PBT, Nilon-6 blends were characterized by Beery et al. (1991). PC/LCP shows a non-Newtonian behavior and their viscosities are intermediate between the LCP and PC. Below 300 $s^{-1}$ the viscosity increases with LCP concentration but above this shear rate, the viscosity decreases with LCP content. The PBT containing blends shows a different rheological behavior. In this case the viscosities of the blends lie below those of the LCP in the whole shear rate range. For Nylon-6 the viscosities of the blends lie below to that of the pure constituents and the blend viscosity decreases with increasing LCP content.

Isayev and Swaminathan (1989) blended PEI with VECTRA 950 and K161 (HBA/Isophthalic acid (IA)/TA/HQ/BP) respectively. In all the cases the viscosities of the blends were reduced by addition of LCP. At high shear rates an addition of about 5 % LCP reduced the blend viscosity by about 30 %.

## 5.3 Morphology of Liquid Crystalline Polymer in Blends

A liquid crystalline polymer contains rigid rod like structure as discussed earlier which forms the liquid crystal phases. This rod like molecular conformations and chain stiffness give LCPs their most important self-reinforcing properties that are close to that of glass fiber reinforced composites.

Siegmann et al. (1985) were the first reported the formation of fibrillar structure of the LCP phases in a polyamide/Vectra type LCP blends. It was observed that the fibrils mainly form at the skin rather than the core of the injection moulded samples. There was definitely some improvement of tensile strength and elastic modulus.

Polymer blends during melt mixing experiences a complex flow field with a mixture of shear and extensional strain. In between the screw and the barrel of mixing or extrusion devices, the polymer/LCP blend melt is compressed and disintegrated. This involves chain scission and break-up, similar to the mixing of low molecular weight liquids (Christiansen et al. 1987). As the LCP phase deforms under the flow field, the surface tensional force accelerates the break-up process. Here blend composition is important because at a high LCP content, the LCP domains tend to coalesce into a sheet-like structure, whilst at a low content, they break up into small particles. In order to achieve a better dispersion of LCP in the matrix polymer, the LCP phase should have a lower viscosity than the matrix polymer. Many researchers have found that the formation of the morphology of the LCP phase takes place at the die entrance, where the converging force of the die generates a strong elongational flow field (Kurtz et al. 1999; Zhou and Yan 2005).

Recent work showed that for a Vectra A950/polypropylene blend, the LCP phase exists mainly in spherical particles of various sizes before the die entrance. Inside the die capillary, the elongational strain is transformed into a simple shear flow. Kohli et al. (1989) found that in the shear region of the capillary, the elongated phase either relaxed back to a spherical shape, or broke up into smaller spherical droplets, mainly as a consequence of competition between viscous and interfacial forces. At a high capillary length/width (L/D) ratio, a low deformation rate takes place and the LCP forms a spherical structure. At a low L/D ratio, relaxation time is short and the elongated structure of the LCP is preserved. At the die exit, an elongated strain is applied to the polymer melt, which is related to the draw-down ratio (ratio of haul-off speed to exit speed). Many researchers have found that in order to obtain a high aspect ratio for the LCP phase, it is essential to apply a high draw-down ratio. At a low draw-down ratio, the LCP phase either relaxes back to ellipsoids or breaks up into particles (Fig. 5.7b). At high drawdown ratios, thin and long fibrils can be obtained (Fig. 5.7a).

The morphology of tensile fractured PP/LCP blends in the transverse direction of melt flow was shown in Fig. 5.7.

From Fig. 5.7a, it is clear that there is some elongated structure formation during blending of LCP at high temperature. The average length of the LCP fibrils was found to be around 68 μm. The widths of the fibrils are around 4.5 μm. This phenomenon happens when the relaxation time within the die is very less. The

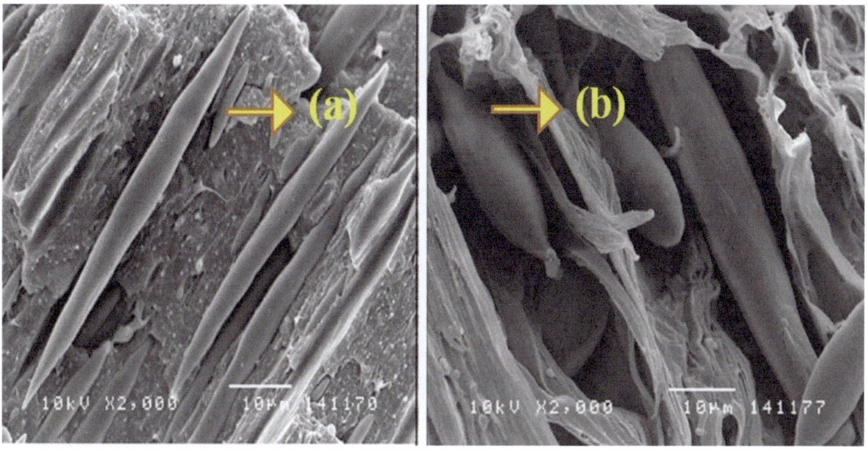


**Fig. 5.7** Core and skin morphology formation of LCPs during processing. (**a**) Elongated microstructure. (**b**) Ellipsoidal microstructure

shear force is very high and so the LCP takes the shape of long fibrils. Figure 5.7b clearly shows that there is some ellipsoidal structure formation during blending of LCP at high temperature. The average length of the both side stretched ellipsoidal LCP fibrils was found to be around 32 μm. The widths of the fibrils are around 9 μm. This phenomenon happens when the relaxation time within the die is high enough. Actually the above two picture for morphology development was studied in presence of nanosilica as nanofiller. The nanosilica was found as white and tiny particles distributed over the entire area. The phases observed in between the two long or short LCP microfibrils (on both the pictures) were for the polypropylene phases. It has been recognized that the molecular chains of LCP tend to follow the flow direction of the melt. LCPs processed by injection moulding would invite structural distribution since the process involved complex shear and extensional stress profile with non-isothermal conditions during mold filling (Narh et al. 2000b). However, the skin–core difference, typical of moulded parts, was not obvious in this set of specimens. This was probably attributed to the small sample thickness. The proportion of skin tends to increase with decreasing specimen thickness because the shear rate at the mold wall is very high. This high shear rate would deform the LCP particles. Furthermore, effects contributed by thermal relaxation would not be as great compared to thicker samples. Thus, the shear stress and cooling rate experienced by the 1 mm thick tensile bars would be fairly uniform and the shape of the discrete LCP phase was similar in the skin and core regions. LCP fibrils were obtained for all composites. However, the aspect ratio (ratio of length-to-diameter of the LCP particle) of the fibrillar LCP increased with the quantity of nanosilica added. The addition of nanosize fillers could improve shear-induced fibrillation of the LCP in LCP/PP composites.

The Capillary number (Ca) is a dimensionless number that predicts fibrillation originated from the hydrodynamic theorem. The basis of fibrillation was apparently embedded in the droplet deformation and break up theory (Karam and Bellinger 1968). The capillary number represents the effect of viscous force versus surface tension acting across the interface between two immiscible liquids or in case of polymer blend we can say that the effect of viscous force versus surface tension acting across the interface in between two immiscible polymer blends at melting condition.

$$Ca = \frac{\mu V}{\gamma} \tag{5.1}$$

Where $\mu$ is the dynamic viscosity of the polymer during melt flow, $V$ is the velocity during melt flow and $\gamma$ is the interfacial tension. The expression of Ca in (5.1) shows that Ca is a ratio of the viscous stress to the interfacial force. Deformation of the droplet is dependent on the interaction between the viscous force and interfacial force. When the disruptive stress due to viscous drag of the medium is capable of overcoming the cohesive effect of surface tension, the suspended phase would be deformed and elongated into cylindrical shape. The magnitude of Ca was affected by many factors such as shear rate, interfacial tension, % LCP content, material properties, viscosity ratio and processing conditions which in turn influence the droplet-fibril transition (Tjong et al. 2000a; Sukhadia et al. 1990; Bretas and Baird 1992; Shah et al. 1986; Christiansen et al. 1987; Brannock and Paul 1990; Harris and Robeson 1988; Manson and Seferis 1992). Effective fibrillation would occur when Ca exceeds the critical value by many times (Bucknall 1977).

Often, the development of good quality LCP fibrils by melt processing techniques such as extrusion and injection moulding was difficult because the fibrillation conditions were not as favorable as those in an optimized spinning process. The results obtained was interesting as this method of preparing in-situ composites was able to achieve high aspect ratio LCP fibrils especially in PP/PET-PHB systems where poor fibrillation was observed even with spun fibres (Landel and Nielsen 1993).

## 5.4 Processing Conditions for Liquid Crystalline Polymers

Owing to the poly-domain and extended chain structures, LCPs can readily form fibers. Pure LCPs form self-reinforced structures with fracture surfaces similar to that of wood. Kiss (1987) reported that for blends of various thermoplastic resins containing Vectra LCPs at a concentration of ca. 30 %, all blends contained well-developed microfibril of the LCP. However, processing conditions do have a significant effect on the LCP phase morphology, and as such, on the properties of the resulting blends. The most important processing parameters are the blend composition, the processing temperature, the viscosity ratio, the shear as well as elongational strain and the die head design, as well as the type and grade of the LCPs and the matrix polymers. Blend composition is one of the most important factors affecting the LCP phase morphology (García et al. 2003; Chung 1986; Kumar and Das 2005; Blizard and Baird 1987b; La Mantia et al. 1992; Dutta et al. 1993). Siegmann et al. (Beery et al. 1991) found that as the LCP concentration increased, the LCP phase changed from a particle-like structure to elongated ellipsoids and then to micro fibrils at high concentrations. For Vectra A900/polypropylene blends, Qin et al. (1993) found that the LCP exists as particles and short droplets at a PP/LCP wt/wt ratio of 100/2.5. At a ratio of 100/15, the LCP fibrils formed a network structure. At an intermediate ratio of around 100/10, the LCP existed in elongated micro fibrils. The thickness of the LCP fibrils is also affected by the LCP concentration. For a LCP/polycarbonate blend, Malik et al. (1989) found that the fibril diameter was roughly 5 pm when the LCP concentration was about 5–10 %. It increased when the concentration was increased to 25 %. Extrusion conditions also have a significant effect on the LCP phase morphology. At a high extrusion rate, there is a high converging force at the die entrance, and a high degree of shearing is applied to the LCP phase. Thinner LCP fibrils were found at a high extrusion rate. Draw-down ratio is another important factor. Many researchers have found that at a high draw-down ratio, the LCP is capable of deforming into fibrillar structures (Chung 1986; Collings 1990; Tjong 2003; Vollrath and Knight 2001). Kohli et al. (1989) found that at a relatively low draw-down ratio of 50, the LCP phase was only slightly elongated. At a draw-down ratio of 500, well-oriented micro fibrils were formed, which were in the draw direction.

A low capillary L/D ratio was found essential in order to generate a fibrillar structure for the LCPs. Blizard et al. 1987 reported that at a low L/D ratio of 7.82, oriented fibrils were generated, which oriented in the direction of flow. At a higher L/D ratio of 21.4, no fibrils were found in the blend. The poor performance at the high L/D ratio may be attributed to the fact that the effect of any extensional forces at the converging entrance to the die may have been lost because of the relaxation of the structure in the shear field of the capillary itself. Siegmann et al. found that in an LCP/polyamide blend, the skin region of the extruded product contained relatively more fibrils than the core region. This was accompanied by the difference in the mechanical properties of the skin and the core. After the removal of the skin, the core showed a lower tensile strength than the original product. Similar results were reported by Silverstein et al. for an LCP/PET blend, where the LCP fibrils were found to be more oriented in the skin than in the core region. The LCP phase morphology is also been affected by the compatibility or interphase tension between the two phases. Some researchers found that for the same LCP, thinner fibrils can be produced in a matrix The LCP phase morphology is also affected by the polymer more compatible to the LCP (Narh et al. 2000b). For example, Blizard and Baird (1987a) found that a more compatible LCP/polycarbonate blend contained fibrils of smaller diameters than an incompatible LCP/nylon blend. It is obvious that the final structure of the blend is the result of a complex processing history involving many experimental parameters, and, therefore, it is not easy to see the effect of each individual parameter. This principle applies when comparing the same LCP in different matrix polymers and the same matrix polymer containing different LCPs. In summary, a low viscosity ratio between LCP and the matrix polymer, a low capillary L/D ratio, a high extrusion rate and a high draw-down ratio are essential in order to develop thin and long LCP fibrils for an effective reinforcement of the matrix polymer.

## 5.5 Processing of Liquid Crystal Polymer Blends

It has been shown that the improvement in the mechanical properties of thermoplastics when these are blended with LCP is associated with the fibrillar structure of the LCP. A number of processing and mixing techniques have been developed for the production of self-reinforcing LCP blends. These include fiber spinning, extrusion, film casting, thermoforming, and injection, compression and blow molding. Injection molding and extrusion have been the two dominant processing operations in the search for suitable methods and conditions for optimal LCP fibrillation. The high orientability and the shear thinning characteristics of thermoplastics LCPs are unique and beneficial in the polymer processing. Orientation of mesogenic moieties of LCPs occurs readily under a flow field during melt processing.

Easy molecular alignment is characteristics for liquid crystals and it often leads to highly anisotropic materials with poor transverse properties. Accordingly, thermoplastic blends typically show a marked increase in both tensile strength and modulus after processing, but simultaneously the elongation at break is drastically reduced. A further problem in blending LCPs and thermoplastics is that the majority of thermotropic LCPs have much higher melting points than thermoplastics. The blending is then difficult to carry out owing to the risk of degradation of the thermoplastics at high temperature.

The mechanical properties achieved in the blends of thermotropic conventional polymers and LCPs are closely related to the fibrillar blend morphology, which in turn depends on processing methods, and also on several other factors including miscibility between the LCP and thermoplastics, LCP concentration, blending time, viscosity ratio and shear rate.

Most thermoplastic blends and polymer composites are commonly fabricated by extrusion and injection molding routes. Similarly, both processing routes are also frequently used for forming LCP/thermoplastic blends. During extrusion of thermotropic LCPs and their blends, high molecular orientations can be achieved by using a drawing step immediately after the extrudate leaves the extruder die, followed by rapid cooling.

In certain cases, the LCP fibers can be formed by extruding the material through the die equipped with a small hole spinneret. The extensional flow associated with the fiber spinning process leads to relatively highly molecular orientation along the extrusion direction in LC copolyesters. On the contrary, the melt flow behavior of injection molding process is rather complicated. The materials are subjected to both elongational and shear flows. During molding, the molten polymer flows faster in the center of the mold. The flow pattern of the material from the center towards the surfaces is called the 'fountain flow'. The fountain melt front is in contact with cooled mold wall will freeze immediately. Accordingly, the mesogen molecules of LCPs tend to align with the flow to form the skin layer of molded product. Less oriented material is located in the center of the molding where shear flow predominates, and the solidification occurs more slowly. Following melt solidification; molded articles exhibit so-called 'skin-core' morphology. Such morphology is similar to that of injection molded glass fiber-reinforced thermoplastics. Similar skin-core structure can be found in the LCP/thermoplastic blends as in pure LCPs. Several operating parameters such as injection speed, barrel and nozzle temperature, screw dwell time, mold temperature, injection temperature can affect the melt viscosity, morphology and mechanical properties of pure LCP injection molded articles. Ophir and Ide have investigated the effects of barrel temperature, packing pressure and injection speed (fast or slow) on the mechanical properties of a wholly co polyester based on 60 mol% p-acetoxy-benzoid acid, 20 mol% Terephthalic acid and 20 mol% naphthalene diacetate It can be seen that the tensile strength and stiffness of such aromatic co polyester increases with decreasing injection speed or with increasing barrel temperature. This is attributed to the higher orientation of mesophases molecules along the flow direction in the skin section.

## 5.6 Compatibility of Liquid Crystalline Polymer Blends

Polymer blend is the intimate mixture of two or more polymer and/or copolymers resulting from common processing steps. Throughout the last decades scientific and technical literatures in this area have expanded remarkably as evidenced by appearance of several books and proceedings of various conferences (Beery et al. 1991). The word compatibility has been used by many investigators to describe single phase behavior. However the term compatible and incompatible refers to the degree of intimacy of the blends, which depends on the measurement procedure during examination. A blend could be considered as a compatible blend, if it does not exhibit the gross symptoms of component polymer segregation, whereas a heterogeneous blends in a macroscopic level is incompatible. Compatible system of blends containing high molecular weight polymers have been identified usually when a favorable specific interaction such as hydrogen bonding, dipole interaction or ionic interaction exists between two components. Although the majority of thermoplastics/elastomer blends share heterogeneous, the component may be referred to as compatible if some technically advantageous combination of properties can be realized from the blends. Partial compatibility implies that above a particular level either the minor or the major various components remain as a dispersed phase.

There are some technical problems, which are frequently the results of some type of mutual incompatibility, which provides an inferior set of properties when dissimilar polymers are blended together. The blending process and the quality of the blends can be improved by adjusting the individual raw polymer viscosity. As a result the effective viscosities of the phase will no longer mismatch. The thermodynamic incompatibility can be overcome if the surface energy differences between polymers are small enough to permit the formation of very small micro domains of the individual polymer phases and there is sufficient adhesion between the phases by formation of crosslink across the interface during blending. The compatibility of various components and the generation of single phase from multiphase system play a major role in influencing the physical properties of the polymer blends. The crying need of designer of polymer blends is how to assure good stress transfer between the components of the multicomponent system which can only vouch for the efficient utilization of component physical properties of the blends with desired properties. Numerous techniques have been utilized to determine the compatibility of the blends but few predict good results.

## 5.7 Factors for Liquid Crystalline Polymer Fibrillation in Blends

Thermotropic liquid crystalline polymers (TLCPs) have gained increased commercial attention because of their unique properties. These include their low coefficients of thermal expansion, low viscosity, and high modulus, low permeability to gases, low dielectric constants, and chemical resistance. As the demand for these characteristics increases, it is anticipated that the use of TLCPs will grow, rising at a projected annual growth rate of 25 % from an estimated use of ten million pounds in recent years. In expanding the potential uses for TLCPs, it has been found that TLCP/TLCP blends can possess characteristics which are better than those of either individual TLCP (Utracki and Favis 1989). But the better result is only possible if the LCP fibrillation is prominent in the blend system.

The materials and processing parameters are the most vital aptitude of LCP to become fiber during dispensation in molten state. Some of the most important parameters in influencing fibrillation of LCP in blends include viscosity ratio, shear rate, LCP content, and interfacial properties (Seppälä et al. 1992). Elongational flow has been recognized to be more effective than shear flow in forming LCP fibers during processing. To ensure the fibrillation process or to study the mechanism of LCP fibrillation it is recommended to go for injection molding.

The LCP content also plays an important role in the factors affecting the fibrillation of the LCP phase in the blend system. As we go on increasing or sometimes decreasing the LCP content the mechanical properties, dynamical properties and the morphological properties changes which indicates the adhesion of the LCP with matrix in the blend system. If we get improved properties in every parameter then only we can conclude that the LCP fibrillation occurs well with the matrix.

The interfacial force acting between the matrix and filler components in the polymer blend is another major parameter that controls the development of fibrillar morphology.

Since injection molding is the most widely used processing technique for the production of plastic products, it is important to establish the conditions for LCP fibrillation so that strong injection-molded products can readily be obtained from LCP-based polymer blends. Moreover, as shear flow is inferior to elongational flow in developing LCP fibrillation, the identification of the conditions for LCP fiber formation under shear flow (such as in injection molding) would specify the minimum condition that is required for LCP fiber formation.

The structure and orientation of thermotropic liquid crystalline polymers (TLCPs) can significantly affect the mechanical performance of TLCP. The flow during processing plays a central role in defining the orientation on the macroscopic level. Extensional flows have been shown to be much more effective than shear flow in producing a high degree of molecular orientation. In view of the importance of extensional flow in processing of LCPs, the extensional rheological properties of LCPs, such as extensional viscosity, are crucial to understand the micro structural variation in extensional flow.

While the shear rheology of liquid crystalline polymer (LCPs) has been studied extensively, little attention was paid to their extensional behavior. Extensional viscosity of LCPs is important in processing of polymers, such as extrusion, blow molding, and fiber spinning. However, it is much more difficult to measure the extensional viscosity than the shear viscosity. Several techniques have been used to measure the extensional viscosity of LCP melts, including fiber spinning, rotary clamp method, lubricated squeezing technique, and the rheometrics Melt Extensional rheometer (RME) and entry flow. So the extent of fibrillation of the LCP phase that was developed during processing could be determined by studying the morphology of the LCP phase in the extrudate both in shear force as well as transient force.

## 5.8 Effect of Fillers on Liquid Crystal Polymers

Particulate inorganic fillers are widely used in thermoplastic industries to reinforce the polymer matrix and to improve the heat-resistance characteristics of materials (Qin et al. 1993; Malik et al. 1989). The high anisotropy of the short-fiber-reinforced thermoplastic can also be avoided by using particulate fillers. Generally, the viscosity of the thermoplastic will dramatically increase with the addition of fillers. When a particulate-filled thermoplastic is blended with TLCP, the TLCP will reduce the viscosity of the blend and enhance its processability. Furthermore, the reduced viscosity ratio of TLCP to matrix will favor fibril formation of the TLCP phase, providing further reinforcement of the blend. Previous researchers (Blizard and Baird 1987a; Meng et al. 1998; Acierno and Naddeo 1994) have used this type of ternary system to bring about a viscosity reduction in the blend.

However, there is little or no understanding how the particulate fillers interact with the LCP phase and whether the filler has any influence on the capacity for the LCP phase to fibrillate. This is the focus of the current review. Most LCPs and thermoplastics are immiscible at the molecular level. The incompatibility between the matrix polymer and dispersed LCP phase leads to poor interfacial adhesion. Accordingly the mechanical strength of the thermoplastic/LCP or elastomer blend is considerably lower than that expected from theoretical calculations. Therefore, compatibilisation is required to improve the adhesion between the LCP fibrils and matrix interface. The nature of the interface and the processing conditions has a significant effect on the morphology and structure of the thermoplastic/LCP blends. The extent of fibril formation and orientation and their control are of major concern in the processing of in-situ composites (Acierno and Naddeo 1994). In the current review, the role and contribution of particulate fillers on fibrillation of the LCP phase in LCP-based blends will also be examined through what is available in published literature.

### *5.8.1 Nanofillers*

Fillers are particles added to material (plastics, Composite material, Concrete) to lower the consumption of more expensive binder material or to better some properties of the mixture material. Worldwide more than 50 million tons of fillers with a total sum of ca. 25 billion Euro for different application areas, such as paper, plastics, rubber, paints and adhesives, are consumed annually. Formerly, fillers were used predominantly to cheapen end products. Today, it has been proven that fillers are also able to enhance technical properties of the products. As a result, a number of optimized types of fillers, Nano-fillers or surface treated goods have been developed.

Nanofillers have for many years had a high significance in the plastics industry. Nanofillers are basically additives in solid form, which differ from the polymer matrix in terms of their composition and structure. They generally comprise inorganic materials, more rarely organic materials. Inactive fillers or extenders raise the overall bulk volume and lower the prices, while active fillers bring about targeted improvements in certain mechanical or physical properties. The activity of active fillers may have a variety of causes, such as the formation of a chemical bond (e.g. cross linking by carbon black in elastomers) or filling of a certain volume and disruption of the conformational position of a polymer matrix, and also the immobilization of adjacent molecule groups and possible orientation of the polymer material (Dutta et al. 1993).

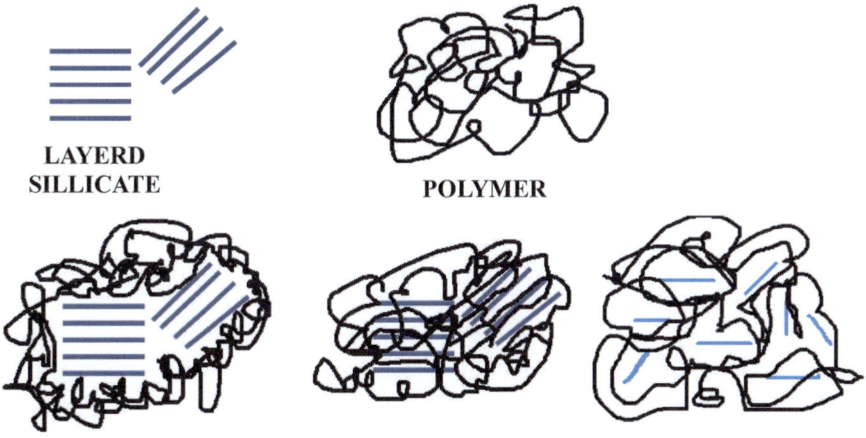


**Fig. 5.8** Possible morphology formation of the nanofiller in the LCP matrix phases

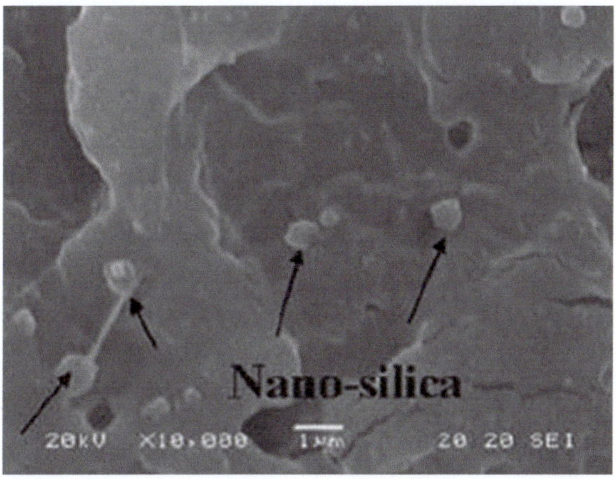


**Fig. 5.9** Nanosilica as observed from TEM in presence of LCP phase

## *5.8.2 Effect of Nanofiller*

Nano filler are expected to improve the properties of materials significantly more, even at lower loading than conventional/micro-fillers. There are some reasons for that (Dutta et al. 1993).

The size effects, large number of particulate with smaller inter particle distance and high specific surface area affects the properties of the ultimate composites. However the requisite properties are achievable for low filler content.

Figure 5.8 shows the different types of morphology formation of the nanofiller during mixing. Figure 5.9 shows TEM images of a nanosilica filled LCP specimen where nanosilica forms a mixed morphology in presence of LCP phase. Here the phase separated structure is more prominent because nanosilica agglomerates to form spherical droplets and comes out from the LCP phase.

Nanosilica is widely used in thermoplastic industries to reinforce the polymer matrix and to improve heat-resistance characteristics of materials. The high anisotropy of the short-fiber-reinforced thermoplastic can also be avoided by using particulate fillers. Generally, the viscosity of the thermoplastic will dramatically increase after the addition of nanosilica (Zhou and Yan 2005). When nanosilica is blended with TLCP, the TLCP will reduce the viscosity of the blend and enhance the processability of the TLCP to matrix will favor the fibril formation of TLCP, providing further reinforcement of the blend. The nanosilica as it has hydroxyl group generally get dispersed to the phase which have less polarity or less viscosity. In most of the cases it dispersed to the LCP phase. But if we take the example of PEI then in case the nanosilica will disperse in the PEI phase rather than LCP. By dispersing to the less viscosity phase it helps to increase the viscosity of the whole system and thus helps in processing. The nanosilica fillers used are powder in nature. The dispersion of nanoparticles is a critical factor in the development of nanocomposites. A composite with improved properties at low filler loading is always the optimum choice. The properties of nanocomposites strongly depend on the organic matrix, the nanoparticles, and the way in which they were prepared. The studied organic matrices include epoxy resin, polystyrene, Polyacrylate, and nylon, whereas the inorganic phases are usually clay, layered silicates, and nanoparticles such as $SiO_2$, $TiO_2$, ZnO, and $CaCO_3$ (Karam and Bellinger 1968; Kiss 1987; Qin et al. 1993; Malik et al. 1989; Blizard and Baird 1987a) among which nanosilica is the first produced nanoparticles. It has been reported that nanosilica can increase the hardness and scratch resistance of a coating and keep the coating clear at the same time (Meng et al. 1998; Acierno and Naddeo 1994). Nanosilica has enhanced the tensile strength and elongation of polyurethane elastomers; however, the modulus and hardness are lower than those of corresponding microsize filled polyurethane (Rath et al. 2006). Because the surface properties, particle size, and dispersability of nanoparticles are so diversified and nanoparticles tend to aggregate in the organic matrix on account of their high surface energy, it is very difficult to predict the properties of newly prepared nanocomposites.

Liquid crystalline polymers (LCPs) are attractive high performance engineering materials, because of their high strength and stiffness, high chemical resistance, good dimensional stability, and low linear thermal expansion coefficient. Particularly, thermotropic liquid crystalline polymers (TLCPs) have received special concerns during last decades, because of their viscosity-reducing function and the reinforcing ability of in-situ generated fibrils in thermoplastics (TPs) blends during melt processing (Shibaev et al. 1984).

Polymer blending is a useful and attractive approach to new polymeric materials with more balanced properties than their components. Most polymers are immiscible so that during the processing these component polymers form a multi-phase system in their blends. Several factors such as the composition, viscosity ratio, elasticity ratio, interfacial tension, shear rate and mixing time have their own influences on the final morphology of the phases.

The LCP acts as a 'processing aid' to reduce the whole melt viscosity and its in situ fibrillation results in a 'self-reinforced' blend. These excellent properties strongly depend on the LCP morphology, which is greatly affected by LCP content viscosity ratio rheological properties (Popa-Nita et al. 2009) and processing conditions. But sometime the mechanical property of the blend system is not so effecting due to the immiscible character of the individual substances. So recently researchers have tried to introduce some filler to the system to increase the effect of fibrillation of the LCP to increase the mechanical properties and some other criteria as well. When additional hydrodynamic effects arise due to the presence of fillers added into in situ composites, the morphological evolution of LCP becomes more attractive, which sometimes is beyond the scope of the understanding of conventional binary blends. Induced by different fillers, various hydrodynamic factors affect the morphological evolution of LCP in different ways. Considering the hydrodynamic effects induced by the added fillers, He and coworkers suggested that LCP melt droplets underwent elongational flow, which favored the LCP fibrillation more effectively than simple shear flow. However, these traditional fillers or fibers made in situ composites much heavier due to their own large density and high loading needed. Nano-size particulates filled polymers (nanocomposites) are of great interest due to their lower density, excellent performances and attractive prospect, compared to conventionally filled polymers (Zhou and Yan 2005). Nano-size fillers such as carbon black (CB) and nano-$SiO_2$ have been reported to have unique influence on the morphology of immiscible blends such as phase inversion, especially in conductive materials. Recently, it has been reported that nano-fillers such as nano-$SiO_2$ and nano-clay can also act as a compatibilizers in immiscible blends during mixing process. The introduction of Nano-sized fillers into in-situ composites provides a novel method to prepare a group of filled in-situ composites. These composites possess much lower density than those filled by conventional fillers, because of their much smaller size and much lower filling needed for a significant effect. Also, different new effects could be endowed to the composites, depending on the type of fillers used.

### *5.8.3 Interaction of Nanosilica with Liquid Crystalline Polymer*

Thermotropic LCPs have received much attention accredited to the melt processability using conventional compounding equipment. Due to the shear-thinning effect of the LCP melts, these polymers exhibited reduced viscosities and good processability. Furthermore, under appropriate processing conditions, the LCP could extend into fibrous structures and orient in the flow direction. This fibrillar morphology retained in the solidified blend was able to render self-reinforcement to the matrix. Currently, the issue of nanocomposite has been an uprising topic of research. Nano-composites are a new class of composites for which at least one dimensions of the dispersed phase is in the nanometer range (Zhou and Yan 2005). Seemingly, the incorporation of nanosized fillers into polymers could render greater reinforcement efficiency than conventional composites with microfillers due to the larger surface area. Thermoplastic polymer/clay nanocomposites have been widely investigated. It was often reported that the presence of nanoscale fillers could enhance the tensile and flexural modulus, thermal stability and barrier properties. The morphology of a polymer blend is dependent on the material characteristics, blend composition, rheological properties and processing conditions. When a thermotropic LCP is dispersed in an isotropic thermoplastic, the LCP phase can exist in the form of spherical droplets, ellipsoids, fibrils or lamellae structures (Shah et al. 1986). While in-situ composites have been a well-established field over the years, nanocomposites, on the other hand, has drawn considerable attention recently. It would be interesting to investigate the effect of adding nanosize fillers as their combination of the above disciplines might unveil additional knowledge to the polymer community.

### 5.8.4 *Nanofiller Reinforcement*

The difference between the behaviors of micro and Nano-reinforced polymers can be analyzed by observing the specific changes in properties in nanoscale, in which polymer chain lengths approach the filler dimensions so that they might display particular interaction influencing the macroscopic behavior of the materials and by the large numbers of parameters involved that can have a synergistic action. Indeed, many parameters have been taken in to account for the reinforcement efficiency of filler into given medium. The parameters are-

1. chemical nature of fillers
2. shape and orientation of the fillers
3. average size, size distribution, specific area of the particle
4. volume fraction
5. dispersion state
6. interfacial area

### 5.8.5 *Other Types of Nanofillers*

There are many types of nanofillers including: carbon black nanofillers, carbon nanotube nanofillers, carbon fiber nanofillers, activated clay nanofillers, natural clay nanofillers (mined, refined, and treated), clay (synthetic) nanofillers, natural fiber nanofillers, zinc oxide nanofillers and silica nanofillers. Carbon nanotubes (CNTs) are allotropes of carbon with a cylindrical nanostructure. Nanotubes have been constructed with length-to-diameter ratio which is significantly larger than any other material. These cylindrical carbon molecules have novel properties which make them potentially useful in many applications in nanotechnology, electronics, optics and other fields of materials science, as well as potential uses in architectural fields. They exhibit extraordinary strength and unique electrical properties, and are efficient thermal conductors.

All different classes of nanofillers have their distinct way of action. It solely depends on the end product requirement to choose among these nanofillers. In this study we will concise our focus to effect of nanosilica on LCP and its blend system.

## 5.9 In-Situ Reinforcement of Liquid Crystalline Polymers

An in situ composite consists of an immiscible blend of thermotropic liquid crystalline polymer (TLCP) with a thermoplastic in which reinforcement of the mechanical properties is achieved through the existence of LCP fibers that are formed during processing. Such composites arise due to the formation of a fibrillar TLCP phase during extensional melt flow. They are interesting because they have several outstanding features. In addition to the improvement on physical and dimensional stability of the matrix, the incorporation of TLCP also enhances the ease of processing, resulting from the reduced overall melt viscosity of the system. Processing conditions including temperature and the shear and elongational forces strongly effect the molecular orientation of the TLCP phase as well as the fiber aspect ratio which determines the final physical properties of the composite. There are several important factors affecting the morphology and distribution of TLCP domain in the finished product. Good fibrillation was found to be achieved when the viscosity of the TLCP phase is lower than that of matrix.

It is known that the thermoplastics reinforced with either fibers or fillers have higher mechanical properties than the matrix, but the reinforcements worsen process ability leads to greater equipment abrasion. May be the additional motion imparted to the machine due to more frictional force achieved from the filler reinforced system cause more abrasion to the equipment. One of the possible way to ameliorate these negative characteristics is the addition of thermotropic liquid crystalline polymers (TLCPs) (Narh et al. 2000b; Datta et al. 1993b; Kelar and Jurkowski 2000; Tjong et al. 2000a; Sukhadia et al. 1990). This is because of their low melt viscosity when processed in conventional processing machines, which should both improve the processability of the system and also minimize the breakage of the inorganic fiber. Moreover, LCPs should reinforce the system at a smaller scale than the mineral reinforcement does. These systems, in which the thermoplastic matrix is reinforced with LCP is called 'in-situ' composites.

Blending LCPs with thermoplastics permits the design of new materials with high strength and stiffness capable of performing determined functions. The mesogenic moieties of LCPs promote a high degree of molecular alignment in the molten state. Their long relaxation times allow the orientation of the chains to be easily frozen in the solid-state, giving rise to the formation of fibrillar structure under certain processing conditions. In situ reinforcement of thermoplastics, using minor LCP component additions, is attractive in terms of low melt viscosity, ease of orientation, low die swell and low shrinkage (Bretas and Baird 1992; Shah et al. 1986). Most commercial LCPs are incompatible with thermoplastics, leading to the properties of the LCP/thermoplastic blends to fall short of expectations. Fiber pullout is the dominant mechanism responsible for the failure of incompatible LCP/thermoplastic blends. Poor interfacial adhesion remains a major obstacle in achieving high performance in situ composites. Thus, suitable compatibilizers additions are needed to improve the interfacial adhesion of blend components. The effects of blending commercial LC copolyesters with a wide variety of both crystalline and amorphous thermoplastics, e.g. polypropylene (PP), polyethersulfone (PES), polycarbonate (PC), polyethylene terephthalate (PET), polytetrafluoroethylene (PTFE), and ultra-high molecular weight polyethylene (UHMPE) have been widely investigated over the past decades. These blends were produced by conventional compounding techniques i.e. either extruded into strands or injection molded.

## 5.10 Conclusion

The liquid crystalline polymers find more applications in smart composite materials. The popularity of LCP is growing very fast because of its ease of processing. It also helps other engineering thermoplastics to process very smoothly even at higher temperature. This happens mainly due to the fibril formations of LCP in presence of other polymer matrix. From the above mentioned areas of LCPs, it is clear that new results arise due to the incorporation of polymeric LCP in to the other polymer systems. It is very much clear that in future it will create its own vast area in the field of polymer blends and nanocomposites.

## References

Acierno D, Naddeo C (1994) Blends of PEEK and PET-PHB 60: a preliminary study on thermal and morphological aspects. Polymer 35(9):1994–1996

Barón M, Stepto RFT (2002) Definitions of basic terms relating to polymer liquid crystals, IUPAC commission on macromolecular nomenclature. IUPAC recommendations. Pure Appl Chem 74 (3):493–509

Bates FS (1991) Polymer-polymer phase behavior. Science 251(4996):898–905

Beery D, Kenig S, Siegmann A (1991) Structure development during flow of polyblends containing liquid crystalline polymers. Polym Eng Sci 31(6):451–458

Blizard KG, Baird DG (1987) The morphology and rheology of polymer blends containing a liquid crystalline copolyester. Polym Eng Sci 27(9):653–662

Brannock GR, Paul DR (1990) Phase behavior of ternary polymer blends composed of three miscible binaries. Macromolecules 23(25):5240–5250

Brehmer M, de Jeu WH (2012) Liquid crystal elastomers: materials and applications. Springer, Heidelberg

Bretas RE, Baird DG (1992) Miscibility and mechanical properties of poly (ether imide)/poly (ether ether ketone)/liquid crystalline polymer ternary blends. Polymer 33(24):5233–5244

Brown CS, Alder PT (1993). In: Folkes MJ, Hope PS (eds) Polymer blends and alloys. Blackie Academic & Professional, Glasgow, pp 193–227

Bucknall CB (1977) Toughened plastics. Applied Science, London

Christiansen WH, Paul DR, Barlow JW (1987) The phase behavior of ternary blends containing polycarbonate, phenoxy, and polycaprolactone. J Appl Polym Sci 34(2):537–548

Chung TS (1986) The recent developments of thermotropic liquid crystalline polymers. Polym Eng Sci 26(13):901–919

Collings PJ (1990) Nature's delicate phase of matter. Adam Hilger IOP, Bristol, p 196

Datta A, Chen HH, Baird DG (1993) The effect of compatibilization on blends of polypropylene with a liquid-crystalline polymer. Polymer 34(4):759–766

Davidson P (1999) Selected topics in X-ray scattering by liquid-crystalline polymers. In: Mingos DMP (ed) Liquid crystals II. Springer, Berlin, pp 1–39

Dowell F (1988) New theories for smectic and nematic liquid-crystal polymers: backbone LCPs and their mixtures and side-chain LCPs. Mol Cryst Liq Cryst Inc Nonlinear Opt 157 (1):203–227

Dutta D, Weiss RA, Kristal K (1993) Liquid crystalline polymer/fluoropolymer blends: preparation and properties of unidirectional "prepregs" and composite laminates. Polym Eng Sci 33 (13):838–844

García M, Eguiazabal JI, Nazabal J (2003) Two scale reinforcement in hybrid composites based on poly (ether sulfone), glass fiber and liquid crystalline polymer. Compos Sci Technol 63 (15):2163–2170

Han H, Bhowmik PK (1997) Wholly aromatic liquid-crystalline polyesters. Prog Polym Sci 22 (7):1431–1502

Harris JE, Robeson LM (1988) Miscible blends of poly (aryl ether ketone)s and polyetherimides. J Appl Polym Sci 35(7):1877–1891

Isayev A, Swaminathan S (1989) US Patent 4,835,047. U.S. Patent and Trademark Office, Washington, DC

Jaarsma FC, Chen JC, Conley DO (1985) A new chemically resistant liquid crystal polymer. Manag Corros Plast 7:71–76

Karam HJ, Bellinger JC (1968) Deformation and breakup of liquid droplets in a simple shear field. Ind Eng Chem Fundam 7(4):576–581

Kelar K, Jurkowski B (2000) Preparation of functionalised low-density polyethylene by reactive extrusion and its blend with polyamide 6. Polymer 41(3):1055–1062

Kiss G (1987) In situ composites: blends of isotropic polymers and thermotropic liquid crystalline polymers. Polym Eng Sci 27(6):410–423

Kohli A, Chung N, Weiss RA (1989) The effect of deformation history on the morphology and properties of blends of polycarbonate and a thermotropic liquid crystalline polymer. Polym Eng Sci 29(9):573–580

Koning C, Van Duin M, Pagnoulle C, Jerome R (1998) Strategies for compatibilization of polymer blends. Prog Polym Sci 23(4):707–757

Kulichikhin VG, Plate NA (1991) Blend composites based on liquid crystal thermoplast. Rev Polym Sci USSR 33(1):1–37

Kumar ES, Das CK (2005) Mechanical, dynamic mechanical properties and thermal stability of fluorocarbon elastomer–liquid crystalline polymer blends. Polym Compos 26(3):306–315

Kurtz SM, Muratoglu OK, Evans M, Edidin AA (1999) Advances in the processing, sterilization, and crosslinking of ultra-high molecular weight polyethylene for total joint arthroplasty. Biomaterials 20(18):1659–1688

La Mantia FP, Valenza A, Magagnini PL (1992) Liquid crystal polymer-based blends: "Universal grade" polymers? J Appl Polym Sci 44(7):1257–1265

La Mantia FP, Paci M, Magagnini PL (1997) Isothermal elongational behavior of liquid-crystalline polymers and LCPs based blends. Rheol Acta 36(2):152–159
Landel RF, Nielsen LE (1993) Mechanical properties of polymers and composites. CRC, Boca Raton
Malik TM, Carreau PJ, Chapleau N (1989) Characterization of liquid crystalline polyester polycarbonate blends. Polym Eng Sci 29(9):600–608
Manson JAE, Seferis JC (1992) Process simulated laminate (PSL): a methodology to internal stress characterization in advanced composite materials. J Compos Mater 26(3):405–431
Meng YZ, Tjong SC, Hay AS (1998) Morphology, rheological and thermal properties of the melt blends of poly (phthalazinone ether ketone sulfone) with liquid crystalline copolyester. Polymer 39(10):1845–1850
Narh KA, Zhang Q, Li Z (2000a) The effect of mixing on the properties of liquid crystalline polymer/polyethylene (terephthalate) blend. J Appl Polym Sci 75:1783–1787
Narh KA, Zhang Q, Li Z (2000b) The effect of mixing shear rate on the properties of liquid crystalline polymer/polyethylene terephthalate blends. J Appl Polym Sci 75(14):1783–1787
National Materials Advisory Board (1990) Liquid crystalline polymers. National Academy Press, Washington, DC. http://plc.cwru.edu/tutorial/enhanced/files/lc/phase/phase.htm
Plummer CJG (1993) Advanced thermoplastic composites: characterization and processing. Hanser, New York (Chap 8)
Plummer CJG, Zülle B, Demarmels A, Kausch HH (1993) The structure of filled and unfilled thermotropic liquid crystalline polymer injection moldings. J Appl Polym Sci 48(5):751–766
Popa-Nita V, Gerlič I, Kralj S (2009) The influence of disorder on thermotropic nematic liquid crystals phase behavior. Int J Mol Sci 10(9):3971–4008
Qin Y, Brydon DL, Mather RR, Wardman RH (1993) Fibres from polypropylene and liquid crystal polymer (LCP) blends: 1. Effect of LCP concentration. Polymer 34(6):1196–1201
Rath T, Kumar S, Mahaling RN, Mukherjee M, Das CK, Pandey KN, Saxena AK (2006) Flexible composite of PEEK and liquid crystalline polymer in presence of polyphosphazene. J Appl Polym Sci 104:3758–3765
Roetting O, Hinrichsen G (1994) Blends of thermotropic liquid crystalline and thermoplastic polymers: a short review. Adv Polym Technol 13(1):57–64
Roth D, Thomas L (1989) Why LCP film. In: Abstracts of papers of the American Chemical Society, vol 198, p 3-CMEC. American Chemical Society, Washington, DC
Rudko O (2002) Liquid crystalline polymers. Uniaxial-biaxial nematic phase transition
Scott CE, Macosko CW (1995) Morphology development during the initial stages of polymer-polymer blending. Polymer 36(3):461–470
Seo YP, Seo Y (2004) Thermal properties of biaxially deformed in situ composites. Polym Eng Sci 44(8):1419–1428
Seppälä J, Heino M, Kapanen C (1992) Injection-moulded blends of a thermotropic liquid crystalline polymer with polyethylene terephthalate, polypropylene, and polyphenylene sulfide. J Appl Polym Sci 44(6):1051–1060
Shah VS, Keitz JD, Paul DR, Barlow JW (1986) Miscible ternary blends containing polycarbonate, SAN, and aliphatic polyesters. J Appl Polym Sci 32(3):3863–3879
Shibaev VP, Kozlovsky MV, Beresnev LA, Blinov LM, Plate NA (1984) Thermotropic liquid crystalline polymers. Polym Bull 12(4):299–301
Siegmann A, Dagan A, Kenig S (1985) Polyblends containing a liquid crystalline polymer. Polymer 26(9):1325–1330
Sukhadia AM, Done D, Baird DG (1990) Characterization and processing of blends of polyethylene terephthalate with several liquid crystalline polymers. Polym Eng Sci 30(9):519–526
Thomas LD, Roth DD (1990) Films from liquid-crystals. Chemtech 20(9):546–550
Tjong SC (2003) Structure, morphology, mechanical and thermal characteristics of the in situ composites based on liquid crystalline polymers and thermoplastics. Mater Sci Eng: R: Rep 41 (1):1–60

Tjong SC, Li RKY, Xie X (2000) Properties of in situ composites based on semiflexible thermotropic liquid crystalline copolyesteramide and polyamide 66 blends. Polym J 32(11):907–914

Utracki LA, Favis BD (1989) Handbook of polymer science and technology. In: Cheremisinoff NP (ed) Polymer alloys and blends. Marcel Dekker, Inc., New York, pp 121–201

Vollrath F, Knight DP (2001) Liquid crystalline spinning of spider silk. Nature 410 (6828):541–548

Weiss RA, Huh W, Nicolais L (1987) Novel reinforced polymers based on blends of polystyrene and a thermotropic liquid crystalline polymer. Polym Eng Sci 27(9):684–691

Yoon HN, Charbonneau LF, Calundann GW (1992) Synthesis, processing sand properties of thermotropic liquid-crystal polymers. Adv Mater 4(3):206–214

Zhou J, Yan F (2005) Improvement of the tribological behavior of ultra-high-molecular-weight polyethylene by incorporation of poly (phenyl p-hydroxyzoate). J Appl Polym Sci 96 (6):2336–2343

# Chapter 6
# Effect of Polymer Network in Polymer Dispersed Ferroelectric Liquid Crystals (PSFLC)

**Tapas Pal Majumder, Tanmoy Lahiri, and Prabir Kumar Mukherjee**

## 6.1 Introduction

Polymer stabilized liquid crystals (PSLC) (Broer et al. 1988, 1989a, b, c, 1990, 1991; Crawford and Žumer 1996) have been attracted among the researchers because of their potential applications in reflective displays, heat-repelling sheets, optical shutters and dynamic holography. Ferroelectric liquid crystals (FLC) show excellent electro-optic properties (Hikmet et al. 1995; Furue et al. 1997; Miyazaki et al. 2001; Petit et al. 2006) having switching times of the order of microsecond and are producing faster devices than conventional nematics. The polymer network may enhance the mechanical stability without affecting their favorable electro-optic properties (Broer et al. 1990; Li et al. 2002; Ma and Yang 2000; Kossyrev et al. 2002). Since polymer stabilized ferroelectric liquid crystals (PSFLC) exhibit faster switching mechanism (Clark and Lagerwall 1980) than pure FLCs, they are more attractive over the common PSLCs in high performance display activities (Kitzerow et al. 1992; Molsen and Kitzerow 1994; Lee et al. 1994a, b). In recent years, researchers found further advanced performances of PSFLCs based on grayscale capability (Lee et al. 2005), V-shaped switching (Furue et al. 2002; Tsuda et al. 2008) and strong mechanical stability (Archer et al. 2008). Fujikake et al. (1997) and Sato et al. (2002) realized the spatial grayscale memory states in PSFLCs by forming enormous number of small liquid crystal (LC) domains with different anchoring of polymer networks. This grayscale capability in PSFLCs at the memory state finds unique applications in flexible displays (Fujikake

T. Pal Majumder, Ph.D. (✉) • T. Lahiri, Ph.D.
Department of Physics, University of Kalyani, Kalyani 741235, West Bengal, India
e-mail: tpmajumder1966@gmail.com; tpm@klyuniv.ac.in; lahiri.tanmoy@yahoo.co.in

P.K. Mukherjee, Ph.D.
Department of Physics, Government College Engineering and Textile Technology, 12 William Carey Road, Hooghly, Serampore 712201, West Bengal, India
e-mail: pkmuk1966@gmail.com



V.K. Thakur, M.R. Kessler (eds.), *Liquid Crystalline Polymers*,
DOI 10.1007/978-3-319-20270-9_6

et al. 2004), particularly in smart card with very low power consumption (Li et al. 2002; Dierking et al. 1999, 2000; Molsen and Kitzerow 1994; Lueder et al. 1998). Experimental findings (Furue et al. 2001; Lee et al. 2006; Murashige et al. 2004; Fujikake et al. 1999) suggest crucial role of suitable morphology of the polymer networks in the liquid crystal (LC) medium for the existence of such grayscale memory states in the PSFLCs. The same morphology is equally responsible in finding the material with favorable electro-optic properties and resistibility towards mechanical stress. Therefore, a detail understanding of this morphology demands special importance both from the research point of view in physicochemical area and also from their potential applications in display industry.

Earlier theories suggested by several research groups (Crawford et al. 1995; Chiccoli et al. 2002; Furue et al. 1998) highlighted the importance of layer-to-layer molecular alignment surrounding the polymer chains, which could describe various novel properties exhibited by PSFLCs. Brehmer et al. (1996) described the effect of crosslinkings of polymer chains on FLC layer to be considered as Hook's like elastic energy. Li et al. (2002) introduced the concept of rigid polymer network that exerted a bulk anchoring force to the liquid crystal directors. This anchoring force is taken as a field like effect in the composite system and expressed as an elastic term with a suitable interaction co-efficient connecting the polymer network and the liquid crystal molecules. Thus, there are several ideas published (Li et al. 2002; Guymon et al. 1995, 1998; Furue et al. 1999; Takahashi et al. 1999; Sato et al. 2003) over the years in order to understand the basic physical functionalities of this modern hi-tech material. However, the importance of the internal structure of a polymer and its possible influence on the FLC layer has never been noticed. It is an established observation that semicrystalline or amorphous polymers contain some nanoscale free volume elements having dimension of the order of (2–10) nm (Simha and Somcynsky 1969; Jean 1990; Bartoš 1996; Dlubek et al. 2004). The presence of such nano-dimensional volumes is confirmed experimentally in many types of polymer chains with the help of Positron Annihilation Lifetime Spectroscopy (PALS) (Dupasquier and Mills 1995; Mogensen 1995; Eldrup et al. 1981; Schrader and Jean 1988; Jean et al. 2003). When a polymer network stabilizes the liquid crystal system, these free volumes bound to play some effective role in determining some of the parameters of the composite medium more effectively than the theoretical models (Brehmer et al. 1996; Li et al. 2002; Fung et al. 1997) have predicted so far. We believe the presence of free volumes and the role of cross-link chains together determine the basic features of a typical PSFLC system.

## 6.2 Free Volume Model and Free Energy of the System

The Landau type free energy expression can be supplemented by the free volume contribution to produce a model free energy for the composite system. In general, the standard Landau free energy density for SmC*-SmA transition can be described by taking the tilt angle $\theta$ as the primary order parameter (Musevic et al. 2002) and the spontaneous polarization $P_S$ as the secondary order parameter (Dvorak 1974). Therefore, the free energy density (F) can be expressed in terms of both the primary order parameter ($\theta$) and the secondary order parameter ($P_S$) as given below:

$$F = F_0 + \frac{1}{2}\alpha(T - T_0)\theta^2 + \frac{1}{4}b\theta^4 - CP_S\theta + \frac{P_S^2}{2\varepsilon_0\chi_0} + \frac{1}{4}\eta P_S^4 - \frac{1}{2}\Omega P_S^2\theta^2 \quad (6.1)$$

Here $F_0$ represents the free energy density for the undistorted SmecticC* phase. $\alpha$ is a temperature dependent mean-field co-efficient. The second term described by b is the standard Landau co-efficient determining the second order phase transition and C is the bilinear coupling co-efficient between polarization ($P_S$) and tilt angle ($\theta$). The next term $\frac{P_S^2}{2\varepsilon_0\chi_0}$ is the dipolar ordering term with $\chi_0$ indicating the dielectric susceptibility along the spontaneous polarization of the medium. The term $\frac{1}{4}\eta P_S^4$ is inserted in the free energy just for stability reasons. The last term proportional to $\Omega$ determines the biquadratic coupling between $P_S$ and $\theta$. Additional terms for PSFLC are introduced in (6.1) to incorporate the effect of polymer network. By taking into the consideration of Li et al. (2002) model with the inclusion of elastic like coupling in the form of $\left(\frac{1}{2}W_P Sin^2\theta \approx \frac{1}{2}W_P\theta^2\right)$ to take into account the interaction between the liquid crystal (LC) molecules and the polymer network as a whole. Here, $W_P$ is the coupling coefficient that determines the strength of this interaction. This elastic term is held responsible for restoring the orientation of the local molecular director to that of the polymer network. In view of this new elastic term, the free energy expression in (6.1) now takes the following form:

$$F = F_0 + \frac{1}{2}\alpha(T - T_0)\theta^2 + \frac{1}{4}b\theta^4 - CP_S\theta + \frac{P_S^2}{2\varepsilon_0\chi_0} + \frac{1}{4}\eta P_S^4 - \frac{1}{2}\Omega P_S^2\theta^2 + \frac{1}{2}W_P\theta^2 \quad (6.2)$$

Equation (6.2) is the effective free energy expression to study PSFLCs till now. The elastic like bulk anchoring provided by the polymer network is not enough to hold the smectic layers against permanent deformations, but, it is the combination of bulk anchoring provided by the rigid polymer network and the director anchoring over the free volume surfaces that determine the overall stability of the composite medium (Lahiri and Pal Majumder 2011). The cause of such concept lies in the property of grayscale memory states observed in PSFLCs (Fujikake et al. 1997; Sato et al. 2002), which occurs mainly due to the strong existence of the director against the thermal energy and the guiding effects of the walls of the container. Such strong existence is achievable only when the composite system has anchoring voids with sizes in molecular dimension to trap the FLC director. Voids can not be realized in the realm of rigid polymer network as suggested in Li et al. (2002). Such voids can be explored by considering the incorporating free volume defect structures within cross-link chains of the polymer network. A detailed analysis on the trapping phenomena may be considered as a polymer structure consisting of large molecular chains extending over few thousand angstroms constantly twisting and wriggling around one another in a very complex fashion. The free volumes can be considered to be existing voids in between those molecular chains and their sizes are defined as the difference in volume between the total volume ($V_t$) of the system and the occupied volume ($V_0$) (Jean 1990). The occupancy of free volumes in polymer structure is mainly investigated by Positron Annihilation Lifetime Spectroscopy (PALS) with the precise determination of shapes, sizes and concentrations (Jean et al. 2003) of these nanovolumes. When such a polymeric network is inserted into a liquid crystal medium, the molecular directors become anchored with a stabilized orientation over the surfaces of the free volumes. For simplicity, we assume the shapes of the free volumes to be spherical in nature. However, free volumes may exhibit anisotropy in shapes due to elastic distortion that they experience from the host system. It has been theoretically (Jean et al. 1990) observed that such a distorted polymeric network occupies free volumes elliptical in shapes (Fig. 6.1), but for simplicity we restricted to a simpler spherical configuration of the free volumes for convenience.

When a polymer network goes inside an LC medium, the free volumes play an important role in determining the true microscopical interaction between the polymer chains and the host molecules. Some nano dimensional trapping of the layer director could take place over volumetric holes. If this is the situation as depicted in Fig. 6.1, then the surface of the free volumes is subjected to some potential barrier of the form $\Gamma \text{Sin}^2 \frac{\theta}{2}$ (Lahiri and Pal Majumder 2011) with $\theta$ representing the surface stabilized tilt angle between the layer normal and the rigidly fixed director and $\Gamma$ denotes the strength of this potential barrier. One way to look into the origin of this surface barrier is by considering the orientational disorder of the molecular director over the free volume boundaries.

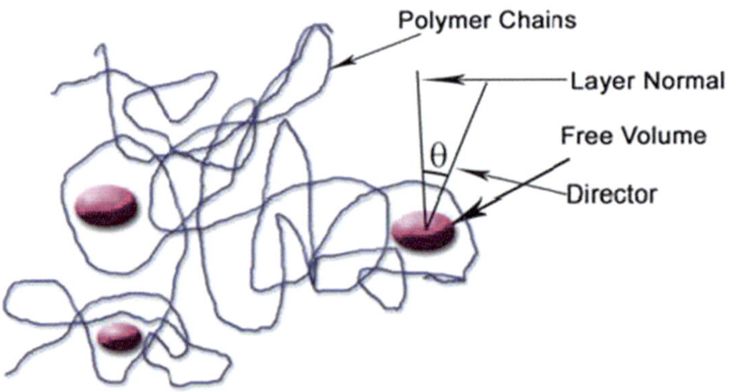


**Fig. 6.1** Pictorial representation of the polymer chains inserted into the liquid crystal layers. The free volumes are represented by the dark ellipses inside the polymeric chains. This figure demonstrates the trapped director mode over the surface of the nano volumes

It is assumed here that the anchoring potential is proportional to the difference in orientational order of the FLC director represented as $(P_0(\cos\theta) \sim P_1(\cos\theta))$, where $P_0(\cos\theta)$ is the zeroth order Legendre Polynomial signifying the parallel orientation of the director with the layer normal and $P_1(\cos\theta)$ is the first order Legendre Polynomial signifying the tilt of the director relative to the layer normal. Therefore, the introduction of the barrier potential $\Gamma\sin^2\frac{\theta}{2}$ can be justified. The orientational order may be considered to calculate the barrier potential which has the advantage of realizing the microscopical origin of the mechanical stability provided by the polymer network. This consideration may be interpreted as some kind of switching mechanism between two nematic states of the director. Then the parallel state denoted by $P_0(\cos\theta)$ is the desired self-healing state of the FLC and the tilted state $P_1(\cos\theta)$ determines the deformed state of the smectic layer. Since, the director is constrained to flip in between these two nematic states, so it may be concluded that the self-healing effect is mainly induced by the free volume trapping mechanism further advanced the FLC medium with the desired mechanical stability. We restricted our assumption only by considering zeroth and first order Legendre Polynomials to describe the orientational order in the molecular director and ignored higher order Legendre Polynomials in describing the self-healing mechanism because of the fact that any higher order Legendre Polynomials may create anomalous fluctuations in the molecular director with the increase of the asymmetry of the layer structure.

Now for a spherically symmetric hollow sphere, we can evaluate the bulk self-energy of these free volumes as $\frac{1}{2}\Gamma\left(1-\frac{r}{R}\cos\theta\right)$ with r denoting the distance from the boundary to the point of evaluation and R being the radius of the spherical volumes. Therefore, the exact free energy of a PSFLC can be written as (Lahiri and Pal Majumder 2011)

$$F = F_0 + \frac{1}{2}\alpha(T-T_0)\theta^2 + \frac{1}{4}b\theta^4 - CP_S\theta + \frac{P_S^2}{2\varepsilon_0\chi_0} + \frac{1}{4}\eta P_S^4 - \frac{1}{2}\Omega P_S^2\theta^2 + \frac{1}{2}W_P\theta^2 + \frac{1}{2}\Gamma\left(1-\frac{r}{R}cos\theta\right) \tag{6.3}$$

The free energy in (6.3) is thermodynamically stable only when $\Gamma < 0$, which is reasonable, since, $\Gamma$ determines the energetic contribution in F due to free volumes present in the medium.

To obtain a thermodynamically stable condition of the system, the minimization of F with respect to $P_S$ provides the following relation:

$$\frac{\partial F}{\partial P_S} = -C\theta + \frac{P_S}{\varepsilon_0\chi_0} + \eta P_S^3 - \Omega\theta^2 P_S = 0 \tag{6.4}$$

From, (6.4), we have

$$-\frac{1}{2}\Omega P_S^2\theta^2 = \frac{1}{2}CP_S\theta - \frac{P_S^2}{2\varepsilon_0\chi_0} - \frac{1}{2}\eta P_S^4 \tag{6.5}$$

Hence, in view of (6.5), we have from (6.3)

$$F = F_0 + \frac{1}{2}\alpha(T - T_0)\theta^2 + \frac{1}{4}b\theta^4 - \frac{1}{4}\eta P_S^4 - \frac{1}{2}CP_S\theta + \frac{1}{2}W_P\theta^2 + \frac{1}{2}\Gamma\left(1 - \frac{r}{R}\text{Cos}\theta\right) \tag{6.6}$$

Again, it is also important to minimize the free energy with respect to θ because large scale fluctuations in θ not only destabilizes the system but also increases the asymmetric behavior. After minimizing the free energy in (6.6) with respect to θ, we have

$$\theta\left\{\alpha(T - T_0) + W_P + \frac{\Gamma r}{2R}\right\} + b\theta^3 - \frac{1}{2}CP_S = 0 \tag{6.7}$$

Equation (6.7) then yields

$$P_S = \frac{2a}{C}\theta + \frac{2b}{C}\theta^3, \text{ where } a = \alpha(T - T_0) + W_P + \frac{\Gamma r}{2R} \tag{6.8}$$

In view of (6.8), we get

$$F = F_0 + \frac{1}{2}a^*\theta^2 + \frac{1}{4}b^*\theta^4 + \frac{1}{2}\Gamma\left(1 - \frac{r}{R}\text{Cos}\theta\right) \tag{6.9}$$

$$\text{Here, } a^* = \alpha\left(T - T_C'\right) \text{ with } T_C' = T_C + \frac{W_P}{\alpha} \text{ and } T_C = T_0 + \frac{\Gamma r}{R\alpha} \tag{6.10}$$

And $b^* = 3b + \frac{16a^4\eta}{C^4}$.

Therefore, the transition point $T_C' = T_0 + \frac{\Gamma r}{R\alpha} + \frac{W_P}{\alpha}$ is now shifted due to two terms, one is associated with the interaction between polymer network and LC molecules, behaving as a composite rigid network with contribution $\frac{W_P}{\alpha}$ and the other one is associated with the existence of nano-sized free volumes interacting with the tilt of the molecular director with contribution $\frac{\Gamma r}{R\alpha}$. Earlier studies also predicted the same shifting phenomenon in PSFLCs (Petit et al. 2006) apart from the fact that a better part of this shift must be attributed to the self-healing effect induced by the free volumes. The overall mechanism of the shifting process may be explained as follow: Once we begin to polymerize the sample, dimers, trimers and oligomers are formed, which are uniformly dispersed in the FLC matrix, well below their solubility limit. As the polymer content increases, the growing polymer becomes insoluble in the matrix and phase-separates from the FLC, quickly increasing its transition temperature to the value expected for the neat matrix. During the polymerization process, the free volumes in between the grown polymer network stabilize the director by strongly anchoring it over the boundary surfaces. As the grown polymer reaches its insolubility limit, a discontinuous system is formed, where the polymer network stabilizes the liquid crystal molecules, while the liquid crystal recovers most of its original properties of the neat material, gaining the benefit of the self-healing property from the nano-confined geometry of the free volumes. Therefore, introduction of polymer network into FLC medium largely retains the favorable electro-optic properties of the neat FLC while gaining the benefit of desired mechanical stability. The polymerization kinetics discussed so far is consistent with SEM studies on different systems (Dierking et al. 1998a, b).

Very close to the transition point, we have from (6.8)

$$\theta = (\gamma_{PSFLC})P_S$$
$$\text{Where,}\quad \gamma_{PSFLC} = \frac{C}{2\alpha(T - T_0) + 2W_P + \frac{\Gamma r}{R}} \tag{6.11}$$

Here, $\gamma_{PSFLC}$ denotes the coupling coefficient between the spontaneous polarization ($P_S$) and the tilt angle ($\theta$) for a polymer stabilized FLC system (Blinov et al. 1995).

Now for a pure FLC sample, we have

$$\gamma_{FLC} = \frac{C}{2\alpha(T - T_0)} \tag{6.12}$$

As expected, $W_P$ and $\Gamma$ both are equal to zero for $\gamma_{FLC}$. Hence, from (6.11) and (6.12), we have

$$\frac{\gamma_{PSFLC}}{\gamma_{FLC}} = \frac{1}{1 + \frac{W_P + \frac{\Gamma r}{R}}{\alpha(T - T_0)}} \tag{6.13}$$

Equation (6.13) clearly reveals that if we can make $\alpha$ significantly large in the composite LC-polymer medium, then we may have $\gamma_{PSFLC} \approx \gamma_{FLC}$, which means the polarization-tilt ratio $\frac{P_S}{\theta}$ shows no improvement due to polymer stabilization. This condition is somewhat disadvantageous from the application point of views because the polarization-tilt ratio determines the electro-optic response of the medium. Therefore, it is important to identify the possibilities, which can lead $\alpha$ to that critical value for which $\gamma_{PSFLC} \approx \gamma_{FLC}$. One, however, notes that for PSFLCs, $\alpha$ does not change with polymer stabilization (Archer et al. 2008). Therefore, it is expected that PSFLCs may show superior electro-optic properties compared to pure FLCs as $\alpha$ remains fixed.

There is one more possibility from (6.13), which further shows that we may again have the condition

$$\gamma_{PSFLC} \approx \gamma_{FLC}, \quad \text{when} \quad \Gamma = -W_P \frac{R}{r}.$$

This condition implies that when the free volume energetic contribution ($\Gamma$) balances the entropic part or the elastic interaction ($W_P$) between the LC molecules and the polymer network, then PSFLCs may not exhibit any improvements in its electro-optic properties. However, it has been found that the difference in magnitude between these two energetic contributions is in general so large that they rarely balance out. Therefore, introduction of free volume geometry does not hamper the favorable electro-optic response of these materials.

In view of (6.11), we have from (6.9)

$$F = F_0 + \frac{1}{2}\mu P_S^2 + \frac{1}{4}\lambda P_S^4 + \frac{1}{2}\Gamma\left(1 - \frac{r}{R}\right) \tag{6.14}$$

The normalized Landau co-efficients are now given by

$$\mu = \frac{C^2\alpha(T - T_0) + C^2 W_P + \frac{C^2\Gamma r}{R}}{\left[2\alpha(T - T_0) + 2W_P + \frac{\Gamma r}{R}\right]^2} \quad \text{and}$$

$$\lambda = \frac{C^4}{\left[2\alpha(T - T_0) + 2W_P + \frac{\Gamma r}{R}\right]^4}\left(3b + \frac{16a^4\eta}{C^4}\right)$$

Equation (6.14) completes the free energy formalism that describes almost every novel properties exhibited by PSFLCs.

## 6.3 Rotational Viscosity

The electro-optic properties of PSFLCs depend mainly on the variation of rotational viscosity and spontaneous polarization of the medium. The dielectric variation may also have some important implications in case of optical properties. We investigate all these three parameters that seem to govern the superiorities of PSFLCs over conventional FLCs. Unlike isotropic liquids, the flow behavior in liquid crystal is extremely complex and difficult to study experimentally due to their inherent anisotropy (Skarp 1988). This is because the translational motions are now coupled to the inner orientational motions of the molecules. In most cases, the flow disturbs the orientational alignment and conversely a change in the alignment created by the walls of the container or by the external field induces a flow in the liquid crystal medium. In general, we distinguish between two possible sources of viscosity in liquid crystals. One is due to a velocity field resulting in the flow of matter and the other due to the local state of alignment of the molecular director. The flow of matter induces a shear viscosity due to pressure gradient and velocity gradient of the flowing matter. Apart from the shear viscosity, there is another viscosity called the rotational viscosity, which arises due to viscous torques being exerted on the director during rotation or during shear flow when the director orientation is held fixed. Therefore, the local orientation of the director in any direction is now accompanied by a dissipation of energy which is described by the rotational viscosity. In case of ferroelectric smectic phases, we distinguish between two viscosity co-efficients $\gamma_\theta$ and $\gamma_\varphi$, respectively. The term $\gamma_\theta$ is associated with the tilt angle variation of the director and $\gamma_\varphi$ corresponds to the azimuthal orientation of the molecular director and both of these terms are referred as the soft mode and Goldstone mode viscosity co-efficients, respectively. Here, we mainly investigate the tilt fluctuation i.e. the soft mode rotational viscosity of the system. This can be done by considering Landau-Khalatnikov (L-G) equation of the form (Lahiri and Pal Majumder 2011)

$$\gamma_\theta \frac{\partial \theta}{\partial t} = -\frac{\partial F}{\partial \theta} \tag{6.15}$$

Here, $\gamma_\theta$ now describes the rotational viscosity related with soft mode relaxation of the PSFLC system. From (6.9), we have

$$\gamma_\theta \frac{\partial \theta}{\partial t} = -a^*\theta - b^*\theta^3 - \frac{1}{2}\Gamma\frac{r}{R}\sin\theta \tag{6.16}$$

It has been found that a simple hydrodynamic relaxation description of the tilt angle $\theta$ as governed by the L-G equation is valid only when $\theta$ is small. In that case, (6.16) can be modified into the following form

$$\gamma_\theta \frac{\partial \theta}{\partial t} = -a^*\theta - b^*\theta^3 - \frac{1}{2}\Gamma\frac{r}{R}\left(\theta - \frac{\theta^3}{3}\right) \tag{6.17}$$

After some simple algebraic transformation, (6.17) yields

$$\int \gamma_\theta \frac{d\theta}{-A\theta - B\theta^3} = \int dt \tag{6.18}$$

Where $A = a^* + \frac{1}{2}\Gamma\frac{r}{R}$ and $B = b^* - \frac{1}{6}\Gamma\frac{r}{R}$

Performing the integration in (6.18) and with some rearrangement of terms, we have

$$\gamma_\theta = \frac{A\tau_R}{\ln\left(\theta^{-1} + \frac{2B}{A}\theta + \frac{B^2}{A^2}\theta^3\right)} \tag{6.19}$$

With the condition $t + D/A = \tau_R$ or $A\ (t + D/A) = A\tau_R$ or $A\ (t + t_0) = A\tau_R = A/2\pi f_R$; $\tau_R$ and $f_R$ are the relaxation time and relaxation frequency, respectively. Here, D is the integration constant manipulated at will to get the relaxation time $\tau_R$. Figure 6.2 shows the variation in rotational viscosity $\gamma_\theta$ with the temperature difference $\left(T - T_C'\right)$.

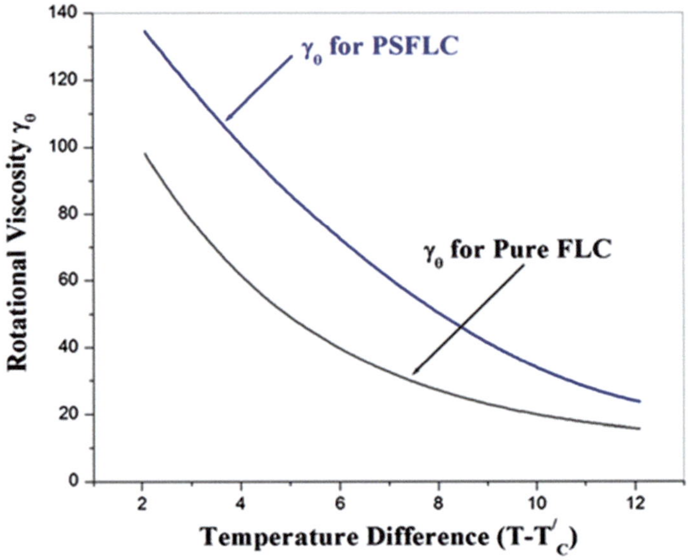


**Fig. 6.2** The variation of rotational viscosity ($\gamma_\theta$) with temperature difference $\left(T - T_C'\right)$ for polymer-stabilized ferroelectric liquid crystal and pure ferroelectric liquid crystal. The units are taken arbitrarily

The experimentally determined material (PSFLCs and FLCs) viscosity obeys typical Arrhenius behavior (6.19) with increasing temperature. This means with increase in temperature $\gamma_\theta$ decreases in a typical exponential manner for both PSFLCs and FLCs. We also observe that addition of polymer network increases the rotational viscosity in case of PSFLCs compared to that of pure FLCs (Fig. 6.2). This is a direct consequence of (6.19) and remarkably, experimental observations made by Pal Majumder et al. (1994) strongly supports our stand. Increase in $\gamma_\theta$ for PSFLCs implies the incorporation of polymer network which reduces the rotational mobility of the molecular director and otherwise it has large rotational freedom in case of neat FLCs.

This behavior of $\gamma_\theta$ in case of PSFLCs may be due to four specific reasons: First, during the onset of the polymerization process, the viscosity of the FLC–monomer dispersion is reduced as monomers are now used up to form the polymer network; second, the phase-separated polymer network acts as the source of elastic interactions with the LC molecules, which can also be held responsible for an effective viscosity observed in Fig. 6.2; third, the polymer network influences the tilt angle and the spontaneous polarization, which creates cascading effect on $\gamma_\theta$; finally, the free volumes present in between the polymer chains restrain the director mobility, which further reduces $\gamma_\theta$. Therefore, it seems combination of effects resulting from a complex picture of interaction mechanism in case of PSFLCs produces the effective viscosity of the medium.

## 6.4 Polarization Profile and Dielectric Constant

The investigation of the polarization profile and the dielectric constant of the medium are given below:

By putting the applied electric field, $E = \frac{\delta F}{\delta P_S}$ in (6.14), we have

$$E = \mu P_S + \lambda P_S^3 - \frac{\Gamma}{2R}\frac{\delta r}{\delta P_S} \tag{6.20}$$

The cubic equation for spontaneous polarization can be written as:

$$P_S^3 + A^* P_S - B^* = 0 \tag{6.21}$$

By normalizing some of the co-efficients in (6.21) we get

$$A^* = \frac{\mu}{\lambda} \quad \text{and} \quad B^* = \frac{1}{\lambda}\left(E + \frac{\Gamma}{2R}\frac{\delta r}{\delta P_S}\right)$$

When (6.21) is solved analytically, then one of the three solutions of the cubic equation can be written in the following expression for $P_S$

$$P_S = \frac{3\left[\frac{B^*}{2} + \sqrt{\frac{B^{*2}}{4} + \frac{A^{*3}}{27}}\right]^{2/3} - A^*}{3\left[\frac{B^*}{2} + \sqrt{\frac{B^{*2}}{4} + \frac{A^{*3}}{27}}\right]^{1/3}} \tag{6.22}$$

Since, $A^{*3} = \frac{\mu^3}{\lambda^3} \approx \alpha^{-3}(T - T_C')^9$ and both $(T - T_C')^9$ and $\alpha^{-3}$ are very small (Petit et al. 2006) compared to $B^*$, hence $A^{*3}$ term can be neglected in (6.22). With this condition, we may well get a physically acceptable solution for $P_S$ only when $E \neq 0$ (In accordance with the expression for $B^*$). Otherwise, the whole system would diverge to a nonlinear singular response mode at zero-external field, which would further dilute the whole theoretical predictions for both PSFLCs and neat FLCs, respectively. In view of this approximation, the second root of (6.21) results in a zero value of spontaneous polarization ($P_S$), which may be regarded as the trivial solution for the paraelectric phase. The third root is physically unacceptable as it provides untenable behavior of the spontaneous polarization and hence, discarded altogether. With these considerations, (6.22) can be utilized to get the desired solution for spontaneous polarization $P_S$ in the form of $P_S = B^{*1/3} - \frac{A^*}{3B^{*1/3}}$. In terms of actual variables, we now have

$$P_{PSFLC} = \lambda^{-1/3}\left[E + \frac{\Gamma}{2R\left(\frac{\delta P_S}{\delta r}\right)}\right]^{1/3} - \mu\lambda^{-2/3}\left[E + \frac{\Gamma}{2R\left(\frac{\delta P_S}{\delta r}\right)}\right]^{-1/3} \tag{6.23}$$

Here, in (6.23)

$$\frac{\mu}{\lambda^{2/3}} = \frac{1}{3}\left(\frac{4\alpha^5}{C^2}\right)^{1/3}\left[\frac{\left(T - T_C'\right)^5}{\left(3b + \frac{16a^4\eta}{C^4}\right)^2}\right]^{1/3} \quad \text{with } a = \alpha\left(T - T_C'\right).$$

When, $\Gamma = 0$ or $\frac{\delta P_S}{\delta r} \to \infty (i.e.\, \Delta r \to 0)$

Then, we recover the polarization profile of the pure FLC medium and it is given by

$$P_{FLC} = \frac{\lambda^{1/3}E^{2/3} - \mu}{\lambda^{2/3}E^{1/3}} \tag{6.24}$$

Equation (6.24) is valid only for the condition $E \neq 0$, as already discussed.

When the nano-sized free volumes contain infinite charge density i.e. $\frac{\delta P_S}{\delta r} \to \infty$. Equation (6.24) shows the same behavior as similar for pure FLCs. This infinite charge density may produce a large electrostatic field due to the presence of charge carriers inside the free volumes. Such field then produces electron-hole pairs inside the free volume states. Therefore, the negative charges are now repelled away from the electron, which are surrounded by holes that act to reduce the electron's true charge. The situation is highly similar to that of a charged body being inserted inside a polarizable medium. The charge of the nano-volumes is now less than that of the actual charge and consequently the system behaves like a polarized medium. Thus inside the composite system we now have some "vacuum polarization" due to the infinite charge density of the nano-volumes. This polarization being dominant enough corresponding to an infinite bare charge density prevails all-over the composite system. Therefore, it becomes hard to distinguish between the actual polarization of the medium and the vacuum polarization that resides inside the material. As a result, the whole system now behaves entirely like a neat FLC. However, these ideas are true only when we have large enough bare charge density inside the free volumes and to achieve this, a large external field has to be applied over the system that can lead to the situation mentioned above. Therefore, the effect of polymer network in the liquid crystal matrix can be best realized if we assume the existence of such nano-volumes and their effective interaction with the LC medium. Introducing the free volume contribution to the Landau free energy in our theory leads to an entirely new concept about the interactive nature of polymer network and the LC molecules and thus lends further interest in exploring the area of PSFLCs. Figure 6.3 shows the variation of the polarization for PSFLCs and the neat FLCs with temperature difference $\left(T - T_C'\right)$.

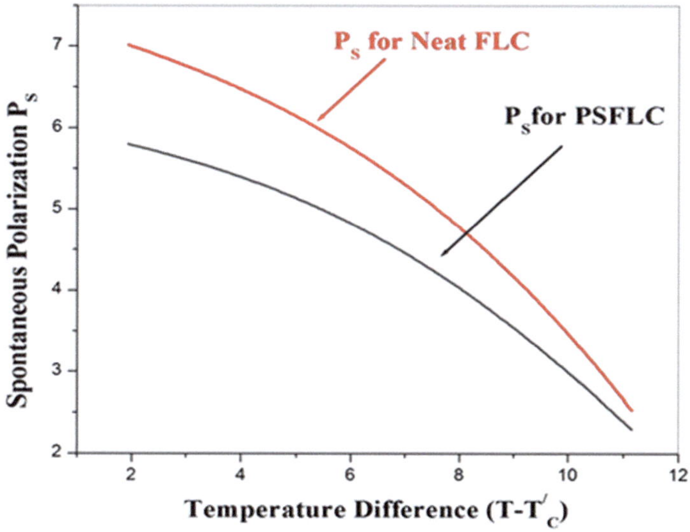


**Fig. 6.3** Variation of spontaneous polarization with temperature difference $\left(T - T_C'\right)$ for both PSFLC and the pure FLC with a fixed applied field. The units are taken arbitrarily

The spontaneous polarization decreases with increase in temperature for both pure FLCs and PSFLCs (Fig. 6.3). The magnitude of the polarization is slightly reduced for the polymerized ferroelectric system compared to that of a pure FLC system. This may be due to the dilution effect induced by the achiral monomer in FLCs. As the temperature increases, the interaction between liquid crystal molecules and polymer network becomes appreciably strong with coherence length several hundred of nanometers approaching to the bulk of the sample. The motion of the trapped molecular director over the free volume surfaces is now severely constrained that reduces the tilt angle $\theta$. The underlying mechanism that contributes to the reduction in tilt angle $\theta$ will be clear when we will introduce the combined geometry of free volumes and cross-link conformations in Sect. 5. As a result of the trapped motion, the tilt angle is reduced compared to that of the bulk value $\theta(T)$ with the liquid crystal molecules are now very close to that of the polymer network. The reduction in tilt angle then reduces the overall spontaneous polarization in PSFLCs. This conjecture finds strong experimental support in a series of work (Kundu et al. 2000; Karapinar et al. 2002; Archer and Dierking 2005, 2008, 2009) undertaken by several groups.

The reduction in polarization due to the introduction of polymer network at all temperatures is described in Kundu et al. (2000) and Karapinar et al. (2002), in which the phase-separation mechanism and the superior electro-optic effects that depends on the polarization profile is worked out using a photo polymerization technique. Recently, Archer et al. (Archer and Dierking 2005, 2008, 2009) working on FELIX M4851/050 experimentally described the impact of tilt angle reduction in PSFLCs. Here also, the reduction in $P_S$ is mainly attributed to the combined effect of tilt angle reduction and dilution effect. All these experimental findings find the theoretical counterpart in our work. We believe apart from the dilution effect, the nano confinement effect due to the presence of free volumes also contribute in determining the polarization profile of the medium.

An essential part of the electro-optic effect in PSFLCs depends on the dielectric response of the medium. Assuming the composite medium to be subjected to a constant isotropic dielectric constant, we have

$$\varepsilon_{PSFLC} = \frac{1}{\varepsilon_0}\frac{\partial P_S}{\partial E} = \frac{1}{\varepsilon_0}\frac{\lambda^{1/3}\left[E + \frac{\Gamma}{2R\left(\frac{\delta P_S}{\delta r}\right)}\right]^{2/3} + \mu}{3\lambda^{2/3}\left[E + \frac{\Gamma}{2R\left(\frac{\delta P_S}{\delta r}\right)}\right]^{4/3}} \tag{6.25}$$

Since, we know that the liquid crystals are strongly anisotropic and therefore, a static dielectric constant may preclude with the ideal case of dielectric anisotropy. However, the surface stabilization of the molecular director over the free volume boundaries enable us to work with this isotropic approximation because the surface stabilization produces strong network correlation in polymer chains when they template the LC order. In this context our assumption is valid, however, in a more general case we have to relax our approximation considering the anisotropy in dielectric constant.

Again, from (6.25) when $\frac{\delta P_S}{\delta r} \to \infty (\text{i.e.}\, \Delta r \to 0)$ or $\Gamma = 0$ we have

$$\varepsilon_{FLC} = \frac{1}{\varepsilon_0}\frac{\lambda^{1/3}E^{2/3} + \mu}{3\lambda^{2/3}E^{4/3}} \tag{6.26}$$

In (6.25), $\frac{\delta P_S}{\delta r}$ describes the charge density of the nano-structured free volumes present in the system. Figure 6.4 shows the variation in dielectric constant with the applied electric field E. We observe a significant difference between the two curves of dielectric constant for PSFLCs and pure FLCs, respectively (Fig. 6.4).

The magnitude of ε is quite small in case of PSFLCs compared to that of a pure FLC in accordance with the observation made by Dierking et al. (2000). Most importantly, this theory predicts a small contribution to the dielectric constant in PSFLC at zero applied field. This may be due to the vacuum polarization arising out of the bare charges inside the nano-volumes present in the system. Moreover, for PSFLCs, the decrease in the dielectric constant with applied bias can be interpreted as due to the formation of some irregularly scattered domains with higher polymer concentrations. Therefore, some local stabilization of the molecular director within the polymer dominated regions counter the vacuum polarization resulted from the bare charges of the free volumes at zero bias. This is clearly understandable from the decreasing nature of the dielectric curve for both PSFLCs and pure FLCs, respectively. However, these conjectures are appreciable only when we are very close to the transition region between SmecticC*-SmecticA phases and the polymer network introduced into the liquid crystal matrix contains significant amount of free

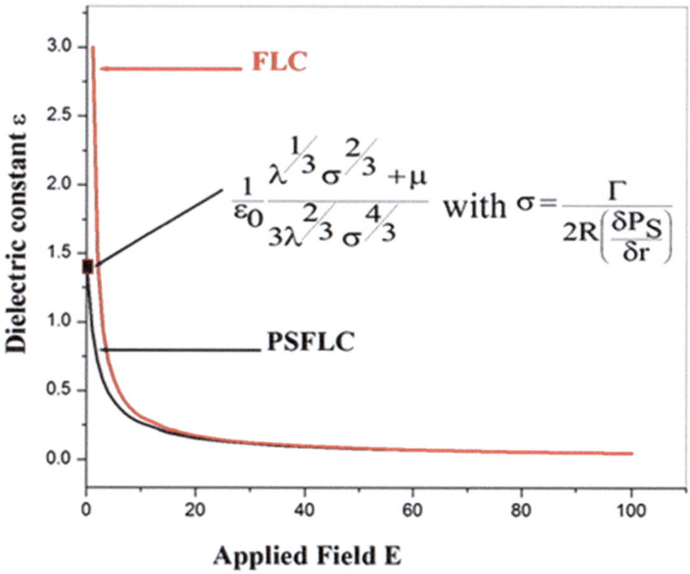


**Fig. 6.4** Variation of the dielectric constant (ε) for PSFLC and pure FLC with the applied field (E). The units are taken arbitrarily

volumes. Since free volumes are observed almost in any polymeric samples, therefore, when polymer network is dispersed within the LC matrix we could observe such modulation in the value of polarization even when the applied field is zero. In that case, we can think of devices that may have superior electro-optic properties compared to that of pure or even conventional PSFLCs.

## 6.5 Origin of Memory States and Cross-link Modeling

Apart from the superior electro-optic properties, PSFLCs also exhibit excellent grayscale memory states. It has potential application in high-resolution multistable matrix panel displays. Therefore, it is of primary importance for the theorists to explore the origin of this memory features in PSFLCs. In order to identify the underlying mechanism that works between the polymer chains and the FLC molecules to create the memory states, we propose the following microscopic model.

When a polymer network goes inside an FLC medium, we may consider the random networks of cross-link chains to be stretched between two smectic planes spaced L apart (Fig. 6.5). Gaussian like cross-links with orientationally uncorrelated chains, freely joined by spherical particles and connected by harmonic springs can be considered as shown in Fig. 6.6. $r(z) = [x(z), y(z)]$, describes the path of cross links conformations where "z" stands for the direction of cross-link alignment without any loops and overhangs. Further, we assume the cross-link

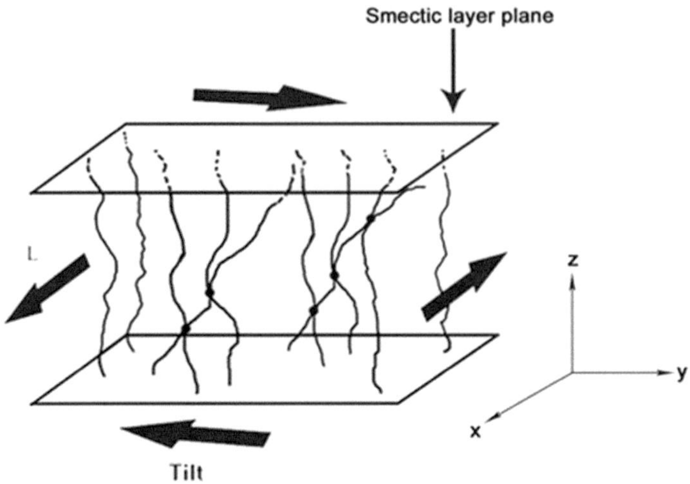


**Fig. 6.5** Schematic diagram of polymer cross-links stretched between two smectic planes L apart

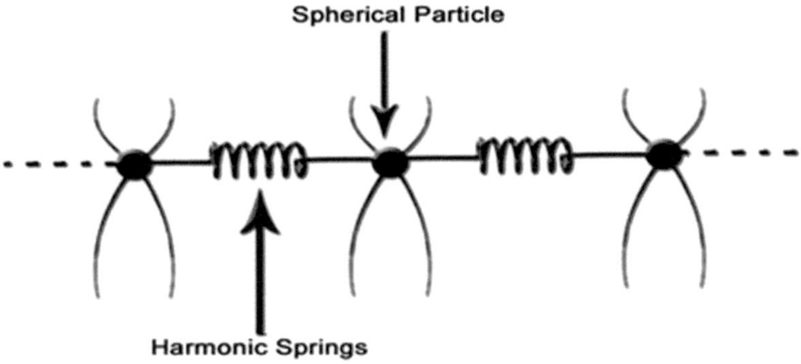


**Fig. 6.6** One-dimensional schematic diagram of spring-ball model

chains to have arbitrary positions both at $z = 0$ and $z = L$, respectively. As the LC molecules get strongly attached in between the cross-link interstices, the assumed Gaussian chains rotate with the FLC director when an electric field is applied over the system. However, once the field is removed, the chains provide enough anchoring strength to the FLC director to overcome the thermal energy and to occupy a specific orientation. Such phenomenon may provide information about the memory characteristics of the PSFLCs (Lahiri and Pal Majumder 2012) because of the cross-link chains restriction to the cone motion of the liquid crystal molecules. The reduction of the rotational viscosity occur due to the existence of cross link chains which restrict the motion of the cone of FLCs. After turning off the external field, this reduced viscosity is then supposed to create an elastic torque on the memory state that promotes both the multistability and greater resolution. The tilt angle ($\theta$) and the spontaneous polarization ($P_S$) is also expected to be reduced due to the reduction of the rotational viscosity. Recent studies (Li et al. 1998; Dierking et al. 1998a; Suresh and Chien 2003; Zheng 2007; Shinkawa et al. 2008; Inoue et al. 2008; Singh 2011) strongly demonstrate this behavior of the tilt angle and spontaneous polarization in PSFLCs.

## 6.6 Construction of Free Energy and Tilt Angle Variation

The effective interaction between cross-links (Fig. 6.6) is described by the following equation (Ulrich et al. 2006, 2010):

$$\varphi = \frac{1}{2L_c^2}\sum_{e=1}^{M}\left[r_{ie}(z) - r_{je}(z)\right]^2 \tag{6.27}$$

Here, we consider N identical particles positioned at $\left(\vec{r}_1 \ldots \vec{r}_N\right)$ with M pairs of monomers. The parameter $L_C$ can be considered as the typical length of a cross-link. The interaction (φ) is measured in units of $K_BT$, where $K_B$ stands for Boltzmann constant. For convenience the cross-link interaction may be assumed to be local in z direction and it depends only on the in-plane distance of the polymer segments. The inter-layer harmonic field may also influence the polarization vectors in the respective smectic layers. So the free energy density with an energetic contribution resulting from the coupling between the harmonic field and the spontaneous polarization can be added as more. As a result this coupling is expected to match the experimentally verified polarization profile observed in PSFLCs. The polymer and FLC composite may be assumed as some elastic continuum subjected to three elastic deformations such as shear, bulk and tilt. The first two are associated with the shear deformations of the smectic structure, while the last one is associated with the tilt elastic deformation induced by the tilting of the cross-link chains from the preferred axis. The elastic energy for such a composite system as according to de Gennes and Matricon (1964) in context of vortex lines analysis for type-II superconductors can be written as:

$$F_{el} = \frac{1}{2}\int\frac{d^2q}{(2\pi)^2}\int\frac{dq_z}{2\pi}\left[\left(\kappa q_z^2 + Gq^2\right)|u(q, q_z)|^2 + Bq_{\mu'}q_{\nu'}u_{\mu'}(q, q_z)u_{\nu'}(-q, -q^z)\right] \tag{6.28}$$

Where, κ, G and B are the tilt, shear and the bulk modulus, respectively. μ′ and ν′ are Cartesian indices in the xy plane. $u(q, q_z)$ is the Fourier transform of the two component displacement field. It is worth to mention that the shear deformation determines the elastic response of the medium on the longest length scales (Ulrich et al. 2006) compared to any other deformations introduced in (6.28). Therefore, the wavelength of the shear deformations has to be larger expectedly in comparison to all microscopic-length scales associated with the composite matrix. However, as far as the in-plane localization and shear deformations are concerned, our system can be effectively viewed as a two-dimensional model in absence of any long-range positional order (Nelson 2002). Therefore, the only possible elastic energy contribution into our free energy comes from the tilt elasticity of the cross-links, which are intrinsically connected with the existence of free volumes in context of negligible shear deformations. The correlation between cross-link and free volumes may be realized as the free motion of cross-link chains is an essential feature that endows the polymer network with novel mechanical and optical properties. The incorporation of wriggling and jumping of large molecular segments allows the flexibility and hence help to uncoil of the cross-link networks associated with the elastic deformations. The existence of intangible voids or empty space can lead to such twisting and wriggling. These voids are called free volumes and the incorporation of free volumes and cross-link conformations induce the tilt elastic deformation in case of PSFLCs. Therefore, the free energy has to be supplemented accordingly both by the tilt elastic energy and also by the free volumes contribution. The free volumes contribution was already demonstrated in (6.3) and the required tilt elastic energy may be obtained from (6.28). The tilt elastic energy may be written in the form $F_{el} = \frac{1}{2}\kappa\left(\frac{\partial\theta}{\partial z}\right)^2$, where κ is the tilt modulus of the polymeric segments. The two-component displacement field (u) in (6.28) is replaced by the term "relative cross-link displacement" $\left(\frac{\partial\theta}{\partial z}\right)$ in case of PSFLCs which is representing the variation in tilt angle θ with cross link alignment length z as the FLC director rigidly attached between the cross-link interstices and rotate accordingly with the cross-link chains. By considering the excluded volume contribution $\left[\frac{1}{2}\Gamma\left(1-\frac{r}{R}\,Cos\theta\right)\right]$ in (6.3) (Simha and Somcynsky 1969; Jean 1990; Bartoš 1996; Dlubek et al. 2004; Dupasquier and Mills 1995; Mogensen 1995; Eldrup et al. 1995; Schrader and Jean 1988; Jean et al. 2003; Lahiri and Pal Majumder 2011), the free energy of PSFLCs can be proposed as given below (Lahiri and Pal Majumder 2012):

$$F = F_0 + \frac{1}{2}\alpha(T-T_0)\theta^2 + \frac{1}{4}b\theta^4 + \frac{P_S^2}{2\varepsilon_0\chi_0} - CP_S\theta + \frac{1}{2}W_p\theta^2 + \frac{1}{2}\Gamma\left(1-\frac{r}{R}Cos\theta\right) - P_SE + \frac{1}{2}\kappa\left(\frac{\partial\theta}{\partial z}\right)^2 - \varsigma P_S^2\left(\frac{\partial\nabla\varphi}{\partial P_S}\right) \tag{6.29}$$

Here, $(F\text{-}F_0)$ is the difference in free energy density between the high temperature SmA phase and the low temperature SmC* phase. Although order parameters ($\theta$, $P_S$) change abruptly from layer to layer (Fig. 6.5) but the sum or difference of them varies slowly along the layer normal. Therefore θ and $P_S$ can be considered as continuous variables representing average tilt angle and spontaneous polarization of the medium (6.29). α is the temperature dependent mean-field co-efficient and b is as usual Landau co-efficient for second order phase transition. The $\frac{P_S^2}{2\varepsilon_0\chi_0}$ term is the dipolar ordering term with $\chi_0$ indicating the dielectric susceptibility along the direction of the electric field. For simplicity the term proportional to $P_S^4$ in (6.29) can be neglected as it is originally introduced in (6.3) only for stability reasons (Gieβelmann and Zugenmaier 1995; Gieβelmann et al. 1997). C is the bilinear coupling co-efficient between θ and $P_S$, which is intrinsically chiral in nature. The biquadratic coupling (achiral in nature) between θ and $P_S$ in (6.29) can be ignored. By assuming a highly polarized surface stabilized PSFLC compound (Brehmer et al. 1996; Bahr and Heppke 1990), terms proportional to $W_P$ and Γ are already discussed earlier in (6.3). The coupling between the spontaneous polarization ($P_S$) and the applied electric field (E) describes the ferroelectric torque acting over the sample. The co-efficient $\varsigma$ signifies the coupling between the harmonic field ($\nabla\varphi$) and the spontaneous polarization ($P_S$). The specific nature of this coupling is considered to be $P_S^2\left(\frac{\partial\nabla\varphi}{\partial P_S}\right)$ instead of $\varsigma(\nabla\varphi)P_S$ because it may destabilize the free energy of the system with appropriate increase of it along the direction "z" (local in nature) makes saddle point for SmC* phase and does not provide any contradictory for achieving zero spontaneous polarization of the disordered paraelectric phase (Lahiri and Pal Majumder 2012). The $\left(\frac{\partial\nabla\varphi}{\partial P_S}\right)$ term signifies a smoothly varying function of harmonic field with the spontaneous polarization with any kind of rapid phase fluctuations being suppressed by the term $P_S^2$. This kind of coupling not only stabilizes the free energy but also produces a sound mathematical formulation for the polarization profile of the system in accordance with the established experimental observations (Kundu et al. 2000; Lahiri and Pal Majumder 2012).

To obtain a thermodynamically stable condition of the system, the minimization of F in (6.28) with respect to polarization $P_S$ provides the following relation:

$$\frac{\partial F}{\partial P_S} = \frac{P_S}{\varepsilon_0\chi_0} - C\theta - E - 2\varsigma P_S\left(\frac{\partial\nabla\varphi}{\partial P_S}\right) - \varsigma P_S^2\left(\frac{\partial^2\nabla\varphi}{\partial P_S^2}\right) = 0 \qquad (6.30)$$

Again, it is also important to minimize the system as a whole with respect to θ because large fluctuations in θ not only destabilizes the system but also increases the asymmetric behavior. Therefore, after minimizing the free energy in (6.29) with respect to θ, we have

$$\frac{\partial F}{\partial \theta} = \kappa\left(\frac{\partial^2\theta}{\partial z^2}\right) + \alpha(T - T_0)\theta + b\theta^3 - CP_S + W_P\theta + \frac{1}{2}\Gamma\frac{r}{R}\sin\theta = 0 \qquad (6.31)$$

The linear variation in the spontaneous polarization ($P_S$) with the tilt angle (θ) (Bahr and Heppke 1990) yields the following simplified relation from (6.30)

$$P_S = C\varepsilon_0\chi_0\theta + \varepsilon_0\chi_0 E \qquad (6.32)$$

Comparing (6.31) and (6.32), we get

$$\frac{\partial F}{\partial \theta} = \kappa\left(\frac{\partial^2\theta}{\partial z^2}\right) + \theta\{\alpha(T - T_0) - C^2\varepsilon_0\chi_0 + W_P\} + b\theta^3 + \frac{1}{2}\Gamma\frac{r}{R}\sin\theta - C\varepsilon_0\chi_0 E = 0 \tag{6.33}$$

Rearranging terms in (6.33), we obtain a second order non-linear differential equation of the form:

$$\frac{\partial^2\theta}{\partial z^2} + \Omega^2\theta + \xi_0\theta^3 = \gamma E \tag{6.34}$$

Here, the different parameters in (6.34) are denoted as

$$\Omega^2 = \frac{\alpha(T - T_C')}{\sigma}, \quad T_C' = T_C + \frac{(C^2\varepsilon_0\chi_0 - W_P)}{\alpha} \quad \text{and} \quad T_C = T_0 + \frac{\Gamma r}{2R\alpha},$$

$$\xi_0 = \frac{b - \frac{\Gamma r}{12R}}{\kappa} \quad \text{and} \quad \gamma = \frac{C\varepsilon_0\chi_0}{\kappa}$$

In order to obtain a suitable solution for (6.34), we consider a family of differential equations with a spatial variation in E (assumed to be applied along the "z" direction) as

$$\frac{\partial^2\theta}{\partial z^2} + \Omega^2\theta + \xi\theta^3 = \gamma E_0\cos(Kz) \tag{6.35}$$

Here, $E_0$ is the amplitude of the applied field directed along the cross-link alignment axis z and K being the wave vector. Further, $\xi$ is a parameter occupying an interval ($I_\xi$) which includes $\xi = 0$. When $\xi = \xi_0$ we recover (6.34) and when $\xi = 0$, we obtain the linearize equation corresponding to the family of equations represented in (6.35). Equation (6.35) can be further rearranged by considering $Kz = Z$ and absorbing the $K^2$ term both in $\Omega^2$ and $\xi$, respectively with the amplitude of the applied electric field being absorbed in $\gamma$.

In that case, we rewrite (6.35) in a more compact form as

$$\frac{\partial^2\theta}{\partial Z^2} + \Omega^2\theta + \xi\theta^3 = \gamma \mathrm{Cos} Z \tag{6.36}$$

The solution of (6.36) is of fundamental importance as it relates the desired variation of the tilt angle $\theta$ with the "effective" cross-link alignment length z (now being absorbed in Z). In view of (6.36), we attempt its solution as the function of both $\xi$ and Z. Therefore, we represent $\theta = \theta\ (\xi,\ Z)$. Now, utilizing the most elementary version of the perturbation method (Lahiri and Pal Majumder 2012), we attempt a representation of the solutions of (6.36) in the form of a power series in $\xi$.

$$\theta(\xi, Z) = \theta_0(Z) + \xi\theta_1(Z) + \xi^2\theta_2(Z) + \cdots \tag{6.37}$$

Here, the co-efficients $\theta_i(Z)$ are the functions of Z alone. To form equations for $\theta_i(Z)$, where $i = 0, 1, 2, 3, \ldots$, we insert the series in (6.37) back into (6.36). This gives

$$\begin{aligned}&\left(\theta_0'' + \xi\theta_1'' + \xi^2\theta_2'' + \cdots\right) + \Omega^2\left(\theta_0 + \xi\theta_1 + \xi^2\theta_2 + \cdots\right) + \\ &\xi\left(\theta_0 + \xi\theta_1 + \xi^2\theta_2 + \cdots\right)^3 = \gamma\cos Z\end{aligned} \tag{6.38}$$

Since, (6.38) is supposed to hold for every member of the family of (6.36) i.e. for every $\xi$ on $I_\xi$ , the co-efficients of the powers of $\xi$ must balance in (6.38) and we obtain

$$\begin{aligned}\theta_0'' + \Omega^2\theta_0 &= \gamma\cos Z \\ \theta_1'' + \Omega^2\theta_1 &= -\theta_0^3 \\ \theta_2'' + \Omega^2\theta_2 &= -3\theta_0^2\theta_1 \\ &\textit{and so on}\end{aligned} \tag{6.39}$$

Here, we restrict ourselves to the periodic solutions in (6.39) having period $2\pi$ of the forcing term. Then, for all $\xi$ on $I_\xi$ and for all Z we have

$$\theta(\xi, Z + 2\pi) = \theta(\xi, Z) \tag{6.40}$$

From (6.37) it is possible to write for all Z

$$\theta_i(Z + 2\pi) = \theta_i(Z), \quad \text{where } i = 0, 1, 2, 3\ldots \tag{6.41}$$

Therefore, (6.39) together with the condition in (6.41) gives the required solutions for $\theta$. From the detailed investigation of the first equation for the set of (6.39), we see that it is same as the linearize equation $\theta_0'' + \Omega^2\theta_0 = 0$. This is because by putting $\xi_0 = 0$ in (6.34) as equivalent of putting $\xi = 0$ in (6.37). Therefore, our solution for $\theta$ will be restricted to the solutions that are closely associated with the linearize equation as suggested above. Now, the solution of the first equation in the set of (6.39) is the following $c_1\cos\Omega Z + c_2\sin\Omega Z + \frac{\gamma}{\Omega^2-1}\cos Z$. This solution is physically appreciable only when $\Omega \neq$ an integer. Here, $c_1$ and $c_2$ are arbitrary constants. Since, $\Omega$ is not an integer the only solution that has the periodicity of $2\pi$ is the following

$$\theta_0(Z) = \frac{\gamma}{\Omega^2 - 1}\cos Z \tag{6.42}$$

Inserting (6.42) back into the second equation in the set of (6.39), we get

$$\theta_1'' + \Omega^2\theta_1 = -\left(\frac{\gamma}{\Omega^2 - 1}\right)^3 \cos^3 Z = \frac{\gamma^3}{\left(\Omega^2 - 1\right)^3}\left[-\frac{3}{4}\cos Z + \frac{1}{4}\cos 3Z\right] \tag{6.43}$$

The solution of (6.43) having the periodicity of $2\pi$ is the following

$$\theta_1(Z) = \frac{3}{4}\frac{\gamma^3}{\left(\Omega^2 - 1\right)^4}\cos Z - \frac{1}{4}\frac{\gamma^3}{\left(\Omega^2 - 1\right)^3\left(\Omega^2 - 9\right)}\cos 3Z \tag{6.44}$$

Therefore, we have complete solution for $\theta$ as

$$\theta = \frac{\gamma}{\Omega^2 - 1}\cos Z + \xi\left(\frac{3}{4}\frac{\gamma^3}{\left(\Omega^2 - 1\right)^4}\cos Z - \frac{1}{4}\frac{\gamma^3}{\left(\Omega^2 - 1\right)^3\left(\Omega^2 - 9\right)}\cos 3Z\right) + O(\xi^2) \tag{6.45}$$

Equation (6.45) represents the variation in the tilt angle ($\theta$) with the effective cross-link alignment length "z". Although, it puts serious restrictions to the behavior of $\theta$ when the values of $\Omega^2$ are equal to 1, 9, 25...etc, but one way out of this problem is achieved when we assume the condition $\Omega \neq$ an integer. Even if $\Omega^2$ terms take values close to 1, 9, 25, etc., the series in (6.45) will not converge and therefore, higher order terms in (6.45) will describe $\theta$ ($\xi$, Z) only poorly. These terms may be regarded as some "feeding back" into the linear equation with large non-linear disturbances like the forcing terms and hence, can be discarded for smooth variation in $\theta$. Even if $\Omega$ takes even integers, the solution to the first equation of the set of (6.38) also contains the $2\pi$ periodicity for all values of $c_1$ and $c_2$. This condition modifies (6.45) a bit but the actual variation in the tilt angle ($\theta$) with z remains almost unaffected. We, therefore, restrict our solution in the form of (6.45), which provides a reasonable microscopical view of the role of cross-link chains with stabilization of $\theta$. Figure 6.7 shows the variation in the tilt angle ($\theta$) with the cross-link alignment length (z). The creation of field induced molecular tilt near SmC*-SmA transition point is investigated experimentally by several researchers (Petit et al. 2006, 2009). The typical variation may be summarized as follow: The tilt angle ($\theta$) first increases linearly with the increase in applied electric field at the low field region, and then it deviates significantly from the linear behavior at the moderate field region. But at larger field regions, it shows a tendency towards saturation. The behavior of $\theta$ with z (since E is a function of z) is depicted in Fig. 6.7. The curve in Fig. 6.7 shows the variation in $\theta$ both at the small field (increasing z) regions and also at the moderate field (intermediate value of z) regions. It is interesting to note from (6.45) that when the field is turned off, the average molecular tilt tends to go down to zero for certain values of z. The minimum value in $\theta$ may be allowed to have zero value once the field is removed, however, the perturbation approach which leads to such variation in tilt breaks

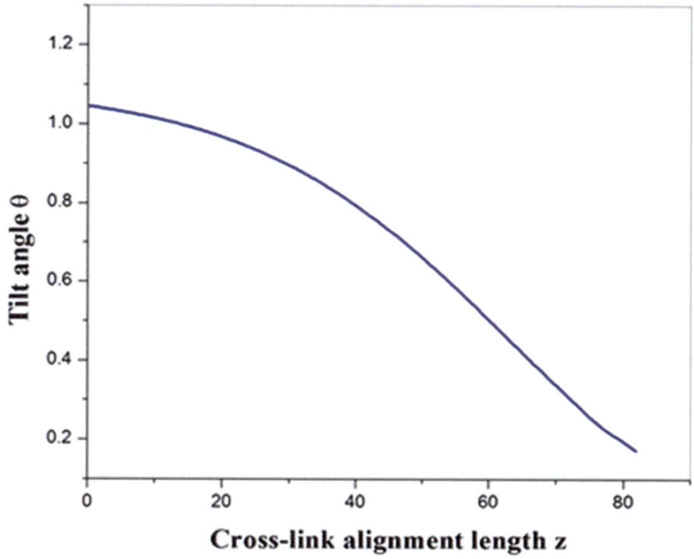


**Fig. 6.7** The variation in the tilt angle θ with the effective cross-link alignment length (z) for a PSFLC system. The units are taken arbitrarily

down as θ achieves the zero value. Therefore, we get a finite but reduced (consequence of the perturbation approach) value for the director tilt (θ) in our theory. This is precisely the state of the material that provides memory characteristics to a PSFLC system. Let us now look into the scenario from the cross-link alignment perspective. With increasing z the cross-link chains get strongly intersticed with the FLC director (Fig. 6.5). This situation seriously restricts the free movement of the FLC director resulting in the reduction of the tilt angle (θ). Once the applied field is turned off, the cross-links are now at their most preferred alignment direction suppressing almost any kind of rotational freedom for the FLC director. This is clearly observed in the decreasing nature of the tilt angle curve (Fig. 6.7) with increasing z. Archer and Dierking (2005, 2008) worked out almost similar behavior in case of tilt angle (θ) with increasing polymer concentration for polymer-FLC matrix. They observed an almost linear decrease in tilt angle with increasing polymer concentration for PSFLCs. This experimental finding is largely dominated by the ξ-coefficient in our solution for θ in (6.45). The decrease in tilt angle with increase in polymer concentration leads to the conclusion that the ξ-coefficient must increase with increase in polymer concentration and understandably, the same is observed in the variation of θ with the effective cross-link alignment length (z) as found in Fig. 6.7. The periodicity of 2π in our solution is profoundly important for the multistability and resolution of the memory states and will be taken up in the rotational viscosity section for its proper understanding.

## 6.7 Multistability and Resolution of Memory States and Rotational Viscosity

The Landau-Khalatnikov (LG) equation associated with the rotational viscosity has the generalized form (Lahiri and Pal Majumder 2011, 2012)

$$\gamma_\theta \frac{\partial \theta}{\partial t} = -\frac{\partial F}{\partial \theta} \tag{6.46}$$

In view of (6.33), we have

$$\gamma_\theta \frac{\partial \theta}{\partial t} + \kappa\left(\frac{\partial^2 \theta}{\partial z^2}\right) = -A\theta - B\theta^3 + C\varepsilon_0\chi_0 E \tag{6.47}$$

By normalizing (6.47), the obtained values of A, B and λ parameters are given below those slightly different compared to earlier in (6.18):

Here, $A = \alpha\left(T - T_C'\right)$; $T_C' = T_C + \frac{\left(C^2\varepsilon_0\chi_0 - W_P\right)}{\alpha}$ and $T_C = T_0 + \frac{\Gamma r}{2R\alpha}$, and $B = b - \frac{\Gamma r}{12R}$ with $\lambda = C\varepsilon_0\chi_0$.

Equation (6.47) may be restated in the following form with the applied field (E) containing both spatial and temporal parts:

$$\gamma_\theta \frac{\partial \theta}{\partial t} + \kappa\left(\frac{\partial^2 \theta}{\partial z^2}\right) + A\theta + B\theta^3 = \lambda E_0 \cos(\omega t - Kz) \tag{6.48}$$

In order to solve (6.48), we make the following substitutions (Lahiri and Pal Majumder 2012)

$$y = \omega t - Kz \quad \text{and} \quad \theta = f(y) \tag{6.49}$$

When (6.49) is inserted back into (6.48) and after carrying out some derivatives and simple algebra, we obtain

$$f''(y) + \ell f'(y) + \Omega^2 f(y) + \xi_0 f^3(y) = \gamma \cos y \tag{6.50}$$

Here, $\Omega^2$, $\xi_0$ and $\gamma$ are already given in (6.33) and the new parameter l can be defined as $\ell = \frac{\gamma_\theta \omega}{\kappa K^2}$.

By considering a periodic solution of f with period $\frac{2\pi}{p}$, f (y) can be represented as the Fourier series for all y in terms of p as follow

$$f(y) = a_0 + a_1 \cos py + b_1 \sin py + a_2 \cos 2py + \cdots \tag{6.51}$$

Substituting (6.51) back into (6.50), we obtain

$$\begin{aligned}&-p^2[a_1\cos py+b_1\sin py+4a_2\cos py+\cdots]+\\&\ell p[-a_1\sin py+b_1\cos py-2a_2\sin 2py+\cdots]+\\&\Omega^2[a_0+a_1\cos py+b_1\sin py+a_2\cos 2py\cdots]+\\&\xi_0[a_0+a_1\cos py+b_1\sin py+a_2\cos 2py+\cdots]^3=\gamma\cos y\end{aligned} \tag{6.52}$$

Equation (6.52) may be rearranged in the following form

$$A_0+A_1\cos py+A_2\sin py+\cdots=\gamma\cos y \tag{6.53}$$

Here, in (6.52), $A_0=\Omega^2 a_0+\xi_0 a_0^3$

$$A_1=-p^2a_1+\ell pb_1+\Omega^2a_1+3\xi_0a_0^2a_1$$

$$A_2=-p^2b_1-\ell pa_1+\Omega^2b_1+3\xi_0a_0^2b_1$$

and so on.

Now, we utilize the Harmonic balance method (Lahiri and Pal Majumder 2012) to obtain a suitable solution for f(y). Setting $p=1$ in (6.52), we have

$$A_1=\gamma \quad \text{and} \quad A_0=A_2=A_3=\cdots=0 \tag{6.54}$$

Utilizing (6.53), we obtain the Fourier components for the solution of f(y) as

$$\begin{aligned}a_0&=0\\a_1&=\frac{\gamma(\Omega^2-1)}{\ell^2+(\Omega^2-1)^2}\\b_1&=\frac{\gamma\ell}{\ell^2+(\Omega^2-1)^2}\\&\text{and so on.}\end{aligned}$$

Here, $a_0$ is set equal to zero just to maintain the $2\pi$ periodicity of the function f(y) and hence for θ. Equation (6.49) can now be written in terms of the leading order variation in f(y) as

$$\theta(z,t)=\frac{\gamma(\Omega^2-1)}{\ell^2+(\Omega^2-1)^2}\cos(\omega t-Kz) \tag{6.55}$$

Extracting the time part from (6.54) and comparing it with (6.45), we have the following relation

$$\frac{\gamma}{\Omega^2-1}=\frac{\gamma(\Omega^2-1)}{\ell^2+(\Omega^2-1)^2}\exp(i\omega t) \tag{6.56}$$

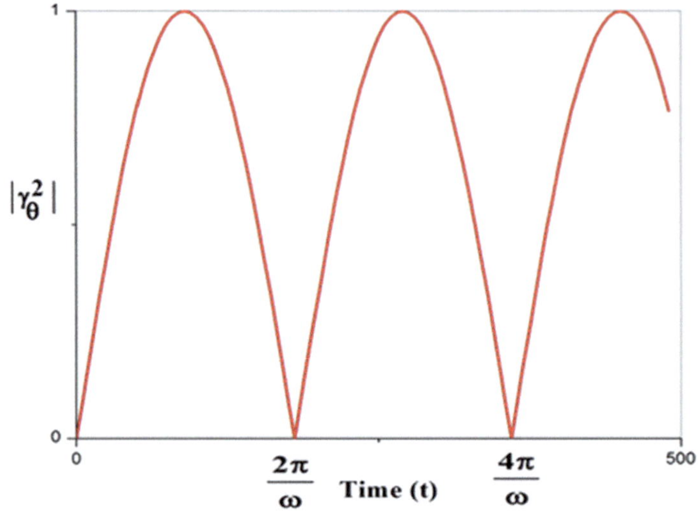


**Fig. 6.8** Sinusoidal variation of the amplitude of the rotational viscosity $|\gamma_\theta{}^2|$ with time t

Utilizing the expressions for γ, $\Omega^2$ and $\ell$ respectively, we obtain

$$|\gamma_\theta^2| = 2\left[\alpha\left(T - T_C'\right) - \kappa K^2\right]^2 \frac{\sin\frac{\omega t}{2}}{\omega^2} \tag{6.57}$$

Figure 6.8 shows the variation in rotational viscosity ($|\gamma_\theta^2|$) with time t. The most remarkable feature of the result in (6.56) is that, as a function of time, the square of the amplitude of the rotational viscosity ($|\gamma_\theta^2|$) oscillates sinusoidally (Fig. 6.8). After rising to a maximum of $^2\left[\alpha\left(T-T_C'\right)-\kappa K^2\right]^2/_{\omega^2}$, it drops back to zero. Since, (6.56) is obtained by comparing (6.55) with (6.45), the maximum value allotted to ($|\gamma_\theta^2|$) must not exceed one, else the perturbative method will be invalid and hence, (6.56). This result is highly surprising. At times $t_n = \frac{2n\pi}{\omega}$, where n = 1, 2, 3... the FLC molecules certain to have almost infinite rotational mobility resulting into zero rotational viscosity as evident from Fig. 6.8. Therefore, the applied field should not keep for a longer period of time but it should turn off after an interval of time $\frac{\pi}{\omega}$ for maximizing the chances to produce finite rotational torque on the FLC director and as a result to obtain the system with finite rotational viscosity. Once the field is turned off, this reduced viscosity seems to produce a finite but weaker torque at the memory state and thereby, promotes multistability and resolution in the memory states as proposed earlier.

## 6.8 Cross-link Conformations and Polarization Profile

We now investigate the nature of the polarization profile that is expected to be influenced by the cross-link harmonic interaction introduced in (6.29). The dynamics of the polarization in response to a time-dependent driving field can be obtained utilizing the Landau-Khalatnikov (Lahiri and Pal Majumder 2011, 2012; Lo 2003) equation of the form

$$\gamma_P \frac{\partial P_S}{\partial t} = -\frac{\partial F}{\partial P_S} \tag{6.58}$$

Here, in (6.57), $\gamma_P$ denotes the dynamic response parameter for the polarization profile in a PSFLC system.

In view of (6.30), we have

$$\gamma_P \frac{\partial P_S}{\partial t} = -\frac{P_S}{\varepsilon_0 \chi_0} + C\theta + E + 2\varsigma P_S \left(\frac{\partial \nabla \varphi}{\partial P_S}\right) + \varsigma P_S^2 \left(\frac{\partial^2 \nabla \varphi}{\partial P_S^2}\right) \tag{6.59}$$

Now, from (6.27), for a pair of monomers, the interaction potential between the cross-link chains takes the following form

$$\varphi = \frac{1}{2L_C^2}[\Delta r(z)]^2 \tag{6.60}$$

Here, $\Delta r(z)$ denotes the separation between the intersticed particles with $\Delta r(z) = r_i(z) - r_j(z)$; where, i and j denote the monomer pairs (Fig. 6.6).

Therefore, $\dfrac{\partial \nabla \varphi}{\partial P_S} = \Delta \left(\dfrac{\partial z}{\partial P_S}\right)$ and $P_S \left(\dfrac{\partial^2 \nabla \varphi}{\partial P_S^2}\right) = \Delta P_S \dfrac{\partial^2 z}{\partial P_S^2}$,

where

$$\Delta = \frac{1}{L_C^2}\left(\frac{\partial \Delta r}{\partial z}\right)^2 \tag{6.61}$$

In evaluating (6.61), we invoke the following approximation. For Gaussian like polymeric chains, the bending stiffness of the wormlike cross-links with a finite persistence length $L_b$ can be neglected as $L_b >> L$ (de Gennes et al. 1992). In that case, (6.61) is obtained by dropping terms involving $\dfrac{\partial^2 \Delta r}{\partial z^2}$. With this approximation and the variation in $\theta$ as given in (6.55), we simplify (6.59) into the following form

$$\gamma_P \frac{\partial P_S}{\partial t} + \frac{P_S}{\varepsilon_0 \chi_0} = \left\{\frac{C\gamma(\Omega^2 - 1)}{\ell^2 + (\Omega^2 - 1)^2} + E\right\} \cos(\omega t - Kz) + \varsigma \Delta P_S \left(2\frac{\partial z}{\partial P_S} + P_S \frac{\partial^2 z}{\partial P_S^2}\right) \tag{6.62}$$

Now, (6.62) can be regarded as the real part of the following generalized equation

$$\gamma_P \frac{\partial P_S}{\partial t} + \frac{P_S}{\varepsilon_0 \chi_0} = \left\{ \frac{C\gamma(\Omega^2 - 1)}{\ell^2 + (\Omega^2 - 1)^2} + E \right\} \exp i(\omega t - Kz) + \varsigma \Delta P_S \left( 2 \frac{\partial z}{\partial P_S} + P_S \frac{\partial^2 z}{\partial P_S^2} \right) \tag{6.63}$$

In order to solve (6.63), we make some suitable transformations (Polyanin and Zaitsev 2003) as follow

$$W = P_S^2 \left( \frac{\partial z}{\partial P_S} \right) \quad \text{and} \quad \beta = P_S^2 \exp i(\omega t - Kz) \tag{6.64}$$

In view of (6.64), dimensional analysis introduces some normalized parameters for (6.63) as

$$KW = \frac{\gamma P}{\varsigma \Delta}, \quad \omega = \frac{1}{\varepsilon_0 \chi_0 \varsigma \Delta} \quad \text{with the substitution } E' = E_t + E \tag{6.65}$$

Here, in (6.65) $E_t = \frac{C\gamma(\Omega^2 - 1)}{\ell^2 + (\Omega^2 - 1)^2}$, which denotes the average tilt induced field of the cross-linked polymeric chains (Fig. 6.5). With the transformations in (6.65) and suitable parameters introduced in (6.64), we finally reduce (6.63) into the following simplified form

$$\frac{\partial W}{\partial \beta} (iKW - 1) + \nu = 0 \ \text{ with } \ \varsigma \nu \Delta = E' \tag{6.66}$$

Therefore, (6.66) is now an autonomous equation (Polyanin and Zaitsev 2003) with one of the roots of the particular solution taking the following form

$$W = \frac{1 - \sqrt{2iKI_1 + 1 + \frac{2iKE' P_S^2 \exp i(\omega t - Kz)}{\varsigma \Delta}}}{iK} \tag{6.67}$$

The other root of the particular solution of (6.66) comes out to be physically unacceptable as it provides untenable behavior of the spontaneous polarization $P_S$ and hence, we discard it altogether. Further, $I_1$ in (6.67) is an arbitrary integration constant, to be determined from the suitable parameters introduced in (6.65). Here, we utilize the first parameter in (6.65) to achieve a spatial variation in the spontaneous polarization ($P_S$) with the cross-link alignment length "z" as follow

$$\mathrm{KW} = \frac{\gamma \mathrm{P}}{\varsigma \Delta} \tag{6.68}$$

Inserting (6.68) back into (6.67), we obtain the arbitrary constant $I_1$, which eventually eliminates the temporal part of the spontaneous polarization. $I_1$ is given by

$$I_1 = -\frac{\gamma_P}{\varsigma \Delta K} - \frac{E' P_S^2}{\varsigma \Delta} \exp i(\omega t - Kz) \tag{6.69}$$

In evaluating (6.69), we ignored terms involving $\gamma_P^2, \gamma_P^3$, etc. Inserting (6.68) back into (6.67) we achieve a suitable simplified spontaneous polarization profile for PSFLC as given below

$$P_S = P_{SO} \exp\left(-\frac{z}{\delta}\right) \tag{6.70}$$

In (6.70), $P_{SO}$ indicates the spontaneous polarization at $z = 0$ in smectic layer structure as shown in Fig. 6.5. Further, $\delta = \frac{\gamma_P}{\varsigma \Delta K}$ signifies the depth of penetration for the spontaneous polarization $P_S$ into the bulk of the sample as "z" increases. Figure 6.9 shows the variation in spontaneous polarization ($P_S$) with cross-link alignment length z. Petit et al. (2009) experimentally observed suitable dielectric relaxation pattern for PSFLCs that could explain the behavior of the spontaneous polarization ($P_S$) as obtained in Fig. 6.9. They realized the qualitative ineffectiveness

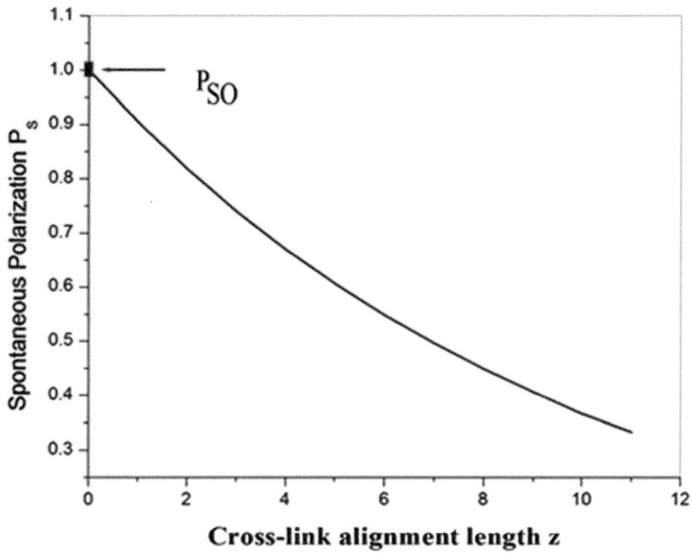


**Fig. 6.9** Variation in spontaneous polarization ($P_S$) with the cross-link alignment length z for a PSFLC system. The units are taken arbitrarily

of the thermal behavior of the Goldstone mode by the polymer network. However, they observed higher degree of quantitative difference in the dielectric strength of the Goldstone mode with increasing polymer concentration. The reduction in strength is as high as 23 % for PSFLCs with the increase in polymer concentration from 0 to 7 %. This result is remarkable and reflects strong light on the molecular morphology for the reduction of the spontaneous polarization in the composite medium. Since, Goldstone mode dielectric relaxation can be reasonably attributed to the polarization profile of the medium (Levstik et al. 1987), we propose mainly two reasons for the typical behavior of $P_S$ in Fig. 6.9. First, as the cross-link alignment length (z) increases with increasing polymer concentration, the elastic torque created over the composite system essentially acts to distort the helical structure of the FLC network. This situation is expected to enhance the polymer network-FLC interactions resulting in the enhancement of the effective elasticity of the PSFLC medium. As a consequence of this, the dielectric strength and hence, the polarization of the medium reduces. Second, with increasing z, the cross-link harmonic field expected to strain the molecular dipolar ordering by attaching the cores of the FLC molecules tidily in between the intersticed chains of the cross-links. This also acts to further reduce the effective polarization of the medium. This reduction of the spontaneous polarization with z summarizes the proposals for the origination of distinct memory states in PSFLCs.

## 6.9 Concluding Remarks

A lot has been told about the morphology of polymer network in stabilizing the ferroelectric liquid crystals. To summarize, presence of defects such as free volumes into PSFLCs create profound implications in controlling some of their important physical and mechanical properties. The free volume model captures the underlying mechanism that stabilizes the FLC medium against any mechanical stress and also highlights the role of geometry in electro-optic performance of the material. It is demonstrated that polymer network not only creates the desired self-healing effect in FLCs but it also retains their superior electro-optic properties. The material properties which, to a large extent, determine the electro-optic performance of PSFLCs are spontaneous polarization ($P_S$), rotational viscosity ($\gamma_\theta$) and response time ($\tau$), respectively. The response time $\tau$ is associated with both the spontaneous polarization ($P_S$) and the rotational viscosity ($\gamma_\theta$) via the relation $\tau = \frac{\gamma_\theta}{P_S E}$, where E is the electric field. The free volume model reveals that with increase in temperature, the rotational viscosity in case of PSFLCs increase considerably compared to pure FLCs and on the contrary, the spontaneous polarization decreases. The net effect is an increase in the response time of the medium subjected to an external field E. The generation of grayscale memory states in PSFLCs is also explained with the combination of effects resulting from the elastic like interaction between the cross-link chains of polymer network and the FLC director and the surface stabilization created by the free volumes in between those polymeric chains. The variation in rotational viscosity ($\gamma_\theta$) with time gives an entirely new concept of realizing the memory states in these substances with greater resolution and multistability. One interesting point is that the present model predicts the idea of electron-hole pair creation inside free volume surfaces when electric field is applied over the system. This concept is new and studied extensively by Positron Annihilation Lifetime Spectroscopy, Doppler Broadening Spectroscopy and by Angular Correlation measurements. Therefore, it seems volume stabilization conjectured in our theoretical model may create new avenues in preparing FLC flat panel displays with high contrast ratio. This conjecture, remarkably, finds experimental counterpart in the work of Pirs et al. (1995). They showed that volume stabilized LC cells not only show internal memory states but also exhibit high contrast shuttering phenomena. Moreover, recently, it is demonstrated by Fujisawa et al. (2008) that FLCs when doped with photo-curable monomers, then polymer stabilized FLCDs exhibit high quality moving images without any blurring and break. Therefore, the role of free volume geometry seems to be an extremely important theoretical knowledge that enables us not only to achieve the suitable background in commercializing FLCDs but also amplifies the present state of affairs in making flat panel displays with exceedingly high contrast ratio and electro-optic performance.

## References


Archer P, Dierking I (2005) Quantitative experimental determination of the Landau-potential of chiral enantiomer doped ferroelectric liquid crystals. Eur Phys J E 18(4):373–381

Archer P, Dierking I (2008) Elastic coupling in polymer stabilized ferroelectric liquid crystals. J Phys D: Appl Phys 41(15):155422

Archer P, Dierking I (2009) Electro-optic properties of polymer-stabilized ferroelectric liquid crystals before, during and after photo-polymerization. J Opt A: Pure Appl Opt 11(2):024022

Archer P, Dierking I, Osipov MA (2008) Landau model for polymer-stabilized ferroelectric liquid crystals: experiment and theory. Phys Rev E 78(5):051703

Bahr CH, Heppke G (1990) Influence of electric field on a first-order smectic-A-ferroelectric-smectic-C liquid-crystal phase transition: a field-induced critical point. Phys Rev A 41 (8):4335–4342

Bartoš J (1996) Free volume microstructure of amorphous polymers at glass transition temperatures from positron annihilation spectroscopy data. Colloid Polym Sci 274(1):14–19

Blinov LM, Beresnev LA, Haase W (1995) Tilt angle, polarization and susceptibility and Landau expansion coefficients for multicomponent ferroelectric liquid crystal mixtures. Ferroelectrics 174:221–239

Brehmer M, Zentel R, Gießelmann F, Germer R, Zugenmaier P (1996) Coupling of liquid crystalline and polymer network properties in LC-elastomers. Liq Cryst 21(4):589–596

Broer DJ, Finkelmann H, Kondo K (1988) In-situ photopolarization of an oriented liquid-crystalline acrylate. Makromol Chem 189:185–194

Broer DJ, Boven J, Mol GN, Challa G (1989a) In-situ photopolymerization of oriented liquid crystalline acrylates. Makromol Chem 190:2255–2268

Broer DJ, Hikmet RAM, Challa G (1989b) Influence of a lateral methyl substituent on monomer and oriented network properties of a mesogenic diacrylate. Makromol Chem 190:3201–3215
Broer DJ, Mol GN, Challa G (1989c) In-situ oriented polymer networks from a mesogenic diacrylate. Makromol Chem 190:19–30
Broer DJ, Gossink RG, Hikmet RAM (1990) Oriented polymer networks obtained by photopolymerization liquid crystal-crystalline monomers. Angew Makromol Chem 183:45–66
Broer DJ, Mol GN, Challa G (1991) Influence of alkylene spacer on the properties of the mesogenic monomers and the formation and properties of oriented polymer networks. Makromol Chem 192:59–74
Chiccoli C, Pasini P, Skačej G, Zannoni C, Žumer S (2002) Polymer network-induced ordering in a nematogenic liquid: a Monte Carlo study. Phys Rev E 65:051703
Clark NA, Lagerwall ST (1980) Submicrosecond bistable electro-optic switching liquid crystals. Appl Phys Lett 36:899
Crawford GP, Žumer S (1996) Liquid crystals in complex geometries formed by polymer and porous networks. Taylor and Francis, London
Crawford GP, Scharkowski A, Fung YK, Doane JW, Žumer S (1995) Internal surface, orientational order, and distribution of a polymer network in a liquid crystal matrix. Phys Rev E 52: R1273
de Gennes PG, Matricon J (1964) Collective modes of vortex lines in superconductors of the second kind. Rev Mod Phys 36(1):45–49
de Gennes PG, Cifferi A, Kringbaum WR, Meyer RB (1992) Polymer liquid crystals. Academic, New York
Dierking I, Kosbar LL, Lowe AC, Held GA (1998a) Polymer network structure and electro-optic performance of polymer stabilized cholesteric textures-II: the effect of curing conditions. Liq Cryst 24:397–406
Dierking I, Kosbar LL, Lowe AC, Held GA (1998b) Polymer network structure and electro-optic performance of polymer stabilized cholesteric textures-I: the influence of curing temperature. Liq Cryst 24(3):387–395
Dierking I, Komitov L, Lagerwall ST, Wittig T, Zentel R (1999) Horizontal chevron domain formation and smectic layer reorientation in SmC* liquid crystals stabilized by polymer networks. Liq Cryst 26(10):1511–1519
Dierking I, Osipov MA, Lagerwall ST (2000) The effect of a polymer network on smectic phase structure as probed by polarization measurements on a ferroelectric liquid crystal. Eur Phys J E 2(4):303–309
Dlubek G, Sen Gupta A, Pointeck J, Krause-Rehberg R, Kaspar H, Lochhaas KH (2004) Temperature dependence of the free volume in fluoroelastomers from positron lifetime and PVT experiments. Macromolecules 37(17):6606–6618
Dupasquier A, Mills AP Jr (1995) Positron spectroscopy of solids. In: Proceedings of the international school of physics Enrico Fermi, Course CXXV, Amsterdam and SIF Bologna, IOS Press
Dvorak V (1974) Improper ferroelectrics. Ferroelectrics 7(1):1–9
Eldrup M, Lightbody D, Sherwood JN (1981) The temperature dependence of positron lifetimes in solid pivalic acid. Chem Phys 63(1–2):51–58
Fujikake H, Takizawa K, Kikuchi H, Fujii T, Kawakita M, Aida T (1997) Polymer stabilized ferroelectric liquid crystals devices with grayscale memory. Jpn J Appl Phys 36:6449–6454
Fujikake H, Aida T, Takizawa K, Kikuchi H, Fujii T, Kawakita M (1999) Study of smectic layer structure of polymer-stabilized ferroelectric liquid crystal with grayscale memory. Electron Commun Jpn 82(8):1–8
Fujikake H, Sato H, Murashige T (2004) Polymer-stabilized ferrielectric liquid crystal for flexible displays. Displays 25(1):3–8
Fujisawa T, Nishiyama I, Hatsusaka K, Takeuchi K, Takatsu H, Kobayashi S (2008) Field sequential full color LCDs using polymer-stabilized V-shaped ferroelectric liquid crystals. Ferroelectrics 364(1):78–85

Fung YK, Borštnik A, Žumer S, Yang DK, Doane JW (1997) Pretransitional nematic ordering in liquid crystals with dispersed polymer networks. Phys Rev E 55:1637–1645
Furue H, Miyama T, Limura Y, Hasebe H, Takatsu H, Kobayashi S (1997) Mesogenic polymer stabilized ferroelectric liquid crystal display exhibiting monostability with high contrast ratio and grayscale capability. J Appl Phys 36:L1517–L1519
Furue H, Iimura Y, Hasebe H, Takatsu H, Kobayashi S (1998) The effect of polymer stabilization on the alignment structure of surface-stabilized ferroelectric liquid crystals. Mol Cryst Liq Cryst Sci Technol A 317(1):259–271
Furue H, Takahashi T, Kobayashi S (1999) Monostabilization of surface-stabilized ferroelectric liquid crystal using polymer stabilization. Jpn J Appl Phys 38(1):5660
Furue H, Yokoyama H, Kobayashi S (2001) Newly developed polymer-stabilized ferroelectric liquid crystals: microsized bistable domains and monostable V-shaped switching. Jpn J Appl Phys 40:5790–5794
Furue H, Takahashi T, Kobayashi S, Yokoyama H (2002) Models of molecular alignment structure in polymer-stabilized ferroelectric liquid crystals. Jpn J Appl Phys 40(1):7230–7233
Gieβelmann F, Zugenmaier P (1995) Mean-field coefficients and the electroclinic effect of a ferroelectric liquid crystal. Phys Rev E 52(2):1762–1772
Gieβelmann F, Heimann A, Zugenmaier P (1997) Experimental determination of Landau-expansion coefficients in ferroelectric liquid crystals. Ferroelectrics 200:237–256
Guymon CA, Hoggan EN, Walba DM, Clark NA, Bowman CN (1995) Phase behaviour and electro-optic characteristics of a polymer stabilized ferroelectric liquid crystal. Liq Cryst 19:719–727
Guymon CA, Dougan LA, Martens PJ, Clark NA, Walba DM, Bowman CN (1998) Polymerization conditions and electrooptic properties of polymer-stabilized ferroelectric liquid crystals. Chem Mater 10:2378–2388
Hikmet RAM, Boots HMJ, Michielsen M (1995) Ferroelectric liquid crystal gels—network stabilized ferroelectric display. Liq Cryst 19:65–74
Inoue T, Higuchi N, Furue H (2008) The effect of polymer doping on the formation of helical structure in ferroelectric liquid crystals. Ferroelectrics 364:113–120
Jean YC (1990) Positron annihilation spectroscopy for chemical analysis: a novel probe for microstructural analysis of polymers. Microchem J 42:72–102
Jean YC, Nakanishi H, Hao LY, Sandreczki TC (1990) Anisotropy of free-volume hole dimensions in polymers probed by positron annihilation spectroscopy. Phys Rev B 42:9705–9708
Jean YC, Mallon PE, Schrader DM (2003) Principles and application of positron and positronium chemistry. World Scientific, Singapore
Karapinar R, Neill MO, Hird M (2002) Polymer dispersed ferroelectric liquid crystal films with high electro-optic quality. J Phys D Appl Phys 35(9):900–903
Kitzerow HS, Molsen H, Heppke G (1992) Linear electro-optic effects in polymer-dispersed ferroelectric liquid crystals. Appl Phys Lett 60:3093
Kossyrev PA, Qi J, Priezjev NV, Pelcovits RA, Crawford GP (2002) Virtual surface, director domain and the Frèedericksz transition in polymer-stabilized liquid crystals. Appl Phys Lett 81:2986
Kundu S, Roy SS, Pal Majumder T, Roy SK (2000) Spontaneous polarization and response time of a polymer dispersed ferroelectric liquid crystal (PDFLC). Ferroelectrics 243:197–206
Lahiri T, Pal Majumder T (2011) Theoretical approach to study the effect of free volumes on the physical behavior of polymer stabilized ferroelectric liquid crystal molecules. J Appl Phys 109:114109
Lahiri T, Pal Majumder T (2012) The effect of cross-linked chains of polymer network on the memory states of polymer stabilized ferroelectric molecules. Polymer 53:2121–2127
Lee K, Suh SW, Lee SD (1994a) Fast linear electro-optical switching properties of polymer-dispersed ferroelectric liquid crystals. Appl Phys Lett 64:718
Lee K, Suh SW, Lee SD, Kim CY (1994b) Ferroelectric response of polymer-dispersed chiral smectic C* liquid crystal composites. J Korean Phys Soc 27(1):86

Lee JH, Lim TK, Kwon YW, Jin J (2005) Memory effects in polymer stabilized ferroelectric liquid crystals, and their dependence on the morphology of the constituent molecules. J Appl Phys 97 (8):084907

Lee JH, Lim TK, Kwon YW, Jin J, Kwon SB, Shin ST (2006) Realization of grayscale memory operation in s step-growth based polymer-stabilized ferroelectric liquid crystal system. Jpn J Appl Phys 45(1):5872

Levstik A, Carlsson T, Filipic C, Levstik I, Zeks B (1987) Goldstone and soft mode at the smectic-A-smectic-C* phase transition studied by dielectric relaxation. Phys Rev A 35:3527–3534

Li J, Wang Z, Cai Y, Huang X (1998) Study of EO properties of polymer network stabilized of ferroelectric liquid crystal in smectic C* phase. Ferroelectrics 213:91–98

Li J, Zhu X, Xuan L, Huang X (2002) V-shaped electro-optic characteristics in FLC gels. Ferroelectrics 277:85–105

Lo VC (2003) Simulation of thickness effect in thin ferroelectric films using Landau-Khalatkinov theory. J Appl Phys 94(5):3353

Lueder E, Puecke M, Polach S (1998) Proceedings of the 18th IDRC, Asia Display 98', Seoul Korea, pp 173

Ma RQ, Yang DK (2000) Frèedericksz transition in polymer-stabilized nematic liquid crystals. Phys Rev E 61:1567–1573

Miyazaki Y, Furue H, Takahashai T, Shikada M, Kobayashi S (2001) Mesogenic polymer-stabilized FLCDs exhibiting asymmetric and symmetric (V-shape) electro-optical characteristics. Mol Cryst Liq Cryst 364:491–499

Mogensen OE (1995) Positron annihilation in chemistry. Springer, Berlin

Molsen H, Kitzerow HS (1994) Bistability in polymer-dispersed ferroelectric liquid crystals. J Appl Phys 75:710–716

Murashige T, Fujikake H, Ikehata S, Sato F (2004) Memory effect of ferroelectric liquid crystal stabilized by polymer fibers. Electron Commun Jpn 87(4):16–24

Musevic I, Blinc R, Zeks B (2002) The physics of ferroelectric and antiferroelectric liquid crystals. World Scientific, Singapore

Nelson DR (2002) Defects and geometry in condensed matter physics. University Press, Cambridge

Pal Majumder T, Mitra M, Roy SK (1994) Dielectric relaxation and rotational viscosity of a ferroelectric liquid crystal mixture. Phys Rev E 50(6):4976–4800

Petit M, Daoudi A, Ismaili M, Buisine JM (2006) Electroclinic effect in a chiral smectic-A liquid crystal stabilized by an anisotropic polymer network. Phys Rev E 74:061707

Petit M, Hemine J, Daoudi A, Ismaili M, Buisine JM, Da Costa A (2009) Effect of the network density on dynamics of the soft mode and the Goldstone modes in short-pitch ferroelectric liquid crystals stabilized by an anisotropic polymer network. Phys Rev E 79:031705

Pirs J, Blinc R, Marin B, Pirs S, Doane JW (1995) Polymer network volume stabilized ferroelectric liquid crystal displays. Mol Cryst Liq Cryst 264:155–163

Polyanin AD, Zaitsev VF (2003) Handbook of exact solutions for ordinary differential equations, 2nd edn. Chapman & Hall, Boca Raton

Sato H, Fujikake H, Lino Y, Kawakita M, Kikuchi H (2002) Flexible grayscale ferroelectric liquid crystal device containing polymer walls and networks. Jpn J Appl Phys 41:5302–5306

Sato H, Fujikake H, Kikuchi H, Kurita T (2003) Rollable polymer stabilized ferroelectric liquid crystal device using thin plastic substrates. Opt Rev 10(5):352–356

Schrader DM, Jean YC (1988) Positron and positronium chemistry. Elsevier, Amsterdam

Shinkawa K, Takahashi H, Furue H (2008) Ferroelectric liquid crystal cell with phase separated composite organic film. Ferroelectrics 364:107–112

Simha R, Somcynsky T (1969) On the statistical thermodynamics of spherical and chain molecule fluids. Macromolecules 2:342–350

Singh U (2011) Effect of polymer networks on the smectic-C* phase of a ferroelectric liquid crystal: high-resolution x-ray studies. Phys Rev E 83:061707

Skarp K (1988) Rotational viscosities in ferroelectric smectic liquid crystals. Ferroelectrics 84:119–142

Suresh S, Chien LC (2003) Electro-optical properties of polymer-stabilized ferroelectric liquid crystal. Ferroelectrics 287:1–6

Takahashi T, Umeda T, Furue H, Kobayashi S (1999) Modelling and computer simulation of the electrooptic response of polymer-stabilized ferroelectric liquid crystal cells. Jpn J Appl Phys 38:5991–5995

Tsuda H, Waki N, Furue H (2008) Response time of polymer-stabilized ferroelectric liquid crystals. Ferroelectrics 365:108–114

Ulrich S, Mao X, Goldbart PM, Zippelius A (2006) Elasticity of highly cross-linked random networks. Eur Phys Lett 76(4):677

Ulrich S, Zippelius A, Benetatos P (2010) Random network of cross-linked directed polymers. Phys Rev E 81:021802

Zheng W (2007) Electrooptical properties of diacrylate polymer based network stabilized ferroelectric liquid crystals. Mol Cryst Liq Cryst 475:173–181

# Chapter 7
# Electro-optic and Dielectric Responses in PDLC Composite Systems

**R.R. Deshmukh**

## 7.1 Introduction

Though liquid crystalline materials have been known for several decades, it was only in the late 1970s that they attracted the attention of researchers to create polymer dispersed liquid crystal (PDLC) composite films for possible generation of new display technology. PDLC hold potential for a variety of electro-optic applications such as light shutters, diffractive optics, Fresnel lenses, spatial light modulators, large area flexible displays (Fuh et al. 1999; Crawford 2005; Im et al. 1995; Yang et al. 2010; Ren et al. 2003; Fontecchio et al. 2001). To cite some examples of early work in this area, electrically controlled optical properties of encapsulated cholesteric liquid crystals were reported by Churchill and Cartmell (1971a, b). Shanks (1977) explored the use of polymer binder in holding liquid crystals for large-area displays. Liquid Crystal (LC) has a thermodynamic phase whose degree of order parameter is between that of a crystal and a liquid. Its molecules have a definite crystalline arrangement while it flows as a liquid. The two common types of structures that LC molecules possess are calamatic (elongated) and discotic (disc-like) (Bahadur 1990). The molecules which demonstrate liquid crystalline states over a certain temperature range are called as thermotropic. Another class of LCs, having two distinct parts/building blocks; hydrophilic and hydrophobic are called lyotropic LCs. All soaps and detergents are in fact lyotropic LCs and their properties depend on the concentration in the solvent and the shape of the molecule. Friedel (1922) classified thermotropic LCs broadly into nematic, cholesteric and smectic. There are several sub-classifications of smectic LCs in accordance with the positional and directional arrangements of the molecules.

R.R. Deshmukh (✉)
Department of Physics, Institute of Chemical Technology, Matunga, Mumbai 400 019, India
e-mail: rajedeshmukh@rediffmail.com; rr.deshmukh@ictmumbai.edu.in



V.K. Thakur, M.R. Kessler (eds.), *Liquid Crystalline Polymers*,
DOI 10.1007/978-3-319-20270-9_7

Nematic LCs are basically one-dimensional (rod like) ordered structures, with unit vector $\vec{n}$ called as director pointing along long molecular axis. LCs exhibit uniaxial symmetry around the director, which gives them shape anisotropy. This shape anisotropy manifests itself in other physical properties such as optical and dielectric anisotropy. The optical anisotropy–birefringence is the heart of LC science and technology. It has two refractive indices; the ordinary refractive index (RI) ($n_o$) and the extra-ordinary ($n_e$). The difference between $n_e$ and $n_o$ is termed as birefringence ($\Delta n$); when $\Delta n > 0$, it is termed as positive birefringence and when $\Delta n < 0$, as negative birefringence. This birefringence decreases with increase in temperature and at clearing temperature it becomes zero (Bahadur 1990; Deshmukh and Jain 2014). The random orientation of LC droplets in the polymer matrix results in the average of the above two and is considered as $n_{lc}$. To quantify the amount of orientational order in the nematic LC phase, the scalar order parameter $S$ is used and expressed by the following equation (Chandrasekhar 1977):

$$S = \frac{1}{2}\left(3\left\langle \cos^2\theta \right\rangle - 1\right) \tag{7.1}$$

where $\theta$ is the angle between the axis of an individual molecule and LC director and the average is taken over the complete ensemble. In LC phases, $S$ can range from 0.3 to 0.9 (Collings 1990; Choudhary et al. 2014) and for typical nematic LC is ~0.5 to 0.7. Response of LC molecules to an electric field is the predominant characteristic utilized for applications in industrial use. LC possesses two components of dielectric permittivity; the component parallel to the director ($\varepsilon_{\|}$) and that perpendicular to the director ($\varepsilon_{\perp}$). The difference in these two dielectric constants is called the dielectric anisotropy ($\Delta\varepsilon$):

$$\Delta\varepsilon = \varepsilon_{\|} - \varepsilon_{\perp} \tag{7.2}$$

## 7.2 Polymer Dispersed Liquid Crystal (PDLC)

PDLC films are composed of phase separated micron-sized birefringent LC droplets dispersed in an isotropic polymer matrix, having a spatially varying refractive index with good light-scattering properties in the OFF state (Lee et al. 2002; Qin et al. 2012; Liu et al. 2012; Manohar et al. 2006). Taylor (1978) proposed phase separation methods for forming PDLC composite films exhibiting dynamic light scattering. In PDLC, the polymer and LC are mixed homogeneously and then allowed to phase separate to form three dimensional structures on a transparent conducting substrate (usually an ITO coated glass plate). Another ITO coated glass plate is used to 'sandwitch' this film to form a PDLC cell. Subsequently several methods of preparation/synthesis including addition of fillers (such as nano particles, nano tubes, etc.) were adopted/investigated to induce optical non-linearity (Filpo et al. 2012; Wang et al. 2012; Jamil et al. 2011). Novel electro-optical properties continue to be revealed with changes in physical and chemical properties of polymer and LC. PDLC-based display devices have an edge over conventional liquid crystal displays because in PDLC systems light can be modulated without the use of polarizers and alignment layers, thus enhancing optical brightness, contrast and giving wide viewing angle. The reduction in the cost of the device is an added advantage. Since droplets of LC are embedded in a polymer matrix, self supporting large area display devices with flexibility and processability like polymer are made possible. The electro-optical (EO) properties of PDLC films are relatively insensitive to the slight variations in the cell gap, which makes construction of large area devices easier (Drzaic 1995). A PDLC system consists of a large number of micron sized LC droplets dispersed in a polymer matrix and the large surface-to-volume ratio in the composite film leads to many features being dominated by the interface, rather than the bulk properties of the LC. Therefore, the morphology, texture, droplet shape, size and distribution play a vital role in electro-optical properties of PDLC films. The isolated/discrete droplets of uniform size or highly interconnected LC losing the character of distinct droplets may be obtained depending upon the concentration of LC in a polymer. Sometimes intermediate morphologies with varying interconnection of the nematic cavities are obtained. The LC and polymer may form interpenetrating three dimensional structures. Many different textures (configurations) of LCs in droplets have been obtained by varying film forming parameters (Liu et al. 2012; Ohta et al. 2012). There are many parameters affecting the droplet morphology such as, type of LC and polymer, phase separation method, viscosity, miscibility of LC in polymer and vice versa, concentration of LC, addition of filler or dye molecules etc. (Lucchetta et al. 2004; Kalkar and Kunte 2002; Parab et al. 2012; Meng et al. 2010; Ahmad et al. 2012a; Li et al. 2008; Liu et al. 2012; Lu et al. 2013; Han 2000, 2002). The ability to control and to reproduce the droplet size, shape and distribution is crucial in developing reproducible PDLC devices.

### *7.2.1 Methods of Preparation of PDLC Films*

Historically, Fergason (1984, 1985) introduced preparation of PDLC films using emulsification of the LC into an aqueous solution of a film forming polymer. In his work, LCs were mixed with the aqueous solution of PVA and stirred rapidly to form emulsion. This emulsion was coated on conducting transparent substrate and dried to induce phase separated structure. Doane and co-workers are the pioneers in the phase separations methods (Doane et al. 1986). The phase separation of polymer and LC is very crucial in the preparation of PDLC films with desired properties. Starting with a homogeneous solution of a monomer, oligomer or pre-polymer and LC, the polymeric components of the system eventually become sufficiently incompatible with the LC causing phase separation. The induced morphology as a result of phase separation ultimately controls the electro-optical properties of the PDLC films.

The morphology of PDLC films depends on the chemical nature of the constituents (LC and polymer) and on the kinetics of phase separation. Phase separation methods can be broadly classified into three categories: SIPS, PIPS and TIPS.

#### 7.2.1.1 Solvent-Induced Phase Separation (SIPS)

In this method, a polymer and LC are mixed in a common solvent to form a single-phase homogeneous solution. This solution is then poured on an optically flat surface (ITO coated glass which acts as a transparent conducting electrode). When the solvent evaporates, the LC phase separates out as droplets in polymer matrix causing polymer-rich and LC-rich phases. After the solvent evaporation, a second ITO coated glass plate is laminated to the film using heat and pressure to form a PDLC cell. The morphology and droplet size can be controlled by adjusting the solvent evaporation rate. Since thermoplastic polymers are used in SIPS, it is possible to "manipulate" the droplet size by heating the cell up to the clearing temperature of LC ($T_{NI}$) and totally dissolving it in the polymer; upon cooling; the LC once again phase-separates from the polymer, with the LC droplet size now depending on the cooling rate. Deshmukh and Malik (2008a) have prepared PDLC composite of films of polymer PMMA and LC E7 using SIPS technique and have studied electro-optical and thermo-optical properties for different LC loading. Rout and Jain (1992) have prepared PDLC composite films of polymer PMMA and LC K24 using SIPS technique and have studied dielectric properties as a function of frequency and temperature.

#### 7.2.1.2 Polymerization Induced Phase Separation (PIPS)

In this method, the LC is homogeneously mixed with monomer, oligomer or pre-polymer. On application of suitable energy in the form of heat, light or radiation, polymerization is induced. The growing polymer chains phase separate from the LC to form a polymer matrix surrounding LC droplets. The inherent flexibility of the polymerization process allows us to use a wide range of materials which are readily available. The phase separation and hence droplet morphology can be controlled by temperature and light intensity. Of the three phase separation techniques, PIPS method has proved to be the most efficient in forming durable PDLC films with good electro-optical properties. The effect of UV intensity on cure kinetics of pre-polymer NOA-65 and the liquid crystal E7 and on morphology and electro-optical properties has been reported (Smith 1991; Lovinger et al. 1994; Cho et al. 2013). In case of dichroic PDLC, the addition of dye molecules in LC alters the kinetics of photo-polymerization. Deshmukh and Malik (2013a) have studied photo-polymerization kinetics and electro-optical properties of dichroic dye-doped PDLC (DPDLC) system. They have used commercially available photo-curable pre-polymer (SAM 114) as a polymer binder and LC BL038. The addition of dichroic dye molecules reduces polymerization rate leading to increased droplet size and finally tailoring the electro-optical properties of DPDLC films prepared by PIPS technique. Another advantage of photo-polymerization technique used in PIPS method is that the mask pattern can be impregnated into the PDLC films by modulating the intensity (by keeping mask pattern between the light source and the film) of UV light across the film during curing. The morphology of the film will be different in the masked portion as compared to the rest of the film. The different portions of such composite films will require different biasing voltages for orientation while the mask pattern can be made to appear or disappear in the film by controlling the intensity of electric field across the film.

#### 7.2.1.3 Thermally Induced Phase Separation (TIPS)

In this method, a homogeneous mixture of LC and thermoplastic polymer is obtained at relatively high temperatures. When this homogeneous mixture is allowed to cool, phase separation takes place with the solidification of polymer. The droplet size can be controlled by the rate of cooling. Rapid cooling leads to the formation of large number of small droplets. Slower cooling rate allows droplets to grow and coalescence to form larger droplets. Thus suitable droplet size can be obtained by controlling the cooling rate. Chen et al. (2012) have prepared scattering mode light shutters in PDLC composite of films of photoconductive polymer PVK and LC K24 using TIPS technique. They have studied effect of different switching temperatures, cooling rates, and cell gaps on the formation of LC light shutter. Though the method seems to be relatively simple, it is less preferred because at high temperatures the polymer and LC phases have high mutual solubility, which leads to poor electro-optical properties of PDLC device.

## 7.3 LC Solubility in Polymer Matrix

Maximum ON-state transmittance occurs when the refractive index of polymer ($n_p$) matches with the ordinary refractive index ($n_o$) of LC. During the film formation, it is possible that some fraction of LC dissolved in polymer matrix can have a profound effect on the PDLC film properties. Therefore, determination of partition of LC between the polymer and LC phases is an important factor in evaluating the performance of PDLC films. The reorientation of LC portion in a PDLC composite film is responsible for the optical non-linearity and electro-optical properties of the device. The absorption of LC into an isotropic polymer results in the LC becoming a part of the polymer phase. In this state, reorientation of LC does not happen with an applied electric field, leaving less amount of LC behind for scattering of light. Therefore, selection of suitable concentration of LC in PDLC films is crucial in optimizing film properties. LC dissolved in polymer matrix alters refractive index, dielectric constant, viscosity etc. of the host polymer. As explained earlier, for best electro-optical responses, a polymer and LC material are chosen on the basis of their physical and chemical properties. One such property is refractive index. When the LC is dissolved in a polymer matrix, the average RI of LC ($n_{lc}$) (which is quite different from $n_p$) increases the effective RI of polymer causing mismatch in $n_o$ and $n_p$. A refractive index difference of 0.01 between $n_o$ and $n_p$ is enough to induce significant haziness into the PDLC film in the high field (ON) state (Drzaic 1995), leading to a poor contrast ratio. Liquid crystalline materials are eutectic mixtures of several components (Nwabunma and Kyu 2001). These components may have different solubility characteristics in the host polymer matrix, causing change in the composition of LC that have remained in the form of droplets. This may change droplet configuration, elastic constants, birefringence, dielectric anisotropy etc. Similarly, low molecular weight monomers, oligomers may get dissolved into the LC; changing the above mentioned properties of LC. Thus an understanding of solubility characteristics of LC and polymer in each other is very important. There are methods to estimate solubility of one component in the other and vice versa. Selecting a proper combination of a polymer and LC may help in reducing this effect. The extraction of LC from PDLC composite films and comparing the wt% of extracted LC with LC loading can give some rudimentary ideas about the solubility effect. In order to study droplet size, shape and distribution of LC in PMMA/E7 and E8 composite films using SEM, films were soaked in methanol (methanol is a solvent for LC (E7 and E8) and non solvent for PMMA polymer matrix) for extraction of LC as reported elsewhere (Deshmukh and Malik 2008a, b; Kalkar et al. 2008). It was observed that for PMMA/E7 composite films with very low ($<$10 %) concentration of LC, there was no significant change in morphology indicating that the LC components remained molecularly dissolved in the polymer matrix. However for higher concentration of LC in PMMA/E7 and PMMA/E8 composite films, clear phase separation was observed. The weight change of the composite films after the extraction of LC largely corresponds to the loaded LC weight fraction, this clearly indicates that most of the LC fraction was extracted in the case of films with higher concentration of LC.

## 7.4 Droplet Morphology

The electro-optical properties of PDLC films strongly depend upon droplet shape, size, distribution and defects. Therefore, it is important to understand the effect of the film forming process and parameters on the droplet configuration.

### *7.4.1 Theoretical Aspects of Droplet Configurations*

LC droplets are usually spherical due to the effect of surface tension, but due to continued photo-polymerization reaction, the texture changes significantly to adopt a different configuration during the remaining process. Finally, in the polymer cavity the LC molecules arrange themselves in a way so as to minimize elastic free energy and acquire axial, radial, toroidal and/or bipolar structure depending upon several factors such as the droplet size, relative values of elastic constants of the LC droplet, orientation, their contact angle and anchoring energy at the droplet wall etc. (Wang et al. 2009; Deshmukh and Malik 2014). Anchoring effects are pronounced in small droplets. When LC molecules are confined in spheroidal geometry, the curvature induces three basic Frank elastic deformations namely: splay, bend and twist (Kumar 2007). The contribution of each deformation to the overall energy density F is given by (Drzaic 1995):

$$F = \frac{1}{2}\left[k_{11}\left(\nabla \cdot \vec{n}\right)^2 + k_{22}\left(\vec{n} \cdot \nabla \times \vec{n}\right)^2 + k_{33}\left(\vec{n} \times \nabla \times \vec{n}\right)^2\right] \tag{7.3}$$

where the proportionality constants $k_{11}$, $k_{22}$ and $k_{33}$ are associated with splay, twist and bend deformations, respectively. The resistance of nematic directors to the surface forces induces these deformations in LC droplets. When LCs are confined to small cavities; curved surfaces contort the director field, including significant elastic deformations in the director configuration. When nematic directors are packed in spherical cavities, one or more defects are created. Point or line defects can exist either on the surface or within the droplet (Brinkman and Cladis 1982). These defects can influence the ability of the droplet to transform from one configuration to another. The transformation and/or variation in these elastic deformation forces is a result of varying molecular anchoring at the droplet wall. Application of electric field may transform one type of defect into another, changing the light scattering and dielectric properties of the composite film, which is an important factor in the development of PDLC devices. Hence it is worthwhile to understand the droplet configuration from the scientific and technological viewpoints.

The alignment of nematic LC directors at the LC-polymer interface is very important as it determines the droplet configuration. Generally the anchoring energy which causes this preferred alignment is relatively strong as compared to other elastic forces within the droplets. The anchoring energy is more for smaller droplets because of increased surface to volume ratio for smaller droplets. Large droplets possess a lower elastic free energy density. Defects arise when the elastic energy density of a nematic grow sufficiently large. Defects are high energy structures (Drzaic 1995), and naturally the nematic will adopt the configuration which will minimize both the number and the strength of these defects. Volovik and Lavrentovich (1983) have classified defects in nematic droplets as *hedgehoges* or *boojams*. The point defects at opposite ends of bipolar droplets are called as *boojams*, while the point defects in the centre of the radial droplet are a *hedgehoge*. The most common configuration observed in nematic LC is bipolar, in which the director field is anchored parallel at the interface. The director field shows cylindrical symmetry, with the symmetry axis passing through the two point defects which lie at the opposite ends (poles) of the droplet (Kumar and Raina 2007). The major elastic deformation present in bipolar droplets is a bend deformation (the curvature of the director as it follows the droplet surface from one pole to the other).

### 7.4.2 In-Situ Measurement of Droplet Morphology

The phase separation kinetics can be studied by optical microscopy, light scattering, and calorimetry. Polarization optical microscopy (POM) has been extensively used for studying evolution of droplet morphology during phase separation. Similarly the effect of electric field on droplet configuration can be readily probed using optical microscopy. The size, shape and distribution of the droplets (domains) in PDLC films strongly depend on the phase separation method and the kinetics of phase separation. The final droplet morphology of PDLC composite films is governed by numerous factors such as, the nucleation and growth, diffusion, coalescence and coarsening processes and the freezing of the droplet structures by the formation of a solid polymer matrix. Rapid curing/polymerization process usually result in the formation of large number of smaller droplets and a sharp increase in viscosity of polymer-rich phase. Conversely, slow cure rate allows enough time for the growth, diffusion and coalescence of the droplets resulting in bigger droplet size. The gelation of polymer also takes place at a reduced rate and the viscosity of polymer increases slowly. This enables diffusion and coalescence of LC droplets. The addition of dye molecules can slow down the cure process (Deshmukh and Malik 2013a; Deshmukh and Jain 2014; Cho et al. 2013; Niiyama and Kikuchi 2013; Xia et al. 2006; Benmouna et al. 2013).

### 7.4.3 Morphological Evolution of PDLC Composite Films

We have investigated the morphological evolution of pure NOA65/E7 mixture and dichroic dye doped NOA65/E7 mixture (filled in a prefabricated cell having gap of 23 μ) during the course of photo-polymerization, when exposed to 8 mW UV light of wavelength 354 nm. While preparing such DPDLC films, dye molecules are first dissolved in LC and then such dye-loaded LCs are mixed with either monomer, oligomer or pre-polymer to form a homogeneous mixture. This is then allowed to phase-separate as mentioned above. Figure 7.1 shows the kinetics of polymerization (evolution of morphology) in pure NOA65/E7 mixture. NOA65 and E7 formed a highly miscible mixture with no pre-existing structure formation seen at reaction time $t = 0$ s. Nwabunma and Kyu (2001) have observed similar effects using polymer NOA65 and LC K21. Lin et al. (2005) have studied phase separation kinetics of NOA65 and E48 filled in a conventional cell and a polyimide coated unaligned cell. Upon exposure to UV light, the polymer/LC system undergoes phase separation, driven by the increasing molecular weight of growing polymer. Since photo-polymerization is inherently a very fast process (Crawford et al. 2006; Nwabunma et al. 2000), the phase separation starts almost instantaneously and at 5 s, the diffusion of the LC molecules begins in the viscous polymer medium. As a consequence, the polymer-rich region is formed and the LC with the unreacted monomers (LC-rich region) gets segregated within the network. The expulsion of

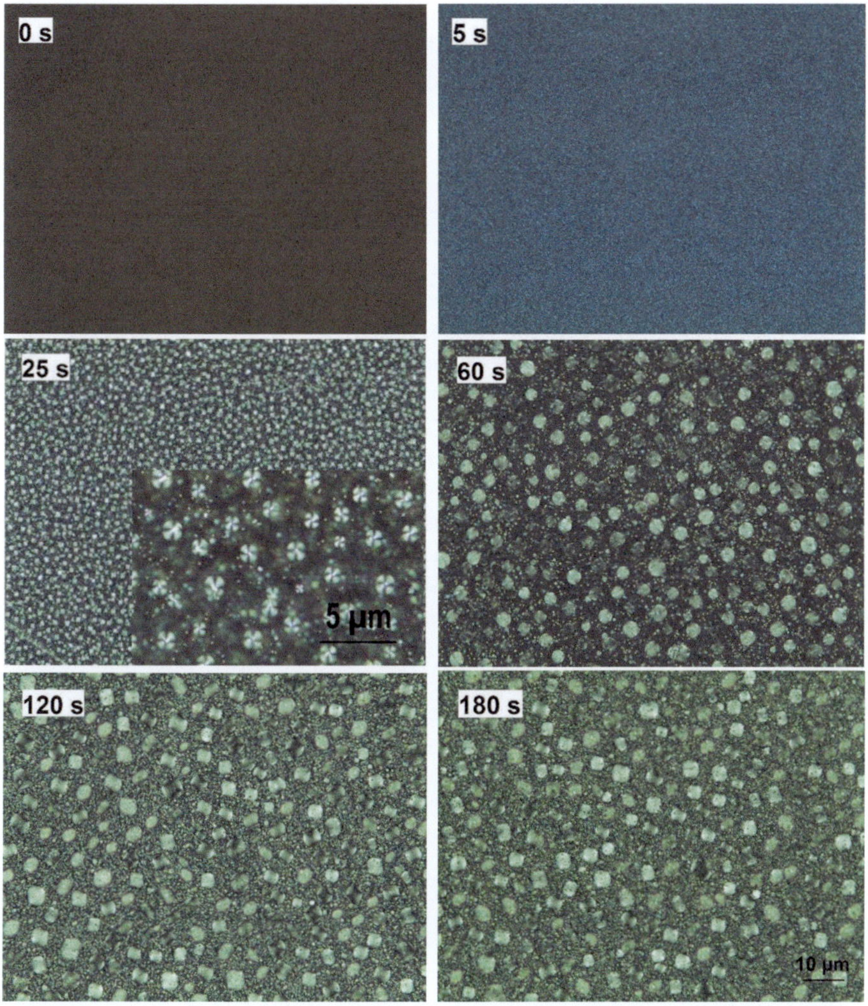


**Fig. 7.1** The evolution of phase separation morphology of NOA65/E7 under UV exposure for different duration of time, observed through a POM at a magnification of ×200 (Deshmukh and Malik 2013b)

LCs from the polymer phase continues with further irradiation of UV energy, and at around 25 s, the morphology emerges with clearly identifiable birefringent LC droplets (domains). At around 30 s (image not shown), spherical droplets emerge out with a distribution of large and small domains simultaneously. The droplets rapidly flow and grow with time through coalescence as LC molecules have high mobility. At this stage, the viscosity of polymer does not increase much to prevent any movement of LC domains. The coarsening process gets pronounced at around 60 s of UV exposure time. As the polymerization advances, polymer gelation gradually occurs; which resists the growth of the moving and nucleating LC droplets. The coarser droplets continue to grow without changing their position during the remaining period of photo-polymerization process and at 120 s, the LC boundaries are eventually frozen as the polymer matrix reaches its gelation point. The LC directors reorganize and the primary LC droplet configuration seen is of a bipolar type at 180 s.

Emergence of LC domains with a texture similar to nematic directors in a radial configuration was observed at around 25 s (a part of magnified image taken at $\times 400$ is shown in the inset). The observed radial configuration (texture), is due to homeotropic anchoring of nematic directors at the LC-polymer interface. The increasing molecular weight of the growing polymer during the course of polymerization could have possibly impinged upon the nematic ordering in LC domains. Polymerization causes separation of mixture into LC-rich and polymer gel-rich domains. The increasing molecular weight of polymer exerts surface forces on the coalescing LC droplets. The nematic nucleation results in sufficient elastic energy density giving rise to radial textures as seen in Fig. 7.1 at around 25 s, with a point defect in the centre; indicative of splay deformation. The splay ordering increases till the reaction time of 30 s (image not shown here), and then gets perturbed by the increasing polymer molecular weight. The orientation of nematic director changes and the point defect present at the centre moves away towards the edges (poles) in a process to maintain the balance between the competing elastic and surface forces. The energy inside the LC droplets rises to form two point defects on the edges resulting in the formation of bipolar structures with bend deformation dominating over the splay elastic constant. As mentioned above, these defects are high energy structures; the nematics finally adopt a configuration that minimizes the number and the strength of these defects. With polymerization time of 120 s and beyond, the LC domain growth practically ceases with no further change in the resultant morphology of bipolar configuration seen in the composite film.

### 7.4.4 *Morphological Evolution of Dye Doped PDLC Composite Films*

The addition of dichroic azo red dye in NOA65/E7 was found to have a considerable influence on the UV-curing polymerization (photo-polymerization). The method of preparation of dye doped PDLC (DPDLC) is described elsewhere (Deshmukh et al. 2012a; Jain and Deshmukh 2014). The dye in polymer-LC mixture impedes/slowdowns the polymerization process, with almost no appearance of separation of LC-rich and polymer-rich regions for a comparatively long period of photo-polymerization. The 0.5 wt% dye-doped PDLC mixture showed start of phase separation of LC material at around 60 s, as shown in Fig. 7.2. The slower rate of photo-polymerization facilitates the growth of LC droplets by diffusion, coalescence process; as the increase in polymer viscosity is not much.

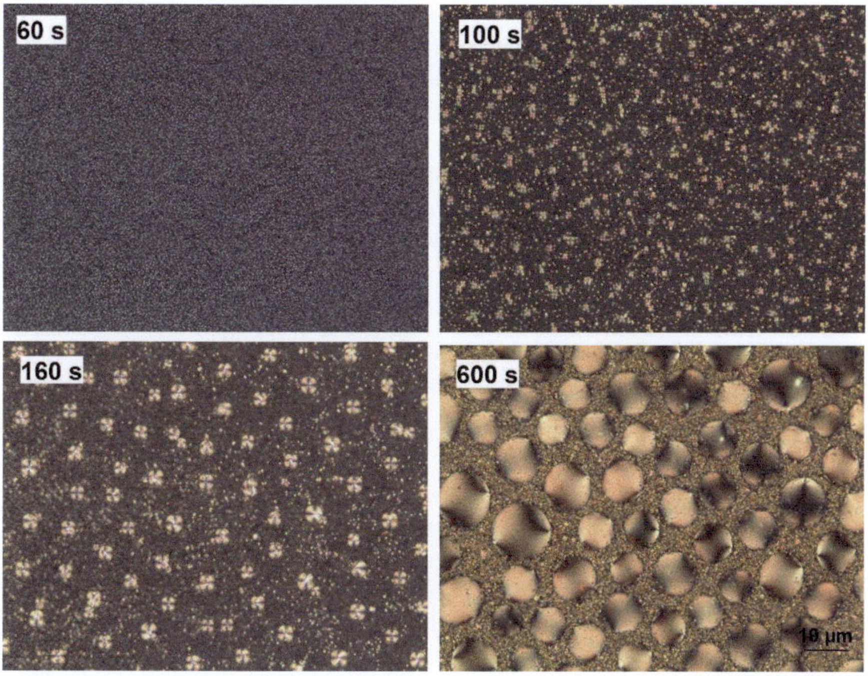


**Fig. 7.2** Morphological evolution of LC domains in 0.5 wt% dichroic azo red dye-doped NOA65/E7 mixture during photo-polymerization, observed through POM at ×200 (Deshmukh and Malik 2013b)

The LC nucleation and domain growth are seen to continue with the progression of the cross-linking reaction leading to formation of splay structures at around 100 s. It was noticed that the time interval between the incipience of phase separation and the occurrence of splay deformations was greater when compared to that of the pure PDLC mixture in Fig. 7.1. Also, the time period for which the splay structures are preserved during the course of polymerization was found to be considerably more with the addition of dye. This could be due to the consequence of a lower UV-curing energy (as part of the UV energy is absorbed by dye molecules) in PDLC mixtures containing dye which slowdowns the cross-linking of matrix polymer. This prevents the polymer from exerting strong anchoring forces at the droplet surfaces and the splay formation continues for most of the reaction time. The LC domains continue to grow and the coarsening process gets pronounced at around 290 s (image not shown) of reaction time.

The *hedgehogs* in the centre of splay structures move to the edges (poles) with growing polymer network. With further polymerization, the polymer solidification quenches the growth and mobility of LC domains. The morphology remains more or less the same and at around 600 s, exhibits random optical axis orientations. It can be seen that the LC domains do not exhibit preferential alignment or orientation in the composite films. In fact, this observed structural heterogeneity is an origin of the optical heterogeneity accountable for light scattering in PDLC devices, rather than polarization of light as is the case with other LC technologies. We have demonstrated that with higher dye content, the polymerization kinetics proceeded at an even much slower rate giving rise to bigger droplet size (Deshmukh and Malik 2013a, b; Deshmukh et al. 2012a). A uniform distribution of phase separated LC droplets can be seen in all composite films exhibiting bipolar and radial configurations; with bipolar configuration being more dominant.

## 7.5 Electro-optical (EO) Properties

The basis of most of the PDLC based devices is the high scattering of light in the OFF state. Light scattering properties of PDLC films which can be controlled by electric field, has generated considerable interest for scientific and technological reasons. The light scattering effects are insensitive to the initial polarization of light and hence PDLC films do not require the use of polarizer films unlike normal LC devices. The light-scattering property of a PDLC film depends on many parameters such as, droplet size and shape, droplet density, refractive indices of polymer and LC, wavelength etc. However, the nematic director orientation/configuration within the droplet dominate the light scattering properties of the films. With the application of electric field, directors can be oriented along the direction of electric field causing transformation from highly scattering to highly transparent film. The mechanism of EO properties can be explained point wise with the help of following Fig. 7.3

(a) At zero field, the symmetry axis of each droplet varies randomly with respect to the adjacent droplet. Because of this random orientation and birefringence nature of the LC, refractive index varies from droplet to droplet. Also RI of polymer is different from that of average RI of LC. Hence light experiences a total mismatch of RI across the film and its path changes from point to point as shown in Fig. 7.3a. Therefore, a PDLC film scatters light strongly and the film appears opaque.

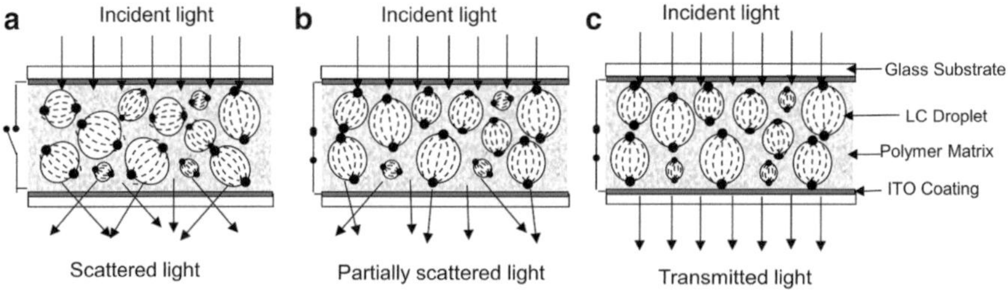


**Fig. 7.3** Schematic of OFF state and ON state transmission mechanism. (**a**) Zero field, (**b**) intermediate field and (**c**) high field

(b) At intermediate fields, some of the droplets are completely oriented along the direction of applied field, some are oriented partially and some are still unchanged as shown in Fig. 7.3b. Droplets having smaller size have more anchoring energy, due to high surface to volume ratio as mentioned above and hence their director configuration is not changed with this intermediate field. In this case light encounters less change in the RI and scattering is reduced resulting in a semi-transparent film.
(c) At sufficiently high fields, symmetry axis of each droplet is aligned along the direction of applied field as shown in Fig. 7.3c. For incident light normal to the film, each droplet possesses a uniform RI across the film thickness. Light encounters only ordinary RI of LC. If this ordinary RI of LC is very close to the RI of polymer, the film appears transparent. It is also termed as matching of RI. Therefore, while selecting the pair of LC and polymer matrix, it is necessary to take into account their refractive indices. Substantial difference in the values of refractive indices would scatter light weakly and the film would appear hazy.

Further, a single droplet model can be considered to explain the scattering and/or propagation of light at different voltage levels, as shown in Fig. 7.4 below.

When no voltage is applied, different droplets will have different orientation. The droplet under investigation (Fig. 7.4a) will scatter light at the polymer-LC interface. When a small voltage (below $V_{th}$, defined later) (Fig. 7.4b) is applied, due to the strong anchoring at the interface, the alignment of LC directors doesn't change much except in the interior part of the droplet. The interior part experiences the electric field effect and aligns directors along the direction of the field. When the field (Fig. 7.4c) is further increased to the intermediate level, the bulk of the LC droplet is oriented along the direction of applied electric field except the directors which are on the polymer-LC interface, experiencing sufficient anchoring forces. At sufficiently high electric field (Fig. 7.4d) i.e. above $V_{on}$ (defined later) all the directors are aligned along the direction of electric field and light encounters only ordinary refractive index of LC, which is very close to the RI of the polymer. Therefore a clear transparent film is observed at sufficiently high voltages.

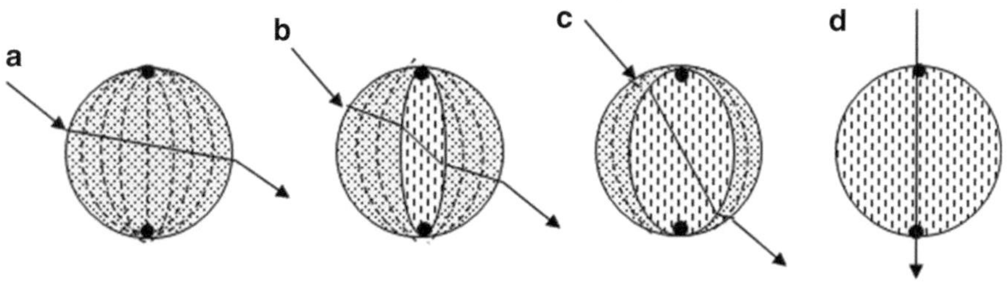


**Fig. 7.4** Single LC droplet model. (**a**) No voltage, (**b**) below $V_{th}$, (**c**) intermediate voltage, (**d**) above $V_{on}$

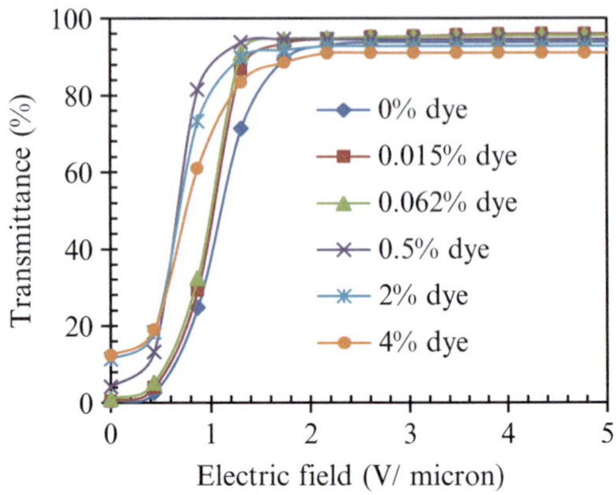


**Fig. 7.5** Electric field dependent optical properties of NOA65/E7 DPDLC films with different dichroic azo red dye content

**Table 7.1** Voltage-transmittance data of dichotic red dye-doped NOA65/E7 DPDLC films

| Dye concentration (wt%) | $T_0$ (%) | $T_{sat}$ (%) | ΔT | CR | $V_{th}$ (V/μ) | $V_{on}$ (V/μ) |
|---|---|---|---|---|---|---|
| 0.0 | 0.50 | 93.80 | 93.3 | 187.6 | 0.63 | 1.56 |
| 0.015 | 0.41 | 96.14 | 95.73 | 234.49 | 0.59 | 1.29 |
| 0.06 | 1.01 | 95.52 | 94.91 | 94.58 | 0.56 | 1.24 |
| 0.5 | 4.16 | 94.52 | 90.36 | 22.73 | 0.43 | 0.93 |
| 2.0 | 11.19 | 92.77 | 81.58 | 8.29 | 0.45 | 1.10 |
| 4.0 | 12.19 | 90.97 | 78.78 | 7.46 | 0.45 | 1.30 |

Electric field dependent optical transmittance of NOA65/E7 DPDLC films studied at a fixed frequency (200 Hz) is shown in Fig. 7.5. EO phenomenon observed in PDLC is based on the orientation mechanism of LC droplets in electric field (Blinov and Cigrinov 1994). However EO response is not a single valued function of voltage but it also depends on various film forming parameters (Liu et al. 2012; Deshmukh and Jain 2014). Various parameters such as minimum transmittance ($T_0$), maximum transmittance ($T_{sat}$), transmittance difference (ΔT), contrast ratio (CR), threshold voltage ($V_{th}$) and driving voltage ($V_{on}$) etc. were calculated from the above figure and the data is given in Table 7.1.

In addition to this, the ON and OFF state transmittance difference ($\Delta T$), rise time, decay time and hysteresis effect need to be considered while discussing the electro-optical properties of PDLC film. In the absence of an electric field, PDLC film is opaque because the preferred alignment direction of LC droplets are randomly oriented in the polymer matrix resulting in the refractive index difference (a) between an LC droplet and a polymer, (b) between two adjacent droplets and (c) within an LC droplet (Malik et al. 2012). On increasing the voltage at a slow and steady rate, under the action of the director torque, the LC directors start reorienting and aligning themselves parallel to the direction of applied field, due to which transmittance reaches the 10 % of the value of OFF state transmittance ($T_{OFF}$); the voltage corresponding to this is termed as the threshold voltage ($V_{th}$). At an electric field of 0.7–0.8 V/μ, we observe intermediate transmittance state. It can be explained from Fig. 7.4 which shows that below $V_{th}$ and slightly above it, the applied electric field has an intermediate value and is not enough to align all the droplets along its direction. Smaller droplets in particular have a high anchoring energy and cannot be aligned easily (as depicted in Fig. 7.3b) as explained above, hence the film appears hazy at such fields. When sufficient voltage is applied across the PDLC film which overcomes the anchoring at polymer-LC interface (Deshmukh and Malik 2013b), the LC droplet attains minimum free energy by completely aligning itself parallel to the field direction. At this stage the film becomes fully transparent and is termed as ON state transmittance ($T_{ON}$). Two terms namely, transmittance difference ($\Delta T$) and contrast ratio (CR) have been calculated to evaluate the optimized PDLC film. Transmittance difference is difference between ON and OFF state transmittance whereas contrast ratio is the ratio of these two.

$$\Delta T(\%) = T_{ON} - T_{OFF} \tag{7.4}$$

$$CR = \frac{T_{ON}(\%)}{T_{OFF}(\%)} \tag{7.5}$$

The minimum value of voltage above which transmittance saturates is termed as saturation voltage ($V_{on}$) (Blinov and Cigrinov 1994; Drzaic 1995; Deshmukh and Jain 2014). Wu et al. (1987) gave the mathematical formulae of threshold and saturation voltage (for bipolar droplet) by the balance between the elastic force, surface interaction and applied electric force and is estimated as (Kalkar et al. 2008; Ahmad et al. 2012b):

$$V_{th} = \frac{d}{3a}\left[\frac{\rho_P}{\rho_{LC}} + 2\right]\left[\frac{k(l^2 - 1)}{\Delta\in}\right]^{\frac{1}{2}} \tag{7.6}$$

$$V_{on} = \frac{d}{3a}\left[\frac{k(l^2 - 1)}{\Delta\in}\right]^{\frac{1}{2}} \tag{7.7}$$

where d is the film thickness, $\rho_p$ and $\rho_{LC}$ are the resistivities of polymer and LC respectively, the aspect ratio $l = a/b$, where $a$ and $b$, are the length of the major and minor axes of LC droplet respectively, $k$ is the elastic constant, and $\Delta\in$ is the dielectric anisotropy of the LC.

Table 7.1 shows decrease in the value of minimum transmittance ($T_0$) and the increase in $T_{sat}$ with the addition of a low amount of dye. As the dye concentration is increased, the $T_0$ increases and $T_{sat}$ decreases, resulting in a low CR for films with a higher dye content. The CR for 0.015 % dye doped sample is ~25 % times more and it is less than that of the undoped sample for the other concentrations of dye doped samples. The value of $T_0$ has a high impact on the value of CR as compared to $T_{sat}$; for example, when 0.06 wt% of dye was added, the increase in $T_0$ is ~0.5 % where as $T_{sat}$ is also increased by 1.7 %. The CR in this case was found to be less as compared to the undoped sample. Therefore it would be worthwhile to state that ΔT (difference between $T_0$ and $T_{sat}$) should also be considered while determining the optimum concentration of dye in the PDLC composite film. Therefore 0.015 and 0.06 wt% dye doped samples could be considered as the optimum loading of dye concentrations for better EO properties. From Fig. 7.5, one finds that switching from an opaque to a transparent state occurs in a short electric field span *i.e.* threshold ($V_{th}$) and saturation ($V_{on}$) fields are closer. Moreover, in comparing the transmission, it was found that the maximum transmission shifted to lower values with the increase of dye concentration because the dichroic dye lowers the polymerization rate causing an increase in the LC droplet size. The bigger droplets are easy to orient because of less boundary/anchoring effect. However, the increase in OFF-state transmittance with an increase in dye content is indicative of the low scattering efficiency of larger LC droplets due to fewer optical boundaries which override the mismatching of RI preposition. Also, during phase separation, the dye remains dissolved (for low concentration of dye) in the LCs as the system phase separates into droplets and a polymer matrix. With the increase in dye concentration, a large amount of dye can be left behind in the polymer matrix during the phase separation. In the ON-state, the dye dissolved in the polymer matrix is no longer able to change its orientation in response to an electric field, thus causing transmission losses by cell scattering and reflection from the film substrate (Deshmukh and Malik 2013a).

In addition to the threshold and saturation voltage, response time is also an important parameter in evaluating the performance of a PDLC film. A minimum time is necessary for the LC molecules to align along the electric field or to relax to their initial orientation when the electric field is removed. Response time is the sum of rise time ($\tau_r$) and decay time ($\tau_d$). Rise time is the time in which transmittance of film reaches from 10 to 90 % on application of electric field. Similarly decay time is the time in which transmittance falls from 90 to 10 % upon removal of applied electric field and are computed as (Doane et al. 1988; Song et al. 2012).

$$\frac{1}{\tau_r} = \frac{1}{\gamma_1}\left[\Delta\varepsilon \times V^2 + \frac{k\left(l^2 - 1\right)}{a^2}\right] \tag{7.8}$$

and

$$\tau_d = \frac{\gamma_1 \times a^2}{k\left(l^2 - 1\right)} \tag{7.9}$$

where, $\gamma_1$ is the rotational viscosity of LC, other symbols bear the same meaning defined earlier.

**Table 7.2** Rise time and decay time of DPDLC films

| | Rise time (ms) at | | Decay time (ms) at | |
|---|---|---|---|---|
| Dye concentration (wt %) | $V_{th}$ | $V_{on}$ | $V_{th}$ | $V_{on}$ |
| 0 | 2.25 | 1.25 | 7.5 | 7.5 |
| 0.015 | 1.75 | 0.5 | 6.5 | 6.8 |
| 2 | 3 | 2.15 | 8.1 | 8.2 |

For higher electric fields,

$$\tau_r = \frac{\gamma_1}{\Delta\varepsilon \times V^2} \tag{7.10}$$

An analysis of above equation indicates that $\tau_r$ is predominantly a function of the applied voltage, where $\tau_d$ mostly depends on the LC domain size and shape and the anchoring energy with polymer wall (Kalkar et al. 2008). The variation in rise time and decay time studied for undoped NOA 65/LC E7 and dye doped PDLC composite film is given in Table 7.2.

It is clear from Table 7.2 that, addition of very small amount of dye in LC while preparing DPDLC films, helps in reducing rise time and decay time; thus making the device faster as compared to the undoped sample. It indicates that such dye doped film shows fast response to the electrical signal. Addition of dye molecules reduces photo-polymerization rate due to which droplet size increases as seen from Figs. 7.1 and 7.2. The increase in droplet size reduces anchoring effect and the reorientation of LC directors becomes easy on application or on removal of electric field. However, if dye concentration is increased further, excess dye molecules are expelled at the LC-polymer interface and in the polymer matrix. The presence of dye molecules at the interface creates additional anchoring and hence higher dye content films show increased response time. Recently, we have demonstrated the method of estimating work of adhesion and interfacial tension between the dye doped LC and polymer using sessile drop contact angle method (Deshmukh and Malik 2014). It was found that the contact angle of LC with respect to the polymer decreases with addition of dye in LC; indicating an improvement in work of adhesion as the dye content increases. Thus for best EO properties, it is necessary to optimize the concentration of dye in the system. Our observations on rise time and decay time for various samples are in good agreement with (7.8) and (7.9). As the driving field increases, the rise time is found to be decreased whereas the decay time is independent.

We have reported the variation in response time as a function of the applied electric field and dye concentration for dye doped NOA 65/E7 DPDLC films elsewhere (Deshmukh and Malik 2013b). The overall trend in all the systems that we have investigated show similar behaviour with respect to applied field, with and without dye and other conditions (Deshmukh and Malik 2008a, b, 2013b; Malik et al. 2012; Deshmukh and Jain 2014; Malik and Deshmukh 2014).

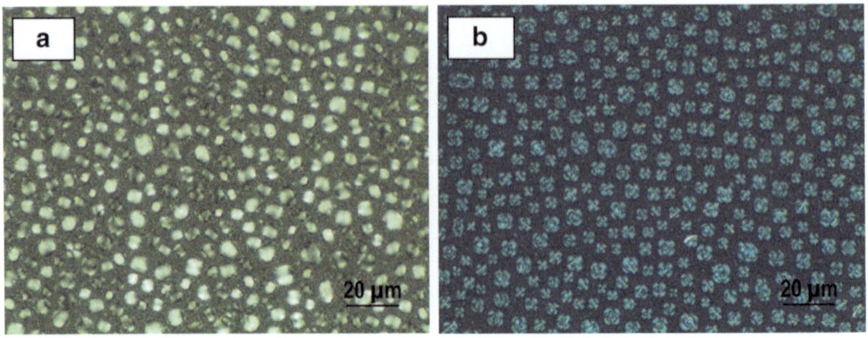


**Fig. 7.6** Droplet orientation of NOA65/E7 DPDLC composite films doped with 0.015 wt% of dichroic azo red dye (**a**) without electric field and (**b**) 2.2 V/μ

The textural changes in the droplet morphology under the influence of applied electric field of 2.2 V/μ for DPDLC films doped with optimum concentration of 0.015 wt% dye are shown in Fig. 7.6. It is observed that the randomly oriented LC droplets in the polymer matrix demonstrate maltese type crosses as a consequence of high order alignment of nematic directors along the direction of electric field.

## 7.6 Hysteresis Effect

During the study of electro-optical properties, a well-known hysteresis phenomenon is observed in PDLC films (Drzaic 1988). Hysteresis has been a problem, which must be tackled for perfect practical applications as a display material. It has been found that the transmittance values obtained at various voltages during scan-down cycle do not follow the same path as that of the scan–up cycle. A measure of hysteresis is given by the voltage width at half of maximum transmittance ($\Delta V_{50}$). The transmittance lags behind and may trace any of the three paths as shown in Fig. 7.7:

**Hysteresis**: The phenomenon when a higher transmission at a given voltage is observed during the scan down cycle as compared to the transmission at the same voltage during the scan up cycle in intermediate voltage range is termed as hysteresis. Here the transmission at a given voltage depends on the previous voltage state.

**Persistence:** Hysteresis is found only in intermediate states but if the film does not return to its complete scattering state immediate after removal of field then this situation is termed as persistence. Persistence arises if the LC domains are highly interconnected and there is a defect structure in any droplet. Then the other connected droplets may get trapped in the high field even after the removal of field until the defect escapes from the trap.

**Memory Effect:** This can be called a semi-permanent persistence because the film remains permanently in the transparent state even after removal of field. The reason

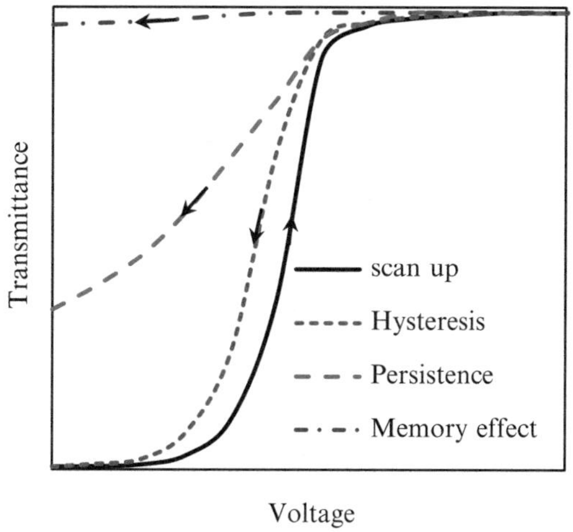


**Fig. 7.7** Schematic of hysteresis, persistence and memory effect in a PDLC film

behind this may be a weak anchoring force which develops between the boundaries of LC domain and polymer network during the period when a field is applied/ON state.

There are some common factors in hysteresis, persistence and memory effect in PDLC films. Persistence and memory effects are caused because of predominance of some specific factors.

Various factors/sources are known to cause hysteresis (Drzaic 1988; Reamey et al. 1992; Kalkar and Kunte 2002; Deshmukh and Malik 2008a). It has been suggested that hysteresis could be due to possible defect movement in a droplet (Drzaic 1988; Reamey et al. 1992) which may ultimately depend on the polymer-LC compatibility induced interfacial polarization influencing the relaxation time distribution in the droplets. Kalkar and Kunte (2002) have observed that interfacial structural changes induced by dielectric loss energy may be one of the main reasons for hysteresis where the original dielectric-structure always lags behind the applied field. Deshmukh and Malik (2008b) and Malik et al. (2012) have suggested that hysteresis could be due to the residual electric charge (on field removal) which serves as a capacitor during the scan-down cycle. When a PDLC film is driven to ON-state, the actual field that appearing across the LC droplet gets strongly masked by the space-charge build-up at the polymer–LC interface due to differences in their electrical conductivities (Jain et al. 1993; Wu et al. 1989). On field removal, the LC molecules orient to their initial state in the presence of an additional electric field due to space-charges and thus take a long time to decay. In addition to these factors, morphology and anchoring energy also play an important role in hysteresis. For example, on addition of a small amount of dye molecules, hysteresis effect may reduce. As seen from POM images, it is clear that the addition of dye molecules increases the size of droplets. The bigger sized droplets will have less anchoring energy and LC directors inside the droplets will get reoriented more easily upon field removal. If LC droplets are too small, anchoring energy would be considerable which will lead to enhanced hysteresis. Addition of dye molecules may not be always favourable to reduce this effect. Particularly, for high dye content films, the expelled dye molecules may create additional anchoring effect at the interface leading to increase in this effect. Therefore, desired concentration of dye molecules is necessary to reduce the hysteresis effect.

## 7.7 Dielectric Properties

LC material has a unique set of electrical characteristics that are dependent on their dielectric properties such as complex dielectric constant and dielectric loss. Accurate measurement of these properties can provide valuable information about their molecular arrangement, molecular dynamics, phase transitions and specific intermolecular interactions to suitably incorporate that material into its intended application with improved performance (Parab et al. 2012; Dixit et al. 2013). The dielectric properties of LC are anisotropic; it has two components of dielectric constant $\varepsilon_{\parallel}$ & $\varepsilon_{\perp}$, along and perpendicular to the director respectively and the dielectric anisotropy is given as $\Delta\varepsilon = \varepsilon_{\parallel} - \varepsilon_{\perp}$. It can have positive or negative value depending upon the permanent dipole moments and polarizability of the molecules present. If the component of the net dipole moment of the LC molecules is greater along the long axis of the molecule than that along the transverse (short) axis; the sign of $\Delta\varepsilon$ will be positive and vice versa (Itatani and Shikata 2001). PDLC composite film is considered as a heterogeneous dielectric system. For an in-depth knowledge of intermolecular interaction and co-operative process between the LC molecules and polymer matrix; a time dependent electric field dielectric relaxation spectroscopy (DRS) has been employed as a powerful tool over wide frequency and temperature range (Jonscher 1983). The migration of charge carriers to the boundaries between components of the heterogeneous system under the influence of applied electric field leads to the interfacial polarization and is referred to as Maxwell–Wagner effect (Wertheimer et al. 1976; Deshmukh et al. 2012b). In such a system LC organization within the droplets is governed primarily by polymer-LC interfacial interactions and LC–LC interactions (Mei and Higgins 1998). Dielectric response is based on the concept of “energy storage” and resulting “relaxation” per release of this energy by the systems individual components (Lvovich 2012). In a DRS the effect of alternating electric field on a system is determined by obtaining the term complex dielectric constant using Debye and Cole-Cole model for single and secondary relaxations as a function of frequency, given as: (Ishida 2010; Schlosser and Schonhals 1989)

$$\varepsilon^{*}(\omega) = \varepsilon^{'}(\omega) - i\varepsilon^{''}(\omega) \tag{7.11}$$

where, $\varepsilon'(\omega)$ and $\varepsilon''(\omega)$ are real and imaginary parts of the complex dielectric constant. The spectra of real and imaginary parts are called as dispersion and absorption curves respectively. The experimental relaxation spectra were fitted with the Debye model and Cole-Cole model for interpretation.

Debye Model: Parab et al. (2013, 2014a, b) and Jain and Deshmukh (2014).
The real part of complex dielectric constant, related to stored energy is given as:

$$\varepsilon^{'}(\omega) = \varepsilon_{\infty} + \frac{(\delta\varepsilon^{'})}{1 + \omega^2\tau^2} \tag{7.12}$$

and the imaginary part of complex dielectric constant, related to dissipated energy is given as:

$$\varepsilon^{''}(\omega) = \frac{(\delta\varepsilon^{'})\omega\tau}{1 + \omega^2\tau^2} \tag{7.13}$$

The loss tangent is given by:

$$tan\ \delta = \frac{\varepsilon^{''}(\omega)}{\varepsilon^{'}(\omega)} \tag{7.14}$$

Cole-Cole model: The Cole–Cole equation is a relaxation model used to describe secondary relaxations given by the equation (Cole and Cole 1941)

$$\varepsilon^{*}(\omega) = \varepsilon^{'}(\omega) - i\varepsilon^{''}(\omega) = \varepsilon_{\infty} + \frac{(\delta\varepsilon^{'})}{1 + (i\omega\tau)^{1-\alpha}} \tag{7.15}$$

where, $\delta\varepsilon^{'} = \varepsilon_s - \varepsilon_{\infty}$, is a measure of the voltage, a material can withstand before breakdown occurs, termed as dielectric strength. $\varepsilon_s$ and $\varepsilon_{\infty}$ are dielectric constants at the static and infinite frequency respectively, ω is the angular frequency and τ is an average relaxation time. The exponent α, known as the distribution parameter ranges from 0 to 1, explains different spectral shapes. The distribution parameter α is expressed as:

$$\varphi = \frac{\alpha\pi}{2} \tag{7.16}$$

Where, φ is angle between arc radius and ε′-axis. If the centre of the semicircle lie on the ε′-axis then the distribution parameter α = 0 (Debye type) and if the centre is below the ε′-axis then α ≠ 0 (non-Debye type) (Kent 2007; Jonscher 1983; De Gennes and Prost 1993; Parab et al. 2012; Rahman et al. 2008; Cole and Cole 1941; Raju 2003).

For the qualitative evaluation of distribution of relaxation time, Cole-Cole plot is drawn between the real part (ε′) and imaginary part (ε″) of the complex dielectric constant (Parab et al. 2013). How well the ε′ and ε″ are fitted to form a semicircle, is an indication of the nature of relaxation behavior. The highest point of semi-circle curve of gives the information about the relaxation frequency $f_r$ of the orientational

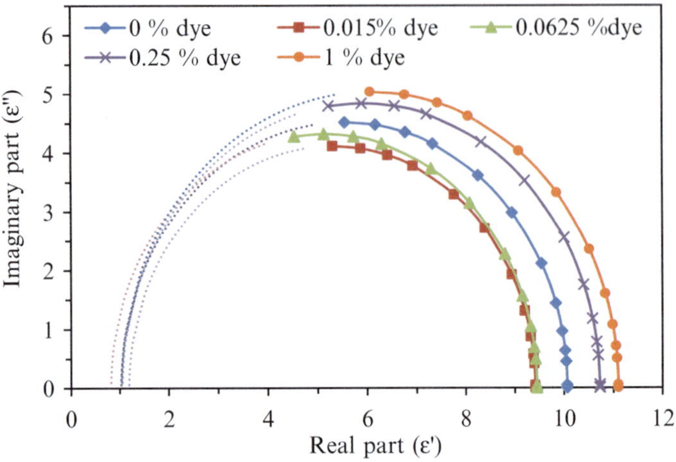


**Fig. 7.8** Cole–Cole plot of SAM-114 and BL036 composite system with varying dye concentration

polarization. At the relaxation frequency ($f_r$), dielectric hitting is maximum. Similarly the frequency corresponding to the point of intersection of ordinate axis (real part) and the semicircle curve indicates dielectric losses. The dielectric losses are minimum for low dye doped samples as compared to the undoped samples, thus addition of small amount of dye is favourable. Figure 7.8 shows Cole-Cole plot for dichroic azo orange dye doped polymer SAM-114 and LC BL036 composite films. The value of $\alpha$ was calculated for all the composite films and was found in the range of 0.02–0.05, clearly indicating mono dispersive non-Debye type relaxation (Jain and Deshmukh 2014).

## 7.8 Conclusion and Future Perspective

It can be concluded that for best EO responses, the morphological characteristics of PDLC composite films play an important role. The morphology in general and droplet shape, size, distribution in specific is influenced by the physical and chemical properties of both the constituents of PDLC composite film. Therefore, primarily LC and polymer are suitably selected on the basis of their properties. Subsequently, the method of preparation, film forming parameters, kinetics/rate of polymerization, the addition of dichroic dye etc. are varied to obtain desired morphology to get the best EO responses. We have demonstrated that the addition of very small amount of dichroic dye can tailor the properties of PDLC devices. It can be further concluded that the addition of a small amount of dye have suitably improved the CR, $\Delta T$ and response time along with a lowering of operating voltages. The incorporation of dye in LC impedes photo-polymerization rate in turn allows enough time for the growth of the LC droplets. PDLC is a heterogeneous system in which LC-rich and polymer-rich regions are present with different electrical conductivities. The dielectric relaxation spectroscopy (DRS) helps us in understanding the intermolecular interaction and co-operative process between the LC molecules and polymer matrix. For such heterogeneous systems, the Maxwell-Wagner effect is profound because of the interfacial polarization at the boundaries in the low frequency region. The model obtained with real and imaginary parts of the relative dielectric constant can provide valuable information about the type of relaxation behaviour. When the value of distribution parameter $\alpha$ is zero the Cole-Cole model reduces to the Debye model. In the present case, $\alpha$ has a non-zero value and therefore it could be concluded that the relaxation behaviour is non-Debye type. A proper understanding of EO and dielectric properties of PDLC films could find suitable applications. Efforts to improve CR, wide viewing angle, reduction in operating voltages, development of new materials and overall cost reduction of the device would be possible using nano-technology. The incorporation of carbon nano-particles and CNT may improve above mentioned properties. Few groups have initiated work on guest–host PDLC films wherein fillers/nano-particles of various types have been incorporated. Use of ITO coated polyester films in lieu of ITO coated glass plates would replace the present liquid crystal displays in favour of the flexible display devices.

**Acknowledgement** I would like to thank Prof. A.K. Kalkar for his valuable suggestions during the preparation of this article. I am also thankful to my Ph.D. student Ms. Anuja Katariya-Jain for helping me in getting references, drawing schematic diagrams and type setting.

## References

Ahmad F, Jamil M, Woo LJ, Jeon YJ (2012a) The investigation of molecular affinity involved in poly(ethylene glycol)-based polymer-dispersed liquid crystal display. Colloid Polym Sci 290:599–606

Ahmad F, Jamil M, Jeon YJ, Woo LJ, Jung JE, Jang JE (2012b) Investigation of nonionic diazo dye doped polymer dispersed liquid crystal film. Bull Mater Sci 35: 221–231

Bahadur B (1990) Liquid crystals applications and uses, vol I. World Scientific, Singapore

Benmouna F, Dembahri ZB, Benmouna M (2013) Polymerization-induced phase separation: phase behavior developments and hydrodynamic interaction. J Macromol Sci Part B: Phys 52:998–1008

Blinov LM, Cigrinov VG (1994) Electro-optic effects in liquid crystal materials. Springer, New York

Brinkman WF, Cladis PE (1982) Defects in liquid crystals. Phys Today 35(5):48–54

Chandrasekhar S (1977) Liquid crystals. Cambridge University Press, Cambridge

Chen YD, Fuh AYG, Cheng KT (2012) Particular thermally induced phase separation of liquid crystal and poly(N-vinyl carbazole) films and its application. Opt Express 20:16777–16785

Cho JD, Lee SS, Park SC, Kim YB, Hong JW (2013) Optimization of LC droplet size and electro-optical properties of acrylate-based polymer-dispersed liquid crystal by controlling photocure rate. J Appl Polym Sci 130:3098–3105

Choudhary AS, Malik MK, Patil SR, Prabhu KH, Deshmukh RR, Sekar N (2014) Phenazines and thiazine: green synthesis, photophysical properties and dichroic behavior in nematic host. Can Chem Trans 2:365–380
Churchill D, Cartmell JV (1971a) Radiation sensitive display device containing encapsulated cholesteric liquid crystals. US Patent 3,578,844
Churchill D, Cartmell JV (1971b) Display device containing minute droplets of cholesteric liquid crystals in a substantially continuous polymeric matrix. US Patent 3,600,060
Cole KS, Cole RH (1941) Dispersion and absorption in dielectrics-I alternating current characteristics. J Chem Phys 9:341–351
Collings PJ (1990) Liquid crystals: nature's delicate phase of matter. Princeton University Press, Princeton
Crawford GP (2005) Flexible flat panel displays. Wiley, Chichester
Crawford NJ, Dadmun ND, Bunning TJ, Natarajan LV (2006) Time-resolved light scattering of the phase separation in polymer-dispersed liquid crystals formed by photo-polymerization induced phase separation. Polymer 47:6311–6321
De Gennes PG, Prost J (1993) The physics of liquid crystals. Clarendon, Oxford
Deshmukh RR, Malik MK (2008a) Effect of temperature on the optical and electro-optical properties of poly(methyl methacrylate)/E7 polymer-dispersed liquid crystal composites. J Appl Polym Sci 109:627–637
Deshmukh RR, Malik MK (2008b) Effects of the composition and nematic-isotropic phase transition on the electro-optical responses of unaligned polymer-dispersed liquid crystals. I. Composites of poly(methyl methacrylate) and E8. J Appl Polym Sci 108:3063–3072
Deshmukh RR, Malik MK, Parab SS (2012a) Dichroic dye induced nonlinearity in polymer dispersed liquid crystal materials for display devices. Adv Mater Res 584:79–83
Deshmukh RR, Parab SS, Malik MK (2012b) Effect of host polymer matrices on electro optical and dielectric behavior of polymer dispersed liquid crystal system. Adv Mater Res 584:531–535
Deshmukh RR, Malik MK (2013a) Effect of dichroic dye on the phase separation kinetics and electro-optical characteristics of polymer dispersed liquid crystals. J Phys Chem Solids 74:215–224
Deshmukh RR, Malik MK (2013b) Photo-polymerization kinetics and electro-optical properties in mixtures of dichroic dye-doped nematic liquid crystal and photocurable polymer. Liq Cryst 40:1050–1059
Deshmukh RR, Jain AK (2014) The complete morphological, electro-optical and dielectric study of dichroic dye doped polymer dispersed liquid crystal. Liq Cryst 41:960–975
Deshmukh RR, Malik MK (2014) A method for estimating interfacial tension of liquid crystal embedded in polymer matrix forming PDLC. J Appl Polym Sci. doi:10.1002/app.41137
Dixit S, Misra A, Manohar R, Arora AK (2013) Enhancement in dielectric properties of nematic liquid crystal by gamma irradiation. Mol Cryst Liq Cryst 571:77–85
Doane JW, Vaz NA, Wu BG, Zumer S (1986) Field controlled light scattering from nematic micro-droplets. Appl Phys Lett 48:269–271
Doane JW, Golemme A, West JL, Whitehead JB Jr, Wu BG (1988) Polymer dispersed liquid crystals for display application. Mol Cryst Liq Cryst 165:511–532
Drzaic PS (1988) Reorientation dynamics of polymer dispersed nematic liquid crystal films. Liq Cryst 3:1543–1559
Drzaic PS (1995) Liquid crystal dispersions. World Scientific, Singapore
Fergason JL (1984) Encapsulated liquid crystal and method. US patent 4,435,047
Fergason JL (1985) Polymer encapsulated nematic liquid crystals for display and light control applications. In: SID digest, pp 68–70
Filpo GD, Siprova S, Chidichimo G, Mashin AI, Nicoletta FP, Cupelli D (2012) Alignment of single-walled carbon nanotubes in polymer dispersed liquid crystals. Liq Cryst 39:359–364
Fontecchio AK, Bowley CC, Chmura SM, Le L, Faris S, Crawford GP (2001) Multiplexed holographic polymer dispersed liquid crystals. J Opt Technol 68:652–656

Friedel G (1922) The mesomorphic states of matter. Ann Phys 18:273–474
Fuh AYG, Tsai MS, Huang LJ, Liu TC (1999) Optically switchable gratings based on polymer-dispersed liquid crystal films doped with a guest–host dye. Appl Phys Lett 74:2572–2574
Han JW (2000) Effects of composition, curing-time, and temperature on the electro-optical characteristics of polymer-dispersed liquid crystal films. J Korean Phys Soc 36:156–163
Han JW (2002) Effect of the nematic-isotropic phase transition on the electro-optical characteristics of polymer-dispersed liquid crystal films. J Korean Phys Soc 40:849–855
Im SJ, Jin YW, Sung JH, Park WY, Sakong DS (1995) A study of electro-optical property and VHR of PDLC depending on the molecular structure of monomer and preparation process. Synth Met 71:2203–2204
Ishida H (2010) Characterisation of composite materials. Momentum Press, LLC, New York
Itatani S, Shikata T (2001) Dielectric relaxation behavior of aqueous dodecyldimethyl-amineoxide solutions. Langmuir 17:6841–6850
Jain AK, Deshmukh RR (2014) Influence of a guest dichroic azo dye on a host liquid crystal dispersed in polymer matrix. Int J ChemTech Res 6:1813–1816
Jain SC, Thakur RS, Lakshmikumar ST (1993) Switching response of a polymer dispersed liquid-crystal composite. J Appl Phys 73:3744–3748
Jamil M, Ahmad F, Rhee JT, Jeon YJ (2011) Nanoparticle-doped polymer-dispersed liquid crystal display. Curr Sci 101:1544–1553
Jonscher AK (1983) Dielectric relaxation in solids. Chelsea Dielectric Press, London
Kalkar AK, Kunte VV (2002) Electro-optical studies on polymer dispersed liquid crystal composite films. II. Composites of PVB/E44 and PMMABA/E44. Mol Cryst Liq Cryst 383:1–25
Kalkar AK, Kunte VV, Bhamare SA (2008) Electrooptic studies on polymer-dispersed liquid-crystal composite films. III. Poly(methyl methacrylate-co-butyl acrylate)/E7 and poly(methyl methacrylate-co-butyl acrylate)/E8 composites. J Appl Polym Sci 107:689–699
Kent JA (2007) Kent & Riegel's handbook of industrial chemistry and biotechnology-I. Springer, New York
Kumar P (2007) Preparation and characterization of dichroic polymer dispersed liquid crystals. PhD thesis, Thapar University, Patiala, India
Kumar P, Raina KK (2007) Morphological and electro-optical responses of dichroic polymer dispersed liquid crystal films. Curr Appl Phys 7:636–642
Lee SH, Lim TK, Shin ST, Park KS (2002) A method for improving contrast ratio of polymer dispersed liquid crystal film using the oriented azo-dye molecules in polymer matrix. Jpn J Appl Phys 41:208–210
Li W, Cao Y, Cao H, Kashima M, Kong L, Yang H (2008) Effects of the structures of polymerizable monomers on the electro-optical properties of UV cured polymer dispersed liquid crystal films. J Polym Sci Part B: Polym Phys 46:1369–1375
Lin YH, Ren H, Wu YH, Liang X, Wu ST (2005) Surface anchoring effect on the morphology and performance of polymer-dispersed liquid crystal. Proc SPIE 5741:74–82
Liu F, Cao H, Mao Q, Song P, Yang H (2012) Effects of monomer structure on the morphology of polymer networks and the electro-optical properties of polymer-dispersed liquid crystal films. Liq Cryst 39:419–424
Lovinger AJ, Amundson KR, Davis DD (1994) Morphological investigation of UV curable polymer dispersed liquid crystal (PDLC) materials. Chem Mater 6:1726–1736
Lu Y, Wei J, Shi Y, Jin O, Guo J (2013) Effects of fabrication condition on the network morphology and electro-optical characteristics of polymer-dispersed bistable smectic A liquid crystal device. Liq Cryst 40:581–588
Lucchetta DE, Criante L, Francescangeli O, Simoni F (2004) Light amplification by dye-doped holographic polymer dispersed liquid crystals. Appl Phys Lett 84:4893–4895
Lvovich VF (2012) Impedance spectroscopy applications to electrochemical and dielectric phenomena. Wiley, New Jersey
Malik MK, Bhatia PG, Deshmukh RR (2012) Effect of nematic liquid crystals on optical properties of solvent induced phase separated PDLC composite films. Ind J Sci Technol 5:3440–3452

Malik MK, Deshmukh RR (2014) Electro-optics of homogeneously aligned nematic liquid crystals stabilized by a polymer network. Int J ChemTech Res 6:1833–1835
Manohar R, Tripathi G, Singh AK, Srivastava AK, Shukla JP, Prajapati AK (2006) Dielectric and optical properties of polymer–liquid crystal composite. J Phys Chem Solids 67:2300–2304
Mei E, Higgins DA (1998) Polymer-dispersed liquid crystal films studied by near-field scanning optical microscopy. Langmuir 14:1945–1950
Meng Q, Cao H, Kashima M, Liu H, Yang H (2010) Effects of the structures of epoxy monomers on the electro-optical properties of heat-cured polymer dispersed liquid crystal films. Liq Cryst 37:189–193
Niiyama S, Kikuchi H (2013) Optical microscopic observation of morphology in liquid crystal/ polymer composites and their electro-optical properties. Mol Cryst Liq Cryst 577:1–14
Nwabunma D, Chiu HW, Kyu T (2000) Morphology development and dynamics of photopolymerization induced phase separation in mixtures of a nematic liquid crystal and photocuratives. Macromolecules 33:1416–1424
Nwabunma D, Kyu T (2001) Phase behavior, photopolymerization and morphology development in mixtures of eutectic nematic liquid crystal and photocurable monomer. Polymer 42:801–806
Ohta S, Inasawa S, Yamaguchi Y (2012) Size control of phase-separated liquid crystal droplets in a polymer matrix based on the phase diagram. J Polym Sci Part B: Polym Phys 50:863–869
Parab SS, Malik MK, Deshmukh RR (2012) Dielectric relaxation and electro-optical switching behavior of nematic liquid crystal dispersed in poly (methyl methacrylate). J NonCryst Solids 358:2713–2722
Parab SS, Malik MK, Dabrowski R, Deshmukh RR (2013) Thermodynamic and bias field characterization of quickly operating antiferroelectric liquid crystal. J Mol Liq 183:20–25
Parab SS, Malik MK, Deshmukh RR (2014a) Investigation of dielectric properties of poly(methyl methacrylate)-E7 composite films. Int J ChemTech Res 6:1836–1839
Parab SS, Malik MK, Bhatia PG, Deshmukh RR (2014b) Investigation of liquid crystal dispersion and dielectric relaxation behavior in polymer dispersed liquid crystal composite films. J Mol Liq 199:287–293
Qin A, Zhang Y, Wang Y (2012) Electro-optical properties of polymer-dispersed liquid crystal prepared by controlled graft living radical polymerization. J Appl Polym Sci 124:2200–2208
Rahman M, Hsieh CW, Wang CT, Jian BR, Lee W (2008) Dielectric relaxation dynamics in liquid crystal-dye composites. Dyes Pigments 84:128–133
Raju GG (2003) Dielectrics in electric fields. Marcel Dekker, New York
Reamey RH, Montoya W, Wong A (1992) Video-microscopy of NCAP films: the observation of LC droplets in real time. Proc SPIE 1665:2–7
Ren H, Fan YH, Wu ST (2003) Tunable Fresnel lens using nanoscale polymer-dispersed liquid crystals. Appl Phys Lett 83:1515–1517
Rout DK, Jain SC (1992) Dielectric properties of polymer-liquid crystal composites. Jpn J Appl Phys 31:1396–1398
Schlosser E, Schonhals A (1989) Recent development in dielectric relaxation spectroscopy of polymer. Colloid Polym Sci 267:963–969
Shanks IA (1977) Liquid crystal devices. US Patent 4,048,358
Smith GW (1991) Cure parameters and phase behaviour of an ultraviolet cured polymer dispersed liquid crystal. Mol Cryst Liq Cryst 196:89–102
Song P, Cao H, Wang F, Liu F, Yang H (2012) The influence of the structure of curable epoxy monomers on the electro-optical properties of polymer dispersed liquid crystal devices prepared by UV-initiated cationic polymerisation. Liq Cryst 39:433–440
Taylor LJ (1978) Preparation of liquid crystal containing polymeric structure. US Patent 4,101,207
Volovik GE, Lavrentovich OD (1983) Topological dynamics of defects: boojams in nematic drops. Sov Phys JETP 58:1159–1166
Wang JH, Zhang BY, Qu WZ, Chu HS, Li H (2009) The kinetics of photo-polymerisation in the fabrication of polymer-dispersed liquid crystals doped with nano-graphite. Liq Cryst 37:1–11
Wang J, Shi S, Han X, Zhou S, Zhang B (2012) The effect of nanographites dispersed in Undec-10-enic acid on optical property and morphology of PDLC films. Colloid Polym Sci 290:1659–1664

Wertheimer MR, Paquin L, Schreiber HP (1976) Polymer composites containing plasma-treated mica. II. Dielectric properties. J Appl Polym Sci 20:2675–2690
Wu BG, Erdmann JH, Doane JW (1989) Response times and voltages for PDLC light shutters. Liq Cryst 5:1453–1465
Wu BG, West JL, Doane JW (1987) Angular discrimination of light transmission through polymer-dispersed-liquid-crystal films. J Appl Phys 62:3925–3931
Xia J, Wang J, Lin Z, Qiu F, Yang Y (2006) Phase separation kinetics of polymer dispersed liquid crystals confined between two parallel walls. Macromolecules 39:2247–2253
Yang KJ, Lee SC, Choi BD (2010) Dye-doped polymer dispersed liquid crystal films for flexible displays. Jpn J Appl Phys 49:05EA05–05EA08

# Chapter 8
# UV-Cured Networks Containing Liquid Crystalline Phases: State of the Art and Perspectives

**Giulio Malucelli**

## 8.1 Photocrosslinkable Liquid-Crystalline Polymers

The investigation on liquid-crystalline polymers (LCPs) started from the work concerning concentrated polymer solutions published by Flory in 1956, which can be considered the first pioneering paper that has given rise to the study of these mesogenic systems: in particular, the author described the possibility that rigid rod-like polymer chains may become liquid-crystalline (LC) when a critical concentration is achieved within the polymer solution (Flory 1956). About 20 years later, de Gennes suggested that the combination of rigid and flexible units could promote the formation of thermotropic main chain LCPs (de Gennes 1975); in the same year, Roviello and Sirigu were able to synthesize polyalcanoates of p,p′-dihydroxy-α,α′ dimethylbenzalazine, which represent the first example of thermotropic LC main chain polymers (Roviello and Sirigu 1975). Since then, a huge number of very interesting peer-reviewed papers mainly referring to thermotropic LC polyethers, LC polyesters, LC poly(ester-amide)s and LC poly(ether-ester)s appeared in the open-literature (Percec et al. 1984, 1987; Shaffer and Percec 1985; Percec and Yourd 1989; Percec and Tsuda 1990; Percec and Kawasumi 1991; Muthusamy et al. 2013).

Meanwhile, the set-up and industrial availability of photocrosslinkable systems (i.e. *UV-curable resins*), opened a new pathway for their wide exploitation in different high-tech fields (i.e., printed circuits, printing inks, printing plates, energy exchange systems, photocurable protective coatings on a variety of substrates, including paper, metal, plastic, textile and wood). This was possible because of the outstanding characteristics of the photocrosslinkable resin systems, such as high

G. Malucelli (✉)
Department of Applied Science and Technology, Politecnico di Torino,
Viale T. Michel 5, Alessandria 15121, Italy
e-mail: giulio.malucelli@polito.it



V.K. Thakur, M.R. Kessler (eds.), *Liquid Crystalline Polymers*,
DOI 10.1007/978-3-319-20270-9_8

curing rate, broad chemical resistance and high thermal stability, as reported by several reviews (Decker 1996, 2002; Crivello 1983).

Indeed, UV-curing is one of the fastest and most efficient methods available for producing highly crosslinked three-dimensional polymer networks in a very short time (usually within tenths of seconds, which is a "negligible" time as compared to traditional thermal curing processes), with a reduced energy consumption and the absence of Volatile Organic Compounds (VOC) emissions. Furthermore, as compared to traditional thermal polymerization methods, UV-curing allows the spatial control of initiation, as well as the tuning of the properties of the resulting materials (e.g. glass transition temperature, mechanical properties, swelling, barrier properties).

From an overall point of view, the UV-curing reactions can be schematized as depicted in Fig. 8.1.

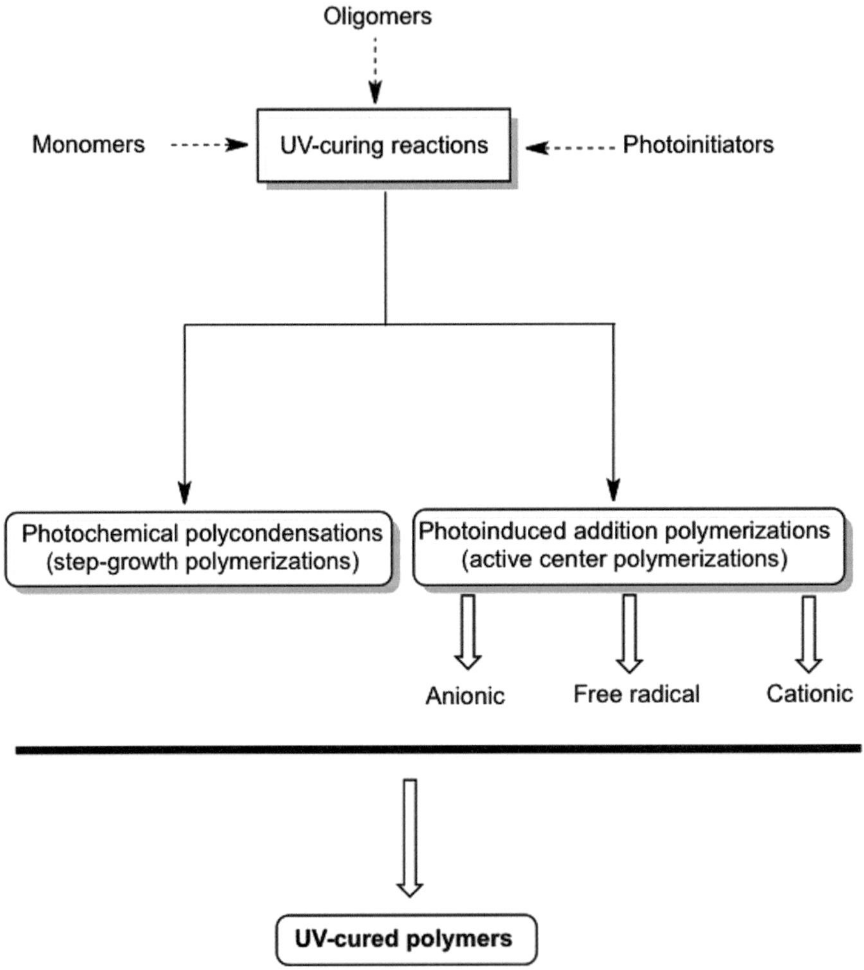


**Fig. 8.1** UV-curing reactions

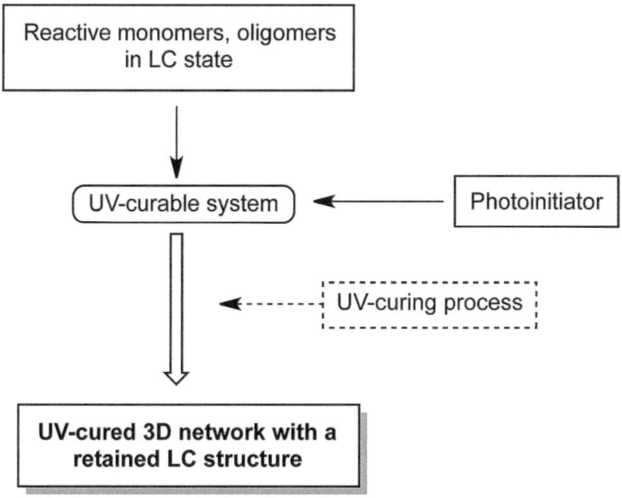


**Fig. 8.2** Scheme of the formation of UV-cured 3D liquid crystalline networks

The synthesis of liquid-crystalline monomers bearing reactive functionalities that can undergo the UV-curing process allows obtaining useful photocrosslinkable liquid-crystalline polymers: indeed, these latter can be exploited for the design of anisotropic network systems like LC thermosets and LC elastomers, information storage devices, and thin films (to be used for microelectronics and optoelectronics), retaining a controlled orientation of functional groups (Muthusamy et al. 2013).

Taking into account the phase behavior of liquid crystalline monomers, the UV-curing technique has the additional advantage that it can be performed over a range of temperatures for most monomer/initiator systems (Pogue et al. 1999). This finding is very important, since it becomes possible to obtain highly crosslinked polymers from LC monomers while they are in an LC phase and to assess the possibility of retaining the liquid crystalline form in the UV-cured final material, as depicted in Fig. 8.2.

More specifically, the presence of dual functionalities (i.e. mesogenic groups and UV-curable functionalities) in main chain photocrosslinkable LCPs allows to create novel materials exhibiting interesting features. Indeed, while the mesogenic groups take part to the definition of the LC properties, the UV-curable functionalities, located in the polymer backbone, can promote such different phenomena as photoisomerization, photocrosslinking, photodimerization, photodissociation, and photoassociation of the polymer chains upon UV exposure (Ravikrishnan et al. 2010).

As an example, the photo-triggered phenomena are responsible for the possibility of using LC polymers bearing α,β-unsaturated carbonyl functionalities (mainly cinnomoyl or benzylidene groups) as negative type photoresists (Kawatsuki et al. 2001; Sakthivel and Kannan 2006; Kaniappan and Murugavel 2009): their

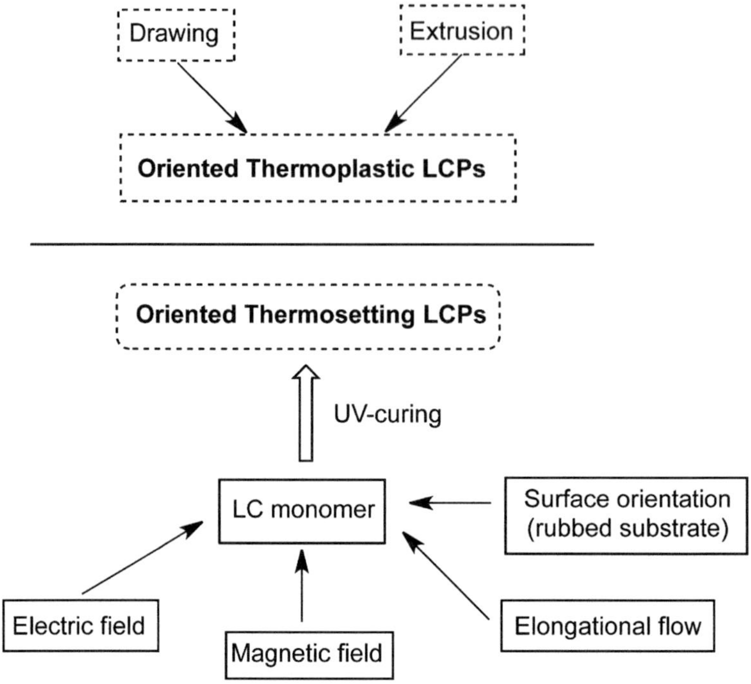


**Fig. 8.3** Orientation phenomena in thermoplastic and thermosetting LCPs

photosensitivity is due to the π-electron density of the chromophores located in the polymer backbone (Allcock and Cameron 1994).

In addition, it has been demonstrated that such molecularly designed polymers with narrow polydispersities can be used for further functionalization or crosslinking reactions and exploited for fabricating optical data storage, electro-optic displays and nonlinear optical systems (Percec and Tomazos 1992).

Notwithstanding with the aforementioned functional (optical and electrical) applications of main chain photocrosslinkable LCPs, another reason that pushed the research toward the investigation of the effect of orientation of liquid crystal polymers refers to the possibility of fabricating materials that exhibit anisotropic mechanical properties analogous to composites (Schultz and Chartoff 1998). According to the scheme depicted in Fig. 8.3, as far as thermoplastic liquid crystal polymers are considered, orientation can be induced by applying shear forces such those promoted from drawing or extrusion. In thermosets, however, the network locks the molecular order in place: as a consequence, the desired alignment must be accomplished before polymerization of the monomer. More specifically, surface interactions with rubbed substrates (i.e. the so-called *surface orientation*) can be successfully exploited in order to align liquid crystal monomers: such a procedure is quite effective as far as thin films are considered, as opposed to those applications that involve large structural parts, for which alternative means of orientation are required to align the bulk of the liquid crystal monomer. Thus, other possibilities, which can be used for aligning liquid crystal monomers have been developed and include electric field, magnetic field and elongational flow. Orientation can be done on nematic or smectic LC monomers, although the latter usually exhibit higher viscosity values as compared to nematic counterparts.

Once the monomer is aligned, photopolymerization can then take place, using a UV-source or a laser: this way, it is possible to freeze the induced orientation of the liquid crystal monomer, as the UV-curing process occurs while the monomer is still in the liquid crystalline state. Depending on the type of external stimulus adopted for the orientation of the liquid crystal monomer (as an example, the use of magnetic fields is very effective for inducing alignment over much greater distances than rubbed substrates), it is possible to create a wide range of anisotropic parts. Stereolithography, a commercially available well-known layer-by-layer technology for rapidly fabricating prototype parts of arbitrary shape from a liquid UV-curable monomer, represents one potential application for these aligned systems (Ullett et al. 1996).

Thus, thanks to the anisotropy that can be achieved in LC photocrosslinkable polymers, specific mechanical properties can be engineered directly into structural parts by selectively UV-curing different portions under different orientations. The achieved LC monomer organization may influence the photopolymerization kinetics, the polymer structure and the microstructure of the macromolecules (He et al. 1995; Hoyle et al. 1993, 1996; Guymon et al. 1997; Broer et al. 1991; Hoyle and Watanabe 1994; Hellermark et al. 1992; Andersson et al. 1992).

The anisotropy achieved in these materials can be better controlled by taking into consideration how the processing conditions may affect the orientation of the liquid crystal monomer: in particular, two of the processing variables, time and temperature, have been found to directly affect the extent of order in a liquid crystal monomer. It is worthy to note that, though the order of a liquid crystal cannot be directly evaluated, it can be estimated on the basis of the anisotropy of a macroscopic material property.

Broer and co-workers (Broer et al. 1989, 1991; Broer 1996) clearly demonstrated that both the curing temperature and the LC phase significantly influence the curing rate and the final properties of polymers obtained by UV-curing LC monomers: indeed, both curing rate and degree of monomer conversion increase with temperature up to the nematic-to-isotropic transition. This behavior can be ascribed to the molecular ordering and the high vitrification point as well. Above the transition temperature, both polymerization rate and conversion decrease.

As a consequence, the reaction processes undergone by the LC monomers are worthy to investigate, aiming to exploit the UV-curing approach in the most efficient way. More specifically, the optimization of the curing behavior of the resin can lead to higher conversions at the required temperatures and to higher polymerization rates. Furthermore, the tuning of the photoinitiator concentration may affect the penetration depth of the UV radiation, and hence can provide control over the energy needed to start the polymerization reaction.

## 8.2 Synthesis of Photocrosslinkable LC Polymers

Table 8.1 collects the main results referring to the synthesis of photocrosslinkable LC polymers, which will be discussed in the followings.

Gangadhara and Kaushal Kishore synthesized a series of photocrosslinkable main chain LC nematic polymers bearing bis(benzylidene)cycloalkanone (namely -cyclopentanone, -cyclohexanone and -cycloheptanone) units, which are known to act both as mesogens and photoactive centres (Kishore and Kishore 1995). Three dicarboxylic acids with ether linkages, which were obtained from oligoethylene oxides (namely triethylene glycol, tetraethylene glycol and pentaethylene glycol), were selected as spacers. The polymerization of these LC polymers was performed in two different ways, i.e. solution or interfacial polycondensation (this latter gave rise to the formation of high molecular weight polymers). The LC features of the obtained polymer were investigated by using differential scanning calorimetry and polarized-light optical microscopy tests. First of all, the LC transition temperatures were correlated with the polymer structure: in particular, they were found to decrease with increasing the size of the cycloalkanone ring, due to the change in geometrical anisotropy of bis(benzylidene)cycloalkanone units. Furthermore, two kinds of photoreactions, namely photoisomerization and photo-crosslinking, were found to occur in these systems, according to the mobility of the polymer chains: in particular, photo-crosslinking dominated over photoisomerization (which disrupted the parallel stacking of the chromophores) when the photolysis was carried out below the glass transition temperature. Finally, as for the LC transition temperatures, also the photosensitivity of the polymers turned out to decrease with increasing the size of the cycloalkanone ring.

Hoyle and co-workers investigated the photopolymerization kinetics of three different cholesteric liquid-crystalline methacrylate monomers synthesized on purpose, differing for the spacer length and for the possible presence of a vinylic unsaturation in the mesogen unit (Hoyle and Watanabe 1994). Differential scanning calorimetry (DSC) and polarization microscopy were used for measuring the phase behavior and phase diagrams of the different products. Photo-DSC was further exploited for studying the photopolymerization kinetics, which was performed either completely in smectic phase or initially in an isotropic phase followed by conversion to a biphasic medium, by simply selecting suitable temperatures: thus it was possible to correlate the effect that LC organization and phase behavior exert on the propagation and termination rate constants. In particular, it was found that the values of termination rate constants were much lower in the smectic phase at low temperatures (45 °C) than in the isotropic phase at higher temperatures (80 °C). Furthermore, when photopolymerization started in an isotropic phase, all the three chosen LC monomers, at different temperatures, showed two rate acceleration processes that can be correlated with distinct changes in the medium composition and order; at the same time, the termination rate processes exhibited a strong decrease.

**Table 8.1** Synthesis of photocrosslinkable LC polymers

| Synthesized LC polymer or type of LC monomer | Adopted polymerization technique | Main findings | References |
|---|---|---|---|
| Photocrosslinkable main chain LC nematic polymers bearing bis (benzylidene) cycloalkanone units | Solution or interfacial polycondensation | – LC transition temperatures decrease with increasing the size of the cycloalkanone ring<br>– Photoisomerization and photo-crosslinking reactions occur<br>– Photo-crosslinking dominates over photoisomerization when photolysis is carried out below the glass transition temperature<br>– The photosensitivity of the polymers decreases with increasing the size of the cycloalkanone ring | Kishore and Kishore (1995) |
| Cholesteric liquid crystalline methacrylate monomers | Photopolymerization performed in smectic phase or initially in an isotropic phase, followed by conversion to a biphasic medium | – The values of termination rate constants are much lower in the smectic phase at low temperatures (45 °C) than in the isotropic phase at higher temperatures (80 °C)<br>– When photopolymerization started in an isotropic phase, all the three chosen LC monomers, at different temperatures, show two rate acceleration processes | Hoyle and Watanabe (1994) |
| Cholesteric liquid crystalline methacrylate monomers | Photopolymerization processes using high repetition rate laser pulses | – High molecular weight polymer primary radicals initiated by pulses previously applied with regard to the given pulse are not able to combine with polymer radical chains in the smectic phase<br>– The use of the laser pulse at a high repetition rate in the smectic phase does not promote the lowering of the molecular weight of the obtained polymer<br>– Enhanced termination is achieved when high repetition rates are applied in the isotropic phase | Hoyle and Chawla (1995) |

(continued)

**Table 8.1** (continued)

| Synthesized LC polymer or type of LC monomer | Adopted polymerization technique | Main findings | References |
|---|---|---|---|
| Styryloxy liquid crystalline monomer | Free radical and cationic photopolymerization performed with continuous or pulsed light sources | – No phase changes or segregation phenomena occurred up to ca. 20 % conversion of the UV-curable system, in a wide temperature range<br>– The LC phase does not affect the free radical polymerization rate<br>– Unlike cationic photopolymerization, the free-radical mechanism promotes rate accelerations at conversions related to compositions, for which changes in medium opacity are observed | Williamson et al. (1996) |
| Semifluorinated liquid crystalline diacrylate monomer | Radical photopolymerization | – It is possible to retain the smectic order originally shown by the monomer prior the UV-curing process<br>– Both termination and propagation rate processes are diffusion-controlled<br>– The polymer chain kinetic lifetime is significantly longer in the smectic phase as compared to the isotropic one | Hoyle et al. (1996) |
| Poly(benzylidene-ether)esters containing a photoreactive benzylidene chromophore in the main chain | Interfacial polycondensation | – The obtained homopolymer systems are stable up to 300 °C<br>– The copolymers start degrading at a slightly lower temperature (281 °C)<br>– The thermal stability of the homopolymers decreases as the acid methylene spacer length increases<br>– The obtained photocrosslinkable polymers show birefringence and stirred opalescence properties | Muthusamy et al. (2013) |

Pursuing this research, Hoyle and Chawla used one of the three cholesteryl methacrylate monomers previously tested for photopolymerization processes, using high repetition rate laser pulses: thus, it was possible to assess whether a smectic liquid crystalline phase may affect termination of primary radicals generated by a given laser pulse in a sequence of pulses with high molecular weight polymer radicals initiated by pulses previously applied with regard to the given pulse (Hoyle and Chawla 1995). It was demonstrated that these primary radicals are not able to combine with polymer radical chains in the smectic phase; in addition, the use of the laser pulse at a high repetition rate in the smectic phase did not promote the lowering of the molecular weight of the obtained polymer. On the other hand, enhanced termination was achieved when high repetition rates were applied in the isotropic phase.

Williamson and co-workers synthesized a styryloxy liquid crystalline monomer, which was photopolymerized with continuous or pulsed light sources, exploiting a free radical and cationic mechanism (Williamson et al. 1996). This work was aimed to study the effect of the cholesteric nematic mesophase on the photopolymerization kinetics: more specifically, the goal was to assess whether photopolymerization in a pure cholesteric nematic mesophase has higher or lower rates as compared to that proceeding in an isotropic phase. In addition, the effect of phase changes taking place during photopolymerization from initially isotropic or cholesteric nematic phases was investigated as far as a possible rate acceleration for cationic or free radical polymerization is considered. To this aim, a liquid crystalline styryloxy monomer bearing a cholesteric mesogens linked to the styryloxy monomer unit via a methylene chain was synthesized and used. On the basis of the chosen temperature, a homogeneous isotropic or cholesteric nematic phase was stable (i.e. no phase changes or segregation phenomena occurred) up to ca. 20 % conversion, in a wide temperature range. Furthermore, the LC phase did not affect the free radical polymerization rate. Finally, unlike cationic photopolymerization, the free-radical mechanism promoted rate accelerations at conversions related to compositions, for which changes in medium opacity were observed.

The same research group also studied the photopolymerization of a semifluorinated liquid crystalline diacrylate monomer, sterically hindered, bearing four ethylene oxide units and two fluorinated ether groups, as a function of temperature (Hoyle et al. 1996). The LC monomer was synthesized on purpose and investigated by exotherm measurements in both isotropic and smectic phases. The obtained results clearly indicated that it was possible to retain the smectic order originally shown by the monomer prior the UV-curing process. As far as the photopolymerization kinetics is considered, both termination and propagation rate processes were diffusion-controlled; in addition, the polymer chain kinetic lifetime was significantly longer in the smectic phase as compared to the isotropic one, presumably because of a strong decrease in termination rate with respect to propagation.

Very recently, Muthusamy and co-workers synthesized a series of poly (benzylidene-ether)esters containing a photoreactive benzylidene chromophore in the main chain, starting from 2,6-bis(4-hydroxy-3-methoxybenzylidene) cyclohexanone in the presence of different aliphatic and aromatic diacid chlorides, exploiting an interfacial polycondensation technique (Muthusamy et al. 2013). These photocrosslinkable polymers can be applied for the fabrication of negative photoresists. The thermal stability and the photochemical properties were thoroughly investigated. More specifically, the obtained homopolymer systems were stable up to 300 °C, while the copolymers started degrading at a slightly lower temperature (281 °C) because of the flexibility of bisphenol-A unit. Furthermore, the thermal stability of the homopolymers was lowered as the acid methylene spacer length increased. As revealed by differential scanning calorimetry, the aliphatic polyesters exhibited odd-even effect (Roviello and Sirigu 1982) and the aromatic polyesters containing terephthaloyl and isophthaloyl units showed higher isotropization temperature as compared to the aliphatic counterparts; in addition, most of the obtained photocrosslinkable polymers exhibited birefringence and stirred opalescence properties. Referring to the UV-curing process, UV spectral studies demonstrated that the polymers undergo photocrosslinking upon exposure to UV-radiation, without using any photoinitiator; in addition, the curing rate was found to strongly decrease with increasing the methylene chain length of the used acid spacer.

## 8.3 Effect of the LC Structure on the Photopolymerization Reaction, Morphology and Final Properties of the Obtained UV-Cured Networks

Table 8.2 summarizes the possible effects of the different liquid crystalline structures on the photopolymerization reaction, as well as on the morphology and final properties of the obtained UV-cured systems: this issues will be discussed in the followings.

Pogue and co-workers investigated the UV-curing process of two commercially-available diacrylate liquid crystalline monomers, having a similar chemical structure, but with a different MW and flexibility (Pogue et al. 1999). In particular, the effects of several cure conditions, including temperature, on the conversion and reaction rates of the two monomers were thoroughly investigated.

To this aim, differential photocalorimetry was used: it was found that each monomer exhibited a nematic mesophase at high temperature, which affected both the reaction rates and conversion. More specifically, the UV-curing of the liquid crystalline diacrylates showed a behavior, in terms of fractional conversion and heat flow, similar to traditional isotropic counterparts: indeed, the reactions entered a region of autoacceleration, but at very low monomer conversion, because of the rigidity of the networks, which was increased by the mesogenic structure of the monomers. After having achieved the maximum rate, the reactions entered a region of autodecelaration, which was ascribed to the inhibition of propagation exerted by the forming networks. Furthermore, the photopolymerization kinetics

**Table 8.2** Effects of the different liquid crystalline structures on the photopolymerization reaction, the morphology and final properties of the obtained UV-cured systems

| Investigated systems | Main findings | References |
|---|---|---|
| Commercially-available diacrylate liquid crystalline monomers | – The fractional conversion and heat flow of the photopolymerization process is not affected by the presence of the LC phases<br>– The photopolymerization kinetics strongly depends on the adopted temperature and is affected by the photoinitiator concentration | Pogue et al. (1999) |
| Methacrylate monomers having a Schiff-base structure | – Unlike in the crystalline phase, the polymerization processes occurs in both nematic and isotropic phases<br>– Irrespective of the used monomer, the conversion increases in polymerization processes performed in nematic phases, while it decreases in the isotropic phases with increasing the temperature<br>– The monomer conversion in photopolymerization reactions increases with time and is saturated at a later stage, as compared with the thermally-induced reactions<br>– The polymers obtained in the nematic phase show a very stable birefringence, regardless of the adopted polymerization method<br>– The crosslinked polymers obtained by UV-curing show a better orientation of mesogens with respect to those prepared by thermal polymerization | Ogiri et al. (1999) |
| Acrylate monomer bearing a cyanobiphenyl mesogen and an alkoxycyanobiphenyl compound bearing the same mesogen | – When the UV-curing process is performed in the LC phase, specific anisotropic phase-separated structures of LC polymer networks are observed for the blends having dissimilar mesogens<br>– When the UV-curing process is performed in the isotropic phase conditions, polygonal or continuous phase-separated structures of liquid crystalline polymer networks are obtained for the dissimilar mesogenic blends<br>– The ordering field of LC molecules promotes the formation of LC polymer networks having anisotropic morphologies and the long range as the same length scale of LC textures | Kihara et al. (2002) |

(continued)

**Table 8.2** (continued)

| Investigated systems | Main findings | References |
|---|---|---|
| Liquid crystalline blend composed of 4-(6-acryloyloxyhexyloxy) benzoic acid and 4-hexyloxy-4′-cyanobiphenyl | – Photopolymerization induces phase separation in (LC monomer/LC) blends, leading to the formation of fine fibers of LC polymers<br>– The obtained fibers are oriented and well-ordered according to the LC textures present when photopolymerization is carried out | Kihara and Miura (2005) |
| Lattice-patterned LC-polymer composites | – The extent of the phase separation between the used LC and prepolymers is strictly correlated with the difference in solubility between these two components<br>– The polymer wall thickness is directly related to the relative composition of the acrylate monomers in the multicomponent prepolymer blends | Sung et al. (2013) |
| Commercially-available liquid crystalline diacrylate monomers | – The UV-curing process is able to freeze the molecular anisotropy induced by the magnetic field<br>– The change in monomer orientation angle as a function of time follows an exponential law with two distinct time constants | Schultz and Chartoff (1998) |
| LC multifunctional dendrimers containing mesogenic azobenzene groups | – The presence of the crosslinking agent and the photoinitiator in the reactive mixture does not preclude the formation of the mesophase | Cervera-Procas et al. (2013) |
| Commercially-available monomer mixture containing a diacrylate azobenzene monomer and a diacrylate monomer | – The different topographical characteristics of the obtained LC networks are very stable to shape adaptations and topographical surface manipulation | McConney et al. (2013) |

was found to strongly depend on the adopted temperature: in particular, an increase in polymerization rate as the temperature increased below the nematic-to-isotropic transition temperature and a quick decrease in rate and conversion above this temperature were observed. These findings were ascribed to different factors including a reduced molecular order at high temperatures, as well as gelation phenomena below the glass transition temperature. Finally, the effects of photoinitiator type and concentration on the kinetic parameters were assessed: it was found that the polymerization rate is affected by the photoinitiator concentration (more specifically, the maximum reaction rates increased as the photoinitiator concentration was increased). A variance, no significant differences in the curing extent were found in samples containing different types of photoinitiators when they were UV-cured at the same temperature.

Ogiri et al. studied the effect of the structure of the nematic and isotropic phases, on the thermal and photo-induced polymerization behavior of monomers having a Schiff-base structure; furthermore, the orientation of the obtained polymers was thoroughly investigated (Ogiri et al. 1999). To this aim, methacrylate monomers differing for the end group only, but with different phase behaviors, were considered, in order to eliminate the effect of the molecular structure on the polymerization process. Unlike in the crystalline phase, it was possible to carry out the polymerization processes in both nematic and isotropic phases. In particular, irrespective of the used monomer, the conversion was found to increase in polymerization processes performed in nematic phases and to decrease in the isotropic phases with increasing the temperature. The former finding was ascribed to the viscosity decrease as temperature increases, which favors the addition of monomer molecules to the propagating polymer radicals, thus resulting in a higher conversion in the nematic phases. As far as photopolymerization in considered, the monomer conversion was found to increase with time and was saturated at a later stage, as compared with the thermally-induced reactions. The molecular alignment of the polymers was assessed through optical polarizing microscopy measurements. As expected, the polymers produced in the isotropic phase showed no birefringence irrespective of the type of reaction (i.e. thermal or photo-induced polymerization); surprisingly, the polymers obtained in the nematic phase exhibited birefringence (very stable and unchanged up to 200 °C), regardless of the adopted polymerization method. Finally, the crosslinked polymers obtained by UV-curing showed a better orientation of mesogens with respect to those prepared by thermal polymerization: this finding was ascribed to the crosslinking, which favors the alignment of the mesogens in a more stable way.

Kihara and co-workers investigated the morphology of liquid crystalline polymer networks prepared by photopolymerization of (LC monomer/LC) blends containing a suitable photoinitiator and a crosslinker (Kihara et al. 2002). One of the blends consisted of an acrylate monomer bearing a cyanobiphenyl mesogen (namely, 4-Cyano-4′-acryloyloxyhexyloxybiphenyl) and an alkoxycyanobiphenyl compound (namely, 4-cyano-4′-hexyloxybiphenyl), which bears the same mesogen. The other blend contained the same cyanobiphenyl mesogen and 4-alkoxybenzoic acid, which, in turn, has dissimilar mesogens.

As far as the blends having the same mesogen are concerned, photopolymerization was exploited for freezing the orientation order of the liquid crystalline textures. In particular, specific anisotropic phase-separated structures of liquid crystalline polymer networks were detected for the blends having dissimilar mesogens, when the UV-curing process was performed in the LC phase. On the contrary, by performing the photopolymerization in isotropic phase conditions, polygonal or continuous phase-separated structures of liquid crystalline polymer networks were obtained for the dissimilar mesogenic blends. Finally, the ordering field of LC molecules was able to promote the formation of LC polymer networks having anisotropic morphologies and the long range as the same length scale of LC textures.

Pursuing this research, Kihara and Miura used a UV-curing process in order to obtain a hydrogen-bonded liquid crystalline polymer, exploiting the phase separation induced by photopolymerization (Kihara and Miura 2005). Indeed, since a liquid crystalline monomer is generally miscible with a low-molar-mass LC solvent, their blends can show an LC phase. However, once the monomer is photopolymerized, the corresponding polymer is often insoluble in the blend and hence phase separates from the LC solvent (Kihara et al. 2002; Kihara and Miura 2004). As a consequence, the photopolymerization-induced phase separation in (LC monomer/LC) blends leads to the formation of fine fibers of LC polymers and networks. Because of the fabrication process, the obtained fibers are oriented and well-ordered according to the LC textures present when photopolymerization is carried out. Following this approach, Kihara and Miura photopolymerized a liquid crystalline blend composed of 4-(6-acryloyloxyhexyloxy)benzoic acid and 4-hexyloxy-4′-cyanobiphenyl, in the presence of small amounts of an inhibitor and photoinitiator. Aiming to clarify the factors determining the morphology of the obtained polymer, two irradiation temperatures were selected: one was in the LC temperature range of the polymer (70 °C), the other in the isotropic range (110 °C). The structures of the obtained polymers were investigated by optical microscopy and scanning electron microscopy (SEM). SEM images showed that the film obtained at the lower temperature consisted of randomly-extended fibers (diameter: about 1.0 μm) having some branches. On the contrary, the film prepared at the higher temperature was made of polymer particles (diameter: about 1.5 μm). These findings clearly indicate that for the obtainment of a fibrous morphology, the polymer needs to exhibit the LC phase immediately after the photopolymerization process. On the other hand, the LC ordering field existing before photopolymerization did not play a crucial role in forming a fibrous structure, although favored the macroscopic orientation and reduced the fiber branches.

Very recently, Sung and co-workers fabricated lattice-patterned LC-polymer composites, starting from different acrylate multicomponent prepolymers and exploiting the photoinduced phase separation mechanism (Sung et al. 2013). The obtained materials could be used in the fabrication of high-quality flexible displays. First of all, the relationship between the phase separation behavior of the lattice-patterned LC-polymer composites and the prepolymer composition was investigated; then, the interfacial properties between the liquid crystalline and the polymer walls were considered, assessing the surface anchoring effect of the lattice-structured polymer walls on the movement of the liquid crystalline molecules. It was found that the extent of the phase separation between LC and prepolymers is strictly correlated with the difference in solubility between LC and prepolymers. Furthermore, the polymer wall thickness turned out to be directly related to the relative composition of the acrylate monomers in the multicomponent prepolymer blends: in particular, the presence of a low surface energy fluorinated acrylate monomer in the blends induced a decrease of the interactions taking place in between the LC and polymer walls.

Schultz and Chartoff investigated the molecular orientation in two commercially-available liquid crystalline diacrylate monomers and the retention of the magnetically-induced orientation in photocrosslinked polymers obtained by a stereolithography process (Schultz and Chartoff 1998). Dielectric permittivity measurements were performed in order to evaluate the retention of orientation upon exposure to UV-radiation: the estimation of the order parameter as a function of temperature was made using an extrapolation method based on the isotropic transition temperature. Dielectric permittivity measurements were also carried out as a function of time after reorientation induced by an external magnetic field. It was found that the calculated change in monomer orientation angle as a function of time followed an exponential law with two distinct time constants. The faster of the two time regimes was ascribed to bulk reorientation and obeyed an Arrhenius type of temperature dependence. The slower time regime was attributed to inhibition phenomena due to either the presence of impurities or surface interactions with the electrodes. Finally, dynamic mechanical analyses clearly showed that the UV-curing process was able to freeze the molecular anisotropy induced by the magnetic field.

Very recently, Cervera-Procas and co-workers prepared and characterized polymer networks based on LC multifunctional dendrimers containing mesogenic azobenzene groups and obtained by thiol-yne photopolymerization (Cervera-Procas et al. 2013). Dendrimers are very interesting structures that have found potential applications in several fields related to nanoscience and nanotechnology, such as catalysis, sensing, drug delivery, gene therapy, light-amplifying systems, tissue engineering, ... (Fréchet and Tomalia 2001). The use of these dendritic structures in LC systems allowed avoiding problems derived from the macrosegregation of the component moieties. Furthermore, it was found that the presence of the crosslinking agent and the photoinitiator in the reactive mixture does not preclude the formation of the mesophase. Finally, free-standing films of the obtained LC polymer networks were further exposed to UV radiation, in order to assess their photoinduced deformation (*trans–cis* photoisomerization): the relationship between the deformation and the LC director orientation was found consistent in all the samples investigated.

Very recently, McConney and co-workers prepared complexly patterned photoresponsive liquid crystal polymer networks, starting from a monomer mixture containing a diacrylate azobenzene monomer and a diacrylate monomer (both were commercially-available), together with a suitable photoinitiator (McConney et al. 2013). Then, their photochemically-initiated mechanical responses, subsumed with the director field of a wide range of topological defects, was thoroughly investigated. In particular, the different topographical characteristics of the investigated LC networks were found to be very stable to shape adaptations and topographical surface manipulation and therefore they could be useful for the fabrication of flow control components, haptic displays, devices for optics/photonics and catalysts.

## 8.4 Recent Applications of UV-Cured Networks Containing Liquid Crystalline Phases

Some of the recent applications of UV-cured networks containing liquid crystalline phases are schematized in Fig. 8.4 and listed in Table 8.3: they will be detailed in this paragraph.

Malucelli and co-workers succeeded in preparing *shape memory main-chain nematic coatings*, by photopolymerizing liquid crystalline elastomers eventually subjected to uniaxial stress. To this aim, a mesogenic diglycidyl-terminated

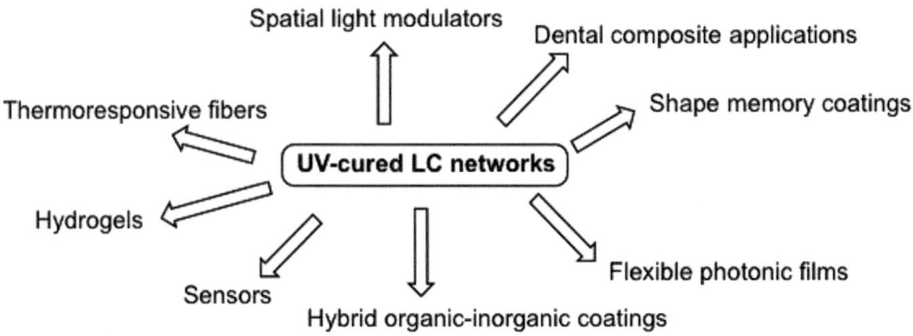


**Fig. 8.4** Some applications of UV-cured networks containing liquid crystalline phases

**Table 8.3** Recent applications of UV-cured networks containing liquid crystalline phases

| LC monomer system | Applications | References |
|---|---|---|
| Mesogenic diglycidyl-terminated monomer blended with its corresponding allyl homologous | Shape memory main-chain nematic coatings | Malucelli et al. (2009) |
| *N*-isopropylacrylamide and *N*,*N*-′-methylenebisacrylamide with a surfactant (polyoxyethylene cetyl ether) | Hydrogels | Forney and Guymon (2011) |
| Commercially-available nematic LC crosslinked with a UV-curable nematic diacrylate monomer, in the presence of an acrylate crown ether | Optical sensor films | Stroganov et al. (2012)) |
| LC oligomers bearing methacrylic groups | Hybrid organic–inorganic UV-cured coatings | Giamberini and Malucelli (2013) |
| UV-curable liquid crystalline urethane methacrylates | Dental resin-based composites | Buruiana et al. (2011) |
| LC biphenyl epoxy added to a conventional cycloaliphatic epoxy resin | Dental restoration materials | Hsu et al. (2012) |
| LC monomer bearing side-on attached mesogens and crosslinkable units | Thermoresponsive fibers | Fleischmann et al. (2014) |
| LC hosts mixed with a small amount of UV-curable difunctional monomer | Spatial light modulator | Sun and Wu (2014) |
| Induced chiral nematic mesophase mixed with a bifunctional UV-curable monomer | Imprinted flexible photonic cell | Chien and Liu (2014) |

monomer (namely, 4,4′-diglycidyloxybiphenyl) was blended with its corresponding allyl homologous and the obtained blends were thermally cured with decandioic acid, giving rise to the formation of LC nematic networks. Then, the networks containing 30 mol% of allyl-terminated monomer were put in contact with a liquid radical photoinitiator (namely, 2-hydroxy-2-methyl-1-phenyl propan-1-one) for different times and photocrosslinked either as such or during the application of an uniaxial stress (Malucelli et al. 2009). Finally, the UV-cured products were quenched to the glassy state, in order to retain the achieved orientation and their elastic response was assessed by thermo-mechanical analysis. First of all, the presence of the allyl-terminated monomer, acting as a "diluent", induced the perturbation of the ordered structure of the LC network: indeed, both clearing enthalpy and entropy decreased with increasing the amount of allyl-terminated monomer. Furthermore, only the liquid crystalline elastomers also subjected to photocrosslinking in the oriented state were found to fully recover their orientation after having been subjected to heating/cooling cycles. On the other hand, they also showed limited dimensional changes because of their high crosslinking degree. At variance, the materials not exposed to UV radiation during the application of an uniaxial stress were unable to retain the orientation achieved upon stretching and showed shape memory behavior only when high stresses were applied.

Forney and Guymon studied the thermal response and mechanical behavior of poly(*N*-isopropylacrylamide) *hydrogels* having a controlled nanostructure achieved by means of a photopolymerization process performed in a bicontinous cubic lyotropic liquid crystal template (Forney and Guymon 2011). This way, it was possible to tune the water uptake, the deswelling kinetics, the volume transition and the mechanical features of the obtained hydrogels. The lyotropic liquid crystalline template samples were synthesized by mixing the monomers (i.e. *N*-isopropylacrylamide and *N*,*N*′-methylenebisacrylamide) with a surfactant (polyoxyethylene cetyl ether) and a radical photoinitiator and exposing the mixtures to UV radiation. The obtained nanostructured hydrogels showed an increased water uptake at temperatures below LCST (lower critical solution temperature) and an increased rate and degree of deswelling above the LCST, as compared to the isotropic counterparts; it is worthy to note that all these enhancements occurred without compromising the mechanical integrity of the obtained networks.

Recently, Stroganov and co-workers prepared simple and reliable *optical sensor films* based on photocured polymer-stabilized cholesteric composites embedding crown ether fragments, which turned out to be optically sensitive to the complexation with potassium and barium ions (Stroganov et al. 2012). More specifically, a commercially-available nematic LC, which, in combination with a chiral dopant gave rise to the cholesteric structure, was crosslinked with a UV-curable nematic diacrylate monomer, in the presence of an acrylate crown ether and a suitable radical photoinitiator. The obtained UV-cured polymer-stabilized cholesteric LC composite films containing crown ether moieties exhibited interesting features as optical sensors: indeed, their complexation with potassium and barium ions led to blue shift of the selected light reflection. In addition, the increase of the concentrations of these ions promoted an increase of rate of change in the spectral features, as well as an enhancement in maximal values of the selective light reflection shift. The obtained sensor films were easy to handle and exhibited a good stability at room temperature for several days.

Recently, Giamberini and Malucelli succeeded in preparing hybrid organic–inorganic UV-cured films containing liquid crystalline nematic units, by exploiting a dual-cure process, which involves a UV-curing reaction of a methacrylic oligomer containing the LC phases and a subsequent thermal treatment that allows the formation of silica phases through hydrolysis/condensation of alkoxysilanes (sol–gel treatment). These *hybrid materials* containing LC units could be used for different high-tech applications, such as in electronic and biomedical fields, exploiting their mechanical, thermal and optical features (Giamberini and Malucelli 2013). More specifically, three different LC oligomers bearing methacrylic reactive groups were synthesized on purpose. Then, each LC oligomer was mixed with tetraethoxysilane (the silica precursor), in the presence of a coupling agent (namely, methacryloyl-oxypropyl-trimethoxysilane, which is capable to covalently link the inorganic phase with the polymer network) and a suitable radical photoinitiator. The mixtures were coated on glass substrates and exposed to the UV-radiation. Subsequently, the coatings were thermally-treated in order to promote the formation of the silica phase. All the obtained products showed a nematic liquid crystalline phase; their morphology was found to be strongly affected by the composition of the hybrid formulations and the presence of the LC units as well. Scanning electron microscopy and atomic force microscopy observations demonstrated that the distribution of the sol–gel derived silica on the two sides of the coating is asymmetric and promotes the formation of conical shapes, preferably on the side directly exposed to the UV-radiation. This peculiar topology was ascribed to the occurrence of interactions, taking place during the sol–gel reactions, between the liquid crystalline unit of the oligomers and the alkoxysilanes: as a consequence, silica was forced to grow in an acicular morphology. Furthermore, it is worthy to note that the presence of silica domains promoted the increase of the glass transition temperature of the polymer network, as well as the occurrence of segregation phenomena of the liquid crystalline phases.

Liquid crystalline urethane methacrylates were synthesized and characterized as co-monomers in UV-curable *dental resin-based composites* by Buruiana and coworkers (2011). The goal of this work was to demonstrate that the UV-curable liquid crystalline urethane methacrylates remain in a LC state under processing conditions of the dental composites, thus inducing a low volume shrinkage upon polymerization. To this aim, selected formulations based on an alkylammonium or cholesteryl urethane methacrylate and 2,2-bis[4-(2-hydroxy-3-methacryloyloxypropyl)-phenyl]propane or its derivative modified with urethane methacrylate groups, further diluted with triethyleneglycol dimethacrylate, were prepared and reinforced with 70 % zirconium silicate nanopowder. Differential scanning photo calorimetry was exploited for investigating the relationship between the LC monomer structure and the photopolymerization kinetics; furthermore, some other specific properties of the dental resin composites, like water sorption, water solubility, hydrophobicity and volume shrinkage were also considered. In the case of alkylammonium monomers, the LC mesophases turned out to be stable at room temperature, while the cholesteryl urethane methacrylates at a higher temperature (40 °C). Furthermore, the viscosity of the comonomers mixture affected both the rate and the degree of conversion of the UV-curing reaction: more specifically, the photopolymerization rate for liquid crystalline urethane methacrylates added to the standard dental formulation was higher as compared to that of mesogenic monomers alone because of the lower viscosity of the former formulations. Furthermore, the polymerization shrinkage and the water sorption were found to decrease in the presence of the liquid crystalline urethane methacrylates added to the standard dental formulation.

Pursuing this research topic, quite recently, Hsu and co-workers prepared a low-shrinkage resin-based UV-curable liquid crystalline epoxy nanocomposite, reinforced with silica nanoparticles, to be used as a *dental restoration material* (Hsu et al. 2012). To this aim, a liquid crystalline biphenyl epoxy synthesized on purpose was added to a conventional epoxy resin (namely cyclohexylmethyl-3,4-epoxycyclohexanecarboxylate), in the presence of a suitable cationic photoinitiator (namely, 4-octylphenyl phenyliodonium hexafluoroantimonate) and camphorquinone as photosensitizer. Before use, the silica nanoparticles were treated with a sizing agent (namely γ-glycidoxypropyltrimethoxysilane), in order to improve their dispersibility in the epoxy system and to create a strong interface with the polymer network after the UV-curing process. First of all, the presence of the liquid crystalline biphenyl epoxy resin in the dental formulation reduced the post-gelation shrinkage of this latter and lowered its water sorption as well. Furthermore, thanks to the LC domains, the obtained UV-cured nanocomposite system exhibited an improved hardness and flexural modulus, as well as a reduced coefficient of thermal expansion with regard to the conventional UV-cured epoxy resin. Finally, the restorative nanocomposites showed a good biocompatibility as assessed by cytotoxicity tests.

Very recently, Fleischmann and co-workers prepared *thermoresponsive fibers* made of a crosslinked liquid crystalline side-chain acrylate polymer: to this aim, a liquid crystalline monomer bearing side-on attached mesogens and crosslinkable units (namely, 4″-acryloyloxybutyl) 2,5-di(4-butyloxybenzoyloxy) benzoate) was synthesized and processed in solution in a microfluidic co-flowing device (Fleischmann et al. 2014). By exploiting the combination of the low interfacial tension between the dispersed polymer solution and the immiscible ambient fluid with the high viscosity of the polymer, it was possible to produce and stabilize a liquid jet without forcing its breaking into droplets, even at low shear rates. After the photopolymerization process, well-oriented crosslinked LC elastomeric fibers were obtained: they exhibited suitable features as actuators, since their orientability can be tuned by photocrosslinking at low temperatures within the liquid crystalline phase. In addition, the actuation could be optimized by exploiting two different strategies, i.e. using different copolymers bearing various amounts of crosslinkable groups and/or changing the degree of conversion of the UV-curable functionalities (by varying the UV exposure time). In particular, contractions of several hundred percent (up to 500 %) were achieved by the LC elastomeric fibers upon the nematic-isotropic phase transition. In addition, a high crosslinking promoted the formation of non-contracting fibers, because of the poor flexibility of the obtained polymer networks to rearrange upon the nematic to isotropic phase transition; in a similar way, contraction was found to lower when the polymers were exposed to UV-radiation for a very long time.

Referring to photonic and display applications, very recently Sun and Wu succeeded in fabricating a polymer network liquid crystal *spatial light modulator*, which is a useful electro-optic device that allows modulating the intensity, phase and polarization state of an incident radiation (Sun and Wu 2014). This is possible because this device shows such good characteristics as broad bandwidth, low operation voltage and high birefringence. To this aim, the authors mixed liquid crystalline hosts with a small amount of UV-curable difunctional monomer, in the presence of a suitable photoinitiator. Then, the mixture was filled into a LC cell with homogeneous alignment; upon exposure to UV-radiation, polymer network crosslinking occurred. The UV-curable monomer used for the fabrication of the polymer network liquid crystal device was found to play an important role in determining the LC domain size and thus the achieved low operating voltage.

Very recently, Chien and Liu fabricated an imprinted *flexible photonic cell* used as a new photonic bandgap, by carrying out multiple photocrosslinking processes in an induced chiral nematic mesophase, utilizing flexible poly(ethylene terephtalate) films as substrates (Chien and Liu 2014). More specifically, a homogeneous mixture containing a nematic LC, a chiral dopant, a bifunctional UV-curable monomer (namely, 4,4′-bis-(6-(acryloyloxy)-hexyloxy)biphenyl) and a suitable photoinitiator (1-Hydroxy-cyclohexyl-phenyl-ketone) was prepared in order to have a cholesteric liquid crystalline phase showing helical molecular orientations. The mixture was filled into a poly(ethylene terephthalate) cell (thickness: 12 μm) at room temperature by capillarity. Before the UV-curing process, a photomask was placed on the top substrate, thus ensuring a gradient profile of the monomer concentration, moving from the non-irradiated area to the irradiated one, so that a diffusion process of the monomer was promoted. The UV-curing/diffusion process was replicated eight times and then the LC matrix and unreacted monomers were extracted, using appropriate solvents: this way, after drying, it was possible to obtain a well-imprinted cholesteric LC polymer matrix. Thanks to its flexibility, the fabricated cell could undergo uniaxial thermal stretching: by stretching in different lengths, it was possible to tune the reflection color of the cell.

## 8.5 Conclusions and Future Perspective

In this chapter, the feasibility of performing UV-curing reactions on suitable liquid crystalline monomers and oligomers has been thoroughly discussed. Indeed, the synthesis of liquid-crystalline monomers bearing reactive groups that can undergo the UV-curing process allows obtaining useful photocrosslinked polymers, which still contain LC phases and therefore can give rise to anisotropic networks (like LC thermosets and LC elastomers). These latter can be successfully exploited for the building up of both structural and functional materials, having a well defined and controlled orientation and therefore suitable for high-tech applications.

The significant number of peer-reviewed papers that have been published in the literature so far undoubtedly suggests that this specific research field still deserves further investigation. In addition, it seems that the gap between the research performed at a laboratory scale on this topic and the desirable scale-up of the most promising and valuable fabricated prototypical components/devices may be reduced in the forthcoming years, thus promoting a real industrialization of LC-based materials and technologies.

## References

Allcock HR, Cameron GC (1994) Synthesis of photocrosslinkable chalcone-bearing polyphosphazenes. Macromolecules 27:3131–3135

Andersson H, Gedde UW, Hult A (1992) Preparation of ordered, crosslinked and thermally stable liquid crystalline poly(vinylether) films. Polymer 33:4014–4018

Broer DJ (1996) Networks formed by photoinitiated chain cross-linking. In: Crawford GP, Zumer S (eds) Liquid crystals in complex geometries, Taylor and Francis, London, pp 239–255

Broer DJ, Boven J, Mol GN, Challa G (1989) In-situ photopolymerization of oriented liquid-crystalline acrylates, 3. Oriented polymer networks from a mesogenic diacrylate. Makromol Chem 190:2255–2268

Broer DJ, Mol GN, Challa G (1991) In-situ photopolymerization of oriented liquid-crystalline acrylates, 5. Influence of the alkylene spacer on the properties of the mesogenic monomers and the formation and properties of oriented polymer networks. Makromol Chem 192:59–74

Buruiana T, Melinte V, Costin G, Buruiana EC (2011) Synthesis and properties of liquid crystalline urethane methacrylates for dental composite applications. J Polym Sci A Polym Chem 49:2615–2626

Cervera-Procas R, Sànchez-Somolinos C, Serrano JL, Omenat A (2013) A polymer network prepared by the tyol-yne photocrosslinking of a liquid crystalline dendrimer. Macromol Rapid Commun 34:498–503

Chien CC, Liu JH (2014) Optical behaviors of flexible photonic films via the developed multiple UV-exposed fabrications. Macromol Rapid Commun 35:1185–1190

Crivello JV (1983) Photoinitiated cationic polymerization. Annu Rev Mater Sci 13:173–190

de Gennes PG (1975) The physics of liquid crystals, Clarendon, Oxford

Decker C (1996) Photoinitiated crosslinking polymerization. Prog Polym Sci 21:593–650

Decker C (2002) Kinetic study and new applications of UV radiation curing. Macromol Rapid Commun 23:1067–1093

Fleischmann EK, Forst FR, Zentel R (2014) Liquid-crystalline elastomer fibers prepared in a microfluidic device. Macromol Chem Phys 215:1004–1011

Flory PJ (1956) Phase equilibria in solutions of rod-like particles. Proc R Soc Lond A 234:73–89

Forney BS, Guymon CA (2011) Fast deswelling kinetics of nanostructured poly (N-isopropylacrylamide) photopolymerized in a lyotropic liquid crystal template. Macromol Rapid Commun 32:765–769

Fréchet JMJ, Tomalia DA (2001) Dendrimers and other dendritic polymers. Wiley, Chichester

Giamberini M, Malucelli G (2013) Hybrid organic-inorganic UV-cured films containing liquid-crystalline units. Thin Solid Films 548:150–156

Guymon CA, Hoggan EN, Clark NA, Rieker TP, Walba DM, Bowman CN (1997) Effect of monomers structure on their organization and polymerisation in a smectic liquid crystal. Science 275:57–59

He L, Zhang S, Jin S, Qi Z (1995) In-situ photopolymerization of an oriented chiral liquid crystal acrylate in an electric field. Polym Bull 34:7–12

Hellermark C, Gedde UW, Hult A (1992) Synthesis and characterization of poly(11-(4′-cyano-trans-4-stilbenyloxy)undecanyl vinyl ether). Polym Bull 28:267–274

Hoyle CE, Chawla CP (1995) Effect of repetition rate on the pulsed-laser-initiated polymerization of a liquid crystalline monomer. Macromolecules 28:1946–1951

Hoyle CE, Watanabe T (1994) Kinetics of polymerization of liquid-crystalline monomers: an exotherm and light scattering analysis. Macromolecules 27:3790–3796

Hoyle CE, Kang D, Jariwala C, Griffin AC (1993) Efficient polymerization of a semi-fluorinated liquid crystalline methacrylate. Polymer 34:3070–3075

Hoyle CE, Mathias LJ, Jariwala C, Sheng D (1996) Photopolymerization of a semifluorinated difunctional liquid crystalline monomer in a smectic phase. Macromolecules 29:3182–3187

Hsu SH, Chen RS, Chang YL, Chen MH, Cheng KC, Su WF (2012) Biphenyl liquid crystalline epoxy resina s a low-shrinkage resin-based dental restorative nanocomposite. Acta Biomater 8:4151–4161

Kaniappan K, Murugavel SC (2009) Photocrosslinkable phosphorus containing homo- and copolyesters: synthesis, characterization, and photosensitive properties. J Appl Polym Sci 111:1606–1614

Kawatsuki N, Goto K, Yamamoto T (2001) Photoinduced anisotropy and photoalignment of nematic liquid crystals by a novel polymer liquid crystal with a coumarin-containing side group. Liq Cryst 28:1171–1176

Kihara H, Miura T (2004) Morphological self-control of a phase-separated polymer during photopolymerization in a liquid-crystalline medium. Polymer 45:6357–6363

Kihara H, Miura T (2005) Morphology of a hydrogen-bonded LC polymer prepared by photopolymerization-induced phase separation under an isotropic phase. Polymer 46:10378–10382

Kihara H, Miura T, Kishi R (2002) Morphological studies of LC polymer networks prepared by photopolymerization of (LC monomer/LC) blends. Polymer 43:4523–4530

Kishore G, Kishore K (1995) Synthesis and characterization of photo-crosslinkable main-chain liquid-crystalline polymers containing bis(benzylidene)cycloalkanone units. Polymer 36:1903–1910

Malucelli G, Iannelli P, Giamberini M (2009) Freezing the orientation of a nematic stretched elastomer by photocrosslinking. Polymer 50:1948–1956

McConney ME, Martinez A, Tondiglia VP, Lee KM, Langley D, Smalyukh II, White TJ (2013) Topography from topology: photoinduced surface features generated in liquid crystal polymer networks. Adv Mater 25:5880–5885

Muthusamy A, Balaji K, Murugavel SC (2013) Synthesis, thermal, and photocrosslinking studies of thermotropic liquid crystalline poly(benzylidene-ether)esters containing α, β-unsaturated ketone moiety in the main chain. J Polym Sci A Polym Chem 51:1707–1715

Ogiri S, Ikeda M, Kanazawa A, Shiono T, Ikeda T (1999) Polymerization of liquid-crystalline monomers having a Schiff-base structure. Polymer 40:2145–2150

Percec V, Kawasumi M (1991) Liquid crystalline polyethers based on conformational isomerism. 18. Polyethers based on a combined mesogenic unit containing rigid and flexible groups: 1-(4-hydroxy-4′-biphenyl)-2-(4-hydroxyphenyl)butane. Macromolecules 24:6318–6324

Percec V, Tomazos D (1992) Molecular engineering of side-chain liquid-crystalline polymers by living cationic polymerization. Adv Mater 4:548–561

Percec V, Tsuda Y (1990) Liquid-crystalline polyethers based on conformational isomerism. 10. Synthesis and determination of the virtual mesophases of polyethers based on 1-(4-hydroxyphenyl)-2-(2-methyl-4-hydroxyphenyl)ethane and .alpha.,.omega.-dibromoalkanes containing from 17 to 20 methylene unit. Macromolecules 23:3509–3520

Percec V, Yourd R (1989) Liquid crystalline polyethers based on conformational isomerism. 2. Thermotropic polyethers and copolyethers based on 1-(4-hydroxyphenyl)-2-(2-methyl-4-hydroxyphenyl)ethane and flexible spacers containing odd number of methylene units. Macromolecules 22:524–537

Percec V, Shaffer TD, Nava H (1984) Functional polymers and sequential copolymers by phase transfer catalysis. 10. Polyethers of mesogenic bisphenols: a new class of main-chain liquid crystalline polymers. J Polym Sci Polym Lett Ed 22:637–647

Percec V, Nava H, Jonsson H (1987) Functional polymers and sequential copolymers by phase transfer catalysis. 24. The influence of molecular weight on the thermotropic properties of a random copolyether based on 1,5-dibromopentane, 1,7-dibromoheptane, and 4,4′-dihydroxy-α-methylstilbene. J Polym Sci A Polym Chem 25:1943–1965

Pogue RT, Ullet JS, Chartoff RP (1999) Determination of the effects of cure conditions on the photopolymerization of liquid crystalline monomers using differential photo-calorimetry. Thermochim Acta 339:21–27

Ravikrishnan A, Sudhakar P, Kannan P (2010) Stilbene-based liquid crystalline and photocrosslinkable polynaphthylphosphate esters. J Mater Sci 45:435–442

Roviello A, Sirigu A (1975) Mesophasic structures in polymers. A preliminary account on the mesophases of some poly-alkanoates of p, p′-di-hydroxy-α, α′-di-methyl benzalazine. J Polym Sci C Polym Lett 13:455–463

Roviello A, Sirigu A (1982) Odd–even effect in polymeric liquid crystals. Makromol Chem 183:895–904

Sakthivel P, Kannan P (2006) Thermotropic liquid crystalline-cum-photocrosslinkable poly (benzylidene arylphosphate ester)s containing the cyclopentanone moiety. Liq Cryst 33:341–351

Schultz JW, Chartoff RP (1998) Photopolymerization of nematic liquid crystal monomers for structural applications: molecular order and orientation dynamics. Polymer 2:319–325

Shaffer TD, Percec V (1985) Functional polymers and sequential copolymers by phase transfer catalysis, 13. Thermotropic copolyethers from 4,4′-dihydroxy-α-methylstilbene and α, ω-dibromoalkanes. Makromol Chem Rapid Commun 6:97–104

Stroganov V, Ryabchun A, Bobrovsky A, Shibaev V (2012) A novel type of crown ether-containing metal ions optical sensors based on polymer-stabilized cholesteric liquid crystalline films. Macromol Rapid Commun 33:1875–1881

Sun J, Wu TS (2014) Recent advances in polymer network liquid crystal spatial light modulators. J Polym Sci B Polym Phys 52:183–192

Sung SJ, Jung EA, Sim K, Kim DH, Cho KY (2013) Phase separation structure and interfacial properties of lattice-patterned liquid crystal-polymer composites prepared from multicomponent prepolymers. Polym Int 63:214–220

Ullett JS, Chartoff RP, Schultz JW, Bhatt JC, Dotrong M, Pogue RT (1996) Solid freeform fabrication proceedings, pp 471–479

Williamson SE, Kang D, Hoyle CE (1996) Phase effects on the polymerization of a styloxy cholesteric liquid crystalline monomer. Macromolecules 29:8656–8665

# Chapter 9
# Liquid Crystal Diffraction Gratings Using Photocrosslinkable Liquid Crystalline Polymer Films as Alignment Layers

**Tomoyuki Sasaki, Kohei Noda, Hiroshi Ono, and Nobuhiro Kawatsuki**

## 9.1 Introduction

Liquid crystals (LCs) are excellent materials for various optical applications because they are highly transparent in the visible region, have a large optical anisotropy, and respond to external fields (Chandrasekhar 1992). Currently, LC displays are widely used in diverse information display devices (Yeh and Gu 1999). Additionally, with the increasing need for advanced optical signal processing technologies, LCs are of interest for use in for diffractive optical elements (Scharf 2007). Sato (1979) investigated LC optical lenses fabricated from rubbed glass substrates and glass lenses and demonstrated that their focal lengths can be controlled by applying electric or magnetic fields to the cells. Chen et al. (1995) proposed an electro-optically controlled LC diffraction grating consisting of two alternating stripes. Subacius et al. (1997a, b) fabricated Raman–Nath-type diffraction gratings formed from cholesteric LCs and uni-directionally rubbed alignment layers. They demonstrated that the periods of the gratings were varied by applying an electric field. The alignment treatment of LC molecules is very important for high functionalization of their properties for use in diffractive optical elements. Various alignment methods have been investigated to form orientation patterns of the LC molecules. Honma and Nose (2003, 2004, 2012a, b, c) fabricated various LC diffraction gratings using microrubbing and characterized their unique diffraction properties based on the director distribution in the cells. Gwag et al. (2007) focused

T. Sasaki • K. Noda • H. Ono (✉)
Department of Electrical Engineering, Nagaoka University of Technology,
1603-1, Kamitomioka, Nagaoka, Niigata 940-2188, Japan
e-mail: onoh@nagaokaut.ac.jp

N. Kawatsuki
Department of Applied Chemistry, Graduate School of Engineering,
University of Hyogo, 2167 Shosha, Himeji, Hyogo 671-2280, Japan

 
V.K. Thakur, M.R. Kessler (eds.), *Liquid Crystalline Polymers*,
DOI 10.1007/978-3-319-20270-9_9

on the use of groove-induced surface anchoring energy and reported bistable LC devices fabricated using nanoimprinting lithography.

Photoalignment is a useful technique for fabricating LC diffractive optical elements because periodic orientation distributions of LC molecules can be easily obtained using optical methods such as exposure through a photomask, laser beam drawing, and holographic recording. Poly(vinyl cinnamate) (PVCi) is a well-known photoalignment material (Schadt et al. 1992). In PVCi films, the dominant photoreaction that is caused by exposure to ultraviolet (UV) light is the (2 + 2) cycloaddition of the cinnamoyl groups on the cinnamic acid side chains. Therefore, anisotropic photocrosslinking reactions are induced in the PVCi films by irradiating them with linearly polarized (LP) UV light. The exposed PVCi films exhibit negative optical anisotropy and the slow axis is perpendicular to the polarization direction (Barachevsky 1991; Schadt et al. 1992). While the typical birefringence of PVCi films exposed with LP UV light is relatively small (~0.01), the exposed PVCi films can be used in alignment layers for low-molecular-weight LCs. Schadt et al. investigated the photoalignment effect of PVCi films and also fabricated twisted nematic (TN) and supertwisted nematic LC displays using the PVCi films for the alignment layers. Subsequently, various photocrosslinkable polymers have been developed for use in the photoalignment of LCs (Ichimura et al. 1997; Kawatsuki et al. 1997a; Obi et al. 1999; Duffy et al. 2001; Seki et al. 2013).

Azobenzene-containing polymers (azopolymers) also exhibit an axis-selective photoreaction. Azobenzene molecules orient when exposed to LP UV light through a *trans–cis–trans* photoisomerization reaction (Kumar and Neckers 1989; Natansohn and Rochon 2002). This molecular orientation changes the dielectric tensor of the film. This means that information about the polarization of the light that the molecules were exposed to can be obtained. Therefore, azopolymers have been widely studied as recording media for polarization holography (Nikolova and Todorov 1984; Todorov et al. 1985; Ebralidze and Ebralidze 1992; Labarthet et al. 1999; Nikolova and Ramanujam 2009; Fabbri et al. 2011; Kawatsuki et al. 2012; Yadavalli and Santer 2013). Alternatively, an azopolymer (a silicone polyimide copolymer doped with a diazodiamine dye) was used by Gibbons et al. (1991) to photoalign a low-molecular-weight nematic LC. The LC molecules in the cell that were irradiated with a linearly polarized (LP) green laser beam were oriented perpendicular to the polarization azimuth of the laser beam. An LC grating with a period of 10 μm by holographic recording was also fabricated. Ichimura et al. (1988) and Ichimura (2000) reported a reversible change in the alignment mode of nematic LCs that was photochemically regulated using command surfaces modified with an azobenzene monolayer. They also showed that the pretilt angles of azobenzene-containing photoalignment layers can be controlled using the incident angle of the LP light. These studies demonstrated that azopolymers are useful for photoalignment of LCs. However, azobenzene molecules absorb in the visible region, which could be problematic for optical applications.

We have previously studied photoreactions of photocrosslinkable liquid crystalline polymers (PLCPs) comprised of cinnamate side groups (Kawatsuki et al. 1997a, b, 1998, 1999, 2000, 2001, 2002). A biphenyl mesogenic group and a photoreactive cinnamoyl group are linked with a polymethacrylate main chain with alkylene spacers. PLCPs show nematic liquid crystalline behavior at a particular transition temperature and exhibit an axis-selective photocycloaddition because the absorption axis is parallel to the mesogenic side chains. When a PLCP film is exposed to LP UV light, the side chains are crosslinked axis-selectively and a small optical anisotropy is induced in the film because of the photocycloaddition. Because the mobility of the crosslinked groups is much lower than that of the non-reacted mesogenic side chains, annealing the exposed PLCP film at the LC temperature induces a particular molecular orientation. The optical anisotropy of the exposed PLCP film can be enhanced by annealing. The enhanced optical anisotropy is larger than 0.2. A three-dimensional molecular orientation is also achieved in PLCP films by irradiating with LP UV light at oblique incidence. Additionally, PLCP films are highly transparent in the visible region and are thermally stabile. Therefore, we believe that our PLCPs are suitable for various birefringent optical applications.

We have previously studied polarization holographic gratings induced in PLCP films (Ono et al. 2003a, b). The polarization gratings were thermally stable and exhibited high diffraction efficiencies. This study demonstrated that PLCPs are also suitable as recording media for polarization holography. One of the most important characteristics of PLCP films is that the direction of the molecular orientation can be controlled using the polarization state and the exposure energy of the incident light beam. When the exposure energy is relatively low, the mesogenic molecules are oriented perpendicular to the polarization azimuth of LP UV light. Alternatively, the molecules are oriented parallel to the polarization azimuth when the exposure energy is sufficiently high. This self-organized molecular orientation originates in the liquid crystalline nature of PLCPs. We fabricated polarization diffraction gratings consisting of periodic orientation distributions of the PLCP molecules with a two-step LP UV exposure (Ono et al. 2004; Emoto et al. 2005). Additionally, we recorded angle-multiplexed polarization holograms in PLCP films (Ono et al. 2003a, b; Sasaki et al. 2006, 2014). Because optical anisotropy is based on the molecular orientation in exposed PLCP films after annealing, there is a low amount of crosstalk between multiplexed holograms. These studies demonstrated that various patterns of molecular orientation distribution are formed in PLCP films.

PLCP films with uniaxial molecular orientation can also function as alignment layers for low-molecular-weight nematic LCs. We fabricated planar and TN LC cells using photoaligned PLCP films for the alignment layers (Kawatsuki et al. 1997a). We also confirmed that the pretilt angles of LC cells were controlled by the incident angle of LP UV light (Kawatsuki et al. 2007). We believe our PLCP films are suitable media for fabricating LC diffractive optical elements because of their good optical properties mentioned above. LC diffraction gratings made from photoalignment layers have attracted attention for many years because photoalignment techniques realize manipulation of molecular orientation in fine regions (Park et al. 2003; Yu et al. 2005; Crawford et al. 2005; Batalioto et al. 2005; Wu et al. 2006; Choi et al. 2006; Provenzano et al. 2007; Hu et al. 2012; Sun et al. 2013). We aim to demonstrate the effectiveness of PLCP photoalignment layers for use in LC diffraction gratings.

Here, we report LC diffraction gratings fabricated using photoregulated PLCP films for alignment layers. The LC gratings were formed using three different methods, a two-step exposure with a photomask (Sasaki et al. 2005, 2007), polarization holographic recording (Ono et al. 2007, 2009), and a one-step polarizer-rotation exposure (Kuzuwata et al. 2012; Sasaki et al. 2013). These fabrication procedures and diffraction properties of the resultant LC gratings demonstrate the utility of PLCP films and photoalignment techniques. The diffraction properties of the LC gratings, especially polarization conversion properties, are theoretically characterized based on the alignment patterns in the PLCP films and the resultant orientation distributions of the LC molecules in the cells.

## 9.2 Two-Step Exposure Method Using a Photomask

We fabricated an LC grating using a two-step exposure method with a line-and-space patterned photomask. The PLCP consisted of a methacrylate group with mesogenic 4-(4-methoxycinnamoyloxy)biphenyl side groups and nonmesogenic *N*-phenylmaleimide (**P1**). The chemical structure and synthesis procedure of **P1** was previously reported (Kawatsuki et al. 2004). **P1** has a nematic LC phase between 89 and 298 °C. Figure 9.1a shows the fabrication process and the structure

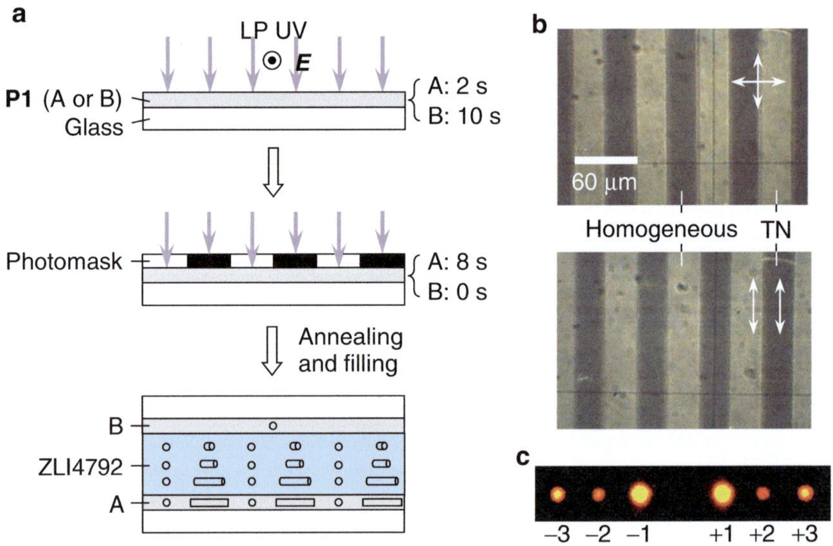


**Fig. 9.1** The structure of the LC grating that was fabricated using a two-step exposure method and a diffraction pattern of the LC grating: (**a**) the fabrication process. ***E*** is the polarization direction of the LP UV light. The *rods* and *circles* show the direction of the orientation; (**b**) POM images. The *arrows* show the transmission axes of the polarizer and analyzer; (**c**) a diffraction pattern. The diffraction pattern was observed by removing the 0th order beam

of the LC grating. The LC grating consisted of two quartz substrates that were coated on the inside with **P1** films and low-molecular-weight nematic LC (ZLI4792, Merck). **P1** films were prepared by spin coating a methylene chloride solution of **P1** onto cleaned quartz substrates. The thickness of the resultant films was 0.2 μm. An ultrahigh-pressure Hg lamp was used as a light source for LP UV exposure in combination with a Glan-Taylor prism to act as a linear polarizer and a wavelength filter with a cutoff wavelength below 290 nm. The intensity of the LP UV light was 150 mW/cm$^2$ at a wavelength of 365 nm. One **P1** film (film A) was exposed to the LP UV light for 2 s and then for 8 s through a photomask with a line width of 30 μm and a space width of 30 μm. The polarization azimuth of the LP UV light was set to be perpendicular to the grating vector of the photomask. The other **P1** film (film B) was exposed to the LP UV light for 10 s without using the photomask. The two films were annealed at the LC temperature of 180 °C for 10 min and then cooled to room temperature. The molecular reorientation was induced perpendicular or parallel to the polarization azimuth of the LP UV light for the regions in which the total irradiation time was either 2 or 10 s. An empty cell was fabricated using the two substrates and 12-μm-thick spacers. The orientation directions of films were set to be mutually perpendicular at the line regions and parallel at the space regions. The nematic LC doped with a small amount of chiral dopant was filled into the empty cell using capillary action. Figure 9.1b shows polarizing optical microscope (POM) images of the resultant LC cell. From the POM images, we infer that the LC cell consists of homogeneously aligned regions and TN alignment regions as seen in Fig. 9.1b. Diffraction properties of the LC grating were investigated using a LP He–Ne laser with a wavelength of 633 nm at an incidence that was normal to the substrates. The intensities and polarization states of the diffracted beams were measured using an optical power meter and a Glan-Thompson prism. Figure 9.1c shows a typical diffraction pattern that was observed by removing the 0th order beam to improve the contrast. As seen in Fig. 9.1c, the intensities of the even (second) order diffraction beams were relatively low. This indicates that the phase distribution in the LC grating is close to a rectangle.

We characterized the diffraction properties of the LC grating based on the Jones matrix method and Fraunhofer diffraction theory. By considering that the nematic LC layer consists of $M$ plates with a thickness of $d_m = d/M$, the grating vector is parallel to the $x$-direction, the substrates are parallel to the $xy$-plane, and that the substrate normal is parallel to $z$-direction, the Jones matrix of the LC grating is

$$\boldsymbol{W}(x) = \boldsymbol{W}_{\mathrm{o}} \cdot \left( \prod_{m=1}^{M} \boldsymbol{W}_m \right) \cdot \boldsymbol{W}_{\mathrm{i}}, \tag{9.1}$$

where $d$ is the cell thickness, $M$ is a substantially large number, $\boldsymbol{W}_m$ is the Jones matrix of the $m$th thin plate, and $\boldsymbol{W}_{\mathrm{i}}$ and $\boldsymbol{W}_{\mathrm{o}}$ are the Jones matrices for the PLCP films at the input and output sides, respectively. Defining the rotation matrix in the $xy$-plane as $\boldsymbol{R}$, these Jones matrices are

$$\boldsymbol{W}_m(x,z) = \boldsymbol{R}(-\phi)\cdot\begin{bmatrix} e^{-i\Gamma_m} & 0 \\ 0 & e^{i\Gamma_m}\end{bmatrix}\cdot\boldsymbol{R}(\phi), \tag{9.2}$$

$$\boldsymbol{W}_\mathrm{i}(x) = \boldsymbol{R}(-\phi_\mathrm{i})\cdot\begin{bmatrix} e^{-i\gamma} & 0 \\ 0 & e^{i\gamma}\end{bmatrix}\cdot\boldsymbol{R}(\phi_\mathrm{i}), \tag{9.3}$$

$$\boldsymbol{W}_\mathrm{o}(x) = \boldsymbol{R}(-\phi_\mathrm{o})\cdot\begin{bmatrix} e^{-i\gamma} & 0 \\ 0 & e^{i\gamma}\end{bmatrix}\cdot\boldsymbol{R}(\phi_\mathrm{o}), \tag{9.4}$$

$$\Gamma_m = \frac{\pi d_m \Delta n}{\lambda}, \tag{9.5}$$

$$\gamma = \frac{\pi d_\mathrm{P} \Delta n_\mathrm{P}}{\lambda}, \tag{9.6}$$

where $\phi(x,z)$ is the angle between the $x$-axis and the director of the nematic LC, $z$ is defined in $[0,d]$, $\phi_\mathrm{i}(x)$ and $\phi_\mathrm{o}(x)$ are the angles between the $x$-axis and the director of the PLCP films of input and output sides, respectively, $\Delta n$ is the optical anisotropy of the nematic LC, $\lambda$ is the wavelength of the probe beam, $d_\mathrm{P}$ is the thickness of the PLCP film, and $\Delta n_\mathrm{P}$ is the optical anisotropy of the PLCP film. Here, we assumed that there were no pretilt angles in the LC layer and the PLCP films. If the grating is a thin grating (i.e., a Raman–Nath-type grating), the Jones vector of the probe beam transmitted through the LC grating is

$$\boldsymbol{E}_\mathrm{o}(x) = \boldsymbol{W}\cdot\boldsymbol{E}_\mathrm{i} \equiv \begin{bmatrix} E_x \\ E_y \end{bmatrix}, \tag{9.7}$$

where $\boldsymbol{E}_\mathrm{i}$ is the Jones vector of the probe beam. Because there is no interaction between $E_x$ and $E_y$ in isotropic free space, the diffracted beams in the far field can be calculated based on Fraunhofer diffraction theory. The Jones vector of the diffracted beams is given by

$$\widetilde{\boldsymbol{E}}(\nu_x) = \begin{bmatrix} F(E_x) \\ F(E_y) \end{bmatrix}, \tag{9.8}$$

where $\nu_x$ is the spatial frequency component for $x$ and $F(g)$ represents the Fourier transformation of $g(x)$.

For simplicity, we assumed that the LC cell consisted of homogeneous alignment regions and 90° TN alignment regions as shown in Fig. 9.1b. We applied the Mauguin condition for the TN alignment regions because the retardation in the LC layer ($2M\Gamma_m$) was larger than the twist angle ($\pi/2$) in this experiment. When the probe beam was incident on film A, the Jones matrix of the LC grating is

$$W(x) = \begin{bmatrix} e^{i\gamma} & 0 \\ 0 & e^{-i\gamma} \end{bmatrix} \begin{bmatrix} e^{iM\Gamma_m} & 0 \\ 0 & e^{-iM\Gamma_m} \end{bmatrix} \begin{bmatrix} e^{i\gamma} & 0 \\ 0 & e^{-i\gamma} \end{bmatrix} \quad (0 \leq x < \Lambda/2), \quad (9.9a)$$

$$W(x) = \begin{bmatrix} e^{i\gamma} & 0 \\ 0 & e^{-i\gamma} \end{bmatrix} \left( R\left(-\frac{\pi}{2}\right) \begin{bmatrix} e^{-iM\Gamma_m} & 0 \\ 0 & e^{iM\Gamma_m} \end{bmatrix} \right) \begin{bmatrix} e^{-i\gamma} & 0 \\ 0 & e^{i\gamma} \end{bmatrix} \quad (\Lambda/2 \leq x < \Lambda), \quad (9.9b)$$

where $\Lambda$ is the grating pitch. When the probe beam was incident on film B, the Jones matrix of the LC grating is

$$W(x) = \begin{bmatrix} e^{i\gamma} & 0 \\ 0 & e^{-i\gamma} \end{bmatrix} \begin{bmatrix} e^{iM\Gamma_m} & 0 \\ 0 & e^{-iM\Gamma_m} \end{bmatrix} \begin{bmatrix} e^{i\gamma} & 0 \\ 0 & e^{-i\gamma} \end{bmatrix} \quad (0 \leq x < \Lambda/2), \quad (9.10a)$$

$$W(x) = \begin{bmatrix} e^{-i\gamma} & 0 \\ 0 & e^{i\gamma} \end{bmatrix} \left( R\left(-\frac{\pi}{2}\right) \begin{bmatrix} e^{iM\Gamma_m} & 0 \\ 0 & e^{-iM\Gamma_m} \end{bmatrix} \right) \begin{bmatrix} e^{i\gamma} & 0 \\ 0 & e^{-i\gamma} \end{bmatrix} \quad (\Lambda/2 \leq x < \Lambda). \quad (9.10b)$$

The observed and calculated polarization states of the 0th and ±1st order beams diffracted from the LC grating are shown in Fig. 9.2 as polar plots. The parameters used in this calculation were $d = 11\,\mu\text{m}$, $d_P = 0.2\,\mu\text{m}$, $\Delta n = 0.1$, $\Delta n_P = 0.2$, and $\lambda = 633\,\text{nm}$. Here, the values of $\Delta n$ and $\Delta n_P$ were assigned in accordance with literatures values (Stallinga 1999; Kawatsuki et al. 2004) and $d$ was determined by fitting the measured and calculated polarization states of the diffracted beams.

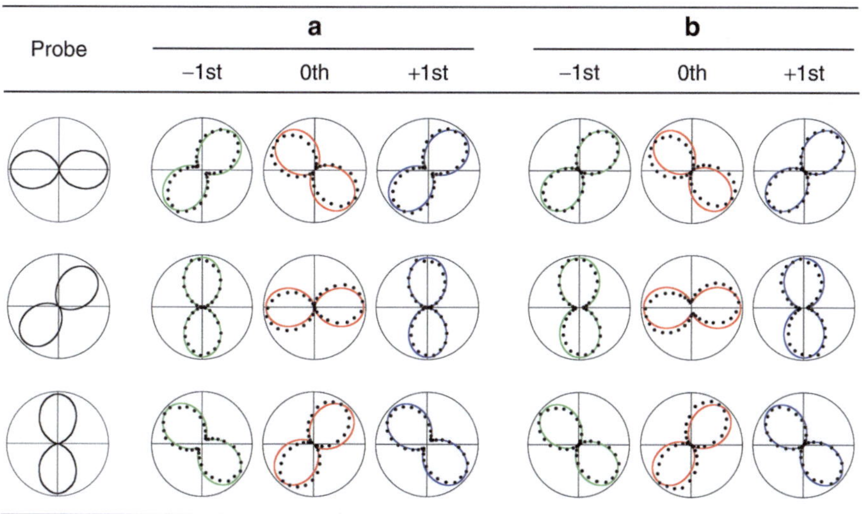


**Fig. 9.2** The polarization states of the beams diffracted from the LC grating that was fabricated using the two-step exposure method. The probe beam was incident on (**a**) film A or (**b**) film B. The polarization azimuth angle of the LP probe beam was 0°, 45°, or 90°. The *filled circles* show the experimental results and *solid lines* show the calculated results

As seen in Fig. 9.2, the observed polarization conversion properties, including the difference in polarization in the incident direction, were described on the basis of our simple binary model involving homogeneous and TN alignments. When the probe beam was incident on film A, the ellipticity of the diffracted beams was almost zero (the diffracted beams were LP). However, the diffracted beams were elliptically polarized when the probe beam was incident on film B. The measured diffraction efficiencies of the 0th and ±1st order beams were approximately 51 % and 20 %, respectively, for any polarization state of the probe beam. This was also in reasonable agreement with the calculated diffraction efficiencies, which are 50 and 20 % for the 0th and ±1st order diffracted beams.

In summary, using a two-step exposure method with a line-and-space photomask we fabricated an LC grating inner coated with PLCP photoalignment films. The diffraction efficiencies and the unique polarization conversion properties of the LC grating were well explained based on the simple binary model involving planar and TN alignments, the Jones matrix method, and diffraction theory.

## 9.3 Polarization Holographic Recording

### *9.3.1 One-Dimensional Grating*

We fabricated LC diffraction gratings using polarization holographic recording. In this study, a PLCP with 4-(4-methoxycinnamoyloxy)biphenyl side groups (**P2**) was used. The chemical structure and synthesis procedure of **P2** was previously reported (Kawatsuki et al. 2002). The experimental setup for polarization holographic recording is illustrated in Fig. 9.3a. A LP He–Cd laser with an operating wavelength of 325 nm was used as the light source for the recording process. The LP UV laser beam was divided into two beams with intensities of 0.85 W/cm$^2$. The two beams overlapped on the **P2** film. The **P2** film was prepared by spin coating a methylene chloride solution of **P2** onto a cleaned glass substrate. The incident angles of the two beams were ±0.37°. The irradiation time was 20 s, which forms a molecular orientation parallel to the polarization azimuth. After irradiation, the **P2** film was annealed at 150 °C for 10 min and then cooled to room temperature.

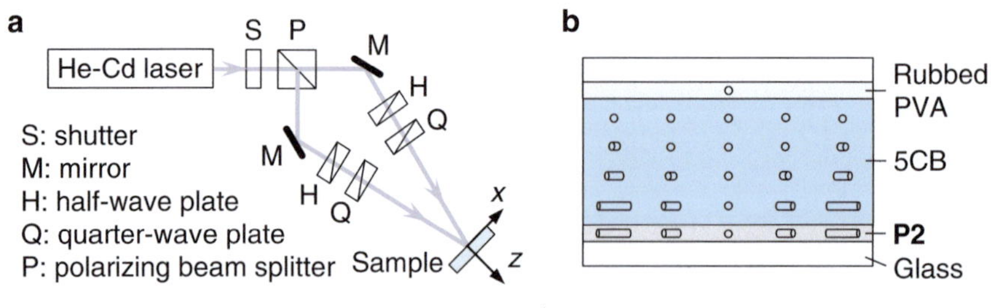


**Fig. 9.3** The schematic illustrations of the LC grating fabricated using polarization holographic recording: (**a**) the experimental setup for recording; (**b**) the structure of the resultant LC grating

The polarization states of the two recording beams were set to be right- and left-hand circular polarized using wave plates. When the plane of incidence is the $xz$-plane, the Jones vectors of the two recording beams are

$$\boldsymbol{E}_{\pm} = \sqrt{\frac{I}{2}}\begin{bmatrix} 1 \\ \pm i \end{bmatrix} \exp[-ik(\mp x \sin\theta + z\cos\theta)], \tag{9.11}$$

where $I$ is the intensity of the light, $k$ is the wavenumber, and $\pm\theta$ are the incident angles. For simplicity, we assumed that $|\theta| \ll 1$ in accordance with the experimental conditions. From (9.11), the Jones vector of the resulting interference field of the two beams is

$$\boldsymbol{E}(x) = \boldsymbol{E}_{+} + \boldsymbol{E}_{-} = \sqrt{2I}\begin{bmatrix} \cos(\xi/2) \\ -\sin(\xi/2) \end{bmatrix} \exp(-ikz\cos\theta), \tag{9.12}$$

where the pitch of the interference pattern is given by $\Lambda = \lambda'/|2\sin\theta|$ where $\lambda'$ is the wavelength of the two beams and $\xi = 2\pi x/\Lambda$. In this experiment, $\lambda' = 325\,\text{nm}$ and $\theta = 0.37°$ so we obtain $\Lambda = 25\,\mu\text{m}$. Equation (9.12) indicates that the interference field is LP and the polarization azimuth is rotated linearly to the $x$-direction. Because the orientation of the annealed **P2** film is parallel (or perpendicular) to the polarization azimuth, the periodic distribution of the easy axis is realized by polarization holographic recording. By using this **P2**-coated glass substrate, a poly(vinyl alcohol) (PVA)-coated glass substrate, nematic LC (4-pentyl-4′-cyanobiphenyl; 5CB, Merck), and 10-μm-thick spacers, we fabricated an LC cell. The PVA film was rubbed uniaxially using a silk cloth. The structure of the LC cell is illustrated in Fig. 9.3b. The diffraction properties of the resultant LC grating cell were probed using an LP He–Ne laser with a wavelength of 633 nm that was incident normal to the substrates. The intensities of the ±1st order diffraction efficiencies were then measured using an optical power meter.

To characterize the diffraction properties of the LC grating, we estimated the director distribution in the cell based on the elastic continuum theory of nematic LCs. The elastic energy density of a deformed nematic LC, $u$, is given by

$$u = \frac{1}{2}K_1(\nabla\cdot\boldsymbol{n})^2 + \frac{1}{2}K_2[\boldsymbol{n}\cdot(\nabla\times\boldsymbol{n})]^2 + \frac{1}{2}K_3|\boldsymbol{n}\times(\nabla\times\boldsymbol{n})|^2, \tag{9.13}$$

where $\boldsymbol{n}$ is the director in the $xyz$-coordinate system and $K_1, K_2$, and $K_3$ are the splay, twist, and bend elastic constants, respectively. We assumed that the angle between the $x$-axis and $\boldsymbol{n}$ (i.e., the twist angle) was given by $\phi = \phi(x, z)$ and the angle between the $xy$-plane and $\boldsymbol{n}$ (the tilt angle) was 0. The grating vector is parallel to the $x$-axis and the normal vector of the substrates is parallel to the $z$-axis. By substituting $\boldsymbol{n}(x, y) = (\cos\phi, \sin\phi, 0)$ into (9.13), we obtain

$$u = \frac{1}{2}(K_1 \sin^2\phi + K_3 \cos^2\phi)\left(\frac{\partial\phi}{\partial x}\right)^2 + \frac{1}{2}K_2\left(\frac{\partial\phi}{\partial z}\right)^2. \tag{9.14}$$

Using a common approximation $\left(K_1 = K_2 \equiv \overline{K}\right)$, (9.14) can be rewritten as

$$u = \frac{1}{2}\overline{K}\left(\frac{\partial\phi}{\partial x}\right)^2 + \frac{1}{2}K_2\left(\frac{\partial\phi}{\partial z}\right)^2. \tag{9.15}$$

For one cycle of the director distribution, the free energy in the cell, $U$, is

$$U = \int_0^d \int_0^\Lambda u\,dx dz. \tag{9.16}$$

where $z = 0$ and $z = d$ are the boundaries between the PLCP films and the LC. The director distribution satisfies the minimum free energy condition that $\delta U = 0$. This variational problem was solved using a finite element method. For simplicity, the fixed boundary condition was assumed at $z = 0$ and $z = d$. The periodic boundary condition $\phi(x, z) = \phi(x + \Lambda, z)$ was also applied in the calculation. By substituting the estimated $\phi(x, z)$ into (9.2), we calculated the diffraction properties of the grating on the basis of (9.1)–(9.8).

Figure 9.4 shows the measured and calculated diffraction efficiencies. The parameters used in this calculation were $\overline{K} = 8.47$ pN, $K_2 = 4.09$ pN, $d = 13.5\,\mu\text{m}$, $\Lambda = 25\,\mu\text{m}$, $\phi(x, 0) = \phi_\text{i}(x) = \pi x/\Lambda$, $\phi(x, 0) = \phi_\text{o} = \pi/2$, $\lambda = 633$ nm, $\Delta n = 0.18$, $\Delta n_\text{P} = 0.17$, and $d_\text{P} = 0.3\,\mu\text{m}$. The elastic constants were given by literature values (Yeh and Gu 1999) and the thicknesses and optical anisotropy were determined by fitting the measured and calculated diffraction

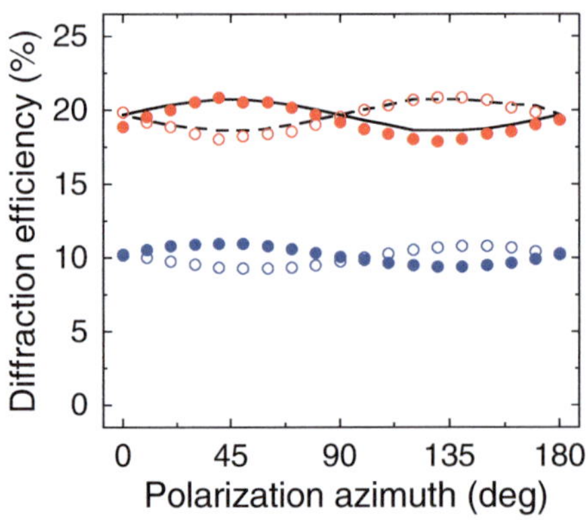


**Fig. 9.4** The diffraction efficiencies for the one-dimensional LC gratings fabricated using polarization holographic recording. The probe beam was LP. The *filled* (*open*) *circles* show the measured +1st (−1st) diffraction efficiencies. The *red* (*blue*) *circles* show the results for the LC gratings with the PLCP (PVCi) film. The *solid* (*dashed*) *lines* show the calculated +1st (−1st) diffraction efficiencies

efficiencies. We assumed that $\gamma = 0$ for $\boldsymbol{W}_o$ because the birefringence of the PVA film is small. As shown in Fig. 9.4, the observed diffraction properties were well explained based on the estimated director distribution in the LC cell. The diffraction efficiencies of an LC cell grating that was fabricated using a PVCi-coated substrate instead of the **P2**-coated substrate are also shown in Fig. 9.4. PVCi is a typical material that is used for the photoalignment of LCs. Figure 9.4 clearly shows that the diffraction efficiencies of the LC grating fabricated using the **P2** film are higher than that fabricated using the PVCi film. The diffraction properties of the LC grating that was fabricated using the PVCi film cannot be explained using the theoretical model, even when taking into account the difference in the optical anisotropy. We attribute this to the difference in the anchoring strengths of the **P2** and PVCi films. We measured the azimuthal anchoring strengths of both films by observing widths of Néel walls (Okubo et al. 2003). The azimuthal anchoring strength of the PVCi film was $\sim 10^{-6}\,\mathrm{J/m^2}$. The azimuthal anchoring strength of the **P2** film was $\sim 10^{-5}\,\mathrm{J/m^2}$ for the exposure energy. This value is comparable to that of the rubbed PVA films, which are regarded as alignment films that satisfy the strong anchoring condition or the fixed boundary condition. Therefore, our PLCPs are also superior from the viewpoint of anchoring energy.

In summary, one-dimensional LC gratings were fabricated using holographically patterned PLCP or PVCi film and a rubbed PVA film. The diffraction properties of the LC gratings were characterized based on the elastic continuum theory of nematic LCs. We found that the anchoring strength of the photoalignment layer affected the diffraction properties. The anchoring strength of the PLCP film was one order of magnitude larger than the PVCi film and the diffraction properties of the LC grating with PLCP film were described assuming the strong anchoring condition. This study showed that our PLCPs are excellent photoalignment materials from the point of view of anchoring energy.

### *9.3.2 Two-Dimensional Grating*

We fabricated two-dimensional LC diffraction gratings by polarization holographic recording of **P2**-coated glass substrates. The procedure for polarization holographic recording and annealing are the same as mentioned above in Sect. 9.3.1. Using the two recorded substrates, 5CB, and spacers with a thickness of 10 μm, LC cells were fabricated. The properties of the LC cells were then investigated. The crossing angle ($\alpha$) of the grating vectors in the two **P2** films was controlled for the cell fabrication. For comparison, we prepared similar LC gratings using PVCi-coated glass substrates and observed their properties. Figure 9.5 shows the POM images of the resultant two-dimensional LC gratings. As seen in Fig. 9.5a–c, the disclination lines that were formed at the boundaries of the right- and left-handed twist alignment regions were observed in the LC cells using **P2** films. In contrast, disclination lines were not observed in the LC cells with PVCi films [Fig. 9.5d–f]. This is

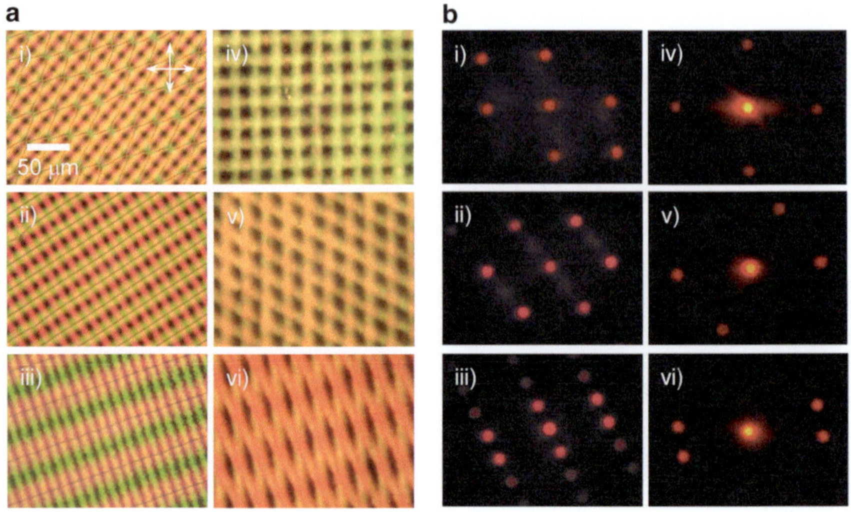


**Fig. 9.5** The observed results for the two-dimensional LC gratings fabricated using polarization holographic recording: (**a**) POM images; (**b**) diffraction patterns. *i*, *ii*, and *iii* (*iv*, *v*, and *vi*) show the results for the LC gratings with the PLCP (PVCi) film. For *i* and *iv* $\alpha = 90°$, for *ii* and *v* $\alpha = 60°$, and for *iii* and *vi* $\alpha = 30°$

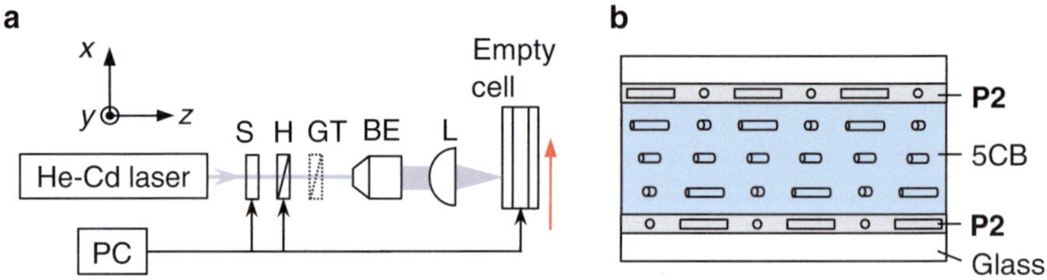


**Fig. 9.6** The schematic illustrations for the LC grating fabricated using the one-step polarizer-rotation exposure method: (**a**) the experimental setup where GT is the Glan-Thompson prism, BE is the beam expander, and L is the cylindrical lens; (**b**) the structure of the LC grating

because the anchoring strength of the PVCi films is lower than that of the **P2** films. Figure 9.6 shows typical diffraction patterns for the two-dimensional LC gratings. The patterns were observed using a LP He–Ne laser with a wavelength of 633 nm at an angle of incidence that was normal to the substrates. Higher-order diffraction beams appeared for the LC gratings with the **P2** films but were almost invisible for the LC gratings with the PVCi films. For the LC gratings with the **P2** films, the diffraction efficiencies and the polarization states of the diffracted beams were also explained based on elastic continuum theory and Jones calculus. Our observations indicate that two-dimensional LC gratings with a fine orientation distribution and high diffraction efficiency can be formed using alignment layers with high anchoring energy such as in our PLCP films.

## 9.4 One-Step Polarizer-Rotation Exposure Method

Using PLCP films as alignment layers, we can fabricate TN cells using one-step exposure method. One of the most important characteristics of our PLCPs is that the mesogenic side chains are oriented perpendicular or parallel to the polarization azimuth in accordance with the exposure energy of the LP UV light. When the exposure energy is relatively low, the director is perpendicular to the polarization azimuth (the order parameter is negative). In contrast, the director is parallel to the polarization azimuth (the order parameter is positive) for relatively high exposure energy. Using this property, we fabricated a TN LC cell using a one-step exposure method on an empty glass cell that was coated on the inside with PLCP films (Kuzuwata et al. 2012). Because the exposure energy differs between the two PLCP films because of the difference in UV light absorption by the front PLCP film, mutually orthogonal easy axes were formed in the front and rear PLCP films by appropriately controlling the exposure energy. In this section, we describe the LC gratings that were fabricated by combining the one-step exposure method with the polarizer-rotation exposure method. The details of the polarizer-rotation exposure method can be found elsewhere (Emoto et al. 2011).

**P2** films with a thickness of 0.3 μm were prepared by spin coating a solution of **P2** in methylene chloride onto cleaned glass substrates. An empty planar cell was fabricated using the two **P2**-coated glass substrates and spacers with a thickness of 10 μm. A one-step polarizer-rotation exposure was performed on the empty cell using a LP He–Cd laser with a wavelength of 325 nm. The experimental setup is illustrated in Fig. 9.6a. The intensity and the polarization azimuth of the LP UV light beam were controlled using a rotating half-wave plate and a Glan-Thompson prism. The laser beam with a Gaussian profile was focused onto the empty cell with a cylindrical lens. The spot dimensions of the laser beam incident on the cell were 50 μm and 4 mm in the $x$- and $y$-directions, respectively. The coordinate system is defined in Fig. 9.6b. The empty cell was moved linearly in the $x$-direction using a linear stage (SIGMA KOKI, SGSP20-35) and the half-wave plate was rotated using a rotational stage (SIGMA KOKI, VWB-201). The stages were controlled using a stage controller (SIGMA KOKI SHOT-204MS) and a computer. Orthogonally parallel and perpendicular molecular orientations can be realized for the front and rear **P2** films, respectively, using the light absorption in the **P2** film on the front substrate. After the one-step polarizer-rotation exposure, the empty cell was annealed at 150 °C for 15 min to induce the molecular orientation in both **P2** films. The cell was cooled to room temperature and then was filled with 5CB by capillary action. The structure of the resultant LC grating is illustrated in Fig. 9.6b.

We attempted to fabricate three LC gratings using the above method. One grating was homogeneous (HOMO), one grating was TN, and the third grating was a mixture of HOMO and TN (MIX). For the HOMO grating, the orientation direction of the **P2** films (the easy axes of the alignment layers) were

$$\phi_{\rm i} = \frac{\pi}{\Lambda}x, \quad (9.17)$$

$$\phi_{\rm o} = \phi_{\rm i}, \quad (9.18)$$

where $\phi_{\rm i}$ and $\phi_{\rm o}$ are the azimuth angle of the director for the front and rear **P2** films. The grating pitch, Λ, was 500 μm. To induce the orientation distributions for the **P2** films, the empty cell that was inner coated with **P2** was moved in the $x$-direction at 187.5 μm/s and the half-wave plate was rotated at 33.75°/s. The exposure energy of the LP UV light beam was set at 2.0 J/cm$^2$ so that $\phi_{\rm i} = \phi_{\rm o}$.

For the TN grating, $\phi_{\rm i}$ and $\phi_{\rm o}$ are

$$\phi_{\rm i} = \frac{\pi}{\Lambda}x, \quad (9.19)$$

$$\phi_{\rm o} = \phi_{\rm i} + \frac{\pi}{2}. \quad (9.20)$$

To fabricate the TN grating with a grating pitch of 500 μm, the empty cell was moved in the $x$-direction at 125 μm/s, the half-wave plate was rotated at 22.5°/s, and the exposure energy was set at 1.2 J/cm$^2$ so that $\phi_{\rm o} = \phi_{\rm i} + \pi/2$.

For the MIX grating, $\phi_{\rm i}$ and $\phi_{\rm o}$ are

$$\phi_{\rm i} = \frac{\pi}{2}, \quad (9.21)$$

$$\phi_{\rm o} = \begin{cases} 0 & (0 \le x < \Lambda/2) \\ \pi/2 & (\Lambda/2 \le x < 0) \end{cases}. \quad (9.22)$$

To fabricate the MIX grating with a grating pitch of 500 μm, the empty cell was moved in the $x$-direction at 125 μm/s and the half-wave plate was rotated at 22.5°/s. Additionally, a Glan-Thompson prism (a linear polarizer) with a transmission axis that was parallel to the $x$-direction was placed behind the half-wave plate as shown in Fig. 9.6a. Therefore, the intensity of the LP UV light beam on the cell was sinusoidally modulated in the $x$-direction. The lower and higher values of the sinusoidally modulated exposure energy were set to 0.6 and 2.2 J/cm$^2$.

Figure 9.7a–c show POM images for the HOMO, TN, and MIX gratings, respectively. Figure 9.7d–f show the director distributions that were estimated based on the elastic continuum theory mentioned above for the HOMO, TN, and MIX gratings. The calculation parameters used here were $\overline{K} = 7.48$ pN, $K_2 = 3.81$ pN, $\Lambda = 500\,\mu$m, and $d = 10\,\mu$m. The observed POM images for the three gratings were consistent with the calculated director distributions. This indicates that the photoalignments described in (9.17)–(9.22) were realized using the one-step polarizer-rotation exposure method. Figure 9.8a–c show the measured and calculated diffraction properties for the HOMO, TN, and MIX gratings. The diffraction properties were probed with a polarized He–Ne laser with a wavelength of 633 nm that was incident normal to the substrates. The intensities and

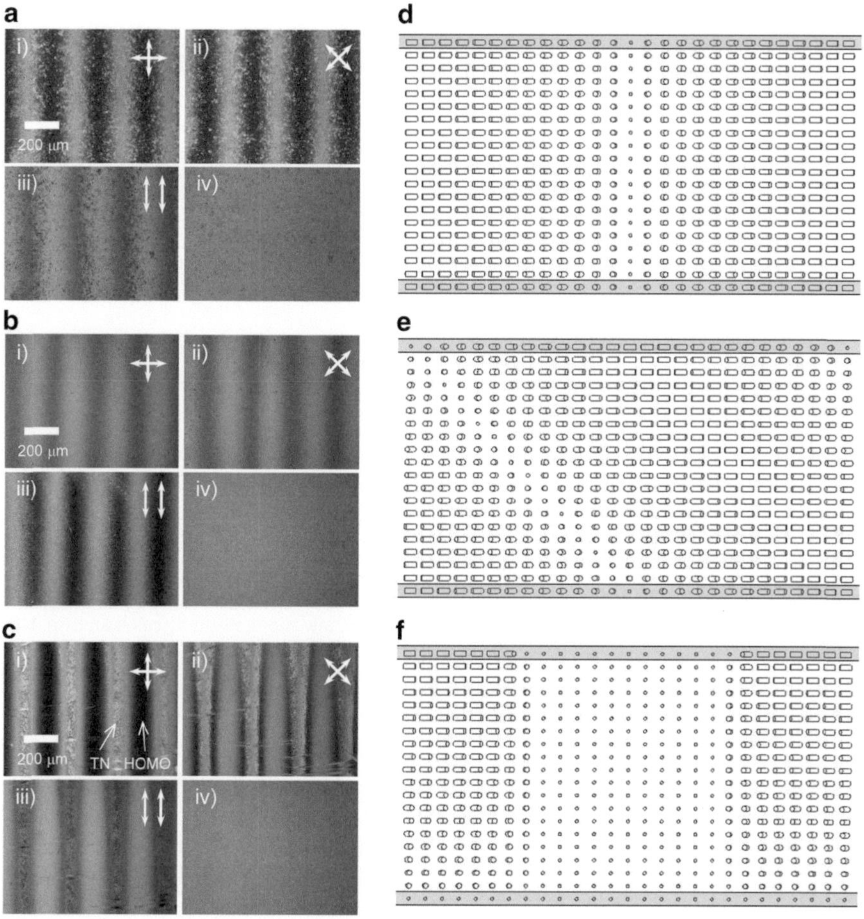


**Fig. 9.7** The orientation distributions where (**a**–**c**) show the POM images (Note that the polarizer and the analyzer were removed in *iv*). (**d**, **f**, **g**) show the calculated director distributions in the *xz*-plane. (**a**, **d**) are the HOMO grating, (**b**, **f**) are the TN grating, and (**c**, **g**) are the MIX grating

polarization states of the ±1st-order diffraction beams were measured using an optical power meter and a Glan-Thompson prism. In the calculations, $d_P = 0.3\,\mu m$, $\Delta n = 0.1776$, $\Delta n_P = 0.2$, and $\lambda = 633\,nm$. The cell thickness $d$ for the HOMO, TN, and MIX gratings was determined to be 10.52, 10.62, and 11.50 µm, respectively, by fitting the measured and calculated diffraction properties. As seen in Fig. 9.8a, b, the observed diffraction efficiencies and the polarization conversion properties were well described for the HOMO and TN gratings based on the estimated director distributions. This also indicates that the theoretical models are valid. In contrast, the observed diffraction properties for the MIX grating cannot be accounted for using the estimated director distribution [Fig. 9.8c]. This was presumably because the fixed boundary condition used in the calculation was not satisfied in the fabricated MIX

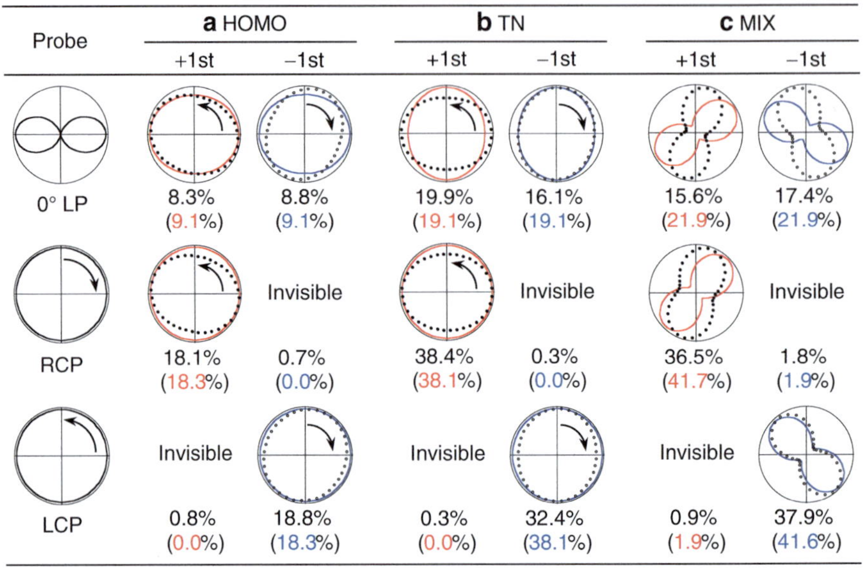


**Fig. 9.8** The polarization states of the diffracted beams and diffraction efficiencies: (**a**) HOMO; (**b**) TN; (**c**) MIX gratings. The *filled circles* (*solid lines*) show the measured (calculated) polarization states. The numerical values are the measured (calculated) diffraction efficiencies

grating because the order parameter (the anchoring strength) of the rear PLCP film was modulated through zero based on the modulation of the exposure energy. We will attempt to consider the anchoring energy of PLCP films in the theoretical model.

In summary, we have demonstrated that LC gratings with twisted alignment can be obtained using one-step polarizer-rotation exposure method. An empty cell was inner coated with PLCP films and was subsequently annealed. The diffraction properties of the fabricated LC gratings were well explained based on the theoretical director distributions that were estimated using the elastic continuum theory of nematic LCs when assuming that there were strong anchoring conditions. Compared with conventional techniques such as polarization holographic recording, our one-step polarizer-rotation exposure method enables flexible design of director distributions because the polarization state and intensity of UV light beam can be controlled. We believe that our proposed method and the PLCP films present a simple yet efficient, competitive and practical procedure for fabricating various LC gratings.

## 9.5 Conclusions and Future Perspective

We investigated various LC diffraction gratings made from PLCP alignment films with mesogenic side chains that were oriented by irradiating them with polarized UV light and subsequent annealing. PLCP films are transparent in the visible region of the spectrum and the orientation direction can be controlled using both the polarization state and the exposure energy of UV light. Additionally, the anchoring strength of PLCP films is one order of magnitude larger than that of common PVCi photoalignment films. Using these characteristics, LC gratings with a twisted alignment were fabricated using photomask exposure, polarization holographic recording, and one-step polarizer-rotation exposure. The diffraction efficiencies and the unique polarization conversion properties of the LC gratings were described based on the elastic continuum theory of LCs, diffraction theory, and Jones calculus. We believe that PLCP films are suitable media for fabricating LC diffractive optical elements because the orientation distribution can be effectively controlled, they are thermally stabile, and are transparent in the visible region of the spectrum.

## References


Barachevsky VA (1991) Photoanisotropic polymeric media and their application in optical devices. Proc SPIE 1559:184–193

Batalioto F, Bechtold IH, Oliveira EA, Evangelista LR (2005) Effects of microtextured substrates on the molecular orientation of a nematic liquid-crystal sample. Phys Rev E 72:031710

Chandrasekhar S (1992) Liquid crystals. Cambridge University Press, New York

Chen J, Bos PJ, Vithana H, Johnson DL (1995) An electro-optical controlled liquid crystal diffraction grating. Appl Phys Lett 67:2588–2590

Choi H, Wu JW, Chang HJ, Park B (2006) Holographically generated twisted nematic liquid crystal gratings. Appl Phys Lett 88:021905

Crawford GP, Eakin JN, Radcliffe MD, Jones AC, Pelcovits RA (2005) Liquid-crystal diffraction gratings using polarization holography alignment techniques. J Appl Phys 98:123102

Duffy WL, Hindmarsh P, Kelly SM, Owen GJ (2001) An investigation of the role of cross-linking and photodegradation of side-chain coumarin polymers in the photoalignment of liquid crystals. Chem Mater 13:694–703

Ebralidze TD, Ebralidze NA (1992) Hologram record by means of film anisotropy photoinduction. Appl Opt 31:4720–4724

Emoto A, Ono H, Kawatsuki N, Uchida E (2005) Polarization gratings in photocrosslinkable polymer liquid crystals prepared using two-step ultraviolet exposure. Jpn J Appl Phys 44:535–538

Emoto A, Wada T, Shioda T, Sasaki T, Manabe S, Kawatsuki N, Ono H (2011) Vector gratings fabricated by polarizer rotation exposure to hydrogen-bonded liquid crystalline polymers. Jpn J Appl Phys 50:032502

Fabbri F, Garrot D, Lahlil K, Boilot JP, Lassailly Y, Peretti J (2011) Evidence of two distinct mechanisms driving photoinduced matter motion in thin films containing azobenzene derivatives. J Phys Chem B 115:1363–1367

Gibbons WM, Shannon PJ, Sun ST, Swetlin J (1991) Surface-mediated alignment of nematic liquid crystals with polarized laser light. Nature 351:49–50

Gwag JS, Fukuda J, Yoneya M, Yokoyama H (2007) In-plane bistable nematic liquid crystal devices based on nanoimprinted surface relief. Appl Phys Lett 91:073504

Honma M, Nose T (2003) Polarization-independent liquid crystal grating fabricated by microrubbing process. Jpn J Appl Phys 42:6992–6997

Honma M, Nose T (2004) Liquid-crystal blazed grating with azimuthally distributed liquid-crystal directors. Appl Opt 43:5193–5197

Honma M, Nose T (2012a) Temperature-independent achromatic liquid-crystal grating with spatially distributed twisted-nematic orientation. Appl Phys Express 5:062501

Honma M, Nose T (2012b) Highly efficient twisted nematic liquid crystal polarization gratings achieved by microrubbing. Appl Phys Lett 101:041107

Honma M, Nose T (2012c) Twisted nematic liquid crystal polarization grating with the handedness conservation of a circularly polarized state. Opt Express 20:18449–18458

Hu W, Srivastava AK, Lin XW, Liang X, Wu ZJ, Sun JT, Zhu G, Chigrinov V, Lu YQ (2012) Polarization independent liquid crystal gratings based on orthogonal photoalignments. Appl Phys Lett 100:111116

Ichimura K (2000) Photoalignment of liquid crystal systems. Chem Rev 100:1847–1873

Ichimura K, Suzuki Y, Seki T, Hosaki A, Aoki K (1988) Reversible change in alignment mode of nematic liquid crystals regulated photochemically by "command surfaces" modified with an azobenzene monolayer. Langmuir 4:1216–1219

Ichimura K, Akita Y, Akiyama H, Kudo K, Hayashi Y (1997) Photoreactivity of polymers with regioisomeric cinnamate side chains and their ability to regulate liquid crystal alignment. Macromolecules 30:903–911

Kawatsuki N, Ono H, Takatsuka H, Yamamoto T, Sangen O (1997a) Liquid crystal alignment on photoreactive side-chain liquid-crystalline polymer generated by linearly polarized UV light. Macromolecules 30:6680–6682

Kawatsuki N, Takatsuka H, Yamamoto T, Ono H (1997b) Photoregulated liquid crystal alignment on photoreactive side-chain liquid-crystalline polymer. Jpn J Appl Phys 36:6464–6469

Kawatsuki N, Suehiro C, Yamamoto T (1998) Photoinduced alignment of photo-cross-linkable side-chain liquid crystalline copolymers comprising cinnamoylethoxybiphenyl and cyanobiphenyl groups. Macromolecules 31:5984–5990

Kawatsuki N, Yamamoto T, Ono H (1999) Photoinduced alignment control of photoreactive side-chain polymer liquid crystal by linearly polarized ultraviolet light. Appl Phys Lett 74:935–937

Kawatsuki N, Matsuyoshi K, Hayashi M, Takatsuka H, Yamamoto T (2000) Photoreaction of photo-cross-linkable methacrylate polymer films comprising 2-cinnamoyloxyethoxybiphenyl side group by linearly polarized ultraviolet light and liquid crystal alignment on the resultant films. Chem Mater 12:1549–1555

Kawatsuki N, Kawakami T, Yamamoto T (2001) A photoinduced birefringent film with a high orientational order obtained from a novel polymer liquid crystal. Adv Mater 13:1337–1339

Kawatsuki N, Goto K, Kawakami T, Yamamoto T (2002) Reversion of alignment direction in the thermally enhanced photoreorientation of photo-cross-linkable polymer liquid crystal films. Macromolecules 35:706–713

Kawatsuki N, Kuwabara M, Matsuura Y, Ono H, Emoto A (2004) Control of thermally enhanced photoinduced reorientation direction of photocrosslinkable copolymer liquid crystals and application to polarization gratings using linearly polarized ultraviolet light. Jpn J Appl Phys 43:5447–5450

Kawatsuki N, Hamano K, Ono H, Sasaki T, Goto K (2007) Molecular-oriented photoalignment layer for liquid crystals. Jpn J Appl Phys 46:339–341

Kawatsuki N, Nishioka E, Emoto A, Ono H, Kondo M (2012) Blazed surface relief formation in azobenzene-containing polymeric films by asymmetric polarization holography. Appl Phys Express 5:041601

Kumar GS, Neckers DC (1989) Photochemistry of azobenzene-containing polymers. Chem Rev 89:1915–1925

Kuzuwata M, Sasaki T, Kawatsuki N, Ono H (2012) Fabrication of twisted nematic structure and vector grating cells by one-step exposure on photocrosslinkable polymer liquid crystals. Opt Lett 37:1115–1117

Labarthet FL, Buffeteau T, Sourisseau C (1999) Azopolymer holographic diffraction gratings: time dependent analysis of the diffraction efficiency, birefringence, and surface modulation induced by two linearly polarized interfering beams. J Phys Chem B 103:6690–6699

Natansohn A, Rochon P (2002) Photoinduced motions in azo-containing polymers. Chem Rev 102:4139–4175

Nikolova L, Ramanujam PS (2009) Polarization holography. Cambridge University Press, Cambridge

Nikolova L, Todorov T (1984) Diffraction efficiency and selectivity of polarization holographic recording. Opt Acta 31:579–588

Obi M, Morino S, Ichimura K (1999) Factors affecting photoalignment of liquid crystals induced by polymethacrylates with coumarin side chains. Chem Mater 11:656–664

Okubo K, Kimura M, Akahane T (2003) Measurement of genuine azimuthal anchoring energy in consideration of liquid crystal molecular adsorption on alignment film. Jpn J Appl Phys 42:6428–6433

Ono H, Emoto A, Takahashi F, Kawatsuki N, Hasegawa T (2003a) Highly stable polarization gratings in photocrosslinkable polymer liquid crystals. J Appl Phys 94:1298–1303

Ono H, Emoto A, Kawatsuki N, Hasegawa T (2003b) Multiplex diffraction from functionalized polymer liquid crystals and polarization conversion. Opt Express 11:2379–2384

Ono H, Emoto A, Kawatsuki N, Uchida E, Kuwabara M (2004) New fabrication method for anisotropic gratings formed in photocrosslinkable polymer liquid crystals. Appl Phys A 79:1725–1727

Ono H, Oikawa S, Kawatsuki N (2007) Effects of anchoring strength on diffraction properties of liquid crystal phase gratings formed on photoalignment polymer films. J Appl Phys 101:123523

Ono H, Hishida M, Emoto A, Shioda T, Kawatsuki N (2009) Elastic continuum analysis and diffraction properties of two-dimensional liquid crystalline grating cells. J Opt Soc Am B 26:1151–1156

Park JH, Yu CJ, Kim J, Chung SY, Lee SD (2003) Concept of a liquid-crystal polarization beamsplitter based on binary phase gratings. Appl Phys Lett 83:1918–1920

Provenzano C, Pagliusi P, Cipparrone G (2007) Electrically tunable two-dimensional liquid crystal gratings induced by polarization holography. Opt Express 15:5872–5878

Sasaki T, Ono H, Kawatsuki N, Kuwabara M (2005) Liquid crystal phase gratings using photoregulated photocrosslinkable polymer liquid crystals. Appl Phys Lett 87:161112

Sasaki T, Hatayama A, Emoto A, Ono H, Kawatsuki N (2006) Simple detection of light polarization by using crossed polarization gratings. J Appl Phys 100:063502

Sasaki T, Ono H, Kawatsuki N, Kuwabara M (2007) Diffraction properties of nematic phase gratings with photoregulated liquid crystal cells. Jpn J Appl Phys 46:698–702

Sasaki T, Kuzuwata M, Noda K, Kawatsuki N, Ono H (2013) Liquid Crystal gratings with twisted alignment produced by one-step polarizer-rotation exposure on photocrosslinkable polymer liquid crystal films. Jpn J Appl Phys 52:042503

Sasaki T, Wada T, Noda K, Kawatsuki N, Ono H (2014) Merged vector gratings recorded in a photocrosslinkable polymer liquid crystal film for polarimetry. J Appl Phys 115:023110

Sato S (1979) Liquid-crystal lens-cells with variable focal length. Jpn J Appl Phys 18:1679–1684

Scharf T (2007) Polarized light in liquid crystals and polymers. Wiley, New Jersey

Seki T, Nagano S, Hara M (2013) Versatility of photoalignment techniques: from nematics to a wide range of functional materials. Polymer 54:6053–6072

Schadt M, Schmitt K, Kozinkov V, Chigrinov V (1992) Surface-induced parallel alignment of liquid crystals by linearly polarized photopolymers. Jpn J Appl Phys 31:2155–2164

Stallinga S (1999) Berreman 4 × 4 matrix method for reflective liquid crystal displays. J Appl Phys 85:3023–3031

Subacius P, Bos PJ, Lavrentovich OD (1997a) Switchable diffractive cholesteric gratings. Appl Phys Lett 71:1350–1352

Subacius P, Shiyanovskii SV, Bos PJ, Lavrentovich OD (1997b) Cholesteric gratings with field-controlled period. Appl Phys Lett 71:3323–3325

Sun J, Srivastava AK, Wang L, Chigrinov VG, Kwok HS (2013) Optically tunable and rewritable diffraction grating with photoaligned liquid crystals. Opt Lett 38:2342–2344

Todorov T, Nikolova L, Stoyanova K, Tomova N (1985) Polarization holograph. 3: some applications of polarization holographic recording. Appl Opt 24:785–788

Wu WY, Mo TS, Fuh AYG (2006) Polarization characteristics of diffracted beams from twisted nematic gratings fabricated by the photoalignment effect in dye-doped liquid crystal films. J Opt Soc Am B 23:1737–1742

Yadavalli NS, Santer S (2013) In-situ atomic force microscopy study of the mechanism of surface relief grating formation in photosensitive polymer films. J Phys Chem B 115:1363–1367

Yeh P, Gu G (1999) Optics of liquid crystal displays. Wiley, New York

Yu CJ, Kim DW, Kim J, Lee SD (2005) Polarization-invariant grating based on a photoaligned liquid crystal in an oppositely twisted binary configuration. Opt Lett 30:1995–1997

# Chapter 10
# Liquid Crystalline Polymer Blends as Fillers for Self-Reinforcing Polymer Composites

**Kwabena A. Narh**

## 10.1 Introduction

A polymer blend is a mixture of at least two polymers or copolymers. The use of polymer blends is for developing materials with a full set of desired properties at a low cost. The disadvantages of one polymer are often compensated by the advantages of other polymers. For example, the weaknesses of polyphenylene ether (PPE) are its poor processability and low impact strength; polyamide (PA) with good processability and high impact strength can be used to blend with PPE to compensate for the disadvantages of PPE.

Most polymer materials are used because they have desirable mechanical properties at an economical cost. For this reason, the mechanical properties should be considered the most important of all the physical and chemical properties of polymers in applications of polymers. The usual and better method of improving the mechanical properties of polymers is to fill polymers with reinforcing materials, or to develop polymer composites.

A polymer composite is a mixture of at least one polymer and fillers (the aspect ratio of fillers is close to one), or is a material in which at least one polymer matrix is combined with reinforcing fibers, which are stronger and stiffer than the polymer matrix. Fillers are used to decrease the cost of materials; reinforcements are used to improve the mechanical properties of polymers. The fiber-reinforced polymer can be classified as long fiber-reinforced polymers and short fiber-reinforced polymers. Recent years have seen rapid growth in the use of short fiber-reinforced polymers. This is because rapid, economical process technology is probably the single most important factor in the growth of the composites, and a major advantage of short fiber-reinforced polymers is that forming of the polymers is possible by normal

K.A. Narh (✉)
Department of Mechanical and Industrial Engineering, New Jersey Institute of Technology, University Heights, Newark, NJ 07102, USA
e-mail: narh@njit.edu



V.K. Thakur, M.R. Kessler (eds.), *Liquid Crystalline Polymers*,
DOI 10.1007/978-3-319-20270-9_10

injection molding or extrusion techniques. These techniques are the most economical processes when cheap and precise manufacture of very large quantities of components is required. In everyday life, one can find many products, which are manufactured with these materials such as tennis racquet frames, storage tanks, pipes, seating for public places, plastic window frames, and components of boats and planes. But the presence of rigid fillers or reinforcing materials in a polymeric matrix entails many problems such as high processing temperature which can easily result in potential degradation of the polymer matrix, high energy required for processing, and the abrasion of processing equipment. From environmental considerations, it is not easy to re-engineer the reinforced materials, thereby possibly discarding them after using, and resulting in environmental problems. The difficulty in re-engineering the materials originates from the significant decrease of mechanical properties after repeated recycling due to the reduction of aspect ratio of reinforcing fibers during reprocessing. Is it possible to find a material that can be used to reinforce polymers as traditional fibers did and at the same time does not entail the problems which occur at the processing of fiber-reinforced polymers? Liquid crystalline polymers (LCPs) provide a solution to the challenge, as described in the following sections.

## 10.2 Liquid Crystalline Polymer as Self-Reinforcing Fillers for Polymeric Materials

A liquid crystalline polymer (LCP) is capable of forming highly oriented crystalline structures when subjected to shear flow above its melting point. The mechanical properties of LCP products are close to those of fiber reinforced composites. However, due to high molecular orientation in the flow direction, their ultimate properties in the direction transverse to flow are usually weak. The addition of LCP to a second thermoplastic component will improve its processability (Baird et al. 1993), reduce its cost, retaining however, to a large extent the exceptional material properties for which LCP's are known. Blending LCP with a flexible coil thermoplastic (TP) can lead to the unique concept that the rigid rod polymer in such blends acts as *self-reinforcing molecular fiber* if efficient dispersions are achieved. The term 'self-reinforcing' or 'in-situ composites' has been used to describe such a blend.

In self-reinforcing composites, the reinforcing species is not actually present before the processing of the resin, but comes into existence during the processing. This arises from the tendency of the thermotropic liquid crystalline polymer to form well-oriented fibrous domains or inclusions within the matrix of the flexible polymer during melt blending. These oriented entities then act as reinforcing agents because of the inherent stiffness of the LCP (Kayaisang et al. 2009; Yang et al. 2014).

## 10.3 Properties of LCP-Based Self-Reinforced Polymer Composites

Self-reinforcing composites acquire most of their rheological characteristics from those of the thermotropic liquid crystalline polymer. In general, LCPs have low melt viscosity and near zero extrudate swell (Jerman and Baird 1981). Hence, in addition to forming a discrete reinforcing phase in the blends, the LCP domains within the blend can decrease the viscosity of the melt, thereby making the processing of high viscosity thermoplastics possible (Baird et al. 1993; Dutta et al. 1990). Furthermore, the orientation of the LCP fibers is shear rate dependent. Therefore, achieving improved mechanical properties from TP/LCP blends does not result from simply blending the polymers but from processing the blends in a manner that forms a reinforcing fiber-like structure within the blends. Most of the studies on TP/LCP blend focused on capillary extrusion and spinning processes (Crevecoeur and Groenickx 1990; Dutta et al. 1990; Isayev 1987; La Mantia et al. 1990b; Lin and Yee 1994; Nobile et al. 1989). These studies showed that draw ratio, concentration of LCP, and viscosity ratio of TP to LCP were key factors for the development of a fibrillar morphology in the TP/LCP blends.

### *10.3.1 Mechanical Properties*

One of the primary objectives of blending LCPs with thermoplastic polymers has been to use the LCP as reinforcement for flexible thermoplastic polymers. Mechanical properties of blends have been reported in a number of studies. Most of the researchers have attempted to explain the changes in mechanical properties in term of the morphology of the LCP domains in the blends.

In nearly all the studies of TP/LCP blends, the two components of the blends were immiscible. The size, shape, and distribution of the LCP phase depended on many factors such as composition, processing conditions, viscosity ratio of the component polymers, and the rheological characteristics of the matrix polymer. The first results of mechanical properties of extruded LCP blends were reported by Kiss (1987) on blends of polyether sulfone with different thermotropic LCPs including two HBA/hydroxy-2-naphthoic acid (HNA) copolyesters, A and B, and one copolyesteramide. All the blends show a remarkable increase of both tensile strength and tensile modulus; the elongation at break is, on the contrary, drastically reduced. Whereas the pure polymer, matrix and LCP, are almost unaffected by shear rate, the blends show a significant dependence on shear rate. Increasing the shear rate, both tensile strength and tensile modulus of the blend with copolyester A show a significant increase; as for the blend with the copolyester B the dependence on shear rate is reversed. Finally, the blend with the LCP copolyesteramide shows the same increase in the tensile strength observed for the blend with the two LCP copolyesters; however the tensile modulus is far higher.

Increased orientation of the dispersed LCP microfibrils formed by drawing can result in greatly improved mechanical properties of the blend. Dutta et al. (1990) observed an order of magnitude increase in modulus as the draw ratio was increased from 1 to 1000 for fibers extruded from blends of an HNA/poly-hydroxybenzoate (PHB)/hydroquinone (HQ)/terephthalic acid (TA) LCP with PC (Dutta et al. 1990). Kim et al. (1998) studied mechanical properties and morphology of fibers of blends of polycarbonate and PET/60PHB considering, in particular, the influence of the draw ratio. The tensile modulus and the ultimate tensile strength of the fibers increased both with the LCP content and with the draw ratio. The elongation at break, however, decreased sharply. The improvement of modulus and strength increased more than linearly with draw ratio. According to the authors, at the highest draw ratio, the LCP droplets were transformed to continuous fibers because of the elongational flow experienced during drawing. Lin et al. (1993) also studied the variations of elastic modulus and tensile strength with draw ratio for PC/Vectra B950 (70/30 %). The elastic modulus of the strand with a draw ratio of 35 is more than three times that of undrawn one. The highest value of elastic modulus is 19 GPa for composite strands. The tensile strength of the strands with 30 wt% LCP rises sharply with increasing draw ratio up to ~15. The highest value of tensile strength was 146 MPa. Again these increments are attributed to microfibril formation of the LCP phase, and more importantly to the LCP molecular orientation. The authors also pointed out that all samples broke in a brittle mode despite the high ductility of the PC matrix. No necking was detected, and the elongation to break was less than 2 %. In a recent study by Mandal et al. (2012), Polypropylene (PP) and Vectra A950, blends were prepared in a single-screw extruder with the variation in Vectra A950 content in the presence of fixed amount (2 %, with respect to PP and LCP mixture as a whole) of ethylene-acrylic acid (EAA) copolymer as a compatibilizer. Mechanical analysis of the compatibilized blends within the range of LCP incorporations under study (2–10 %) indicated pronounced improvement in the moduli, ultimate tensile strength, and hardness.

Obviously, the improvement of tensile strength and tensile modulus due to the shear rate or melt drawing was the result of improved orientation and fibrillation of the dispersed LCP domains. These results demonstrate that the modulus and strength can be markedly improved by varying elongational rate and shear rate.

A number of studies have considered the effect of LCP concentration on the mechanical properties. Zhuang et al. (1988) found that adding small amount of 40PET/60PHB to PET, PC, polystyrene (PS) increased the modulus and tensile strength of compression molded films, extrudates, and melt spun filaments; the elongation at break was, on the contrary, drastically reduced in all cases. The modulus was also found to increase with LCP percentage studied by Crevecoeur and Groenickx (1990).

Mechanical properties of injection molded blends of thermoplastic polymers and thermotropic LCP materials were first reported by Siegmann et al. (1985) with reference to a blend of an amorphous polyamide and Vectra A. In particular, the modulus was found to undergo a 1.5-fold increase by addition of 25 % of LCP, and elongation at break underwent a dramatic decrease with only 5 % of LCP. These two features are quite general with blends with thermotropic LCPs and are related to the morphology that builds up during the processing. In injection molded objects, the change of mechanical properties is mainly related to the structure at the sample skin. Indeed, the same authors found that the core had a lower modulus and a higher elongation. With this system, ultimate strength was found to undergo a 1.3-fold increase by addition of 25 % LCP, however the ultimate strength is very sensitive to adhesion between components and thus its behavior may change with the system considered. Kiss (1987) reported data with reference to many systems having 30 % LCP. He also found that elongation at break had much lower values than the matrix, the modulus increased at least as expected on the basis of the additive rule of mixtures, and the strength often increased, but in some cases decreased by the addition of LCP. Surprisingly, his data showed values of flexural moduli not larger than the tensile ones.

Isayev and Modic (1987) reported mechanical properties both in the flow and in the transverse direction. Along the flow direction, the results are consistent with those described in the preceding sections including the morphology of the system, specifically, the modulus increases and impact strength decreases much faster with LCP content. Along the transverse (tangential) direction the elongation at break decreases even more quickly than along the flow direction; conversely the modulus is higher than in both pure components. In the same paper, the authors reported that a decrease of flow rate reduces the increase of both modulus and strength in the flow direction; an opposite effect of flow rate on the mechanical properties was reported by Brinkmann et al. (1991). The effect of the matrix polymer is shown in the work by Siegmann et al. (1985) also with reference to mechanical properties. The results are in line with their morphological observations mentioned previously. In particular, modulus increases with amorphous nylon (AN), PC and slightly with PBT; tensile strength increases with AN, PC and remains essentially unchanged with PBT matrix; both modulus and strength decrease as LCP content increase in a Nylon 6 matrix.

O'Donnell and Baird (1996) studied the effect of injection molding conditions on the mechanical properties of an *in situ* composite. The variables of filling time, mold temperature, and the thickness of mold cavity were analyzed. The flexural moduli increase from 3.9 to 5.2 GPa, and tensile moduli increase from ~2.9 to 4.7 GPa with filling time from 0.8 to 5 s for 1.5 mm molds. This same property/processing trend occurs for 2.3 mm mold: the flexural moduli increased from 2.62 to 4.08 GPa and the tensile moduli increased from 2.12 to 3.08 GPa while the filling time is increased from 1.3 to 7.8 s. The authors also found that the transverse flexural properties were unaffected by changing filling time. The average transverse modulus for the 1.0 mm thick plaques was approximately 1.5 GPa, and for the 1.5 and 2.3 mm thick plaques was 1.4 GPa. Flexural modulus for the thin molds were larger than that for the thick molds. The authors did not obtain a simple relationship between mold temperature and mechanical properties for the PP/LCP blend, even for pure LCP; these results are different from the results reported by Suokas (1989) with reference to a pure thermotropic LCP. Suokas found that the tensile modulus of pure LCP decreases with increasing mold temperature. Yi et al. (1996) analyzed the tensile modulus and strength of PC/LCP blends at the melt temperatures of 280 and 320 °C and at the volume flow rates of 8 and 80 $cm^3\ s^{-1}$, and found that there are larger tensile modulus and strength at lower melt temperature, and lower volume flow rate. At the injection conditions of lower melt temperature and lower volume flow rate, larger average aspect ratio of LCP fibers was observed.

### *10.3.2 Rheological Properties*

Blizard and Baird (1987) studied the morphological development in shear flow of a blend of 60 % hydroxybenzoic acid (HBA)/poly (ethylene terephthalate) (PET) (as a LCP) and polycarbonate (PC) (as a matrix). The experimental results of samples subjected to shear flow in cone and plate rheometer show that the LCP particles are in spherical form regardless of the shear rate. The diameter of the LCP particles remains constant and uniform with increasing shear rate, and no fibrous structure was found. The reason that no fibrous structure was observed may be attributed to low shear rate (10 $s^{-1}$) in cone and plate rheometer. SEM micrographs of samples extruded through capillaries reveal fibrils of the LCP phase near the skin region for shear rate from 45.7 to 457 $s^{-1}$. The authors did not give the size range of LCP fibrils. Blizard and Baird (1987), and La Mantia et al. (1990a) demonstrated that the fibrils were formed and retained only in short capillaries. By increasing the length of the capillary, no fibrous structure was observed. Blizard and Baird (1987) explained this behavior by considering that the fibrils formed at the entrance of the capillary can be lost because of the relaxation of these structures in the shear flow field of the capillary. Obviously, the shear flow is not a key factor for the formation of LCP fibers according to the authors' explanation. An interesting phenomenon was found in the study of Blizard and Baird (1987), i.e. that the fibers tend to be of smaller diameter with increasing shear rate and the presence of phase inversion can be detected at high shear rate. The higher concentration of the LCP near the edge of the samples appears as a matrix from which near-spherical droplets of PC have been extracted. But the results mean that shear flow can really deform the LCP particles only at very high shear rate or at skin region. La Mantia et al. (1990a) pointed out also that the elongated particles could deform and break in shear flow into small particles if the viscosity of the matrix is much larger than that of the dispersed phase.

Beery et al. (1991a) also studied the shear flow of PC/LCP and polybutyleneterephthalate (PBT)/LCP blends in capillaries. They found in the PC/LCP blends that the LCP structure on the surface of the extruded strand changes with shear rates. At low shear rates up to 54 $s^{-1}$, the LCP appears as spherical particles that change to highly oriented fibrillar structure as the shear rate increases above 270 $s^{-1}$. The relative amount of fibrillar structure as well as the length of the fibrils gradually increases with increasing shear rate. The resulting diameter of fibrils is rather inhomogeneous. The LCP structure development with shear rates in PBT/LCP blends is different. In the case of PBT/LCP blend, even at high shear rates above

2700 $s^{-1}$, the skin of the strand includes spherical particles of the LCP dispersed phase with occasional elongated particles, which is considerably different from the fibrillar appearance of the PC/LCP blends described in the preceding sections. The authors explained these results on the basis of rheological analysis that the PBT viscosities lie below the LCP ones in the whole shear rates studied. Beery et al. (1991b) observed the LCP morphological development at the center region of extruded strands in the PBT/LCP and Nylon 6/LCP blends. No fibrous structures were found in center region of the strands both at low shear rates down to 27 $s^{-1}$ and high shear rates up to 5400 $s^{-1}$. Actually, the shear rates in the center region are very low and the viscosities of matrix polymer are larger than those of the LCP in that region. It is worth noting that only if the viscosities of the two phases are similar or if the viscosity of the matrix polymer is moderately higher than that of the LCP, can fibrous structure develop (Beery et al. 1991a).

The skin-core morphology in the extrudates from capillary or extruder die mostly supports the effect of shear flow on the LCP fibrillar formation.

Seo et al. (1995) investigated the structure development of ethylene-propylene-diene terpolymer (EPDM) (as a matrix)/Vectra B950 (as a LCP) blends in a straight die with no converging section, and in converging dies with different converging angles: 7.5°, 10°, 30°, and 45°. SEM photographs reveal that the strand from the straight die has relatively large particles in the core region that appeared to be deformed very little; more oriented materials exists at the outer region, which has undergone mostly shear flow. In the strands from converging dies, SEM photographs did not show any evidence of particle deformation in the core region even though extensional flow exists in this region. It seems that the extensional flow is not large enough to deform LCP particles, but the orientation of deformed particles near the outer region in converging dies with large converging angle is higher than that with small converging angle. Seo et al. (1995) also tested the effect of screw speeds on particle deformation in the die. As expected, more macrofibrils and fibers appear in the outer region, because of more severe shearing action, and overall particle size is reduced. The conclusion drawn from these results is that shear flow can deform dispersed LCP particle only at the outer region.

Other typical studies related to the shear flow induced morphology of LCP blends include Isayev and Modic (1987), Weiss et al. (1987), and James et al. (1987). All of them reported on fibrous structure formations at high shear rate.

Recently, Li et al. investigated the rheological behavior of polyamide-66/liquid crystalline blend, and found that rheological behavior of the blend strongly depended on the LCP content (Li et al. 2013).

From all the above reported results, one observes that fibrous structures of TP/LCP blends subjected to shear flow could be formed only at very high shear rates and mostly near the skin region. It could be concluded that shear flow is not a key factor in the formation of fibrous morphology of TP/LCP blends.

In general, in elongational flow, droplets of the dispersed phase are deformed into fibrils. Evidence of this statement has been found for blends containing LCPs both in convergent flow at the entrance of the capillary (Blizard and Baird 1987; Kohli et al. 1989; Weiss et al. 1987) and in non-isothermal elongational flow (spinning processes) (Carfagna et al. 1991; La Mantia et al. 1990b; Lin et al. 1993; Mehta and Deopura 1993; Qin et al. 1993). In this latter case, La Mantia et al. (1990b) observed that relatively high draw ratios are needed in order to produce fibrils of the liquid crystal phase. Indeed, nylon/Vectra B950 blends show fibrils only for draw ratios larger than 50, but the author did not explain the reasons of this behavior. Lin et al. (1993) reported that the sample of PC/LCP blends with draw ratio of 25 shows uniform and well-defined LCP microfibrils. The diameter of the microfibrils is in the range of 0.05–0.3 μm. The experimental results of La Mantia et al. and Lin et al. show that the components of the TP/LCP blends are crucial to the generation of an LCP fibrillar structure. Lin et al. explained the morphology of PC/LCP *in situ* composites using Taylor's theory on the deformation of Newtonian droplets suspended in another Newtonian fluid. They concluded that drawing increases the deforming force, thus leading to a higher aspect ratio. Their experimental results also show that for given sufficiently large LCP droplets, the LCP phase can be readily stretched into the fibrillar form by drawing.

The viscosity ratio is also very important in the spinning process for the formation of LCP fibers. Baird and Ramanathan (1990) showed that the LCP phase of the two blends, with the same matrix and different LCP samples having similar viscosity, has different size (droplets or fibrils) depending on the temperature dependence of the viscosity of the two components. Fibrils are formed and retained when the solidification behavior of the LCP component is similar to that of the thermoplastic matrix. Carfagna et al. (1991) studied the spinning process of the blends of polyetherimide Ultem 1000 as a matrix and the fully aromatic polyester K161 as a LCP, which have different glass transition temperatures ($T_g$). Since the K161 has a $T_g$ at 280 °C, over 70 °C higher than that of Ultem 1000, during spinning of the extrudate at room temperature, K161 quickly cools to a temperature close to its $T_g$, freezing any motion of the rods and not allowing further macromolecular orientation, by spinning. Therefore, the authors pointed out that the maximum level of orientation and deformation for LCP fibers in the blend is reached only inside the capillary. During fiber spinning, Ultem 1000 is drawn after the die exit, but the LCP fibers quickly stiffen due to the cooling. Thus, the LCP is almost unaffected by the fiber spinning, retaining the degree of molecular orientation reached inside the capillary. They also stressed the importance of shear viscosity ratio. To obtain a liquid crystalline phase, in the form of fibers with a high aspect ratio, LCP viscosity should be lower than that of matrix.

Heino and Seppala (1992) studied the effects of LCP content and different types of matrices on morphology developed during spinning process. At low LCP contents, the LCP phases can form only a few small fibers and no significant reinforcement is achieved, while at larger LCP contents there are enough oriented LCP fibers in the matrix (see Fig. 10.1) to carry a significant amount of the load and reinforce the matrix. A clear difference in morphology of PET/LCP blends can be seen in Fig. 10.1. The left and right in Fig. 10.1 are the morphologies of the core and skin regions, respectively. The thickness of the skin layer was generally about 100–200 μm. Although the length of the LCP fibrils could not be measured accurately at the transversally fractured surface, they ranged from about 5–10 μm

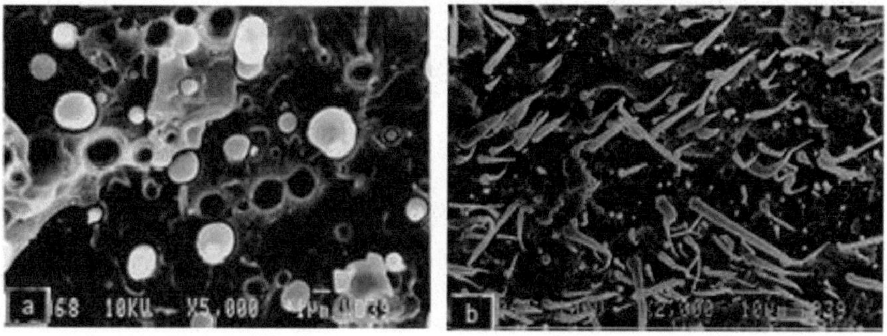

**Fig. 10.1** SEM studies of PET/ 10 wt% LCP blend (Heino and Seppala 1992)

at lower LCP contents (5 and 10 wt%) to at least tens of microns at higher LCP contents. The diameter of the LCP fibers and spheres was mostly 1–3 μm in the PET/LCP blends. The amount of fibrillar LCP domains increased with increasing LCP contents and increasing draw ratio. The effect of draw ratio could be seen from Fig. 10.1, where the spherical particles in the core region at lower LCP contents are seen to be slightly deformed at even higher draw ratio.

The rheological characteristics of the pure components, PET and LCP, and PET/LCP blends at various temperatures are shown in Fig. 10.2. The viscosity of LCP shows a continuous shear-thinning behavior. This is due to the orientation of nematic domains in LCP in the direction of flow. The oriented nematic domains smoothly glide past each other, thus lubricating LCP melt and lowering the melt viscosity. The viscosity of the PET/LCP blend also showed the same behavior as the 100 % LCP, resulting in the formation of LCP fibers. Thus, the reduction of PET/LCP blend viscosity is more significant at high shear rate because of the presence of the fibers. Figure 10.2 shows that the viscosity of PET/LCP blend is apparently lower than that of the original components of the blend, PET and LCP, which, as noted by several authors, will improve the processability of the blend.

It is noted that there exists a crossover point between the flow curves of PET melt and that of the LCP melt. This point has a special significance because, according to the literature (Levitt et al. 1996; Meijer and Janssen 1994), maximum deformation during flow in a capillary occurs when the viscosity ratio of the original components of the blend is close to unity, or larger than unity. This suggests that the shear rate during processing should be equal to or higher than the value at the crossover point, in order to achieve maximum fibrillation during processing.

There have been some concerns expressed regarding transesterification reaction that might occur during the processing of the TP/LCP blends. According to the available literature, transesterification reaction could lead to enhanced compatibility between the polymer blends (Heino and Seppala 1992; Porter and Wang 1992). The fact that the blends in our studies remain incompatible through all the processing variables used indicates that no transesterification reaction has taken place. Furthermore, according to Heino and Sepalla (1993), transesterification reaction between Vectra A950 and PET can occur only in the presence of certain catalysts, such as $Sb_2O_2$, stannous octoate or zinc acetate.

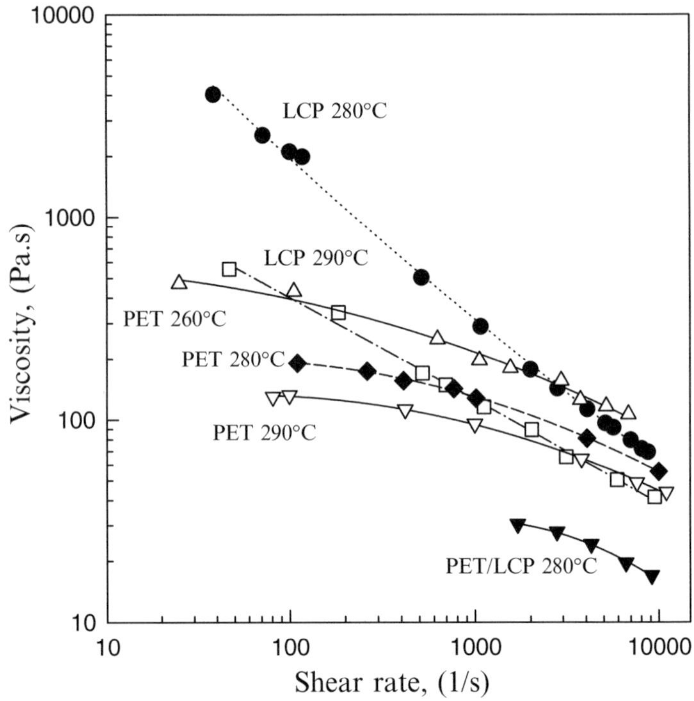


**Fig. 10.2** Shear viscosity as a function of shear rate for PET, LCP, and PET/LCP blend

## 10.4 Experimental Procedures for Preparation of TP/LCP Self-Reinforced Composites

In this section two experimental procedures are described, one for preparing a blend of LCP and a thermoplastic polymer that did not result in self-reinforcing composite, the other for preparing self-reinforcing composite. The two experimental procedures were chosen from the author's own work to illustrate the importance of the preparation method in achieving the self-reinforcing effect.

### *10.4.1 Materials*

The materials used in this study were poly (ethylene terephthalate) (Shell Cleartuf 7202C) with a 0.72 intrinsic viscosity, and a liquid crystalline polymer, produced by Ticona under the trade name Vectra A950. The LCP is a co-polyester of 73 mol% HBA (4-hydroxybenzoic acid) and 27 mol% HNA (6-hydroxy-2-naphthoic acid).

### 10.4.2 Batch Mixing of PET/LCP Blends

As discussed Sect. 10.3.2, in the molten state, and with no effect of elongational shear, the morphology of blends of LCP and thermoplastics consist of near-spherical droplets of PC in the matrix polymer (see Fig. 10.1).

In order to get a reliable measure of the effect of shear rates on LCP droplet size formation in the TP matrix, a Brabender batch mixer was used to mix about 300 $cm^3$ of the LCP and PET. The concentration of LCP for all blends was fixed at 15 % by weight, with no addition of a compatibilizer. Mixing was conducted at 280 °C over a time period of 3 min. The mixing temperature of 280 °C was determined from preliminary tests, and was found to be the lowest possible temperature needed to ensure minimum degradation, but high enough to melt the LCP. The components were thoroughly dried in a vacuum oven at around 60 °C for 24 h before mixing. Five blends were prepared by varying mixing speeds from 30 to 90 rpm in 15 rpm increments.

The speed of the mixer was calibrated using a Mitutoyo digital tachometer before and during the mixing experiments to ensure accurate and reproducible mixing speeds from batch to batch. The prepared blends were dried and kept in separate sealed plastic bags for further analyses.

Samples obtained from the batcher mixer were subsequently compression molded for mechanical testing. A simple plaque mold was designed and constructed for this purpose. The mold was a two piece combination, one piece is a metal sheet (3.18 cm by 14.65 cm by 0.5 cm) having a rectangular cut in the center with the dimensions of 1.27 cm by 10.16 cm by 0.01 cm, the other is a solid metal sheet cover.

### 10.4.3 Injection Molding Procedure to Produce PET/LCP Self-Reinforced Composite

The PET pellets were manually mixed with different concentrations of LCP pellets, and loaded into a confined hopper with a dryer installed directly to feed a slot of the injection molding machine (Model: TOYO PLASTAR Ti-90G). The mixture of PET/LCP pellets was dried for 3 h at 165 °C in the hopper prior to molding. The specimens were molded to ASTM code D638 Type I. The most important parameters in injection molding, which have significant effects on the melt flow in mold cavity, are the injection speed (or filling time), mold temperature, and melt temperature. In this study, the three parameters were varied to determine their effects on the morphology development of the LCP blends. Injection speed, expressed by the percentage of the maximum injection rate of injection molding machine, were changed from 10 to 16 %, giving injection speeds of between 13 and 21 mm $s^{-1}$. The adjustment of the mold temperature was completed by changing the temperature of coolant, and the temperatures were set to be 18, 35, and 50 °C. Melt temperatures were set to be 304, 310, and 320 °C. All other parameters, listed in Table 10.1, were kept constant for all runs.

**Table 10.1** Injection molding conditions

| Packing pressure (psi) | Holding time (s) | Screw rotation speed[a] (rpm) | | Back pressure[a] (psi) | |
|---|---|---|---|---|---|
| 700 | 15–20 | 50 | 5 | 0 | 2418 |

[a]The first value applies to the screw's axial position from 0 to 0.623 in. The second value applies to this position from 0.623 to 0.625 in.

### *10.4.4 Self-Reinforced Composite from Liquid Crystalline Oligomers via Reactive Extrusion Process*

In this two-step process (Narh 2009), first a reactive extrusion procedure is utilized to randomly place reactive sites along a flexible thermoplastic chain in the melt. Subsequently, in the same pass, an end-functionalized liquid crystalline oligomer (LCO) is added to the TP and melt mixed vigorously in a co-rotating reactive extruder, thereby grafting the LCO onto the reactive site on the thermoplastic chain. Since the LCO is still immiscible with the thermoplastic, upon cooling the LCO will phase separate in a self-assembly fashion to form domains along the chain. Having covalent bonding to the thermoplastic chain, the nano-sized LCO domains result in a liquid crystalline domain for a self-reinforcing composite. The thermoplastic polymer is randomly functionalized with a variable number of epoxy species at up to 4 wt% epoxy groups. The LCO for these initial studies are analogous to the Ticona polyester LCP that was used in the studies described in this chapter—just much lower in molecular weight.

### *10.4.5 Mechanical Property Measurement*

All the mechanical properties of the samples were measured in the tensile mode using the Instron universal-testing instrument, Model 1125. All tensile specimens were tested at a crosshead speed of 5 mm/min. The gage length was set to 50 mm. Each data point in this work was the average of at least four measurements.

## 10.5 Morphology Analysis of Self-Reinforced Composites

In this section experimental procedures to analyze the morphological characteristics of PET/LCP blend are described.

### 10.5.1 Sample Preparation for Morphology Examination

The samples for the morphological studies were cut from the middle section of the injection molded specimens. The cut surface was polished starting with 400 emery sand cloth and finishing with Billiard cloth. The polished samples were subsequently etched with 40 % by weight aqueous methylamine solution for 24 h to remove the PET. The etched samples were washed with distilled water, and then dried in air at room temperature.

### 10.5.2 Optical and Scanning Electron Microscopy Analysis of Morphology

The morphology of the etched samples was examined with both optical microscope (OM), and the Electro Scan Environmental Scanning Electron Microscope (ESEM), without any surface coating. The former helps to examine the entire surface while the latter reveals fine details of the morphology.

The ESEM is integrated with Kevex Sigma 3 Energy Dispersive X-ray (EDX) detector. Compared with a conventional Scanning Electron Microscope (SEM) which requires all samples to be observed and analyzed in a vacuum, the ESEM offers the advantage of viewing samples at much higher pressures (up to 50 Torr) and under a variety of environmental conditions. Samples can be analyzed with little or no sample preparation at all. The optical micrographs were taken with a stereomicroscope. The samples in all the micrographs in this paper were etched under identical conditions.

For the SEM, magnifications of between 270 and 450 were used. The electron beam scanned the sample in the thickness direction, starting from the sample's edge to its center. This enables the gradual changes of morphology across the thickness of samples to be viewed. For a more detailed structure of the morphology, a higher magnification of about ×450 was used.

#### 10.5.2.1 Effects of Mixing Shear Rates on Morphology

In theoretical studies (Manas-Zloczower and Feke 1988; Meijer and Janssen 1994), the term "uniform particle distribution" is generally accepted as an indication that a mixture is perfectly random. Prior to the start of the mixing process, the components are assumed to be completely separated and the particles of the individual components are indistinguishable from one another. As mixing progresses, these agglomerates are first ruptured, followed by separation of the fragments from each other, and finally the distribution of the ultimate particles throughout the melt. This is referred to in the literature as dispersive mixing (Meijer and Janssen 1994) as opposed to distributive mixing (Porter and Wang 1992) which does not involve

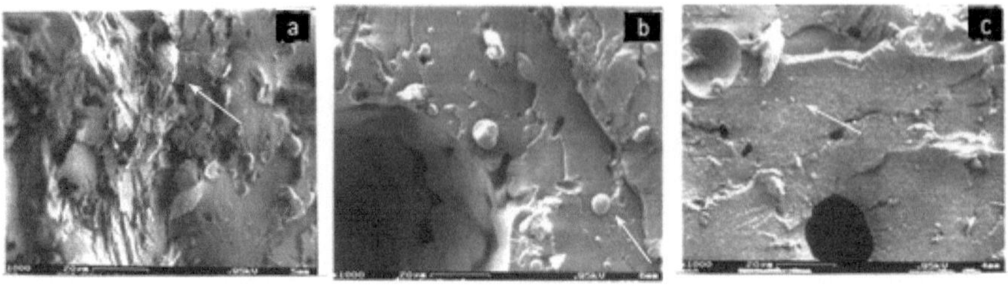


**Fig. 10.3** SEM photo of 15 % PET/LCP blend mixed in a Brabender batch mixer at (**a**) 30 rpm, (**b**) 60 rpm, and (**c**) 90 rpm. The spherical particles in (**a**, **b**) are LCP dispersed phase. The average sizes of the LCP "particles" in (**c**) are less than 2 μm. The *arrows* show the LCP particles; Note the reduction in size with increasing mixing speed, from a macro scale (**a**) to a micro scale (**c**)

breaking up of particles. Several optical methods have been used to capture various stages of the degree of mixing and miscibility of polymer blends (Yu et al. 1995), including light scattering (Narh 1996), high power optical microscopy, and scanning electron microscopy. Scanning electron microscopy technique was used in the present studies to determine, qualitatively, the degree of mixing of the blends, and, hence, the effects of mixing speed on the degree of mixing through changes in morphology of the blend. Analyses were conducted for three different samples prepared at 30 rpm, 60 rpm and 90 rpm respectively. The SEM microphotographs are presented in Fig. 10.3a–c respectively.

Figure 10.3a shows rather large, nearly spherical particles of the minor LCP phase dispersed in the PET matrix. As the mixing speed is increased, one sees further evidence of dispersive mixing, where the relatively large spherical particles seen in Fig. 10.3a are now broken into much smaller sizes (Fig. 10.3b). As the mixing speed is increased from 60 to 90 rpm, it is observed that the dispersed particles have been further reduced to barely discernible sizes which are uniformly distributed throughout the matrix (Fig. 10.3c). These observations correlate well with the results from the mechanical property tests, as discussed in the next section.

#### 10.5.2.2 Tensile Modulus Dependence of Mixing Speed

Figure 10.4 shows a plot of tensile modulus as a function of mixing speed. The figure also shows two data points corresponding to the pure PET and LCP. This figure clearly shows that the modulus increases with the mixing speed. Furthermore, although intensive mixing is not expected to result in the formation of fibers from the LCP components which could lead to optimal properties as one would from extrusion processes, nevertheless Fig. 10.4 shows a synergetic effect of the LCP component in the form of increased tensile modulus of the blend over the pure PET sample.

An alternative method of assessing the effect of the degree of mixing on the mechanical properties of the blend is by measuring the toughness of the blend. This result is also shown in Fig. 10.4. Here, the toughness is the area under the tensile stress–strain curve. In a similar study, but using the number of passes through a twin

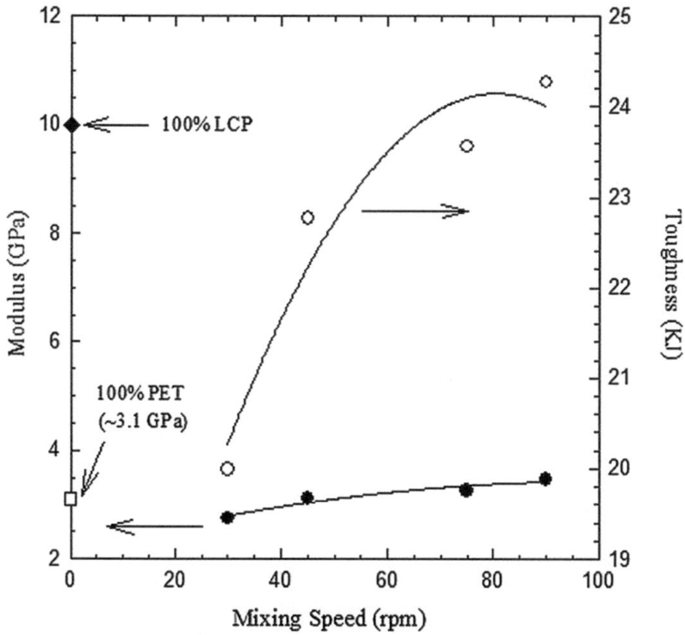


**Fig. 10.4** Tensile modulus and toughness versus mixing speed for 15 % PET/LCP blends. *Filled circle* tensile modulus; *open circle* toughness (area under stress–strain curve); *filled diamond* tensile modulus for 100 % LCP

screw extruder as a measure of degree of mixing, Baird et al. (1993) reported that the toughness of PEI/LCP (Ultem/LCP) blends increased with the number of passes. Figure 10.4 shows that increasing the degree of mixing of the blends also increases the toughness of the material, thus corroborating the results of Baird et al.

## *10.5.3 Morphological Development During Injection Molding of PET/LCP Blends*

Figure 10.5 is a montage of SEM photographs clearly showing morphological hierarchy in the skin-core structure. The LCP domains in the picture are easily distinguished from the dark PET matrix. The left side in the picture is the edge of the sample and the arrow indicates the flow direction. The individual fibers can be clearly seen in the skin zone and appear to be highly aligned in the direction of flow; non-deformed LCP domains exist mainly in core region. Three sublayers can be approximately identified in the micrograph through differences in the size, shape, and orientation of the LCP domains. The first layer is the top skin, approximately 390 μm thick close to the edge of the sample, and consists of longer and highly oriented fibers.

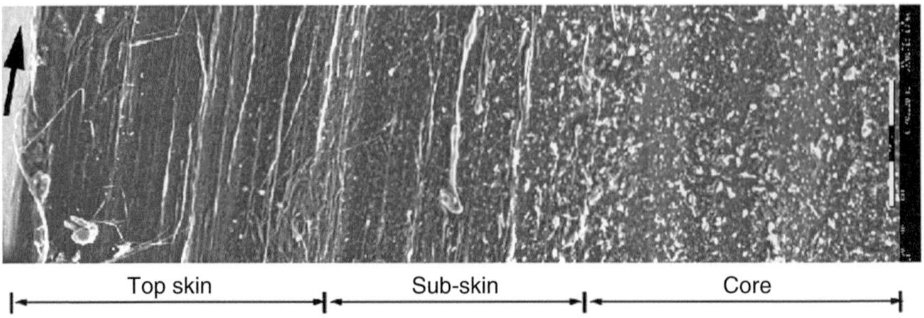


**Fig. 10.5** A SEM micrograph of in situ formed LCP fibers, following extraction of PET matrix from the PET/LCP blend. The injection speed and melt temperature were 13 mm $s^{-1}$ and 304 °C, respectively. The *left* of the picture is sample and the *right* is the sample core. The *arrow* indicates the flow direction

The diameters of the fibers range from 1.6 to 3.2 μm. The second layer is the subskin, approximately 310 μm thick. Compared with fibers in the top skin, the fibers in the subskin are shorter and less oriented, and the diameters of the fibers in the subskin range from 3.5 to 6.0 μm. But almost all of the fibers in the skin area (top and sub skins) are parallel to each other. The third layer is the core with non-deformed LCP domains. The aspect ratio of these non-deformed LCP domains is approximately unity.

#### 10.5.3.1 Mechanical Properties of Injection Molded PET/LCP Blends

Several mechanical tests were conducted on specimens with morphologies similar to that in Fig. 10.5, for several processing conditions. From the results of such tests, it has now been established that improvement in mechanical properties of TP/LCP blends can be correlated with the processing parameters. In particular, these studies have established that the relative thickness of the skin/core structure in the cross section of injection molded samples correlates with the magnitude of the mechanical properties. The latter, in turn correlates with the magnitude of three of the most critical process parameters in injection molding process, namely, injection speed, mold temperature and melt temperature.

## 10.6 Micromechanics Model for Self-Reinforcing Composites

As a result of the findings described in Sect. 10.5, a new phenomenological model was developed that enables us to predict the correlation between the mechanical properties of PET/LCP blends and process parameters. The predictions are in good agreement with experimental results that we obtained, as described below (Li and Narh 2000, 2001; Narh and Li 2000; Narh et al. 2000).

This model can be summarized as follows: Assuming a perfect bonding and absence of physical or chemical interactions of constituents at the interface, and perfect oriented continuous fibers, the well-known Halpin–Tsai (Halpin and Tsai 1969) composite model can be used to predict the modulus of the blends, provided it incorporates a variable fiber aspect ratio. The model is recast in the following form:

$$\frac{E}{E_m} = \frac{1 - ABV_f}{1 - BV_f} \tag{10.1}$$

with

$$B = \frac{E_f/E_m - 1}{E_f/E_m + A} \quad \text{and} \quad A = 2\frac{L}{D}$$

where E, $E_f$ and $E_m$ are the moduli of composite, the reinforcing phase, and the matrix, respectively; L/D is the aspect ratio of fibers. $V_f$ is the volume fraction of dispersed phase.

The fiber aspect ratio is related to strain ε in the form:

$$\frac{L}{D} = \sqrt{\frac{3}{2}}(\varepsilon + 1)^{\frac{3}{2}}\left(\frac{L_o}{D_o}\right)^{\frac{3}{2}} \tag{10.2}$$

where $L_o/D_o$ ($L_o/D_o = 1$ for a sphere) and L/D are the aspect ratios of the LCP domains before and after deformation.

Figure 10.6 shows a sample result based on (10.2). The figure shows the variation of the aspect ratio of LCP fibers across the thickness of the mold cavity, at different injection speeds, melt temperatures (similar results were obtained for different mold temperatures). Here, it is assumed that the original aspect ratio of the fibers is unity, a fact borne out by the results shown in Fig. 10.5. Based on prediction results such as those in Fig. 10.6 and, showing that the aspect ratio of the LCP fibers varies from large, near the skin, to small, in the core, and the fact that most of the LCP fibers near the surface are parallel to each other, we have developed a composite model that incorporates a variable fiber aspect ratio, the latter being dependent on process parameters such as injection speed and mold temperature. The model was based on the assumption of an affine deformation of the LCP domains, without interfacial reactions.

Several studies (Yi et al. 1996; Zachariades and Porter 1983) have reported that the modulus of LCP fibers is a function of the fiber aspect ratio that increases monotonically with fiber length, and more rapidly as the fiber diameter decreases. Yi et al. (1996) gave an approximate linear function to describe the relationship between the modulus and aspect ratio of Vectra A950 fiber in the form:

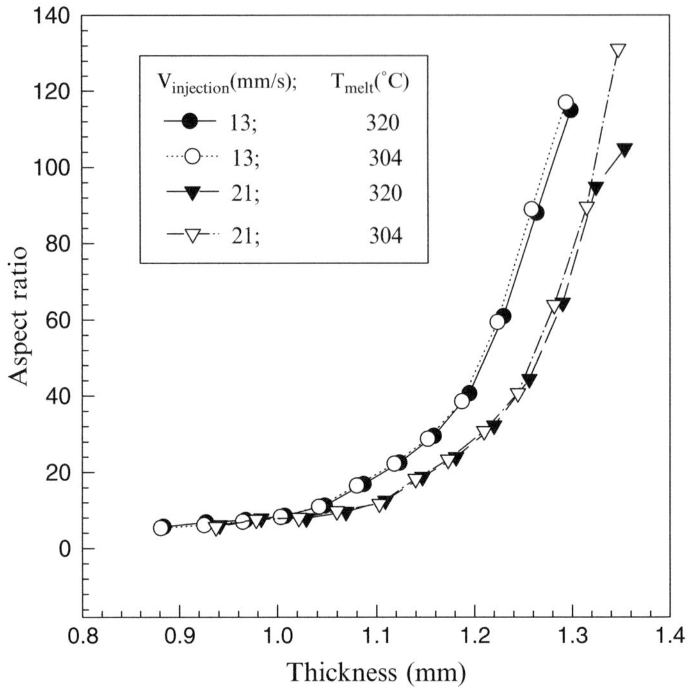


**Fig. 10.6** Effect of injection speed and melt temperature on aspect ratio at a mold temperature of 18 °C for PET/LCP (85/15 wt%) blend. The thickness is measured from the center of mold cavity (*left* of the figure)

$$E_f = 17.4 + 0.33\frac{L}{D}\,\text{GPa} \tag{10.3}$$

Equation (10.3) suggests that the modulus of the LCP has a minimum of 17.73 GPa when the aspect ratio is unity. But some studies (Acierno and Nobile 1993) have reported that the modulus could be as low as 9.7 GPa. Equation (10.3) also shows that there is no limiting value for the modulus of Vectra A950 if the aspect ratio of the LCP fiber increases infinitely. Obviously, this is not reasonable. We used a logarithmic polynomial to model the relationship between the modulus of the LCP and its fiber aspect ratio, as suggested by Zachariades and Porter (1983):

$$\log\left(\frac{1}{E_f}\right) = C_1 + C_2\log(X) + C_3\log(X)^2 \tag{10.4}$$

where $X = (L/D)^{-2}$, $C_1$, $C_2$, and $C_3$ are material-specific constants.

Substituting (10.1) and (10.2) into (10.4), replacing the constants $C_1$, $C_2$, and $C_3$ by their numerical values (Li and Narh 2001), and integrating the modulus with respect to the thickness of mold cavity, we have:

$$E_{total} = \frac{2}{H}\int_0^{H/2} E\,dy = \frac{2E_m}{H}\int_0^{H/2} \frac{1 - \sqrt{6}B(\varepsilon+1)^{3/2}V_f}{1 - BV_f}dy \tag{10.5}$$

A numerical integration was used to solve (10.5).

Figure 10.7 shows the modulus of Vectra A950 as a function of aspect ratio. It is seen that the modulus increases more rapidly at small aspect ratio, and slowly as the aspect ratio exceeds 40. The results from (10.4) are comparable to those given by others (Blizard et al. 1990; O'Donnell and Baird 1996). Blizard et al. (1990) reported a maximum value of E = 28 GPa for Vectra A950. This corresponds to our data, when the aspect ratio is above 40. O'Donnell and Baird (1996) presented a range of modulus, from 10.3 to 18.2, for Vectra A950, at different mold temperatures and plaque thickness, which corresponds to our calculations with aspect ratio ranging from 2 to 10.

The measured PET modulus ($E_m$) in the range of 2.75–2.9 GPa and three different volume fractions of the LCP ($V_f$) were used to calculate the modulus of the blends.

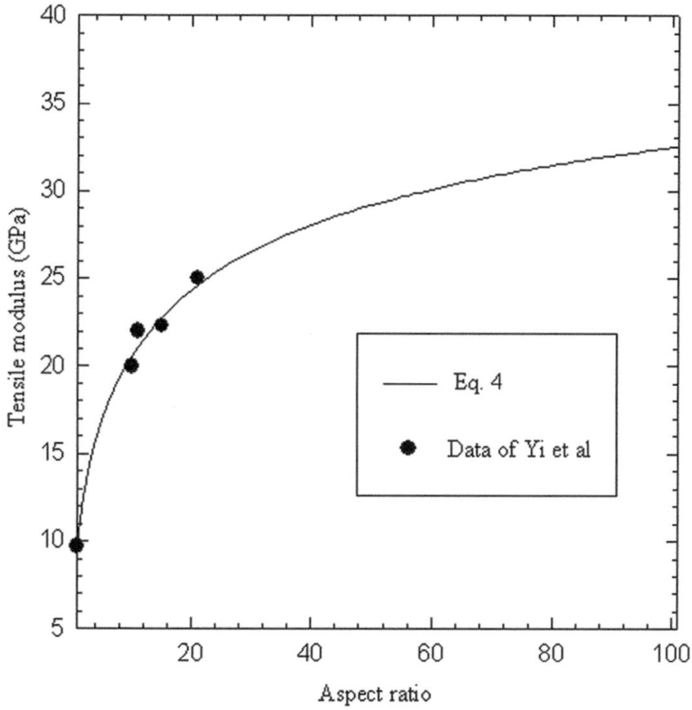


**Fig. 10.7** Tensile modulus of Vectra A950 as a function of LCP fiber aspect ratio

Figure 10.8 shows a sample result of measured, as well as predicted tensile modulus for a PET/LCP (85/15) blend, as a function of injection speed, at three different mold temperatures. As can be seen, the agreement between predicted and experimental results is reasonable in view of assumptions used. In particular, the assumption that the LCP fibers are perfectly oriented and parallel in the blends may not be valid for all the processing conditions. Consequently, the true strength of this model can be rigorously evaluated only after several measurements involving different processing conditions, different blend compositions, processing temperatures etc. In the present model, like most classical micromechanics models for two component composites (Christensen 1979), we have assumed perfect interfacial adhesion and absence of physical or chemical interactions of constituents at the interface. Additionally, we assumed perfect oriented fibers in the matrix.

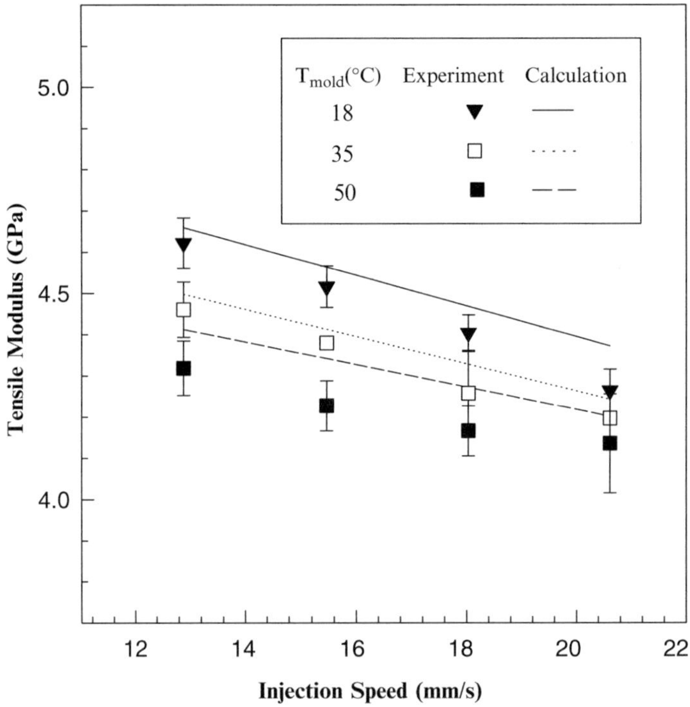


**Fig. 10.8** Tensile modulus of PET/LCP (85/15 wt%) blend vs. injection speed at a melt temperature of 304 °C. The data are shown with ±1δ

## 10.7 Conclusions and Future Perspective

Based on a combined scanning electron and optical microscopy studies of injection molded specimens, the existence of distinct morphological hierarchy in injection molded samples of PET/LCP blends, was demonstrated. A skin layer, and a core layer, can be distinguished according to the shape of LCP domains. In the core layer, LCP domains exist mainly in spherical forms, while in the skin layer, fibril-like LCP domains are found. The skin layer can be subdivided into topskin and subskin layers, the former consisting of very long LCP fibers, while the latter contains short LCP fibers. The diameters of the fibers are about 1.6–6.0 μm in the 15 % LCP blend. The thickness of the skin layer, one of the characteristics identified in the morphology of the blends was measured. Its magnitude ranged from approximately 0.4–0.6 mm in the 15 % LCP blends, depending on the processing conditions. The morphology of the blends is very sensitive to the mold temperature, and injection speed, but not to the melt temperature.

The results have also shown that the tensile modulus of PET/LCP blends is sensitive to both injection speed, and melt temperature of the blends. The modulus of the blends increases with decreasing injection speed, and decreasing melt temperature. The variation of modulus is related to the change of morphology caused by the changes of these two processing parameters.

In most of experiments that result in self-reinforcing composite, the LCP fibrous phase is not permanent and disappears upon reprocessing. It is, however, possible to use a reactive extrusion approach to randomly place reactive sites along a thermoplastic chain in the melt before mixing with the LCP, in order to form a reprocessable self-reinforcing TP/LCP composite.

The modulus of pure LCP fibers is basic data for studying mechanical properties of LCP blends. For future studies based on the studies described in this chapter, it would help if a technique to measure the LCP fiber size in the matrix, and to determine the distribution across the thickness of the mold cavity could be developed, in order to confirm the theoretical predictions of morphology of LCP blends. One suggested method for experimental measurement of the effect of LCP fiber aspect ratio on the mechanical properties of LCP blends is to construct a mold with several inserts. Using these inserts one can prepare ASTM D638 standard specimens with several thickness ranging from 3 mm to 100 μm or less under identical processing conditions. One can then compare the measured modulus versus thickness, with predictions.

In the micromechanical model presented, like most classical micromechanics models for two component composites (Christensen 1979), we have assumed perfect interfacial adhesion and absence of physical or chemical interactions of constituents at the interface. Additionally, we assumed perfect oriented fibers in the matrix. The validity of these assumptions in our model will be one of the tasks to be undertaken for future studies on this subject. For example, we have not taken into account the fact that transesterification reaction might occur during the processing of TP/LCP blends, as well as non-perfect fiber orientation. According to the literature, transesterification reaction between the LCP and TP molecules could lead to enhanced compatibility between the polymer blend constituents.

# References

Acierno D, Nobile MR (1993) Processing and properties of thermotropic liquid crystalline polymers. In: Mantia FPL (ed) Thermoplastic liquid crystal polymer blends. Technomic, New York, p 76

Baird DG, Ramanathan R (1990) The in-situ generation of liquid crystalline reinforcements in engineering thermoplastics. In: Culbertson BM (ed) Contemporary topics in polymer science, vol 6. Plenum, New York, pp 73–93

Baird DG, Bafna SS, de Souza JP, Sun T (1993) Mechanical properties of in-situ composites based on partially miscible blends of polyetherimide and liquid crystalline polymers. Polym Compos 14:214

Beery D, Kenig S, Siegman A (1991a) Structure development during flow of polyblends containing liquid crystalline polymers. Polym Sci Eng 31:451

Beery D, Kenig S, Siegmann A (1991b) Structure and properties of molded polyblends containing liquid crystalline polymers. Polym Eng Sci 31:459

Blizard KG, Baird DG (1987) The morphology and rheology of polymer blends containing a liquid crystalline copolyester. Polym Eng Sci 27(9):653–662

Blizard KG, Federici C, Federico O, Chapoy LL (1990) The morphology of extruded blends containing a thermotropic liquid crystalline polymer. Polym Eng Sci 30:1442

Brinkmann T, Hoeck P, Michaeli W (1991) Proceedings of compalloys Europe'91, Loughborough

Carfagna C, Amendola E, Nicolais L, Acierno D, Francescangeli O, Yang B, Rustichelli F (1991) Blends of a polyetherimide and a liquid crystalline polymer: fiber orientation and mechanical properties. J Appl Polym Sci 43(5):839–844

Christensen RM (1979) Mechanics of composite materials. Wiley, New York

Crevecoeur G, Groenickx G (1990) Morphology and mechanical properties of thermoplastic composites containing a thermotropic liquid crystalline polymer. Polym Eng Sci 30:1442

Dutta D, Fruitwala H, Kohli A, Weiss RA (1990) Polymer blends containing liquid crystals: a review. Polym Eng Sci 30:1005

Halpin JC, Tsai SW (1969) Effects of environmental factors on composite materials. U.S. Government Report, Washington

Heino MT, Seppala JV (1992) Extruded blends of a thermotropic liquid crystalline polymer with polyethylene terephthalate, polypropylene and polyphenylene sulfide. J Appl Polym Sci 44:2185

Heino MT, Seppala JV (1993) Effect of mixing time and catalyst addition on the compatibility of PET/LCP. Acta Polytech Scand Chem Technol Metall Ser 25:214

Isayev AI (1987) Injection and compression molding fundamentals, vol 15. Marcel Dekker, New York

Isayev AI, Modic M (1987) Self-reinforced melt processible polymer composites: extrusion, compression, and injection molding. Polym Compos 8:158

James SG, Donald AM, MacDonald WA (1987) Blends of a liquid crystalline copolyester with polyethersulphone. Mol Cryst Liq Cryst 153:491

Jerman RE, Baird DG (1981) Rheological properties of copolyester liquid crystalline melts. I. Capillary rheometry. J Rheol 25:275

Kayaisang S, Amornsakchai T, Saikrasun S (2009) Influence of liquid crystalline polymer and recycled PET as minor blending components on rheological behavior, morphology, and thermal properties of thermoplastic blends. Polym Adv Technol 20:1136–1145

Kim GM, Michler GH, Reichert P, Kressler J, Mulhaupt R (1998) Micromechanical deformation processes in PA12-layered silicate nanocomposite. Polym Mater Sci Eng 79:178

Kiss G (1987) In situ composites: blends of isotropic polymers and thermotropic liquid crystalline polymers. J Polym Eng Sci 27:410

Kohli A, Chung N, Weiss RA (1989) The effect of deformation history on the morphology and properties of polycarbonate and a thermotropic liquid crystalline polymer. Polym Eng Sci 29:573

La Mantia FP, Saiu M, Valenza A, Paci M, Magagnini PL (1990a) Relationships between mechanical properties and structure for blends of nylon 6 with a liquid crystal polymer. Eur Polym J 26:323
La Mantia FP, Valenza A, Paci M, Magagnini PL (1990b) Rheology-morphology relationships in nylon 6/liquid crystalline polymer blends. Polym Eng Sci 30:7
Levitt L, Macosko CW, Pearson SD (1996) Influence of normal stress difference on polymer drop deformation. Polym Eng Sci 36:1647–1655
Li Z, Narh KA (2000) Morphological development during injection molding of self-reinforcing composites: II. Numerical simulations and analytical predictions. Polym Compos 21:978–987
Li Z, Narh KA (2001) Experimental determination and numerical prediction of mechanical properties of injection molded self-reinforcing polymer composites. Compos B Eng 32:103–109
Li Y, Pan Y, Sun Z, Sheng X, Shi W, Li J, Huang H (2013) The rheological behavior of polyamide-66/liquid crystal polymer blends (conference paper). Adv Mater Res 631–632:100–103
Lin Q, Yee A (1994) Elastic modulus of in-situ composites of a liquid crystalline polymer and polycarbonate. Polym Compos 15:156
Lin Q, Jho J, Yee AF (1993) Effect of drawing on structures and properties of a liquid crystalline polymer and polycarbonate in-situ composite. Polym Eng Sci 33:789
Manas-Zloczower I, Feke DL (1988) Analysis of agglomerate separation in linear flow fields. Int Polym Proc 2:182
Mandal PK, Siddhanta SK, Chakraborty D (2012) Engineering properties of compatibilized polypropylene/liquid crystalline polymer blends. J Appl Polym Sci 124:5279–5285
Mehta S, Deopura BL (1993) Fibers from blends of PET and thermotropic liquid crystalline polymer. Polym Eng Sci 33:931
Meijer HEH, Janssen JMH (1994) Mixing of immiscible liquids. In: Manas-Zloczower I (ed) Mixing and compounding of polymers. Hanser/Gardner Publications, New York, p 85
Narh KA (1996) The effect of shear flow on phase behavior of polybutadiene/polyisoprene blends. Adv Polym Technol 15:245
Narh KA (2009) Liquid crystalline blends as route to self-reinforcing nanocomposites. J Reinforc Plast Compos 28:1957–1973
Narh KA, Li Z (2000) Morphological development during injection molding of self-reinforcing composites: I. Experimental results. Polym Compos 21:751–761
Narh KA, Zhang Q, Li Z (2000) The effect of mixing on the properties of liquid crystalline polymer/polyethylene (terephthalate) blend. J Appl Polym Sci 75:1783–1787
Nobile MR, Amendola E, Nicolais L (1989) Physical properties of blends of polycarbonate and a liquid crystalline copolyester. Polym Eng Sci 29:244
O'Donnell HJ, Baird DG (1996) The effect of injection molding conditions on the mechanical properties of an in-situ composite. I: Polypropylene and a liquid crystalline copolyester based on hydroxynaphthoic acid. Polym Eng Sci 36:963
Porter RS, Wang L-H (1992) Compatibility and transesterification in binary polymer blends. Polym Rev 33:2019
Qin Y, Brydon DL, Mather RR, Wardman RH (1993) Fibers from polypropylene and liquid crystal polymer blends: 2. Effect of extrusion and drawing conditions. Polymer 34:1202
Seo Y, Hwang SS, Hong SM, Park TS, Kim KU (1995) Effect of die geometry on the structural development of a thermotropic liquid crystalline polymer in a thermoplastic elastomer matrix. Polym Eng Sci 35:1621
Siegmann A, Dagan A, Kenig S (1985) Polyblends containing a liquid crystalline polymer. Polymer 26:1325–1330
Suokas E (1989) The microstructure of injection-moulded thermotropic copolyester: 1. Optical and scanning electron microscope studies. Polymer 30:1105
Weiss RA, Hu W, Nicolais L (1987) Novel reinforced polymers based on blends of polystyrene and a thermotropic liquid crystalline polymer. Polym Eng Sci 27:684

Yang Q, Hirata M, Hsu Y-I, Lu D, Kimura Y (2014) Improved thermal and mechanical properties of poly(butylene succinate) by polymer blending with a thermotropic liquid crystalline polyester. J Appl Polym Sci 131:39952

Yi XS, Shi F, Michaeli W (1996) The modulus of shear-induced fibers from a thermotropic liquid crystalline polymer in a thermoplastic matrix. Polym Compos 17:423

Yu D-W, Esseghir M, Gogos CG (1995) The use of on-line optical microscopy for monitoring compounding and other polymer processing operations. In: Proceedings of SPE ANTEC, vol 41. Boston, p 136

Zachariades AE, Porter RS (1983) The strength and stiffness of polymers. In: Plastics engineering, vol 4. Marcel Dekker, New York, p 382

Zhuang P, Kyu T, White JL (1988) Characteristics of hydroxybenzonic acid-ethylene terephthalate copolymers and their blends with polystyrene, polycarbonate, and polyethylene terephthalate. In: SPE ANTEC tech. papers, vol 34, p 1237

# Chapter 11
# Optical Fredericks Transition in a Nematic Liquid Crystal Layer

**Hong Zhou, Eric P. Choate, and Hongyun Wang**

## 11.1 Introduction and Overview

The term "liquid crystals" was first introduced by Otto Lehmann in 1889 and it is quite self-explanatory. It describes a matter state where materials exhibit many properties of solid crystals such as anisotropy and symmetry, but flow like fluids or viscous fluids. This combination of properties makes liquid crystals absolutely unique and complex materials. There are many different types of liquid crystal phases. One of the most common phases is the nematic phase where the rod-like molecules tend to be aligned parallel to each other along some common preferred direction ("anisotropic axis"). Examples of nematic liquid crystals include p-azoxyanisole (PAA), 4-methoxybenzylidene-4′-butylaniline (MBBA) and 4-pentyl-4′-cyanobiphenyl (5CB). In a typical nematic such as PAA the molecules closely resemble rigid rods of length 20 Å and width 5 Å (de Gennes and Prost 1993) where 1 angstrom (Å) equals $10^{-10}$m. The state of PAA depends on the temperature. In particular, PAA is a solid below 118 °C, a nematic liquid crystal between 118 and 135.5 °C, and an isotropic liquid above 135.5 °C (de Jeu 1980). The local direction of the average molecular alignment in liquid crystals is usually

H. Zhou, Ph.D. (✉)
Department of Applied Mathematics, Naval Postgraduate School, 833 Dyer Road, Bldg. 232, SP-250, Monterey, CA 93943, USA
e-mail: hzhou@nps.edu

E.P. Choate, Ph.D.
Department of Mathematics and Statistics, Radford University, 801 East Main Street, Radford, VA 24142, USA
e-mail: echoate2@radford.edu

H. Wang, Ph.D.
Department of Applied Mathematics and Statistics, University of California, Santa Cruz, 1156 High Street, Santa Cruz, CA 95064, USA
e-mail: hongwang@soe.ucsc.edu



V.K. Thakur, M.R. Kessler (eds.), *Liquid Crystalline Polymers*,
DOI 10.1007/978-3-319-20270-9_11

described by a unit vector $\vec{n}$, called the "director". Due to the lack of polarity of liquid crystals the two vectors $\vec{n}$ and $-\vec{n}$ are indistinguishable.

Liquid crystals have been used in a wide variety of applications because external perturbation can cause significant changes in the macroscopic properties of the liquid crystal system. Both electric and magnetic fields can be applied to induce these changes. In fact, one of the most intensively studied phenomena in the physics of liquid crystals is the Fredericks transition which has been known for over 80 years. It refers to a transition from a uniform director configuration to a deformed director configuration of a homeotropically or planarly aligned nematic liquid crystal (NLC) layer upon application of a magnetic or electric field of sufficiently large magnitude. The behavior of a nematic liquid crystal is determined by two competing torques: (1) the elastic torque within the material that tries to align the NLC molecules along the easy direction as determined by the anchoring boundary conditions; and (2) the external-field-induced torque that tends to align the NLC molecules either parallel or orthogonal to the applied field. A uniform director distribution occurs when only one of these torques is present. When both torques exist, there is a competition between them. Once the applied field strength exceeds a threshold value, the uniform NLC orientation will deform and the Fredericks transition takes place.

The description of how the director is aligned on the boundary surfaces is called anchoring. The main types of anchoring include strong, weak and conical anchoring. Strong anchoring occurs when the surface energy is sufficiently large and the director orientation on the boundary of the sample can be considered simply fixed in a preferred orientation determined by an alignment technique. In the case of weak anchoring, the director alignment on the boundary is no longer rigidly fixed to the surface and it is allowed to vary. Conical anchoring refers to the situation where the director at a boundary makes a fixed angle with the tangent plane of the boundary. Strong anchoring conditions are commonly used due to their simplicity. There are several kinds of strong anchoring boundary conditions. For the planar alignment, the molecules lie parallel to the cell surface. It can be obtained by either rubbing the alignment layer (usually a polyimide thin film) or by UV photo-alignment. For the homeotropic alignment, the director is perpendicular to the cell surface. It is usually obtained by means of detergents or by applying a surfactant to the alignment surface. It is also possible to obtain any pretilt angle by evaporating silicon dioxide at different angles of incidence. Other alignment techniques, such as micron rubbing by microsphere, micro-contact printing, ion beam bombardment, are available as well (Yilmaz et al. 2008).

The Fredericks transition plays a fundamental role in the operation of many liquid crystal displays because the director orientation can be controlled easily by applying an external magnetic or electric field. One striking feature is that the magnitude of the Fredericks threshold voltage is essential to the operation of many liquid-crystal devices. If the Fredericks threshold voltage could be reduced reliably, this would make it possible for the manufacture of very low-power liquid crystal displays and other liquid-crystalline devices. The Fredericks transition can also be employed to achieve tunable all-optical switching and diode operation in the photonic devices (Miroshnichenko et al. 2006).

There have been many studies on the Fredericks transition (Khoo 1981, 1982a, b; Shoarinejad and Shahzamanian 2008; Torrent et al. 1988; Vena et al. 2007; Yang 1983; Zakhlevnykh 2004). For example, experimental observations include the first observation of the optical-field-induced Fredericks transition in the nematic liquid crystal 5CB (Durbin et al. 1981), nonlinear light amplification in MBBA above the Fredericks transition by a magnetic field (Khoo and Zhuang 1980). A quantitative theory to describe the reorientation of the director of a nematic liquid crystal induced by the field of a light wave was developed by Zel'dovich and his co-workers using the geometrical optics approximation (Zel'dovich and Tabiryan 1980; Zel'dovich et al. 1981). They calculated the solution for the spatial distribution of the director direction for intensities above and below threshold. They noted that a Fredericks transition induced by light fields in an NLC with a positive dielectric anisotropy $(\varepsilon_a > 0)$ can occur only when the light wave is normally incident on an NLC cell with homeotropic orientation. When an ordinary light wave is normally incident on an NLC cell with planar orientation, Fredericks transition will not occur. They also predicted that near the threshold the maximum angle of deviation of the director from the unperturbed direction is proportional to the square root of the excess of the intensity above the threshold intensity and that for certain nematic liquid crystals the Fredericks transition in a light field is accompanied by hysteresis in broad beams, which they believed has no analog in the case of static fields. In fact, in 1974 Deuling and Helfrich (Deuling and Helfrich 1974) had shown that a homeotropic nematic layer with negative dielectric (or conductive) anisotropy enclosed between two metallized glass plates undergoes a Fredericks transition. In 1981 Durbin et al. (Durbin et al. 1981) expressed the total electromagnetic energy density as the ratio of the magnitude of the Poynting vector to the ray velocity and then used the infinite-plane-wave approximation in which the time average of the magnitude of the Poynting vector of the optical field was assumed to be a constant throughout the medium. Minimization of the free energy led to the Euler equation. The solution of the Euler equation with proper boundary conditions gave the average molecular reorientation of the liquid crystal. Their theoretical predictions were in quantitative agreement with their experimental results. These earlier theoretical studies only took the large anisotropies of nematic liquid crystals into account but they lack the ability to properly describe the associated nonlinear effects. The ground-breaking treatment of the optically induced Fredericks transition was given by Ong (1983). In this classical work Ong provided a detailed study which includes the interaction between the optical field and the NLC. An Euler equation, which is consistent with Maxwell's equations, is derived for describing the optically induced spatial reorientation of the director of a homeotropically oriented nematic liquid crystal for the case of normal incidence. The solution of the Euler equation describes the deformation angle of the NLC. By examining the maximum deformation angle near the threshold, Ong obtained the threshold intensity and the criterion for the physical parameters that indicate whether the Fredericks transition is first order or second order at the threshold. In contrast to Durbin's assumption, Ong showed that the time average of the magnitude of the Poynting vector is not a constant, but instead the time average of the z-component of the Poynting vector is a constant. Ong's observation is consistent with the geometrical optics approximation (Zel'dovich and Tabiryan 1980; Zel'dovich et al. 1981). In 1995 Sugimura et al. (Sugimura et al. 1995) generalized the Rapini–Papoular formula (Rapini and Papoular 1969) for the anchoring energy and provided rigorous expressions for the threshold and saturation fields for a field-controlled twisted chiral nematic slab with weak boundary coupling. Self et al. (2002) considered the situation where the electromagnetic field interacts with the liquid crystal deformation. Assuming strong anchoring at solid surfaces and a one-dimensional distortion, they identified three dimensionless parameters, which are related to the elastic anisotropy of the liquid crystal, the anisotropy of the electrical permittivity of the liquid crystal, and the applied voltage, respectively. They applied systematic asymptotic methods to determine approximate solutions in five various regimes defined by physically relevant limiting cases of the parameters. Shi (2002) summarized the main results in his two papers (Shi and Yue 2000; Yang et al. 2000) to give the exact criteria for the first order Fredericks transition to occur in magnetically, electrically or optically driven liquid crystal cells. All these criteria are expressed in terms of the material parameters, cell geometrical parameters and surface anchoring parameters where the anchoring strengths at the surfaces are described by the Rapini–Papoular surface anchoring energy. It was noted that the first order magnetic or electrical Fredericks transition occurs at surface anchoring conditions that current technology is not mature enough to achieve. The first order transition cannot occur for a rigid anchoring condition and this explains why first order transitions in the magnetic and electrical cases are so difficult to observe. However, the first order transition can be easily realized for a three-dimensional Fredericks transition, as in a twist liquid crystal cell.

Colloidal suspensions—dispersions of small solid particles in a host liquid consisting of small molecules—in nematic liquid crystals can also undergo Fredericks transition. For a general review on the physics of colloidal dispersions in nematic liquid crystals, see the article by Stark (2001). Ferronematics is the name for suspensions of monodomain ferro or ferromagnetic particles in nematic liquid crystals. Dispersed ferromagnetic particles greatly enhance the magnetic properties of the liquid crystals. In 2003 Bena and Petrescu (Bena and Petrescu 2003) considered an infinite flat layer of ferronematics (FN) between two identical substrates to study the magnetic field-induced Fredericks transition in a homeotropically aligned FN with weak anchoring molecules to the walls. They have found a correction to the Burylov–Raikher expression (Burylov and Raikher 1993) for the magnetic threshold field, together with an expression for the saturation magnetic field as a function of material parameters and anchoring conditions. The saturation transition does not occur in the case of rigid anchoring molecules to the cell walls. In 2009 they extended their work to analyze the influence of anchoring conditions on the Fredericks transition in ferronematics subjected to a superposed magnetic static field and an optical field (Petrescu and Bena 2009). Using the Rapini–Papoular expression (Rapini and Papoular 1969) for the anchoring energy and the Euler–Lagrange analytical method, they obtained a correlation between the threshold intensities of the magnetic and the laser fields, the anchoring strength, the material and device parameters. They found that the fields needed for the transition are smaller than those in the rigid anchoring case.

Another system analogous to ferronematics involves ferroelectric rather than ferromagnetic colloidal particles. The colloidal particles are extremely small and at low concentrations these submicron colloids macroscopically appear similar to a pure liquid crystal with no evidence of dissolved particles. These dilute suspensions are stable because the nanoparticles do not significantly perturb the director field in the liquid crystals, and interaction between the particles is weak. In these colloidal suspensions the dielectric response is strongly enhanced and the values of the effective dielectric functions are higher than those of the analogous pure liquid crystals. Such systems were first fabricated by Reznikov et al. (2003) using particles of the ferroelectric thiohypodiphosphate ($Sn_2P_2S_6$) for the suspension and have been the subject of many researches (e.g. Li et al. 2006; Shelestiuk et al. 2011). Shelestiuk et al. (2011) explored the Fredericks transition in ferroelectric liquid-crystal nanosuspensions. Their study suggested that doping a nematic liquid crystal with ferroelectric particles can significantly decrease the electric Fredericks threshold field.

A computational approach to the optical Fredericks transition in a homeotropic nematic liquid crystal cell has been provided by Ilyina et al. (2006). They combined the popular finite difference time domain (FDTD) method to solve the Maxwell's equations with a relaxation algorithm to solve the Ginzburg–Landau equations for the liquid crystal director.

As one can see, it is impossible for us to provide a comprehensive list of all relevant publications on the Fredericks transition in liquid crystals. For more details on the theory and applications of liquid crystals, please refer to the classical books by de Gennes and Prost (1993), Chandrasekhar (1992), Khoo (1995), Virga (1994), and Stewart (2004).

## 11.2 The Classical Fredericks Transitions in a Nematic Liquid Crystal Cell

In general, the classical Fredericks transition in nematics can be fairly well-explained using continuum theory of nematic liquid crystals developed by Frank, Ericksen and Leslie. Before we present a detailed analysis on the optical Fredericks transition, which couples the interaction between the applied electromagnetic field of a light wave and the orientation of liquid crystals, we would like to briefly review the classical results (de Gennes and Prost 1993; Virga 1994; Stewart 2004). Many of our ideas here are borrowed from Stewart (2004).

Conventional analytic study of this problem normally considers an electromagnetic field in which the magnetic component is dominant. Generally the effect of the nematic orientation does not affect the magnetic field, and so the electromagnetic field structure decouples from the orientation and the electric field components can be ignored. To get started, let us consider a nematic liquid crystal sample confined between two parallel glass plates at $z = 0$ and at $z = L$. We assume strong anchoring and impose the same director at each plate. We also assume the anchoring is uniform in both the $x$- and $y$-directions so that the orientation is a function of $z$ only.

We will consider three common geometries concerning the relationship between the anchoring conditions and imposed magnetic field: splay, bend, and twist. For the splay geometry, the director is parallel to the bounding plates where $\mathbf{n}(z) = (1, 0, 0)$ is the uniform equilibrium solution. In this case, the natural choice for the magnetic field to apply in order to observe the Fredericks transition is $\mathbf{H} = H(0, 0, 1)$, $H = |\mathbf{H}|$. Assuming that the director attempts to reorient itself to be parallel to the magnetic field and any reorientation of the director only depends on $z$, we seek solutions for the director of the form $\mathbf{n} = (\cos\phi(z), 0, \sin\phi(z))$, where $\phi(z)$ is called the director angle. The bend geometry reverses these roles with the director anchored perpendicularly to the plates while the magnetic field is parallel to the plates. We set the magnetic field $\mathbf{H} = H(1, 0, 0)$ and seek solutions of the form $\mathbf{n} = (\sin\phi(z), 0, \cos\phi(z))$ in the case of the bend geometry. Figure 11.1 illustrates these three geometries.

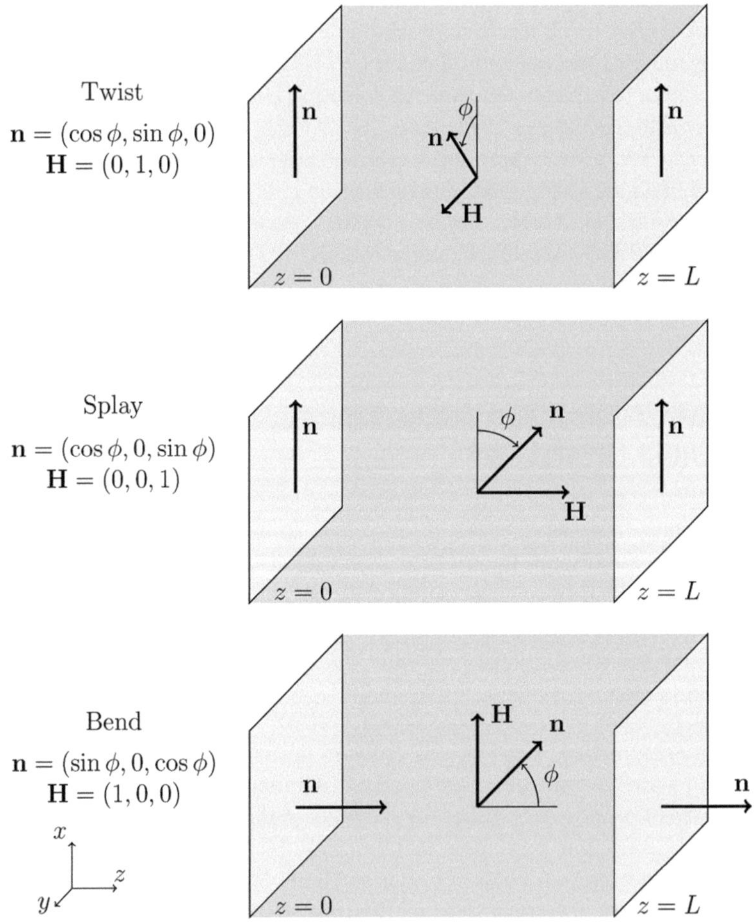


**Fig. 11.1** A simple illustration of the three common geometries

We begin by focusing on the twist geometry, in which both the initial director alignment and the magnetic field are parallel to the glass plates but are mutually orthogonal to each other. We impose a magnetic field of the form $\mathbf{H} = H(0, 1, 0)$ and search for directors of the form $\mathbf{n} = (\cos\phi(z), \sin\phi(z), 0)$. The analysis is easier in the twist geometry, and we present a detailed derivation of its solution. Afterwards, we consider the splay and bend geometries, which are very similar to each other and reduce to a form like the twist geometry under a simplifying assumption.

For boundary conditions in all three geometries, we use

$$\phi(0) = \phi(L) = 0 \tag{11.1}$$

so that all three cases have the uniform solution $\phi(z) = 0$ for all $0 \leq z \leq L$. Below the critical field strength $H_c$ of the Fredericks transition, this uniform solution is stable, but above the critical value, a stable non-uniform solution appears.

To find the stable solution, we need to minimize the free energy of the system, which balances contributions from elastic and magnetic parts. The Frank elastic free energy density for nematics is given by

$$f_{Frank} = \frac{1}{2}\left[K_1(\nabla \cdot \mathbf{n})^2 + K_2(\mathbf{n} \cdot \nabla \times \mathbf{n})^2 + K_3(\mathbf{n} \times \nabla \times \mathbf{n})^2\right] \tag{11.2}$$

where the Frank elasticity constants $K_1$, $K_2$, and $K_3$ represent the splay, twist, and bend moduli, respectively (Virga 1994). The elastic constants depend on temperature and are commonly of the order $10^{-7}$–$10^{-6}$ dyn with $K_3$ often being two or three times larger than $K_1$ and $K_2$. For example, for PAA, at 120 °C the Frank elastic constants are $K_1 = 5.0 \times 10^{-7}$ dyn, $K_2 = 3.8 \times 10^{-7}$ dyn, and $K_3 = 10.1 \times 10^{-7}$ dyn. At 125 °C, these values are changed to $K_1 = 4.5 \times 10^{-7}$ dyn, $K_2 = 2.9 \times 10^{-7}$ dyn, and $K_3 = 9.5 \times 10^{-7}$ dyn.

For analytical purposes, it is often common to use the simplified single-constant assumption with $K_1 = K_2 = K_3 = K_{all}$. Under this assumption, the Frank elastic free energy density simplifies to

$$f_{Frank} = \frac{1}{2}K_{all}\nabla\mathbf{n} : \nabla\mathbf{n} = \frac{1}{2}K_{all}\mathrm{tr}\left(\nabla\mathbf{n} \cdot (\nabla\mathbf{n})^T\right). \tag{11.3}$$

On the other hand, the free energy density contribution from the magnetic field takes the form

$$f_{magnetic} = -\frac{1}{2}\chi_a(\mathbf{n} \cdot \mathbf{H})^2 \tag{11.4}$$

where $\chi_a$ is the anisotropic magnetic susceptibility in Gaussian cgs units. We will assume $\chi_a > 0$. Putting these together, we obtain the total free energy density: $f = f_{Frank} + f_{magnetic}$.

### 11.2.1 Twist Geometry and Single-Constant Assumption

Inserting the form of the twist geometry $\mathbf{n} = (\cos\phi(z), \sin\phi(z), 0)$ and $\mathbf{H} = H(0,1,0)$ into (11.2) and (11.4), we get the total free energy density

$$f = \frac{1}{2}K_2\phi'^2 - \frac{1}{2}\chi_a H^2 \sin^2\phi. \tag{11.5}$$

The total energy functional of $\phi(z)$ is obtained by integrating the free energy density over the thickness of the sample layer:

$$F[\phi] = \int_0^L f\left(\phi(z), \phi'(z)\right)dz \tag{11.6}$$

To look for functions $\phi(z)$ that minimize $F$, we consider the perturbed solution $\phi(z) + \delta\eta(z)$ where $\eta(0) = \eta(L) = 0$, and write

$$\begin{aligned} F[\phi] = \int_0^L \Bigg[ & f(\phi, \phi') + \delta\left(\frac{\partial f}{\partial \phi}\eta + \frac{\partial f}{\partial \phi'}\eta'\right) \\ & + \frac{\delta^2}{2}\left(\frac{\partial^2 f}{\partial \phi^2}\eta^2 + 2\frac{\partial^2 f}{\partial \phi \partial \phi'}\eta\eta' + \frac{\partial^2 f}{\partial \phi'^2}\eta'^2\right)\Bigg] dz + O(\delta^3) \end{aligned} \tag{11.7}$$

For $\phi(z)$ to minimize $F[\phi]$, the first variation must be zero:

$$\int_0^L \frac{\partial f}{\partial \phi}\eta + \frac{\partial f}{\partial \phi'}\eta'\, dz = 0. \tag{11.8}$$

Integration by parts using the boundary conditions $\eta(0) = \eta(L) = 0$ yields

$$\int_0^L \left(\frac{\partial f}{\partial \phi} - \frac{d}{dz}\frac{\partial f}{\partial \phi'}\right)\eta\, dz = 0. \tag{11.9}$$

Since $\eta$ is arbitrary, we obtain the Euler–Lagrange equation

$$\frac{\partial f}{\partial \phi} - \frac{d}{dz}\frac{\partial f}{\partial \phi'} = 0 \tag{11.10}$$

Now, upon substituting (11.5) we arrive at the equilibrium equation for the director angle:

$$K_2\phi'' + \chi_a H^2 \frac{\sin 2\phi}{2} = 0 \tag{11.11}$$

This equation is of the form that has been considered by Freedericksz and Zolina (1933) and Zocher (1933).

For any value of $H$ and the boundary conditions described in (11.1), $\phi(z) = 0$ is a trivial but important solution to (11.11). It represents a uniform solution in which the nematic orientation is homogeneous. Its stability can be seen from the second variation in (11.7). In order for $\phi(z) = 0$ to be a minimum, we need

$$\int_0^L \frac{\partial^2 f}{\partial \phi^2}\eta^2 + 2\frac{\partial^2 f}{\partial \phi \partial \phi'}\eta\eta' + \frac{\partial^2 f}{\partial \phi'^2}\eta'^2 dz > 0. \tag{11.12}$$

Integrating by parts and inserting (11.5) yields

$$\int_0^L -\chi_a H^2 \cos 2\phi\eta^2 + K_2\eta'^2 dz > 0, \tag{11.13}$$

or once we inserting $\phi(z) = 0$,

$$\int_0^L -\chi_a H^2\eta^2 + K_2\eta'^2 dz > 0. \tag{11.14}$$

If we assume a representation of $\eta(z) = \sum_{j=1}^{\infty} \eta_j \sin \frac{\pi jz}{L}$, we find that we must have

$$-\chi_a H^2 + \frac{\pi^2 j^2}{L^2} K_2 > 0. \tag{11.15}$$

for each $j \geq 1$ for stability. The most restrictive condition of these is when $j = 1$. Therefore, the homogeneous solution $\phi(z) = 0$ minimizes the free energy if $H < H_c$ where

$$H_{c,\text{uniform}} = \frac{\pi}{L}\sqrt{\frac{K_2}{\chi_a}} \tag{11.16}$$

The question remains as to whether or not there are other solutions.

Since the free energy density (11.5) has an even dependence on $\phi$, if $\phi(z)$ minimizes the free energy, then $-\phi(z)$ also minimizes the free energy. Also, due to the boundary conditions (11.1), we expect nontrivial solutions to take the symmetric form

$$\phi(z) = \phi(L - z), \quad 0 \leq z \leq L \tag{11.17}$$

which implies that

$$\phi'\left(\frac{L}{2}\right) = 0. \tag{11.18}$$

From (11.11), if $\phi' = 0$ and $0 < \phi < \frac{\pi}{2}$, then $\phi''\left(\frac{L}{2}\right) < 0$, and so $\phi\left(\frac{L}{2}\right)$ must be a local maximum, which we will denote as $\phi_{\max}$.

We now multiply (11.11) by $\phi'$ and integrate to get

$$K_2\phi'^2 + \chi_a H^2 \sin^2\phi = C. \qquad (11.19)$$

The constant of integration $C$ may be found by evaluating at $z = \frac{L}{2}$:

$$C = \left[K_2\phi'^2 + \chi_a H^2 \sin^2\phi\right]\Big|_{z=\frac{L}{2}} = \chi_a H^2 \sin^2\phi_{\max}. \qquad (11.20)$$

Thus, (11.19) can be written as

$$K_2\phi'^2 = \chi_a H^2\left[\sin^2\phi_{\max} - \sin^2\phi\right]. \qquad (11.21)$$

Since the left side is positive, we can conclude from the right side that $0 \leq \phi \leq \phi_{\max}$ on the interval $0 \leq z \leq \frac{L}{2}$ with $\phi$ increasing monotonically from 0 to $\phi_{\max}$. Therefore,

$$\sqrt{K_2}\phi' = \sqrt{\chi_a} H \sqrt{\sin^2\phi_{\max} - \sin^2\phi}. \qquad (11.22)$$

Upon integration, we can define $\phi(z)$ on $0 \leq z \leq \frac{L}{2}$ implicitly through the equation

$$\sqrt{\chi_a} H z = \sqrt{K_2}\int_0^{\phi} \frac{d\tilde{\phi}}{\sqrt{\sin^2\phi_{\max} - \sin^2\tilde{\phi}}}, \qquad (11.23)$$

once we have found the value of $\phi_{\max}$, which we may find by first solving the implicit equation coming from (11.23) evaluated at $z = \frac{L}{2}$ and $\phi = \phi_{\max}$:

$$\sqrt{\chi_a} H \frac{L}{2} = \sqrt{K_2}\int_0^{\phi_{\max}} \frac{d\tilde{\phi}}{\sqrt{\sin^2\phi_{\max} - \sin^2\tilde{\phi}}}. \qquad (11.24)$$

We can then extend the solution to the interval $\frac{L}{2} \leq z \leq L$ by using the symmetry (11.17).

To aid us both in analysis and computation, we can rewrite these using elliptic functions if we first make a change of variables similar to the one used by Zocher (1933)

$$\sin\psi = \frac{\sin\tilde{\phi}}{\sin\phi_{\max}}, \quad \cos\psi\, d\psi = \frac{\cos\tilde{\phi}}{\sin\phi_{\max}} d\tilde{\phi}. \qquad (11.25)$$

Equation (11.24) can be rewritten as

$$\frac{HL}{2}\sqrt{\frac{\chi_a}{K_2}} = \int_0^{\frac{\pi}{2}} \frac{d\psi}{\sqrt{1 - \sin^2\psi \sin^2\phi_{\max}}} = K(\sin\phi_{\max}) \tag{11.26}$$

where $K(k)$ is the complete elliptic integral of the first kind with modulus $k = \sin\phi_{\max}$. MATLAB provides a built-in function *ellipke* to compute complete elliptic integrals of the first and second kinds. This also allows us to see that in the limit of strong magnetic fields ($H \to \infty$), we must have $\phi_{\max} \to \frac{\pi}{2}^-$.

Once $\phi_{\max}$ is found, the distorted solution $\phi(z)$ can be found on $0 \le \frac{z}{L} \le \frac{1}{2}$ by rewriting (11.23) as

$$\begin{aligned} Hz\sqrt{\frac{\chi_a}{K_2}} &= \int_0^{\sin^{-1}\left(\frac{\sin\phi}{\sin\phi_{\max}}\right)} \frac{d\psi}{\sqrt{1 - \sin^2\psi \sin^2\phi_{\max}}} \\ &= F_{elliptic}\left(\sin^{-1}\left(\frac{\sin\phi}{\sin\phi_{\max}}\right), \sin\phi_{\max}\right), \end{aligned} \tag{11.27}$$

where $F_{elliptic}(\psi, k)$ is the incomplete elliptic function of the first kind with modulus $k = \sin\phi_{\max}$, which can be computed in MATLAB with the built-in function *ellipticF*. The solution for $\frac{1}{2} \le \frac{z}{L} \le 1$ can then be obtained via the symmetry (11.17).

This distorted solution, however, does not exist for small values of $H$. To see this and find the critical threshold $H_{c,\text{distorted}}$ above which the distorted solution does exist, we define the function

$$G(\psi, \phi_{\max}) \equiv \frac{1}{\sqrt{1 - \sin^2\psi \sin^2\phi_{\max}}}. \tag{11.28}$$

Note that $G$ has the symmetry $G(\psi, \phi_{\max}) = G(\phi_{\max}, \psi)$. It is straightforward to check that

$$\frac{\partial G(\psi, \phi_{\max})}{\partial \phi_{\max}} = \frac{\sin^2\psi \sin 2\phi_{\max}}{2(1 - \sin^2\psi \sin^2\phi_{\max})^{3/2}}. \tag{11.29}$$

Thus, for $0 < \phi_{\max} < \frac{\pi}{2}$ and $0 < \psi < \frac{\pi}{2}$,

$$\frac{\partial G(\psi, \phi_{\max})}{\partial \phi_{\max}} > 0. \tag{11.30}$$

Consequently, $G(\psi, \phi_{\max})$ is a monotonically increasing function of $\phi_{\max}$ for $0 < \phi_{\max} < \frac{\pi}{2}$. Furthermore, note that

$$\lim_{\phi_{\max}\to 0^+} G(\psi,\phi_{\max}) = \lim_{\phi_{\max}\to 0^+} \frac{1}{\sqrt{1-\sin^2\psi\,\sin^2\phi_{\max}}} = 1. \tag{11.31}$$

Taking the limit $\phi_{\max} \to 0^+$ on both sides of (11.26), we find that the critical field strength $H_{c,\text{distorted}}$ obeys

$$\frac{H_{c,\text{distorted}}L}{2}\sqrt{\frac{\chi_a}{K_2}} = \frac{\pi}{2}. \tag{11.32}$$

Therefore, we find the same critical value as $H_{c,\text{uniform}}$ from (11.16), and so we have only one critical value

$$H_c = \frac{\pi}{L}\sqrt{\frac{K_2}{\chi_a}} \tag{11.33}$$

such that the uniform solution $\phi(z) = 0$ is the only solution and is stable if $H < H_c$, but for $H > H_c$, the uniform solution becomes unstable and the distorted solution given by (11.26) and (11.27) exists.

Also, when $H > H_c$, we can prove that the distorted solution given by (11.26) and (11.27) is energetically favorable to the uniform solution $\phi(z) = 0$ by showing that the difference in the free energies of the two solutions

$$\Delta F = F[\phi] - F[0] \tag{11.34}$$

is negative. From the energy expression (11.6) and using (11.21), we obtain

$$\begin{aligned}
\Delta F &= \frac{1}{2}\int_0^L K_2\phi'^2 - \chi_a H^2 \sin^2\phi\, dz \\
&= \frac{1}{2}\chi_a H^2 \int_0^L \sin^2\phi_{\max} - 2\sin^2\phi\, dz \\
&= \frac{1}{2}\chi_a H^2 \left[\int_0^{\frac{L}{2}} \sin^2\phi_{\max} - 2\sin^2\phi\, dz + \int_{\frac{L}{2}}^L \sin^2\phi_{\max} - 2\sin^2\phi\, dz\right]
\end{aligned} \tag{11.35}$$

For the second integral, we make a change of variables $\tilde{z} = L - z$ so that

$$\begin{aligned}
\int_{\frac{L}{2}}^L \sin^2\phi_{\max} - 2\sin^2\phi\, dz &= \int_{\frac{L}{2}}^0 \left[\sin^2\phi_{\max} - 2\sin^2\phi(L-\tilde{z})\right](-d\tilde{z}) \\
&= \int_0^{\frac{L}{2}} \left[\sin^2\phi_{\max} - 2\sin^2\phi(\tilde{z})\right] d\tilde{z} \\
&= \int_0^{\frac{L}{2}} \sin^2\phi_{\max} - 2\sin^2\phi(z)\, dz
\end{aligned} \tag{11.36}$$

where the symmetry condition (11.17) is employed. Thus,

$$\begin{aligned}
\Delta F &= \chi_a H^2 \int_0^{\frac{L}{2}} \sin^2\phi_{\max} - 2\sin^2\phi \, dz \\
&= \chi_a H^2 \int_0^{\phi_{\max}} \left(\sin^2\phi_{\max} - 2\sin^2\tilde{\phi}\right) \frac{dz}{d\tilde{\phi}} d\tilde{\phi} \\
&= \sqrt{\chi_a} H \int_0^{\phi_{\max}} \left(\sin^2\phi_{\max} - 2\sin^2\tilde{\phi}\right) \frac{\sqrt{K_2}}{\sqrt{\sin^2\phi_{\max} - \sin^2\tilde{\phi}}} d\tilde{\phi}
\end{aligned} \tag{11.37}$$

where we have differentiated (11.23) to get the expression of $\frac{dz}{d\tilde{\phi}}$. Using the substitution (11.25), we can rewrite this as

$$\begin{aligned}
\Delta F &= \sqrt{\chi_a K_2} H \sin^2\phi_{\max} \int_0^{\frac{\pi}{2}} \frac{\cos 2\psi}{\sqrt{1 - \sin^2\phi_{\max}\sin^2\psi}} d\psi \\
&= \sqrt{\chi_a K_2} H \sin^2\phi_{\max} \int_0^{\frac{\pi}{2}} \cos 2\psi \, G(\psi, \phi_{\max}) d\psi \\
&= -\frac{1}{2}\sqrt{\chi_a K_2} H \sin^2\phi_{\max} \int_0^{\frac{\pi}{2}} \sin 2\psi \frac{\partial G(\psi, \phi_{\max})}{\partial \psi} d\psi
\end{aligned} \tag{11.38}$$

where we have again used (11.28) and integrated by parts in the last step. Thus, we use (11.30) and the fact that $\sin 2\psi > 0$ on $0 < \psi < \frac{\pi}{2}$ to conclude that $\Delta F < 0$ for $H > H_c$.

Therefore, the distorted solution is indeed energetically preferred over the unstable uniform solution, which verifies that the Fredericks transition does occur in the twist geometry and that

$$H_{c,\text{twist}} = \frac{\pi}{L}\sqrt{\frac{K_2}{\chi_a}} \tag{11.39}$$

is the critical threshold.

Under the single-constant approximation $K_1 = K_2 = K_3 = K_{all}$, the free energy density takes the same form (11.5) except that $K_2$ is replaced with $K_{all}$. Therefore all the above analysis can be repeated so that the critical threshold is

$$H_c = \frac{\pi}{L}\sqrt{\frac{K_{all}}{\chi_a}}. \tag{11.40}$$

When $H < H_c$ the uniform solution $\phi(z) = 0$ is the only stable solution, but for $H > H_c$, the stable solution is the distorted solution given by

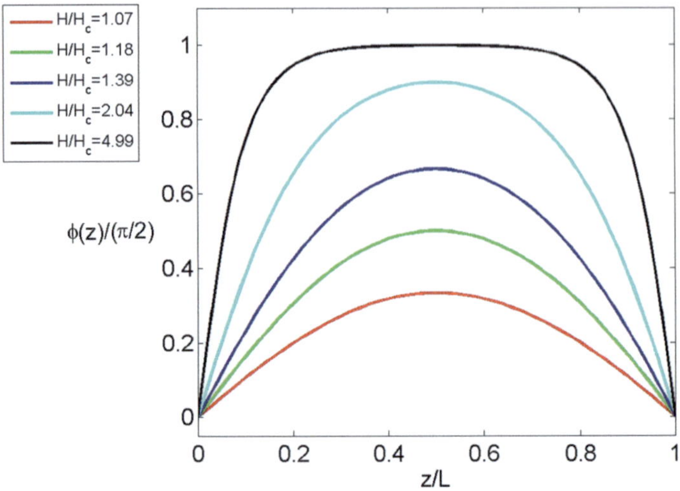


**Fig. 11.2** The solution $\phi(z)$ with respect to $\pi/2$ for different values of the dimensionless magnetic field strength $H/H_c$ using the one-constant approximation

$$\frac{H}{H_c} = \frac{2}{\pi} K(\sin \phi_{\max}), \tag{11.41}$$

$$\frac{Hz}{H_c L} = \frac{1}{\pi} F_{elliptic}\left( \sin^{-1}\left( \frac{\sin \phi}{\sin \phi_{\max}} \right), \sin \phi_{\max} \right), \tag{11.42}$$

where $K$ and $F_{elliptic}$ are the complete and incomplete elliptic functions of the first kind, respectively.

Figure 11.2 plots the distorted solution $\phi(z)$ from (11.41) and (11.42) for various values of the dimensionless magnetic field strength $H/H_c$. As $H/H_c$ increases, the values of $\phi(z)$ in the interior also increase. The transition that we have so far described analytically can be illustrated by the bifurcation diagram of Fig. 11.3, where the solid line denotes the stable solution and the dashed line represents the unstable one. For $H \leq H_c$, the trivial solution $\phi_{\max} = \phi = 0$ is the only solution of the equilibrium equation, and it is necessarily stable. For $H > H_c$, the trivial solution becomes unstable and a new distorted solution arises, which is stable.

The formulation of the solution in terms of elliptic integrals from (11.41) and (11.42) can also be helpful in the asymptotic analysis of the solution. As $k \to 0$, the complete elliptic integral of the first kind has the asymptotic behavior

$$K(k) = \frac{\pi}{2}\left( 1 + \frac{1}{4}k^2 + \frac{9}{64}k^4 \right) + o(k^4). \tag{11.43}$$

Therefore, we have an asymptotic expression for how $\phi_{\max}$ emerges from the uniform solution as $\phi_{\max} \to 0^+$ in terms of the normalized field $H/H_c$ from the asymptotic result

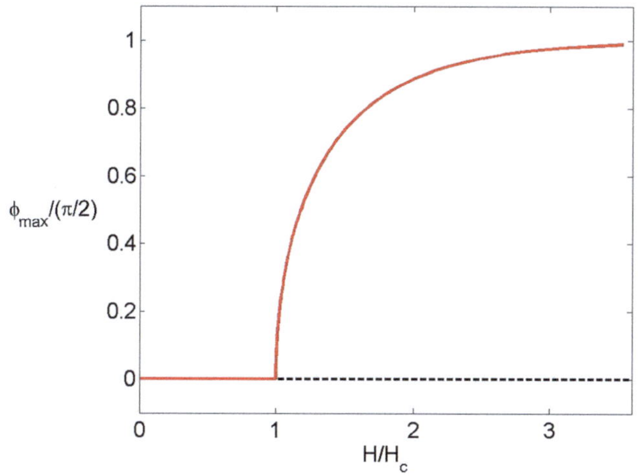


**Fig. 11.3** Phase diagram of the classical Fredericks transition in the one-constant approximation

$$\frac{H}{H_c} = 1 + \frac{1}{4}\phi_{\max}^2 + o(\phi_{\max}^2) \tag{11.44}$$

so that as $\phi_{\max} \to 0^+$

$$\phi_{\max} \approx 2\sqrt{\frac{H}{H_c} - 1}. \tag{11.45}$$

As $\phi_{\max} \to \frac{\pi}{2}$, the asymptotic analysis gives (Stewart 2004):

$$\frac{H}{H_c} \approx \frac{2}{\pi} \ln\left[\frac{4}{\sqrt{1 - \sin^2\phi_{\max}}}\right] \tag{11.46}$$

which in turn yields

$$\phi_{\max} \approx \frac{\pi}{2} - 4\exp\left[-\frac{\pi}{2}\frac{H}{H_c}\right]. \tag{11.47}$$

Thus, $\phi_{\max}$ is close to $\frac{\pi}{2}$ when the ratio $\frac{H}{H_c}$ is big enough (for example, $\frac{H}{H_c} > 3$).

The $\frac{1}{L}$ dependence in these critical values corresponds with the dependence found experimentally by Fredericks. Also, from an experimental point of view, these analytical results also provide a way to use the detection of the critical threshold to measure the Frank elasticity constants if the magnetic anisotropy $\chi_a$ is known.

### 11.2.2 Bend and Splay Geometries

A similar strategy may be followed in the bend and splay geometries, but each case depends on the two different Frank constants $K_1$ and $K_3$ instead of just one. We will highlight the differences for the bend geometry, and the splay geometry can then be found by swapping the roles of $K_1$ and $K_3$.

Inserting the bend geometry form of $\mathbf{n} = (\sin\phi(z), 0, \cos\phi(z))$ and $\mathbf{H} = H(1,0,0)$ into (11.2) and (11.4), we find the total free energy density given by

$$f = \frac{1}{2}[K_1 \sin^2\phi + K_3 \cos^2\phi]\phi'^2 - \frac{1}{2}\chi_a H^2 \sin^2\phi. \tag{11.48}$$

The Euler–Lagrange equation for the director angle (11.10) then becomes

$$(K_1 \sin^2\phi + K_3 \cos^2\phi)\phi'' + (K_1 - K_3)\frac{\sin 2\phi}{2}\phi'^2 + \chi_a H^2 \frac{\sin 2\phi}{2} = 0. \tag{11.49}$$

Again, for any value of $H$ and the boundary conditions (11.1), we have the uniform solution $\phi(z) = 0$. The second variation takes a similar form as (11.14)

$$\int_0^L -\chi_a H^2 \eta^2 + K_3 \eta'^2 \, dz > 0, \tag{11.50}$$

so that the uniform solution minimizes the free energy if $H < H_c$ where

$$H_{c,\mathrm{uniform}} = \frac{\pi}{L}\sqrt{\frac{K_3}{\chi_a}}. \tag{11.51}$$

For the distorted solution, we also assume the symmetry (11.17) and then solve integral equations first for $\phi_{\max}$ and then for $\phi(z)$ on $0 < z < \frac{L}{2}$. Unfortunately, the forms that these integrals take are more complicated and cannot be easily represented with elliptic functions.

We find these integrals by multiplying (11.49) by $\phi'$ and integrating with the same constant of integration as (11.20) so that

$$(K_1 \sin^2\phi + K_3 \cos^2\phi)\phi'^2 = \chi_a H^2 \left[\sin^2\phi_{\max} - \sin^2\phi\right]. \tag{11.52}$$

Taking the square root and integrating, we can define the distorted solution with

$$\sqrt{\chi_a} H z = \int_0^\phi \sqrt{\frac{K_1 \sin^2\tilde{\phi} + K_3 \cos^2\tilde{\phi}}{\sin^2\phi_{\max} - \sin^2\tilde{\phi}}}\, d\tilde{\phi} = \int_0^\phi \sqrt{\frac{K_3 + (K_1 - K_3)\sin^2\tilde{\phi}}{\sin^2\phi_{\max} - \sin^2\tilde{\phi}}}\, d\tilde{\phi}, \tag{11.53}$$

with $\phi_{\max}$ having been found from

$$\sqrt{\chi_a}\,H\frac{L}{2}=\int_0^{\phi_{\max}}\sqrt{\frac{K_3+(K_1-K_3)\sin^2\tilde{\phi}}{\sin^2\phi_{\max}-\sin^2\tilde{\phi}}}\,d\tilde{\phi}. \tag{11.54}$$

Or, using the change of variables (11.25), we have

$$\sqrt{\chi_a}Hz=\int_0^{\sin^{-1}\left(\frac{\sin\phi}{\sin\phi_{\max}}\right)}\sqrt{\frac{K_3+(K_1-K_3)\sin^2\psi\sin^2\phi_{\max}}{1-\sin^2\psi\sin^2\phi_{\max}}}\,d\psi \tag{11.55}$$

$$\sqrt{\chi_a}\,H\frac{L}{2}=\int_0^{\frac{\pi}{2}}\sqrt{\frac{K_3+(K_1-K_3)\sin^2\psi\sin^2\phi_{\max}}{1-\sin^2\psi\sin^2\phi_{\max}}}\,d\psi \tag{11.56}$$

We can find the critical field strength above which this distorted solution exists through a similar argument as the twist geometry by observing that the integrand in (11.55) and (11.56) is an increasing function of $\phi_{\max}$ so that

$$\begin{aligned}\lim_{\phi_{\max}\to 0^+}G(\psi,\phi_{\max}) &= \lim_{\phi_{\max}\to 0^+}\sqrt{\frac{K_3+(K_1-K_3)\sin^2\psi\sin^2\phi_{\max}}{1-\sin^2\psi\sin^2\phi_{\max}}}\\ &=\sqrt{K_3}\end{aligned} \tag{11.57}$$

Thus, taking the limit $\phi_{\max}\to 0^+$ of (11.56) shows that

$$\sqrt{\chi_a}H_{c,\text{distorted}}\frac{L}{2}=\frac{\pi}{2}\sqrt{K_3}, \tag{11.58}$$

and again we have only one critical value

$$H_{c,\text{bend}}=\frac{\pi}{L}\sqrt{\frac{K_3}{\chi_a}}, \tag{11.59}$$

which has the same dependence on the Frank constant as the twist (11.33) and single-constant approximations (11.40). A similar argument to the one above shows that when $H>H_{c,\text{bend}}$, the distorted solution is energetically preferable to the uniform solution.

Similarly, the critical value for the splay geometry is

$$H_{c,splay}=\frac{\pi}{L}\sqrt{\frac{K_1}{\chi_a}}, \tag{11.60}$$

and the distorted solution is given by

$$\sqrt{\chi_a} Hz = \int_0^{\sin^{-1}\left(\frac{\sin\phi}{\sin\phi_{\max}}\right)} \sqrt{\frac{K_1 + (K_3 - K_1)\sin^2\psi \sin^2\phi_{\max}}{1 - \sin^2\psi \sin^2\phi_{\max}}}\, d\psi \tag{11.61}$$

$$\sqrt{\chi_a}\, H \frac{L}{2} = \int_0^{\frac{\pi}{2}} \sqrt{\frac{K_1 + (K_3 - K_1)\sin^2\psi \sin^2\phi_{\max}}{1 - \sin^2\psi \sin^2\phi_{\max}}}\, d\psi. \tag{11.62}$$

### *11.2.3 Extension to Electric Fields*

The above discussion of the Fredericks transition can be extended to include (1) pretilt at the boundaries where the boundary conditions $\phi(0) = \phi(L) = 0$ are replaced by $\phi(0) = \phi(L) = \phi_0$; (2) the tilted fields where the magnetic field is applied across the sample at a fixed angle; (3) weak anchoring where the director is weakly anchored to both boundary plates. A more detailed discussion can be found in Stewart (2004) and Virga (1994).

The Fredericks transition also occurs under the influence of an electric field **E**. From a mathematical perspective, the above analysis can be repeated with the magnetic free energy density (11.4) replaced by an electric free energy density

$$f_{electric} = -\frac{1}{2}\varepsilon_a(\mathbf{n}\cdot\mathbf{E})^2, \tag{11.63}$$

In the above expression $\varepsilon_a = \varepsilon_{\|} - \varepsilon_{\perp}$ is the dielectric anisotropy of the liquid crystal where $\varepsilon_{\|}$ and $\varepsilon_{\perp}$ are the relative permittivities or dielectric constants of the liquid crystal when the field and director are parallel and perpendicular, respectively. The analogous critical field strengths for the splay, twist and bend geometries are (Stewart 2004)

$$E_{c,splay} = \frac{\pi}{L}\sqrt{\frac{K_1}{\varepsilon_a}}, \quad E_{c,twist} = \frac{\pi}{L}\sqrt{\frac{K_2}{\varepsilon_a}}, \quad E_{c,bend} = \frac{\pi}{L}\sqrt{\frac{K_3}{\varepsilon_a}}. \tag{11.64}$$

However, this formulation is known to be less accurate than the magnetic field approximation. So far, we have assumed that the electromagnetic fields are unaffected by the orientation of the liquid crystal medium. For magnetic fields, this assumption works well, but for electric fields, the coupling of the field and the liquid crystal orientation cannot be ignored.

## 11.3 Optical Fredericks Transition with Coupled Orientation and Electromagnetic Fields

We now extend our analysis in two ways to better model the optical Fredericks transition. That is, the reorientation of a nematic liquid crystalline cell by a light wave passing the layer allows for the coupling of the electromagnetic field and the

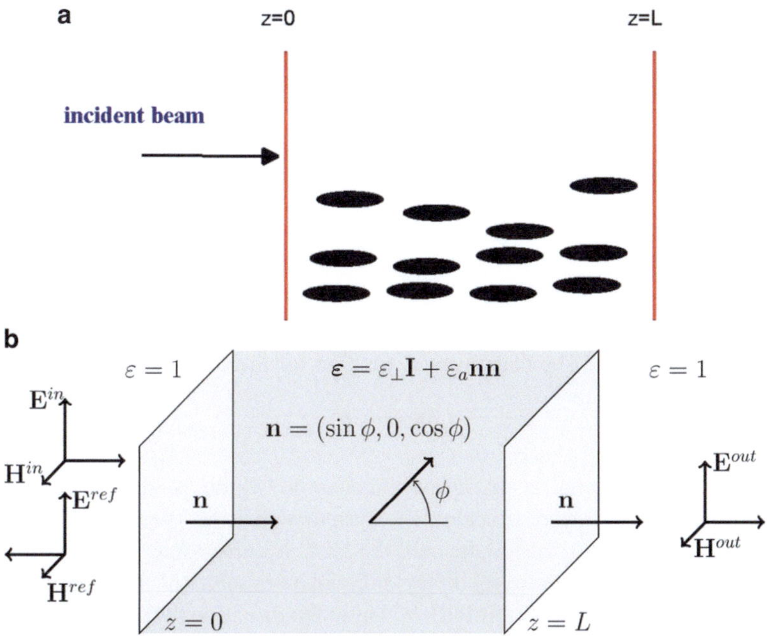


**Fig. 11.4** A schematic diagram of an incident beam on a nematic liquid crystal layer. (**a**) An incident beam entering a liquid crystal layer. (**b**) Mathematical representations of electric field, magnetic field, and orientation distribution

nematic orientation field. We use the bend geometry from Sect. 11.2 with the director $\mathbf{n} = (\sin\phi(z), 0, \cos\phi(z))$, but the electromagnetic field consists of a normally incident beam of light with the polarization parallel to the $xz$-plane, as illustrated in Fig. 11.4. In order to focus on the effect of the NLC, we assume the supporting plates have the same optical properties as the medium outside the cell, which we treat as vacuum.

### 11.3.1 Maxwell Equations

The governing equations of the electromagnetic field are given by Maxwell's equations (see, for example, Yeh 2005) which in Gaussian units are

$$\nabla \times \mathbf{E} + \frac{1}{c}\frac{\partial \mathbf{B}}{\partial t} = \mathbf{0} \tag{11.65}$$

$$\nabla \times \mathbf{H} - \frac{1}{c}\frac{\partial \mathbf{D}}{\partial t} = \frac{4\pi}{c}\mathbf{J} \tag{11.66}$$

$$\nabla \cdot \mathbf{D} = 4\pi\rho \tag{11.67}$$

$$\nabla \cdot \mathbf{B} = 0 \tag{11.68}$$

for the electric field $\mathbf{E}$, the magnetic field $\mathbf{H}$, the electric displacement $\mathbf{D}$, the magnetic induction $\mathbf{B}$, and the speed of light $c$. We include the electric current density $\mathbf{J}$ and the charge density $\rho$ for the moment, but we will soon assume they are zero.

In order to close the system, it is necessary to specify constitutive relations that describe the macroscopic properties of the medium. The magnetic constitutive equation is

$$\mathbf{B} = \mu \mathbf{H}, \tag{11.69}$$

where the relative permeability $\mu$ is assumed to be 1 so that the NLC has no effect on the magnetic field. The constitutive equation for the electric field is

$$\mathbf{D} = \boldsymbol{\varepsilon} \cdot \mathbf{E} \tag{11.70}$$

for the symmetric relative permittivity tensor $\boldsymbol{\varepsilon}$, which is also known as the dielectric tensor. If $\boldsymbol{\varepsilon}$ were a scalar, the system would be isotropic. Outside the NLC cell, we use $\varepsilon = 1$. Inside the cell, the NLC is anisotropic with two different values of the relative permittivity. In the direction parallel to the director $\mathbf{n}$, we have the extraordinary relative permittivity $\varepsilon_{\parallel}$, but in the direction perpendicular to $\mathbf{n}$, we have the ordinary relative permittivity $\varepsilon_{\perp}$. Typical values for these constants for the nematic PAA at 125 °C are $\varepsilon_{\parallel} \approx 3.3$ and $\varepsilon_{\perp} \approx 2.3$. The relative permittivity tensor is defined by

$$\boldsymbol{\varepsilon} = \varepsilon_{\perp}\mathbf{I} + \varepsilon_a \mathbf{nn} = \begin{bmatrix} \varepsilon_{\perp} + \varepsilon_a \sin^2\phi & 0 & \varepsilon_a \sin\phi \cos\phi \\ 0 & \varepsilon_{\perp} & 0 \\ \varepsilon_a \sin\phi \cos\phi & 0 & \varepsilon_{\perp} + \varepsilon_a \cos^2\phi \end{bmatrix} \tag{11.71}$$

where $\mathbf{I}$ is the identity tensor and $\varepsilon_a = \varepsilon_{\parallel} - \varepsilon_{\perp}$ measures the dielectric anisotropy.

Since these are linear equations, they apply to both real and complex electromagnetic vectors. However, since the real part of the product of two complex vectors is not equal to the product of their real parts, care must be taken to distinguish between the real and complex electromagnetic vectors in their products. The simplest form of light is a monochromatic, linearly polarized plane wave. So we assume a monochromatic complex form with angular frequency $\omega$, and for the rest of this section, we use a tilde to denote the real, time-varying vectors. Vectors without tildes represent complex vectors that may depend on $z$ but not $t$. That is,

$$\widetilde{\mathbf{E}} = \mathrm{Re}\left(\mathbf{E}e^{-i\omega t}\right), \quad \widetilde{\mathbf{D}} = \mathrm{Re}\left(\mathbf{D}e^{-i\omega t}\right), \quad \widetilde{\mathbf{H}} = \mathrm{Re}\left(\mathbf{H}e^{-i\omega t}\right), \quad \widetilde{\mathbf{B}} = \mathrm{Re}\left(\mathbf{B}e^{-i\omega t}\right). \tag{11.72}$$

We can now use (11.69) and (11.70) to rewrite Faraday's Law (11.65) and Ampere's Law (11.66) as

$$-\frac{i\omega}{c}\boldsymbol{\varepsilon}\cdot\mathbf{E} = \nabla\times\mathbf{H}, \tag{11.73}$$

$$\frac{i\omega}{c}\mathbf{H} = \nabla\times\mathbf{E}. \tag{11.74}$$

These allow us to condense the electromagnetic field equations into a single vector equation with $\mathbf{E}$ and $\mathbf{n}$, which appears through $\boldsymbol{\varepsilon}$, as

$$\nabla\times\left(\frac{c}{i\omega}\nabla\times\mathbf{E}\right) = -\frac{i\omega}{c}\boldsymbol{\varepsilon}\cdot\mathbf{E} \tag{11.75}$$

or

$$\nabla\times(\nabla\times\mathbf{E}) = \frac{\omega^2}{c^2}\boldsymbol{\varepsilon}\cdot\mathbf{E} \tag{11.76}$$

which, with the help of the vector identity $\nabla\times(\nabla\times\mathbf{E}) = \nabla(\nabla\cdot\mathbf{E}) - \nabla^2\mathbf{E}$, can be expressed as

$$\nabla^2\mathbf{E} - \nabla(\nabla\cdot\mathbf{E}) + k_0^2\boldsymbol{\varepsilon}\cdot\mathbf{E} = \mathbf{0} \tag{11.77}$$

for the wavenumber $k_0 = \frac{\omega}{c} = \frac{2\pi}{\lambda_0}$, where $\lambda_0$ is the free space wavelength of the incident beam.

This can be simplified further under our assumptions that the NLC and electromagnetic fields are homogeneous in both the $x$- and $y$-directions and the polarization of the incident beam has the electric field in the $x$-direction. From the $z$-component of (11.73), we observe that

$$-\frac{i\omega}{c}(\varepsilon_{13}E_1 + \varepsilon_{33}E_3) = \frac{\partial H_2}{\partial x} - \frac{\partial H_1}{\partial y} = 0 \tag{11.78}$$

which implies that $\varepsilon_{13}E_1 + \varepsilon_{33}E_3 = 0$. This allows us to write the $z$-component of the electric field as a function of the $x$-component as

$$E_3 = -\frac{\varepsilon_{13}}{\varepsilon_{33}}E_1 = -\frac{\varepsilon_a\sin\phi\cos\phi}{\varepsilon_\perp + \varepsilon_a\cos^2\phi}E_1 \tag{11.79}$$

A similar argument from (11.74) shows that $H_3 = 0$.

Also, the $y$-component of (11.73) is

$$\frac{\partial^2 E_2}{\partial z^2} + k_0^2\varepsilon_\perp E_2 = 0, \tag{11.80}$$

and given the polarization of the incident beam, we can conclude that $E_2 = 0$. Furthermore, since $\frac{i\omega}{c}H_1 = \frac{\partial E_3}{\partial y} - \frac{\partial E_2}{\partial z} = -\frac{\partial E_2}{\partial z}$ from (11.74), it follows immediately that $H_1 = 0$.

Returning to (11.73), we observe that the $x$-component is

$$\frac{d^2E_1}{dz^2} + k_0^2(\varepsilon_{11}E_1 + \varepsilon_{13}E_3) = 0 \tag{11.81}$$

Substituting (11.79), this yields an equation connecting the $x$-component of the electric field and the director angle as

$$\frac{d^2E_1}{dz^2} + k_0^2\frac{\varepsilon_\perp\varepsilon_{||}}{\varepsilon_\perp + \varepsilon_a\cos^2\phi}E_1 = 0. \tag{11.82}$$

Once $E_1$ is known, we can find the $y$-component of the magnetic field from (11.74) to be

$$H_2 = \frac{1}{ik_0}\frac{dE_1}{dz}. \tag{11.83}$$

## 11.3.2 Free Energy Minimization

Now we need to minimize the free energy of the system to determine the preferred orientation of the system. The elastic free energy density $f_{Frank}$ is the same as (11.2). However, the electromagnetic contributions are more complicated due to the time-varying fields. The electric contribution is given by

$$\begin{aligned}\widetilde{f}_{electric} &= \frac{1}{8\pi}\widetilde{\mathbf{E}}\cdot\widetilde{\mathbf{D}} = \frac{1}{32\pi}\left(\mathbf{E}e^{-i\omega t} + \mathbf{E}^*e^{i\omega t}\right)\cdot\left(\mathbf{D}e^{-i\omega t} + \mathbf{D}^*e^{i\omega t}\right)\\ &= \frac{1}{32\pi}\left(\mathbf{E}\cdot\mathbf{D}e^{-i2\omega t} + \mathbf{E}\cdot\mathbf{D}^* + \mathbf{E}^*\cdot\mathbf{D} + \mathbf{E}^*\cdot\mathbf{D}^*e^{i2\omega t}\right)\\ &= \frac{1}{16\pi}\left[\mathrm{Re}\left(\mathbf{E}\cdot\mathbf{D}e^{-i2\omega t}\right) + \mathbf{E}\cdot\mathbf{D}^*\right]\end{aligned} \tag{11.84}$$

because $\mathbf{E}\cdot\mathbf{D}^*$ is real. It follows in a similar way that the magnetic contribution has the form

$$\widetilde{f}_{magnetic} = \frac{1}{8\pi}\widetilde{\mathbf{B}}\cdot\widetilde{\mathbf{H}} = \frac{1}{16\pi}\left[\mathrm{Re}\left(\mathbf{B}\cdot\mathbf{H}e^{-i2\omega t}\right) + |\mathbf{H}|^2\right]. \tag{11.85}$$

The time-varying Poynting vector, which represents the directional energy flux density of the electromagnetic field, is thus given by

$$\widetilde{\mathbf{S}} = \frac{c}{4\pi}\widetilde{\mathbf{E}} \times \widetilde{\mathbf{H}} = \frac{c}{8\pi}\left[\mathrm{Re}\left(\mathbf{E} \times \mathbf{H}e^{-i2\omega t}\right) + \mathrm{Re}\left(\mathbf{E} \times \mathbf{H}^*\right)\right]. \tag{11.86}$$

The magnitude of the Poynting vector is the irradiance.

According to the energy conservation law Poynting's theorem, the work done by the electric field is

$$\widetilde{\mathbf{J}} \cdot \widetilde{\mathbf{E}} = \frac{1}{4\pi}\left(-\frac{\partial \widetilde{\mathbf{D}}}{\partial t} + c\nabla \times \widetilde{\mathbf{H}}\right) \cdot \widetilde{\mathbf{E}} \tag{11.87}$$

with the insertion of the Ampere's law (11.66). The first term can be written using the identity

$$\frac{\partial}{\partial t}\left(\widetilde{\mathbf{E}} \cdot \widetilde{\mathbf{D}}\right) = \frac{\partial \widetilde{\mathbf{E}}}{\partial t} \cdot \widetilde{\mathbf{D}} + \widetilde{\mathbf{E}} \cdot \frac{\partial \widetilde{\mathbf{D}}}{\partial t} = \frac{\partial \widetilde{E}_i}{\partial t}\varepsilon_{ij}\widetilde{E}_j + \widetilde{E}_i\varepsilon_{ij}\frac{\partial \widetilde{E}_j}{\partial t} = 2\frac{\partial \widetilde{E}_i}{\partial t}\varepsilon_{ij}\widetilde{E}_j = 2\widetilde{\mathbf{E}} \cdot \frac{\partial \widetilde{\mathbf{D}}}{\partial t} \tag{11.88}$$

where we have used Einstein summation convention and the fact that $\boldsymbol{\varepsilon}$ is symmetric. The second term can be rewritten with the vector identity $\nabla \cdot \left(\widetilde{\mathbf{E}} \times \widetilde{\mathbf{H}}\right) = \widetilde{\mathbf{H}} \cdot \nabla \times \widetilde{\mathbf{E}} - \widetilde{\mathbf{E}} \cdot \nabla \times \widetilde{\mathbf{H}}$. Putting everything together, we can rewrite (11.87) as

$$\begin{aligned}
\widetilde{\mathbf{J}} \cdot \widetilde{\mathbf{E}} &= \frac{1}{4\pi}\left[-\frac{1}{2}\frac{\partial}{\partial t}\left(\widetilde{\mathbf{E}} \cdot \widetilde{\mathbf{D}}\right) + c\left(\widetilde{\mathbf{H}} \cdot \nabla \times \widetilde{\mathbf{E}} - \nabla \cdot \left(\widetilde{\mathbf{E}} \times \widetilde{\mathbf{H}}\right)\right)\right] \\
&= \frac{1}{4\pi}\left[-\frac{1}{2}\frac{\partial}{\partial t}\left(\widetilde{\mathbf{E}} \cdot \hat{D}\right) - \widetilde{\mathbf{H}} \cdot \frac{\partial \widetilde{\mathbf{B}}}{\partial t} - c\nabla \cdot \left(\widetilde{\mathbf{E}} \times \widetilde{\mathbf{H}}\right)\right] \\
&= -\frac{\partial}{\partial t}\left(\widetilde{f}_{electric} + \widetilde{f}_{magnetic}\right) - \nabla \cdot \widetilde{\mathbf{S}}.
\end{aligned} \tag{11.89}$$

Under our assumption that the current density $\mathbf{J}$ is zero, then we have

$$\frac{\partial}{\partial t}\left(\widetilde{f}_{electric} + \widetilde{f}_{magnetic}\right) + \nabla \cdot \widetilde{\mathbf{S}} = 0. \tag{11.90}$$

For time-periodic sinusoidal electromagnetic fields, the average power flow per unit time is often more useful. We assume the timescale of the reorientation of the liquid crystal is much slower than the electromagnetic timescale defined by the frequency $\omega$ of our monochromatic incident beam. We therefore approach the liquid crystal part as a steady-state by averaging the appropriate electromagnetic quantities over one period. Therefore we need the time-average free energy densities over one period for the free energy minimization for the $\phi$ equation:

$$f_{electric} = \left\langle \widetilde{f}_{electric} \right\rangle = \frac{1}{16\pi}\mathbf{E}\cdot\mathbf{D}^*, \quad f_{magnetic} = \left\langle \widetilde{f}_{magnetic} \right\rangle = \frac{1}{16\pi}|\mathbf{H}|^2 \quad (11.91)$$

Also, to define the intensity below, we will also need the time-average of (11.90), namely,

$$\begin{aligned} 0 &= \left\langle \frac{\partial}{\partial t}\left(\widetilde{f}_{electric} + \widetilde{f}_{magnetic}\right) + \nabla\cdot\widetilde{\mathbf{S}} \right\rangle \\ &= \frac{\partial}{\partial t}\left\langle \widetilde{f}_{electric} \right\rangle + \frac{\partial}{\partial t}\left\langle \widetilde{f}_{magnetic} \right\rangle + \\ \nabla\cdot &\left\langle \frac{c}{8\pi}\left[\mathrm{Re}\left(\mathbf{E}\times\mathbf{H}e^{-i2\omega t}\right) + \mathrm{Re}\left(\mathbf{E}\times\mathbf{H}^*\right)\right] \right\rangle \\ &= \nabla\cdot\frac{c}{8\pi}\mathrm{Re}\left(\mathbf{E}\times\mathbf{H}^*\right) = \nabla\cdot\langle\mathbf{S}\rangle. \end{aligned} \quad (11.92)$$

From the perspective of the free energy density of the orientation, the optical energies are treated as negative quantities so that the free energy density is

$$f = f_{Frank} - f_{electric} - f_{magnetic}. \quad (11.93)$$

However, since $f_{magnetic}$ does not contain the director angle $\phi$, its contribution to the free energy does not affect the minimization process, and so we omit it.

We now proceed to minimize the free energy function in a manner similar to that of Sect. 11.2, except that it will contain the unknown $E_1$. The free energy density can be expressed in an explicit way as

$$\begin{aligned} f\left(\phi(z), \phi'(z)\right) &= \frac{1}{2}\left[K_1\sin^2\phi(z) + K_3\cos^2\phi(z)\right]\left[\phi'(z)\right]^2 \\ &- \frac{|E_1(z)|^2}{16\pi}\frac{\varepsilon_\perp\varepsilon_\parallel}{\varepsilon_\perp + \varepsilon_a\cos^2\phi(z)} \end{aligned} \quad (11.94)$$

Substituting this into the Euler–Lagrange equation (11.10), we find the following relation:

$$\begin{aligned} \left(K_1\sin^2\phi + K_3\cos^2\phi\right)\frac{d^2\phi}{dz^2} &+ (K_1 - K_3)\sin\phi\cos\phi\left(\frac{d\phi}{dz}\right)^2 \\ &+ \frac{\varepsilon_a\varepsilon_\perp\varepsilon_\parallel\sin 2\phi}{16\pi(\varepsilon_\perp + \varepsilon_a\cos^2\phi)^2}|E_1|^2 = 0. \end{aligned} \quad (11.95)$$

### 11.3.3 Boundary Conditions and Intensity

We now have a coupled second-order system (11.82) and (11.95) for the director angle $\phi$ and the $x$-component of the electric field $E_1$. We need boundary conditions for this system and a way to quantify the intensity of the electromagnetic field that will drive the optical Fredericks transition. For the anchoring conditions, we use

$$\phi(0) = 0, \quad \phi(L) = 0. \tag{11.96}$$

However, the electromagnetic boundary conditions are more complicated.

When the middle layer $(0 < z < L)$ is homogeneous and isotropic, boundary conditions have been given in (Yeh 2005). In our study here, the middle layer is an NLC so we cannot directly use the boundary conditions in (Yeh 2005). Nevertheless, we can borrow ideas from there. Outside the liquid crystal layer, $E_1$ is assumed to take the forms

$$E_1(z) = \begin{cases} E_{in}e^{ik_0 z} + E_{ref}e^{-ik_0 z}, & z < 0 \\ E_{out}e^{ik_0 z}, & z > L \end{cases} \tag{11.97}$$

where the unknown complex constants $E_{in}$, $E_{ref}$, and $E_{out}$ represent, respectively, the incident beam, the beam that reflects off the front of the NLC cell, and the transmitted beam that passes through the cell. From (11.83), the corresponding magnetic field outside the layer is given by

$$H_2(z) = \begin{cases} E_{in}e^{ik_0 z} - E_{ref}e^{-ik_0 z}, & z < 0 \\ E_{out}e^{ik_0 z}, & z > L \end{cases} \tag{11.98}$$

Ideally, we would like to impose $E_{in}$ and then solve for $E_{ref}$ and $E_{out}$. However, it is not possible to do this since it would not assign a boundary condition to the system (11.82) and (11.95). We can get this around in the following way. From the time-average Poynting vector (11.92), it follows that

$$\frac{d}{dz}\langle S_3 \rangle = \frac{c}{8\pi}\frac{d}{dz}\mathrm{Re}\left(E_1 H_2{}^*\right) = 0. \tag{11.99}$$

Thus, $\langle S_3 \rangle = \frac{c}{8\pi}\mathrm{Re}\left(E_1 H_2{}^*\right)$ is constant throughout the domain, both inside and outside the NLC cell. Note that it is only the time-average of the z-component of the Poynting vector that is constant throughout the domain, not the overall Poynting vector, its magnitude, or $S_3$ itself. We define the intensity $I$ as this constant so that

$$I = \frac{c}{8\pi}\mathrm{Re}\left(E_1 H_2{}^*\right) = \begin{cases} \frac{c}{8\pi}\left(|E_{in}|^2 - \left|E_{ref}\right|^2\right), & z < 0 \\ \frac{c}{8\pi}|E_{out}|^2, & z > L \end{cases} \tag{11.100}$$

Thus, we can conclude that $|E_{in}|^2 = \left|E_{ref}\right|^2 + |E_{out}|^2$, as is required to conserve energy. We shall choose $E_{out} = \sqrt{\frac{8\pi I}{c}}$ but choose $I$ as our control variable.

For discontinuous media, the components of the electromagnetic fields that are tangential to the surfaces of the discontinuities must be continuous. For our NLC cell, this means that $E_1$ and $H_2$ must be continuous at both $z = 0$ and $z = L$. This gives us both boundary conditions for (11.82) and way to solve for the unknowns $E_{in}$ and $E_{ref}$. For $E_1$, we have

$$
\begin{aligned}
E_{in} + E_{ref} &= E_1(z=0) \\
E_{out} e^{ik_0 L} &= E_1(z=L)
\end{aligned} \tag{11.101}
$$

For $H_2$, (11.83) gives us

$$
\begin{aligned}
E_{in} - E_{ref} = H_2(z=0) &= \frac{1}{ik_0}\frac{dE_1}{dz}\bigg|_{z=0} \\
E_{out} e^{ik_0 L} = H_2(z=L) &= \frac{1}{ik_0}\frac{dE_1}{dz}\bigg|_{z=L}
\end{aligned} \tag{11.102}
$$

Since (11.82) is linear in $E_1$, we use the rescaling

$$
\begin{aligned}
E_1(z) &= E_{out} e^{ik_0 L} u(z) \\
|E_1(z)|^2 &= \frac{8\pi I}{c}|u(z)|^2
\end{aligned} \tag{11.103}
$$

so that our solution proceeds as follows.

For a given intensity $I$, we first define $E_{out} = \sqrt{\frac{8\pi I}{c}}$. Then we solve

$$
\begin{aligned}
&\frac{d^2u}{dz^2} + k_0^2 \frac{\varepsilon_\perp \varepsilon_\|}{\varepsilon_\perp + \varepsilon_a \cos^2\phi} u = 0, \\
&(K_1 \sin^2\phi + K_3 \cos^2\phi)\frac{d^2\phi}{dz^2} + \frac{1}{2}(K_1 - K_3)\sin 2\phi \left(\frac{d\phi}{dz}\right)^2 \\
&\qquad + \frac{I\varepsilon_a \varepsilon_\perp \varepsilon_\| \sin 2\phi}{2c(\varepsilon_\perp + \varepsilon_a \cos^2\phi)^2}|u|^2 = 0,
\end{aligned} \tag{11.104}
$$

subject to the four boundary conditions

$$
\begin{aligned}
&\phi(0) = 0, \\
&\phi(L) = 0, \quad u(L) = 1, \quad \frac{du}{dz}\bigg|_{z=L} = ik_0.
\end{aligned} \tag{11.105}
$$

Then, we may write

$$
\begin{aligned}
E_{in} &= \frac{1}{2} E_{out} e^{ik_0 L}\left[u(0) + \frac{1}{ik_0}\frac{du}{dz}\bigg|_{z=0}\right], \\
E_{ref} &= \frac{1}{2} E_{out} e^{ik_0 L}\left[u(0) - \frac{1}{ik_0}\frac{du}{dz}\bigg|_{z=0}\right].
\end{aligned} \tag{11.106}
$$

As in Sect. 11.2, for any value of $I$, we have the solution with the uniform orientation $\phi(z) = 0$. The corresponding electric field is the same as in the case of an isotropic layer with permittivity $\varepsilon_\perp$. This solution is only stable for values of

$I$ less than some critical intensity $I_2$. There is a second non-uniform solution that only exists for values of $I$ greater than a critical intensity $I_1$. Unlike the problem in Sect. 11.2, these critical values are not the same, and we have hysteresis with bi-stable solutions in the region $I_1 < I < I_2$.

## 11.4 Numerical Method and Results

We approximate the orientation angle profile $\phi(z)$ using a truncated Fourier sine expansion:

$$\phi(z) = \phi_0 + (\phi_L - \phi_0)\frac{z}{L} + \sum_{k=1}^{K} w_k \sin\frac{k\pi z}{L} \tag{11.107}$$

In this formulation, the orientation angle profile is represented by a vector of $K$ components $\vec{w} = \langle w_1, w_2, \cdots, w_K \rangle$. As a result, the free energy is a function of the vector $\vec{w}$.

To compute the free energy $F\left(\vec{w}\right)$ defined as $F[\phi] = \int_0^L f\left(\phi(z), \phi'(z)\right)dz$ with the free energy density $f(\phi(z), \phi'(z))$ given in (11.94), we first calculate functions $\phi(z)$ and $\phi'(z)$ from vector $\vec{w} = \langle w_1, w_2, \cdots, w_K \rangle$ defined on a fine numerical grid.

Next, $u(z)$ is solved from (11.104) and (11.105) on the same numerical grid using the fourth order Runge–Kutta method. The free energy density $f(\phi(z), \phi'(z))$ is evaluated on the numerical grid from functions $\phi(z), \phi'(z)$ and $u(z)$. Finally, the free energy $F\left(\vec{w}\right)$ is calculated by integrating the free energy density $f(\phi(z), \phi'(z))$ given in (11.94) on the numerical grid.

Once we have a numerical method for calculating $F\left(\vec{w}\right)$ at each value of the beam intensity $I$ and for each coefficient vector $\vec{w}$, we solve numerically $\nabla F\left(\vec{w}\right) = \vec{0}$ for steady states. The stability of each steady state is determined by examining whether or not the Hessian matrix $H(F)\left(\vec{w}\right)$ is positive definite. In numerical simulations, both the gradient $\nabla F\left(\vec{w}\right)$ and the Hessian matrix $H(F)\left(\vec{w}\right)$ are calculated using numerical differentiation.

The problem is specified by parameters below.

- the anchoring conditions at the two boundaries, $\phi_0, \phi_L$
- the ordinary relative permittivity, $\varepsilon_\perp = 2.45$
- the extraordinary relative permittivity, $\varepsilon_\parallel = 3.346$
- two Frank elasticity constants, $K_1 = 4.5 \times 10^{-7}\,\text{dyn}$, $K_3 = 9.5 \times 10^{-7}\,\text{dyn}$
- thickness of the layer, $L = 200\,\text{nm}$

- wavelength of the beam, $\lambda_0 = 1.5\,\mu\text{m}$
- light speed, $c = 3 \times 10^{10}\,\text{cm/s}$
- intensity of the beam, $I = 0.1 \sim 0.7\,\text{GW/cm}^2 = 0.1 \sim 0.7 \times 10^{16}\,\text{dyn/(cm s)}$

The derived parameters are

- $\varepsilon_a = \varepsilon_\parallel - \varepsilon_\perp$
- $k_0 = \frac{2\pi}{\lambda_0}$ wavenumber of the beam
- $E_{out} = \sqrt{\frac{8\pi I}{c}}$

In the following we study the steady states and their stabilities as a function of the beam intensity $I$.

We consider the simplest anchoring conditions: $\phi_0 = \phi_L = 0$. For this anchoring condition, $\phi(z) = 0$ $\left(i.e.\ \vec{w} = \vec{0}\right)$ is always a steady state solution (trivial branch) for any beam intensity $I$. Another branch of steady state solution appears at $I = 0.210578$ and is stable. The trivial branch $\left(\vec{w} = \vec{0}\right)$ remains stable beyond $I = 0.210578$; it becomes unstable at $I = 0.214293$. The phase diagram is shown in Fig. 11.5. In the figure, the vertical coordinate corresponds to the maximum

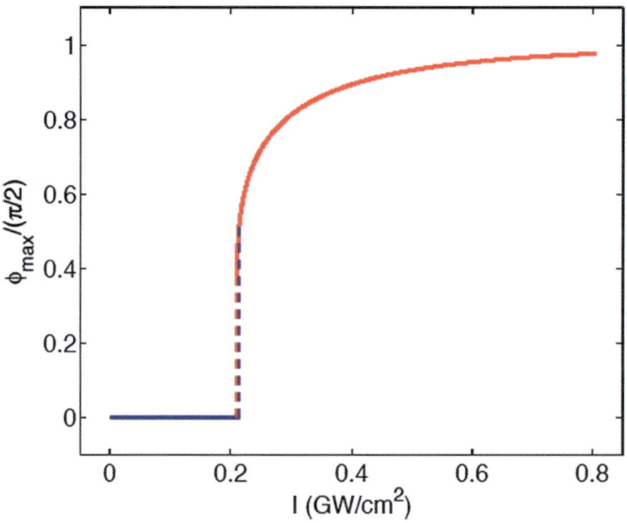


**Fig. 11.5** Phase diagram for the anchoring condition $\phi_0 = \phi_L = 0$. Here $\phi(z) = 0$ $\left(i.e. \vec{w} = \vec{0}\right)$ is always a steady state solution. A second branch of steady solution appears at $I = 0.210578$ and is stable. The trivial branch $\left(\vec{w} = \vec{0}\right)$ remains stable beyond $I = 0.210578$; it becomes unstable at $I = 0.214293$

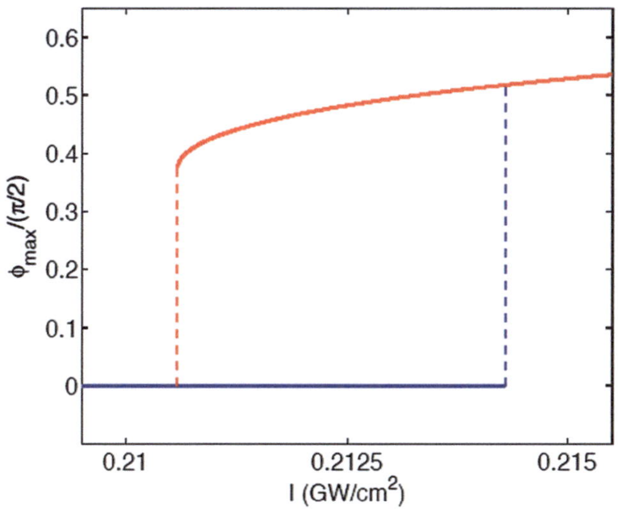


**Fig. 11.6** Details of the phase diagram in the small interval ("bi-stable region") where both the trivial branch and the non-trivial branch are stable

orientation angle: $\max_z \phi(z)/(\pi/2)$. There is a small interval for the beam intensity $I$ in which both the trivial branch and the non-trivial branch are stable. This interval is usually called the bi-stable region. The system exhibits a hysteresis behavior: when the beam intensity is increased from a low value, the transition from the trivial branch to the non-trivial branch occurs at $I = 0.214293$; when the beam intensity is reduced from a high value, the transition from the non-trivial branch to the trivial branch occurs at $I = 0.210578$. Figure 11.6 gives the details of the phase diagram in the small interval for the beam intensity $I$ where both trivial branch and the non-trivial branch are stable..

## 11.5 Conclusion and Future Perspective

The Fredericks transition plays an important role in nonlinear optics and it can occur under magnetic or electric effects. Even though it has been studied extensively both experimentally and theoretically, there are still many problems that remain to be solved. For example, a rigorous mathematical proof on the hysteresis of the Fredericks transition is still unavailable; the transition for the mismatched boundary conditions needs further investigation. To conclude, the Fredericks transition will continue to be a fascinating research subject.

**Acknowledgment and Disclaimer** Hong Zhou would like to thank the Office of Naval Research (ONR) for supporting this work. The views expressed in this document are those of the authors and do not reflect the official policy or position of the Department of Defense or the U.S. Government.

# References


Bena R, Petrescu E (2003) Surface effects on magnetic Freedericksz transition in ferronematics with soft particle anchoring. J Magn Magn Mater 263:353–359

Burylov SV, Raikher YL (1993) Magnetic Fredericksz transition in a ferronematic. J Magn Magn Mater 122:62–65

Chandrasekhar S (1992) Liquid crystals. Cambridge University Press, Cambridge

de Gennes PG, Prost J (1993) The physics of liquid crystals. Clarendon, Oxford

de Jeu WH (1980) Physical properties of liquid crystalline materials. Gordon and Breach, New York

Deuling HJ, Helfrich W (1974) Hysteresis in the deformation of nematic liquid crystal layers with homeotropic orientation. Appl Phys Lett 25:129–130

Durbin SD, Arakelian SM, Shen YR (1981) Optical-field-induced birefringence and Freedericksz transition in a nematic liquid crystal. Phys Rev Lett 47(19):1411–1414

Freedericksz V, Zolina V (1933) Forces causing the orientation of an anisotropic liquid. Trans Faraday Soc 29:919–930

Ilyina V, Cox SJ, Sluckin TJ (2006) A computational approach to the optical Freedericksz. Opt Commun 260:474–480

Khoo IC (1981) Optically induced molecular reorientation and third-order nonlinear optical processes in nematic liquid crystals. Phys Rev A 23(4):2077–2081

Khoo IC (1982a) Nonlinear light scattering by laser- and dc-field-induced molecular reorientations in nematic-liquid-crystal films. Phys Rev A 25(2):1040–1048

Khoo IC (1982b) Theory of optically induced molecular reorientations and quantitative experiments on wave mixing and the self-focusing of light. Phys Rev A 25(3):1636–1644

Khoo IC (1995) Liquid crystals: physical properties and nonlinear optical phenomena. Wiley, New York

Khoo IC, Zhuang S (1980) Nonlinear optical amplification in a nematic liquid crystal above the Frederiks transition. Appl Phys Lett 37:3–4

Li F, Buchnev O, Cheon C, Glushchenko A, Reshetnyak V, Reznikov Y, Sluckin TJ, West JL (2006) Orientational coupling amplification in ferroelectric nematic colloids. Phys Rev Lett 97:147801

Miroshnichenko AE, Pinkevych I, Kivshar YS (2006) Tunable all-optical switching in periodic structures with liquid-crystal defects. Opt Express 14(7):2839–2844

Ong HL (1983) Optically induced Freedericksz transition and bistability in a nematic liquid crystal. Phys Rev A 28(4):2393–2407

Petrescu E, Bena E (2009) Surface anchoring energy and the Freedericksz transitions in ferronematics. J Magn Magn Mater 321:2757–2762

Rapini A, Papoular M (1969) Distortion d'une lamelle nematique sous champ magnetique conditions d'ancrage aux parois. J Phys Colloq 30:C4-54–C4-56

Reznikov Y, Buchnev O, Tereshchenko O, Reshetnyak V, Glushchenko A, West J (2003) Ferroelectric nematic suspension. Appl Phys Lett 82(12):1917–1919

Self RH, Please EP, Sluckin TJ (2002) Deformation of nematic liquid crystals in an electric field. Eur J Appl Math 13:1–23

Shelestiuk SM, Reshetnyak V, Sluckin T (2011) Frederiks transition in ferroelectric liquid-crystal nanosuspensions. Phys Rev E 83:041705

Shi J (2002) Criteria for the first order Freedericksz transitions. Liq Cryst 29(1):121–125

Shi J, Yue H (2000) Surface- and optical-field-induced Freedericksz transitions and hysteresis in a nematic cell. Phys Rev E 62(1):689–698

Shoarinejad S, Shahzamanian MA (2008) On the numerical study of Frederick transition in nematic liquid crystals. J Mol Liq 138:14–19

Stark H (2001) Physics of colloidal dispersions in nematic liquid crystals. Phys Rep 351:387–474

Stewart IW (2004) The static and dynamic continuum theory of liquid crystals: a mathematical introduction. Taylor & Francis, London

Sugimura A, Luckhurst GR, Ou-Yang Z (1995) Director deformation of a twisted chiral nematic liquid crystal cell with weak anchoring boundaries. Phys Rev E 52(1):681–689

Torrent MC, Sagues F, Arias F, San Miguel M (1988) Freedericksz transition in a periodic magnetic field. Phys Rev A 38(5):2641–2649

Vena C, Versace C, Strangi G, D'Elia S, Bartolino R (2007) Light depolarization effects during the Freedericksz transition in nematic liquid crystals. Opt Express 15(25):17063–17071

Virga EG (1994) Variational theories for liquid crystals. Chapman and Hall, London

Yang KH (1983) Freedericksz transition of twisted nematic cells. Appl Phys Lett 43:171–173

Yang G, Shi J, Liang Y (2000) Surface anchoring energy and the first order Freedericksz transition of a NLC cell. Liq Cryst 27(7):875–882

Yeh P (2005) Optical waves in layered media. Wiley, Hoboken

Yilmaz S, Melik H, Angay F, Emek M, Yildirim A (2008) Optical properties of aligned nematic liquid crystals in electric field. J Mod Phys 2:248–255

Zakhlevnykh AN (2004) Threshold magnetic fields and Freedericksz transition in a ferronematic. J Magn Magn Mater 269:238–244

Zel'dovich BY, Tabiryan NV (1980) Light self-focussing in nematic liquid crystals as a method of investigation of the free surface orienting influence. Sov Phys JETP 52:1210

Zel'dovich BY, Tabiryan NV, Chilingaryan YS (1981) Fredericks transitions induced by light fields. Sov Phys JETP 54(1):32–37

Zocher H (1933) The effect of a magnetic field on the nematic state. Trans Faraday Soc 29:945–957

# Chapter 12
# New Liquid Crystalline Poly(azomethine esters) Derived from PET Waste Bottles

**Ahmed Mohammed Issam and A.K. Nurul Khizrien**

## 12.1 Liquid Crystal

The existence of liquid crystal was known when the Austrian chemist Friedrich Reinitzer discovered a strange phenomenon of two distinct melting points on a cholesterol substance back in 1888. The solid crystal melted into cloudy liquid at 145.5 °C until the cloudiness disappeared and changed to a clear transparent liquid at 178.5 °C. The term "liquid crystal" was introduced by Otto Lehmann when he discovered in 1890 that ammonium oleate and *p*-azoxyphenetole showed turbid states between the extremely crystalline and the truly isotropic fluid state. This phenomenon was challenged for almost half of a century by some scientists because they previously knew of three states of matter, but the concept was finally accepted and supported through conclusive experiments and theories.

Pierre-Gilles de Gennes, a French physicist found attractive agreement between liquid crystals, superconductors and magnetic materials in the 1960s and in 1991. He was awarded the Nobel Prize in Physics for discovering and developing methods for studying order phenomena in liquid crystals. George William Gray, a British chemist thereafter made some fundamental contributions, which then profoundly influenced the modern development of liquid crystal science. Until the early 1970s, new types of liquid crystalline states were discovered and applied especially in the display industry.

An erratum to this chapter can be found at DOI 10.1007/978-3-319-20270-9_22

A.M. Issam (✉)
Department of Chemistry, Faculty of Science, University of Malaya,
Kuala Lumpur 50603, Malaysia
e-mail: issam_usm@yahoo.com

A.K. Nurul Khizrien
School of Industrial Technology, Universiti Sains Malaysia, Penang 11800, Malaysia



V.K. Thakur, M.R. Kessler (eds.), *Liquid Crystalline Polymers*,
DOI 10.1007/978-3-319-20270-9_12

Liquid crystals are phenomenal organic materials with regard to the molecules exhibiting intermediate phases, which flow like liquids as well as possess the characteristic of ordered crystals. The materials form regions of highly ordered structure while in the liquid phase yet the degree of order is rather less than a regular solid crystal and possessing unique direction-dependent properties because the fact that the molecules are anisotropically oriented.

Tremendous progress in liquid crystal research has resulted in the expansion of new discoveries, knowledge and applications, for instance, the market for light-weight and low power electronic devices was rising exponentially, fueled up by the application of displays, electro-optical devices, organic light emitting diodes, photovoltaic devices, organic field effect transistors, biosensor, temperature and pressure sensors (Berdague et al. 1993).

Liquid crystalline polymers (LCP) comprises of aromatic rigid mesogenic groups such as benzene ring or biphenyl group with a flexible spacer of a range of linkages such as ester, ether, azo or imine groups and are known to exhibit liquid-crystalline (LC) behavior with positional and orientational orders in LC phase (Issam et al. 2010) upon heating or in concentrated solution. These structures if built into polymer chains can uphold mesophase formation with changes in temperature (thermotropic LC) or at certain concentrations in a suitable solvent. Liquid-crystalline mesophase can be formed by polymers with mesogenic moieties in the main chain (Issam et al. 2012) or mesogenic units bonded to main chain known as side groups (Tomasz and Włodzimierz 2009). As it is known, mesophase formation is generally depends on the aliphatic spacer length and substituent on the mesogen ring.

In general, LCP with completely aromatic polymers had attractive mechanical properties at high temperatures, exceptional chemical resistance and inherent flame retardant as well as good weatherability. Among successful polymers around the world, aromatic polyesters and polyazomethines are acknowledged as a substantial class of high-performance materials because of their excellent thermal stability, good mechanical properties, and environmental resistance but more particularly, as promising materials for coating, electrical, optical, photonic, and magnetic application (Atta et al. 2005).

## 12.2 Liquid Crystal Polyesters

Polyesters are made either by reaction of a dicarboxylic acid with a diol or by the self-condensation of a hydroxyl acid. Polyesters have gained significant attention in the past decade as their various adaptable structures intensify their wide applications such as coating materials, microfibers for outdoor wear and sportswear, industrial fibers, clear casting, polyester concrete and composites (Issam et al. 2011a, b). Polyesters represent the majority of the prepared liquid crystal polymers (LCP) from the concise history of the development of the first-generation LCP. The basic structures in liquid crystal polyesters are benzene rings attached at para positions through ester groups, which also represent the most important class of aromatic LCP. Liquid crystal polyesters are generally called melt liquid crystal type (thermotropic liquid crystal) polymers, and have been remarkably excellent to melt fluidity due to their peculiar behavior.

Melt solution and slurry polycondensation are the usual synthetic routes used for the synthesis of aromatic LC polyesters and the methods utilized are (1) the polycondensation of terephthalic acid diesters and aromatic diols, (2) the polycondensation of terephthalic acid and acetate of aromatic diols with the addition of transesterification catalysts, and (3) the polycondensation of aromatic diols and aromatic diacid dichlorides (Atta et al. 2005, 2007).

The next development in liquid crystal polyesters was the preparation by polycondensation based on terephthalic acid (TPA) and hydroquinone (HQ) or *p*-hydroxybenzoic acid (HBA). The polyesters are insoluble with very high melting temperatures of 600 °C for poly (TPA/HQ) and 610 °C for poly (HBA), which are by far too high to obtain stable liquid crystalline phases for melt processing. In 1972, Economy and coworkers patented several copolyester compositions, and one of these are the copolymerization of poly (4-hydroxybenzoic acid) (PHB) with 4,4′-dihydroxybiphenyl (BP) and terephthalic acid (TPA) due to the need for lower melting, melt-processable polymers. Considerable synthetic efforts have been attempted in order to decrease the melting temperatures of aromatic LC polyesters while retaining LC properties. The copolyester structure was tailored by partial substitution of TPA with isophthalic acid to produce a melt-spinnable material.

The formation of thermotropic LC phases reduces the melt viscosities during processing and melt-processable LCP could be obtained by lowering the chain regularity of the main chain while maintaining the chain stiffness by random copolymerization or the use of other aromatic comonomeric units. Jackson and Kuhfuss (1976) produced X7G fibers at Eastman-Kodak by progressively increasing the rigidity of polyethylene terephthalate (PET), a commercial, thermoplastic polymer by the introduction of increasing amount of HBA units in the flexible macromolecule. Majority of aromatic thermotropic liquid crystal polyesters are known to be excellent in small-thickness moldability, heat resistance, mechanical strength, dimensional stability and the like; since the polyesters can be adapted to soldering temperatures about 240–260 °C. The thermal stability of liquid crystal polymer with low branching coefficient is better than that of the dumbbell-shaped. As there are adamant concerns for biodegradable polymeric materials relating to environmental awareness, several biodegradable aliphatic polyesters have been developed and some of them are now commercially handy because of their good biodegradability but still constitute serious obstacles due to their poor thermal and mechanical properties (Tserki et al. 2006).

## 12.3 Polyethylene Terephthalate (PET)

Polyethylene terephthalate (PET) is a semicrystalline thermoplastic polyester used in the preparation of variety of products such as fibres, filaments, sheets and soft drink bottles (Shukla et al. 2003, 2006). PET is one of the best-known consumer plastic for high softening-point blow-moulded bottles and the world demand for PET bottles have been increasing due to its high mechanical strength, light weight, low permeability to gases, relatively low manufacturing cost and its non-toxicity (Karayannidis et al. 2005; David and Medhat 2005).

Almost 104 million tones of post-consumer PET bottles are being disposed and dumped indiscriminately onto bare lands and water, which thus constitute a serious impediment to the environment. PET does not create any direct hazard in the environment, but due to the increasing plastic's waste and its high resistance to the atmosphere and biological agents, it may be considered as pollutant material (Sinha et al. 2010). Thus, recycling of PET bottles is a common environmentally friendly procedure used to reduce plastic waste and also contribute to the conservation of raw petrochemical products and energy. As there is urgent demand for PET waste conversion, many scientists have converted PET waste to other valuable for raw materials (Issam 2014) and besides react with other monomers to produce new polymers such as unsaturated poly (ester-urethane) (Issam et al. 2011a, b). As terephthalic acid (an aromatic diacid) is used for the preparation of various types of polyesters, including their liquid crystals, it is worth noting that used is predominantly obtained from petroleum resources that are valuable and non-renewable (Atta et al. 2005). The ultimate aim of this research is to regenerate TPA from PET waste bottles to be used as a new raw material in the preparation of other polymers. This will lead to minimizing dependency on the petrochemical resources as well as to reduce the amounts of waste. This work thus uses TPA as one of the main reactants in producing liquid crystalline poly(azomethine esters).

## 12.4 Liquid Crystal Poly(azomethine esters)

Liquid crystal poly(azomethine esters) consists of a rigid mesogenic moiety (e.g., benzene ring, biphenyl group) and an azomethine linkage (–CH=N–) in the main chain or side chain. They are of the interest due to their unique performance properties namely, high mechanical strength, high-thermal stability, environmental resistance, and are promising materials in photonic and opto-electronic applications (Vasanthi and Ravikumar 2013).

This statement is substantiated by the fact that the incorporation of azomethine groups in the main chain LC improves the thermal properties, and the polymer formed mesomorphic phase on heating (thermotropic liquid crystal polymer). However, the drawbacks from their high melting points and poor solubility minimize their practical applications by conventional methods. In order to enhance their processability, several approaches have been attempted such as introduction the flexible aliphatic segments into the main chain, or via the bulky lateral substituents. Issam et al. (2010) reported that the azomethine group with a flexible spacer is very useful in exhibiting LC properties. They can be synthesized by forming the linking groups between rigid core and the flexible spacer (–O– or –OOC–) in the polymerization step or alternatively; polymers are obtained as polyethers (–O– linkages formed in the polymer synthesis) or polyesters (–OOC– linkages formed in the polymer synthesis). Rapid progress in the field of polymer science creates various modified polyazomethines with the sole aim of decreasing their melting temperature, enhance their solubility and to develop mesomorphism such as poly(azomethine-esters), poly(azomethine-ethers), poly(azomethine-carbonates), poly(amide-azomethine-esters), poly (acrylate-azomethines), thermosetting poly-azomethines and poly(azomethine-sulfones) (Aslan et al. 1997). Among these polymers, thermotropic liquid crystals poly(azomethine esters) has received greater technological importance due to their high thermal stability and good processability. Because of these valuable properties, great efforts have been concentrated on the synthesis of new compounds with enhanced properties. Previous studies have shown that poly(azomethine esters) integrated the beneficial properties of both polyazomethines and polyesters, due to the contribution of the anisotropic azomethine rigid cores and isotropic ester moieties. It has been reported that the ester group increases the electrical conductivity of copolymers in poly(azomethine-esters) in two ways i.e. by increasing the flexibility of the chain and making the charge carrier movement more along the chain (Han and Bhowmik 1997). Marin et al. (2006) investigated the usefulness of aliphatic–aromatic polyazomethines with ester groups for opto-electrical applications by synthesizing polyazomethines with various structures of the dialdehyde moiety. By varying the dialdehyde structure, they managed to control the conjugation along the polymer backbone and consequently, control the optical and electrical properties of the polymers.

In order to investigate the role of the sub-unit of dialdehyde, groups such as thiophene, carbazole, stilbene or phenylene were introduced to the polyazomethine main chain. Poly(azomethine esters) with stilbene bonds showed promising opto-electronic properties due to the lowest value of the optical energy gap ($E_g = 2.28$ eV) and the highest thermoluminescent intensity ($I = 42$). They also suggested there was a new application potential for polyazomethines by exploiting their liquid-crystalline and transport properties toward stable organic opto-electronic devices, for example, in OLEDs and solar cells.

Marin et al. (2006) successfully synthesized novel functional thermotropic liquid crystalline polyazomethine derivatives by incorporating new properties of luminescence; fluorene and/or oxadiazole chromophoric units. The polymers showed an impressive mesomorphic state controlling the order degree of the semicrystalline state while being viscous fluids. The semiflexible polyazomethines have high thermal stability, good mechanical properties and good solubility in polar solvents thus relevant for opto-electronic material design with the aim of thermostable films with good mechanical properties. Afterward (Marin et al. 2013), synthesized thermotropic liquid crystalline dimers containing azomethine and pyrrole, indole, or diphenyl chromophoric moieties.

In this chapter, the new thermotropic liquid crystal poly(azomethine esters) were designed and synthesized as T-shaped azomethine bisphenols containing azomethine linkage (–CH=N) attached to different halogens and were reacted with terephthaloyl chloride that has been obtained from the regenerated terephthalic acid of PET waste bottles. The effect of the azomethine linkages and the type of the substituents (halogens) on the liquid crystalline behavior, thermal stability, as well as the crystalline states of these polymers were further studied and analyzed.

## 12.5 Materials and Methods

### 12.5.1 *Materials*

All the chemicals were commercially obtained and used without further purification. PET bottles of Spritzer (Malaysia) mineral water were shredded into flakes. Sodium hydroxide AR (Chem.AR), Ethylene glycol (Aldrich Chem. Co.) and sulphuric acid (Mallinckrodt Chem. Co.) were used as purchased. Aniline, p-iodoaniline, p-bromoaniline and p-chloroaniline were purchased from (Aldrich Chem. Co.). 2,5-dihydroxybenzaldehyde, absolute ethanol, methanol, 1-butanol, dimethylsulfoxide (DMSO), ethyl methyl ketone and diethyl ether (Fluka Chem. Co.).

### 12.5.2 *Instrumentation*

FTIR spectra for the monomers and polymers were recorded on a Perkin-Elmer 2000 FTIR spectrophotometer using KBr pellet method. $^{1}$H NMR and $^{13}$C NMR spectra were recorded on a Bruker 400 MHz NMR spectrometer, Dimethylsulfoxide (DMSO-$d_6$) was used as the NMR.

Solvent and tetramethylsilane (TMS) used as internal reference. Differential Scanning Calorimetry (DSC) was performed on a Perkin-Elmer DSC7 series in nitrogen atmosphere at a heating rate of 10 °C/min. Polarizing optical microscope (POM) was measured in Nikon Eclipse E 600 connected to a Linkam THMS 600 heating stage. The X-ray diffractograms were recorded using Siemens Model D-5000 diffractometer at room temperature with nickel-filtered CuKα radiation. Thermogravimetric analysis (TGA) was performed on a Perkin-Elmer TGA7 series in nitrogen atmosphere at a heating rate of 20 °C/min. Inherent viscosities of polymer solutions (0.2 dL/g in DMSO) were determined at 30 °C in an Ubbelohde viscometer.

## 12.6 Monomer Synthesis

### 12.6.1 *Synthesis of Azomethine Bisphenol 1 (a)*

A solution of 1.38 g (0.1 mol) 2,5-dihydroxybenzaldehyde in 30 mL of absolute ethanol was added dropwise into a mixture of 1.07 g (0.1 mol) of aniline in 30 mL absolute ethanol containing few drops of acetic acid in a 500 mL flask. The mixture was refluxed for 6 h with continuous stirring. 2,5-dihydroxybenzelidene aniline was separated as a precipitate, washed several times with diethylether, and dried in a vacuum oven at 70 °C for 24 h. Final purification was carried out by re-crystallization from 1-butanol to give reddish crystals; Melting point (°C): 150.5–153. Yield: 88 %. FTIR (KBr) ($cm^{-1}$): 3407 (–OH stretching), 1622 (–C=N stretching). $^{1}$H NMR (DMSO-$d_6$) (ppm): 8.36 (1H, –CH=N–), 7.46–6.79 (8H, aromatic).

$^{13}C$ NMR (DMSO-$d_6$) (ppm): 163.7 (–CH=N), 135.0–115.8 (aromatic C=C), 152.8 (–C–N, aromatic). Elemental analysis: Calc. for $C_{13}H_{11}NO_2$: C, 73.23; H, 5.20; N, 6.57; Found: C, 73.18; H, 5.33; N, 6.59.

### 12.6.2 *Synthesis of Azomethine Bisphenol 1 (b)*

The monomer was synthesized using the same procedure as in the above except that aniline was substituted by *p*-chloroaniline. Melting point (°C): 181–183. Yield was 79 %. FTIR (KBr) ($cm^{-1}$): 3350 (–OH stretching), 1620 (–C=N stretching). $^1H$ NMR (DMSO-$d_6$) (ppm): 8.0 (1H, –CH=N–), 7.26–7.83 (7H, aromatic). $^{13}C$ NMR (DMSO-$d_6$) (ppm): 163.9 (–CH=N), 151.9–112.3 (aromatic C=C). Elemental analysis: Calc. for $C_{13}H_{10}NO_2Cl$: C: 63.03, H: 4.04, N: 5.65; Found: C, 63.10; H, 4.05; N, 5.58.

### 12.6.3 *Synthesis of Azomethine Bisphenol 1 (c)*

The same procedure as in the above was adopted except that aniline was substituted by *p*-bromoaniline. Melting point (°C): 192–193.5. Yield was 82 %. FTIR (KBr) ($cm^{-1}$): 3345 (–OH stretching), 1621 (–C=N stretching). $^1H$ NMR (DMSO-$d_6$) (ppm): 8.86 (1H, –CH=N–), 7.25–7.93 (7H, aromatic). $^{13}C$ NMR (DMSO-$d_6$) (ppm): 164.9 (–CH=N), 152.3–110.8 (aromatic C=C). Elemental analysis: Calc. for $C_{13}H_{10}NO_2Br$: C, 53.44; H, 3.42; N, 4.79; Found: C, 53.40; H, 3.46; N, 4.73.

### 12.6.4 *Synthesis of Azomethine Bisphenol 1 (d)*

The same procedure as in the above was adopted except that aniline was substituted by *p*-iodoaniline. Melting point (°C): 197–198. Yield was 78 %. FTIR (KBr) ($cm^{-1}$): 3355 (–OH stretching), 1622 (–C=N stretching). $^1H$ NMR (DMSO-d6) (ppm): 8.53 (1H, –CH=N–), 7.21–7.89 (7H, aromatic). $^{13}C$ NMR (DMSO-$d_6$) (ppm): 163.9 (–CH=N), 152.0–114.4 (aromatic C=C). Elemental analysis: Calc. for $C_{13}H_{10}NO_2I$: C, 46.03; H, 2.95; N, 4.13; Found: C, 45.99; H, 2.98; N, 4.17.

### 12.6.5 *Regeneration of Terephthalic Acid (TPA) from PET Waste Bottles*

Sodium Hydroxide pellets were first dissolved in ethylene glycol at a temperature of between 60 and 100 °C. PET flakes were added to the mixture when NaOH pellets were partially dissolved. The ratio of PET flakes to NaOH pellets to ethylene glycol was 1:2:1.5, respectively. The mixture was well mixed and heated at 190–200 °C in a stainless steel container and the temperature was maintained for about 20 min until the mixture turned milky. The mixture was subsequently stirred continuously and after 20 min, the mixture was cooled to room temperature. Distilled water was afterwards added to dissolve all white precipitates. It was then stirred and filtered through a coffee-filter. When the mixture turned clear yellow, a few drops of 5 M $H_2SO_4$ were added to this solution until pH 2. A white powder appeared and was filtered using Buchner funneled (under suction) while ethylene glycol was collected as the filtrate. The white precipitate (TPA) was collected and washed several times with water and finally dried in a vacuum oven for 24 h at 60 °C.

## 12.7 Preparation of LC Poly(azomethine esters) 3 (a–d)

Terephthalic acid (TPA) was converted into terephthaloyl chloride by using the same procedure as described by Li Zifa et al. (1996). 0.01 mol of terephthaloyl chloride was dissolved in 300 mL THF, an equimolar quantity of compound, I (a) was dissolved in 200 mL THF followed by the addition of 5 mL pyridine. The terephthaloyl chloride solution was taken in a dropping funnel and added dropwise to a mixture of 0.01 mol azomethine bisphenol in 200 mL THF and 5 mL of pyridine at 0 °C for 2 h. The reaction mixture was stirred for 3 h at room temperature and then refluxed for 3 h. The reaction mixture was poured into methanol to precipitate and afterwards filtered. All the obtained poly(azomethine esters) were washed with diethyl ether and methanol and finally dried in a vacuum oven at 75 °C for 24 h. The polymers (b–d) were synthesized using the same method as described in this procedure. The results were summarized in Table 12.1.

## 12.8 Results and Discussion

New thermotropic liquid crystalline poly(azomethine esters) containing azomethine linkages were prepared by polycondensation reaction of azomethine bisphenols with terephthaloyl chloride, which was produced from terephthalic acid. Terephthalic

**Table 12.1** Inherent viscosity, yields and CHN data for LC poly(azomethine esters) 3(a–d)

| Compound | CHN Calculated (%) | Found (%) | Yield (%) | Inherent viscosity, $\eta_{inh}$ (dL/g) |
|---|---|---|---|---|
| 3a | C:73.46, H:3.79, N:4.08 | C:72.52, H:3.65, N:4.10 | 80 | 0.43 |
| 3b | C:66.75, H:3.17, N:3.70 | C:66.19, H:3.18, N:3.54 | 76 | 0.50 |
| 3c | C:61.03, H:2.90, N:3.39 | C:59.99, H:2.98, N:3.23 | 82 | 0.52 |
| 3d | C:53.74, H:2.56, N:2.98 | C:52.69, H:2.47, N:2.91 | 72 | 0.49 |

acid (TPA) was regenerated from PET waste bottles via saponification process, and the chemical structure of the TPA was confirmed by FTIR and $^{1}H$ NMR spectroscopy. FTIR spectrum of TPA showed the presence of C=O band at 1685 $cm^{-1}$ and the –C=C– stretching of the aromatic ring appeared at 1574 and 1510 $cm^{-1}$. The asymmetrical stretching of –OH group appeared at 3064 $cm^{-1}$ while the symmetrical stretching band appeared at 2551 $cm^{-1}$. The absorption bands at 1424 and 937 $cm^{-1}$ were attributed to the C–O stretching and C–O–H deformation, respectively. Furthermore, $^{1}H$ NMR spectrum of terephthalic acid (Fig. 12.1) showed a broad peak centered at 13.3 ppm, corresponding to the proton of hydroxyl group and a strong singlet peak at 8.1 ppm due to the protons of aromatic ring.

The preparation of T-shaped azomethine bisphenols 1 (a–d) from 2,5-dihydroxybenzaldehyde and *p*-substituted aniline was outlined in Scheme 12.1. The structures of azomethine bisphenols were confirmed by FTIR, $^{1}H$ and $^{13}C$ NMR spectroscopy and elemental analysis (CHN). The FTIR spectra of the monomers I (a–d) showed sharp absorption bands at 1620–1622 $cm^{-1}$ which was assigned to the –CH=N– stretching and confirms the formation of azomethine linkage. The absorption bands at 3345–3407 $cm^{-1}$ were due to the stretching vibrations of the hydroxyl groups. Further confirmation was carried out by $^{1}H$ NMR spectroscopy of monomer 1 (b) (Fig. 12.2) which showed the presence of azomethine proton (–C<u>H</u>=N–) at 8.86 ppm. The characteristic singlet peak which centered at 9.12 ppm was attributed to the proton of the phenolic (Ph-O<u>H</u>) group.

$^{13}C$ NMR spectrum of monomer 1 (b) showed peak at 163.11 ppm, which was due to the azomethine carbon (–CH=N–) and thus confirms the formation of azomethine linkage. In addition, the carbons of the aromatic rings appeared at 112.3–131.2 and 148.7, 149.8 and 151.9 for C–N and C–O, respectively. Elemental analysis (CHN) data also supported the formation of the expected monomers and confirmed the calculated data.

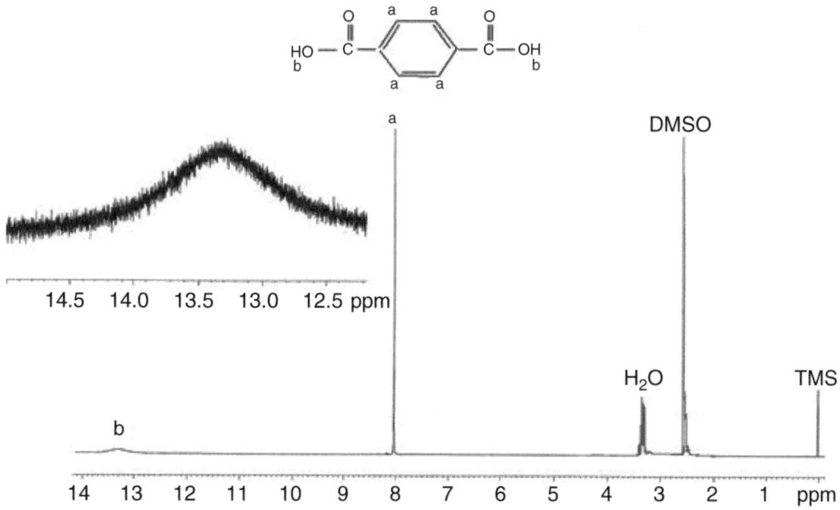


**Fig. 12.1** $^{1}H$ NMR spectrum of terephthalic acid

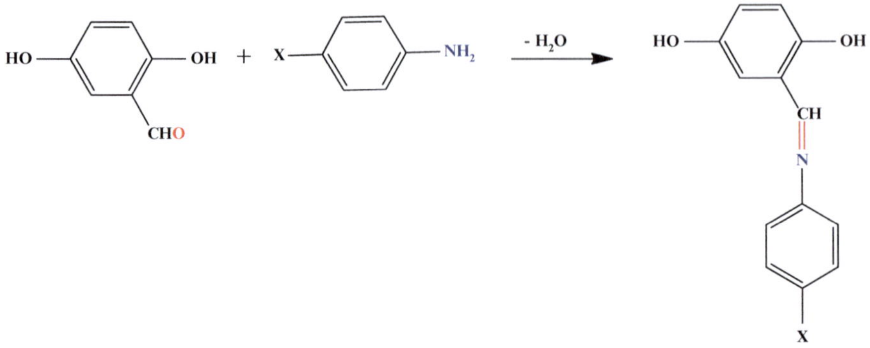


1 (a-d)

Ia: X = -H

Ib: X = -Cl

Ic: X = -Br

Id: X = -I

1 (a-d)

**Scheme 12.1** Synthesis of T-shaped azomethine bisphenols 1 (a–d)

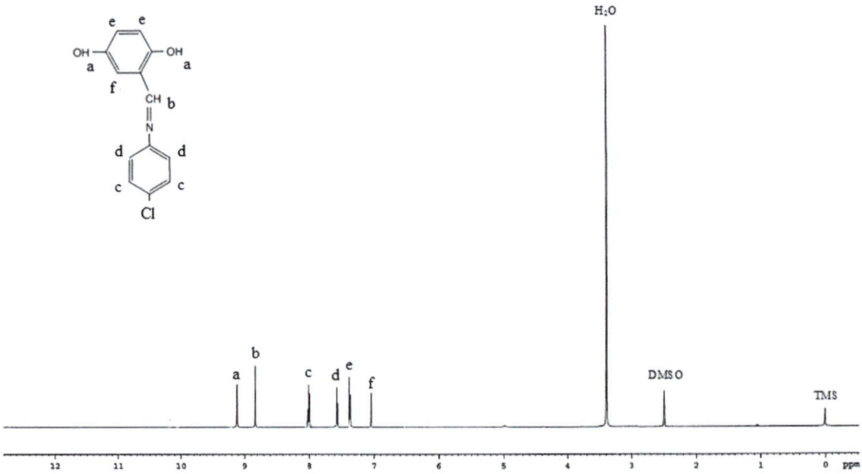


**Fig. 12.2** $^{1}$H NMR spectrum of monomer 1 (a)

The general polymerization reactions of the LC polymers 3(a–d) were as outlined in Scheme 12.2. The structures were confirmed by CHN, FTIR and $^{1}$H and $^{13}$C NMR spectroscopy. The FTIR spectra of polymers showed characteristic absorption bands in the range 1681–1711 cm$^{-1}$ due to the carbonyl stretching (–C=O) of ester linkage, and in the range 1619–1622 cm$^{-1}$ due to the azomethine (–CH=N–) linkage. The other absorption bands which represent the aromatic ring appeared in range of 1508–1603 cm$^{-1}$.

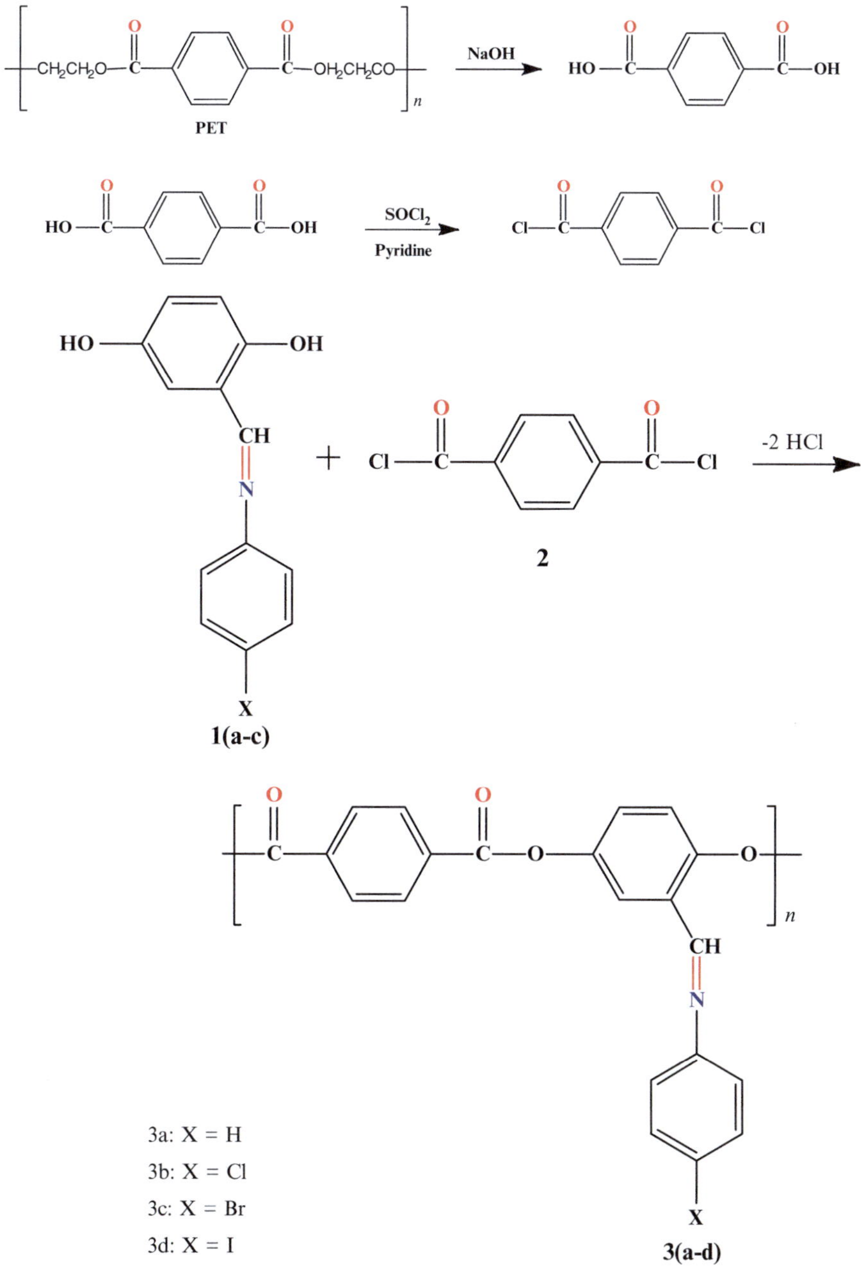


3a: X = H
3b: X = Cl
3c: X = Br
3d: X = I

**Scheme 12.2** Synthesis of poly(azomethine esters) 3 (a–d)

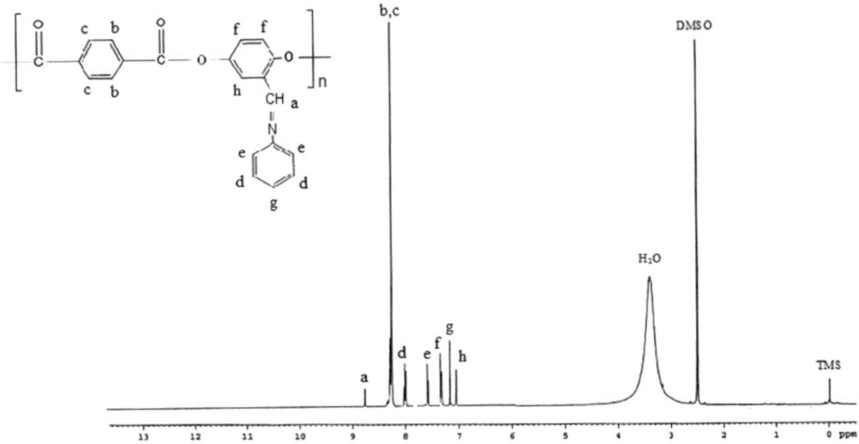


**Fig. 12.3** [1]H NMR spectrum of polymer 1 (a)

In the [1]H NMR spectrum of polymer 3(a) shown in Fig. 12.3, we notice the disappearance of the singlet peak at 9.12 ppm for –OH group and the appearance of two new peaks at 8.09 and 8.1 ppm were due to the protons in aromatic ring of the ester groups. Moreover, the multiplet peaks found at 7.51–8.21 ppm were attributed to the aromatic protons of the monomer. Furthermore, the proton of azomethine group appeared at 8.75 ppm. In the [13]C NMR spectrum, the new peak which is centered at 182 ppm is assigned to carbon of the carbonyl groups of the ester linkages. Other peaks for carbons of azomethine at 176.07 and aromatic rings at the range of 121.11–152.13 confirmed the chemical formula of polymer 3(b).

In addition to the structural confirmation via the spectroscopy analysis, the elemental analysis data (Table 12.1) also supported the formation of the expected polymers 3 (a–d).

Solubility tests carried out on the polymers revealed that all the LC poly (azomethine esters) were insoluble in common organic solvents such as chloroform, THF and acetone. This leads to the difficulty in their molecular weight determination by gel permeation chromatography (GPC). The inherent viscosities ($\eta_{inh}$) of the polymers 3 (a–d) were as given in Table 12.1. The low inherent viscosities may be attributed to the poor solubility of the polymers in the polymerization solution. Precipitate was also formed during the polymerization which hinders further propagation of the polymer chain and thus leads to lower molecular weight.

### *12.8.1 Thermotropic LC Properties of the Polymers 3 (a–d)*

Differential Scanning Calorimetry (DSC) and polarizing optical microscope (POM) were used to study the mesophases of the liquid crystalline poly(azomethine esters). The mesophase confirmation and the temperature data of poly(azomethine esters) were summarized in Table 12.2. The thermal behaviors of the polymers 3 (a–d)

**Table 12.2** DSC and POM data of LC poly(azomethine esters) 3(a–d)

| | DSC | | | | |
|---|---|---|---|---|---|
| Polyesters | $T_m$(°C) | $T_i$ (°C) | ΔT (°C) | Cr–N (°C) | N–I (°C) |
| 3a | 364 | – | – | – | – |
| 3b | 186 | 300 | 114 | 188 | 300 |
| 3c | 174 | 300 | 126 | 174 | 303 |
| 3d | 167 | 296 | 129 | 169 | 298 |

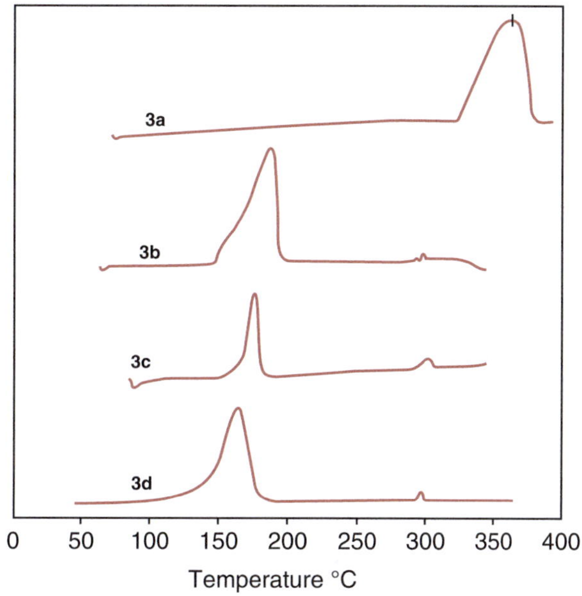


**Fig. 12.4** DSC curves of poly(azomethine esters) 3 (a–d)

were investigated using DSC (Fig. 12.4). The samples were heated from room temperature to 400 °C at the rate of 10 °C/min followed by cooling to room temperature and reheating the samples to 320 °C also at the rate of 10 °C/min. It is apparent in the DSC curve that $T_g$ for polymer 3(b) was 147 °C while the $T_m$ and $T_i$ were centered at 187 °C and 309 °C, respectively. The sample was held for 1 min and then cooled. On cooling, one exothermic peak was observed at 200 °C arising from crystallization and the second run was carried out by reheating of the polymer at the same condition. The melting points ($T_m$) for polymers 3(c) and 3(d) appeared at 174 °C and 167 °C respectively, whereas the $T_i$ for these polymers found at 300 °C and 296 °C, respectively. However, the DSC curve for polymer 3(a) showed only melting point at 364 °C without any sign for $T_i$. The high $T_m$ reflects the high thermal stability of the LC poly(azomethine esters) obtained.

It may be observed from Table 12.2 that polymer 3(d) with the iodine substituent group gives lower melting point ($T_m$) compared to the polymers with chlorine and bromine [3(b) and 3(c)] and also with polymer 3 (a), which is based on aniline. This may be attributed to the size difference which causes lack of intermolecular interaction as the size increases.

Polarized optical microscopy (POM) was performed to confirm the transition temperatures obtained by DSC studies and to find out the type of mesophases present in the synthesized polymers, the crystal phase to nematic phase transition temperature (Cr–N), and nematic phase to isotropic phase (N–I) temperature. It was found that the transition temperatures of the polymers are almost the same as those obtained in DSC. The POM results revealed that the poly(azomethine esters) 3(b–d) exhibit nematic liquid crystal phases and the images were shown in Fig. 12.5.

XRD diffraction studies was also carried out to analyze the crystallinity of the polymers 3(a–d) and was performed at room temperature. To avoid the trace of the solvents, powder samples were placed on a zero-background quartz plate, placed for 20 s. in an oven preheated to 110 °C, quenched to room temperature using a $N_2$ purge, and then analyzed in the diffractometer. The XRD patterns are shown in Fig. 12.6. All the polymers show similar multiple sharp diffraction peaks in the range of $2\theta = 11.0°–30.0°$ which indicates the crystalline to semi-crystalline nature of the polymers. This may be due to the slight difference in the chemical structures of these polymers. Furthermore, the results of the x-ray diffraction were in agreement with the POM textures, and this confirmed that the synthesized polymers 3(b–d) were in liquid crystal phases (nematic mesophase).

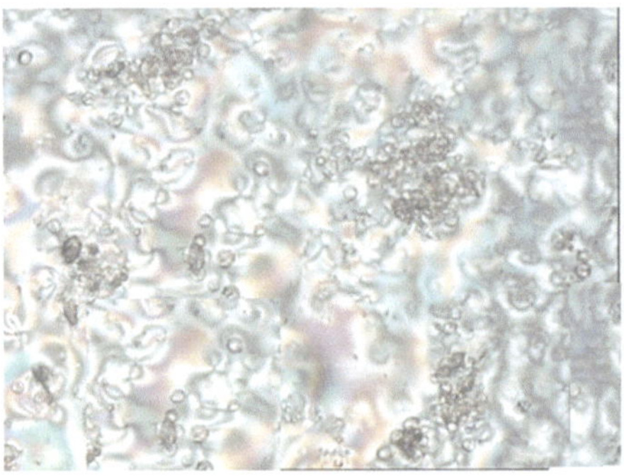

3 (b)

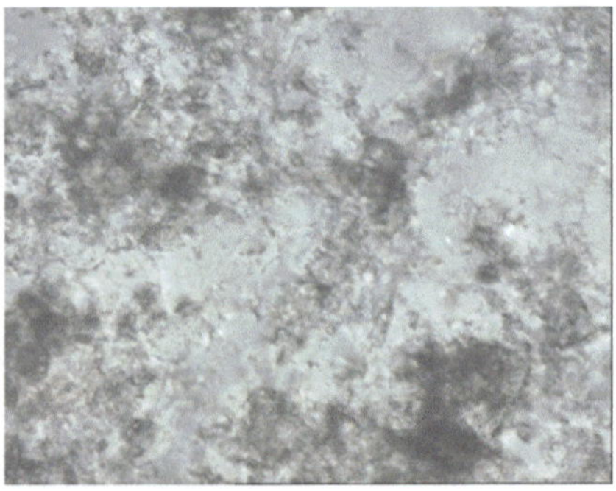

3 (c)

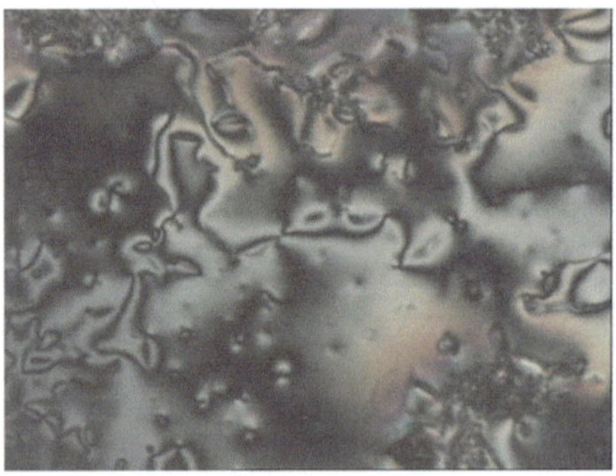

3 (d)

**Fig. 12.5** Polarizing photomicrograph of polymer 3 (b–d)

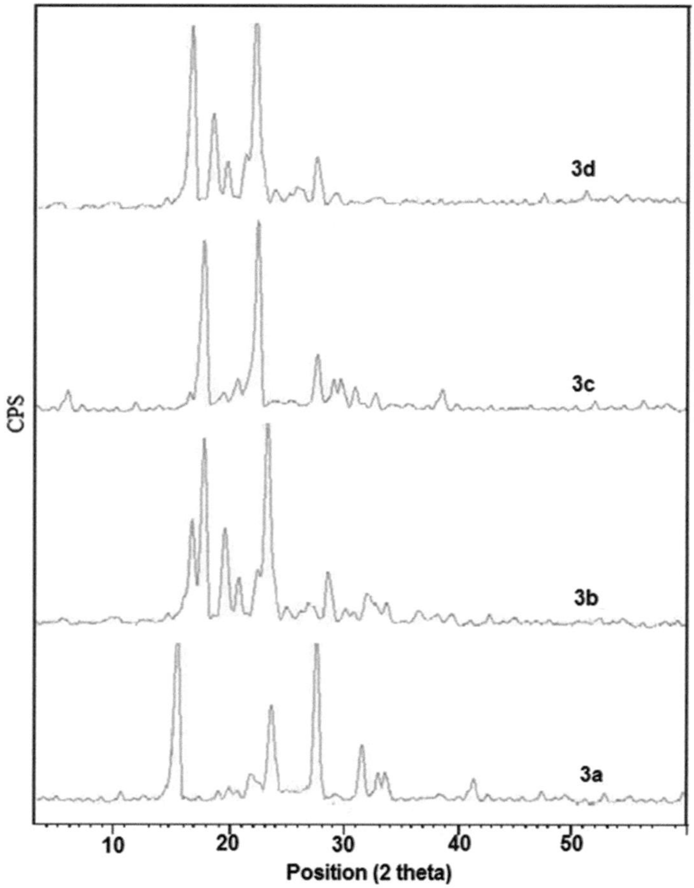


**Fig. 12.6** XRD pattern of poly(azomethine esters) 3 (a–d)

Thermal stability and degradation of the polymers 3(a–d) were analyzed based on the thermograms obtained from TGA at constant heating rate of 20 °C $min^{-1}$ in the temperature range of 30–800 °C under nitrogen atmosphere.

Equal weights of all the polymer samples were used to eliminate the mass effect. Enhancement of thermal stability of the polyesters observed in the present studies can be attributed to the delocalization characteristic of azomethine linkage. The results imply that the presence of azomethine linkage in the backbone of polyester is ideally suited to produce good thermal stability and high yield. The temperature required for 10 % degradation ($T_{10}$) and the amount of chars formed at 800 °C were summarized in Table 12.3. It is evident to the TGA curves that two steps of degradation have occurred. The first step at the range of 320–329 °C was assigned to the cleavage of ester groups, whereas the second step at the range of 450–480 °C, which is attributed to the degradation of azomethine and aromatic rings.

**Table 12.3** TGA data of LC poly(azomethine esters) 3(a–d)

| LC polymers | Degradation temperature at weight loss (°C) | | |
|---|---|---|---|
| | 10 % | 20 % | Residue at 800 °C |
| 3a | 380 | 410 | 25 |
| 3b | 335 | 363 | 15 |
| 3c | 343 | 375 | 17 |
| 3d | 330 | 358 | 13 |

## 12.9 Conclusion

In the present work, terephthalic acid was successfully regenerated from PET waste bottles via saponification process in high yield and purity. The new liquid crystalline poly(azomethine esters) based on terephthalic acid were prepared and characterized successfully using important and vital spectroscopic methods. The combining results that obtained from DSC, POM and XRD studies were confirmed the liquid crystallinity of the poly(azomethine esters). All the polymers exhibit nematic mesophases except that for polymer 3(a). Furthermore, TGA results also confirmed that the new synthesized polymers possess high thermal stability.

**Acknowledgement** The authors would like to thank University of Malaya for their support throughout this work.

## References

Aslan S, Immirzi B, Laurienzo P (1997) Unsaturated polyester resin from glycolized polyethyleneterephthalate, synthesis and comparison of properties and performance with virgin resin. J Mater Sci 32:2329–2336

Atta AM, Elnagdy SI, Abdel-Raouf ME, Elsaeed SM, Abdel-Azim AA (2005) Comprehensive properties and curing behavior of unsaturated polyester resin in the presence of vinyl ester resins derived from recycled poly (ethylene terephthalate). J Polym Res 12:373–383

Atta AM, El-kafrawy AF, Aly MH, Abdel-Azim AA (2007) A new epoxy resins Based on recycled poly (ethylene terephthalate) as organic coatings. Prog Org Coat 58:13–22

Berdague P, Bayle JP, Ho MS, Fung BM (1993) New laterally aromatic branched liquid crystal materials with large nematic ranges. Liq Cryst 14:667–674

David EN, Medhat SF (2005) New motivation for the depolymerization products derived from poly (ethylene terephthalate) (PET) waste: a review. Macromol Mater Eng 209:13–30

Han H, Bhowmik PK (1997) Wholly aromatic liquid-crystalline polyesters. Prog Polym Sci 22:1431–1502

Issam AM (2014) A new approach to obtain Kevlar-49 from PET waste bottles. Res Chem Intermediat 40:3033–3044

Issam AM, Sankar G, Melati K, Abu Bakar M (2010) Synthesis and liquid crystalline properties of new diols containing azomethine groups. Molecules 15:3260–3269

Issam AM, Shilla MN, Abdul Khalil HPS (2011a) Effect of methylene spacers of unsaturated resins on the properties of composites based on oil palm empty fruit bunches and fiberglass. Adv Compos Mater 20:1–12

Issam AM, Sufia H, Nurul Khizrien AK (2011b) A new unsaturated poly (ester-urethane) based on terephthalic acid derived from polyethylene terephthalate (PET) of waste bottles. J Polym Environ 20:469–476

Issam AM, Mohd Fadhley A, Wan Rosli WD (2012) New class of liquid crystalline epoxy resins: synthesis and properties. J Ind Eng Chem 18:364–372

Jackson WJ Jr, Kuhfuss HF (1976) Liquid crystal polymers. I. Preparation and properties of p-hydroxybenzoic acid copolyesters. J Polym Sci Polym Chem Ed 14(8):2043–2058

Karayannidis GP, Archilias DS, Sideridou ID, Bikiaris DN (2005) Alkyd resins derived from glycolized waste poly (ethylene terephthalate). Eur Polym J 41:201–210

Marin L, Cozan V, Bruma M (2006) Comparative study of new thermotropic polyazines. Polym Adv Technol 17:664–672

Marin L, Zabulica A, Sava M (2013) Symmetric liquid crystal dimmers containing a luminescent mesogen: synthesis, mesomorphic behavior, and optical properties. Soft Mater 11:32–39

Shukla U, Rao KV, Rakshit AK (2003) Thermotropic liquid-crystalline polymers: synthesis, characterization and properties of poly(azomethine esters). J Appl Polym Sci 88:153–160

Shukla SR, Harad AM, Mahato D (2006) Depolymerization of nylon 6 waste fibers. J Appl Polym Sci 100(1):186–190

Sinha V, Patel MR, Patel JV (2010) PET waste management by chemical recycling: a review. J Polym Environ 18:8–25

Tomasz G, Włodzimierz S (2009) Side-chain liquid crystal polymers (SCLCP): methods and materials. An overview. Materials 2(1):95–128

Tserki V, Matzinos P, Pavlidou E, Vachliotis D, Panayiotou C (2006) Biodegradable aliphatic polyesters. Part I. Properties and biodegradation of poly(butylenes succinate-co-butylene adipate). Polym Degrad Stab 91:367–376

Vasanthi BJ, Ravikumar L (2013) Synthesis and characterization of poly(azomethine ester)s with a pendent dimethoxy benzylidene group. Open J Polym Chem 3:70–77

Zifa L, Zhang S, Shaokui CAO, Chen Z, Zhou Q (1996) Synthesis and characterization of liquid crystal polymers with T-shaped two-dimensional mesogenic units. Chin J Polym Sci 14 (1):71–79

# Chapter 13
# Liquid Crystalline Polymer Composites for Optoelectronics

**Amrit Puzari**

Composite materials (or composition materials) comprise of two or more constituent materials having significant differences in their physical or chemical properties. The constituent materials are separated by a distinct interface and retain their separate identity in the finished structure. Composite materials are known to possess two distinct phases namely Matrix phase and Dispersed phase and have distinct properties different from their individual component. Matrix phase is the primary phase and is more ductile and less hazardous phase. It holds the dispersed phase which is embedded in the matrix in a discontinuous manner. Dispersed phase or secondary phase is somewhat harder than the primary phase and therefore it is also known as reinforcing phase (Jose et al. 2012). Composite in fact is a multipurpose material and are suitable for photoelectric conversion and electrochromic application. Natural composites are also known which are usually polymeric in nature. Wide range of applications of composite materials is known ranging from automobile industry, medical equipment, space technology, electronic packaging and many more. Although history reveals the use of composite materials date back in 1500 BC, but modern era of composites began after scientists developed plastics in the early 1900s. Introduction of Fiber Reinforced Polymer (FRP) added newer dimensions to this development.

Formation of composite materials is the effective route to improve the performances of polymeric materials through appropriate structural modifications. Remarkable improvement can be obtained in mechanical, thermal, electrical, optical and catalytic properties of the composite materials. Therefore it can be stated that synthesis of composite materials is an efficient way to synthesize newer smart materials (Sun et al. 2013). Liquid crystalline polymers (LCPs) are considered as an important class of high performance engineering materials because of their

A. Puzari (✉)
Department of Science and Humanities, National Institute of Technology Nagaland, Chumukedima, Dimapur 797 103, Nagaland, India
e-mail: amrit09us@yahoo.com



V.K. Thakur, M.R. Kessler (eds.), *Liquid Crystalline Polymers*,
DOI 10.1007/978-3-319-20270-9_13

mechanical strength, stiffness, low linear thermal expansion coefficient and other similar characteristics (Pegoretti and Traina 2009). They have excellent chemical resistance and inherent flame retardancy. LCPs are positionally disordered crystals or orientationally ordered liquids (Godovskii and Papkov 1992).

Liquid crystalline polymeric materials also hold great promise for tunable opto-electronic materials. Of course these materials have pronounced practical applications. This is because of their specialty in structural behavior and also because of their intriguing physics. Functionalized LC Polymers are finding applications in electroluminescent display devices, non-linear optical materials, tunable lasers etc. The optical birefringence and dielectric anisotropy of LC materials as well as molecular level self ordering coupled with polymeric structural pattern, help tuning of the electro-optical performances of liquid crystalline polymers. Because of the specialty in their properties (Cheng et al. 2012), compared to other polymers, liquid crystalline polymer compounds are also relatively easy to process and nano-fillers like Carbon Nano Tubes, $TiO_2$ etc. can be incorporated into the matrix to develop newer series of nanocomposite materials. This special feature has created hope for commercialization of newer series of nanocomposites for application in opto-electronics. Thermotropic liquid crystalline polymers are considered as still better candidates for the purpose (Thomas and Wood 1985). Recently sensing and surface morphological properties of poly[(9,9-dioctylfluorenyl-2,7-diyl)-co-bithiophene] liquid crystalline polymer was investigated for exploiting its potential utility for optoelectronic applications like for fabrication of sensors. Good solubility, sensitivity and responsive behavior of the LC Polymer are promising for fabrication of such devices (Gunduz 2015). Although LC polymers are rarely known to make any direct impact as LC display devices, the optical properties of LC Polymers can be carefully tuned to improve their performances in electro-optic devices. LC polymers are also finding applications in aligning and mechanically stabilizing small molecular displays (Van Boxtel et al. 2000).

## 13.1 Structural Diversity in Liquid Crystalline Polymer

These polymers have the capacity to form highly ordered structure while in the liquid phase. Rigid mesogenic repeating units are attached to the main chain of polymer backbone and these rigid monomer units confers high strength and high melting temperatures to the polymeric materials (Collyer 1989). It is possible for both polymer melts and polymer solutions to exhibit phases with positional and orientational order under certain circumstances. Similar to general liquid crystalline materials, in case of liquid crystalline polymer materials too, one can observe the two broad category of materials i.e. thermotropic liquid crystalline polymer and lyotropic liquid crystalline materials. The former forms liquid crystalline structures with change in temperature while the later forms liquid crystalline structure with change in concentration (Collings 1990). Examples of Lyotropic and Thermotropic LCP are Kevler based on poly (p-phenyleneterephthalamide) (Kwolek 1971) and

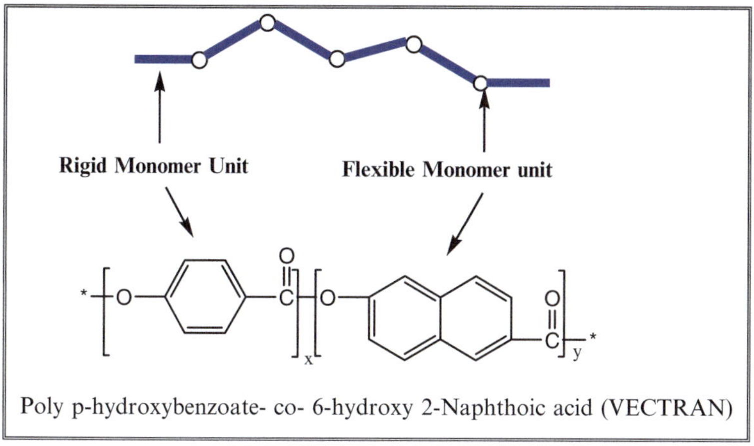


**Fig. 13.1** Molecular structure of aromatic polyester based liquid crystalline polymer, Vectran (Calundann 1980)

Vectra based on aromatic polyesters (Calundann 1980) respectively. Molecular structure of aromatic polyester based liquid crystalline polymer, Vectra is represented in Fig. 13.1.

The thermotropic polymer liquid crystals possess at least one liquid crystalline phase between glass-transition temperature and transition temperature to isotropic liquid. Similarly under certain ranges of temperatures and concentrations, lyotropic polymer liquid crystalline materials possess at least one liquid crystalline phase. Of course some polymers may exhibit both thermotropic and lyotropic behavior.

Again two major groups of thermotropic polymer liquid crystalline materials are known depending on whether the rigid mesogenic unit is incorporated in to the main chain of the polymer or appended as side chain. Accordingly they are named as main-chain liquid crystalline polymer and side-chain liquid crystalline polymer. Three structural units are present in case of side chain liquid crystalline polymers. Apart from the polymer chain these polymer possess some spacer unit and terminal mesogenic group. Depending on the chemical structure of the polymer, diversity in structural properties is observed for such polymers. Thus mesomorphic properties of such polymers can be tuned by varying the chemical nature and structure of the core. Lyotropic polymer liquid crystals are usually resulted when polymer arranges itself in an elongated fashion while in solution (Collins 2005). Two main polymer liquid crystalline phases known as nematic and smectic phase are represented in Fig. 13.2.

In yet another classification, they can be broadly classified into three classes namely ‘Aromatic polyamides’ or ‘Aramides’ like Kevler, Twaron, etc., ‘Aromatic heterocyclic’ and ‘Aromatic copolyesters’ (Pegoretti and Traina 2009). Aromatic heterocyclic LCPs are kind of lyotropic liquid crystal containing wholly aromatic molecular framework with fused heterocyclic ring along the main chain. Aromatic copolyesters exhibit thermotropic behavior and are characterized by high

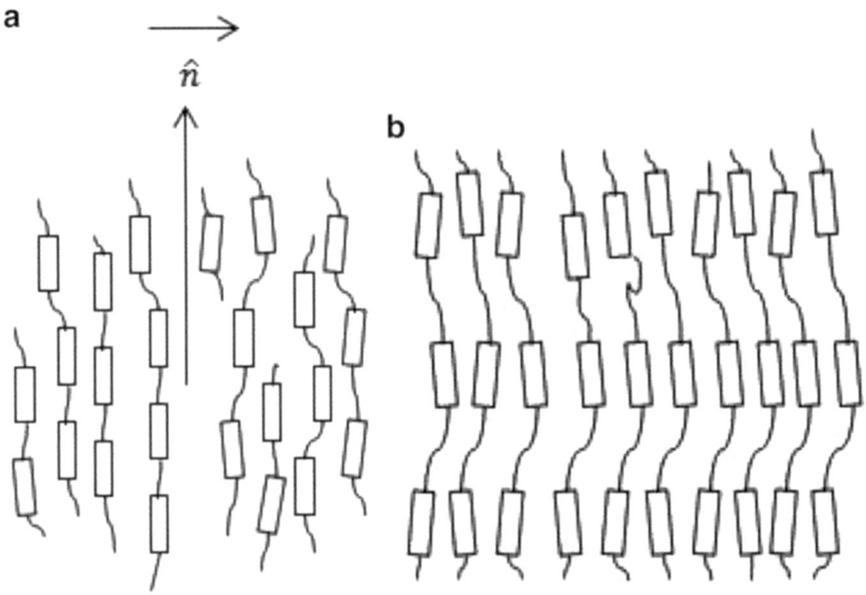


**Fig. 13.2** Main chain polymer liquid crystal phases: (**a**) nematic, (**b**) smectic (Collins 2005) [Reproduced with permission from Wiley VCH Verlag GmBH & Co]

degree of linearity and rigidity in their structure. Polymer Vectra referred above included in this category.

With progress in time newer families of Liquid Crystalline Polymer materials are developed to include newer architectures like liquid crystalline elastomers, dendritic liquid crystalline polymers, block copolymers etc. Conjugated polymers are interesting material for development of numerous optoelectronic applications. As a result new potential applications are emerging in various domains like organic photovoltaic, actuators etc. in which case liquid crystalline polymer composites play a significant role. In general the molecular weight and microcrystalline morphology of the polymer influences the charge carrier mobility (Himmelberger et al. 2014). Poor interconnectivity between the microcrystalline domains might be the possible reason for this. The anisotropic carrier transport properties of liquid crystalline polymers are being exploited for development of Organic Field Effect Transistors (OFETs). Molecular conformation and structural diversities of the polymer coupled with the phenomena of molecular self-assembly, significantly influence the resulting optoelectronic properties of the polymer (Botiz and Stingelin 2014). The Liquid crystalline dendrimers are another class of interesting materials which possess a flexible dendritic network. For such molecular architectures mesogenic or promesogenic groups are attached to the termini of the branches either laterally (side-on) or terminally (end-on) (Puzari 2012).

Liquid crystalline behavior of such molecules is largely governed by the nature of mesogenic or pro-mesogenic promoters. Overall structure of such dendritic architecture depends on several factors such as nature of mesogenic group or pro-mesogenic group attached to the peripheral group of the dendritic network, tendency of dendritic structure to adopt globular isotropic conformation, generation number of the dendritic framework, structural compatibility between the dendritic core and the peripheral group etc. (Frechet and Tomalia 2001). Smectic, nematic or columnar phases can be obtained from such molecular architecture by choosing appropriate mesogenic group and the phase behavior of such dendrimers is also influenced by the generation number. Ferroelectric properties were reported with dendrimers functionalized with chiral mesogenic groups (Busson et al. 1998; Boiko et al. 2000).

Therefore acute structural knowledge about these types of polymeric materials is very much important for tuning them to obtain tailor made properties. These types of molecules possess some interesting physical properties like low viscosity or high solubility, compared to their linear non branched analogue polymers. Such polymers have the capacity to self assemble into supramolecular columns or spheres. Three structural units are present in liquid crystalline polymers. Apart from the polymer chain these polymer possess some spacer unit and terminal mesogenic group. Nature of the mesophase is strongly dependant on the terminal mesogenic group. Numerous types of mesogenic or promesogenic groups are known. Sometimes some other structural concepts are also used to explain the structural organization in Liquid Crystalline dendrimers. Shape-persistent systems (Donio et al. 2007) is one such concept used to elaborate the structure of some main chain dendrimers, where dendritic core itself acts as main mesogenic group and the main chain dendrimers are incorporated as both branching groups (hyperbranched polymers) (Percec and Kawasumi 1992; Bauer et al. 1993) and end-groups (Willow like polymers) (Percec et al. 1995; Li et al. 1996).

## 13.2 Liquid Crystalline Polymer Based Blends

Unusual properties of Liquid Crystalline Polymers such as high thermal stability, flame retardant, high chemical resistance, high mechanical strength etc. attracts greater interest from scientific community for development of commercially important newer smart materials, especially for the cases where chemical inertness and superior strength is a desired condition. Compared to conventional polymeric composites, this class of polymeric materials are projected as superior materials with respect to various intriguing properties such as mechanical strength, tensile strength, structural regularity and the like.

In contrast to conventional composites where fabrication of the composites involve multiple steps, some liquid crystalline composite structures with highly oriented reinforcing fibrous LCP phase can be produced in a single step. Because of this reason these blends are some time referred to as in situ composites (Kiss 1987). Beauty of such composite materials is that the mechanical properties of such composite materials can be tuned by controlling the addition of liquid crystalline composite. Thus these can be considered as promising material for design of high performance material with desired mechanical properties. Even some times the mechanical property of such composites parallels with those of glass fibre reinforced composites (Lin et al. 1993). Another significant difference of such in situ composites from conventional ones is that stretching of the LCP phase in the blend exhibit a marked reinforcement effect and with increasing orientation of the LCP chain along the fibre axis during drawing, the strength of LCP micro fibrils found to increase (Carfagna et al. 1991; Dutta et al. 1992). Theoretical modules are also known which can provide effective guidelines for fabrication and application of such composites (Lin and Yee 1994).

Thermotropic Liquid Crystalline Polymers (TCLP) are frequently been studied for design of speciality materials for commercial application because of their unique properties such as low coefficients of thermal expansion, low viscosity, high modulus, low permeability to gases, low dielectric constants, and chemical resistance (Tjong 2003). Current development in nanotechnology has made it possible the design of particular composite from such Thermotropic Liquid Crystalline Polymers (TCLP) by blending with suitable nanoparticles. Such composite materials have significant improvement in their properties such as barrier properties, electrical properties, mechanical properties and thermal properties.

Thermotropic Liquid Crystalline polymers (TLCPs) have been one of the most interesting developments in the chemistry and technology of polymeric materials during the past two decades. Unique intriguing properties associated with TLCPs have significantly improved their commercial importance and continuously attracting scientific interest. These polymeric materials can be processed easily and can be incorporated with different nanofillers like carbon nanotube (CNT), graphene, graphene oxide etc. to fabricate newer high performance nanocomposites. These high performance nanocomposites are commercially been employed for different applications including aerospace (Vyas et al. 2007; Hearle 2001). These nanocomposites possess proven superiority over conventional CNT/polymer composites when characteristics like mechanical and thermal properties and even electrical properties are considered. It is worth mentioning that specialty of such type of nanocomposites primarily can be attributed to the sophisticated properties of CNT nanofillers and typical structural organization of liquid crystalline polymeric materials. Unique structural arrangement of atoms found in CNTs confers superiority in their properties and makes them ideal reinforcing component for fabrication of polymer nanocomposites (Chung et al. 1989).

TLCPs are also considered for blending with thermoplastics for development of newer smart materials. In such cases the structural organizations of such blends mostly determined by processing conditions (Lin and Winter 1991; Kachidza et al. 1992). Of course such blends possess relatively poor mechanical properties because of the poor interfacial adhesion between the matrix polymer and dispersed LCP phase. Researchers engaged in this field indicate that improvement of this interfacial adhesion might substantially improve the mechanical behavior of such composites. It has been demonstrated that addition of small amount of functionalized CNT can help enhancement of the interfacial adhesion between the LCP and the thermoplastics which in turn significantly improves the mechanical properties of the composite (Cheng et al. 2010).

Although it is anticipated that CNT/LCP composites are promising for functional and structural application, but there are certain limitations on using CNTs as reinforcing fillers. This is mainly because of poor interfacial interaction between CNTs and LCP matrix and problem of dispersion of CNTs during fabrication process. Many researchers are engaged in this field to resolve such problems which lead to development of several methods for improvement of dispersion of CNTs in LCP matrix and to enhance the interfacial interaction. Particularly LCPs with polar functional groups shows better improvement in the process of dispersion of CNTs and also shows enhancement of interfacial interaction, when chemically functionalized CNTs are involved in the fabrication process. In such cases secondary interactions like hydrogen bonding and/or π–π interaction plays significant role in enhancing the interfacial adhesion between LCP phase and CNTs. The effect of chemically functionalized CNTs on their state of dispersion can better be understood from Fig. 13.3, which represents Field emission electron microscope (FESEM) micrographs of TLCP/CNT composites consisting of simple and chemically functionalized CNTs. The representative diagram contains micrographs of composites obtained by dispersing chemically functionalized MWCNT (Multi Walled Carbon NanoTube) in LCP matrix. Functional group involved in this case are carboxylic acid (–COOH) group and hydroxyl benzoic acid (–HBA) group (Sahoo et al. 2009a, b, c).

Other important examples of carbon based nanofillers which attract significant research interest for fabrication of LCP nanocomposites are single and multi layer forms of graphite, graphene nanoplatelets, graphene oxide and graphene. Among these in general graphene is stated to possess extraordinary potential for development of nanocomposites because of their outstanding mechanical properties, high surface area and specific electrical and thermal conductivities (Ramanathan et al. 2008). However research works on graphene polymer nanocomposites primarily focuses on isotropic or random dispersion of the graphene in the polymer matrix.

Like Liquid crystalline Polymers, Liquid Crystalline elastomers (LCE) also possess special characteristics properties which project them as potential candidates for various important commercial applications. Apart from their notable mechanical properties, they possess the important properties like softness, elasticity,

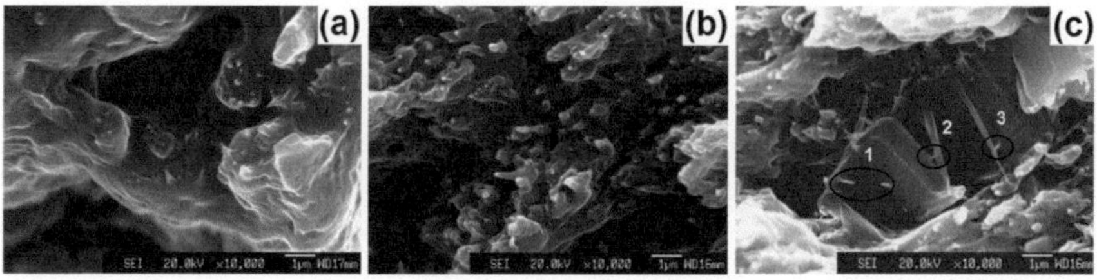


**Fig. 13.3** Field emission electron microscope (FESEM) micrographs for (**a**) raw MWCNT/TLCP, (**b**) MWCNT-COOH/TLCP and (**c**) MWCNT-HBA/TLCP (Sahoo et al. 2009a, b, c) [Reproduced with permission from Wiley VCH Verlag GmbH & Co]

durability, light weight etc. (Warner and Terentjev 2007). The anisotropic unit called the mesogens can exhibit analogous orientational ordering as that of ordinary low molar mass liquid crystals. So with the mesogens phase transition from liquid crystalline state to isotropic state, a significant structural change observed in the macroscopic shape because mesogens are bound to the polymer network. This shape change is fully reversible and is dictated by the thermodynamic equilibrium between the two states. Therefore scientific interest is focused on this type of soft materials, which combines both rubber-like elasticity and anisotropic liquid crystalline ordering, so that potential application of such materials can be explored properly.

Therefore development of LCE nanocomposites using nanoparticle fillers has become an important research objective. LCE nanocomposites devised through careful selection of nanoparticles can lead to better material characteristics. LCEs are consists of the components like mesogen, polymer backbone and the cross linkers. Mesogens which are key element for the formation of liquid crystalline phase are either attached as side group to polymer backbone or incorporated as a portion of the polymer backbone. Cross linkers assist in the formation of polymer network by covalently connecting various polymer chains. Nanoparticles chosen for fabrication of LCE nanocomposites must possess special optical electrical and magnetic properties. Carbon based nanoparticles including CNT are appeared as appropriate candidate for the purpose with suitable commercial implication. Carbon nanotubes has the capacity to absorb light and convert this into heat and therefore deformable elastomers with embedded carbon nanotubes are found to use as conductive layers (Huang and Terentjev 2010) The cylindrical shaped Single walled nanotubes (SWNTs), Multi-walled carbon nanotubes (MWNTs) which have extremely high surface-to-volume ratio and specific thermal, mechanical and electronic properties, are always claimed as superior to other nanoparticles (Ajayan and Tour 2007). Thus introduction of CNTs into LCEs produces kind of composite materials which can respond to light and electric field (Ji et al. 2010; Courty et al. 2003). Since CNTs are also sensitive to magnetic field therefore such material can be tuned by a magnet. Chemical structure of few LCE starting materials with acrylic backbone is represented in Fig. 13.4 (Ji et al. 2012).

Apart from carbon based nanoparticles, other nanoparticles are also used for fabrication of such nanocomposites. Thus Perovskite-based ferroelectric $PbTiO_3$ nanoparticles are employed for fabrication of such nanocomposites with LCEs (Domenici et al. 2010). The high ferroelectricity of such particles can trigger actuation in the resultant composite. Among the non carbon categories, electrically active and magnetically active nanoparticles are deposited onto LC Elastomers. The electrical conductivity of such nanoparticles might influence the actuation properties of such LCE composites. Domenici et al. reported one such LCE composite where a layer of $MoO_{3-x}$ nanowires were deposited onto the LCE surface (Domenici et al. 2011). These nanowires are of similar length with that of carbon nanotube but with larger diameter and are porous. Similarly magnetically active LCEs are also known where superparamagnetic $Fe_3O_4$ nanoparticles are employed (Kaiser et al. 2009). The composite material obtained can be a potential candidate

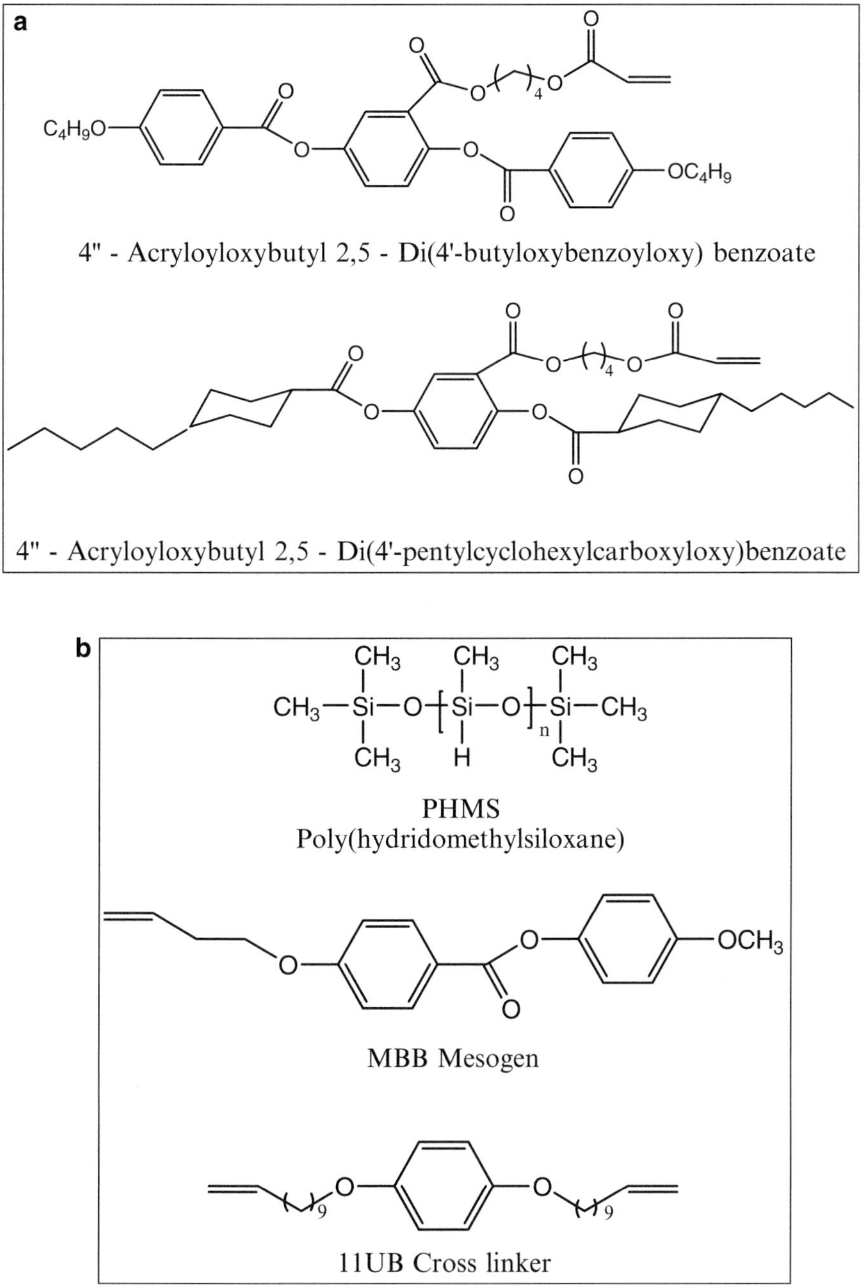


**Fig. 13.4** Chemical structure LCE starting materials. (**a**) LCE with acrylic backbone and (**b**) Polysiloxane based LCE (Ji et al. 2012)

for using as antennae for electromagnetic energy dissipation. Liquid crystalline polymer based blends thus possess many interesting behaviours which make them attractive for several versatile applications.

## 13.3 Liquid Crystalline Polymer in Micro and Opto-Electronics

Liquid Crystalline Polymer can be referred as a thermoplastic polymer that consists in aligned molecular chains with spatial regularity to impart crystal like behaviour. These polymers are suitable materials for numerous electronic and optoelectronic applications. Fabrication of various electronic components usually employ cost effective single crystal silicon because of which scientists are always exploring the possibility of using polymeric species as a substitute for silicon. Apart from this polymer offers certain unique advantageous physical and chemical properties which are not available in silicon. For example we can mention about the properties such as biocompatibility, lower melting temperature and high mechanical fracture limit of polymers. Lower melting temperature of polymers allows one to use the material in situations where temperature stability is not so important. Therefore polymers are important class of materials for such application.

In liquid crystalline polymers rigid and flexible monomers are linked to each other and in liquid crystal state the rigid segments of the molecules align next to one another in the direction of shear flow. This orientation once formed maintains their directions and structure even if the LCP is cooled below the melting temperature (Jayaraj and Farrell 1998). This is not observed in most thermoplastic polymer. These unique features of LCP confer them with electrical, thermal and mechanical properties different from other engineering polymers. These were used as a high-performance thermoplastic material for high-density printed circuit board (PCB) fabrication (Khoo et al. 1997) and semiconductor packaging (Jayaraj and Farrell 1998). Available literature indicates that microfabrication methods developed for liquid crystalline polymer makes them suitable for certain MEMS (microelectromechanical systems) applications (Wang et al. 2003). Two micro sensor applications have also been demonstrated in the article. Liquid crystalline polymer has some more interesting characteristics which make them more attractive for design of materials that can work in adverse environmental conditions. For example LCP has very low moisture absorption and low moisture permeability, excellent chemical resistance, and unaffected by most of the acids, bases and solvents over a wide range of temperature (Culbertson 1995). LCP materials can be used for several microdevice applications and it can serve as a flexible substrate material when fabrication of devices such as planar neuron probes, micro tactile sensors, antennas, capacitors etc. (Wang et al. 2003).

After incorporation of CNT, LCE nanocomposite starts responding to an electric field and when the nanoparticles form a conducting layer on LCE, it exhibit electromechanical actuation. This occurs because of Joule heating when an electric current passes through the material. Chambers et al. have studied in more detail about the electromechanical actuation of carbon processed LCEs and they have also considered the composites where conducting layers in LCEs were prepared by $PbTiO_3$ and $MoO_{3-x}$ nanowires (Chambers et al. 2009). It has been reported that LCE–CNT nanocomposite can actuate well below the nematic–isotropic transition because of the thermo mechanical response induced by pure heating. CNT contractions as well as heating collectively contribute to the observed optical actuation. Available literature data also indicates that CNT contraction occurred after photo absorption is essential for overall actuation of CNT–elastomer composites (Ahir et al. 2006). Whatsoever it is, it can be stated that CNT–LCE material can provide us electrically driven actuation where higher concentration of CNT will provide higher electric charge.

These facts in other way prove the superiority of CNT nanoparticles as these can be used in applications where high temperature is undesirable. CNTs can absorb in various regions such as visible, near-IR, Microwave, terahertz and megahertz frequency ranges (Mizuno et al. 2009). When the concentration of CNT is significantly higher, anisotropic behavior is observed with LCE composites at these frequencies. Such higher loading of CNTs along with its alignment found to induce anisotropic optical properties in LCE composites. Such anisotropic properties in combination with actuating behavior converts the LCE composites to potentially suitable material for preparation of tunable polarizers, wave guides, or other optoelectronic devices (Ji et al. 2012). Of course fabrication of LCE nanocomposites is somewhat difficult job because in doing so one must ensure that even after addition of nanoparticles, the LC properties and thermo-mechanical capability of the polymer should remain intact. In order to achieve useful mono-domain LCE nanocomposites, one needs to properly optimize the concentration of nanoparticles. If the concentration is too low, it may not provide an efficient system for actuation while if the concentration is high it may destroy the LC properties of the polymer. Homogeneous dispersion of nanoparticles in LCEs is also equally important.

CNTs are thus excellent precursor material for fabrication of electrically conductive nanocomposites. High aspect ratio of CNTs is an added advantage for this. However poor solubility of CNTs many times limits their chemical manipulation and potential applications. Therefore chemical and physical modification of the surface of CNTs might enhance their potential applicability. Covalent modification of CNTs are now being investigated on CNT based materials. For individual CNT electrical conductivity is found in the order of $10^6$ S $m^{-1}$ (Baughman et al. 2002) whereas SWCNT films have conductivity in the range of $10^4$–$10^5$ S $m^{-1}$ (Ericson et al. 2004; Sreekumar et al. 2003). This is due to the contact resistance between the individual carbonnanotubes. CNT polymer composites are therefore found to possess a wide range of electrical conductivity. Apart from the concentration of carbon nanotubes, other factors like dispersion, alignment and aspect ration of carbon nanotubes influence the conductivity value. Further to this, processing method employed and the type of functionalization of CNT also determines the electrical conductivity of CNT/polymer composite (Sahoo et al. 2009a, b, c). Exact mechanistic route is still not clear. For TLCP nanocomposites studies carried out with different functionalized MWCNTs indicates that depending on the functionalization, considerable change in electrical conductivity was observed. For example electrical conductivity of raw MWCNT/TLCP composite was measured as $9.80 \times 10^{-12}$ S/cm while for MWCNT$C_6H_4COOH$/TLCP and MWCNT$C_6H_4NH_2$/TLCP, electrical conductivity values were measured as $2.75 \times 10^{-11}$ S/cm and $4.38 \times 10^{-11}$ S/cm respectively (Sahoo et al. 2011a, b). Thus electrical conductivity of CNT/TLCP nanocomposites can be tuned through appropriate selection of nanoparticles i.e. through proper functionalization of CNT. CNT/TLCP nanocomposites have special utility in microwave frequency electronics. Low relative dielectric constant and low dissipation factors makes them suitable for such type of applications (Cheng et al. 2012). The possibility of using such nanocomposites in other important areas like in packaging and microelectromechanical systems (MEMS) is continuously attracting scientific interests.

Photocontrollable materials have significant application in the field of optics and photonics as well as in mass transport and mechanics. Azobenzene containing functionalized liquid crystalline polymer are important example of photocontrollable materials. Molecular alignment in such molecules plays a significant role in controlling optical coefficient, peripheral molecular alignments. It enables us to control optical coefficients, peripheral surface relief structure, and actuation of substances by means of photo irradiation. Exploration of photocontrol mechanism of such materials will open up newer areas of research in this field (Emoto et al. 2012). Photochromic Azobenzene molecule upon irradiation with UV light reversibly changes their shape from a rod-like *trans*-form to a bent *cis*-form and reverse transformation takes place when irradiated with visible light. *cis*- to *trans* transformation can also be induced thermally. So such phenomena can effect macroscopic volume change of the bulk under suitable conditions. Such type of photoinduced contraction was observed in azobenzene liquid crystalline elastomer (Wermter and Finkelmann 2001) where the liquid crystalline nature expected to have control over the response time for macroscopic deformation. Thus photo triggered shrinking or expansion of the surface of the film in such material may lead to bending motion and this was demonstrated by Ikeda and coworkers (Yu and Ikeda 2011).

It is possible to control the bending direction by changing the alignment of the liquid crystalline molecule. Photocontrollable liquid crystalline polymers have potential applicability in the field of holographic memory (Matharu et al. 2007). Azobenzene containing polymer systems are used for active grating coupler controlled by UV and visible light (Bang et al. 2007). Anisotropic grating formed by photocontrollable materials have the application prospects ranging from optics to information technology.

## 13.4 Liquid Crystalline Polymer Nanocomposite for Organic Nanophotonics

Liquid crystalline polymer molecules are usually modified through incorporation of nanoparticles to form nanocomposites. Depending on the type and nature of nanoparticles, the nanocomposites formed show multiple application prospects.

Liquid crystalline polymer composites, those are having potential applicability for photonic applications are discussed under the heading of organic nanophotonics. Nano-phase polymer dispersed liquid crystalline composites (n-PDLCs) are projected as suitable for voltage tunable optical devices (Lewis et al. 2000). The dimensional scale of the liquid crystalline droplets allows control of the optical scattering effects in the visible and near infrared spectrum. Similarly Carbon nanotubes are considered as potential candidate for formation of such nanophotonic devices. It is important to note that carbon nanotubes can enhance the charge mobility in nanocomposites and even these can modulate the charge mobility in nanocomposites. This property can be exploited to tune the colour of a single pixel that has multiple emissive layers through adjustment of the field (Woo et al. 2001). Organic nanophotonics encompasses the important applications like memory devices, Organic light emitting diodes, voltage tunable optical devices etc. It can be stated that each liquid crystalline polymer/nanoparticle nanocomposites can be considered as multifunctional photonic media because each domain/inclusion can be tuned to manifest specific photonic or optoelectronic function.

Tuning of the optical properties during operation is very important for making it suitable for various applications. For example in case of optical switches shifting of band gap is required during operation. Control of the orientation of the optical anisotropy of one materials incorporated in photonic crystal allows one to change the refractive index of at least one material and hence tuning of optical properties (Busch and John 1999). Such type of phenomena is observed in polymer dispersed liquid crystalline (PDLC) nanocomposites. These composites are produced from solution phase to incorporate liquid crystalline domain in a polymer matrix (Drazic 1995; Bunning et al. 2000) and which were used as light valve. Here liquid crystalline droplets produced through evaporation of solvents followed by subsequent annealing, are dispersed in the polymer matrix. It is possible to reorient the direction of a liquid crystalline droplet by applying a field and that changes its refractive index. The anisotropic liquid crystal possesses two refractive indexes, one is ordinary and the other is extraordinary. When randomly oriented, the ordinary refractive index of liquid crystal droplets are fairly close to that of polymer while extraordinary one is very high. This mismatch of refractive index blocks the transmission. When the electric field is applied, liquid crystal droplets reorient and light experience the ordinary refractive index which is fairly matches with that of polymer. Therefore no more blockages are there for transmission of light. Some of the polymer dispersed liquid crystals are used to produce electrically tunable lens and prism grating (Fan et al. 2003).

## 13.5 Self Assembled Liquid Crystalline Polymer Nanocomposite

Self assembled Liquid Crystalline Polymer Nanocomposites constitute another important class of materials having multiple utility. Cellulose nanoparticles (CNCs) constitute one such promising class of material which has the ability to self organize into chiral nematic (cholesteric) liquid crystal phase. The self organized structure has some important characteristics because of which the liquid crystalline polymer holds potential for optoelectronic applications. Self assembly in liquid crystalline polymers allows rational design of macromolecular entity with precisely ordered structure. Because of the presence of internal degrees of freedom observed in such polymeric molecules, they exhibit subtle behaviour of both high polymers and liquid crystals i.e. both liquid crystalline and high polymer conformational entropy are observed in LC Polymers. Therefore unlike conventional polymers, these polymers can spontaneously change their shape on ordering and this spontaneous shape change might impart novel characteristics which are not found in liquid crystals or polymers separately. The shape of liquid crystalline polymers are dependent on the state of nematic order which is again dependent on the process of self assembly in the macromolecular system.

It is possible to synthesise various forms of cellular nanocrystals following different routes and multiple application prospects can be studied (Lagerwall et al. 2014). Apart from biocompatibility other attractive properties observed for such materials are low thermal expansion coefficient, optical transparency and anisotropy, flexible surface chemistry etc. and these are exploited for different applications (Dufresne 2006; Zhang et al. 2012). The chiral nematic phase is characterized by long range orientational order of the nanorods along with a helical internal structure (Lagerwall and Scalia 2012). A representative structural image of CNC nanorods is presented in Fig. 13.5 (Salajkova et al. 2012).

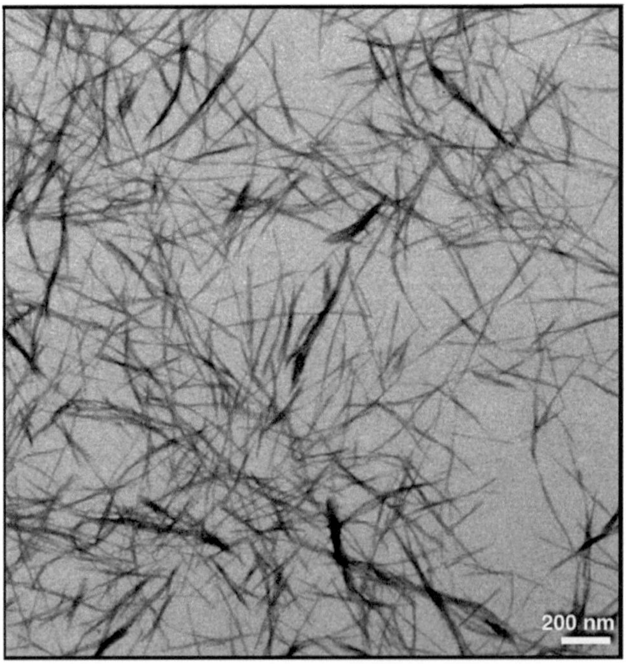


**Fig. 13.5** High-magnification images of CNC goethite nanorods (Salajkova et al. 2012) [Reproduced with permission from The Royal Society of Chemistry]

Because of the combined influence of the alignment and pitch of the helical modulation in different states, the material acquires a photonic band gap. The self assembled bio-derived photonic crystals thus have attractive optical properties for which CNC films are considered for development of newer materials for novel application. Especially the helical arrangement of the nanorods, converts these to materials with higher application potential in the field of optoelectronics. Such kind of films has attractive optical properties and possibly can be used in applications such as security papers and mirrorless lasing. Biological nanomaterials are also biodegradable.

Liquid crystalline dendrimers constitute another important class of compounds for design of intelligent materials. Several of such dendritic molecules can self assemble through various secondary interactions to form bulk molecular architecture. Matrix of the dendritic architecture highly influences the material properties of the complex structure (Puzari 2012). With increasing generation of the dendritic architecture, structural complexity also increases. Generally lower generations possess planar structure while higher generation dendrimers possess globular geometry which can accommodate more surface groups. Overall structure possesses large number of compartments in the interior. Depending on the type and nature of the constituent groups/atoms the overall structure can be flexible or rigid. For flexible molecules more disordered conformation results. Increase in the number of generations also increases the isotropization temperature and hence changes the phase transition behaviour. Structural organization of the liquid crystalline dendrimer depends upon the capacity of the surface functional groups to form network with each other.

The cases where hydrogen bonding is possible with the surface groups, resulting structure possess a very dense periphery and hollow core. Self assembly of such supramolecular structures leads to bulk structural moieties with periodic nanostructures which render possible design of newer materials with multiple utility. In such cases the process of self assembly and formation of quantized bundles of nanoscale building blocks is largely influenced by nature and multiplicity of the interior. Composition of the core and nature of the core group also influence the structural organization of such supramolecular structures. There is bright possibility of using these Bulk nanoscale self-assembly of organic matter as scaffold for photonic materials and for molecular electronics. Process of self assembly in some dendrimers attributes characteristic texture due to liquid crystalline mesophase and for such dendrimers phase transition and spin transition phenomena shows quick response to temperature. Therefore these can be prospective materials for sensor application and similar applications in molecular electronics (Astruc et al. 2010). At the same time these can be considered as promising materials for formation of novel nanocomposites.

## 13.6 Inorganic Organic Polymer Dispersed Liquid Crystalline Nanocomposite

Inorganic organic hybrid polymer dispersed liquid crystalline nanocomposites can be referred as a novel class of compounds and are attracting scientific interests because these can be considered as suitable material for fabrication of polymeric photorefractive (PR) material. Polymeric photorefractive materials are known to have low permeability and large optical nonlinearity which renders them attractive for the people of material science. These are also low cost materials (Steckman et al. 2000; Wurthner et al. 2002). Limitation of using such materials lies in the fact that these materials need relatively large dc electric field. Use of polymer dispersed liquid crystalline nanocomposites (Bunning et al. 2000) and substitution of traditional photosentizers with inorganic nanocrystals (Winiarz et al. 1999) can help to overcome such limitations. Dielectric anisotropy present in Nematic liquid crystals in the mesophase allows liquid crystalline materials to reorient under the influence of relatively moderate electric field. This particular property can be exploited for using them as photorefractive medium.

Inorganic organic hybrid polymer dispersed liquid crystalline nanocomposites are reported as suitable candidate for fabrication of photorefractive medium. Winiarz et al. reported one such photorefractive medium which consists of hole transporting polymer composite matrix, electro-optically active nanodroplets of liquid crystal and Cadmium sulphide quantum dots as photosentizers (Winiarz and Prasad 2002). They employed poly(methylmethacrylate) (PMMA) as polymer matrix, *N*-ethylcarbazole (ECZ) as hole transporting medium and commercial nematic liquid crystal mixture TL 202 for fabrication of the photorefractive medium with Cadmium sulphide as photosentizer. The reported medium exhibits more than 90 % internal diffraction efficiency which is relatively high compared with other PDLC photorefractive medium.

Polymer dispersed ferroelectric liquid crystal (PDFLC) material are also known which can be considered as potential material for optical devices. PDFLCs projected to have several advantages over the surface stabilized liquid crystal devices. More stable and flexible system can be obtained from PDFLCs and are projected as possible alternatives for fast switching devices. Electrooptical switching can be achieved through voltage dependent optical retardation when the cell is placed between crossed polarizers. These materials contain micron sized droplets of ferroelectric symmetric liquid crystal dispersed throughout a polymer (Karapinar et al. 2002).

Scientific community is also exploring the possibility of using holographic polymer dispersed liquid crystal (H-PDLC) technology for design of tunable integrated optical elements, for sensor systems, optical interconnects, optical datastorage etc. (Bunning et al. 2000; Liu and Sun 2008). Capacity of liquid crystalline molecule to reorient under the influence of electric field and efficient optothermo-mechanical response of polymer liquid crystal (PLC) composite is the primary basis of such scientific approach. PLC materials combined with holographic fabrication can be processed to photonic quasi crystals (Zito et al. 2008). Fabrication of a polymer liquid crystal-based fiber Bragg grating (PLC-FBG) with two beam holography was reported with the objective of developing a new platform of integrated fiber system. The system was integrated with hollow channels of a solid core photonic crystal fiber (PCF) (Zito and Pissadakis 2013). A suitable liquid crystalline/photomonomer solution was infiltrated into the cladding channel of a solid core Photonic crystal fiber. This hybrid H-PDLC/PCF system possesses high thermal and bending sensitivities. Microscopic measurement reveals the presence of LC nanodroplets embedded in the polymer matrix.

Low viscous monomers like dipentaerythritol-hydroxy-penta-acrylate (DPHPA) monomer (40–45 % w/w) was employed for the process along with cross linking agents like *N*-vinylpyrrolidone (NVP) monomer (20–25 % w/w) which also enhances fluidity. Rose bengal (RB) has been used as Photoinitiator. Active research works are going on to investigate about the optimization of diffraction efficiency and tuning of the morphology. Possibility of active inclusion of magnetic nanoparticles into the liquid crystal has also been studied (Candiani et al. 2011).

## 13.7 International Trends/Recent Advance in Liquid Crystalline Polymer Composite

Liquid crystalline polymer composites have become an increasingly important material for technological field involving photonic applications. Several attractive features such as dielectric anisotropy present in the mesophase, reasonably wide temperature range, low thermal expansion coefficient, high chemical resistance, specific thermo-mechanical response etc. make them more attractive. Of course several limitations have to overcome in order to avail the total benefit of using such materials for opto-electronic applications.

Recent works in liquid crystalline polymer science, also emphasises more on the use of Thermotropic Liquid Crystalline Polymers (TLCPs) for development of nanocomposites using different nanofillers (Cheng et al. 2012). A discussion has already been made on this in the foregoing discussions. Carbon based nanofillers are more promising in this regards.

Polymer stabilized blue phase liquid crystals (PS-BPLC) are considered for technological innovations, for information display and for photonic applications (Chen and Wu 2014). BPLCs have several promising features like reasonably wide temperature range, sub-millisecond gray-to-gray response time, optically isotropic in the voltage off state, no need for alignment of layers etc. which make them suitable for photonic applications. Most importantly the special feature of fast response time makes it a strong contender for next generation display technologies. The sub-millisecond response time enables it colour sequential display. Therefore there are significant enhancement in their optical efficiency and resolution density. These materials are also projected as attractive candidate for fabricating large panel displays. However limitations like high operation voltage, hysteresis, residual birefringence, charging issues due to the large capacitance etc. restricts their applicability (Yan and Wu 2011).

Blue phases, which are found as a highly twisted nematic liquid crystal cools down from isotropic phase, has the capacity of selective reflection of incident light. Although usually blue phase exhibit blue colour, but it is not always the case. Depending on the chirality of liquid crystal, they might reflect other colours too. The primary limitation of Blue phase is the narrow temperature range. Among several other methods developed to widen the temperature range, Polymer stabilized method developed by Kikuchi et al. has been found to be relatively more successful (Kikuchi et al. 2002). A small fraction of monomers (~8 %) and photoinitiators (~0.5 %) are added to the blue phase system. On UV irradiation the monomers get polymerized and the polymer network formed stabilizes the blue phase. Finally it constitutes a polymer stabilized self-assembly blue phase system where each material component plays a significant role (Yan and Wu 2011). Recently BPLC materials are gaining significant importance for optoelectronic applications. Nanostructured double twist cylinders through the phenomena of self-assembly generates interesting three dimensional cubic lattice structures that increases the importance of BPLCs as electro-optic components, tunable photonic crystals (Etchegoin 2000) and three dimensional lasers (Cao et al. 2002). Polymer stabilized BPLCs are rather more attractive because solid resin of fixed blue phase structure can be obtained with improved characteristics. In fact they are considered as potential precursor for photonic applications (Yan et al. 2011). The process of molecular self-assembly plays a significant role in generating these three dimensional lattice structures. Chemical structure of some nematic LC hosts and some monomer used in PS-BPLCs are given in Fig. 13.6.

In PSBP system polymer network plays important role in controlling material properties. They concentrate on the core in a random coil conformation in the lattice structure leading to stabilization of the blue phase (Kikuchi et al. 2002). Different monomers used collectively stabilize the blue phase and these are also needed to expand the temperature range. However it should be noted that the polymer obtained through polymerization of the monomer, should not readily mix with the liquid crystal. Otherwise there will be no stabilization effect (Kikuchi et al. 2002). Polymer network also influences electrooptic properties and increase in monomer concentration leads to higher operation voltage and smaller hysteresis.

## 13.8 Conclusion

The family of Liquid crystalline polymer composites has unique combinations of several interesting material properties which prove their superiority over other conventional materials of similar kind which are used for design of intelligent functional materials. Design of organic polymer based composite materials has always remains as a thrust area for the peoples of material science since such kind

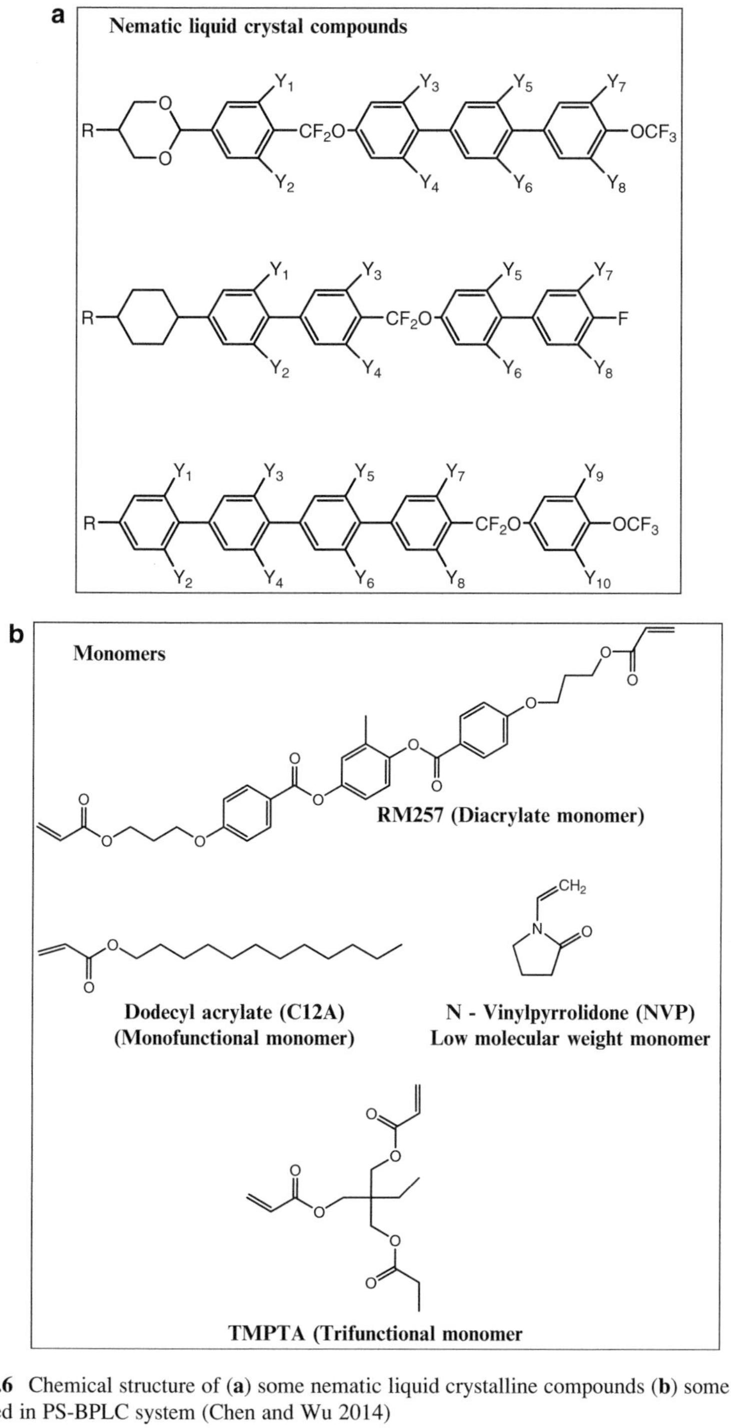


**Fig. 13.6** Chemical structure of (**a**) some nematic liquid crystalline compounds (**b**) some monomer used in PS-BPLC system (Chen and Wu 2014)

of materials can be suitable substitute for the cost effective inorganic counterparts. Scientific community around the world are actively pursuing research activities in this field. Liquid crystalline polymer composites because of their specialty in structural organization can be considered as novel precursors for design of smart materials for electronics and optoelectronics. They can be engineered for generation of nanoscopic devices using their wide range of diversities in structural and functional properties. Proper selection of mesogenic groups for such polymeric structures, the phenomenon of molecular self assembly coupled with other intriguing characteristics of such composites will render it possible the design of newer smart materials for electronics and optoelectronics. Days are not that far when newer innovations with tailor made properties will gradually establish their dominance in the field of molecular electronics.

**Acknowledgements** I am delighted to acknowledge the help and support I received from Dr. Sidananda Sharma, Sr. Scientific Officer, IIT Guwahati and Mr. Alok Ranjan, Research Scholar, Singapore University of Technology and Development (SUTD).

## References

Ahir SV, Squires AM, Tajbakhsh AR, Terentjev EM (2006) Infrared actuation in aligned polymer-nanotube composites. Phys Rev B 73:085420:1–085420:12

Ajayan PM, Tour J (2007) Nanotube composites. Nature 447:1066–1068

Astruc D, Boisselier E, Ornelas C (2010) Dendrimers designed for functions: from physical, photophysical, and supramolecular properties to applications in sensing, catalysis, molecular electronics, photonics, and nanomedicine. Chem Rev 110:1857–1959

Bang CU, Shishido A, Ikeda T (2007) Azobenzene liquid-crystalline polymer for optical switching of grating waveguide couplers with a flat surface. Macromol Rapid Commun 28:1040–1044

Bauer S, Fischer H, Ringsdorf H (1993) Highly branched liquid crystalline polymers with chiral terminal groups. Angew Chem Int Ed Engl 32:1589–1592

Baughman RH, Zakhidov AA, de Heer WA (2002) Carbon nanotubes: the route toward applications. Science 297:787–792

Boiko N, Zhu X, Vinokur R, Rebrov E, Muzafarov A, Shibaev V (2000) New carbosilane ferroelectric liquid crystalline dendrimers. Mol Cryst Liq Cryst 352:342–350

Botiz I, Stingelin N (2014) Influence of molecular conformations and microstructure on the optoelectronic properties of conjugated polymers. Materials 7:2273–2300

Bunning TJ, Natarajan LV, Tondiglia VP, Sutherland RL (2000) Holographic polymer dispersed liquid crystals (HPDLCs). Annu Rev Mater Sci 30:83–115

Busch K, John S (1999) Liquid-crystal photonic-band-gap materials: the tunable electromagnetic vacuum. Phys Rev Lett 83:967–970

Busson P, Ihre H, Hult A (1998) Synthesis of a novel dendritic liquid crystalline polymer showing a ferroelectric SmC* phase. J Am Chem Soc 120:9070–9071

Calundann GW (1980) Melt processable thermotropic wholly aromatic polyester. US Patents 4184996

Candiani A, Margulis W, Sterner C, Konstantaki M, Pissadakis S (2011) Phase-shifted Bragg microstructured optical fiber gratings utilizing infiltrated ferrofluids. Opt Lett 36:2548–2550

Cao WY, Munoz A, Palffy-Muhoray P, Taheri B (2002) Lasing in a three-dimensional photonic crystal of the liquid crystal blue phase II. Nat Mater 1:111–113

Carfagna C, Amendola E, Nicolais L, Acierno D, Francescangeli O, Yang B, Rustichelli F (1991) Blends of a polyetherimide and a liquid crystalline polymer: fiber orientation and mechanical properties. J Appl Polym Sci 43:839–844

Chambers M, Finkelmann H, Remškar M, Sánchez-Ferrer A, Zalar B, Ţumer S (2009) Liquid crystal elastomer–nanoparticle systems for actuation. J Mater Chem 19:1524–1531

Chen Y, Wu ST (2014) Recent advances on polymer-stabilized blue phase liquid crystal materials and devices. J Appl Polym Sci 40556:1–10

Cheng HKF, Sahoo NG, Li L, Chan SH, Zhao J (2010) Molecular interactions in PA6, LCP and their blend incorporated with functionalized carbon nanotubes. Key Eng Mater 447:634–638

Cheng HKF, Basu T, Sahoo NG, Li L, Chan SH (2012) Current advances in the carbon nanotube/ thermotropic main-chain liquid crystalline polymer nanocomposites and their blends. Polymers 4:889–912

Chung TS, Calundann GW, East AJ (1989) Liquid-crystalline polymers and their applications. In: Chermisinoff NP (ed) Encyclopaedia of engineering materials, vol 2. Marcel Dekker, New York, pp 625–675

Collings PJ (1990) Liquid crystals: nature's delicate phase of matter. Princeton University Press, Princeton, pp 149–167

Collins PJ (2005) Liquid crystalline materials. In: Kirk-Othmer encyclopaedia of chemical technology, 5th edn, vol. 15. Wiley, Hoboken, pp 81–120

Collyer AA (1989) Thermotropic liquid crystal polymers for engineering applications. Mater Sci Technol 5:309–322

Courty S, Mine J, Tajbakhsh AR, Terentjev EM (2003) Nematic elastomers with aligned carbon nanotubes: new electromechanical actuators. Europhys Lett 64:654–660

Culbertson EC (1995) A new laminate material for high performance PCBs: liquid crystal polymer copper clad films. In: Proceedings of the 45th electronic components and technology conference. p 520–523

Domenici V, Zupančič B, Laguta VV, Belous AG, V'yunov OI, Remškar M, Zalar B (2010) $PbTiO_3$ nanoparticles embedded in a liquid crystalline elastomer matrix: structural and ordering properties. J Phys Chem C 114:10782–11078

Domenici V, Conradi M, Remškar M, Viršek M, Zupančič B, Mrzel A, Chambers M, Zalar B (2011) New composite films based on $MoO_{3-x}$ nanowires aligned in a liquid single crystal elastomer matrix. J Mater Sci 46:3639–3645

Donio B, Buathong S, Bury I, Gullion D (2007) Liquid crystalline dendrimer. Chem Soc Rev 36 (9):1495–1513

Drazic PS (1995) Liquid crystal dispersions. World Scientific, Singapore

Dufresne A (2006) Comparing the mechanical properties of high performances polymer nanocomposites from biological sources. J Nanosci Nanotechnol 6:322–330

Dutta D, Weiss RA, Kristal K (1992) Blends containing liquid crystalline polymers: preparation and properties of melt drawn fibers, unidirectional prepregs and composite laminates. Polym Compos 13:394–401

Emoto A, Uchida E, Fukuda T (2012) Optical and physical applications of photocontrollable materials: azobenzene-containing and liquid crystalline polymers. Polymers 4:150–186

Ericson LM, Fan H, Peng HQ, Davis VA, Zhou W, Sulpizio J, Wang Y, Booker R, Vavro J, Guthy C et al (2004) Macroscopic, neat, single-walled carbon nanotube fibers. Science 305:1447–1450

Etchegoin P (2000) Blue phases of cholesteric liquid crystals as thermotropic photonic crystals. Phys Rev E 62:1435–1437

Fan Y-H, Ren H, Wu ST (2003) Switchable Fresnel lens using polymer-stabilized liquid crystals. Opt Express 11:3080–3086

Frechet JMJ, Tomalia DA (eds) (2001) Dendrimers and other dendritic polymers, Wiley series in polymeric sciences. Wiley, New York

Godovskii YK, Papkov VS (1992) Chapter 4. In: Plate NA (ed) Liquid crystal polymers. Plenum Press, New York, p 125

Gunduz B (2015) Sensing and surface morphological properties of a poly [(9,9-dioctylfluorenyl-2,7-diyl)-*co*-bithiophene] liquid crystalline polymer for optoelectronic applications. J Appl Polym Sci 132:41659 (1–8)

Hearle JWS (2001) High performance fibers. Woodhead, Philadelphia, p 329

Himmelberger S, Vandewal K, Fei Z, Heeney M, Salleo A (2014) Role of molecular weight distribution on charge transport in semiconducting polymers. Macromolecules 47(20):7151–7157

Huang YY, Terentjev EM (2010) Tailoring the electrical properties of carbon nanotube–polymer composites. Adv Funct Mater 20:4062–4068

Jayaraj K, Farrell B (1998) Liquid crystal polymers and their role in electronic packaging. Adv Microelectron 25:15–18

Ji Y, Huang YY, Rungsawang R, Terentjev EM (2010) Dispersion and alignment of carbon nanotubes in liquid crystalline polymers and elastomers. Adv Mater 22:3436–3440

Ji Y, Marshall JE, Terentjev EM (2012) Nanoparticle-liquid crystalline elastomer composite. Polymers 4:316–340

Jose JP, Malhotra SK, Thomas S, Joseph K, Goda K, Sreekala MS (2012) Advances in polymer composites: macro- and microcomposites—state of the art, new challenges, and opportunities. In: Jose JP, Malhotra SK, Thomas S, Joseph K, Goda K, Sreekala MS (eds) Polymer composites, vol 1, 1st edn. Wiley, Hoboken

Kachidza J, Serpe G, Economy J (1992) Ordering processes in the 2,6-hydroxynaphthoic acid (HNA) rich copolyesters of p-hydroxybenzoic acid and HNA. Makromol Chem Macromol Symp 53:65–75

Kaiser A, Winkler M, Krause S, Finkelmann H, Schmidt AM (2009) Magnetoactive liquid crystal elastomer nanocomposites. J Mater Chem 19:538–543

Karapinar R, O'Neill M, Hird M (2002) Polymer dispersed ferroelectric liquid crystal films with high electro-optic quality. J Phys D Appl Phys 35(9):900–905

Khoo CGL, Brox B, Norrhede R, Maurer FHJ (1997) Effect of copper lamination on the rheological and copper adhesion properties of a thermotropic liquid crystalline polymer used in PCB applications. IEEE Trans Compon Packaging Manuf Technol 20:219–226

Kikuchi H, Yokota M, Hisakado Y, Yang H, Kajiyama T (2002) Polymer-stabilized liquid crystal blue phases. Nat Mater 1:64–68

Kiss G (1987) In situ composites: blends of isotropic polymers and thermotropic liquid crystalline polymers. Polym Eng Sci 27:410–423

Kwolek SL (1971) Poly-(p-benzamide) composition, process and product. US Patent 3600350

Lagerwall JPF, Scalia G (2012) A new era for liquid crystal research: applications of liquid crystals in soft matter nano-, bio- and microtechnology. Curr Appl Phys 12:1387–1412

Lagerwall JPF, Schütz C, Salajkova M, Noh JH, Hyun Park J, Scalia G, Bergström L (2014) Cellulose nanocrystal-based materials: from liquid crystal self-assembly and glass formation to multifunctional thin films. NPG Asia Mater 6:1–12

Lewis KL, Mason IR, Sage I, Lymer KP, Smith GW (2000) Nanodispersed liquid crystal polymer composites for optically tunable filters. In: Proc. SPIE 4104, organic photorefractives, photoreceptors and nanocomposites, 9

Li JL, Crandall KA, Chu P, Percec V, Petschek RG, Rosenblatt C (1996) Dendrimeric liquid crystals: isotropic−nematic pretransitional behavior. Macromolecules 29:7813–7819

Lin YG, Winter HH (1991) High-temperature recrystallization and rheology of a thermotropic liquid crystalline polymer. Macromolecules 24:2877–2882

Lin Q, Yee AF (1994) Elastic modulus of in-situ composites of a liquid crystalline polymer and polycarbonate. Polym Compos 15:156–162

Lin Q, Jho JY, Yee AF (1993) Effect of drawing on structure and properties of a liquid crystalline polymer and polycarbonate in situ composite. Polym Eng Sci 33:789–798

Liu YJ, Sun XW (2008) Holographic polymer-dispersed liquid crystals: materials, formation, and applications. Adv Optoelectron. 2008. Article ID 684349, 52 pages

Matharu AS, Jeeva S, Ramanujam PS (2007) Liquid crystals for holographic optical data storage. Chem Soc Rev 36:1868–1880

Mizuno K, Ishii J, Kishida H, Hayamizu Y, Yasuda S, Futaba DN, Yunura M, Hata KA (2009) A black body absorber from vertically aligned single-walled carbon nanotubes. Proc Natl Acad Sci U S A 106:6044–6047

Pegoretti A, Traina M (2009) Liquid crystalline organic fibres and their mechanical behaviour. In: Bunsell AR (ed) Handbook of tensile properties of textile and technical fibres. Woodhead Publishing, Cambridge, pp 354–436

Percec V, Kawasumi M (1992) Synthesis and characterization of a thermotropic nematic liquid crystalline dendrimeric polymer. Macromolecules 25:3843–3850

Percec V, Chu P, Ungar G, Zhou J (1995) Rational design of the first nonspherical dendrimer which displays calamitic nematic and smectic thermotropic liquid crystalline phases. J Am Chem Soc 117:11441–11454

Puzari A (2012) Chapter 4: liquid crystalline dendrimer: towards intelligent functional materials. In: Iwan A, Schab-Balcerzak E (eds) Liquid crystalline organic compounds and polymers as materials XXI century: from synthesis to application. Transworld Research Network, Trivandrum, pp 95–124

Ramanathan T, Abdala AA, Stankovich S, Dikin DA, Herrera-Alonso M, Piner RD et al (2008) Functionalized graphene sheets for polymer nanocomposites. Nat Nanotechnol 3:327–331

Sahoo NG, Cheng HKF, Cai J, Li L, Chan SH, Zhao J, Yu S (2009a) Improvement of mechanical and thermal properties of carbon nanotube composites through nanotube functionalization and processing methods. Mater Chem Phys 117:313–320

Sahoo NG, Cheng HKF, Li L, Chan SH, Judeh Z, Zhao J (2009b) Specific functionalization of carbon nanotubes for advanced polymer nanocomposites. Adv Funct Mater 19:3962–3971

Sahoo NG, Thet NT, Tan QH, Li L, Chan SH, Zhao J, Yu S (2009c) Effect of carbon nanotubes and processing methods on the properties of carbon nanotube/polypropylene composites. J Nanosci Nanotechnol 9:5910–5919

Sahoo NG, Cheng HKF, Bao H, Pan Y, Li L, Chan SH, Zhao J (2011a) Covalent functionalization of carbon nanotubes for ultimate intermolecular adhesion to liquid crystalline polymer. Soft Matter 7:9505–9514

Sahoo NG, Cheng HKF, Bao H, Li L, Chan SH, Zhao J (2011b) Nitrophenyl functionalization of carbon nanotubes and its effect on properties of MWCNT/LCP composites. Macromol Res 19:660–667

Salajkova M, Berglund LA, Zhou Q (2012) Hydrophobic cellulose nanocrystals modified with quaternary ammonium salts. J Mater Chem 22:19798–19805

Sreekumar TV, Liu T, Kumar S, Ericson LM, Hauge RH, Smalley R (2003) Single-wall carbon nanotube films. Chem Mater 15:175–178

Steckman GJ, Bittner R, Meerholz K, Psaltis D (2000) Holographic multiplexing in photorefractive polymer. Opt Commun 185:13–17

Sun X, Sun H, Peng H (2013) Developing polymer composite materials: carbon nanotubes or graphene? Adv Mater 25(37):5153–5176

Thomas EL, Wood BA (1985) Mesophase texture and defects in thermotropic liquid crystalline polymers. Faraday Discuss Chem Soc 79:229–239

Tjong SC (2003) Structure, morphology, mechanical and thermal characteristics of the in situ composites based on liquid crystalline polymers and thermoplastics. Mater Sci Eng R 41:1–60

Van Boxtel M, Broer D, Bastian Sen C, Baars M, Janssen R (2000) Electro-optical switches based on polymer and dendrimer filled nematics. Macromol Symp 154:25–35

Vyas R, Rida A, Bhattacharya S, Tentzeris MM (2007) Liquid crystal polymer (LCP): the ultimate solution for low-cost RF flexible electronics and antennas. In: Proceedings of the polymers in defence and aerospace applications, Toulouse, France, 18–19 Sept 2007. Smithers Rapra Technology, Shropshire; 2007; Section 6, p 21

Wang X, Engel J, Liu C (2003) Liquid crystal polymer (LCP) for MEMS: processes and applications. J Micromech Microeng 13:628–633

Warner M, Terentjev EM (2007) Liquid crystal elastomers. Oxford University Press, Oxford

Wermter H, Finkelmann H (2001) Liquid crystalline elastomers as artificial muscles. e-Polymers 13:1–13

Winiarz JG, Prasad PN (2002) Photorefractive inorganic–organic polymer-dispersed liquid-crystal nanocomposites photosensitized with cadmium sulfide quantum dots. Opt Lett 27:1330–1332

Winiarz JG, Zhang L, Lal M, Friend CS, Prasad PN (1999) Photogeneration, charge transport, and photoconductivity of a novel PVK/CdS–nanocrystal polymer composite. Chem Phys 245:417–428

Woo HS, Czerw R, Webster S, Carroll DL, Park JW, Lee JH (2001) Organic light emitting diodes fabricated with single walled carbon nanotubes dispersed in hole conducting buffer: the role of carbon nanotubes in a hole conducting polymer. Synth Met 116:369–372

Wurthner F, Wortmann R, Meerholz K (2002) Chromophore design for photorefractive organic materials. Chemphyschem 3:17–31

Yan J, Wu ST (2011) Polymer stabilized blue phase liquid crystals: a tutorial. Opt Mater Express 1 (8):1527–1535

Yan J, Li Y, Wu ST (2011) High-efficiency and fast-response tunable phase grating using a blue phase liquid crystal. Opt Lett 36:1404–1406

Yu H, Ikeda T (2011) Photocontrollable liquid-crystalline actuators. Adv Mater 23:2149–2180

Zhang YP, Chodavarapu VP, Kirk AG, Andrews MP (2012) Nanocrystalline cellulose for covert optical encryption. In: Proc. SPIE 8258, organic photonic materials and devices XIV 8258, 825808

Zito G, Pissadakis S (2013) Holographic polymer-dispersed liquid crystal Bragg grating integrated inside a solid core photonic crystal fiber. Opt Lett 38:3253–3256

Zito G, Piccirillo B, Santamato E, Marino A, Tkachenko V, Abbate G (2008) Two-dimensional photonic quasicrystals by single beam computer-generated holography. Opt Express 16:5164–5170

# Chapter 14
# Functional Materials from Liquid Crystalline Cellulose Derivatives: Synthetic Routes, Characterization and Applications

**S.N. Fernandes, J.P. Canejo, C. Echeverria, and M.H. Godinho**

## 14.1 Introduction

### *14.1.1 Cellulose: Chemical Versatility*

Cellulose is a linear syndiotactic and semi-rigid homopolymer consisting of D-anhydroglucopyranose units (AGU) (Scheme 14.1), where each unit has three hydroxyl (OH) groups at C-2, C-3 and C-6 positions. The AGU units are covalently linked through acetal functions between the equatorial OH group of C4 and the C1 carbon atoms by β-(1 → 4) glycosidic bonds (Credou and Berthelot 2014; Klemm et al. 2005). In the solid state, every second AGU unit is rotated by 180° in the plane due to the constraints of the β-linkage and a $^4C_1$-chair formation of the AGU is assumed. In this conformation the free hydroxyl groups are positioned in the ring plane (equatorial), while the hydrogen atoms are in a vertical position (axial). The chain length of the cellulose polymer varies, depending on the cellulose source and also on the treatment of the raw material (Thakur and Thakur 2014). Terminal groups at the either end of the cellulose molecule are quite different in nature from each other (Scheme 14.1), that is, the cellulose chain consists at one end of a D-glucose unit with an original C4-OH group (the nonreducing end) and the other end is terminated with an original C1-OH group, which is in equilibrium with the aldehyde structure—the reducing end (Klemm et al. 1998b; Kondo 2005; Pérez and Mazeau 2005).

Cellulose is a very versatile starting material for chemical conversions, and its chemical character and reactivity is mainly determined by the presence of the three equatorially positioned hydroxyl (OH) groups in the AGU, one primary and two secondary groups, that can undergo the typical reactions known for primary and

S.N. Fernandes • J.P. Canejo • C. Echeverria • M.H. Godinho (✉)
CENIMAT/I3N, Departamento de Ciência dos Materiais, Faculdade de Ciências e Tecnologia (FCT), Universidade Nova de Lisboa, Caparica 2829-516, Portugal
e-mail: mhg@fct.unl.pt



V.K. Thakur, M.R. Kessler (eds.), *Liquid Crystalline Polymers*,
DOI 10.1007/978-3-319-20270-9_14

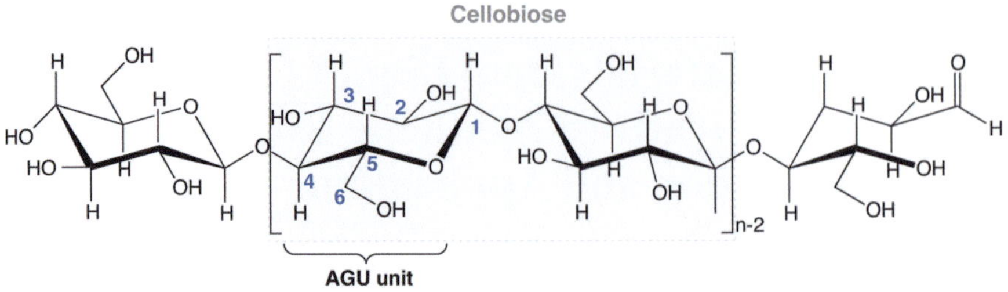


**Scheme 14.1** Molecular structure of cellulose representing the cellobiose unit as a repeating unit showing the non-reducing (*left*) and reducing (*right*) end-groups. This structure also shows that the repeated anydroglucopyranose units (AUG) are rotated of 180° with respect to each other, due to β-linkage constrains (n = DP, degree of polymerization)

secondary alcohols. In addition, the β-glycosidic linkages of cellulose are also susceptible to hydrolytic attack. The reactivity of the hydroxyl groups varies according to the reaction medium in which the functionalization is made. Therefore the hydroxyl groups do not only play a role in the typical reactions of primary and secondary alcohols that are carried out on cellulose, but also play an important role in the solubility of cellulose. For example, the hydroxyl groups are responsible for the extensive hydrogen-bonding network forming both by intra- and intermolecular hydrogen bonding, that induces cellulose insolubility in common organic solvents and in water. In order to dissolve cellulose, the prevailing hydrogen-bonding network must be broken (Isogai et al. 2011; Klemm et al. 1998a, 2011; Varshney and Naithani 2011).

Cellulose can be found mainly in plant cell walls, where the hydrogen bonds contribute to a configuration where the molecules tend to be side-by-side forming crystalline micro fibrils. These show good mechanical properties, with a high tensile strength, giving the cell walls the necessary mechanical support. Cellulose presents very good biodegradability and biocompatibility, is insoluble in water and in many organic solvents. The fact that cellulose is the most abundant natural polymer and, at the same time, that it represents an important natural renewable resource makes it a good candidate for applications in a large range of areas. Cellulosic materials have been used as raw materials for different applications such as textiles manufacturing, building of shelters and production of tools.

One method to simplify the processing of cellulose is through the chemical modification of the cellulose macromolecule. The hydroxyl groups attached to the main chain can be used to promote chemical substitution by ester, ether, or carbamate groups (Nakayama and Azuma 1998), originating different cellulose derivatives depending on the group that is added to the main chain and on the degree of substitution. The possibility to attach diverse lateral substituents to the cellulosic chain, allows for the possibility to produce a wide range of cellulose derivatives that exhibit distinct properties.

### 14.1.2 Cellulose Liquid Crystal

One interesting characteristic of cellulosic materials is the ability to form, under appropriate conditions, liquid crystalline phases. The designation liquid crystal (LC) refers to a phase of the matter, composed of anisotropic molecules, that is an intermediate between the crystalline solid and the conventional liquid, showing mechanical properties of the liquids and optical characteristics of the solids (de Gennes and Frost 1993; Singh and Dunmur 2002). Because the liquid crystalline phases occur between two phases, they are also known as mesophases. One of the main characteristics of liquid crystalline phases is the fact that although the molecules that constitute these materials show no positional order, they exhibit long-range orientational order that extends for distances that are several times higher than the length of the constituent molecules. This is possible because in the mesomorphic state, molecular self-assembly is promoted and by attraction-repulsion interactions, the molecules align spontaneously.

Flory (1956) predicted that, under appropriate conditions, solutions of materials with highly anisotropic molecules that displayed enough stiffness could form a liquid crystal phase. The cellulosic chain fulfills both criteria, is anisotropic due to its length and shows limited flexibility forming a semi-rigid polymer (Gray 1983). Following these principles, one would expect that cellulose could generate liquid crystalline phases.

The value of polymer concentration above which a polymeric solution forms a liquid crystalline phase is known as the critical concentration (C*). Flory's theoretical model takes into account the persistent length (q) of the polymeric chains (Flory 1956). According to this model, the expression (14.1)[1] for the critical concentration in volume fraction, in terms of q, is:

$$C^*(v/v) \simeq \frac{8}{x\left(1-\frac{2}{x}\right)} \tag{14.1}$$

where $x = \frac{2q}{D}$ and $D$ represents the value of the Kuhn segment (Kuhn 1936). Since stiffer polymers have higher persistent lengths, the parameter q, in the model proposed by Flory, is an indicator of the rigidity of the polymeric chains.

Flory's model has the limitation of assuming that all polymeric chains have the same length (same degree of polymerization); however, longer chains have higher hydrodynamic diameters and show increased difficulty for disentanglement and orientation during the formation of the mesophase. Another limitation of this model

[1] The value of the critical concentration, in terms of weight fraction, is given by the expression:

$$C^*(w/w) = \frac{\rho_2 C^*(v/v)}{\rho_1 + (\rho_2 - \rho_1)C^*(v/v)},$$

where $\rho_1$ is the density of the solvent and $\rho_2$ the density of the polymer.

is the fact that it only takes into account the polymer volume fraction and it does not consider the polarity effects that occur in the solvent/polymer system (Bheda et al. 1980).

Due to the stiffness of the main chain, cellulose and cellulose derivatives can be at the origin of chiral nematic phases, which are similar to a nematic phase with a natural twist around an axis normal to the direction defined by the director. While in each plane all the molecules are oriented in the same average direction, from plane to plane this direction is shifted by a few degrees forming a helix (de Gennes and Frost 1993). The cholesteric phases are characterized by the pitch, which can be estimated by the distance necessary for the director to rotate 360°. The first time when a liquid crystalline phase was observed in a cellulosic system was in 1976 (Werbowyj and Gray 1976), when Gray et al. observed bright colours, which changed with the viewing angle, in a solution of hydroxypropylcellulose (HPC) in water. These materials were found to be birefringent and showed optical activity. The birefringence of a cholesteric phase can be related to the pitch by the model proposed by De Vrieș (1951), which assumes a planar orientation of the molecules in each layer and a large number of layers. When the cholesteric is illuminated normally to the planes with white light the reflection is the circularly polarized component of the light in a range of wavelengths, with the maximum wavelength ($\lambda_0$) defined by:

$$\lambda_0 = \overline{\mu} P \tag{14.2}$$

where, μ is the average refractive index of the cholesteric and P represents the value of the helical pitch. If $\lambda_0$ has a value corresponding to the visible light that the cholesteric phase reflects coloured light.

In the case where the incident light reaches the cholesteric with an angle not normal to the cholesteric planes, the reflected light has a wavelength that is dependent on the angle of incidence and the colour is iridescent changing with the this angle (Laivins and Gray 1985c). In this case the value of the wavelength of the reflected light is given by:

$$\lambda = \overline{\mu} P \sin \varphi \tag{14.3}$$

where, φ is the angle of incidence.

According to the de Vries model, the rotatory power of the cholesteric, with units expressed in rad/nm, of light normal to the cholesteric planes with wavelength λ is given by the expression:

$$\theta = \frac{\pi \Delta \mu_c^2 P}{4\lambda^2 \left[1 - (\lambda/\lambda_0)^2\right]} \tag{14.4}$$

where, θ represents the optical rotation and $\Delta\mu_c$ the layer birefringence.

Since the wavelength of the reflected light depends on the value of the cholesteric pitch, any change of the pitch can give rise to changes in the observed colour of the liquid crystal phase (assuming $\lambda_0$ in the visible range) (Bhadani and Gray 1983).

The pitch and handedness of the cholesteric helix can be affected by several factors: temperature, chemical structure of the side chain and degree of polymerization. It was observed that in films casted from HPC cholesteric liquid crystal phase the pitch increases (Charlet and Gray 1987) with temperature.

Most cholesteric cellulose derivatives form the right-handed helical structure. However the occurrence of helical sense inversion, induced by temperature, was also reported for thermotropic oligomeric cellulose derivatives (Yamagishi et al. 1988). The flexible side-chain not only assists in the melting and the orientation of the cellulose backbone, due to an increase in the mobility of the latter, but also plays an important role in the formation of helical structures in the cholesteric mesophases (Yamagishi et al. 2006).

Not only cellulose macromolecules are found to exhibit liquid crystalline phases but also suspensions of cellulose microfibrils can form cholesteric chiral nematic phases (Sugiyama et al. 1990), which are attributed to the chiral screw-like structure of the microfibrils (Fleming et al. 2001).

Chanzy et al. reported the observation of a liquid crystal phase obtained from a cellulose solution in a mixture of *N*-methyl-morpholine-*N*-oxide and water (Chanzy et al. 1980). Subsequent measurements performed by X-ray diffraction showed that fibres obtained from the liquid crystalline solution, heated to 110 °C, exhibited high degree of orientation, which was not observed in fibers produced from the isotropic phase. This observation opened the door to the possibility of production of cellulosic fibers with enhanced mechanical properties similar to Kevlar, since the orientation of the cellulose fibres obtained from the liquid crystal phase was analogous to the orientation that could be achieved in high strength fibers like rayon and Fortisan. After this observation, other cellulosic solutions were reported as being capable of forming oriented films and fibers from liquid crystalline phases (Gray 1983; Laivins and Gray 1985c; Navard and Haudin 1980; Patel and Gilbert 1981; Yim et al. 1992).

Besides being capable of forming lyotropic liquid crystalline phases, cellulose derivatives can also originate thermotropic mesophases in the absence of any solvent.

A list of cellulosic systems that form thermotropic liquid crystalline phases can be found in recent literature (Zugenmaier 2014). Among them cellulosic esters were highlighted because they can be obtained from the easily available HPC ether. Although it was also stressed that, for these cellulose derivatives, experimental data are sometimes difficult to achieve and to understand due to the high clearing temperatures, which can cause degradation and irreversible changes during the measurements. To overcome these problems and better perceive the behavior of cellulosic thermotropic systems fully substituted cellulose ethers and esters were synthesized (Yamagishi et al. 1991).

Despite the drawbacks of thermotropic HPC ether and esters interesting optical properties were observed, e.g., iridescent acetoxypropylcellulose (APC) (Tseng et al. 1981), lower-mass HPC (Shimamura et al. 1981) and HPC trifluoroacetate films (Gray 1983), when heated, can be obtained. The ability that cellulose derivatives show to form thermotropic phases is an indication that the lateral chains can play the role of the solvent; by increasing the distances between the main chains of the molecules they help the mobility of the macromolecules without the presence of a solvent.

As mentioned above, in literature one can find a vast number of cellulose derivatives that form an interesting range of polymer liquid crystals (PLCs). Many of these derivatives form cholesteric liquid-crystalline phases in suitable solvents and some other derivatives, with relatively large side chains, form both lyotropic and thermotropic liquid crystals (Gray 1994). Among them are the cellulosic thermotropic esters, in particular the ones synthesized from HPC, whose characteristics will be revised in this work. A general goal for studying these mesophases is to produce materials with diverse properties by developing better performance fibers and films by new processes to future applications.

## 14.2 Synthesis of Liquid Crystalline Esters Derivatives of Hydroxypropylcellulose

Etherification and esterification, oxidation, grafting, crosslinking and also chain degradation, by various mechanisms, are types of chemical reactions that can take place at the cellulose macromolecule and give rise to a vast number of different products (Atalla and Isogai 2005; Edgar et al. 2001; Goetz et al. 2009; Habibi et al. 2010; Heinze and Liebert 2001; Heinze 2005; Isogai et al. 2011; Klemm et al. 2011; Lagerwall et al. 2014; Revol et al. 1992; Soldi 2005; Thakur et al. 2011, 2013a, b; Varshney and Naithani 2011).

The principal route to obtain cellulose derivatives is the esterification or etherification reactions. These products form a large and well established industry and are used in pharmaceuticals, food, personal care, construction, oil field chemicals, textiles, adhesives, inks, paper and coating, among several others. Additional to their extensive use as polymer materials, cellulose esters and ethers frequently serve as suitable starting materials for consecutive synthetic routes. This leads to one of the trends of modern cellulose chemistry, which is the functionalization of known cellulose derivatives in order to expand their application area (Wüstenberg 2014).

### *14.2.1 Hydroxypropylcellulose*

Hydroxypropylcellulose is a cellulosic ether derivative that can form both lyotropic (Werbowyj and Gray 1976) and thermotropic (Shimamura et al. 1981) liquid

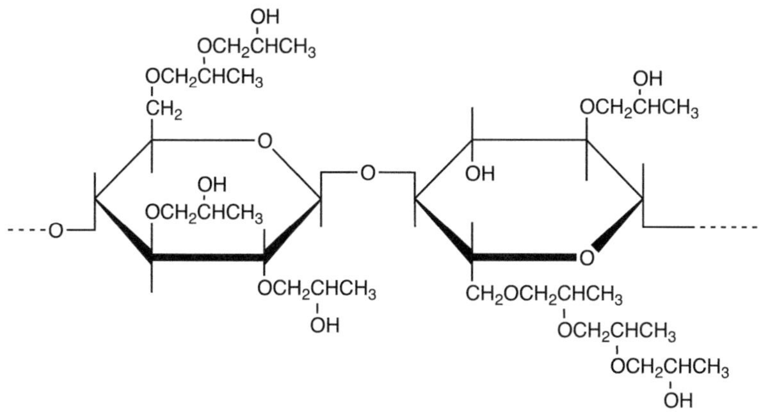


**Scheme 14.2** Possible molecular structure of hydroxypropylcellulose with a molecular substitution of 4 and a degree of substitution of 2.5

crystalline mesophases and is commercially available in a wide range of molecular weights. The production of HPC starts, as for other ethers in general, with the alkalization of cellulose, in order to destroy its intra- and intermolecular hydrogen-bonding network, and subsequent reaction with 1,2-propylene oxide at high temperatures and pressures. The reaction can be carried out in an organic solvent or in an excess of the reagent, and the obtained product can be then purified in hot water (Atalla and Isogai 2005; Wüstenberg 2014). In the reaction with propylene oxide, secondary OH groups are formed, which are capable of undergoing further reaction.

Scheme 14.2 shows one of many possible molecular structures of HPC with a molar and degree of substitution of 4 and 2.5, respectively (Ho et al. 1972). The maximum value of the degree of substitution $\left(\overline{DS}\right)$ is three while there is no theoretical maximum limit of molecular substitution $\left(\overline{MS}\right)$.

The polymer liquid crystal HPC is soluble in water, at temperatures below 38 °C, and soluble in many polar solvents. Bearing in mind its solubility, liquid crystalline properties and the more reachable hydroxyl groups, HPC presents itself as a good starting material that can undergo chemical reactions to give rise to novel and interesting cholesteric materials.

### *14.2.2 Thermotropic Esters HPC-Derivatives*

In this section we focus attention on the preparation of the various ester derivatives of HPC described in the literature, of the type $C_nCP$, where n is the number of methylene units in the side-chain. An idealized chemical structure is represented in Scheme 14.3, where one can see four hydroxypropyl groups and three substituents aliphatoxy groups of the type $–C(=O)(CH_2)_{n-2}–CH_3$. In Sect. 14.3 one can find the structural, thermal and optical characterization of these compounds showing the

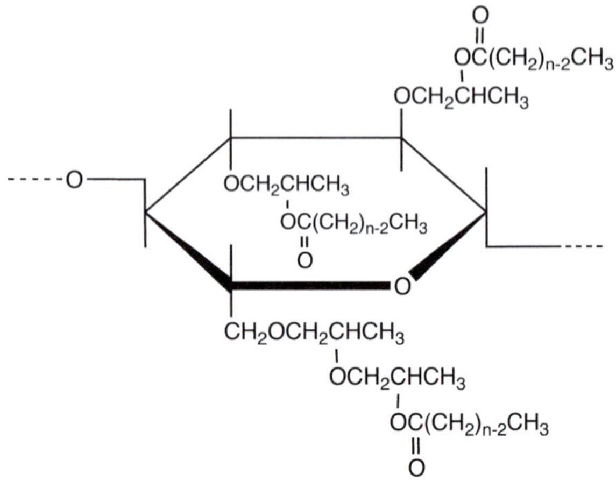


**Scheme 14.3** An idealized chemical structure of an anhydroglucose unit of CnPC with four hydroxypropyl groups and three aliphatoxy groups of the type $–C(=O)(CH_2)_{n-2}–CH_3$ as substituents [based on the schematic representation of Tseng et al. (1981)]

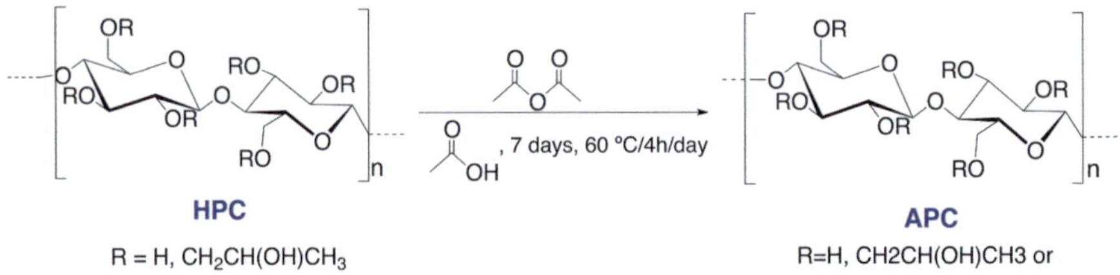


**Scheme 14.4** Esterification reaction of hydroxypropylcellulose with acetic anhydride and acetic acid as initiator to produce acetoxypropylcellulose

influence of the structural modification, as well as the effects of the degree of esterification, length of side-chain, degree of polymerization on the liquid crystalline properties.

One of the reactions that HPC can undergo is esterification. This type of reaction is usually accomplished by reacting the polymer with the corresponding aliphatic acid, either its anhydride or acid chloride form, as the active agent, in a proper solvent with a short heating cycle (Klemm et al. 1998a). In this reaction the hydroxyl groups of HPC are replaced by aliphatoxy groups of the type $–C(=O)(CH_2)_{n-2}–CH_3$ originating its ester derivatives.

In 1981 Tseng et al. reported for the first time that the ester derived from the acetylation of HPC (Mw = 100,000 g/mol), the acetoxypropylcellulose (APC), in the presence of anhydride acetic using pyridine as a catalyst, readily forms cholesteric liquid crystalline solutions with organic solvents and also formed a thermotropic cholesteric phase below 180 °C. The authors additionally presented an alternative (less hazardous) synthetic route where no pyridine was used. This later method requires the use of acetic acid as an initiator of the esterification reaction. Nevertheless a longer time was needed, 1 week vs. 1 day, to achieve a similar degree of esterification $(\overline{DE})$, see Scheme 14.4. The properties of the product obtained by both methods were similar. This ester is insoluble in water but soluble in several organic solvents (Tseng et al. 1981). Further characterization of APC was done, as the determination of chain stiffness (Laivins and Gray 1985a) and the liquid crystal to isotropic transitions (Laivins and Gray 1985b). The effect of chain length and degree of acetylation in the liquid crystalline properties of the APC was also examined (Laivins et al. 1986). Besides this, the group additionally focused its attention on the influence of the molar mass in the cholesteric pitch (Laivins and Gray 1985c).

The acetoxypropylcellulose was further reviewed regarding its liquid crystalline properties in bulk (Cidade et al. 1995; Riti et al. 1997). Aliphatic esters can exhibit cholesteric reflections in the visible range of the spectrum (Costa et al. 2006; Tseng et al. 1981). The selective reflection depends on several parameters including the degree of esterification (Bhadani and Gray 1983) and on the temperature (Rusig et al. 1994). The pitch of the cholesteric helix can also be locked by using different procedures among them by photoinitiated crosslinking of pendant acrylate groups (Müller and Zentel 2000). Taking advantage of these properties, APC was used throughout to produce films, and electrospun fibers: (i) using APC anisotropic solutions (Almeida et al. 2009a; Canejo and Godinho 2013; Canejo et al. 2008; Godinho et al. 2009a, 2010; Kundu et al. 2010); (ii) functionalized APC, via crosslinking reaction with diisocyanate (Costa et al. 2006, 2007; Godinho et al. 2009b; Sena et al. 2011); (iii) with an azo-chromophore, see Scheme 14.5 (Pinto et al. 2011); (iv) blended with ultra-narrow PbS nanowires (Maji et al. 2013).

Gray's group also prepared the propanoate of (2-hydroxypropyl)cellulose (PPC or $C_3PC$) from the esterification reaction of HPC (Mw = 100,000 g/mol) and propionyl chloride or propionic anhydride in acetone (Tseng et al. 1982). The appearance and behavior of the polymer is similar to APC, however this ester displayed iridescent colors at room temperature. In the same year (Bhadani and Gray 1982), the synthesis of the benzoic acid ester of HPC (BzPC) from the reaction of the aromatic benzoyl chloride with HPC in pyridine was described. The polymer showed a thermotropic liquid crystalline phase below 160 °C. Benzoic acid ester of HPC was fractionated from acetone solution by precipitation with methanol-water and polymers with different molecular weights were obtained (Bhadani et al. 1983).

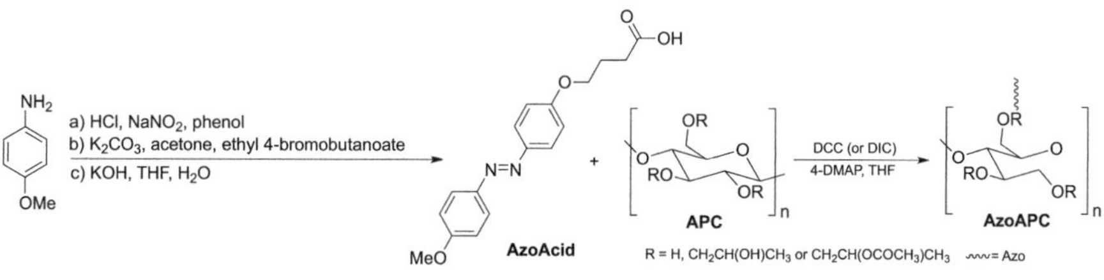


**Scheme 14.5** Functionalization of APC: Schematic synthetic route to yield azoacetoxypropylcellulose (AzoAPC). Reprinted with permission from Pinto LFV, Kundu S, Brogueira P, Cruz C, Fernandes SN, Aluculesei A, & Godinho MH (2011) "Cellulose-Based Liquid Crystalline Photoresponsive Films with Tunable Surface Wettability". Langmuir 27(10):6330–6337. Copyright 2011 American Chemical Society

The authors demonstrated that the fractions of BzPC also exhibited a thermotropic phase, and that the transition temperature increased sharply with increasing the polymers' molar mass (limiting value of 176 °C).

The acrylic acid ester of HPC was prepared from HPC and acryloyl chloride (Bhadani and Gray 1984). The resultant polymer formed a thermotropic cholesteric mesophase with visible reflection bands at temperatures between ambient and 60 ° C. By exposing a thin layer of the mesophase to UV light, the mesophase structure was stabilized to give a crosslinked cholesteric film.

The fairly easy and efficient synthetic routes described above have been used as the basic procedure throughout the years to produce novel esters of HPC. In order to avoid repetition from now on, only new details in the synthetic procedure will be discussed.

Pawloswski et al. described the synthesis of acetoacetoxypropyl cellulose (AAPC), formed by the acetoacetylation of hydroxypropyl cellulose using a diketene/acetone adduct (2,2,6-trimethyl-4-H-1,3-dioxin-4-one) in *N*-methyl-2-pyrrolidinone (NMP) at elevated temperature. The authors showed that as with APC, AAPC also forms both thermotropic and lyotropic liquid-crystalline phases. It is noteworthy to mention that the nematic to isotropic transition occurs at 174 °C and thin films of cholesteric thermotropic AAPC show green reflection colors (Pawlowski et al. 1986, 1987).

Rusig et al. prepared a series of esters of HPC from butyric acid, isobutyric acid and isovaleric acid, BCP, iBCP and iVCP respectively, that presented iridescent colors between 20 and 100 °C and form thermotropic liquid crystalline phases (Rusig et al. 1992). The authors studied the effects of degradation on these esters of HPC presented while heated. The results showed that the temperature accelerates ageing of these cellulosic compounds and at 80 °C, under water-saturated atmosphere, hydrolytic degradation of the main chain occurs. This leads to changes in chemical structure inducing different behavior in its optical properties. In order to prevent this degradation, mainly attributed to the presence of the acid chloride, which exists even after thorough purification procedures, the authors presented three other methods to obtain ester of HPC from the butyric acid chloride in avoid the degradation of the polymer (Rusig et al. 1992). From the three methods explored, the reaction of HPC with dicyclohexylcarbodiimide (DCC), used as decoupling agent, and 4-dimethylaminopyridine (DMAP) as catalyst, at room temperature, was the most successful. However for all the samples prepared, the degree of esterification never exceeded 1.9 (lower in comparison with the one obtained from the original method 2.5), and so changes of the optical properties were observed, for example the absence of color reflection (Rusig et al. 1992).

This group also reviewed the effect of the linear side-chain of the esters derivatives of HPC in its optical properties. For that acetoxypropylcellulose (APC) and propionic (PPC), ***n***-butyric (BPC), isobutyric (IBPc), valeric (VPC), isovaleric (iVPC), hexanionic (HexPC), and heptanionic (HepPC) acid esters of HPC (Mw = 120,000 g/mol) were prepared and characterized (Rusig et al. 1994). The results obtained showed that all these esters form thermotropic cholesteric liquid crystalline phases and that the longer linear-side chains stabilize the mesophase.

In 2000 Wojciechowski (2000) published a very interesting and complete work on the thermotropic mesomorphism of cellulose derivatives. The HPC and three HPC-derivatives: two esters with different lengths of the side-chain (PPC, HxPC) and the cyanoethyl derivative (CEPC) were the macromolecules used. These HPC derivatives were prepared by the esterification reactions of HPC with hexanoic acid, propanoic acid and acrylonitrile (Mays 1988) respectively. With this work the author intended to establish relationships between the thermotropic properties of the cellulose derivatives and their molecular structure, mainly the effect of side-chain length and polarity. Due to the high dipole moment of 2-cyanoethyl groups the polarity of CEPC is much higher than that of the other HPC-esters analysed and the author observed that the transition temperature increases with the increase in the polarity brought by the side-chain group of the HPC derivative (Wojciechowski 2000).

The effect of the degree of esterification on the pitch of a cellulosic derivative was analyzed by Hou et al. (2000). Butyric esters of HPC (BPC) with different degrees of esterification $(\overline{DE})$ were synthesized via control of the esterification time of butyryl chloride and HPC. The authors demonstrated that the pitch of these thermotropic esters of HPC could be tuned by changing the degree of esterification (Hou et al. 2000). Later Greiner's group, presented a series of esters ($C_nPC$ with $n = 3, 4, 5, 7, 9, 23$, BzPC and butyl/ethyl-propylcellulose BPPC) derived from the reaction of HPC with aliphatic and aromatic acyl chlorides (Greiner et al. 2003) in order to exploit its opto-electronic applications, such as light-emitting diodes. Particular emphasis was given to the controlled modulation of the pitch and to the absorption and emission properties as a function of the chemical composition. In this work, the correlation of $\overline{DE}$ and the pitch was attained by a series of butylesters of HPC with different $\overline{DE}$ (obtained by using different esterification reaction times). The authors similarly examined the effect of menthol, as a chiral dopant, for the cellulose ester $C_4PC$ and the effect of composition in blend systems (butryric ester of HPC with different $\overline{DE}$ and also butyric and valeric ester of HPC). Greiner and co-workers (Greiner et al. 2003; Hou et al. 2000) additionally functionalized the HPC esters with coumarins, anthracene and arylenevinylenes as chromophores, to investigate the degree of circular polarization of the emitted light as well as their performance in light-emitting diodes resulting from the introduction of these groups in the polymeric system. Two types of approaches were used to obtain the chromophore ester-HPC derivatives; by chemically linking the chromophores to the HPC backbone through an esterification reaction with the corresponding acid chlorides chromophore, or by using as doping agent. The authors concluded that these cellulose derivatives could be applied in opto-electronics; however a further control of the uniformity of the helix formation and orientation should be accomplished. A similar work, however shorter, has been already presented by the group where a smaller series of esters of HPC were doped with coumarin 334 (Arici et al. 2000).

Huang et al., studied a series of aliphatic esters of HPC-$C_n$PC, where n = 2, 3, 5, 6, 7, 10. The authors observed that as one increases the number of methylene units in the side-chain of the cholesteric liquid crystal polymers, the window of the thermotropic phase transition narrows (Huang et al. 2007). Although the authors presented a similar study to the one published by Kosho et al. in 1999 (Kosho et al. 1999), wide-angle X-ray diffraction (WAXD) studies permitted to detect that the layer spacing of the cholesteric liquid crystals in this series increases linearly with an increase in the methylene units in the side chains.

More recently El-Waki et al. reported the synthesis and characterization of a series of 4-alkyoxybenzoyloxypropyl cellulose (ABPC-*n*) samples, obtained via the esterification of hydroxypropyl cellulose (HPC) with 4-alkoxybenzoic acid bearing different numbers of carbon atoms (n = 1, 2, 3, 4, 5, 7, 8, 12). In this series the authors noted that only for the higher alkoxy chain lengths (7, 8 or 12) a thermotropic liquid crystal behavior was observed; however all the esters of HPC within the series presented lyotropic behavior (El-Wakil et al. 2010).

In 2012 Bagheri et al. presented two new series of amphiphilic cholesteryl-modified HPC derivatives (CHDs), obtained from the reactions involving HPC and cholesterol-based mesogenic dimer, or cholesteryl chloroformate. In order to produce the cholesteryl-terminated dendronized HPC (HPC-G1-Chol) ester, the authors synthesized the 2,2-Bis(cholesteryloxycarbonyloxymethyl)propionic acid (G1-Col). HPC-G1-Chol which was obtained via esterification reaction of HPC with DMAP, 4-(Dimethylamino)pyridinium-4-toluenesulfonate (DPTS), G1-COl and DCC in dichloromethane at room temperature. The cholesteryl-bearing HPC derivative (HPC-Chol) was obtained by the reaction of HPC with cholesteryl chloroformate. All of the synthesized polymers formed thermotropic liquid crystalline phases. The polymers with a mesogenic side chain (i.e., the HPC-G1-Chol series) had wider mesophases than HPC and polymers that were derived from HPC-Chol. The dependence of the side-chain on the thermal properties observed led the authors to suggest that not only the backbone but also the side chains of the CHDs are involved in LC phase formation and phase transitions. The same results also appeared to show that the degree of esterification of each cellulose derivative significantly influences the properties of its thermotropic mesophase. Lyotropic liquid crystallines mesophases in acetone were also obtained (Bagheri and Shateri 2012).

## 14.3 Characterization of HPC and Its Ester Derivatives

In this following section the modification of structure and properties of the derivatives of HPC as a function of their side-chain properties will be presented.

For the sake of the comparison a brief description of the main structural and thermal properties of thermotropic HPC is shown.

### 14.3.1 Thermotropic HPC: Structure, Thermal and Optical Properties

A deep characterization of the solid state of Hydroxypropylcellulose was first carried out by Samuels in 1969 (Samuels 1969). The complex morphological structure of HPC revealed emphasizes the crystalline melt behaviour of the polymer. The narrow diffraction band observed at $2\theta = 8^\circ$ corresponding to the *d*-spacing of 1.04 nm, is attributed to backbone-backbone *d*-spacing of the HPC main chains in the crystalline phase whereas the broad diffraction band at $2\theta = 20^\circ$ is related to slightly ordered amorphous phase of HPC (Wojciechowski 2000). Thus, on the basis of WAXS measurement it was determined that the structure of HPC present a pseudotetragonal unit cell with a chain-axis repeat distance corresponding to three anhydroglucose units oriented at a 120° angle to each other.

In terms of thermal properties, Aspler and Gray (1982) reported the glass transition temperature ($T_g$) of 10 °C for bulk HPC by using inverse chromatography. Later this results was sustained by Suto et al. (1982), by means of Dynamic Mechanical techniques in which a broad endotherm ranging from 160 to 210 °C with a peak at 180 °C was described. This broad endotherm was related to the thermotropic toward isotropic liquid crystal around 200 °C.

Shimamura et al. (1981) characterized optically a series of HPC sample with different molecular weights. These authors observed by POM (Polarized Optical Microscope) that bulk HPC samples showed an isotropization temperature in between 210 and 215 °C. Comparing heating and cooling experiments, they observed that on cooling the samples transmitted 30 % less intensity although they observed that when shearing is applied to the sample this intensity become higher.

For lower molecular weight HPC samples a cholesteric structure can appear between 160 and 205 °C (Shimamura et al. 1981).

### 14.3.2 Effect of the Side-Chain Length on the Structure

The ability of the cellulose derivatives to form liquid crystalline phases without the need of a solvent depends on the substituents side-chain and in its structure. For those polymers, in the absence of solvent, the side-chain controls the average space between quasi-parallel cellulose backbones; thus the d-spacing will be different for ester derivatives with different side-chain. This variation of the structure affects the optical properties and pitch of the cholesteric helix (Tseng et al. 1982).

The local structure within the cholesteric state is accessible to a certain extent from X-ray diffraction (XRD). As a general description, an XRD of an ester of HPC will depict a structured halo at larger scattering angles (due to a short range order within the polymer) and a broad but strong reflection at smaller angles. XRD

**Table 14.1** d-spacing for different ester of HPC

| HPC-ester | d-Spacing (nm)[a] |
|---|---|
| HPC | 1.04[b] |
| APC ($C_2PC$) | 1.19[c] |
| APC ($C_2PC$) Mw = $2.14 \times 10^5$ (g/mol) | 1.22[d] |
| APC ($C_2PC$) Mw = $2.18 \times 10^4$ (g/mol) | 1.25[d] |
| PPC ($C_3PC$) | 1.3[b], 1.28[c] |
| CEPC | 1.2[b] |
| BPC ($C_4PC$) | 1.32[c] |
| VPC ($C_5PC$) | 1,37[c] |
| HexPc ($C_6PC$) | 1.53[b], 1.48[c] |
| HepPc ($C_7PC$) | 1.55[c] |
| $C_9PC$ | 1.65[e] |
| $C_{10}PC$ | 1.86[c] |
| $C_{12}PC$ | 1.9[e] |

[a]d-spacing calculated from the first peak of the XRD pattern measured at room temperature
[b]Wojciechowski (2000)
[c]Huang et al. (2007)
[d]Laivins and Gray (1985c)
[e]Greiner et al. (2003)

position peak and the corresponding intermolecular distance vary with the length of the substituent.

For instance, when examining the structure of different aliphatic acid esters of HPC by means of WAXD, changes in intensity and peak position below $2\theta = 8°$ were observed. Comparison of the diffractograms of HPC and its derivatives indicates that the chemical modification of HPC macromolecules results in a linear increase in *d*-spacing of the polymer main chain. As it has been described (Table 14.1), the d-spacing evolves from 1.04 nm for HPC (Wojciechowski 2000) to 1.9 nm for an aliphatic side-chain containing ten methylene groups ((CH3(CH2) n-2CO-Propylcellulose) (n = 12 in the table).

It is noteworthy that the *d*-spacing of the main chains has an influence on the intensity of the low-angle diffraction band attributed to crystalline phase. Thus, the observed evolution indicates a remarkable change in the packing of the polymer main chain as well as the decreasing contribution of the crystalline phase. The flexible side-chain substituents on the cellulose may facilitate the orientation of the main chain by increasing its mobility and enhancing solubility. The presence of many large substituents may also inhibit crystallization and increase the chain radius (Pawlowski et al. 1987). A structural model, proposed by Arici et al. (2000), described a parallel packing of chain molecules side by side with the lateral groups pointing to each other. They considered that the chain backbones could be arranged in a parallel manner within a plane, and the side-chains are parallel to each other in the plane and angled with respect to the main backbone. This packing model accounted for the observed distances and its increment of about 0.07 nm with the incorporation of each $CH_2$ unit.

Layered structures of the type described by Arici et al. were proposed for several semi-flexible polymers. Lee et al. (1997) used a similar model to describe the structure of several derivatives of HPC with various side-chain length, according to Wegner's bilayer model for substituted aromatic polyesters (Rodriguez-Parada et al. 1989). Gray et al. reported a parallel alignment of the polymer backbone in thermotropic cholesteric phases of HPC esters, in which the backbone's distance is controlled by the length of substituents (Tseng et al. 1982).

Greiner et al. (2003) found that the layer ordering, that considers both the d-spacing and the shape of the small angle reflection, depended on temperature. They observed in an X-ray diagram that the peak intensity decreases and the reflection become broader with temperature. This was accompanied by a variation of the d-spacing. Temperature strongly affects the layer spacing for longer substituents whereas there is no alteration of d-spacing for the short chain substituents. This work also concluded that the conformation of the lateral groups becomes disordered at relatively narrow temperature ranges.

## *14.3.3 Thermal Properties of Esters of HPC*

As will be further presented in this section, there is a correlation between the *d*-spacing of the poly (saccharide) main chain and the thermotropic mesomorphism of the investigated ester of HPC.

There are some determinant factors in the HPC ester derivatives such as, the length of the substituents, degree of esterification, degree of polymerization, that directly affects (i) the glass transition temperatures, $T_g$, of HPC and its derivatives, (ii) the clearing temperature, $T_c$, characterizing the transition from the cholesteric state to the isotropic fluid and (iii) transition enthalpy $\Delta H_i$, as observed from the data collected and summarized in Table 14.2. In this section the effect of the aforementioned factors in the thermal properties will be explored.

### 14.3.3.1 Effect of Substituent Length

The DSC and microscopic studies revealed that the cellulose derivatives are glassy at low temperatures down to −50 °C as shown in Table 14.2. $T_g$ decreases as the number of methylene groups in the side-chain of the aliphatic acid ester of HPC increases as seen in Fig. 14.1. In fact, the $T_g$ of HPC appears at 10–20 °C whereas this $T_g$ decreases up to −63 °C for $C_{10}$PC polymer (Huang et al. 2007; Rusig et al. 1994; Wojciechowski 2000).

The thermotropic LC-phase of HPC was observed above 160 °C with transition temperatures to isotropic state at 205 °C. The increasing length of the side-chain in the polymer series CnPC leads to a stabilization of the isotropic phase (Huang et al. 2007; Rusig et al. 1994; Wojciechowski 2000).

**Table 14.2** Thermal properties of ester of HPC

| HPC-ester | $\overline{DP}$ | $\overline{DE}$ | ΔH (J/g) | $T_g$ (°C) | $T_c$ (°C) | ΔT |
|---|---|---|---|---|---|---|
| HPC | | | | 15 | 205 | 190 |
| APC ($C_2PC$) | 59 | 2.8 | 3.5 | −16[a], −10.2[b] | 164.0[a], 162[b] | 180 |
| PPC ($C_3PC$) | 73 | 2.8 | 2.8 | −25[a] | 153.0[a] | 178 |
| CEPC ($C_3PC$) | | | | −20[c] | 194[c] | 214 |
| BPC ($C_4PC$) | 68 | 2.8 | 2.3 | −40[a], −34[b] | 159.1[a], 143[b] | 177 |
| C ($C_4PC$) | 65 | 2.5 | | −27[a] | 131.7[a] | 158 |
| VPC ($C_5PC$) | 73 | 2.6 | 1.9 | −44.7[b] | 130[b] | 175 |
| iVPC ($C_5PC$) | 79 | 2.8 | | −33[a] | 145.0[a] | 178 |
| HexPc ($C_6PC$) | 71 | 2.8 | 1.8 | −55[a], −48.7[b] | 146.9[a], 124[b] | 174 |
| HepPc ($C_7PC$) | 87 | 2.85 | 1.7 | −68[a], −59.3[b] | 141.0[a], 115[b] | 174 |
| $C_{10}PC$ | 80 | 2.65 | 1.4 | −63.3[b] | 89[b] | 152 |

[a]Rusig et al. (1994)
[b]Huang et al. (2007)
[c]Wojciechowski (2000)
$\overline{DP}$ degree of polymerization, $\overline{DE}$ degree of esterification (determined by NMR), $T_g$ glass transition temperature, $T_c$ clearance transition temperature

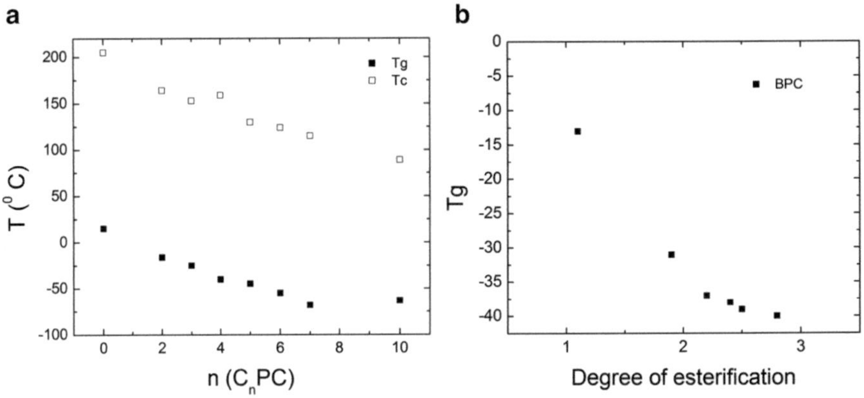


**Fig. 14.1** (**a**) Effect of chain length in glass transition temperature ($T_g$) and clearance temperature ($T_c$) of esters of HPC with different number of methylene groups (n) in the side-chain. (**b**) Effect of degree of esterification in the glass transition temperature of BPC (Butyric propylcellulose, n = 3). Values for both figures were taken from references (Huang et al. 2007; Rusig et al. 1994; Wojciechowski 2000)

For BPC ($C_4PC$) and VPC ($C_5PC$) the structure of the aliphatic side-chain also affects the thermal properties. In fact, for bulky iBPC the $T_g$ appears at −40 °C whereas for linear BPC the $T_g$ is close to −27 °C. A similar result was observed for bulky iVPC, showing a $T_g$ close to −45 °C, and linear VPC with a $T_g$ of −33 °C. The clearing temperature also presents such differences between bulky and linear structures of the aliphatic ester. This effect can be explained by the fact that the bulky chain leads to chemical hindrance that derive in a decrease of chain mobility (Rusig et al. 1994). The bulky side-chains (iBPC and iVPC) lead to a decrease of the mesophase stability, compared to the homologous derivatives with linear side-chains (BPC and VPC).

The clearing temperature depends not only on the length of side-chains but also on their polarity. Two polymers that have the same side-chain length ($C_3PC$) like CEPC and PPC, exhibit different clearing temperatures as summarized in the Table 14.2. CEPC forms lyotropic LC-phases. However thermo-optical characteristics suggest the presence of the thermotropic LC-phase at the elevated but lower temperatures in comparison with HPC. More polar CEPC has higher clearance temperature ($T_i = 194$ °C) than less polar PPC ($T_i = 135$ °C) (Wojciechowski 2000).

#### 14.3.3.2 Effect of Degree of Esterification

Another factor that affects the thermal properties of the thermotropic esters of HPC is the degree of esterification as shown for BPC by Rusig et al. (1994). The BPC, prepared by using different amounts of chloride acid, gave rise to BPC systems with degrees of esterification. It was observed that as the range of $\overline{DE}$ increases the $T_g$ of the polymer decreases quasi exponentially Fig. 14.1b. The values of $T_g$ decrease from −13 to −46 °C when the degree of esterification increases from 1.1 to 2.8.

The degree of esterification is related to the chain stiffness, that is, as the $\overline{DE}$ increases the chain stiffness also increases, which hinders the rotation about glycosidic bonds in the backbone and therefore decreasing the temperature at which the glass transition occurs.

From these results the authors have predicted that not only the backbones but also the side chains, the degree of esterification and their polarity are involved in the formation of the LC phases and phase transitions. The temperature transitions are directly dependent on those factors. The methylene units ($(-CH_2-)_n$) in the side chains not only serve as "diluent" due to the increase of flexibility and volume, these units are also responsible for the phase transitions. The length of the aliphatic chain contributes also to the enthalpy involve with the isotropic to liquid crystal phase transition.

### *14.3.4 Optical Properties: Temperature Dependence of the Pitch*

From the results presented before one can see that the pitch values of the cholesteric structure as well as the optical properties are very much influenced by HPC esterification and length of substituents.

The optical properties are directly related to the polymer and LC structure formation, consequently it is of paramount importance to determine the effect of side chain size and number of substituents, degree of esterification and degree of polymerization in pitch values and the optical properties of the mesophases of thermotropic liquid crystals.

By external appearance, APC ($C_2PC$) (Tseng et al. 1981) and PPC ($C_3PC$) (Tseng et al. 1982) are colorless at room temperature; while BPC ($C_4PC$) (Hou et al. 2000) and VPC ($C_5PC$) present bright blue purple and red prevailing color, respectively. The rest of $C_nPCs$ (with n = 6, 7, and 10) are colorless sticky fluids with high viscosities at room temperature.

An interesting property of the HPC ester derivatives such as APC, PPC, iBPC, VPC, and iVPC, is that the materials can reflect light at a specific light wavelength and in a specific temperature range, which depends on the mentioned variables.

For instance APC ($C_2PC$) (Tseng et al. 1981), PPC ($C_3PC$) (Tseng et al. 1982) and BPC ($C_4PC$) (Hou et al. 2000) display iridescent color at ambient temperature due to the selective reflections of light from the cholesteric LC phase. Under POM all these $C_nPC$ samples displayed the birefringence and typical LC textures from anisotropic LC materials when the samples were assembled between two glass slides as shown in the work done by Huang et al. (2007).

Temperature variations are known to cause in many instances strong modifications of the pitch values as shown by Lin-Liu et al. (1976) and Zugenmaier (1998). Some cases have been reported where a temperature variation could even lead to a sign reversal of the helix (Watanabe and Nagase 1988) but such strong effects have not been observed for cellulose esters but only for cellulose ether derivatives.

The temperature dependence of the thermotropic cholesteric pitch was found to follow the relationship described by the empirical equation (Bhadani and Gray 1983; Rusig et al. 1994; Shimamura et al. 1981; Tseng et al. 1981):

$$\frac{1}{P} = b\left(1 - \frac{T}{T_N}\right) \tag{14.5}$$

where P is the cholesteric pitch, b is a constant for a given polymer-solvent concentration, that increases as the polymer concentration increases, and $T_N$ is the compensation temperature, and is defined as the temperature at which the cholesteric pitch is infinite. At this temperature the molecular arrangement resembles that of the nematic mesophase.

#### 14.3.4.1 Effect of Substituent Length

It is possible to measure the cholesteric maximum reflection peak, $\lambda_0$, of the ester of HPC by means of UV/Vis spectroscopy. This $\lambda_0$ was described to increase with the number of carbons of the aliphatic chain for linear HPC esters going from visible to near infrared region of the spectra, as seen in Fig. 14.2a, b. Thus, the side chain length was also found to affect the cholesteric maximum reflection peak values (Huang et al. 2007).

For APC, PPC, BPC, iBPC, and iVPC (n = 2, 3, 4, 5 respectively) the magnitude of the pitch was comparable to the wavelength of visible light, for a given temperature range, and a linear relationship was also observed between $1/nP$ and T that follows the expression mentioned before (14.5). For example, in the case of APC, the values of the pitch are in the visible range of the spectrum for temperatures between about 100–140 °C. However, for the BPC the corresponding

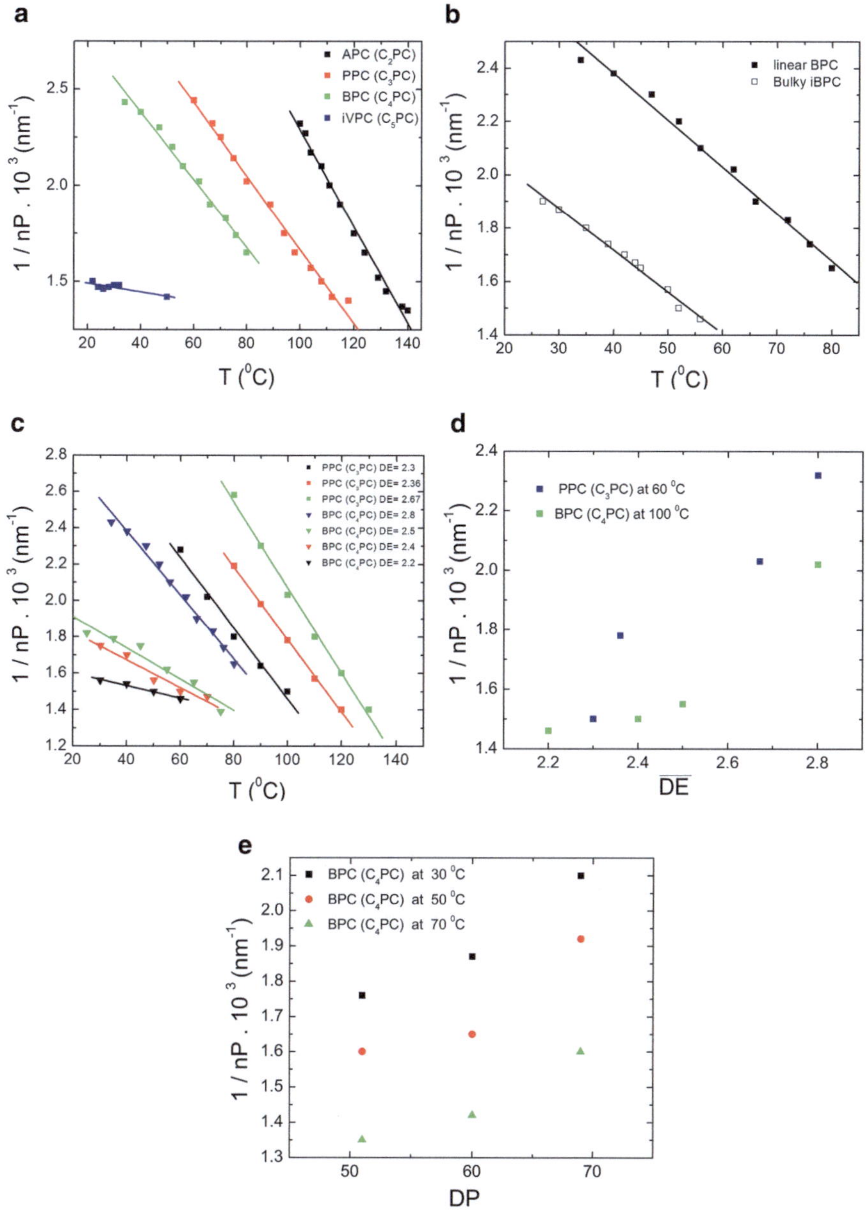


**Fig. 14.2** (**a**) Variation of the pitch with temperature for different side chain length esters of HPC and (**b**) Variation of the pitch with temperature for different size of side chain for BPC. (**c**) Variation of the pitch with temperature for several samples of BPC (values obtained from Rusig et al. (1994)) and PPC (values obtained from Werbowyj and Gray (1984)) with different degree of esterification. (**d**) Variation of the pitch with degree of esterification for the PPC and BPC samples. (**e**) Variation of the pitch for BPC samples with different degree of polymerization at 30, 50 and 70 °C (The values for figure (**a**, **b**, **d**, **e**) were obtained from Rusig et al. (1994))

temperature range is lower (30–90 °C). The increase of the pitch with the size of the side chain substituents can be justified by associating this with the increase in segment separation.

#### 14.3.4.2 Effect of Degree of Esterification

The degree of esterification (DE) of the HPC ester has significant impact on the pitch values. For instance, Hou et al. (2000) reported the dependence of the pitch within the thermotropic cholesteric phase on the $\overline{DE}$ of butylesters of HPC (BPC). A series of BPCs was synthesized with different $\overline{DE}$ (2.20–2.96).

Rusig et al. also studied the effect of $\overline{DE}$ for BPC (DE = 2.2, 2.4, 2.5 and 2.8) and PPC (DE = 2.3, 2.36, and 2.67). The results collected in Fig. 14.2c show a linear relationship between $1/nP$ and T, which can be expressed again as (14.5), from which $T_N$, compensation temperature, can be extrapolated. In the case of BPC samples, the pitch decreases as the values of $\overline{DE}$ increase. The slope of the curves $1/nP$ vs. T are almost constant for BPC samples with different values of the degree of esterification, and the temperatures $T_N$ are all greater than the clearing temperatures where the helicoidal structure does not exist anymore. The values of $T_N$ seem to decrease with the increase of DE. The decrease of the pitch with the increase of $\overline{DE}$ was also observed for PPC (Fig. 14.2d) and also reported for APC samples where complete acetylation of the hydroxyl groups does not further decrease the temperature range within which reflection colours are exhibited but rather shifts this range to higher temperatures (Laivins et al. 1986).

#### 14.3.4.3 Effect of Degree of Polymerization (DP): Effect of Molecular Weight

Laivins et al. studied how the molecular mass of APC (among other parameters such as d-spacing mentioned before) affected the pitch and hence the optical properties of the mesophases (Laivins and Gray 1985b). A dependence of pitch with the molar mass was observed at any studied temperature. The fraction with a higher molar mass ($2.14 \times 10^5$ g/mol) exhibited a lower pitch value than the short chain fractions ($2.18 \times 10^4$ g/mol). Furthermore, this dependence of the pitch was noticeable for lowest fraction (Laivins and Gray 1985c). Rusig et al. found the same trend for BPC ($C_3PC$) samples with different degrees of polymerization (Rusig et al. 1994). As shown in Fig. 14.2e the values of the pitch decrease with the increase of DP. They also found that when DP increases, the value of $T_N$ increases as well.

Laivins et al. proposed a possible explanation of the pitch dependence of the molar mass in which the segment density and hence the chiral forces could be somewhat diluted for a low molar mass polymer.

## 14.4 Recent Advances in Applications of Liquid Crystalline HPC and Its Esters Derivatives

The chemical modification of cellulose can be an effective way to facilitate the processing of cellulosic materials. It was observed that adding lateral chains to the cellulosic main chain increases the number of solvents in which the cellulose derivative is soluble and that a high degree of substitution, or non-uniform substitution, prevents crystallization and improves solubility. Processing cellulose derivatives from the liquid crystal phase opens the door to the production of materials with new properties and several works are reported in literature concerning the processing of cellulosic derivatives from the liquid crystalline phase.

Geng et al. (2013) report the production of a soft motor using a cellulose derivative liquid crystal network. This motor uses a circular loop made from HPC and a difference in humidity, between the two sides of belt, to generate motion. The origin of this motor is the change in the shape of the belt due to changes in the orientational order of the constituents caused by humidity on one side of the HPC network. The film was obtained by shear of a HPC anisotropic solution which induced orientation of the mesogenic units producing a nematic monodomain with the director parallel to the shear direction (Godinho et al. 2009b). It was observed a bending of the film when in presence of water molecules. When a loop is passed over two wheels and one side of the film is exposed to water vapor, the interaction with the water molecules causes the film to bend changing the force equilibrium giving rise to a net torque on the wheel resulting on its rotation of the wheel Fig. 14.3.

As the wheels turn, the area of film exposed to the humidity moves to the other side and unbends as the water molecules evaporate. As dry parts of the film contact

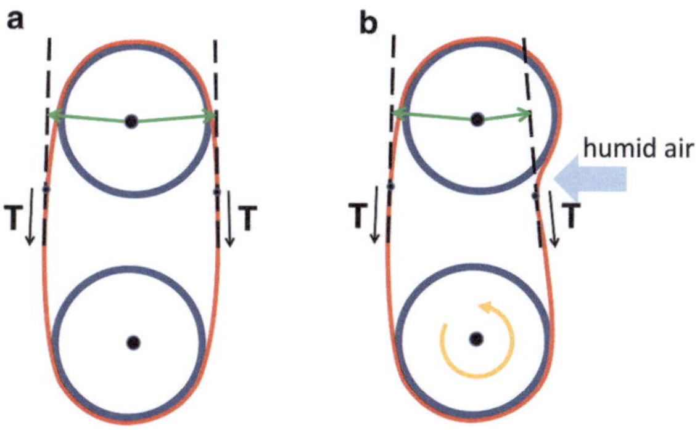


**Fig. 14.3** Schematic of the water-driven cellule motor. (**a**) Before being exposed to humidity the net torque is equal to zero and the lever arms are of equal length. (**b**) In the presence of humidity, the bending of the film causes the configuration to be non-symmetrical and the net torque is not zero. The lever arms are not equal giving rise to the motor rotation (Geng et al. 2013). Reprinted by permission from Macmillan Publishers Ltd: Scientific Reports, Geng Y, Almeida PL, Fernandes SN, Cheng C, Palffy-Muhoray P, & Godinho MH, "A cellulose liquid crystal motor: a steam engine of the second kind". 3:1028, Copyright © 2013

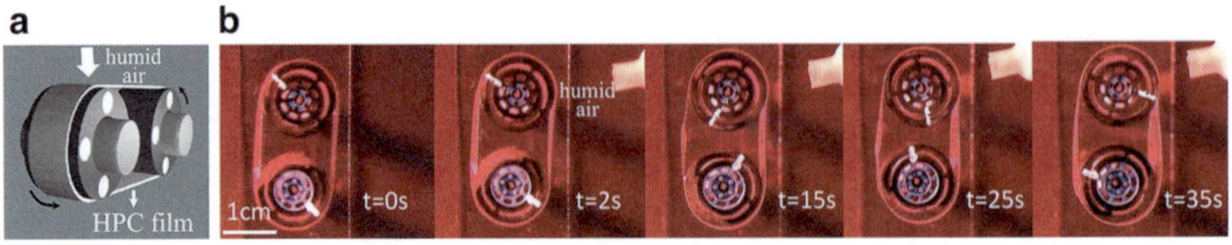


**Fig. 14.4** Moisture-driven liquid crystal cellulose engine. (**a**) Schematic of the motor where it is possible to observe the directions of the moist air and rotation. (**b**) Sequence of video frames showing the motor performing work. Reprinted by permission from Macmillan Publishers Ltd: Scientific Reports, Geng Y, Almeida PL, Fernandes SN, Cheng C, Palffy-Muhoray P, & Godinho MH, "A cellulose liquid crystal motor: a steam engine of the second kind". 3:1028, Copyright © 2013

with the water vapour the motor continues to turn. Since the bending and unbending are reversible, as long as the difference in humidity is maintained the motor will continue running. This is illustrated in Fig. 14.4.

Another way to profit from the LC cellulose derivatives is described in (Pinto et al. 2011). Films were prepared from liquid crystal APC functionalised with azo-containing groups. These groups were introduced through chemical substitution on the hydroxyl groups present on the APC molecule as mention previously. The films produced, when exposed to UV radiation, exhibit reversible wettability. The length and lateral packing for the groups containing the azo units are similar to the packing distances observed for the cellulosic main chain, resulting in minimum disturbance of the liquid crystal order. Due to this, the produced azo-APC forms a liquid crystalline phase at room temperature.

Contact angle measurements showed that UV irradiation promotes an increase in the hydrophobic character of the azo-APC. Irradiation with UV light induces a cis-trans transformation that results in bending of the azo-group attached to the main chain (trans configuration). Further irradiation with visible light promotes the reverse transformation with the corresponding unbending of the azo-group (cis configuration), indicating that the transformations from trans-to-cis and then from cis-to-trans are reversible. This work shows that by means of irradiation with UV light or white light it is possible to control the chemical composition of the surface; one of the two major factors that influence wetting properties, the other factor being the roughness of the surface (Butt et al. 2003; Wenzel 1936). After irradiation with UV light the azo dipole is exposed at the surface, resulting in a less accentuated hydrophobic character (contact angle of approximately 90°). The reverse transformation, after visible light irradiation, induces a cis-to-trans modification and the film/air interface is dominated by the apolar methoxy groups resulting in increased hydrophobicity (contact angle of approximately 120°). This work shows that it is also possible to prepare a stimuli-responsive cellulosic material with a hydrophobic–hydrophilic behavior by chemically attaching azo-groups to the chain of cellulose derivatives Fig. 14.5.

Besides the production of films, cellulose LC derivatives can also be used for the production of fibers and micro/nano fibers. One of the best known techniques

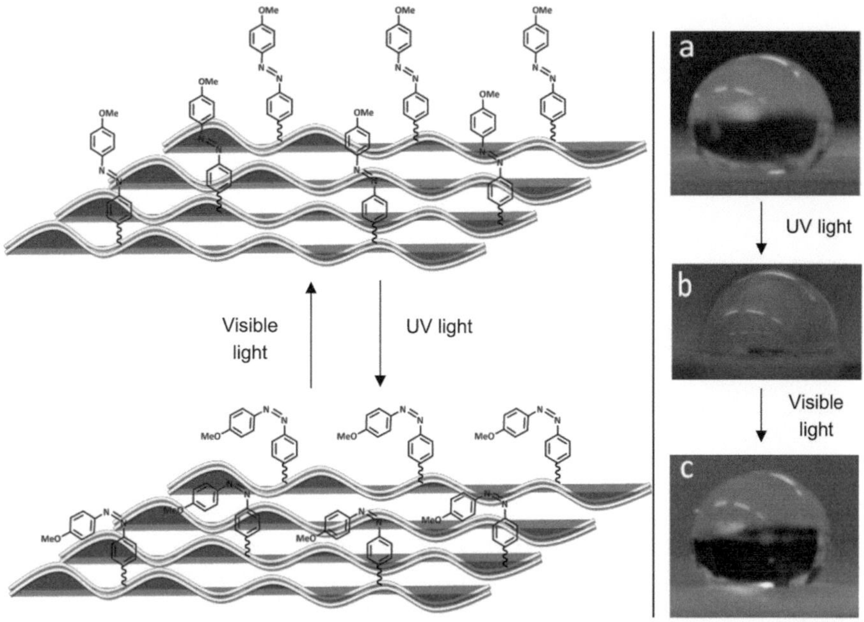


**Fig. 14.5** Schematic of the surface of the azo-APC film and the structural changes that occur upon exposure to visible light and to UV light. (**a**) Water droplet on the surface more hydrophobic, before irradiation with UV light (trans configuration). (**b**) Water droplet after UV irradiation (cis configuration) on the less hydrophobic surface. (**c**) Water droplet after irradiation with visible light (after one cycle of UV-visible light irradiation). Reprinted with permission from Pinto LFV, Kundu S, Brogueira P, Cruz C, Fernandes SN, Aluculesei A, & Godinho MH (2011) "Cellulose-Based Liquid Crystalline Photoresponsive Films with Tunable Surface Wettability". Langmuir 27 (10):6330–6337. Copyright 2011 American Chemical Society

to produce micro fibers is the electrospinning technique (Li and Xia 2004). The advantage of microfibers is that they present a much higher superficial area than bulk material. Several works are reported in literature where micro fibers were electrospun from cellulose derivatives from isotropic solutions (Frenot et al. 2007; Shukla et al. 2005; Wu et al. 2005; Zhao et al. 2003, 2004).

The production of thermotropic APC cellulose derivatives micro fibers is reported (Canejo et al. 2008) for the electrospinning of APC. These were obtained from a lyotropic solution of APC at room temperature. Scanning Electron Microscopy (SEM) observations showed that the APC electrospun fibers exhibit a spontaneous twist along their axis.

When APC electrospun fibres were collected suspended between two electrodes and allowed to release the tension acquired during the spinning process, a twisted rod-to-helix transformation is observed and the fibers exhibit helical structures (Godinho et al. 2009a, 2010). These micro helices (obtained if the fiber is fixed by both extremities) and micro spirals (obtained if the fiber is fixed only by one of the ends) were observed by SEM (see Fig. 14.6).

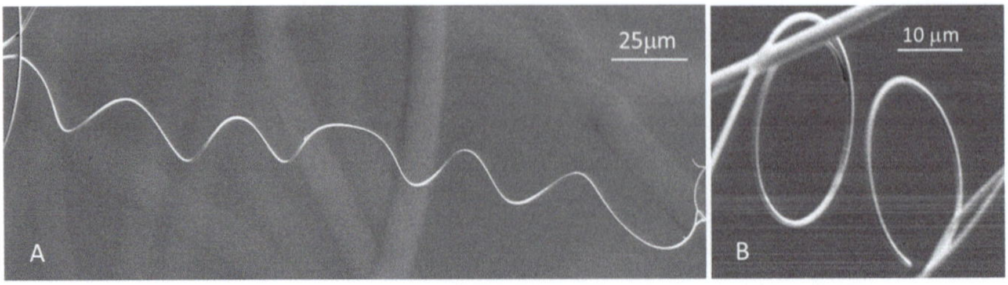


**Fig. 14.6** SEM pictures of helical structures observed in electrospun APC thermotropic fibres. Suspended fibres show helices (**a**) or spirals (**b**) if suspended by two or one end, respectively. Reproduced in part from Godinho MH, Canejo JP, Pinto LFV, Borges JP, & Teixeira PIC (2009a), "How to mimic the shapes of plant tendrils on the nano and microscale: spirals and helices of electrospun liquid crystalline cellulose derivatives". Soft Matter 5:2772, with permission of The Royal Society of Chemistry

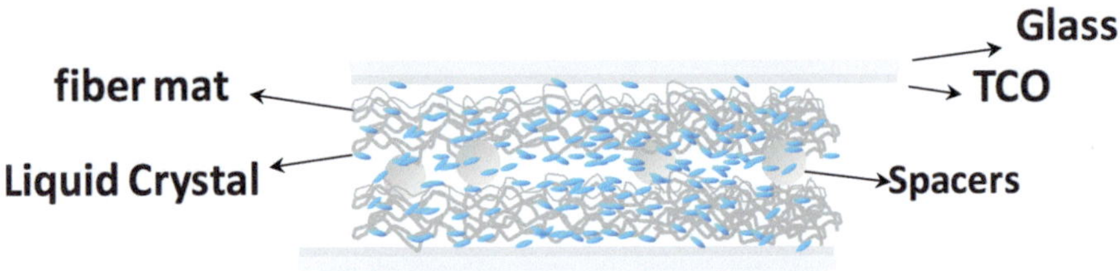


**Fig. 14.7** Schematic of the light scattering device. The cellulose derivate was electrospun on the surfaces of the conductive substrates (Almeida et al. 2009b). Reprinted with permission from Almeida PL, Kundu S, Borges JP, Godinho MH, & Figueirinhas JL (2009), "Electro-optical light scattering shutter using electrospun cellulose-based nano- and microfibers". Applied Physics Letters 95:043501. Copyright © 2009, AIP Publishing LLC

The possibility to produce non-woven mats, by electrospinning, with micro helical structures, like helices and spirals, is an effective way of increasing considerably the superficial area of the mat allowing their use for example as filters.

Non-woven mats produced from HPC can also be incorporated in electro-optical light scattering shutter devices. As described previously in literature cellulose-based polymer dispersed liquid crystal (CPDLC) electro-optical devices use a rough cellulose derivative film surrounded by two layers of nematic liquid crystal placed between two conducting transparent substrates (Godinho et al. 1996, 1998). More recently, Almeida et al. (2009b) reports on the production of an electro-optical shutter where two layers of non-woven mat of HPC and cellulose acetate (CA) were electrospun on the surface of the conductive indium tin oxide-coated substrates. The schematic for this type of fibre device (FD), which can scatter or transmit light depending on the applied voltage (OFF or ON state respectively), is represented in Fig. 14.7.

In this type of device the liquid crystal interpenetrates the fibres of the non-woven mat, in contrast to CPDLCs and polymer dispersed liquid crystals (PDLCs) where it stays confined in layers or droplets, resulting in enhanced electro-optical properties.

**Table 14.3** Comparison between the electro-optical properties for the prepared devices and a device assemble with a polymeric film

| | $T_{max}$ (%) | $T_{min}$ (%) | Eon (V/μm) | $t_{on}$ (ms) | $t_{off}$ (ms) |
|---|---|---|---|---|---|
| FD (CA) | 74 | 0.2 | 2.15 | 1.48 | 6.41 |
| FD (HPC) | 89 | 1.3 | 2.41 | 1.40 | 6.33 |
| Cellulosic film[a] | 82 | 0.4 | 1.70 | −12.04 | 10.54 |

[a]Almeida et al. (2002, 2009b)



In the OFF state, the random orientation of the fibres contributes to a random orientation of the director, which in turn helps improving the light scattering of the device. During the ON state, the similar values of the refractive indexes for the liquid crystal cellulose derivatives result in a very clear state once the electric field is applied. The electro-optical properties of the assembled devices are summarized in Table 14.3.

Another advantage of using electrospun cellulose derivatives non-woven mats is the significant decrease in production costs associated with the fabrication of these type of devices, since the amount of raw materials is much smaller and polarizers and surface treatment are not needed.

## 14.5 Conclusions

Cellulose is the most abundant natural, renewable, biocompatible and ecofriendly polymer on Earth. The research on cellulose due to environmental concerns has become intense in the past decades. The primary and secondary free hydroxyl groups, are responsible for its fascinating chemical versatility and can undergo chemical reactions that are the bases for the high range of cellulose derivatives used worldwide. Cellulose derivatives can furthermore act as a starting material and undergo functionalization reactions, which can expand their field of application.

The cholesteric nature of the cellulose derivatives and its optical, thermal and mechanical properties can be tuned by the chemical modification of the original molecular structure, giving rise to a high brand of thermotropic materials. These new features impart specific properties that make them of high interest for photonics, electronics, stimuli-responsive devices as well as biomedical applications.

**Acknowledgments** The authors would like to acknowledge the Portuguese Science and Technology Foundation (FCT) supported, through contracts PTDC/CTM-POL/1484/2012, PEst-C/CTM/LA0025/2013-14 and Project 441.00 INDIA, also FEDER, COMPETE through contract QREN n. 34169. S. N. Fernandes and C. Echeverria acknowledge FCT for grant SFRH/BPD/78430/2011 and SFRH/BPD/88779/2012, respectively.

## References


Almeida PL, Tavares S, Martins AF, Godinho MH, Cidade MT, Figueirinhas JL (2002) Cross-linked hydroxypropylcellulose films: mechanical behaviour and electro-optical properties of PDLC type cells. Opt Mater 20(2):97–100

Almeida PL, Kundu S, Beja D, Fonseca J, Figueirinhas JL, Godinho MH (2009a) Deformation of isotropic and anisotropic liquid droplets dispersed in a cellulose liquid crystalline derivative. Cellulose 16(3):427–434

Almeida PL, Kundu S, Borges JP, Godinho MH, Figueirinhas JL (2009b) Electro-optical light scattering shutter using electrospun cellulose-based nano- and microfibers. Appl Phys Lett 95:043501

Arici E, Greiner A, Hou H, Reuning A, Wendorff JH (2000) Optical properties of guest host systems based on cellulose derivatives. Macromol Chem Phys 201(15):2083–2090

Aspler JS, Gray DG (1982) Interaction of organic vapours with hydroxypropyl cellulose. Polymer 23(1):43–46

Atalla RH, Isogai A (2005) Recent developments in spectroscopic and chemical characterization of cellulose. In: Dumitriu S (ed) Polysaccharides structural diversity and functional versatility, 2nd edn. Marcel Dekker, New York, pp 123–158

Bagheri M, Shateri S (2012) Synthesis and characterization of novel liquid crystalline cholesteryl-modified hydroxypropyl cellulose derivatives. J Polym Res 19(3):1–13

Bhadani SN, Gray DG (1982) Liquid crystal formation from the benzoic acid ester of hydroxypropylcellulose. Makromol Chem Rapid Commun 3(6):449–455

Bhadani SN, Gray DG (1983) Cellulose-based liquid crystalline polymers; esters of (hydroxypropyl) cellulose. Mol Cryst Liq Cryst 99(1):29–38

Bhadani SN, Gray DG (1984) Crosslinked cholesteric network from the acrylic acid ester of (hydroxypropyl)cellulose. Mol Cryst Liq Cryst 102(8–9):255–260

Bhadani SN, Tseng S-L, Gray DG (1983) Lyotropic and thermotropic liquid-crystalline phase formation from fractions of a semiflexible cellulosic polymer. Makromol Chem 184 (8):1727–1740

Bheda J, Fellers J, White J (1980) Phase behavior and structure of liquid crystalline solutions of cellulose derivatives. Colloid Polym Sci 258:1335–1342

Butt H-J, Graf K, Kappl M (2003) Physics and chemistry of interfaces. Wiley, Weinheim

Canejo J, Godinho M (2013) Cellulose perversions. Materials 6:1377–1390

Canejo JP, Borges JP, Godinho MH, Brogueira P, Teixeira PIC, Terentjev EM (2008) Helical twisting of electrospun liquid crystalline cellulose micro- and nanofibers. Adv Mater 20:4821–4825

Chanzy H, Peguy A, Chaunis S, Monzie P (1980) Oriented cellulose films and fibers from a mesophase system. J Polym Sci Polym Phys Ed 18(5):1137–1144

Charlet G, Gray DG (1987) Solid cholesteric films cast from aqueous (hydroxypropyl)cellulose. Macromolecules 20(1):33–38

Cidade MT, Leal CR, Godinho MH, Martins AF, Navard P (1995) Rheological properties of acetoxypropylcellulose in the thermotropic chiral nematic phase. Mol Cryst Liq Cryst 261 (1):617–625

Costa I, Almeida PL, Filip D, Figueirinhas JL, Godinho MH (2006) Tunable topographical cellulose matrices for electro-optical liquid crystal cells. Opto-Electron Rev 14(4):299–303

Costa I, Filip D, Figueirinhas JL, Godinho MH (2007) New cellulose derivatives composites for electro-optical sensors. Carbohydr Polym 68(1):159–165

Credou J, Berthelot T (2014) Cellulose: from biocompatible to bioactive material. J Mater Chem B 2(30):4767–4788

de Gennes PG, Frost J (1993) The physics of liquid crystals, 2nd edn. Oxford University Press, Oxford

De Vries H (1951) Rotatory power and other optical properties of certain liquid crystals. Acta Crystallogr 4(3):219–226

Edgar KJ, Buchanan CM, Debenham JS, Rundquist PA, Seiler BD, Shelton MC, Tindall D (2001) Advances in cellulose ester performance and application. Prog Polym Sci 26(9):1605–1688
El-Wakil NA, Fahmy Y, Abou-Zeid RE, Dufresne A, El-Sherbiny S (2010) Liquid crystalline behavior of hydroxypropylcellulose esterified with 4-alkoxybenzoic acid. Bioresources 5(3):1834–1845
Fleming K, Gray DG, Matthews S (2001) Cellulose crystallites. Chem Eur J 7:1831–1835
Flory PJ (1956) Statistical thermodynamics of semi-flexible chain molecules. Proc R Soc A Math Phys Eng Sci 234:60–73
Frenot A, Henriksson MW, Walkenström P (2007) Electrospinning of cellulose-based nanofibers. J Appl Polym Sci 103:1473–1482
Geng Y, Almeida PL, Fernandes SN, Cheng C, Palffy-Muhoray P, Godinho MH (2013) A cellulose liquid crystal motor: a steam engine of the second kind. Sci Rep 3:1028
Godinho MH, Martins AF, Figueirinhas JL (1996) Novel PDLC type display based on cellulose derivatives. Liq Cryst 20:373–376
Godinho MH, Martins AF, Figueirinhas JL (1998) Composite systems for display applications from cellulose elastomers and nematic liquid crystals. Opt Mater 9:226–229
Godinho MH, Canejo JP, Pinto LFV, Borges JP, Teixeira PIC (2009a) How to mimic the shapes of plant tendrils on the nano and microscale: spirals and helices of electrospun liquid crystalline cellulose derivatives. Soft Matter 5:2772
Godinho MH, Filip D, Costa I, Carvalho AL, Figueirinhas JL, Terentjev EM (2009b) Liquid crystalline cellulose derivative elastomer films under uniaxial strain. Cellulose 16(2):199–205
Godinho MH, Canejo JP, Feio G, Terentjev EM (2010) Self-winding of helices in plant tendrils and cellulose liquid crystal fibers. Soft Matter 6:5965
Goetz L, Mathew A, Oksman K, Gatenholm P, Ragauskas AJ (2009) A novel nanocomposite film prepared from crosslinked cellulosic whiskers. Carbohydr Polym 75(1):85–89
Gray DG (1983) Liquid crystalline cellulose derivatives. J Appl Polym Sci Appl Polym Symp 37:179–192
Gray DG (1994) Chiral nematic ordering of polysaccharides. Carbohydr Polym 25(4):277–284
Greiner A, Hou H, Reuning A, Thomas A, Wendorff JH, Zimmermann S (2003) Synthesis and opto-electronic properties of cholesteric cellulose esters. Cellulose 10(1):37–52
Habibi Y, Lucia LA, Rojas OJ (2010) Cellulose nanocrystals: chemistry, self-assembly, and applications. Chem Rev 110(6):3479–3500
Heinze T (2005) Chemical functionalization of cellulose. In: Dumitriu S (ed) Polysaccharides structural diversity and functional versatility, 2nd edn. Marcel Dekker, New York, pp 551–554
Heinze T, Liebert T (2001) Unconventional methods in cellulose functionalization. Prog Polym Sci 26(9):1689–1762
Ho FF-L, Kohler RR, Ward GA (1972) Determination of molar substitution and degree of substitution of hydroxypropyl cellulose by nuclear magnetic resonance spectrometry. Anal Chem 44:178–181
Hou H, Reuning A, Wendorff JH, Greiner A (2000) Tuning of the pitch height of thermotropic cellulose esters. Macromol Chem Phys 201(15):2050–2054
Huang B, Ge JJ, Li Y, Hou H (2007) Aliphatic acid esters of (2-hydroxypropyl) cellulose—effect of side chain length on properties of cholesteric liquid crystals. Polymer 48(1):264–269
Isogai A, Saito T, Fukuzumi H (2011) TEMPO-oxidized cellulose nanofibers. Nanoscale 3(1):71–85
Klemm D, Philipp B, Heinze T, Heinze U, Wagenknecht W (1998a) General considerations on structure and reactivity of cellulose. In: Comprehensive cellulose chemistry, vol 1. Wiley, Weinheim, pp 9–154
Klemm D, Philipp B, Heinze T, Heinze U, Wagenknecht W (1998b) General considerations on structure and reactivity of cellulose. In: Fundamentals and analytical methods: comprehensive cellulose chemistry, vol 1. Wiley, Weinheim, pp 9–155
Klemm D, Heublein B, Fink HP, Bohn A (2005) Cellulose: fascinating biopolymer and sustainable raw material. Angew Chem 44(22):3358–3393

Klemm D, Kramer F, Moritz S, Lindstrom T, Ankerfors M, Gray D, Dorris A (2011) Nanocelluloses: a new family of nature-based materials. Angew Chem 50(24):5438–5466
Kondo T (2005) Hydrogen bonds in cellulose and cellulose derivatives. In: Dumitriu S (ed) Polysaccharides structural diversity and functional versatility, 2nd edn. Marcel Dekker, New York, pp 69–98
Kosho H, Hiramatsu S, Nishi T, Tanaka Y, Kawauchi S, Watanabe J (1999) Thermotropic cholesteric liquid crystals in ester derivatives of hydroxypropylcellulose. High Perform Polym 11(1):41–48
Kuhn W (1936) Beziehungen zwischen Molekülgröße, statistischer Molekülgestalt und elastischen Eigenschaften hochpolymerer Stoffe. Kolloid-Zeitschrift 76(3):258–271
Kundu S, Feio G, Pinto LFV, Almeida PL, Figueirinhas JL, Godinho MH (2010) Deuterium NMR study of orientational order in cellulosic network microfibers. Macromolecules 43(13):5749–5755
Lagerwall JPF, Schütz C, Salajkova M, Noh J, Hyun Park J, Scalia G, Bergström L (2014) Cellulose nanocrystal-based materials: from liquid crystal self-assembly and glass formation to multifunctional thin films. NPG Asia Mater 6(1):e80
Laivins GV, Gray DG (1985a) Characterization and chain stiffness of (acetoxypropyl)cellulose. Macromolecules 18(9):1746–1752
Laivins GV, Gray DG (1985b) Liquid crystalline phase transition of a semiflexible polymer: acetoxypropyl cellulose. Macromolecules 18(9):1753–1759
Laivins GV, Gray DG (1985c) Optical properties of (acetoxypropyl)cellulose mesophases: factors influencing the cholesteric pitch. Polymer 26(10):1435–1442
Laivins GV, Sixou P, Gray DG (1986) The liquid-crystalline properties of (acetoxypropyl) cellulose: effect of chain length and degree of acetylation. J Polym Sci B 24(12):2779–2792
Lee JL, Pearce EM, Kwei TK (1997) Morphological development in alkyl-substituted semiflexible polymers. Macromolecules 30(26):8233–8244
Li D, Xia Y (2004) Electrospinning of nanofibers: reinventing the wheel? Adv Mater 16:1151–1170
Lin-Liu YR, Shih YM, Woo C-W, Tan HT (1976) Molecular model for cholesteric liquid crystals. Phys Rev A 14(1):445–450
Maji S, Kundu S, Pinto LFV, Godinho MH, Khan AH, Acharya S (2013) Improved mechanical stability of acetoxypropyl cellulose upon blending with ultranarrow PbS nanowires in Langmuir monolayer matrix. Langmuir 29(49):15231–15239
Mays JW (1988) Solution properties and chain stiffness of cyanoethyl hydroxypropyl cellulose. Macromolecules 21(11):3179–3183
Müller M, Zentel R (2000) Cholesteric phases and films from cellulose derivatives. Macromol Chem Phys 201(15):2055–2063
Nakayama E, Azuma J (1998) Substituent distribution of cyanoethyl cellulose. Cellulose 5:175–185
Navard P, Haudin J-M (1980) Rheology of mesomorphic solutions of cellulose. Br Polym J 12(4):174–178
Patel DL, Gilbert RD (1981) Lyotropic mesomorphic formation of cellulose in trifluoroacetic acid-chlorinated-alkane solvent mixtures at room temperature. J Polym Sci Polym Phys Ed 19 (8):1231–1236
Pawlowski WP, Gilbert RD, Fornes RE, Purrington ST (1986) Acetoacetylation of O-(hydroxypropyl)cellulose by 2,2,6-trimethyl-4H-1,3-dioxin-4-one. Carbohydr Res 156:232–235
Pawlowski WP, Gilbert RD, Fornes RE, Purrington ST (1987) The thermotropic and lyotropic liquid-crystalline properties of acetoacetoxypropyl cellulose. J Polym Sci B 25(11):2293–2301
Pérez S, Mazeau K (2005) Conformations, structures, and morphologies of celluloses. In: Dumitriu S (ed) Polysaccharides structural diversity and functional versatility, 2nd edn. Marcel Dekker, New York, pp 41–68

Pinto LFV, Kundu S, Brogueira P, Cruz C, Fernandes SN, Aluculesei A, Godinho MH (2011) Cellulose-based liquid crystalline photoresponsive films with tunable surface wettability. Langmuir 27(10):6330–6337

Revol JF, Bradford H, Giasson J, Marchessault RH, Gray DG (1992) Helicoidal self-ordering of cellulose microfibrils in aqueous suspension. Int J Biol Macromol 14:170–172

Riti JB, Cidade MT, Godinho MH, Martins AF, Navard P (1997) Shear induced textures of thermotropic acetoxypropylcellulose. J Rheol 41(6):1247–1260

Rodriguez-Parada JM, Duran R, Wegner G (1989) A comparative study of mesophase formation in rigid-chain polyesters with flexible side chains. Macromolecules 22(5):2507–2516

Rusig I, Dedier J, Filliatre C, Godhino MH, Varichon L, Sixou P (1992) Effect of degradation on thermotropic cholesteric optical properties of (2-hydroxypropyl) cellulose (HPC) esters. J Polym Sci A Polym Chem 30(5):895–899

Rusig I, Godinho MH, Varichon L, Sixou P, Dedier J, Filliatre C, Martins AF (1994) Optical properties of cholesteric (2-hydroxypropyl) cellulose (HPC) esters. J Polym Sci B 32(11):1907–1914

Samuels RJ (1969) Solid-state characterization of the structure and deformation behavior of water-soluble hydroxypropylcellulose. J Polym Sci A 7(7):1197–1258

Sena C, Godinho MH, Oliveira CLP, Figueiredo Neto AM (2011) Liquid crystalline cellulosic elastomers: free standing anisotropic films under stretching. Cellulose 18(5):1151–1163

Shimamura K, White JL, Fellers JF (1981) Hydroxypropylcellulose, a thermotropic liquid crystal: characteristics and structure development in continuous extrusion and melt spinning. J Appl Polym Sci 26(7):2165–2180

Shukla S, Brinley E, Cho H, Seal S (2005) Electrospinning of hydroxypropyl cellulose fibers and their application in synthesis of nano and submicron tin oxide fibers. Polymer 46:12130–12145

Singh S, Dunmur DA (2002) Liquid crystals: fundamentals. Liquid crystals: physical properties and nonlinear, Ed. World Scientific Publishing Co. ISBN 9810242506

Soldi V (2005) Stability and degradation of polysaccharides. In: Dumitriu S (ed) Polysaccharides structural diversity and functional versatility, 2nd edn. Marcel Dekker, New York, pp 395–410

Sugiyama J, Okano T, Yamamoto H, Horii F (1990) Transformation of Valonia cellulose crystals by an alkaline hydrothermal treatment. Macromolecules 23(12):3196–3198

Suto S, White JL, Fellers JF (1982) A comparative study of the thermotropic mesomorphic tendencies and rheological characteristics of three cellulose derivates: ethylene and propylene oxide ethers and an acetate butyrate ester. Rheol Acta 21(1):62–71

Thakur VK, Thakur MK (2014) Processing and characterization of natural cellulose fibers/thermoset polymer composites. Carbohydr Polym 109:102–117

Thakur VK, Singha AS, Misra BN (2011) Graft copolymerization of methyl methacrylate onto cellulosic biofibers. J Appl Polym Sci 122(1):532–544

Thakur VK, Thakur MK, Gupta RK (2013a) Rapid synthesis of graft copolymers from natural cellulose fibers. Carbohydr Polym 98(1):820–828

Thakur VK, Thakur MK, Singha AS (2013b) Free radical–induced graft copolymerization onto natural fibers. Int J Polym Anal Char 18(6):430–438

Tseng S-L, Valente A, Gray DG (1981) Cholesteric liquid crystalline phases based on (acetoxypropyl)cellulose. Macromolecules 14(3):715–719

Tseng SL, Laivins GV, Gray DG (1982) The propanoate ester of (2-hydroxypropyl)cellulose: a thermotropic cholesteric polymer that reflects visible light at ambient temperatures. Macromolecules 15(5):1262–1264

Varshney VK, Naithani S (2011) Chemical functionalization of cellulose derived from nonconventional sources. In: Kalia S, Kaith BS, Kaur I (eds) Cellulose fibers: bio- and nano-polymer composites. Springer, Berlin, pp 43–60

Watanabe J, Nagase T (1988) Thermotropic polypeptides. 5. Temperature dependence of cholesteric pitches exhibiting a cholesteric sense inversion. Macromolecules 21(1):171–175

Wenzel RN (1936) Resistance of solid surfaces to wetting by water. Ind Eng Chem 28(8):988–994

Werbowyj RS, Gray DG (1976) Liquid crystalline structure in aqueous hydroxypropyl cellulose solutions. Mol Cryst Liq Cryst 34(4):97–103

Werbowyj RS, Gray DG (1984) Optical properties of hydroxypropyl cellulose liquid crystals. I. Cholesteric pitch and polymer concentration. Macromolecules 17(8):1512–1520
Wojciechowski P (2000) Thermotropic mesomorphism of selected (2-hydroxypropyl)cellulose derivatives. J Appl Polym Sci 76(6):837–844
Wu X, Wang L, Yu H, Huang Y (2005) Effect of solvent on morphology of electrospinning ethyl cellulose fibers. J Appl Polym Sci 97:1292–1297
Wüstenberg T (2014) Hydroxypropylcellulose. In: Wüstenberg T (ed) Cellulose and cellulose derivatives in the food industry. Wiley, Weinheim, pp 319–342
Yamagishi T, Fukuda T, Miyamoto T, Watanabe J (1988) Thermotropic cellulose derivatives with flexible substituents. Polym Bull 20(4):373–377
Yamagishi T, Fukuda T, Miyamoto T, Takashina Y, Yakoh Y, Watanabe J (1991) Thermotropic cellulose derivatives with flexible substituents IV. Columnar liquid crystals from ester-type derivatives of cellulose. Liq Cryst 10(4):467–473
Yamagishi T-A, Nakamoto Y, Sixou P (2006) Preparation and cholesteric mesophase properties of (Butyl-co-pentyl) propylcellulose. Cellulose 13(3):205–211
Yim CT, Gilson DFR, Kondo T, Gray DG (1992) Order parameters and side-chain conformation in ethyl cellulose/chloroform liquid crystal phases. Macromolecules 25:3377–3380
Zhao S, Wu X, Wang L, Huang Y (2003) Electrostatically generated fibers of ethyl-cyanoethyl cellulose. Cellulose 10:405–409
Zhao S, Wu X, Wang L, Huang Y (2004) Electrospinning of ethyl-cyanoethyl cellulose/tetrahydrofuran solutions. J Appl Polym Sci 91:242–246
Zugenmaier P (1998) Cellulosic liquid crystals. Handbook of liquid crystals. Wiley, Weinheim
Zugenmaier P (2014) Cellulosic liquid crystals. In: Goodby JW, Collings PJ, Kato T, Tschierske C, Gleeson H, Raynes P (eds) Handbook of liquid crystals, vol 7:I:3. Wiley, Weinheim, pp 1–39

# Chapter 15
# Liquid Crystal Polymers as Matrices for Arrangement of Inorganic Nanoparticles

**Raisa V. Talroze, Alexey S. Merekalov, Alina M. Shatalova, Olga A. Otmakhova, and Georgiy A. Shandryuk**

## 15.1 Introduction

Preparation of hybrid composites consisting of a polymeric matrix and inorganic nanoparticles (NPs) is considered to be a useful approach for creating new materials. There are different types of such nanocomposites (Shenhar et al. 2005; Paul and Robeson 2008; Lagashetty and Venkataraman 2005; Gerasin et al. 2013) containing, for instance, spherical silicate NPs (Mark 1996), NPs of metals and metal oxides (Pomogailo 2000; Pomogailo et al. 2011; Bronstein et al. 2004), ultradispersed diamonds (detonation nanodiamonds) (Iakoubovskii et al. 2000; Dolmatov 2007), quantum dots (Rossetti et al. 1983; Murray et al. 2000; Dekker 1999; Buhro and Colvin 2003), nanotubes (Martel et al. 2001; Moniruzzaman and Winey 2006; Coleman et al. 2006; Badamshina et al. 2010; Gullapalli and Wong 2011; Saint-Aubin et al. 2013; Chatterjee et al. 2012), nanowhiskers (Okamoto 2004; Palchoudhury et al. 2011; Mashkour et al. 2014) and layered silicates (Okamoto 2004; Paul and Robeson 2008; Špírková et al. 2011).

In general, nanocomposites can be divided by their application into *structural* and *functional* ones (Gerasin et al. 2013). While the incorporation of NPs in *structural* nanocomposites enhances the mechanical properties processability of a polymer matrix, the desired characteristics of functional materials are based on the properties specific to NPs (such as quantum size effect in a semiconductor particle, the excitation of localized surface plasmons in metal particles, superparamagnetism of magnetic particles, etc.) combined with the properties of polymers.

During the last decade the immobilization of NPs by self-assembled and ordered matrices attracted a lot of attention due to the possibility not only to localize, but also to organize NPs in ordered arrays that may find new applications. One group of

R.V. Talroze (✉) • A.S. Merekalov • A.M. Shatalova • O.A. Otmakhova • G.A. Shandryuk
A.V. Topchiev Institute of Petrochemical Synthesis,
Russian Academy of Sciences, Moscow, Russia
e-mail: rtalroze@ips.ac.ru



V.K. Thakur, M.R. Kessler (eds.), *Liquid Crystalline Polymers*,
DOI 10.1007/978-3-319-20270-9_15

systems that has been studied extensively is block copolymers. The self-consistent field theory has been developed (Thompson et al. 2002; Gupta et al. 2006; Lin et al. 2005) that shows the role of NPs on the phase separation in diblock copolymers. The key parameters, namely, NP size, the length of macromolecule blocks, parameters of interactions NP-NP, NP-block and block 1-block 2 strongly influence the location of NPs either in the center or at the interfaces of lamellar or spherical domains. Simultaneously a lot of experimental studies have been carried out (Fogg et al. 1997; Chiu et al. 2005; Thurn-Albrecht et al. 2000; Böker et al. 2007; Li et al. 2008; Zhang et al. 2006; Park et al. 2007; Pavan and Shenhar 2011; Pester and Ruppel 2013; Liedel et al. 2013; Yabu et al. 2011a, b). As an example, we have to mention the introduction of metal NPs as the inorganic phase into the PS-PMMA block copolymer forming a lamellar phase (Misner et al. 2003). NPs are located mainly in the PS domains due to the preferential interactions. When 30 wt.% of gold NPs are introduced, the PS domains become electroconductive, thus converting into nanowires. If the composition of above block copolymers results in the formation of PMMA cylindrical micelles, their solid films can act as templates for CdSe NPs that penetrate into cylindrical structures due to capillary forces and retain their photoluminescence (PL). Microphase separation in block copolymers is also favorable for creation of nanosized periodic structures in which the plasmonic particles are self-assembled to fit the type of domain structure of the polymer matrix, which has an effect on the plasmonic properties (Yabu et al. 2011a, b). Gold and FePt NPs being inserted in PS-PEO block copolymers have experimentally demonstrated the role of their size as well as concentration and polymer composition on the character of NP distribution within the matrix (Aissou et al. 2011; Horechyy et al. 2013).

If anisotropic particles such as gold nanorods (NRs) are placed in a polymer matrix then one can achieve their orientation while stretching a polymer film or fiber above glass transition temperature (Pérez-Juste et al. 2005). The same effect has been observed when different rod-like NPs functionalized with PMMA, PS and PEG macromolecules self-assemble to form LC liotropic phases (Meuer et al. 2007, 2008; Zorn et al. 2008). The other group of self-assembled matrices for NPs is represented by low molecular weight liquid crystals (LCs) (Qi et al. 2008; Khatua et al. 2010; Kobayashi and Toshima 2007; Kobayashi et al. 2008; Kumar et al. 2009; Chandran et al. 2014; Joshi et al. 2010; Sinha et al. 2004; Bellini et al. 2000; Duran et al. 2005; Kopcansky et al. 2010, 2011; Rudzki et al. 2013; Garbovskiy and Glushchenko 2013; Lavrič et al. 2013; Yaroshchuk et al. 2013; Rajh et al. 2013) and liquid crystal polymers (LCPs) (Shandryuk et al. 2005, 2008; Talroze et al. 2007; Barmatov et al. 2005, 2006a, b, 2007; Nikonorova et al. 2007). Such nanocomposites possess novel properties making them different from those of individual LCs and NPs. These include controllable light scattering and photoluminescence (PL), stabilization of blue phases, and the spectral shift of the particle plasmon resonance when an electric field is applied. One very important aspect of the application of these matrices is high sensitivity of LCs and LCPs to the action of external electric, magnetic, and mechanical fields. It means that the properties of nanocomposites may be tuned by the application of the external stimulus. Moreover, the ordered state induced by the external field can be locked in place when LC matrix is cooled below the glass transition temperature ($T_g$).

In the current paper our discussion will focus on the recent advances in new nanocomposites consisting of semiconductor NPs embedded in ordered matrices capable of localization and organization of NPs, namely LCs and LCPs. The priority will be given to NPs composites with LCP matrices.

## 15.2 Phase Separation in Composites of NPs Embedded in LC Systems

One of the major challenges of the nanocomposite preparation is the lack of miscibility of inorganic NPs in organic matrix. The phase separation may happen even in the isotropic phase of LC nanocomposites if the chemical affinity of components is low and it results in the formation of biphasic systems even above the clearing point ($T_{N\text{-}I}$). This issue is resolved by the chemical modification of the surface of NPs with organic ligands. The introduction of strong interactions between NPs and matrix may also contribute in thermodynamic stability of nanocomposites. However, it is much more difficult to control the phase separation at isotropic-nematic (I-N) phase transition. The physical reason for that separation is pretty simple: the homogeneous distribution of isotropic NPs should disrupt the nematic order and decrease the transition temperature. As a result, it is thermodynamically more preferable to concentrate major part of NPs in compact areas in the isotropic phase whereas the rest of the volume may be transformed in the nematic LC state containing small amounts of NPs.

The molecular theory developed in Gorkunov and Osipov (2011), Gorkunov et al. (2013), Lopatina and Selinger (2011), Osipov and Gorkunov (2014) predicts that important parameters such as the temperature interval and the concentration of NPs depend on the constants of the microscopic interaction of molecules and NPs. Both theoretically and experimentally the phase separation is shown to appear in a broad range of parameters and is thermodynamically determined in a broad temperature range below I-N phase transition temperature (Osipov and Gorkunov 2014). It is not easy to suppress such a phase separation by simple chemical methods because the shape and size of NPs play an important role. However, the greater stability may be achieved via replacing the low molecular weight LCs by LC polymer matrix. It results in the transformation of thermodynamically unstable nanocomposites to metastable ones. The latter may have nearly infinite relaxation times. This means that not only the thermodynamic characteristics of components determine the structure of nanocomposites. Methods of preparation and the diffusion rates of NPs also influence the time of the formation of thermodynamically stable phases. The study carried out by Derikov et al. (2014) shows the possibility of preparation of uniform areas with an LC order and concentrations of NPs essentially exceeding the solubility limit of NPs in the LC nematic phase. As mentioned in Derikov et al. (2014) the kinetically stable uniform distribution of NPs may be achieved, for example, under mechanical shaking or shearing, i.e., some supply of mechanical energy, and even fixed if the viscosity of the matrix is sufficiently high. In both cases there are certain constrains related to the size and shape of NPs. As it was shown in Joshi et al. (2010), Sinha et al. (2004), Bellini et al. (2000) the spherical shape of NPs induces the decrease in N-I transition temperature ($T_{N\text{-}I}$) whereas $T_{N\text{-}I}$ increases in nematics embedded with strongly anisotropic NPs (Duran et al. 2005; Kopcansky et al. 2011). These effects were explained by the theoretical studies that lead to general concepts regarding the NPs dispersed in the anisotropic medium (Gorkunov and Osipov 2011; Lopatina and Selinger 2011).

The possible explanation of the maintenance of stability of a non-equilibrium phase observed in Derikov et al. (2014) may be a strong difference between the rate of NPs diffusion and the rate of LC ordering. When the temperature range of the ordered LC state is reached, NPs in amount essentially exceeding a thermodynamically equilibrium value do not serve as a hindrance for the LC ordering. At the same time the slowing of the diffusion of NPs due to the change of low molecular weight matrix by a polymer one prevents the formation of the equilibrium NPs concentrations in phases formed under separation. Thus the LC polymer matrix prevents the NPs displacement out of LC nematic phase of a composite which cannot be accomplished if this matrix is a low molecular weight LC. The results obtained in Derikov et al. (2014) open the way for creating long-lived stable thermodynamically non-equilibrium composites with a LC order in the vicinity of NPs. Note that the discussion above is mainly focused on nematic LC matrices because there is no proper microscopic molecular theory to describe the thermodynamics of nanocomposites with matrices of the smectic type. At the same time, there are some experimental data on the phase transition characteristics of smectic nanocomposites which we will discuss further.

## 15.3 Structure of LC Polymer: NPs Composites

### *15.3.1 Smectic Matrices/QDs*

The first, most detailed study of the structure of nanocomposites composed of CdSe quantum dots (QDs) and LCPs is described in Shandryuk et al. (2005, 2008), Talroze et al. (2007). The chemical structures of LC homo- and copolymers are given in Fig. 15.1. Homopolymers (Fig. 15.1a) form the smectic $C_A$ mesophase due to the hydrogen bond between alkoxybenzoic acid moieties resulted in the formation of mesogen-like dimers (Merekalov et al. 2001; Shatalova et al. 2003). As for the mesophase structure of copolymers (Fig. 15.1b) it depends on the chemical structure and content of comonomer units.

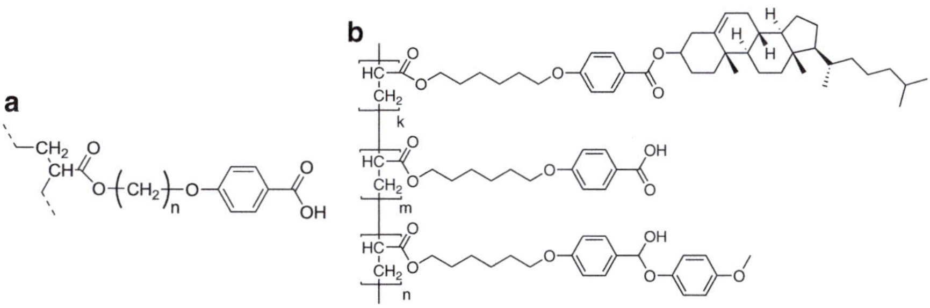


**Fig. 15.1** Structures of the LC polymer (BA-nPA) (**a**) and copolymer (**b**) matrices

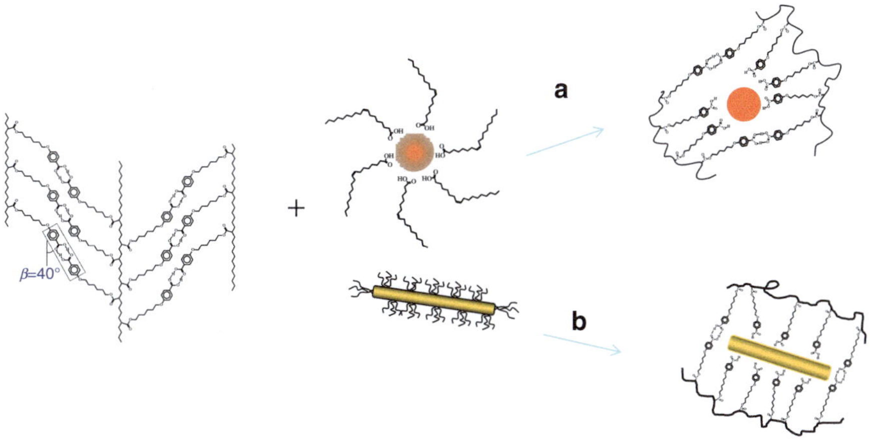


**Fig. 15.2** Schematics of the reaction between H-bonded smectic polymers and QD (**a**) or NR (**b**)

The choice of that group of polymers as matrices has been based on the hypothesis that alkoxybenzoic acid moieties in the side chain of macromolecules substitute the oleic acid molecules that form the stabilizing shell at the surface of NPs (Fig. 15.2).

The essence of the approach under consideration is the application of the anisotropic LCP matrix and chemical bonding of that matrix to the surface of QDs. FTIR spectra have proved the formation of ionic bonds between LC polymers and CdSe QDs (Shandryuk et al. 2008), which impart strong stability to the resulting nanocomposites. The formation of ionic bonds means that H-bonded dimers of alkoxybenzoic moieties are broken and COOH groups are chemically transformed to $COO^-$ anions upon interacting with cadmium ions at the surface of QDs. Figure 15.3a presents the TEM image of the film of the nanocomposite of the smectic polymer containing 20 wt.% of QDs. The smectic structure of the nanocomposites exists in a broad temperature range between $T_g$ and smectic-isotropic transition temperature ($T_{S\text{-}I}$). As an example, the change in the transition temperatures for the nanocomposite of BA-6PA with CdSe QDs is given in

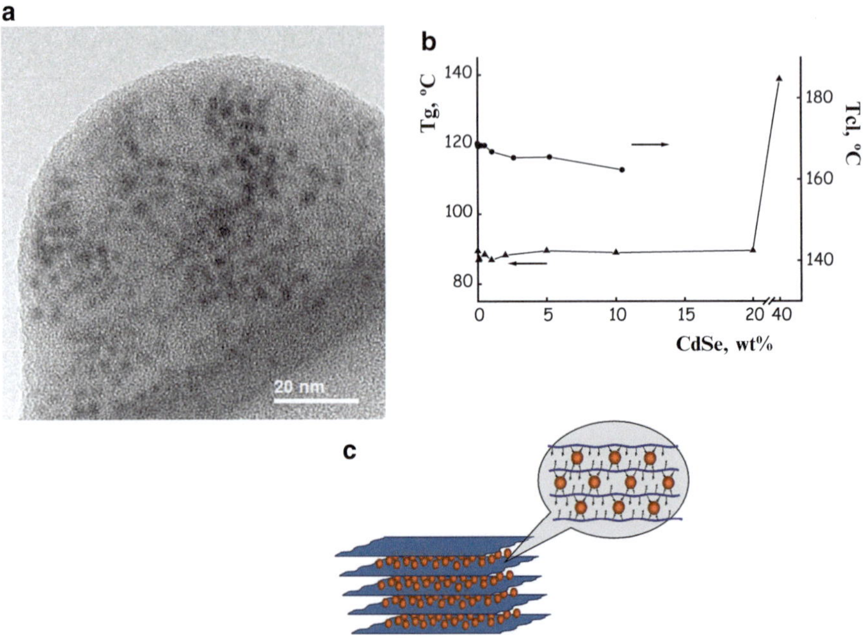


**Fig. 15.3** TEM microphotographs of 2.6 nm CdSe-BA-6PA composite containing 20 wt.% QDs (**a**) (Shandryuk et al. 2008), transition temperature vs. CdSe quantum dots content (**b**) and the model of the smectic nanocomposite structure (**c**) (Shandryuk et al. 2008)

Fig. 15.3b. The temperature $T_{S\text{-}I}$ of the phase transition drops by 8° with the incorporation of 10 wt.% QDs, whereas $T_g$ initially stays stable and then increases as the concentration of QDs rises.

Taking into account the structure of the smectic C polymer one can conclude that NPs are inserted inside the smectic layer where H-bonds are located and it makes quantum dots organized within the smectic layers (Fig. 15.3c) (Shandryuk et al. 2008). The increase in the interlayer distance and its change with the change in the size of QDs has been proved with small angle X-ray diffraction. The estimation of the average distance between QDs at 40 wt.% has shown to be equal to approximately 1–2 diameters. It indicates fairly dense packing of QDs in the layers between carboxylic groups when most of H-bond dimers are broken. It is necessary to mention that this distance is also considered in Denisyuk et al. (2010) as a limit value above which the transformation of "NPs in polymer matrix to polymer in NP matrix" takes place. In our case the LCP phase disappears after being substituted by the polymer-QD layered structure. The rise in $T_g$ observed is in good agreement with the hardening of the composites as a result of polymer cross-linking over QDs. Moreover, the calculations also show that the H-bonded smectic polymer matrix may be filled with a high amount of QDs without any significant aggregation.

Dasgupta et al. (2013) have suggested the very similar model of the nanocomposite based on the same group of H-bonded polymers already considered above, but crosslinked for the preparation of nanoporous smectic LCP network. The latter has been used for in-situ synthesis of silver NPs located in the smectic layers.

## *15.3.2 Smectic Matrix/Nanorods (NRs)*

The next step taken is to analyze the effect of change in the NP chemical structure and shape while fixing the smectic polymer matrix structure. It has been achieved with cadmium sulfide nanorods (NRs) with an average diameter 5 nm and the length ranging from 30 to 100 nm (Fig. 15.4a) (Ezhov et al. 2011). There is one

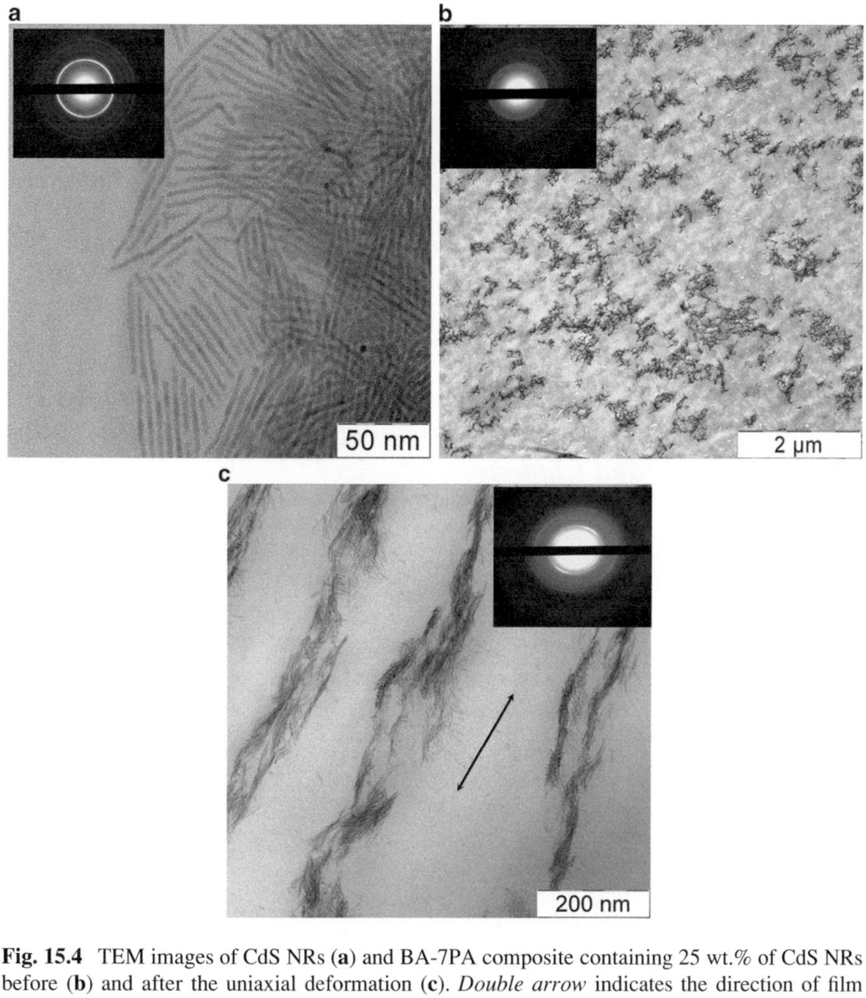


**Fig. 15.4** TEM images of CdS NRs (**a**) and BA-7PA composite containing 25 wt.% of CdS NRs before (**b**) and after the uniaxial deformation (**c**). *Double arrow* indicates the direction of film stretching and the insets show the respective ED patterns (Ezhov et al. 2011)

more feature which makes CdS NRs different from QDs discussed above, namely, the chemical structure of the ligands at the surface. If CdSe QDs are covered with oleic acid and small amounts of TOPO, CdS NRs contain just TOPO molecules at the surface (Fig. 15.2b). FTIR spectra of CdS NRs show an intense band at 1102 $cm^{-1}$ which is assigned to P=O bond strongly coordinated to CdS. As for the composites, contrary to the visible CdSe QDs–smectic polymer interaction there is no specific spectral signature of the bonding between the matrix and CdS except for the change in the relative intensity of the spectral bands at 2552 and 2670 $cm^{-1}$ which are slightly decreased in the spectrum of CdS/BA-7PA. It may be attributed to the decrease in the number of hydrogen-bonded dimers as presented in Shandryuk et al. (2002), Vasilets et al. (2004). The TEM image of the composite film (Fig. 15.4b) shows CdS NRs form small domains (100–200 nm) in the LC polymer matrix. Unlike composites with QDs, the transition temperatures do not change significantly with the insertion of as much as 38 wt.% of CdS NRs. These data confirm that no appreciable disruption of the polymer matrix takes place and that the smectic mesogenic ordering is fully preserved. On the other hand, it also confirms a fairly weak interaction between the NRs and the polymer matrixes. The uniaxial deformation of the composite films above the glass transition temperature results in a partial coalescence of small domains and the formation of elongated wire-like structures. A combination of TEM images and ED patterns clearly demonstrates that the CdS nanocrystals are mainly oriented along the stretching direction (Fig. 15.4c). Thus, NRs which do not have strong interaction with a matrix form a separate phase though the domains of 100–200 nm size are uniformly distributed within the bulk composite. However the stretching and orientation make separation more visible because NRs make long domains.

### *15.3.3 Nematic and Cholesteric Matrices*

In order to follow the H-bonds approach while changing the smectic structure of the matrix polymer by nematic or cholesteric ones, several copolymers have been synthesized and presented in Shatalova et al. (2012), Kutergina (2013), Derikov et al. (2014) (Fig. 15.1b). The major characteristics of copolymers are given in Table 15.1. Note that because the molecular weights of copolymers summarized in the table are quite low ($(5\text{–}20) \times 10^3$), we will apply further the different terminology (co-oligomers).

The FTIR spectra of Cop-D co-oligomers contain the spectral band at 1682 $cm^{-1}$ that represents C=O bond ($\nu_{CO}$) in the cyclic dimers of carboxylic groups shown in smectic systems we have discussed before (Shandryuk et al. 2002; Shatalova et al. 2003; Vasilets et al. 2004). It means that COOH-containing monomer units exist in macromolecules of co-oligomers in the form of cyclic dimers even if their content does not exceed 5 mol.%. With the increase in their content the intensity of $\nu_{CO}$ band increases. We may assume that monomers form cyclic dimers even in

**Table 15.1** Major characteristics of LC co-oligomer

| Sample | n | m | k | $M_w$ ($\times 10^3$ Da) | $M_w/M_n$ | Phase type, T, °C |
|---|---|---|---|---|---|---|
| Cop-D1 | 0.95 | 0.05 | 0 | 12.3 | 1.5 | *g* 22.8 ***SmA*** 84.4 ***N*** 110.4 *I* |
| Cop-D2 | 0.9 | 0.1 | 0 | 5.0 | 2.0 | *g* 27.4 ***SmA*** 71.4 ***N*** 104.5 *I* |
| Cop-D3 | 0.8 | 0.2 | 0 | 14.0 | 1.7 | *g*26.4 ***SmA*** 76.4 ***N*** 113.1 *I* |
| Cop-D4 | 0.7 | 0.3 | 0 | 16.1 | 1.7 | *g* 26.1 ***SmA*** 68.9 ***N*** 113.7 *I* |
| Cop-T1 | 0.6 | 0.05 | 0.35 | 20.7 | 1.5 | *g* 36.3 **Chol** 172.7 *I* |
| Cop-T2 | 0.7 | 0.05 | 0.25 | 12.6 | 1.9 | *g* 26.4 **Chol** 144.3 *I* |
| Cop-T3 | 0.8 | 0.05 | 0.15 | 5.9 | 2.2 | *g* 23.5 **Chol** 135.4 *I* |

solutions before copolymerization and after that they are embedded in a co-oligomer backbones as dimers.

Incorporation of QDs in copolymers results in both the decrease in the intensity of the 1682 $cm^{-1}$ feature and an appearance of the band at 1535 $cm^{-1}$, which indicates the formation of $COO^-$ ions. Interestingly, this band may be visualized at higher contents of QDs only because of the lower extinction in the sample with QDs (Shatalova et al. 2012; Kutergina 2013). Thus, QDs with oleic acid on the surface when embedded in co-oligomers also introduce a change in the shell ligands in favor for interaction with monomer units containing alkoxybenzoic acid groups of macromolecules, thus stabilizing the solid co-oligomer–QDs nanocomposites as was shown in smectic LCP–QDs nanocomposites.

The transition to ternary co-oligomers (Cop-T) containing alkoxybenzoic acid groups in amount of 5 mol.% with variable content of nematic- and chiral (cholesterol) units allows for the formation of cholesteric phase in a broad temperature range of Cop-T co-oligomers (Table 15.1) (Shatalova et al 2012). The formation of the cholesteric phase has been confirmed by polarized light microscopy as oil stripes texture and by UV spectra that show the selective reflection of the light characteristic for cholesterics. The concentration of alkoxybenzoic acid-containing monomer units does not exceed 5 mol.%, but FTIR spectra of these co-oligomers have the same features as discussed above for nematic counterparts. The presence of cyclic dimers has been also proved with $\nu_{CO}$ band at 1682 $cm^{-1}$ of relatively low intensity which disappears from the spectrum after QDs are embedded. Simultaneously, a small band at 1535 $cm^{-1}$ appears confirming the fact of ionic bond formation. QDs are distributed homogeneously across the matrix at least up to 3 wt.% and do not disturb the cholesteric phase. It has been proved by TEM images given in Fig. 15.5.

## 15.4 Photoluminescence of Nanocomposites

Semiconductor NPs are known to be unique because of their quantum confinement properties, strong PL and narrow emission band together with a photostability (Srivastava and Singh 2012). The size of CdSe nanocrystals was successfully

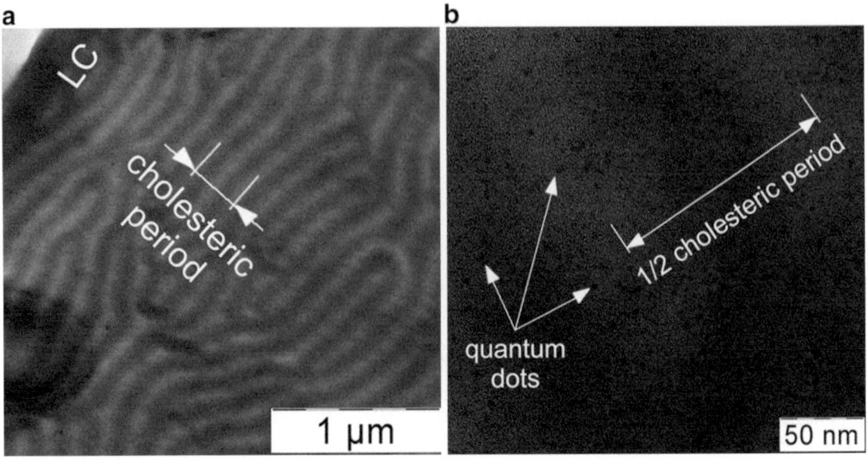


**Fig. 15.5** TEM images of a cholesteric composite of Cop-T2 containing 3 wt.% of CdSe QDs at low (**a**) and high magnification (**b**)

confined within 8 nm and capped with PVA. Composites showed optical properties associated with quantum size effects (Ma et al. 2002). Uniform CdSe–ZnS QDs functionalized with carboxy groups on the surface were used to prepare PVA-QDs nanocomposite in Suo et al. (2010). A stable emission peak for QDs and a shifting peak for PVA were observed. Polarized lights were found to have significant effect on the absorption. PL of CdSe-PVA nanocomposites with small and narrow size distribution obtained by varying the polymer concentrations were reported in Kushwaha et al. (2012). A single emission peak has been obtained at 510 nm with the intensity increasing with the increase in PVA concentration. The concentration and as a result the PL intensity increases due to lowering in particle size. Note that the polymer composites showed absorption at 563 nm, similar to the pure QDs solution. Detailed study of the effect of solvents on the emission of CdSe QDs stabilized by oleic acid ligands has shown that THF affects the desorption of ligands and the formation of surface defects more than toluene (Merekalov et al. 2011). It results in both the red shift of the PL wavelength maximum and the decrease in PL intensity. The addition of polymers of different structures in solutions of QDs changes the effectiveness of PL and the kinetics of its quenching. At the concentration of QDs below 2 wt.% the intensity of PL increases and attains its maximum value and then drops down. This tendency has been described by both the increase in the amount of defects and quenching resulted from the energy transfer on macromolecules. These findings are of great importance because they indicate the conditions of composites preparation if it is related to films cast from solutions.

The role of the structure of LCP matrix on PL properties of CdSe QDs has been studied in details in Tselikov et al. (2013, 2015), Tselikov (2013). Figure 15.6a shows the PL spectra of CdSe QDs embedded in a smectic polymer. These spectra contain a peak caused by the radiative recombination of excitons in CdSe QDs. The position of this peak shifts from 540 to 560 nm as the QD concentration increases

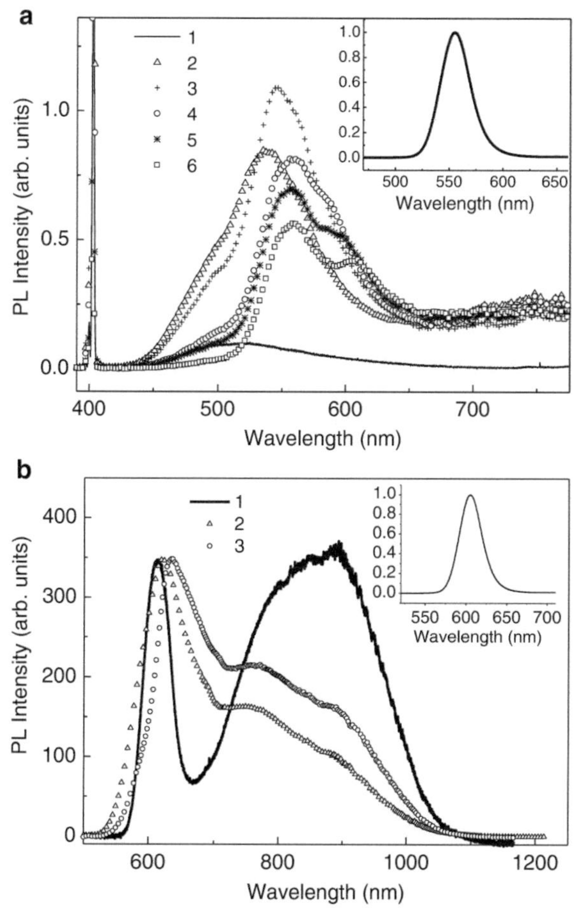


**Fig. 15.6** Photoluminescence spectra of smectic LCP composites with various contents of CdSe QDs (3.2 nm) (**a**) (Tselikov et al. 2013) and of co-oligomers (**b**) Cop-T2 (1) and Cop-D2 (2, 3) containing 10 wt.% content of CdSe QDs (4.2 nm) measured at 25 (1, 2) and 85 °C (3)

from 0.1 to 5 wt.%. The red shift of the exciton PL band can be explained by the effect of light reabsorption over the QD array. It should be noted that the exciton PL band intensity reaches a maximum at a QD weight concentration of 0.5 wt.%. A further increase in the CdSe QD concentration results in a decrease in the intensity of this band. This fact has been explained by the energy transfer from QDs to the smectic LC polymer matrix (Tselikov et al. 2013). One can see the exciton PL peak to exhibit a long wavelength shoulder with maximum changing from 600 to 620 nm. Contrary to the basic peak, its intensity increases with the QD concentration. The appearance of this band is probably caused by the effect of new electronic states at the CdSe/LCP interface.

As mentioned before, the introduction of alien monomer units in macromolecules of the H-bonded polymer results in the change in LC structure of CP-D copolymers (Table 15.1, Fig. 15.1b). The presence of these units in the amount of 70–95 mol.% leads to the appearance of $S_A$ and N phases in the polymer matrix. The shape of the PL spectrum significantly changes (Fig. 15.6b) in comparison with that of the composite having $S_C$ structure (Fig. 15.6a). Contrary to the above mentioned smectic polymer composites, the co-oligomer matrices are filled with CdSe QDs of the different size (4.2 nm). That is why the spectral band observed at 600–650 nm has been attributed to the radiative recombination of excitons in CdSe QDs (Tselikov 2013; Tselikov et al. 2014). The corresponding band in the PL spectrum of the colloidal QDs of that size moves from 605 nm (in solution) (Fig. 15.6b) up to 646 nm when QDs are embedded in a smectic CP-D copolymer (Fig. 15.6b, spectrum 3). The heating of the sample above the temperature of the S-N phase transition leads to the change in the PL spectrum. As an example, the exciton PL peak of Cop-D2 composite with CdSe QD (10 wt.%) is shifted from 647 down to 620 nm (Fig. 15.6b) when the transition of the matrix from the smectic to the nematic phase occurs. This effect may be caused by a decrease in light reabsorption resulted from an increase in the interparticle distance due to changes in LC order.

The incorporation of cholesterol-containing monomer units (Fig. 15.1b, Table 15.1) leads to the change of the mesophase structure in CP-T co-oligomers for the cholesteric LC phase existing in a broad temperature range (Table 15.1). The embedment of QDs in the amount of 2–3 wt.% in cholesterol-containing co-oligomers does not disrupt the cholesteric phase as proved by the presence of the selective reflection stop-zone in composites (Kutergina 2013). However the cholesterol containing composite with 10 wt.% of CdSe QDs does not selectively reflect light, which has been interpreted in terms of the possible destruction of the helical structure characteristic for cholesterics. This may result in the nematic order of the co-oligomer matrix. As for the PL properties, the spectrum of such a composite like Cop-T2 with 10 wt.% of QDs (Fig. 15.6b, spectrum 1) it is very similar to that of the nematic Cop-D co-oligomer composite (spectrum 3) but with a spectral band even more shifted down to 605–610 nm, which is almost the same as that of CdSe QDs in solution.

PL kinetics of nanocomposite systems of different structure with various QD concentrations has been measured at the wavelength corresponding to the exciton PL band maximum peak in Tselikov et al. (2013, 2014), Tselikov (2013). The kinetics curves have been approximated by the sum of the exponents:

$$I = I_1 \exp(-t/\tau_1) + I_2 \exp(-t/\tau_2),$$

where I is the PL intensity, t is the time, $\tau_1$ and $\tau_2$ are the PL lifetimes.

In accordance with (de Mello et al. 2006) the biexponential PL decay of CdSe QDs is generally associated with the recombination of bright and dark excitons in QDs. The lifetimes $\tau_1$ and $\tau_2$ were calculated for the PL of CdSe QDs in smectic and cholesteric matrices yielding 10 and 50 ns and 14 and 70 ns, respectively. The comparison of $\tau_1$ and $\tau_2$ with those of colloidal QDs in solution (15 and 80 ns) indicates that the bright and dark excitons in QDs are sensitive to the matrix properties.

In addition to radiative exciton recombination process discussed above, spectra in Fig. 15.6b contain poorly resolved band in the low-energy spectral region which shows at least two maxima at 750–800 nm and 900 nm. The broad maximum at 750–800 nm results from surface defects of QDs similar to those observed before for selenium vacancies (Babentsov 2006). As for the maximum at 900 nm it may be due to defects at the CdSe/LC matrix interface.

Coming back to CdS NRs in smectic LCP matrix (Fig. 15.4), we will further summarize what is going on with PL. The position of the CdS NRs emission peak in an LCP matrix is nearly the same as it is in a homogeneous solution (Fig. 15.7a). The number of defects is much higher than in CdS NRs in solution and the trap emission is also very high. When the PL of CdS NRs in LCP matrix has been induced at a wavelength longer than the PL emission peak (Ezhov et al. 2011) an

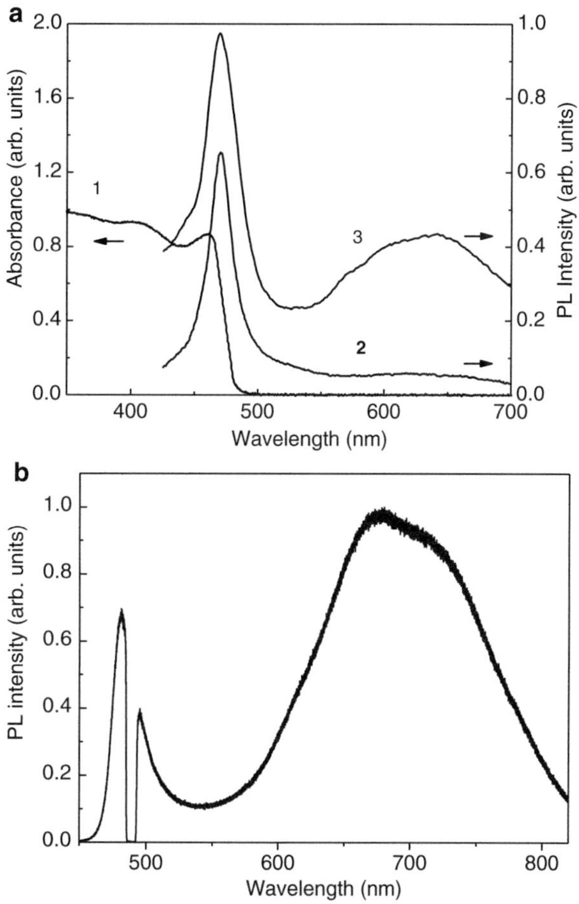


**Fig. 15.7** UV–VIS absorption (1) and PL spectra (excitation wavelength 400 nm) of CdS NRs in heptane solution (2), in LC polymer (BA-10PA) matrix (3) at 2 wt.% content (**a**) and in BA-7PA matrix with 50 wt.% of CdS NRs (**b**) (excitation wavelength 488.5 nm)

anti-Stokes PL (ASPL) has been observed. As an example, the PL spectrum of H-bonded polymer composite containing 33 wt.% of CdS NRs obtained under the excitation with the $Ar^+$- laser (488 nm) is given in Fig. 15.7b. The intensity of the anti-Stokes part of the spectrum is comparable to Stokes luminescence, whereas the PL maximum is located below the wavelength of the excitation source. The nature of ASPL in nanocomposites has been established as involvement of thermally populated states.

## 15.5 Conclusions and Future Perspective

The organic/inorganic hybrid nanocomposites with LC matrix show interesting features resulting in modification of the electro-optical properties of LCs. However these systems cannot hold high contents of NPs and reach thermodynamic stability that leads to macrophase separation. The latter may be avoided if LCPs serve as matrices in which the kinetic stability and homogeneous distribution of NPs is achieved. New approaches for the development of optical materials are already shown for thermodynamically stable nanocomposites-containing NPs of metal chalcogenides and LCP matrices with variable structure and composition. Via modification of the chemical structure of LC polymers one can modify the properties of nanocomposites leading to the change in LC structure type. The nanocomposites of smectic LCPs with QDs demonstrate fairly dense packing of QDs organized into layered structures. The latter provides the red-shift of the exciton PL band due to the radiative recombination of excitons in CdSe QDs of QDs relatively to that of colloidal QDs in solution. The major reason for that is the light reabsorption over the QD array. The transition from smectic to nematic structure of the matrix may lead to the blue shift of PL band, which can be achieved at low packing density of QDs and lower concentration of carboxy groups in co-oligomers. Kinetic characteristics of PL also depend on the type of LC mesophase. Despite the marked progress in this field of research we still do not have qualitative and quantitative answers on the following questions: what surface chemistry of NPs and what type of polymer structure is necessary to provide the thermodynamic stability of nanocomposites; what is the concentration limit of NPs to provide the kinetic stability of composites; should NPs be simply imbedded in the matrix or additionally localized via chemical bonds established between macromolecules and the elements of NP surface; what amount (concentration) of NPs should be embedded in a polymer to improve the properties of composite nanomaterials; what new properties can be expected in nanocomposites with controllable thermodynamic stability and photoluminescence. In the future, it is important to focus on both theoretical and experimental approaches that will provide direct control over the preparation of nanocomposites when the matrix medium is capable of (a) localization of NPs, (b) embedding the optimum amount of NPs, (c) distributing it inside the matrix and providing the kinetic control of stability. Understanding this control will enable a new generation of the nanomaterials with desired structure and functional properties.

## References


Aissou K, Fleury G, Pecastaings G, Alnasser T, Mornet S, Goglio G, Hadziioannou G (2011) Hexagonal-to-cubic phase transformation in composite thin films induced by FePt nanoparticles located at PS/PEO interfaces. Langmuir 27:14481–14488

Babentsov VN (2006) Defects with deep donor and acceptor levels in nanocrystals of CdTe and CdSe. Semiconductor Phys Quantum Electron Optoelectron 9(3):94–98

Badamshina ER, Gafurova MP, Estrin YI (2010) Modification of carbon nanotubes and synthesis of polymeric composites involving the nanotubes. Russ Chem Rev 79:945–980

Barmatov EB, Pebalk DA, Barmatova MV (2005) Influence of silver nanoparticles on the phase behavior of side-chain liquid crystalline polymers. Langmuir 20(25):10868–10871

Barmatov EB, Pebalk DA, Barmatova MV (2006a) Influence of silver nanoparticles on the order parameter of liquid crystalline polymers. Liq Cryst 33:1059–1063

Barmatov EB, Medvedev AS, Pebalk DA, Barmatova MV, Nikonorova NA, Zezin SB, Shibaev VP (2006b) The effect of silver nanoparticles on the phase state of comb-shaped liquid crystalline polymers with cyanobiphenyl mesogenic groups. Polym Sci A 48(7):665–675

Barmatov EB, Pebalk DA, Barmatova MV (2007) New polymer liquid-crystalline CdS nanocomposites forming a chiral nematic phase. Polym Sci B 49(1):47–50

Bellini T, Buscaglia M, Chiccoli C, Mantegazza F, Pasini P, Zannoni C (2000) Nematics with quenched disorder: what is left when long range order is disrupted. Phys Rev Lett 31:1008–1011

Böker A, He J, Emrick T, Russell TP (2007) Self assembly of nanoparticles at interfaces. Soft Matter 3:1231–1247

Bronstein LM, Sidorov SN, Valetsky PM (2004) Nanostructured polymeric systems as nanoreactors for nanoparticle formation. Russ Chem Rev 73(5):501–515

Buhro WE, Colvin VL (2003) Semiconductor nanocrystals: shape matters. Nature Mater 2 (3):138–139

Chandran A, Prakash J, Naik KK, Srivastava AK, Dąbrowski R, Czerwiński M, Biradar AM (2014) Preparation and characterization of MgO nanoparticles/ferroelectric liquid crystal composites for faster display devices with improved contrast. J Mater Chem C 2:1844–1853

Chatterjee T, Mitchell CA, Hadjiev VG, Krishnamoorti R (2012) Oriented single-walled carbon nanotubes–poly(ethylene oxide) nanocomposites. Macromolecules 45:9357–9363

Chiu JJ, Kim BJ, Kramer EJ, Pine DJ (2005) Control of nanoparticle location in block copolymers. J Am Chem Soc 127:5036–5037

Coleman JN, Khan U, Gun'ko YK (2006) Mechanical reinforcement of polymers using carbon nanotubes. Adv Mater 18(6):689–706

Dasgupta D, Shishmanova IK, Ruiz-Carretero A, Lu K, Verhoeven V, van Kuringen HPC, Portale G, Leclère P, Bastiaansen CWM, Broer DJ, Schenning APHJ (2013) Patterned silver nanoparticles embedded in a nanoporous smectic liquid crystalline polymer network. J Am Chem Soc 135(30):10922–10925

de Mello DC, Bode M, Meijering A (2006) Size and temperature dependence of exciton lifetimes in CdSe quantum dots. Phys Rev B 74:085320/1–085320/9

Dekker C (1999) Carbon nanotubes as molecular quantum wires. Phys Today 52(5):22–28

Denisyuk IY, Fokina M (2010) A review of high nanoparticles concentration composites: semiconductor and high refractive index materials, chap 5. In: Masuda Y (ed) Nanocrystals. InTech, Rijeka, pp 109–142

Derikov YI, Kutergina IY, Shandryuk GA, Merekalov AS, Gorkunov MV, Abramchuk SS, Ezhov AA (2014) Stable non-equilibrium composites composed of liquid crystal polymers and CdSe quantum dots. Polym Sci A56(4):488–497

Dolmatov VY (2007) Detonation-synthesis nanodiamonds: synthesis, structure, properties and application. Russ Chem Rev 76(4):339–360

Duran H, Gazdecki B, Yamashita A, Kyu T (2005) Effect of carbon nanotube on phase transitions of nematic liquid crystals. Liq Cryst 32:815–822

Ezhov AA, Shandryuk GA, Bondarenko GN, Merekalov AS, Abramchuk SS, Shatalova AM, Manna ZER, Talroze RV (2011) Liquid crystalline polymer composites with CdS nanorods: structure and optical properties. Langmuir 27(21):13353–13360

Fogg DE, Radzilowski LH, Blanski R, Schrock RR, Thomas E (1997) Fabrication of quantum dot/polymer composites: phosphine-functionalized block copolymers as passivating hosts for cadmium selenide nanoclusters. Macromolecules 30(3):417–426

Garbovskiy Y, Glushchenko A (2013) Optical/ferroelectric characterization of BaTiO3 and PbTiO3 colloidal nanoparticles and their applications in hybrid materials technologies. Appl Opt 52(22):E34–E39

Gerasin VA, Antipov EM, Karbushev VV, Kulichikhin VG, Karpacheva GP, Talroze RV, Kudryavtsev YV (2013) New approaches to the development of hybrid nanocomposites: from structural materials to high-tech applications. Russ Chem Rev 82(4):303–332

Gorkunov MV, Osipov MA (2011) Mean-field theory of a nematic liquid crystal doped with anisotropic nanoparticles. Soft Matter 7:4348–4356

Gorkunov MV, Shandryuk GA, Shatalova AM, Kutergina IY, Merekalov AS, Kudryavtsev YV, Talroze RV, Osipov MA (2013) Phase separation effects and the nematic–isotropic transition in polymers and low molecular weight liquid crystals doped with nanoparticles. Soft Matter 9 (13):3578–3588

Gullapalli S, Wong MS (2011) Nanotechnology: a guide to nano-objects. Chem Eng Prog 107 (5):28–32

Gupta S, Zhang Q, Emrick T, Balazs AC, Russell TP (2006) Entropy-driven segregation of nanoparticles to cracks in multilayered composite polymer structures. Nature Mater 5 (3):229–233

Horechyy A, Nandan B, Zafeiropoulos NE, Formanek P, Oertel U, Bigall NC, Eychmüller A, Stamm M (2013) A step-wise approach for dual nanoparticle patterning via block copolymer self-assembly. Adv Funct Mater 23(4):483–490

Iakoubovskii K, Baidakova MV, Wouters BH, Stesmans A, Adriaenssens GJ, Vul AYa, Grobet PJ (2000) Structure and defects of detonation synthesis nanodiamond. Diamond Relat Mater 9 (3–6):861–865

Joshi T, Kumar A, Prakash J, Biradar AM (2010) Low frequency dielectric relaxations of gold nanoparticles/ferroelectric liquid crystals composites. Liq Cryst 37:1433–1437

Khatua S, Manna P, Chang WS, Tcherniak A, Friedlander E, Zubarev ER, Link S (2010) Plasmonic nanoparticles–liquid crystal composites. J Phys Chem C 114:7251–7257

Kobayashi S, Toshima N (2007) Nanoparticles and LCDs: it's a surprising world SID. Inf Display 23:26–32

Kobayashi S, Sano S, Miyama T, Takatoh K, Toshima N (2008) Enhancement of the characteristics of LCDs by doping nanoparticles reduction of driving voltage, viscosity, and response times. Proc SPIE 5008:23–29

Kopcansky P, Tomašovičová N, Koneracká M, Timko M, Mitróová Z, Závišová V, Éber N, Fodor-Csorba K, Tóth-Katona T, Vajda A, Jadzyn J, Beaugnon E, Chaud X (2010) Structural phase transition in liquid crystal doped with gold nanoparticles. Acta Phys Pol A 118:988–989

Kopcansky P, Tomasovicova N, Koneracka M, Timko M, Zavisova V, Dzarova A, Jadzyn J, Beaugnon E, Chaud X (2011) Phase transitions in liquid crystal doped with magnetic particles of different shapes. Int J Thermophys 32:807–817

Kumar A, Prakash J, Mehta DS, Biradar AM, Haase W (2009) Enhanced photoluminescence in gold nanoparticles doped ferroelectric liquid crystals. Appl Phys Lett 95:023117–023121

Kushwaha K, Gautam N, Singh P, Ramrakhaini M (2012) Synthesis and photoluminescence of CdSe/PVA nanocomposites. J Phys 365, ID12014
Kutergina IYu (2013) Composites based on LC copolymers and CdSe quantum dots. PhD Thesis, D Mendeleyev University of Chemical Technology of Russia, Moscow (Russia)
Lagashetty A, Venkataraman A (2005) Polymer nanocomposites. Resonance 10(7):49–60
Lavrič M, Cordoyiannis G, Kralj S, Tzitzios V, Nounesis G, Kutnjak Z (2013) Effect of anisotropic $MoS_2$ nanoparticles on the blue phase range of a chiral liquid crystal. Appl Opt 52(22):E47–E52
Li QF, He JB, Glogowski E, Li XF, Wang J, Emrick T, Russell TP (2008) Responsive assemblies: gold nanoparticles with mixed ligands in microphase separated block copolymers. Adv Mater 20:1462–1466
Liedel C, Schindler RA, Pavan MJ, Lewin C, Pester CW, Ruppel M, Urban VS, Shenhar R, Böker A (2013) Electric-field-induced alignment of block copolymer/nanoparticle blends. Small 9 (19):3276–3281
Lin Y, Böker A, Sill K, Xiang H, Abetz C, Li X, Wang J, Emrick T, Balazs AC, Russell TP (2005) Self-directed self-assembly of nanoparticle/copolymer mixtures. Nature 434(1):55–59
Lopatina LM, Selinger JM (2011) Maier-Saupe-type theory of ferroelectric nanoparticles in nematic liquid crystals. Phys. Rev E 84:041703/1–041703/7
Lorenz A, Zimmermann N, Kumar S, Evans DR, Cook G, Martínez MF, Kitzerow H-S (2013) X-ray scattering of nematic liquid crystal nanodispersion with negative dielectric anisotropy. Appl Opt 52(22):E1–E5
Ma X-D, Qian X-F, Yin J, Xi H-A, Zhu Z-K (2002) Preparation and characterization of polyvinyl alcohol-capped CdSe nanoparticles at room temperature. J Coll Interface Sci 252(1):77–81
Mark JE (1996) Ceramic-reinforced polymers and polymer-modified ceramics. Polym Eng Sci 36:2905–2920
Martel R, Derycke V, Lavoie C, Appenzeller J, Chan K, Tersoff J, Avouris P (2001) Ambipolar electrical transport in semiconducting single-wall carbon nanotubes. Phys Rev Lett 87 (25):2568051–2568054
Mashkour M, Kimura T, Kimura F, Mashkour M, Tajvidi M (2014) Tunable self-assembly of cellulose nanowhiskers and polyvinyl alcohol chains induced by surface tension torque. Biomacromolecules 15(1):60–65
Merekalov AS, Kuptsov SA, Shandryuk GA, Bezborodov VS, Terentjev EM, Talroze RV (2001) Antiferroelectric alignment and mechanical director rotation in a hydrogen-bonded chiral $SmC_A$* elastomer. Liq Cryst 28(4):495–502
Merekalov AS, Shandryuk GA, Vasil'ev RB, Bykov VI, Bondarenko GN, Shatalova AM, Otmakhova OA, Talroze RV (2011) Photoluminescence of cadmium selenide quantum dots in polymer solutions. Polym Sci B 53(9–10):553–561
Meuer S, Oberle P, Theato P, Tremel W, Zentel R (2007) Liquid crystalline phases from polymer-functionalized $TiO_2$ nanorods. Adv Mater 19:2073–2078
Meuer S, Fischer K, Mey I, Janshoff A, Schmidt M, Zentel R (2008) Liquid crystals from polymer-functionalized $TiO_2$ nanorod mesogens. Macromolecules 41:7946–7952
Misner MJ, Skaff H, Emrick T, Russell TP (2003) Directed deposition of nanoparticles using diblock copolymer templates. Adv Mater 15:221–224
Moniruzzaman M, Winey KI (2006) Polymer nanocomposites containing carbon nanotubes. Macromolecules 39:5194–5205
Murray CB, Kagan CR, Bawendi MG (2000) Synthesis and characterization of monodisperse nanocrystals and close-packed nanocrystal assemblies. Ann Rev Mater Res 30(1):545–610
Nikonorova NA, Barmatov EB, Pebalk DA, Barmatova MV, Domínguez-Espinosa G, Diaz-Calleja R, Pissis P (2007) Electrical properties of nanocomposites based on comb-shaped nematic polymer and silver nanoparticles. J Phys Chem C 111(24):8451–8458
Okamoto M (2004) Encyclopedia of nanoscience and nanotechnology. In: Nalwa HS (ed) Polymer/clay nanocomposites, vol 8. American Scientific, Stevenson Ranch, pp 791–843
Osipov MA, Gorkunov MV (2014) Molecular theory of phase separation in nematic liquid crystals doped with spherical nanoparticles. ChemPhysChem 15(7):1496–1501

Palchoudhury S, An W, Xu Y, Qin Y, Zhang Z, Chopra N, Holler RA, Turner CH, Yu B (2011) Synthesis and growth mechanism of iron oxide nanowhiskers. Nano Lett 11(3):1141–1146

Park SH, Kim BJ, Hawker CJ, Kramer EJ, Bang J, Ha JS (2007) Controlled ordering of block copolymer thin films by the addition of hydrophilic nanoparticles. Macromolecules 40:8119–8124

Paul DR, Robeson LM (2008) Polymer nanotechnology: nanocomposites. Polymer 49 (15):3187–3204

Pavan MJ, Shenhar R (2011) Two-dimensional nanoparticle organization using block copolymer thin films as templates. J Mater Chem 21:2028–2040

Pérez-Juste J, Rodriguez-Gonzales B, Mulvaney P, Liz-Marzan LM (2005) Optical control and patterning of gold nanorod-PVA nanocomposite films. Adv Funct Mater 15:1065–1071

Pester CW, Ruppel M (2013) Block copolymer nanocomposites in electric fields: kinetics of alignment. ACS Macro Lett 2(1):53–58

Pomogailo AD (2000) Hybrid polymer-inorganic nanocomposites. Russ Chem Rev 69(1):53–80

Pomogailo AD, Rozenberg AS, Dzhardimalieva GI (2011) Thermolysis of metallopolymers and their precursors as a method for the preparation of nanocomposites. Russ Chem Rev 80 (3):257–292

Qi H, Kinkead B, Hegmann T (2008) Effects of functionalized metal and semiconductor nanoparticles in nematic liquid crystals. Proc SPIE 6911:691106–691117

Rajh D, Shelestiuk S, Mertelj A, Mrzel A, Umek P, Irusta S, Zak A, Drevenšek-Olenik I (2013) Effect of inorganic 1D nanoparticles on electrooptic properties of 5CB liquid crystal. Phys Status Solidi A 210:2328–2334

Rossetti R, Nakahara S, Brus LE (1983) Quantum size effects in the redox potentials, resonance Raman spectra, and electronic spectra of CdS crystallites in aqueous solutions. J Chem Phys 79:1086–1095

Rudzki A, Evans DR, Cook G, Haase W (2013) Size dependence of harvested $BaTiO_3$ nanoparticles on the electro-optic and dielectric properties of ferroelectric liquid crystal nanocolloids. Appl Opt 52(22):E6–E14

Saint-Aubin K, Poulin P, Jaillet C, Maugey M, Zacri C (2013) Changes of morphology and properties of block copolymers induced by carbon nanotubes. Polymer 54:2285–2291

Shandryuk GA, Koval MV, Kuptsov SA, Sasnovskii GM, Talroze RV, Platé NA (2002) Hydrogen bonding as a method of modifying liquid-crystalline polymers and other compounds. Polym Sci A44:267–274

Shandryuk GA, Rebrov AV, Vasiliev RB, Dorofeev SG, Merekalov AS, Gaskov AM, Talroze RV (2005) Stabilization of cadmium selenide nanoparticles in a liquid-crystalline polymer. Polym Sci B 47(9–10):266–268

Shandryuk GA, Matukhina EV, Vasiliev RB, Rebrov AV, Bondarenko GN, Merekalov AS, Gaskov AM, Talroze RV (2008) Effect of H-bonded liquid crystal polymers on CdSe quantum dot alignment within nanocomposite. Macromolecules 41:2178–2185

Shatalova AM, Shandryuk GA, Bondarenko GN, Kuptsov SA, Talroze RV, Platé NA (2003) Structure and thermal behavior of hydrogen-bonded liquid-crystalline polymers. Polym Sci 45 (2):135–141

Shatalova AM, Kutergina IY, Derikov YI, Shandryuk GA, Talroze RV (2012) Matrices based on acrylic liquid crystalline copolymers for the design of composites with quantum dots. Polym Sci B 54(11–12):533–541

Shenhar R, Norsten TB, Rotello VM (2005) Polymer-mediated nanoparticle assembly: structural control and applications. Adv Mater 17(6):657–669

Sinha G, Glorieux C, Thoen J (2004) Broadband dielectric spectroscopy study of molecular dynamics in the glass-forming liquid crystal isopentylcyanobiphenyl dispersed with aerosils. Phys Rev E 69:031707–031712

Špírková M, Duchek P, Strachota A, Poręba R, Kotek J, Baldrian J, Šlouf M (2011) The role of organic modification of layered nanosilicates on mechanical and surface properties of organic–inorganic coatings. J Coat Technol Res 8(3):311–328

Srivastava P, Singh K (2012) Synthesis of CdSe nanoparticles by solvothermal route: structural, optical and spectroscopic properties. Adv Mater Lett 3(4):340–344

Suo B, Su X, Wu J, Chen D, Wang A, Guo Z (2010) Poly (vinyl alcohol) thin film filled with CdSe-ZnS quantum dots: fabrication, characterization and optical properties. Mater Chem Phys 119 (1–2):237–242

Talroze RV, Shandryuk GA, Matukhina EV, Vasiliev RB, Merekalov AS, Gaskov AM (2007) Liquid crystal polymers as directing medium: creation of organized quantum dots composites. Polym Preprints 48(2):775–776

Thompson KB, Ginzburg VV, Matsen MW, Balazs AC (2002) Block copolymer-directed assembly of nanoparticles: forming mesoscopically ordered hybrid materials. Macromolecules 35 (3):1060–1071

Thurn-Albrecht T, Steiner R, DeRouchey J, Stafford M, Huang E, Bal M, Tuominen M, Hawker CJ, Russell TP (2000) Nanoscopic templates from oriented block copolymer films. Adv Mater 12:787–791

Tselikov GI (2013) Influence of impurities and molecular environment on optical properties of CdSe quantum dots. Ph.D. Thesis, MV Lomonosov Moscow State University, Moscow (Russia)

Tselikov GI, Timoshenko VY, Plenge J, Rühl E, Shatalova AV, Shandryuk GA, Merekalov AS, Talroze RV (2013) Photoluminescence properties of cadmium selenide quantum dots embedded in a liquid-crystal polymer matrix. Semiconductors 47(5):647–649

Tselikov GI, Timoshenko VYu, Golovan' LA, Plenge J, Shatalova AV, Shandryuk GA, Kutergina I, Merekalov AS, Rühl E, Talroze RV (2015) Role of the polymer matrix on the photoluminescence of embedded CdSe quantum dots. ChemPhysChem. 16:1071–1078

Vasilets VN, Shandryuk GA, Savenkov GN, Shatalova AM, Bondarenko GN, Talroze RV, Plate NA (2004) Liquid crystal polymer brush with hydrogen bonds: structure and orientation behavior. Macromolecules 37(10):3685–3688

Yabu H, Jinno T, Koike K, Higuchi T, Shimomura M (2011a) Three-dimensional assembly of gold nanoparticles in spherically confined microphase-separation structures of block-copolymers. Macromolecules 44:5868–5873

Yabu H, Jinno T, Koike K, Higuchi T, Shimomura M (2011b) Nanoparticle arrangements in block-copolymer particles having microphase separated structures. J Polym Sci B 49:1717–1722

Yaroshchuk O, Tomylko S, Gvozdovskyy I, Yamaguchi R (2013) Cholesteric liquid crystal–carbon nanotube composites with photo-settable reversible and memory electro-optic modes. Appl Opt 52(22):E53–E59

Zhang QL, Gupta S, Emrick T, Russell TP (2006) Surface-functionalized CdSe nanorods for assembly in diblock copolymer templates. J Am Chem Soc 128:3898–3899

Zorn M, Meuer S, Nawaz TM, Khalavka Y, Sönnichsen C, Tremel W, Zentel R (2008) Liquid crystalline phases from polymer functionalized semiconducting nanorods. J Mater Chem 18:3050–3056

# Chapter 16
# Side Chain Liquid Crystalline Polymers: Advances and Applications

**G. Siva Mohan Reddy, J. Jayaramudu, S.S. Ray, K. Varaprasad, and E. Rotimi Sadiku**

## 16.1 Liquid Crystals (LCs)

Matters comprise of molecules that exist in three different phases—solids, liquids and gases depending on their arrangement. The arrangement of molecules in a crystalline solid leads to long-range positional and orientational orders. In isotropic liquid, on the other hand, the molecules are arranged randomly and possess no order and flow freely. However, it was observed that a certain class of molecules organise in a different manner in order to exhibit a new phase, popularly known as liquid crystalline phase. Molecules in liquid crystalline phase self-organize, leading to positional/orientation order. Macroscopically, the liquid crystalline phase exhibits flow and appears like an opaque fluid.

---

G. Siva Mohan Reddy (✉) • S.S. Ray
Department of Applied Chemistry, University of Johannesburg, Doornfontein 2028, Johannesburg, South Africa

DST/CSIR-National Centre for Nano-structured Materials, Council for Scientific and Industrial Research, Pretoria 0001, South Africa
e-mail: smreddyg@gmail.com; sgoddeti@csir.co.za

J. Jayaramudu
DST/CSIR-National Centre for Nano-structured Materials, Council for Scientific and Industrial Research, Pretoria 0001, South Africa

K. Varaprasad
Departamento Ingeniería de Materiales-DIMAT, Universidad de Concepcion, Concepcion, Chile

E. Rotimi Sadiku
Department of Chemical, Metallurgical and Materials Engineering (Polymer Division), Tshwane University of Technology, CSIR Campus, Pretoria 0040, South Africa



V.K. Thakur, M.R. Kessler (eds.), *Liquid Crystalline Polymers*,
DOI 10.1007/978-3-319-20270-9_16

## 16.2 Historical Background of Liquid Crystals

In 1888, the Austrian botanical physiologist Reinitzer (1888), while working at the German University of Prague, extracted cholesterol from carrots in order to establish its chemical formula. Reinitzer examined the physico-chemical properties of various derivatives of cholesterol. A number of workers previously observed some distinct color effects on cooling cholesterol derivatives just above the solidification temperature. Reinitzer himself found the same phenomenon in cholesteryl benzoate. However, the colors, near the solidification of cholesteryl benzoate were not the most peculiar feature and he finally detected that cholesteryl benzoate, does not melt like other compounds, as it indicated two melting points. At 145.5 °C, it melted into a cloudy liquid and at 178.5 °C, it melted again and the cloudy liquid suddenly became clear. Furthermore, the phenomenon was reversible.

In 1908, Vorlander started a research group working on LCs and demonstrated the principles of molecular design that highlighted the field. The heroic period of the liquid crystalline state comes to an end around 1920. By this time, a large amount of data on LCs had been collected but LCs was not popular among scientists in the early twentieth century and the material remained as scientific curiosity. Until 1957, all were quiet on the LC front and nothing new could be expected in this area. Brown and Shaw (1957) published an article on the LC phase, which subsequently sparked an international resurgence in LC research that later developed into a mature field of science (Schadt 1989). Later research efforts yielded various interesting developments, which expanded the field of research enormously.

In 1991, when liquid crystal displays were well established in our day-to-day life, while de Gennes (1975, 1984) received the Nobel Prize in Physics. He discovered that "methods developed for studying ordered phenomena in simple systems can be generalized to more complex forms of matter, in particular, to liquid crystals and polymers".

Chandrasekhar et al. (1977) reported that the disc shaped molecules exhibited the liquid crystalline properties. The chronological development from conventional rod and disc shaped to non-conventional bent-shaped, which showed polar order, was incorporated. The concept of supramolecular chemistry has broadened the design of LCs (Lehn 1988; Scherman 2009). A new molecular design from nano to macro scale is important in order to enlarge the functional capabilities of LCs.

## 16.3 Classification of Liquid Crystals

Materials that exhibits mesophases can be broadly classified into two categories: lyotropic and thermotropic liquid crystals (Reddy et al. 2014a). In lyotropics the mesomorphism is influenced by the action of solvents on amphiphiles, while in the case of thermotropic liquid crystals, the transitions are brought about by the action of heat.

### 16.3.1 Lyotropic Liquid Crystals

Lyotropic liquid crystals are obtained when an appropriate concentration of a material is dissolved in a solvent (Collings and Hird 1997; Neto et al. 2005). The most common systems are those formed by water and amphiphilic molecules (molecules that possess a hydrophilic part that interacts strongly with water and a hydrophobic part that is water insoluble), such as: soaps, detergents and lipids. There are a number of phases observed in such water-amphiphilic systems, as the composition and temperature are varied; some appear as spherical micelles, and others possess ordered structures with one-, two-, or three-dimensional positional order. Examples of these kinds of molecules are soaps and various phospholipids, like those present in cell membranes (Lukask and Harden 1985).

### 16.3.2 Thermotropic Liquid Crystals

Thermotropic liquid crystals are those that exhibit mesophase in a certain temperature range. If the temperature is raised too high, thermal motion will destroy the ordering of the liquid crystalline phase, resulting in isotropic (liquid) phase. At sufficiently low temperature, most liquid crystalline materials will form a (anisotropic) crystal. The transition from crystal to mesophase is called melting point, while that from the mesophase to the isotropic liquid is termed the clearing point. Many thermotropic liquid crystals exhibit a variety of phases as the temperature changes. Thermodynamically stable LC phases that occur during heating and cooling processes are called enantiotropic phases, while, thermodynamically unstable mesophases appear during the cooling process due to hysteresis in the crystallization, are referred to as monotropic. These thermotropic LCs can be broadly classified into two categories, determined by the shapes of their molecules. They are: conventional and non-conventional shaped liquid crystals.

## 16.4 Shape of Molecules and Their Mesophases

### 16.4.1 Conventional Shaped Liquid Crystal

The conventional liquid crystals are broadly classified into two types, namely; calamitic liquid crystals and discotic liquid crystals.

#### 16.4.1.1 Calamitic Liquid Crystals

The most commonly encountered LCs is often called calamitic LCs. They consist of rod-like molecules with one molecular axis longer than the other two axes.

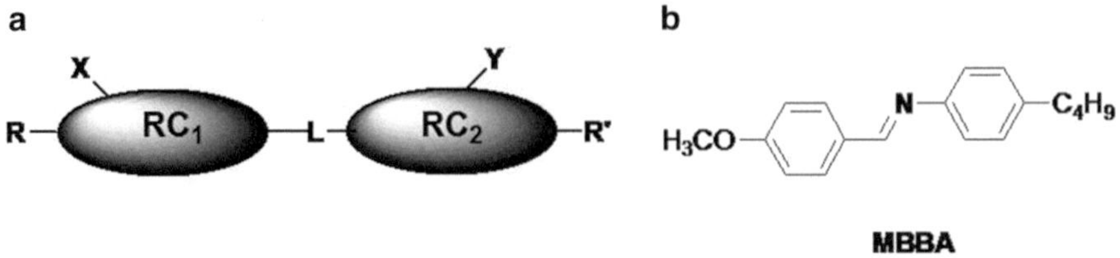


**Fig. 16.1** (**a**) A general template for molecular structure of calamitic LC and (**b**) an example of a calamitic LC

Figure 16.1a shows a general template that describes the structure of calamitic liquid crystal.

As shown in the template, $RC_1$ and $RC_2$ are the rigid cores and are often aromatic in nature (e.g., 1,4-phenyl, 2,5-pyrimidinyl, 2,6-naphthyl etc.) or they can also be alicyclic (e.g., trans-4-cyclohexyl, cholesteryl, etc.,) cores. Generally, the two rigid cores are interconnected through either a covalent bond or linking groups L, such as: –COO–, –CH=N–, $–CH_2–CH_2–$, –N=N– etc. The terminal substituents R and R′ are usually either alkyl or alkoxy chains or the combination of these two. In many cases, one of the terminal units is a polar substituent (e.g., F, CN, Cl, etc.). In some special cases, lateral substituents X and Y (e.g., Cl, CN, F, $CH_3$, etc.,) are incorporated in order to account for a particular special property. A typical example of achiral rod-like mesogen is 4-methoxy-benzylidene-4-n-butylaniline (MBBA), as shown in Fig. 16.1b (Tschierske and Dantlgraber 2003; Ros et al. 2005). The introduction of molecular chirality in rod-like molecules results in optically active LCs and exhibits a variety of interesting mesophases (Meyer et al 1975).

#### 16.4.1.2 Discotic Liquid Crystals

Until 1977, it was believed that only rod-like molecules, having a considerably higher length than width, could show liquid crystalline properties (calamitic mesogens). However, in 1977, Chandrasekhar realized that not only rod-like molecules, but also compounds with a disc-like molecular shape are capable of forming mesophases (Chandrasekhar et al. 1977; Sabine et al. 2007) and these are referred to as discotic liquid crystals (DLCs). The first examples of this kind of mesomorphism were observed in hexa-alkanoyloxybenzenes, hexaalkoxy- and -alkanoyloxy-triphenylenes (Fig. 16.2), and were established by X-ray studies (Kumar 2010). Generally, the disc shaped molecules have central core as benzene or polyaromatic, such as triphenylene or phthalocyanine.

Two basic types of discotic mesophases have been widely recognized; these are columnar and nematic. In general, discotic liquid crystals consist of flat and rigid aromatic cores surrounded by six or eight flexible chains. These flexible chains are connected to the core by different linking units, such as: ether, ester, benzoate, alkyne, etc (Kumar 2010).

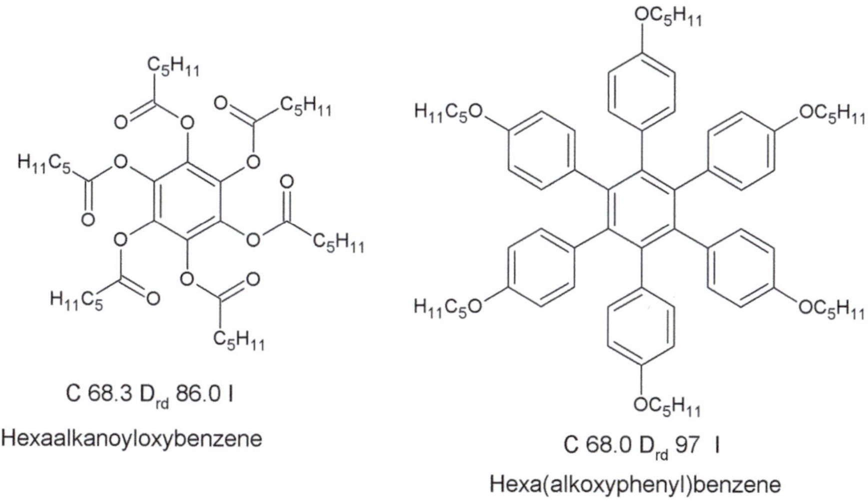


**Fig. 16.2** Example of discotic mesogens (Collings and Hird 1997)

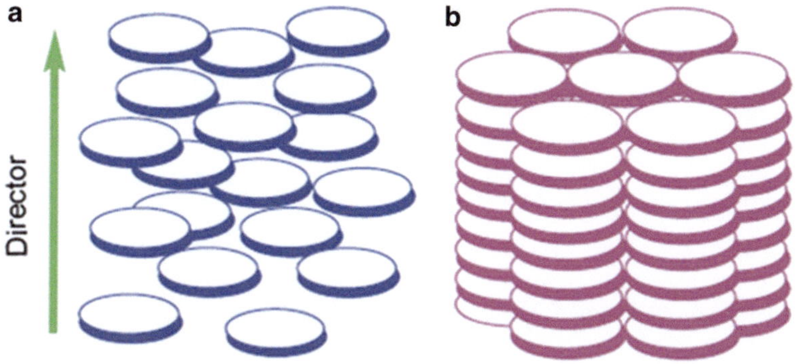


**Fig. 16.3** Schematic representation of: (**a**) a discotic nematic, (**b**) a columnar phase (Kumar 2006)

### 16.4.1.2.1 Nematic Discotic Phase

In the discotic nematic ($N_D$) phase (Fig. 16.3a), molecules have orientationally ordered arrangement of discs with no long-range translational order. This is the least order (usually high temperature) in the disc-like molecules (Kumar 2009). The nematic mesophase can be assimilated to a lamellar nematic liquid crystal, in which the director vector (an optical axis) is perpendicular to the average direction along which the flat molecules are aligned, as illustrated in Fig. 16.3a.

16.4.1.2.2 Columnar Phase

In the columnar phase (Fig. 16.3b); the discs are stacked on top of one another to form columns. There are several different types of columnar mesophases exhibited by discotic molecules. These arise because of the different symmetry classes of the two dimensional lattice of columns and the order or disorder of the molecular stacking within the columns (Kumar 2006). The polymorphism in columnar mesophases is, in many ways, analogous to the polymorphism found in the smectic mesophases generated by calamitic materials.

## *16.4.2 Non-conventional Shaped Liquid Crystals*

Conventionally, mesogens are designed with shape anisotropic rigid cores, such as rods or discs, decorated with flexible chains. The latter contributes fluidity to the material and the former induces the anisotropic character. In contrast, non-conventional mesogens, which have become very popular in recent years, lack such shape-anisotropy of the core unit. Among them, are bent-core mesogens, multi-arm mesogens, cones (Xu and Swager 1993; Komori and Shinkai 1993) shuttle-cocks (Sawamura et al. 2002; Matsuo et al. 2004), dendrons, dendrimers (Percec et al. 2002; Saez et al. 2001; Baars et al. 1998) supramolecular mesogens, such as metallomesogens (Cardinaels et al. 2005) and hydrogen-bonded mesogens (Paleos and Tsiourvas 2001; Saez and Goodby 2005), rod-dendrons (Lecommandoux et al. 2003) and rings (Mindyuk et al. 1998), have gained significant recognitions.

### 16.4.2.1 Banana-Shaped Mesogens

In the mid-1990s, a novel type of mesogenic molecule, the banana LCs or bent-core LCs, entered into the fold of LCs (Pelzl et al. 1999). Since the pioneering work by Niori et al. (1996), extensive research has been carried out on these LCs. The origin of the novelty stems from the compact packing arrangement of bent-shaped molecules (Figs. 16.4 and 16.5), which restricts rotational freedom, thus allowing the molecules to organise into novel types of liquid crystalline phases. These new mesophases are denoted by the letter B, which refers to the characteristic bent or banana molecular shape. Seven mesogenic phases are identified and are simply designated as $B_1$ to $B_7$ (Ros et al. 2005; Hird 2005). All these phases, together with another phase discovered later ($B_8$) are sometimes called *banana phases*, since specific phases differ from the conventional phases found in calamitic (rod-like) mesogens (Dierking 2003). The distinction between phases is made on the basis of the optical textures and the different characteristics of their X-ray diffractograms. The optical textures observed for the $B_1$ and $B_2$ phases are shown in Fig. 16.6 (Srinivasan et al. 2013a; Takezoe and Takanishi 2006).

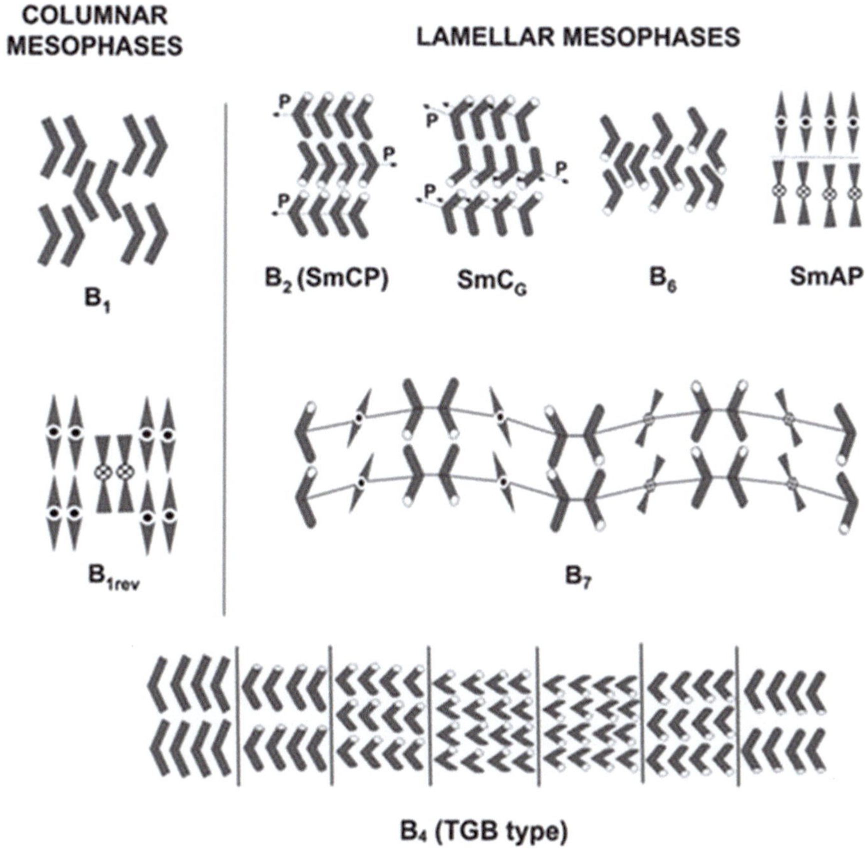


**Fig. 16.4** The basic structural models proposed for some of the mesophases reported for bent-core liquid crystals. Variations of these structures (tilted B1rev, different types of $SmC_G$ and SmCP) have also been proposed (Ros et al. 2005)

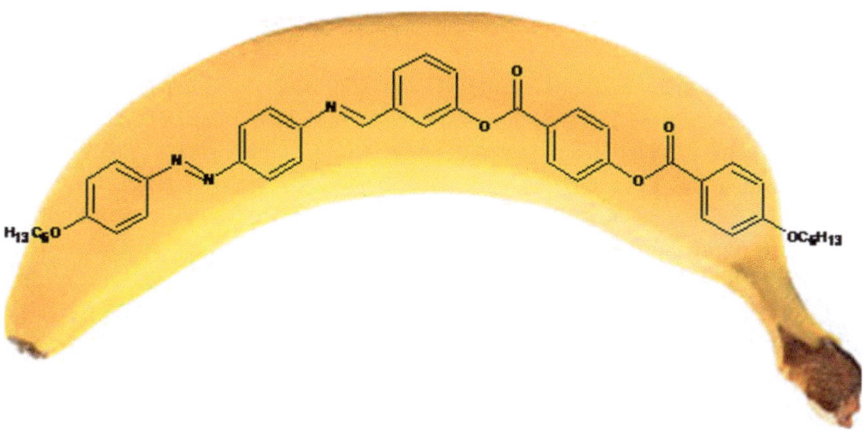


3-(4-(40-hexyloxybenzoyloxy)-benzoyloxy)phenyl4-((4-(4-hexyloxyphenylazo)-phenylimino)methyl)benzoate

**Fig. 16.5** Example of banana mesogen (Srinivasan et al. 2013a)

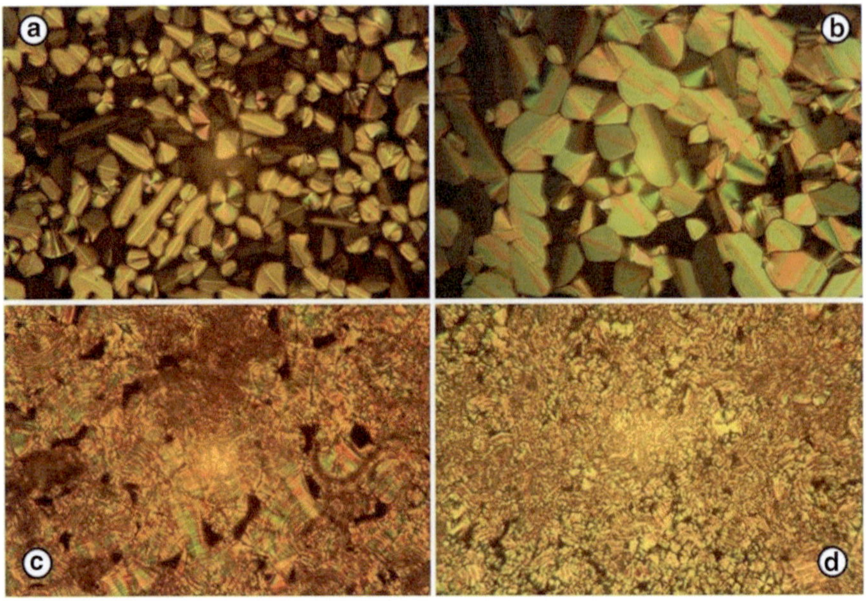


**Fig. 16.6** Microphotographs of B1 (**a**, **b**) and B2 phases (**c**, **d**) (Srinivasan et al. 2013a)

## 16.5 Liquid Crystalline Polymers (LCPs)

LCPs are a special class of polymer that shows liquid crystal phase (Wang and Zhou 2004). When LC meets polymers, the resulting LCPs show anisotropic and isotropic properties, which originate from the mesogenic units and good mechanical properties, which come from long-chain structures (Wang and Zhou 2004). Over the last few decades, LCPs have attracted considerable research attention, owing to their wide applications as engineering plastics, electro-optic or nonlinear optic materials, stationary phases, high-strength and high-modulus fibers and their use as gas separation membranes, etc. (Hsu 1997). The molecular engineering of LCPs is mostly achieved by incorporating mesogenic units, such as rod-like or disc-like mesogens (derived from low molecular weight compounds), into polymer chains, which has been proven to be an efficient way of obtaining LCPs with desired properties. Generally, two typical strategies for the incorporation of mesogenic units into the polymer chain are frequently employed. One, is the embedding of mesogens into the backbone of a polymer chain, and the resulting LCP is known as main-chain liquid crystalline polymer (MCLCP). The other approach is by attaching mesogens to the main chain as side groups and the LCP obtained is named as side-chain liquid crystalline polymer (SCLCP) (McArdle 1989). Based on the position where the mesogens are linked to the backbone, the mesogen can be, either longitudinally or laterally attached into the polymeric system (Table 16.1). Thus, the SCLCPs with a longitudinal connection (end-on)

**Table 16.1** Schematic illustration of structural models of liquid crystalline polymers (Chen et al. 2010)

| | Side-chain LCP | Main-chain LCP |
|---|---|---|
| End-on (longitudinally attached) | 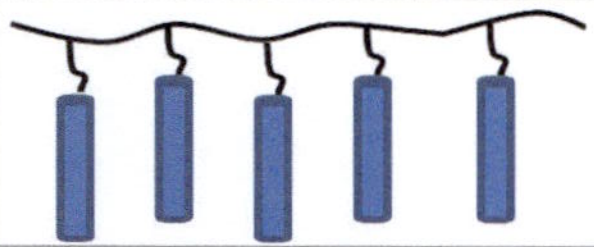 |  |
| Side-on (laterally attached) | 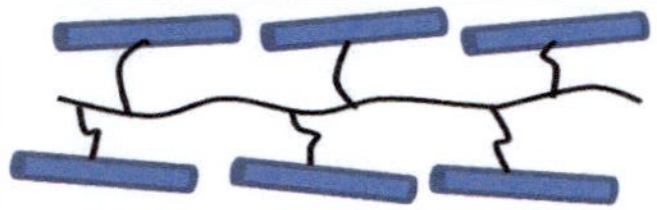 | 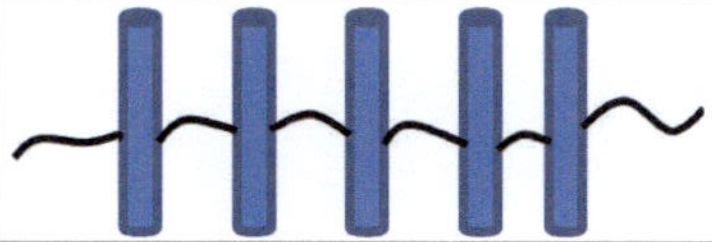 |

of the mesogen are end-on SCLCPs and those with lateral connection (side-on) are called side-on SCLCPs. Similarly, for the MCLCPs, there are also end-on MCLCPs and side-on MCLCPs, as shown in Table 16.1.

### 16.5.1 Main Chain Liquid Crystalline Polymers

In MCLCPs, the mesogenic groups within the polymer backbone are linked to each other through a flexible spacer (Hermay 1987). Each mesogenic group participates in forming the mesophase structure, as in low molecular weight material, but the polymer molecules must adopt a conformation or packing that is compatible with the structure of the mesophase. As a result, mesogenic properties are closely coupled with the polymeric properties. Considering the fact that the polymer chain conformation is dominated by the conformation of the flexible spacer, an alteration of the flexible group may influence the molecular packing into a mesophase and thereby resulting in a structure and properties that will depart from those of the low molecular weight mesogens (Godovsky et al. 1989). A number of liquid crystalline main chain polymers, such as: polyesters (Fig. 16.7), polyamides, poly(p-phenylene)s, polyurethanes, polycarbonates, polyethers, polyimides, poly(esterimide)s, poly(azomethines), poly(azomethine-ester)s etc, have been reported (Rehahn et al. 1990; Percec and Yourd 1998; Ungar et al. 1992). Most of these studies have focused, primarily on the understanding of their structure-property relationships. Many structural modifications have been carried out in order to observe the structure-property relationships of thermotropic aromatic polyesters.

### 16.5.2 Side Chain Liquid Crystal Polymers

Thermotropic SCLCPs came into existence due to the pioneering work of Finkelmann, Ringsdorf, Shaibaev and Plate (Finkelmann et al. 1978a, b; Plate and Shibaev 1984; Shibaev 2009). Prior to their studies, many research groups explored the possibility of inducing mesomorphism in side chain liquid crystals without using flexible methylene spacers (Blumstein and Hsu 1978; Amerik and Krentsel 1967). Amongst them, the contributions of Roviello and Sirigu (1975), Blumstein et al. (1975) and Paleos et al. (1968) are noteworthy. For instance, Blumstein observed birefringence in polymeric acrylates with carboxylic groups (Blumstein et al. 1975). The appearance of birefringence is attributed to the hydrogen bonding arising from the carboxylic group. Paleos et al. (1968) synthesized liquid crystalline monomeric Schiff bases, using 4-amino styrene and 4-alkoxy benzaldehyde. Similarly Perplies et al. (1974) made acrylate-based Schiff bases by reacting acryloxy benzaldehyde with alkyl/alkoxy anilines. In some cases, birefringence disappears without any identifiable mesophase textures are noticed in

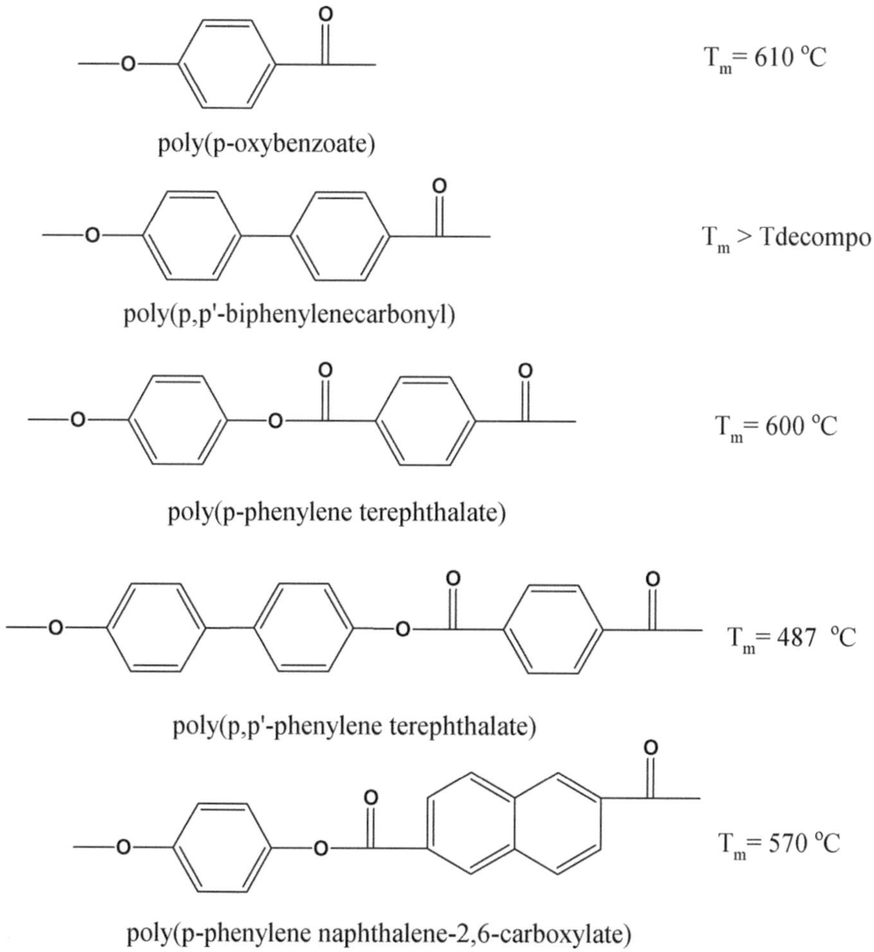


**Fig. 16.7** Structure of the ideal thermotropic polyesters (Hasookhan and Bhowmik 1997)

the monomers. It is interesting to note that olefinic/acrylic comb-like polymers are known, where depending on the length of the pendant group, crystallization is observed (Potemkin and Palyulin 2009). The later works of Finklemann (1982), Shibaev and Plate (1985), Shibaev and Plate (1984) showed unambiguous liquid crystalline phases in SCLCPs. The success of their studies is attributed to a better molecular design wherein the flexible backbone consisting of mostly acrylate/methacrylate chains are well separated by the mesogenic core of the oxymethylene spacers of varying length (Portugall et al. 1982; Rodekirch et al. 1993; Reddy et al. 2013, 2014a, b). Thus, the insertion of spacer between backbone chain and rod-like mesogenic core facilitated the packing of mesogens in parallel fashion, thereby favoring the orientational order. Hence, many SCLCPs with a wide range of molecular architectures have been realized using different polymerization methods (Demus et al. 2008; Reddy et al. 2014a, b). The structural model proposed by Finkelmann,

Ringsdorf, Shibaev and Plate is practiced in contemporary literature for realizing the SCLCPs with different pendant mesogenic core units (Ganicz and Stanczyk 2009).

Mesogenic side group maintains a degree of orientational freedom and dependent on the coupling strength to the backbone and its alignment may be altered by the application of external electric or other fields (Finkelmann and Rehage 1984; Shibaev and Plate 1984). This alignment takes place on a time scale of the orders of magnitude faster for SCLCPs than for MCLCPs. This allows the optical properties of SCLCPs to be readily altered by the use of external electric fields (Finkelmann and Rehage 1984). In SCLCPs, the mesogenic groups are linked to an existing polymer backbone, either directly or via flexible spacer units. The principal concept is to link conventional mesogenic groups to a polymer main chain, via flexible spacer groups. A direct fixation of the mesogenic groups to the polymer main chain causes a coupling of the motions of the main chain and the mesogenic side chain (Finkelmann et al. 1978a, b). Thus, the high mobility of chain segments in the liquid state of the polymer and its tendency towards a statistical chain conformation, consequently prevents an anisotropic orientation of the mesogenic side chains at temperatures above the glass transition temperature ($T_g$) of the polymer.

If the mesogenic groups are fixed to the main chain via flexible spacer, the motions of the main chain and the side chains might be decoupled. In this case, the mesogenic moieties are able to build-up the orientational long range order; although the motions of the main chain tend to prevent an anisotropic order. Direct linkage, except in a few cases, gives only glasses with anisotropic structure that is lost at the glass transition (Amerik and Krentsel 1967). Coupled with steric interactions between the side groups, the tendency towards a statistical distribution of chain conformations hinders the ordered arrangement of the mesogenic groups and LC formation is suppressed. Decoupling of the side groups by using a flexible spacer allows the main chain motion to occur without disturbance of the anisotropic arrangement of the side chains. The polymer may then exhibit LC properties. Figure 16.8 shows the model consideration for the preparation of polymers with LC phases.

Among the polymer backbones, most common backbones used for constructing the SCLCPs are poly (acrylate) (Fig. 16.9a), (methacrylate) (Fig. 16.9b) and siloxane (Fig. 16.9c) systems. Poly (α-chloroacrylates), (itaconates), (phosphazenes) and (ethylene oxides) have also been reported (Noel and Navarad 1991). Typical spacer groups consist of between 2 and 12 methylene units. However, oligooxyethylene or oligosiloxane units may be used to enhance the degree of decoupling through a more flexible spacer (Kostromin et al. 1987; Engel et al. 1985; Percec et al. 1987). SCLCPs of the acrylate, methacrylate or chloroacrylate type have been the most widely synthesized. This is because they can be prepared by free radical polymerization (Decobert et al. 1986; Spassky et al. 1989; Reddy et al. 2013, 2014b). Anionic, group transfer and cationic polymerizations have also been used to obtain different tacticities, molecular weights and poly dispersities and also to vary the nature of the polymer backbone (Froslni et al. 1981; Hahn et al. 1981; Kreuder et al. 1986). Polycondensation reactions have been successfully applied for the preparation of SCLCPs polyesters with flexible, semi-flexible and rigid backbones. Chemical modification of reactive

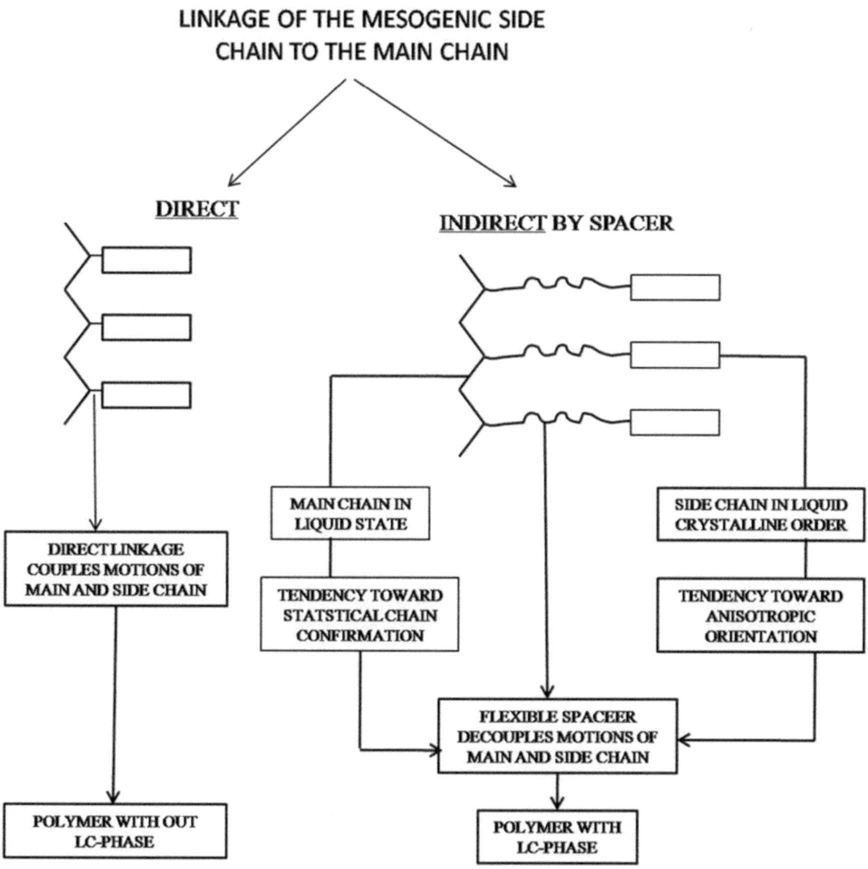


**Fig. 16.8** Model consideration for liquid crystalline polymers (Finkelmann et al. 1978a, b)

polymers, such as sodium polyacrylates or sodium polyitaconates, has also been reported (Keller 1984, 1985). Poly (siloxanes) are usually prepared via the addition process, commencing with a preformed polymer backbone which contains reactive functional groups (Si-H), onto which the mesogenic groups are appended (Keller 1985; Reck and Rangsdorf 1985; Berg et al. 1986).

Non-mesogenic dye side groups may be combined with mesogenic groups to form dyed polymer mesophases (Decobert et al. 1986; Finkelmann and Rehage 1980). Thus, the plethora of structures available in the low molar mass systems is enriched by the copolymerization process that can yield dichroic, fluorescent or optically non-linear mesophases (Spassky et al. 1989). Following these, polymers exhibiting nematic and smectic mesophases can be realised. The cholesteric polymer phase, however, presumes the presence of chiral groups. It is commonly obtained by the copolymerization of: (i) a nematogenic monomer with a chiral comonomer or (ii) a binary mixture of chiral monomers. Cholesteric polymers can

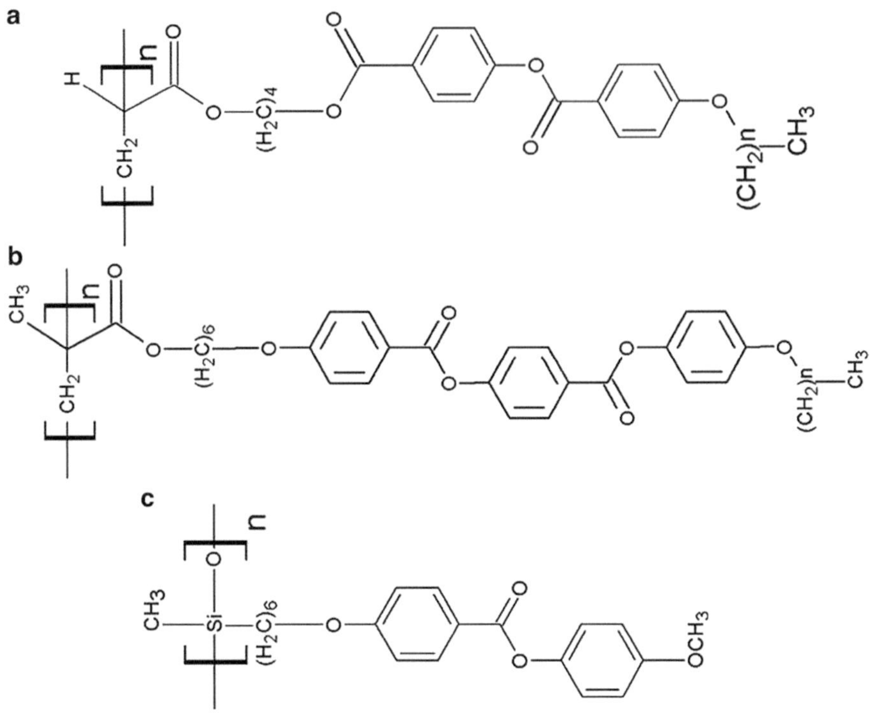

**Fig. 16.9** Example of (**a**) acrylate, (**b**) methacrylate and (**c**) siloxane based SCLCPs (Reddy et al. 2013, 2014b; Wolarz et al. 2005)

also be obtained in the form of mixtures of nematic polymers with chiral polymers or chiral low molar mass compounds.

The structure of SCLCPs can be modified in numerous ways, since a large degree of variation is possible for the chemical structure of the main chain, the nature and the length of the flexible spacer, the molecular structure of the mesogenic group and the size and polarity of the terminal group (Portugall et al. 1982). Systematic studies of a number of SCLCP types have established that certain trends in liquid crystalline behavior accompany particular changes in molecular structure. It has been demonstrated that when mesogenic side groups of the same chemical structure are attached, via flexible spacers of approximately the same length to different polymer backbones, their transition temperatures are affected by the flexibility of the polymer backbone (Cowie and Hunter 1990). It is well known that for conventional polymers, the $T_g$ is lowered as the flexibility of the backbone is increased (Blumstein and Hsu 1978). The bulky mesogenic groups, however, cause severe hindrance to main chain motions. Hence, the glass transition temperatures, $T_g$s of SCLCPs are considerably higher than those of conventional polysiloxanes, polyacrylates and poly (meth) acrylates. Extending the length of the flexible spacer results in, at least, one of the two following effects:

(1) Reduction of the $T_g$: This behaviour is quite general for the poly (acrylate) and poly (methacrylate) series and reflects the increased plasticizing action of the flexible spacer (Noel and Navarad 1991; Shibaev and Plate 1985). It is also quite clear that long flexible spacers move the bulky mesogenic groups farther away from the polymer backbone, which causes less hindrance to the main chain motions. In some series, the glass transition temperatures for the higher members rise, progressively. (2) Purely nematic behaviour tends to give way to a predominantly or purely smectic behaviour in the higher homologues. The change from nematic to smectic with increasing length of the substituent also applies to the alkyl end group of the mesogenic side chains.

Side chain liquid-crystalline poly (meth) acrylates with ester group and turning of the ester group have been reported (Cowie and Hunter 1990). In the turning of the ester group, the smectic layer order is stabilized and mostly, the clearing points show up at higher temperatures. The glass transition temperatures of some polymers show an expected tendency to decrease with increasing number of methylene groups. Unlike the tail group, a long spacer does not lower the $T_g$ in some cases. This phenomenon may be due to a high order in the spacer group, especially in polymers with broad liquid crystalline phases. The spacer length also has such an unexpected influence on the clearing point. Whereas long tail groups broaden the liquid crystalline temperature region, long spacer groups may lower the clearing point. The mainchain of poly (methacrylates) contributes to the liquid crystalline order in the smectic layers (Reddy et al. 2014b). In the case of the new liquid crystalline poly (meth) acrylates (LCPMA), decoupling by a long spacer leads to the lowering clearing points (Reddy et al. 2014b). In most of the new polymers, i.e. methacrylates (LCPMA), their high temperature phase, is the $S_A$ phase with a single layer structure. This is in accordance with the assumption that the methacrylate main chain stabilizes the smectic layer order (Reddy et al. 2014b). The unexpected effects of long spacers on the glass transition temperatures and also on the clearing points may be due to the tendencies of de-mixing of the methylene groups. Furthermore, such segregations of methylene groups may destabilize the liquid-crystalline phase and lower the clearing point. The phase behaviour of liquid crystalline polymers depends on their molecular weights (Rodekirch et al. 1993). This dependence is very strong for low molecular weights, but appears to reach a plateau at a molecular weight of about 15,000, with increasing molecular weight leading to the shifting of all transitions to higher temperatures (Rodekirch et al. 1993). This shift is stronger for the transition from nematic to isotropic than for the glass transition. Therefore, the liquid crystalline phase is widened by increasing the molecular weight.

## 16.6 Bent-Core Side Chain Liquid Crystalline Polymers (BCLCPs)

LCPs are the hottest research topics in the field of liquid crystals because of their unusual properties. Among the LCPs, BCLCPs (Fig. 16.10) possess macroscopic polar order with a variety of useful properties, such as: dielectric, piezo, and

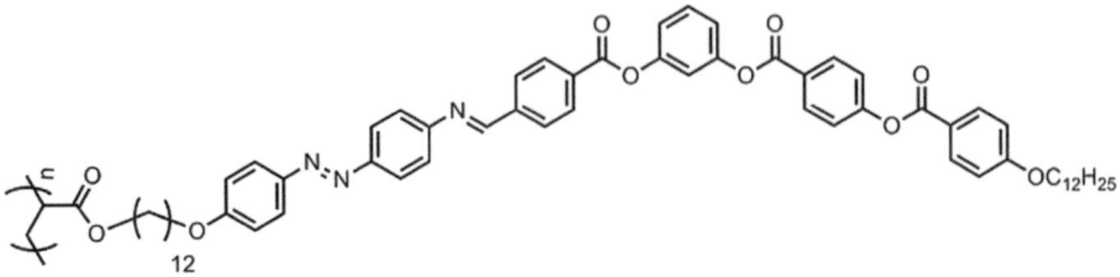


**Fig. 16.10** Example of side chain bent core liquid crystalline polymers (Srinivasan et al. 2013)

pyroelectricity (Balamurugan et al. 2010; Keith et al. 2007; Takezoe and Takanishi 2006). Therefore, the developments of BCLC material into polymeric molecules are interesting research topics that explore potential useful applications. Generally, most of the bent-shaped polymers are reported as main-chain (Gimeno et al. 2011; Choi et al. 2010) as well as side-chain polymers (Chen et al. 2007a, b; Wang et al. 2010; Kishore et al. 2009). The side-chain bent-core polymers are derived from 1,3-phenylene unit with five aromatic rings, which can be symmetric (Wang et al. 2010) as well as asymmetric (Kishore et al. 2009), wherein the polymerizable functional groups are generally used, are 1,3-diene (Sentman and Gin 2003), methacrylates or acrylates group (Chen et al. 2007a, b) and siloxane group (Keith et al. 2005). Interestingly, side-chain bent-core liquid crystalline polymers predominantly demonstrate rich mesomorphism when compared with the main-chain (Saravanan et al. 2008), side-chain (Saravanan and Kannan 2010) and dendronized liquid crystalline polymers (Rudick and Percec 2007, 2008; Rosen et al. 2009; Percec et al. 1998) are classified as follows: (i) side-chain bent-core liquid crystalline polymer (SBCLCPs), (Chen et al. 2007a, b; Srinivasan et al. 2013b) (ii) bent-core mesogen-jacketed liquid crystalline polymers (BMJLCPs) (Chen et al. 2010), (iii) central linked bent-core liquid crystalline polymer (CBCLCPs) (Srinivasan and Kannan 2011) and (iv) hydrogen-bonded side chain bent-cores liquid crystalline polymers (HBCLCPs) (Wang et al. 2010), were reported based on the polymeric back bone attached to the position of bent-core mesogen pendant. Although BCLCs have been extensively investigated, bent-core side-chain polymers have so far, recorded very limited success with respect to their polar switching properties (Sentman and Gin 2003; Chen et al. 2007a, b).

In 2005, Keith et al reported the first bent core side-chain liquid crystalline polymers with a five phenyl ring core, connected with the ester, polysiloxane, as a back bone and terminal alkoxy chain. Moreover, this polymer exhibits fast ferroelectric switching and in addition, the highly interesting phenomenon of spontaneous achiral symmetry breaking, which was previously found in LC systems only for a few low molecular weight bent-core materials, was observed, (Thisayukta et al. 2000; Reddy and Sadashiva 2003; Ortega et al. 2003) Furthermore, Srinivasan et al. (2013b) synthesized poly (meth) acrylate based BCLCPs containing six phenyl rings with different linking groups, such as: azo, azomethine and ester with higher alkyl chain (dodecyl spacer) on both terminal chains as well as the spacer. They reported that their molecules exhibited dual switching properties, namely photo and electrically switchable properties. However, none of these polymers, consisting of the above linkages, have so far been reported to have dual switching properties. The BCLCPs combined with azo and imino functional groups represent an important development in the switchable material and because of the imino linkage molecule, they (BCLCPs) are more conducive to exhibit mesomorphism. The azobenzene is considered to be a prototype molecular switch, thus the BCLCPs behave as dual switch with light and electric fields. Gimeno et al. (2014) prepared photo-responsive BCLCPs because of their wide applications, such as: photo-alignment, light-induced chirality or photo-recording capabilities, which can also be explored and studied for their glassy and liquid crystalline characteristics.

## 16.7 Mesogenic-Jacketed Liquid Crystalline Polymers (MJLCPs)

The decoupling concept (in which flexible spacers are used to decouple the motion of the ordered arrangement of side-chain mesogens from that of the random-coiled main chain), is undoubtedly a landmark for the development of SCLCPs. Zhou et al. (1987, 1988), proposed MJLCPs (Fig. 16.11) with bulky side groups (often mesogens) laterally attached to the main chain of LCPs connecting with a short spacer or a single covalent bond (Chen et al. 2010). Although, MJLCPs have molecular structures similar to SCLCPs, they exhibit properties similar to those of MCLCPs, such as forming banded textures by shearing in the LC state and having large persistent lengths in solutions, (Xu et al. 1993; Wan et al. 1995; Zhang et al. 1997). This is because the backbones of MJLCPs are forced to take the extended-chain

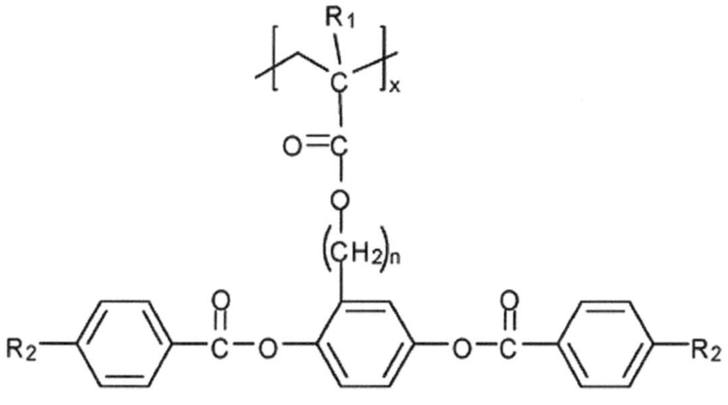


$R_1$ = $-CH_3$; m = 11; $R_2$ = $-OC_nH_{2n+1}$ (n = 1-8)
m = 11; $R_2$ = $-C_3H_7$
m = 6; $R_2$ = $-OC_4H_9$

**Fig. 16.11** Example of mesogenic-jacketed liquid crystalline polymers (Chen et al. 2010)

conformation, owing to the steric effect of the crowded side chains, i.e. the "jacketing" effect (Percec et al. 1998). Many LC phases of MJLCPs, including columnar nematic (ΦN), (Ye et al. 2004; Zhu et al. 2012), hexatic columnar nematic (ΦHN), (Ye et al. 2004), hexagonal columnar (ΦH), (Ye et al. 2004; Chai et al. 2007) rectangular columnar, (Zhu et al. 2012; Chen et al. 2006a, b) smectic A (SmA), (Chai et al. 2007; Chen et al. 2007a, b) and smectic C (Gopalan et al. 2002) phases, in which the MJLCP chain behaves like a cylinder or tablet to form the LC phases, have been reported. With the development of the structural library of MJLCPs, many efforts have been expended and targeted on the molecular engineering of the MJLCP main chains. The first reported series of MJLCPs are based on polyacrylates and poly (meth) acrylates (Hessel and Finkelmann 1985, 1986; Chen et al. 2010). When compared to polyacrylates, the poly (meth) acrylate backbone tends to depress the LC phase formation, indicating the influence of the polymer backbone on liquid crystallinity. The most studied MJLCPs are those with a polystyrene main chain, which can be synthesized via controlled radical polymerizations (Gao et al. 2009). A series of MJLCPs with side-chain cores based on 2-vinylhydroquinone, (Zhou et al. 1989) 2-vinyl- 1,4-phenylenediamine, (Zhang et al. 1997) 2-vinylterephthalic acid (VTA), (Zhang et al. 1999; Ye et al. 2004), vinylbiphenyl, (Zhang et al. 2011; Wu et al. 2012) vinylterphenyl, (Yu et al. 2003a, b) and oxadiazole-containing conjugated structure (Chai et al. 2007; Xu et al. 2009), have been successfully synthesized and characterized (Gao et al. 2009). By hydrosilylation of polymethyl hydrosiloxane with the styrenic monomers, MJLCPs with a polysiloxane main chain were synthesized (Zhang et al. 2010). Besides MJLCPs with flexible backbones, the jacketed structure has also been applied to polymers with rigid backbones, such as polyalkynes (Chen et al. 2006a, b; Peng et al. 2010) and polythiophenes (Yang et al. 2009, 2010). In addition, MJLCPs with a polynorbornene backbone and laterally attached 2,5-bis [(4′-n-alkoxybenzoyl)oxy] mesogens on side chains, display enantiotropic nematic mesophases (Pugh and Schrock 1992). Terminating the hydrocarbon substituents of the mesogen with immiscible components, such as fluorocarbon segments or oligodimethylsiloxane segments, can induce smectic layering with a polynorbornene backbone (Arehart and Pugh 1997; Pugh et al. 1998). Very recently, Yang et al. (2013) reported a new series of MJLCPs with a polynorbornene backbone. When the number of carbon atoms in the alkoxy terminal chains of the side-chain mesogen is more than nine, the polymers exhibit the liquid crystalline property. In the past two decades, Percec and co-workers have intensively synthesized and investigated other type of jacketed polymers, i.e. dendron-jacketed polymers, (Rosen et al. 2009; Percec and Tomazos 1992; Rudick and Percec 2008) by a set of living polymerization methods, such as living cationic polymerization, (Percec et al. 1991, 1992) living polymerization of acetylenes (Percec et al. 2005) and living metathesis polymerization (Percec et al. 1996). With the consideration of the high tolerance of functional groups e.g. fullerene, the convenient synthesis by ring-opening metathesis polymerization (ROMP), (Ball et al. 2006; Bielawski and Grubbs 2007) and the excellent properties of polynorbornenes, (Maier 2001; Piunova et al. 2013), the exploration of MJLCPs with polynorbornene main chain, is of great interest (Zhu et al. 2014).

## 16.8 Liquid Crystal Elastomers (LCEs)

LCEs are unique solid materials that possess mobile directors and rubber elasticity (Urayama 2007). LCEs comprise of rubbery networks of cross-linked polymer chains that have rod-like elements, sufficient to induce a mesophase. LCEs are often referred to as a new class of materials due to the hybrid characteristics between LCs and elastomeric solids. Unlike uncross-linked LC polymers, LCEs behave as solids; i.e., they do not flow because crosslinking prohibits the motion of entire molecules. Nevertheless, the mesogenic groups in LCEs still have considerable mobility because of the network backbone formed by distant and adjacent crosslinks in the rubbery state. In 1991, Finkelmann and Kupfer pioneered a method of producing globally aligned monodomain LCEs (Kupfer and Finkelmann 1991). It proved that LCEs exhibit several novel properties that are not simple extensions of the inherent characteristics of LCs and elastomers (Warner and Terentjev 1996, 2003; Xie and Zhang 2005). The most important characteristic of LCEs is the strong coupling of the orientational order of LCs with the rubber elasticity of the polymer networks. The alignment of the constituent mesogens is always coupled with the macroscopic shape of the LCEs, and vice versa. A remarkable property arises due to this characteristic uniaxial spontaneous deformation in response to various environmental factors (such as temperature and light irradiation), affecting their orientational order parameter. Externally imposed deformation can also realign the director along the stretching direction. Importantly, LCEs exhibit a memory effect with regard to their directors in the preparation state and the mesogen alignment recovers its original state when the external factors are removed.

Various LCEs including nematic, cholesteric, $S_A$, smectic C* (Chiral $S_C$) and discotic LCEs have been synthesized and characterized (Warner and Terentjev 2003; Brand et al. 2006). These LCEs have numerous potential applications, such as: soft actuators, artificial muscles and nonlinear optical devices (Urayama 2007; Jiang et al. 2013). The coupling of the mesogen alignment with elastomeric networks also yields many academically interesting issues, such as: nonsymmetric elasticity, unusual phase transition and significantly, nonlinear mechanical responses. Most of the earlier studies focused on temperature-induced shape changes in the dry state without volume change (Kupfer and Finkelmann 1991; Warner and Terentjev 1996, 2003; Xie and Zhang 2005). LCEs immersed in solvents undergo variations in volume as well as shape and they behave as anisotropic gels having temperature-responsive volume and shape. The rubber elasticity of nematic networks leads to some interesting features in the phase diagrams that are absent in noncross-linked liquid crystal polymer/solvent and non-nematogenic network/solvent systems (Urayama 2007). The electric field effect with response times of 10 ms and induced strains of more than 10 %, demonstrate that LCEs can be promising materials as soft actuators with high actuation speed (Xie and Zhang 2005). This electric field response also reveals an LCE specific deformation pathway during a 90° rotation of the director.

## 16.9 Conclusion and Future Perspective

Thermotropic SCLCPs have become important in view of their interesting material properties. The ability to form films, fibres and sheets, characteristic of synthetic polymers, and their tendency to respond to external stimuli, typical of LCs, are conveniently manifested in SCLCPs. In contrast to their MCLCPs counterparts, SCLCPs are easy to solubilise in common solvents at room temperature and their mesophase temperatures are relatively low (ambient) and they have many interesting applications. Owing to this interesting characteristic, SCLCPs have attracted a lot of attention from researchers around the world. The developments in SCLCPs further paved the way for thermotropic LCEs, in which mesogenic monomers are one of the key components. LC elastomers are often prepared by reacting mesogenic monomer with non-mesogenic cross-linking agent. However, studies have also been reported whereby mesogenic bifunctional monomers serve as cross-linking agents. LCEs are used as actuators which can find other uses in not only biomedical sciences (as artificial muscles and heart valves, for examples), but also in a wide array of other fields (thermal and chemical sensors, microfluidics, touch sensors, etc). The potentials of these materials are vast and they will form the basis for a rich scientific field in coming years. SCLCPs have also entered into new class of polymers, such as bent core SCLCPs, which are mainly used for dual switching devices, photo-responsive, dielectric, piezo and pyroelectricity devices because of their very interesting and unique properties and research is continuing on BCSCLCPs in order to find new applications. MJLCPs are a class of SCLCPs with many interesting applications in several fields, such as: optical data storage, chromatographic stationary phase, separation membranes and nonlinear optics and because of these interesting applications, research on MJLCPs has become even, a more fascinating field. As a result of these many interesting applications, SCLCPs have become more popular in the field of LCPs.

**Acknowledgements** One of the authors (G.S.M.R.) sincerely thanks to the National Research Foundation (NRF) of South Africa for the financial support in the form of NRF-Free standing postdoctoral fellowship (Grant number 88329). G.S.M.R. and S.S.R. would like to thank the University of Johannesburg for financial support.

## References

Amerik YB, Krentsel BA (1967) Polymerization of certain vinyl monomers in liquid crystals. J Polym Sci Part C Polym Symp 16:1383–1392

Arehart SV, Pugh C (1997) Induction of smectic layering in nematic liquid crystals using immiscible components. 1. laterally attached side-chain liquid crystalline poly(norbornene)s and their low molar mass analogs with hydrocarbon/fluorocarbon substituents. J Am Chem Soc 119:3027–3037

Baars MWPL, Suntjens SHM, Fischer HM, Peerlings HWI, Meijer EW (1998) Liquid-crystalline properties of poly(propylene imine) dendrimers functionalized with cyanobiphenyl mesogens at the periphery. Chem Eur J 4:2456–2466

Balamurugan S, Kannan P, Chuang MT, Wu SL (2010) Antiferroelectric bent-core liquid crystals for molecular switching applications. Ind Eng Chem Res 49:7121–7128

Ball ZT, Sivula K, Frechet JMJ (2006) Well-defined fullerene-containing homopolymers and diblock copolymers with high fullerene content and their use for solution-phase and bulk organization. Macromolecules 39:70–72

Berg S, Krone V, Ringsdorf H (1986) Structural variations of liquid-crystalline polymers: cross-shaped and laterally linked mesogens in main chain and side group polymers. Makromol Chem Rapid Commun 7:381–388

Bielawski CW, Grubbs RH (2007) Living ring-opening metathesis polymerization. Prog Polym Sci 32:1–29

Blumstein A, Hsu EC (1978) Liquid crystalline order in polymers. In Blumstein A. (ed) Academic: New York

Blumstein A, Blumstein RB, Clough SB, Hsu EC (1975) Oriented polymer growth in thermotropic mesophases. Macromolecules 8:73–76

Brand HR, Pleiner H, Martinoty P (2006) Selected macroscopic properties of liquid crystalline elastomers. Soft Matter 2:182–189

Brown GH, Shaw WG (1957) The mesomorphic state—liquid crystals. Chem Rev 57:1049–1157

Cardinaels T, Ramaekers J, Guillon D, Donnio B, Binnemans KA (2005) Propeller-like uranyl metallomesogen. J Am Chem Soc 127:17602–17603

Chai CP, Zhu XQ, Wang P, Ren MQ, Chen XF, Xu YD, Fan XH, Ye C, Chen EQ, Zhou QF (2007) Synthesis and phase structures of mesogen-jacketed liquid crystalline polymers containing 1,3,4-oxadiazole based side chains. Macromolecules 40:9361–9370

Chandrasekhar S, Sadashiva BK, Suresh KA (1977) Liquid crystals of disc-like molecules. Pramana J Phys 9:471–480

Chen L, Chen YW, Zha DJ, Yang Y (2006a) Synthesis and properties of polyacetylenes with directly attached bis(4-alkoxyphenyl)terephthalate mesogens as pendants. J Polym Sci Part A Polym Chem 44:2499–2509

Chen XF, Tenneti KK, Li CY, Bai YW, Zhou R, Wan XH, Fan XH, Zhou QF (2006b) Design, synthesis, and characterization of bent-core mesogen-jacketed liquid crystalline polymers. Macromolecules 39:517–527

Chen S, Gao LC, Zhao XD, Chen XF, Fan XH, Xie PY, Zhou QF (2007a) Synthesis and properties of mesogen-jacketed liquid crystalline polymers with asymmetry mesogenic core. Macromolecules 40:5718–5725

Chen X, Kishore KT, Li CY, Bai Y, Wan X, Fan X, Zhou QF, Rong L, Hsiao BS (2007b) Side-chain liquid crystalline poly(meth)acrylates with bent-Core mesogens. Macromolecules 40:840–848

Chen XF, Shen Z, Wan XH, Fan XH, Chen EQ, Ma Y, Zhou QF (2010) Mesogen-jacketed liquid crystalline polymers. Chem Soc Rev 39:3072–3101

Choi EJ, Kim EC, Ohk CW, Zin WC, Lee JH, Lim TK (2010) Synthesis and mesomorphic properties of main-chain polymers containing V-shaped bent-core mesogens with acute-subtended angle. Macromolecules 43:2865–2872

Collings JP, Hird M (1997) Introduction to liquid crystals: chemistry and physics. Taylor and Francis, London

Cowie MG, Hunter W (1990) Synthesis and characterisation of chiral nematic liquid-crystalline copolymers with the chiral centre in the flexible spacer. Makrmol Chem 191:1393–1401

De Gennes PG (1975) The physics of liquid crystals. Clarendon, Oxford

De Gennes PG (1984) "Flexible" nematic polymers: stiffening near the clearing point. Mol Cryst Liq Cryst 102:95–104

Decobert G, Dubots JC, Esselin S, Noel C (1986) Some novel smectic* liquid-crystalline side-chain polymers. Liq Cryst 1:307–317

Demus D, Goodby J, Gray GW, Spiess HW, Vill V (2008) Behavior and properties of side group thermotropic liquid crystal polymers, in Handbook of liquid crystals set. Wiley-VCH Verlag GmbH, Weinheim
Dierking I (2003) Textures of liquid crystals. Wiley-VCH, Weinheim
Engel M, Hisgen B, Keller R, Kreuder W, Reck B, Ringsdorf H, Schmidt HW, Tschmner P (1985) Synthesis, structure and properties of liquid crystalline polymers. Pure Appl Chem 57:1009–1014
Finkelmann H (1982) Polymer Liquid Crystals. Academic Press, New York
Finkelmann H, Rehage G (1980) Investigations on liquid crystalline polysiloxanes, 1. Synthesis and characterization of linear polymers. Makromol Chem Rapid Commun 1:31–34
Finkelmann H, Rehage G (1984) Liquid crystal side chain polymers. Liquid crystal polymers II/III. Adv Polym Sci 60/61:99–172
Finkelmann H, Happ M, Portugall M, Ringsdorf H (1978a) Liquid crystalline polymers with biphenyl-moieties as mesogenic group. Makromol Chem 179:2541–2544
Finkelmann H, Ringsdorf H, Wendorf JH (1978b) Model considerations and examples of enantiotropic liquid crystalline polymers. Polyreactions in ordered systems. Makromol Chem 179:273–276
Froslni A, Levita G, Luplnacci D, Magagnini PL (1981) Mesophase formation in polymers with anisometric side–groups in relation to their chemical and stereochemical structure. Mol Cryst Liq Cryst 66:21–36
Ganicz T, Stanczyk W (2009) Side-chain liquid crystal polymers (SCLCP): methods and materials. An overview. Materials 2:95–128
Gao LC, Fan XH, Shen ZH, Chen X, Zhou QF (2009) Jacketed polymers: controlled synthesis of mesogen-jacketed polymers and block copolymers. J Polym Sci Part A Polym 47:319–330
Gimeno N, Antoni SF, Sebastian N, Mezzenga R, Blanca Ros M (2011) Bent-core based main-chain polymers showing the dark conglomerate liquid crystal phase. Macromolecules 44:9586–9594
Gimeno N, Pintre I, Martinez-Abadia M, Serrano JL, Ros MB (2014) Bent-core liquid crystal phases promoted by azo-containing molecules: from monomers to side-chain polymers. RSC Adv 4:19694–19702
Godovsky YK, Papkov VS, Dusek K (1989) Thermotropic mesophases in element-organic polymers. Adv Polym Sci 88:129–180
Gopalan P, Andruzzi L, Li XF, Ober CK (2002) Fluorinated mesogen-jacketed liquid-crystalline polymers as surface-modifying agents: design, synthesis and characterization. Macromol Chem Phys 203:1573–1583
Hahn B, Wendorff JH, Portugall M, Ringsdorf H (1981) Structure and phase transitions of a liquid crystalline polymer. Colloid Polym Sci 259:875–884
Hasookhan H, Bhowmik PK (1997) Wholly aromatic liquid-crystalline polyesters. Prog Polym Sci 22:1431–1502
Hermay FM (1987) Encyclopedia of polymers and engineering, vol 9, 2nd edn. Wiley, New York
Hessel F, Finkelmann H (1985) A new class of liquid crystal side chain polymers mesogenic groups laterally attached to the polymer backbone. Polym Bull 14:375–378
Hessel F, Finkelmann H (1986) Optical biaxiality of nematic lc-side chain polymers with laterally attached mesogenic groups. Polym Bull 15:349–352
Hird M (2005) Banana-shaped and other bent-core liquid crystals. Liq Cryst Today 14:9–21
Hsu CS (1997) The application of side-chain liquid-crystalline polymers. Prog Polym Sci 22:829–871
Jiang H, Li C, Huang X (2013) Actuators based on liquid crystalline elastomer materials. Nanoscale 5:5225–5240
Keith C, Reddy RA, Tschierske C (2005) The first example of a liquid crystalline side-chain polymer with bent-core mesogenic units: ferroelectric switching and spontaneous achiral symmetry breaking in an achiral polymer. Chem Commun 7:871–873
Keith C, Dantlgraber G, Amaranatha Reddy R, Baumeister U, Tschierske C (2007) Ferroelectric and antiferroelectric smectic and columnar liquid crystalline phases formed by silylated and non-silylated molecules with fluorinated bent Cores. Chem Mater 19:694–710

Keller P (1984) Synthesis of liquid crystalline side-chain polyacrylates via phase-transfer catalysis. Macromolecules 17:2937–2939

Keller P (1985) Synthesis of liquid crystalline side-chain polyitaconates via phase-transfer catalysis. Macromolecules 18:2337–2339

Kishore KT, Chen X, Li CY, Shen Z, Wan FX, Zhou QF, Rong L, Hsiao BS (2009) Influence of lc content on the phase structures of side-chain liquid crystalline block copolymers with bent-core mesogens. Macromolecules 42:3510–3517

Komori T, Shinkai S (1993) Novel columnar liquid crystals designed from cone-shaped calix[4] arenes. the rigid bowl is essential for the formation of the liquid crystal phase. Chem Lett 22:1455–1458

Kostromin SG, Shibaev VP, Plate NA (1987) Thermotropic liquid-crystalline polymers XXVI. Synthesis of comb-like polymers with oxygen containing spacers and a study of their phase transitions. Liq Cryst 2:195–200

Kreuder W, Webster OW, Ringsdorf H (1986) Liquid crystalline polymethacrylates by group-transfer polymerization. Makromol Chem Rapid Commun 7:5–13.5

Kumar S (2006) Critical review self-organization of disc-like molecules chemical aspects. Chem Soc Rev 35:83–109

Kumar S (2009) Hierarchical discs. Liq Cryst Today 18:2–7

Kumar S (2010) Chemistry of discotic liquid crystals. CRC, New York

Kupfer J, Finkelmann H (1991) Nematic liquid single crystal elastomers. Macromol Chem Rapid Commun 12:717–726

Lecommandoux S, Klok HA, Sayar M, Stupp SI (2003) Synthesis and self-organization of rod–dendron and dendron–rod–dendron molecules. J Polym Sci Part A 41:3501–3518

Lehn JM (1988) Supramolecular chemistry: scope and perspectives molecules supramolecular and molecular devices. Angew Chem Int Ed Engl 27:89–112

Lukask KT, Harden MM (1985) Supported phospholipid bilayers. Biophys J 47:105–113

Maier G (2001) Low dielectric constant polymers for microelectronics. Prog Polym Sci 26:3–65

Matsuo Y, Muramatsu A, Hamasaki R, Mizoshita N, Kato T, Nakamura E (2004) Regioselective synthesis of 1,4-Di(organo)[60]fullerenes through DMF-assisted monoaddition of silylmethyl grignard reagents and subsequent alkylation reaction. J Am Chem Soc 126:432–433

McArdle CB (1989) Side chain liquid crystal polymers. Blackie and Son Ltd, Glasgow

Meyer RB, Liebert L, Strzelecki LL, Keller P (1975) Ferroelectric liquid crystals. J Phys Lett 36:69–71

Mindyuk OY, Stetzer MR, Heiney PA, Nelson JC, Moore S (1998) High resolution X-ray diffraction study of a tubular liquid crystal. J Adv Mater 10:1363–1366

Neto F, Antonio M, Silvio RA (2005) The physics of lyotropic liquid crystals. Oxford University Press, New York

Niori T, Sekine T, Watanabe J, Furukawa T, Takezoe H (1996) Distinct ferroelectric smectic liquid crystals consisting of banana shaped achiral molecules. J Mater Chem 6:1231–1233

Noel C, Navarad P (1991) Liquid crystalline polymers. Prog Polym Sci 16:55–110

Ortega J, Folcia CL, Etxebarria J, Gimeno N, Ros MB (2003) Interpretation of unusual textures in the B2 phase of a liquid crystal composed of bent-core molecules. Phys Rev E 68:011707–011710

Paleos CM, Tsiourvas D (2001) Supramolecular hydrogen-bonded liquid crystals. Liq Cryst 28:1127–1161

Paleos CM, Laronge TM, Labes MM (1968) Liquid crystal monomers: *N*-(*p*-alkoxybenzylidene)-*p*-aminostyrene. Chem Commun (London) 18:1115–1116

Pelzl G, Diele S, Weissflog W (1999) Banana-shaped compounds-a new field of liquid crystals. Adv Mater 11:707–724

Peng H, Chen Y, Chen L, He X, Li F (2010) Luminescent mesogen jacketed poly(1-alkyne) bearing lateral terphenyl with hexyloxy tail. J Polym Sci Part A Polym Chem 48:5679–5692

Percec V, Tomazos D (1992) Molecular engineering of side-chain liquid-crystalline polymers by living cationic polymerization. Adv Mater 4:548–561

Percec V, Yourd R (1998) Liquid crystalline polyethers and copolyethers based on conformational isomerism. 3. The influence of thermal history on the phase transitions of the thermotropic polyethers and copolyethers based on 1-(4-hydroxyphenyl)-2-(2-methyl-4-hydroxyphenyl)ethane and flexible spacers containing an odd number of methylene units. Macromolecules 22:3229–3242

Percec V, Rodrguez-Prarada JM, Ericsson C (1987) Synthesis and characterization of liquid crystalline poly (p-vinylbenzyl ether)s. Polym Bull 17:347–352

Percec V, Heck J, Ungar G (1991) Liquid-crystalline polymers containing mesogenic units based on half-disk and rodlike moieties. 5. side-chain liquid-crystalline poly(methylsiloxanes) containing hemiphasmidic mesogens based on 4-[[3,4,5-tris(alkan-l-yloxy) benzoyl]oxy]-4'-[[p-(propan-1-yloxy)-benzoyl] oxy] biphenyl groups. Macromolecules 24:4957–4962

Percec V, Heck J, Lee M, Ungar G, Alvarez-Castillo A (1992) Poly{2-vinyloxyethyl 3,4,5-tris [4-(*n*-dodecanyloxy)benzyloxy]benzoate}: a self-assembled supramolecular polymer similar to tobacco mosaic virus. J Mater Chem 2:1033–1039

Percec V, Schlueter D, Ronda JC, Johansson G, Ungar G, Zhou JP (1996) Tubular architectures from polymers with tapered side groups. assembly of side groups via a rigid helical chain conformation and flexible helical chain conformation induced via assembly of side groups. Macromolecules 29:1464–1472

Percec V, Ahn CH, Ungar G, Yeardley DJP, Moller M, Sheiko SS (1998) Controlling polymer shape through the self-assembly of dendritic side-groups. Nature 391:161–164

Percec V, Holerca MN, Uchida S, Cho WD, Ungar G, Lee Y, Yeardley DJP (2002) Exploring and expanding the three-dimensional structural diversity of supramolecular dendrimers with the aid of libraries of alkali metals of their $AB_3$ minidendritic carboxylates. Chem Eur J 8:1106–1117

Percec V, Rudick JG, Peterca M, Wagner M, Obata M, Mitchell CM, Cho WD, Balagurusamy VSK, Heiney PA (2005) Thermoreversible cis-cisoidal to cis-transoidal isomerization of helical dendronized polyphenylacetylenes. J Am Chem Soc 127:15257–15264

Perplies E, Ringsdorf H, Wendorff JH (1974) Polyreaktionen in orientierten Systemen, 3. Polymerisation ungesättigter Benzylidenaniline mit flüssig-kristallinen Eigenschaften. Macromol Chem 175:553–561

Piunova VA, Miyake GM, Daeffler CS, Weitekamp RA, Grubbs RH (2013) Highly ordered dielectric mirrors via the self-assembly of dendronized block copolymers. J Am Chem Soc 135:15609–15616

Plate NA, Shibaev VP (1984) Thermotropic liquid crystalline polymers-problems and trends. Makromol Chem 6:3–27

Portugall M, Ringsdorf H, Zentel R (1982) Synthesis and phase behaviour of liquid crystalline polyacrylates. Makromol Chem 183:2311–2321

Potemkin II, Palyulin VV (2009) Comblike macromolecules. Poly Sci Ser A 51:123–149

Pugh C, Schrock RR (1992) Synthesis of side-chain liquid crystal polymers by living ring-opening metathesis polymerization. 3. influence of molecular weight, interconnecting unit, and substituent on the mesomorphic behavior of polymers with laterally attached mesogens. Macromolecules 25:6593–6604

Pugh C, Bae JY, Dharia J, Ge JJ, Cheng SZD (1998) Induction of smectic layering in nematic liquid crystals using immiscible components. 2. laterally attached side-chain liquid-crystalline poly(norbornene)s and their low-molar-mass analogues with hydrocarbon/oligodimethylsiloxane substituents. Macromolecules 31:5188–5200

Reck B, Rangsdorf H (1985) Combined liquid crystalline polymers: mesogens in the main chain and as side groups. Makromol Chem Rapid Commun 6:291–299

Reddy RA, Sadashiva BK (2003) Influence of fluorine substituent on the mesomorphic properties of five-ring ester banana-shaped molecules. Liq Cryst 30:1031–1050

Reddy GSM, Narasimhaswamy T, Jayaramudu J, Sadiku ER, Raju KM, Ray SS (2013) A new series of two-ring-based side chain liquid crystalline polymers: synthesis and mesophase characterization. Aust J Chem 66:667–675

Reddy GSM, Jayaramudu J, Varaprasad K, Sadiku R, Jailani SA, Aderibigbe BA (2014a) Chapter 9—Nanostructured liquid crystals. In: Thomas S, Shanks R, Chandrasekharakurup S (eds) Nanostructured polymer blends. William Andrew, Oxford, pp 299–324

Reddy GSM, Narasimhaswamy T, Raju KM (2014b) Synthesis and mesophase characterization of novel methacrylate based thermotropic liquid crystalline monomers and their polymers. New J Chem 38:4357–4364

Rehahn M, Schluter AD, Wegner G (1990) Soluble poly(*para*-phenylene)s, 3. Variation of the length and the density of the solubilizing side chains. Makromol Chem 191:1991–2003

Reinitzer F (1888) Contributions to the knowledge of cholesterol. Monatsh Chem 9:421–441

Rodekirch G, Riibner J, Zschuppe V, Wolf D, Springer J (1993) New side-group liquid-crystalline polymethacrylates. Makromol Chem 194:1125–1135

Ros MB, Luis Serrano J, Rosario de la Fuente M, Ce Sar LF (2005) Banana-shaped liquid crystal: a new field to explore. J Mater Chem 15:5093–5098

Rosen BM, Wilson CJ, Wilson DA, Peterca M, Imam MR, Percec V (2009) Dendron-mediated self-assembly, disassembly, and self-organization of complex systems. Chem Rev 109:6275–6540

Roviello A, Sirigu A (1975) Mesophasic structures in polymers. A preliminary account on the mesophases of some poly-alkanoates of p, p'-di-hydroxy-α, α'-di-methyl benzalazine. J Polym Sci Polym Lett Ed 13:455–463

Rudick JG, Percec V (2007) Helical chirality in dendronized polyarylacetylene. New J Chem 31:1083–1096

Rudick JG, Percec V (2008) Induced helical backbone conformations of self-organizable dendronized polymers. Acc Chem Res 41:1641–1652

Sabine L, Angelika B, Nelli S, Frank G (2007) Discotic liquid crystals: from tailor-made synthesis to plastic electronics. Angew Chem Int Ed 46:4832–4887

Saez IM, Goodby JW (2005) Supermolecular liquid crystals. J Mater Chem 15:26–40

Saez IM, Goodby JW, Richardson RM (2001) A liquid-crystalline silsesquioxane dendrimer exhibiting chiral nematic and columnar mesophases. Chem Eur J 7:2758–2764

Saravanan C, Kannan P (2010) Fluorine-substituted azobenzene destabilizes polar form of optically switchable fulgimide unit in copolymer system. J Polym Sci Part A Polym Chem 48:1565–1578

Saravanan C, Senthil S, Kannan P (2008) Click chemistry-assisted triazole-substituted azobenzene and fulgimide units in the pendant-based copoly(decyloxymethacrylate)s for dual-mode optical switches. J Polym Sci Part A Polym Chem 46:7843–7860

Sawamura M, Kawai K, Matsuo Y, Kanie K, Kato T, Nakamura E (2002) Stacking of conical molecules with a fullerene apex into polar columns in crystals and liquid crystals. Nature 419:702–705

Schadt M (1989) The history of liquid crystal displays and liquid crystalline materials technology. Liq Crsyt 5:57–71

Scherman OA (2009) From leading to function. Nature Chem 1:524–525

Sentman AC, Gin DL (2003) Polymerizable bent-core mesogens: switchable precursors to ordered, polar polymer materials. Angew Chem Int Ed 42:1815–1819

Shibaev VP (2009) Liquid–crystalline polymers: past, present, and future. Poly Sci Series A 51:1131–1193

Shibaev V, Plate NA (1984) Thermotropic liquid-crystalline polymers with mesogenic side groups. Liquid Crystal Polymers II/III. Adv Polym Sci 60/61:173–252

Shibaev VP, Plate NA (1985) Synthesis and structure of liquid-crystalline side-chain polymers. Pure Appl Chem 57:1589–1602

Spassky N, Lacoudre N, Le Borgne A, Varion JP, Jun CL, Friedrich C, Noel C (1989) Liquid crystal polymers with terminally 1- phenyl −2-(4-cyanophenyl)-ethane substituted side chains. Makromol Chem Macromol Symp 24:271–281

Srinivasan MV, Kannan P (2011) Photo-switching and nonlinear optical behaviors of center linked bent-core azobenzene liquid crystalline polymers. J Mater Sci 46:5029–5043

Srinivasan MV, Kannan P, Roy A (2013a) Investigations on photo and electrically switchable asymmetric bent-core liquid crystals. J Mater Sci 48:2433–2446

Srinivasan MV, Kannan P, Roy A (2013b) Photo and electrically switchable behavior of azobenzene containing pendant bent-core liquid crystalline polymers. J Polym Sci Part A Polym Chem 51:936–946

Takezoe H, Takanishi Y (2006) Bent-core liquid crystals: their mysterious and attractive world. Jpn J Appl Phys 45:597–625

Thisayukta J, Nakayama Y, Kawauchi S, Takezoe H, Watanabe J (2000) Distinct formation of a chiral smectic phase in achiral banana-shaped molecules with a central core based on a 2,7-dihydroxynaphthalene unit. J Am Chem Soc 122:7441–7448

Tschierske C, Dantlgraber G (2003) From antiferroelectricity to ferroelectricity in smectic mesophases formed by bent-core molecules. Pramana J Phys 61:455–481

Ungar G, Percec V, Zuber M (1992) Liquid crystalline polyethers based on conformational isomerism. 20. Nematic-nematic transition in polyethers and copolyethers based on 1-(4-hydroxyphenyl)2-(2-R-4-hydroxyphenyl)ethane with R=fluoro, chloro and methyl and flexible spacers containing an odd number of methylene units. Macromolecules 25:75–80

Urayama K (2007) Issues in liquid crystal elastomers and gels. Macromolecules 40:2277–2288

Vorlander D (1908) About transparently clear, crystalline liquids. Rep Ger Chem Soc 41:2033–2052

Wan XH, Zhang F, Wu PQ, Zhang D, Feng XD, Zhou QF (1995) Characterization of the chain stiffness for a mesogen-jacketed liquid crystal polymer- poly (2,5-bis[(4-methoxybenzoyl)oxy] styrene. Macromol Symp 96:207–218

Wang X, Zhou Q (2004) Liquid crystalline polymers. World Scientific Publishing Co., Singapore

Wang LY, Tsai HY, Lin HC (2010) Novel supramolecular side-chain banana-shaped liquid crystalline polymers containing covalent- and hydrogen-bonded bent cores. Macromolecules 43:1277–1288

Warner M, Terentjev EM (1996) Nematic elastomers—a new state of matter? Prog Polym Sci 21:853–891

Warner M, Terentjev EM (2003) Liquid crystal elastomers. Oxford University Press, Oxford

Wolarz E, Fischer TH, Stumpe J (2005) Photoreorientation in liquid crystalline side-group polysiloxane with azobenzene derivative admixture. Mol Cryst Liq Cryst 437:1245–1253

Wu H, Zhang L, Xu Y, Ma Z, Shen Z, Fan X, Zhou Q (2012) Amphiphilic mesogen-jacketed liquid crystalline polymers: design, synthesis, and self-assembly behaviors. J Polym Sci Part A Polym Chem 50:1792–1800

Xie P, Zhang R (2005) Liquid crystal elastomers, networks and gels: advanced smart materials. J Mater Chem 15:2529–2550

Xu B, Swager TM (1993) Rigid bowlic liquid crystals based on tungsten-oxo calix[4]arenes: host-guest effects and head-to-tail organization. J Am Chem Soc 115:1159–1160

Xu GZ, Wu W, Shen DY, Hou JN, Zhang SF, Xu M, Zhou QF (1993) Morphological study of oriented films obtained from side-chain liquid crystalline polymers. Polymer 34:1818–1822

Xu Y, Yang Q, Shen Z, Chen X, Fan X, Qifeng Zhou Q (2009) Effects of mesogenic shape and flexibility on the phase structures of mesogen-jacketed liquid crystalline polymers with bent side groups containing 1,3,4-oxadiazole. Macromolecules 42:2542–2550

Yang Q, Jin H, Xu Y, Wang P, Liang X, Shen Z, Zhou Q (2009) Synthesis, photophysics, and electroluminescence of mesogen-jacketed 2D conjugated copolymers based on fluorene-thiophene- oxadiazole derivative. Macromolecules 42:1037–1046

Yang Q, Xu Y, Jin H, Shen Z, Chen X, Zou D, Zhou Q. (2010) A novel mesogen-jacketed liquid crystalline electroluminescent polymer with both thiophene and oxadiazole in conjugated side chain. J Polym Sci A Polym Chem, 48:1502–1515

Yang H, Zhang F, Lin BP, Keller P, Zhang XQ, Sun Y, Guo LX (2013) Mesogen-jacketed liquid crystalline polymers and elastomers bearing polynorbornene backbone. J Mater Chem C 1:1482–1490

Ye C, Zhang HL, Huang Y, Chen EQ, Lu YL, Shen DY, Wan XH, Shen ZH, Cheng SZD, Zhou QF (2004) Molecular weight dependence of phase structures and transitions of mesogen-jacketed liquid crystalline polymers based on 2-vinylterephthalic acids. Macromolecules 37:7188–7196

Yu Z, Tu H, Wan X, Chen X, Zhou QF (2003a) Synthesis and characterization of mesogen-jacketed liquid-crystal polymers based on 2, 5-bis(4′-alkoxyphenyl)styrene. J Polym Sci Part A Polym Chem 41:1454–1464

Yu ZN, Tu HL, Wan XH, Chen XF, Zhou QF (2003b) Synthesis and properties of liquid crystalline 4,4″-dialkoxy-2″-methyl-p-terphenyls. Mol Cryst Liq Cryst 391:41–55

Zhang D, Zhou QF, Ma YG, Wan XH, Feng XD (1997) Mesogen-jacketed liquid crystal polymers with mesogens of aromatic amide structure. Polym Adv Technol 8:227–233

Zhang D, Liu Y, Wan X, Zhou QF (1999) Synthesis and characterization of a new series of "mesogen-jacketed liquid crystal polymers" based on the newly synthesized vinylterephthalic acid. Macromolecules 32:4494–4496

Zhang L, Chen S, Zhao H, Shen Z, Chen X, Fan X, Zhou Q (2010) Synthesis and properties of a series of mesogen-jacketed liquid crystalline polymers with polysiloxane backbones. Macromolecules 43:6024–6032

Zhang L, Wu H, Shen Z, Fan X, Zhou QJ (2011) Synthesis and properties of mesogen-jacketed liquid crystalline polymers containing biphenyl mesogen with asymmetric substitutions. Polym Sci Part A Polym Chem 49:3207–3217

Zhou Q, Li H, Feng X (1987) Synthesis of liquid-crystalline polyacrylates with laterally substituted mesogens. Macromolecules 20:233–234

Zhou QF, Li HM, Feng XD (1988) Synthesis of a new class of side chain liquid crystal polymers—polymers with mesogens laterally attached via short linkages to polymer backbones. Mol Cryst Liq Cryst 155:73–82

Zhou Q, Zhu X, Wen Z (1989) Liquid-crystalline side-chain polymers without flexible spacer. Macromolecules 22:491–493

Zhu YF, Guan XL, Shen Z, Fan XH, Zhou QF (2012) Molecular engineering of elastic and strong supertough polyurethanes. Macromolecules 45:3346–3355

Zhu YF, Zhang ZY, Zhang QK, Hou PP, Hao DZ, Qiao YY, Shen Z, Fan XH, Zhou QF (2014) Mesogen-jacketed liquid crystalline polymers with a polynorbornene main chain: green synthesis and phase behaviors. Macromolecules 47:2803–2810

# Chapter 17
# Liquid Crystalline Semiconducting Polymers for Organic Field-Effect Transistor Materials

**Qun Ye and Jianwei Xu**

## 17.1 Introduction

Field effect transistor (FET) has been developed as a switching or amplifying component which serves as a key element in modern electronic integrated circuits. The concept of the field-effect transistor was firstly proposed by Lilienfeld in 1930 (Lilienfeld 1930) and about 30 years later Kahng and Atalla successfully fabricated the first silicon-based field-effect transistor (Kahng and Atalla 1960). Along with the discovery of conducting plastics in the 1970s (Shirakawa et al. 1977), scientists started to implement organic based semiconductor materials into the device of field-effect transistors. The first report of organic semiconductor based FET device was reported by Koezuka and co-workers in 1987 using an electrochemically polymerized polythiophene as the semiconducting layer (Koezuka et al. 1987). After that, tremendous effort has been devoted to the development of organic field-effect transistors. Both the organic semiconductor materials and the device physics and engineering have been advanced significantly during the past decades. After the late 1980s, a vast number of new organic molecules and conjugated polymers have been prepared by synthetic chemists for the pursuit of high performance OFET materials for modern applications (Allard et al. 2008; Anthony et al. 2010; Murphy and Fréchet 2007; Takimiya et al. 2007; Wang et al. 2012). Molecules such as pentacene and polymers such as poly(3-hexylthiophene) (P3HT) have been extensively investigated for the fundamental study and elucidation of device physics of organic field-effect transistors.

Compared with inorganic semiconductor materials, organic materials have several attractive features. First of all, the physical properties of organic materials can be conveniently tuned by adjusting building blocks during the synthesis and hence

Q. Ye (✉) • J. Xu (✉)
Institute of Materials Research and Engineering, A*STAR (Agency for Science, Technology and Research), 3 Research Link, Singapore 117602, Republic of Singapore
e-mail: yeq@imre.a-star.edu.sg; jw-xu@imre.a-star.edu.sg



V.K. Thakur, M.R. Kessler (eds.), *Liquid Crystalline Polymers*,
DOI 10.1007/978-3-319-20270-9_17

to satisfy some specific requirements. Secondly, thin films formed by organic semiconductor materials are basically bonded by weak interactions such as van der Waals forces and dipole-dipole interactions between neighboring molecules or polymer threads. Therefore the organic thin films are relatively softer and more flexible compared with inorganic semiconductors. This feature makes organic semiconductor materials suitable for applications on flexible substrates such as plastics, foils, fibres, and papers (Arias et al. 2010). Thirdly, organic semiconductors, especially polymeric materials, are designed to have good solubility in common organic solvents. This makes the organic semiconductor materials compatible with current solution processing methods such as roll-to-roll printing and ink-jet printing technology (Sirringhaus et al. 2000a). Hence, cheap, flexible and large-area electronics application with a low processing cost has been envisaged as a promising industrialization direction for OFET materials. Such organic electronic products include flexible displays, electronic paper, radio-frequency ID tags, smart cards and various kinds of FET based sensors. All these application possibilities have attracted serious industrial interest as more and more organic materials have demonstrated comparable or even better performance than the conventional amorphous Si materials (Würthner 2001).

Conjugated polymers have been widely explored as charge transport materials in organic electronics. For conjugated polymers used in OFET devices, they typically possess coupled aromatic monomers to form a π-conjugated backbone and solubilizing functionalities attached on the periphery of the conjugated backbone. The rigid backbone conformation facilitates close π-stacking and provides the desired charge transport property along and between the polymer chains while the flexible chains surrounding the backbone render solution processability to the polymer material. This "hairy-rod" design strategy has been successfully applied in the synthesis and demonstration of a large number of high performance OFET polymer materials. P3HT (Fig. 17.1d), for example, is the first polymeric semiconductor which exhibits charge mobility higher than 0.1 $cm^2\ V^{-1}\ s^{-1}$(Sirringhaus et al. 1999). This kind of design strategy actually resembles that of liquid crystalline materials. For liquid crystal (LC) materials, e.g., calamitic and discotic liquid crystals, the molecule normally contains a rigid core and some flexible side chain attached (Fig. 17.1). Indeed, as the OFET conjugated polymers follow the same design strategy, the liquid crystalline property has been observed for quite a number of polymers in this research area. The phase transitions of the LC polymers, i.e., the crystal to LC mesophase transition and the LC mesophase to isotropic transition, typically correspond to the side chain melting and main chain melting, respectively. For these LC polymers, the liquid crystalline property is not just an attribute of the material itself. The intrinsic self-assembly properties and various supramolecular phases of liquid crystalline materials also provide extra benefits for the morphology control and improvement of the performance of the polymer thin films in the OFET device. The thermal annealing post-treatment method, which aims at controlling the morphology of the polymer film with thermal energy, is a common practice for performance enhancement of OFET materials. The detailed understanding of the

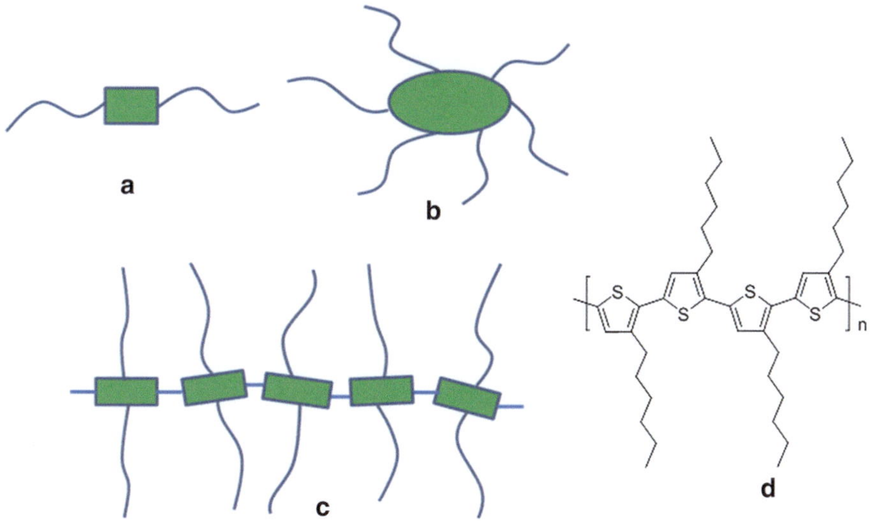


**Fig. 17.1** (**a**) Calamitic liquid crystals; (**b**) discotic liquid crystals; (**c**) hairy-rod type conjugated polymers and (**d**) the chemical structure of poly(3-hexylthiophene) (P3HT)

physics and engineering associated with the thermal annealing process is greatly enhanced by studying the conjugated polymers with liquid crystalline properties.

In this chapter, we mainly focus our discussion on conjugated polymers with liquid crystalline property as organic field-effect transistor materials. We will firstly give a briefly description of the transistor device and related physics. In the next session, the design strategy, property and performance of the LC polymers will be discussed to illustrate how liquid crystalline property of the polymer influences the overall performance of the device. Last but not least, we will end up our discussion by providing a perspective on future research activities in this research area.

## 17.2 Fundamentals of Organic Field Effect Transistors

The device structure of an organic material based FET device is essentially the same as that of a conventional inorganic metal oxide FET device. Figure 17.2 depicts three major types of configurations for OFET devices, that is, top-contact, bottom-gate device (Fig. 17.2a), bottom-contact, bottom-gate device (Fig. 17.2b) and bottom-contact, top-gate device (Fig. 17.2c). An OFET device is mainly built up with three electrodes (source, drain and gate), a dielectric layer and an organic semiconductor layer. The commonly used electrode materials include gold, silver and aluminum. The dielectric layer can be constructed by a variety of materials. Thermally grown silicon dioxide ($SiO_2$) is the most commonly used dielectric material for routine device testing due to its ready availability from conventional

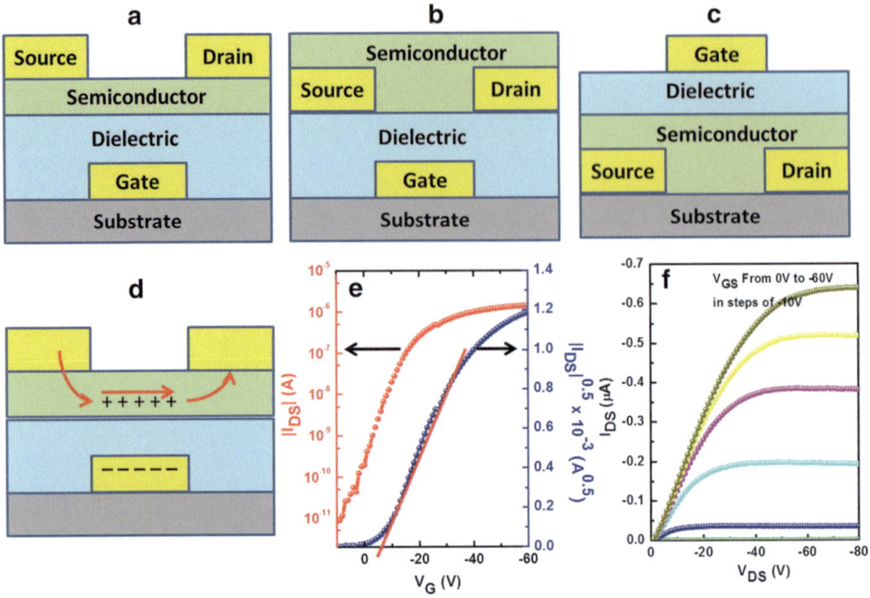


**Fig. 17.2** (**a**) Top-contact, bottom-gate device; (**b**) bottom-contact, bottom-gate device; (**c**) bottom-contact, top-gate device; (**d**) working mechanism of a FET device with p-type materials; (**e**) transfer characteristic curve of FET device and (**f**) output characteristic curve of FET device

Si technology. Other dielectric materials used in OFET devices include inorganic oxides ($Al_2O_3$, $TiO_2$, etc.), organic polymers (polyethylene, polypropylene, CYTOP, poly(vinyl alcohol), poly(vinyl phenol), etc.), hybrid composite materials and self-assembled layers. High capacity and small surface roughness are major requirements for the dielectric materials used in FET devices (Veres et al. 2004). The experimental OFET devices are normally fabricated on an n-doped silicon wafer (gate electrode) with thermally grown $SiO_2$ as the dielectric layer and vapor-deposited gold as the source and drain electrode.

The basic operation principle of OFET device is shown in Fig. 17.2d. As the device is turned off, i.e., no voltage is applied on the gate electrode, the off current ($I_{off}$) should be zero or close to zero. Small current is commonly observed even when the device is at off state mainly due to the presence of charge carriers caused by doping in the semiconducting layer. For *p*-type materials, when a negative voltage is applied on the gate electrode, the device behaves as a capacitor and holes are accumulated at the interface between the semiconducting layer and the dielectric layer. The accumulated charges at the interface serve as a channel for the current flow from the source to the drain and current is modulated by the voltage applied on the gate ($V_G$). The detected current versus the applied voltage can be plotted to the transfer characteristic curve and the output curve as shown in Fig. 17.2e, f. The drain current can be described in the following formula:

$$(I_{SD})_{lin} = (W/L)\mu\ C_i(V_{SG}\text{-}V_T\text{-}V_{SD}/2)\ V_{SD}$$
$$(I_{SD})_{sat} = (W/2L)\mu\ C_i(V_{SG}\text{-}V_T)^2$$

where $(I_{SD})_{lin}$ is the current in the linear region, $(I_{SD})_{sat}$ is the current in the saturated region, μ is the charge carrier mobility, W is the channel width, L is the channel length, $C_i$ is the capacitance per unit area of the insulator layer, $V_T$ is the threshold voltage, $V_{SD}$ is the drain voltage and $V_{SG}$ is the gate voltage.

Critical parameters for the evaluation of an OFET device include charge carrier mobility (μ), current on/off ratio ($I_{on}/I_{off}$) and threshold voltage ($V_T$). For many practical applications, the charge carrier mobility μ is expected to be greater than 0.1 $cm^2\ V^{-1}\ s^{-1}$ while $I_{on}/I_{off} > 10^4$ and ideally the threshold voltage should be close to zero so that the device can be turned on at a very low voltage. Practically the threshold voltage ($V_T$) stability is as important as field-effect mobility, because it is closely associated with the shelf life and operational life of the device.

## 17.3 LC Materials for Organic Field-Effect Transistors

For field-effect transistor materials, several attributes are considered essential to achieve high performance in the device. First of all, the material should pack closely in the thin film so that the charge carrier can be transferred via the closely interacting frontier orbitals. Secondly, the morphology of the thin film is desirable in such a way that large crystalline domains with ordered long range packing and minimal grain boundaries are present. Liquid crystalline materials, due to the presence of both rigid conjugated core and flexible aliphatic chains, can form close π-π stacking for efficient charge transport while at the same time have a high tendency to form homogeneous and continuous films with high quality over the whole channel of the device. All these features of liquid crystalline materials make them as a highly promising class of materials for OFET applications. Indeed, some start-of-the-art OFET devices with LC materials have achieved the mobility values among the best ever reported. In this session, we will first briefly overview some small molecule LC materials used as OFET materials. And the major focus is on conjugated liquid crystalline polymers which have been successfully applied as OFET materials.

### *17.3.1 Small Molecule LC Materials for OFET Materials*

As we have discussed in the previous text, small LC molecules are mainly classified into two types: calamitic (rod-like) and discotic (disc-like) (Fig. 17.1). Calamitic

a

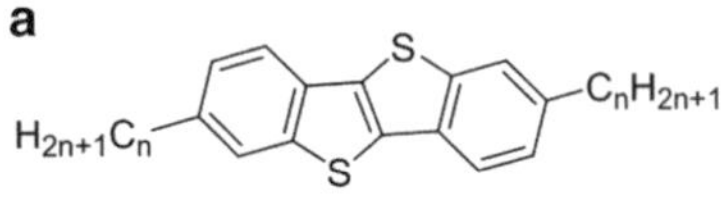

b

n = 6 Cr $\xrightarrow{115.8^{\circ}C}$ LC $\xrightarrow{119.9^{\circ}C}$ I

n = 7 Cr $\xrightarrow{109.3^{\circ}C}$ LC $\xrightarrow{129.5^{\circ}C}$ I

n = 8 Cr $\xrightarrow{111.2^{\circ}C}$ LC $\xrightarrow{126.5^{\circ}C}$ I

n = 9 Cr $\xrightarrow{106.7^{\circ}C}$ LC $\xrightarrow{127.8^{\circ}C}$ I

n = 10 Cr $\xrightarrow{112.2^{\circ}C}$ LC $\xrightarrow{124.2^{\circ}C}$ I

n = 11 Cr $\xrightarrow{109.2^{\circ}C}$ LC $\xrightarrow{122.9^{\circ}C}$ I

n =12 Cr $\xrightarrow{112.7^{\circ}C}$ LC $\xrightarrow{118.3^{\circ}C}$ I

n = 13 Cr $\xrightarrow{112.3^{\circ}C}$ LC $\xrightarrow{116.7^{\circ}C}$ I

c

| Materials | m ($cm^2V^{-1}s^{-1}$) | $I_{on}/I_{off}$ | Remarks |
|---|---|---|---|
| 2Cn-BTBT (n = 5-14) | 2.8 | $10^7$ | Top-contact, spin-coating, annealed at 80 °C |
| $2C_{12}$-BTBT | 3.9 | $10^6$-$10^8$ | Top-contact, ODTS-SAM |
| $2C_8$-BTBT | 5 | $10^6$ | Top-contact, DTS-SAM, directing crystallization |
| $2C_8$-BTBT | 31 | $10^5$-$10^7$ | Top-gate, CYTOP dielectric, ink-jet |

**Fig. 17.3** (**a**) Chemical structure of BTBT; (**b**) the phase transition of the alkylated BTBT; (**c**) Representative OFET performance and relevant device setup of BTBT based molecules

LC molecules are an especially important category of materials which have shown great potential as OFET charge transport materials. 2,7-Dialkylbenzo [*b*]benzo[4,5]thieno[2,3-*d*]thiophenes (BTBT) (Fig. 17.3a) are excellent examples to demonstrate the feasibility of LC materials. Interestingly the derivatives of this aromatic system are firstly synthesized and investigated for the elucidation of their LC properties (Kaszynski and Dougherty 1993). It is Takimiya's research group at Hiroshima who firstly explored the potential of these molecules as the charge transport materials in the OFET device (Takimiya et al. 2014). The rigid core of this series of molecules contains four rings (two benzene rings and two thiophene rings) fused together. The isoelectronic structure of this aromatic system is chrysene instead of tetracene and hence the absorption of this series of materials falls in the UV region ($\lambda_{max} \sim 310$ nm). The thin films formed are thus typically transparent in the visible light. Typically the 2,7-dialkyl-BTBT molecules enter a smectic LC phase (Košata et al. 2003) around 110 °C (Fig. 17.3b). The temperature range for the mesophase is rather narrow, about 10–20 °C. The highest occupied molecular orbital (HOMO) energy level for this series of molecules is ca. −5.7 eV, which is sufficiently deep to ensure the ambient stability of the transistor device. Strong crystallinity and homogeneity of the films are observed. Both out-of-plane and in-plane X-ray diffraction (XRD) study confirm that the molecules in the thin films pack in the same way as in the single crystal. All these features favor the charge transport and hence mobility values up to 2.8 $cm^2 V^{-1} s^{-1}$ have been obtained (Takimiya et al. 2014).

After the continuous development for nearly a decade, more and more research work has demonstrated mobilities higher than 1 $cm^2 V^{-1} s^{-1}$ using BTBT based molecules (Fig. 17.3c). A surprisingly high mobility value of 31.3 $cm^2 V^{-1} s^{-1}$ with average of 16.4 $cm^2 V^{-1} s^{-1}$ can be obtained by 2,7-dioctyl-BTBT following a non-vacuum, room-temperature ink-jet printing process (Minemawari et al. 2011). Due to these exciting experimental works, it is no wonder that some of the BTBT derivatives have already been commercialized by chemical vendors (for C8-BTBT, Sigma Aldrich).

Discotic liquid crystals (DLC) are firstly discovered by Chandrasekhar in 1977 (Chandrasekhar et al. 1977) and the associated chemistry and LC property have been intensively studied over the past decades (Laschat et al. 2007). Although a large library of DLC materials is available in the literatures, the application of DLC materials as organic electronics, especially as charge transport materials in OFET devices, is still in its infancy. Exciting OFET performance data based on this type of materials is still very limited. DLC materials typically exhibit charge carrier mobility in the order of $10^{-3}$ $cm^2 V^{-1} s^{-1}$. Some examples have demonstrated mobility in the order of $10^{-2}$ $cm^2 V^{-1} s^{-1}$ (Dong et al. 2009). Other charge carrier mobility measurement methods such as space charge-limited current (SCLC) technique (Blom et al. 1997) and pulse-radiolysis time-resolved microwave conductivity method (Warman et al. 2004) have exhibited more promising results to demonstrate the charge transport property of DLC materials. The major challenge is to control the alignment of DLC materials in the OFET channel by the development of appropriate processing techniques and to make the materials more compatible with current FET device engineering.

## *17.3.2 Liquid Crystalline Polymers for OFET Materials*

### 17.3.2.1 Fluorene Based LC Polymers for OFET Materials

Fluorene based polymers are especially good candidates for the application of organic light-emitting diode (OLED) materials due to their superior solid-state photoluminescence quantum efficiency, good solubility and film-forming property, and versatile tuning of their physical properties by modification of substituents on the 9-position (Xie et al. 2012). Since the first work on the synthesis of soluble poly (9,9-dialkylfluorene-2,7-diyl) by Fukuda and co-workers (Fukuda et al. 1993), fluorene based polymers have been intensively investigated as a novel blue LED material. Given the popularity of this type of material in the research field of OLED, polyfluorene (PF) and related polymers have been tested in OFET devices to study their charge transport properties. Another attribute of fluorene based polymers is their liquid crystalline properties due to their "hairy-rod" like chemical structures.

**Table 17.1** Summary of fluorene based LC polymers (**P1**–**P6**)

| | Structure | Phase transition | Mobility ($cm^2\ V^{-1}\ s^{-1}$) |
|---|---|---|---|
| **P1** | 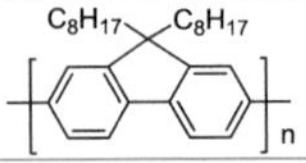 | Cr $\xrightarrow{170^\circ C}$ N $\xrightarrow{270^\circ C}$ I | $9 \times 10^{-3}$ (TOF) |
| **P2** | 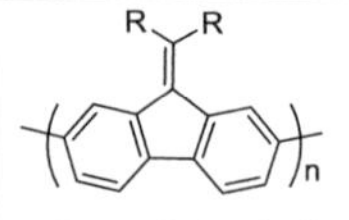 R= $C_6H_{13}$, $C_8H_{17}$, $C_{10}H_{21}$ | Cr $\longrightarrow$ N $\longrightarrow$ I | $2 \times 10^{-3}$ |
| **P3** | 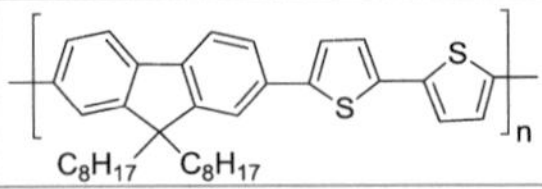 | Cr $\xrightarrow{120^\circ C}$ LC $\xrightarrow{263^\circ C}$ I | $0.4 \times 10^{-3}$ |
| **P4** | 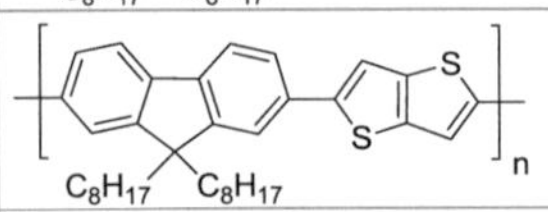 | Cr $\xrightarrow{160^\circ C}$ LC $\xrightarrow{260^\circ C}$ I | $1.1 \times 10^{-3}$ |
| **P5** | 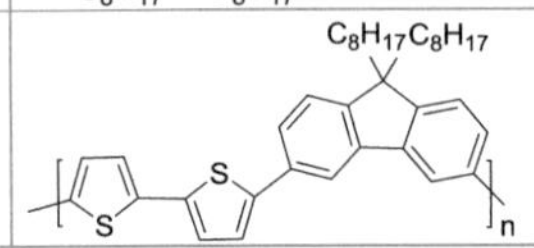 | Cr $\xrightarrow{184^\circ C}$ $LC_1$ $\xrightarrow{237^\circ C}$ $LC_2$ $\xrightarrow{311^\circ C}$ I | 0.02 (aligned) |
| **P6** | 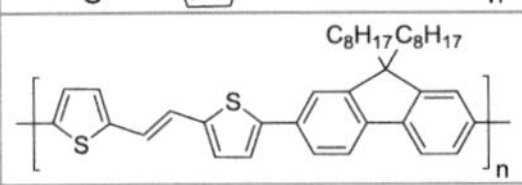 | Amorphous | 0.02 |

The LC property of the fluorene containing polymers, as will be described later, plays a crucial role in improving the charge mobility of the polymers.

Poly(9,9-dioctylfluorene) (**P1** in Table 17.1) (Redecker et al. 1998, 1999) is a solution-processable polymer that displays nematic mesophase above 160 °C. For a uniformly aligned nematic glass of **P1** on a rubbed polyimide substrate, the hole mobility can be improved from $4 \times 10^{-4}\ cm^2\ V^{-1}\ s^{-1}$ to $9 \times 10^{-3}\ cm^2\ V^{-1}\ s^{-1}$ as determined by time-of-flight (TOF) measurement. To further improve the charge carrier mobility of PFs, Heeney and co-workers prepared a series of polyalkylidene fluorene polymers (**P2** in Table 17.1) (Heeney et al. 2004). For this series of polymers, the *sp*$^3$ carbon of fluorene is replaced with a *sp*$^2$ carbon, which allows the solubilizing alkyl chains to adopt a co-planar instead of a perpendicular conformation relative to the planar backbone. This design strategy encourages stronger π-π interactions between neighbouring polymer chains and hence a higher mobility value compared with normal PF polymers. This series of polymers shows a nematic liquid crystalline mesophase while detailed analysis on the LC mesophase is not carried out due to multiple overlapping processes. The absorption of **P2** exhibits a ca. 30 nm red shift compared with that of **P1**, indicating an enhanced planarization due to the sp$^2$-hybridized linkage. The OFET devices are fabricated for **P2** and field-effect mobility values from $1 \times 10^{-4}$ to $9 \times 10^{-4}\ cm^2\ V^{-1}\ s^{-1}$ are obtained from *as-spun* thin films. To optimize the device performance, the polymer films were annealed in the nematic phase under nitrogen atmosphere. The annealing post-treatment aims to optimize the morphology of the film and render enhanced order and π-π stacking within the film. After annealing treatment, the hole mobility increases to about $2 \times 10^{-3}$ cm$^2$ V$^{-1}$ s$^{-1}$. Optimal annealing temperature was found at 170–190 °C for this series of polymers and higher or lower annealing temperature does not achieve such improvement of the mobility.

Fluorene moiety is also implemented to form copolymers to explore its potential as OFET materials (**P3–P6**). **P3** and **P4** are copolymers between fluorene and bithiophene and thieno[2,3-*b*]thiophene, respectively via the 2,7-linkage (Lim et al. 2005). For **P3**, it exhibits a crystal to liquid crystal transition at 120 °C and a liquid crystal to isotropic phase at 263 °C. While for **P4**, the crystal to liquid crystal transition temperature is at 160 °C and liquid crystal to isotropic phase transition is at 260 °C. Typical nematic liquid crystal textures are observed for both polymers. To study their charge transport properties, the films are annealed at 285 °C and then quenched to room temperature to maintain the nematic alignment. **P4** exhibits a hole mobility of $1.1 \times 10^{-3}$ cm$^2$ V$^{-1}$ s$^{-1}$, which is about 3 times higher than that of **P3** ($0.4 \times 10^{-3}$ cm$^2$ V$^{-1}$ s$^{-1}$). The much higher mobility of **P4** than **P3** is also reflected by its better long distance packing order as revealed by XRD study. For **P3**, the XRD pattern is essentially featureless while for **P4** multiple pronounced diffraction peaks can be observed. The more planar and rigid backbone of **P4** would account for this observation. This result indicates that the FET mobility is closely related to the order of packing in the solid state.

Sirringhaus and co-workers studied a 3,6-linked copolymer poly-9,9′-dioctyl-fluorene-co-bithiophene (**P5**) (Sirringhaus et al. 2000a, b). This polymer exhibits two mesophase transitions at 184 °C and 237 °C and the mesophase to isotropic phase at 311 °C (Kinder et al. 2004). Compared with **P3**, all the transition temperatures are much higher. In this study, the polymer chains of **P5** are aligned by a rubbed polyimide substrate. The aligned thin film is then heated to 275–285 °C and quenched to room temperature to maintain the nematic packing state. Top-gate OFET devices are fabricated on the aligned nematic glass. High mobility up to 0.02 cm$^2$ V$^{-1}$ s$^{-1}$ with $I_{on}/I_{off} > 10^5$ and good ambient stability are obtained for **P5** when measurement is done along the alignment direction. If measurement is done perpendicular to the alignment direction, much lower mobility (0.001–0.002 cm$^2$ V$^{-1}$ s$^{-1}$) is observed. For the isotropic devices with no aligned substrate, mobilities of 0.003–0.005 cm$^2$ V$^{-1}$ s$^{-1}$ are obtained. The annealing process also has a significant effect on the overall performance of the device of **P5**. For the *as-spun* thin films, much lower mobility ($<10^{-3}$ cm$^2$ V$^{-1}$ s$^{-1}$) is obtained.

To further explore the potential of fluorene based polymer for OFET, Chung and co-workers prepared a copolymer between fluorene and 1,2-bis(2′-thienyl)vinyl moiety (**P6** in Table 17.1) (Chung et al. 2008). Compared with **P3**, there is one extra vinyl linkage in each repeating unit for **P6**. Due to the extra vinyl moiety, the conjugated polymer backbone of **P6** is more rigid and extended than **P3**, and **P6** shows an amorphous nature without any liquid crystalline mesophase observed for **P3**. Despite its amorphous nature, top-contact OFET device for **P6** with PDMS treated $SiO_2$ as the dielectric layer shows a promising a hole mobility of 0.02 cm$^2$ V$^{-1}$ s$^{-1}$. This mobility value is among the highest reported for fluorene based polymers.

#### 17.3.2.2 Thiophene Based Liquid Crystal Polymers for OFET Materials

Thiophene based polymers have been widely explored as organic electronics materials (McCulloch et al. 2009; Lu and Liu 2010). Poly(3-hexyl-thiophene) (P3HT), for example, is a benchmark material to establish the foundation of OFET research. A lot of thiophene and fused thiophene based polymers have the liquid crystalline attribute and have been demonstrated as high performance OFET materials. Some of them have been listed in Table 17.2.

Ong and co-workers prepared a solution processable regioregular polythiophene, poly(3,3‴-didodecyl-quaterthiophene) (**P7** in Table 17.2) via ferric chloride mediated

**Table 17.2** Summary of thiophene based LC polymers and their OFET mobility (**P7–P14**)

| | Structure | Phase transition | Mobility ($cm^2\ V^{-1}\ s^{-1}$) |
|---|---|---|---|
| **P7** | 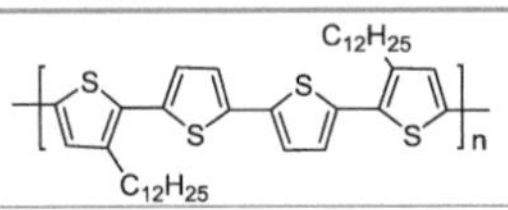 | Cr $\xrightarrow{110°C}$ LC $\xrightarrow{140°C}$ I | 0.14 |
| **P8** | 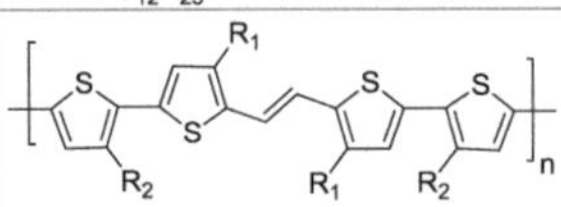 **P8a** $R_1 = C_6H_{13}$, $R_2 = C_{12}H_{25}$<br>**P8b** $R_1 = R_2 = C_{12}H_{25}$ | **P8a** Cr $\xrightarrow{130°C}$ LC $\xrightarrow{185°C}$ I<br>**P8b** Cr $\xrightarrow{80°C}$ LC $\xrightarrow{170°C}$ I | **P8a**: 0.47<br>**P8b**: 0.94 |
| **P9** | 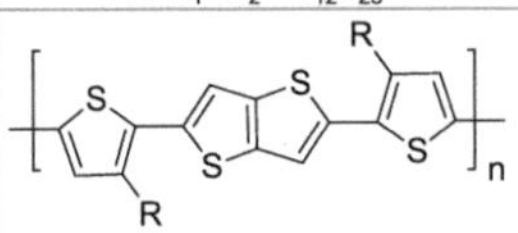 **P9a** R = $C_{10}H_{21}$ **P9b** R = $C_{12}$H25<br>**P9c** R = $C_{12}H_{29}$ **P9d** R = $C_{16}H_{33}$ | **P9a** Cr $\xrightarrow{171°C}$ LC $\xrightarrow{251°C}$ I<br>**P9b** Cr $\xrightarrow{143°C}$ LC $\xrightarrow{244°C}$ I<br>**P9c** Cr $\xrightarrow{141°C}$ LC $\xrightarrow{248°C}$ I | **P9a**: 0.3<br>**P9b**: 0.3<br>**P9c**: 0.39<br>**P9d**: 0.72 |
| **P10** | 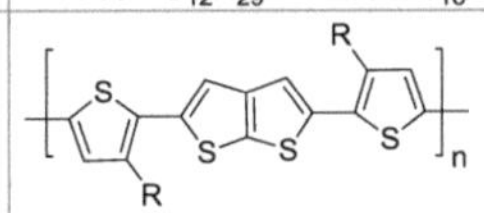 **P10a** R = $C_{10}H_{21}$<br>**P10b** R = $C_{12}H_{25}$ | **P10** Cr $\xrightarrow{rt}$ LC $\xrightarrow{180°C}$ I | **P10a**: 0.15<br>**P10b**: 0.12 |
| **P11** | 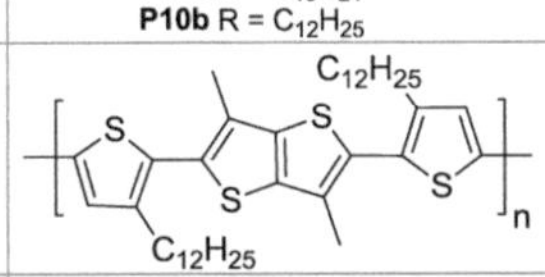 | Cr $\xrightarrow{85°C}$ LC $\xrightarrow{131°C}$ I | $8.1 \times 10^{-7}$ |
| **P12** | 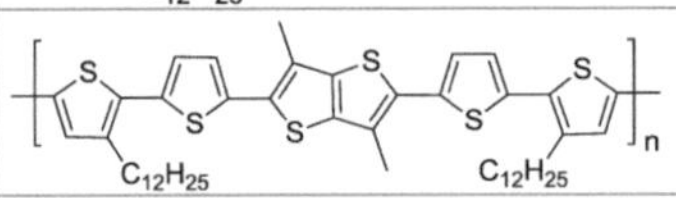 | Cr $\xrightarrow{72°C}$ LC $\xrightarrow{221°C}$ I | 0.03 |
| **P13** | 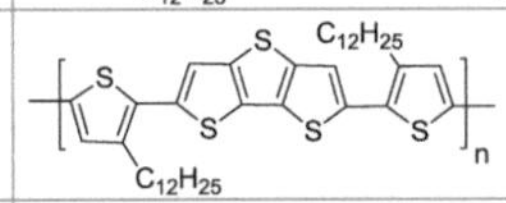 | No transition | 0.3 |
| **P14** | 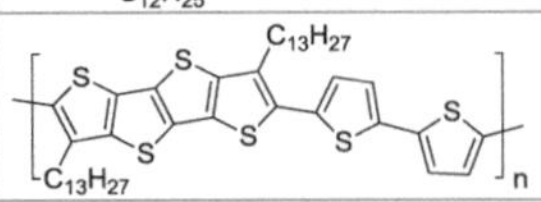 | No transition | 0.33 |

oxidative coupling polymerization (Ong et al. 2004; Zhao et al. 2004). **P7** exhibits two endotherms at 110 °C and 140 °C, corresponding to the crystal to LC mesophase transition and LC mesophase to isotropic transition, respectively. Annealing at the mesophase is found to be crucial in determining the morphology and the long range ordered packing of the polymer in the thin film. Typically the polymer is annealed in the LC mesophase. According to XRD analysis, much stronger diffraction peaks have been observed for the annealed sample while for the *as-spun* thin films featureless diffraction pattern is observed. AFM study on the annealed thin film of **P7** clearly shows a well-developed crystalline domain structure. This indicates that **P7** has an intrinsic tendency to self-organize into a highly ordered lamellar packing motif, which is very beneficial for the charge transport in the OFET device. For the *as-spun* thin film of **P7**, it exhibits a hole mobility of 0.02–0.05 $cm^2 V^{-1} s^{-1}$ with $I_{on}/I_{off}$ ratio of $10^6$. After annealing, the mobility is increased up to 0.14 $cm^2 V^{-1} s^{-1}$ with $I_{on}/I_{off}$ ratio $>10^7$. This result clearly demonstrates the importance of annealing as well as morphology control in order to enhance the OFET performance.

A vinylene moiety is inserted into the polythiophene backbone to increase the effective conjugation and planarity of the backbone and **P8** is prepared by Kim and co-workers to function as OFET materials (Kim et al. 2011). The absorption maximum of **P8** ($\lambda_{max} \sim 510$ nm) is about 40 nm red shifted compared with that of **P7** ($\lambda_{max} \sim 470$ nm), indicating a more conjugated backbone due to the presence of the vinylene moiety. **P8b** with four dodecyl side chains shows lower endotherms than **P8a** with two hexyl and two dodecyl chains as a result of the increased flexibility of the side chains. Both polymers pack in a lamellar order and exhibit very strong X-ray diffractions with the fifth diffraction order (500) observed. Nano ribbon-type morphology is observed for these two polymers. For the *as-spun* thin films of **P8a** and **P8b**, OFET mobility values as high as 0.29 and 0.24 $cm^2 V^{-1} s^{-1}$ are obtained, respectively. After annealing, the mobility of the two polymers is improved to 0.47 and 1.05 $cm^2 V^{-1} s^{-1}$, respectively. The improved mobility by thermal annealing is mainly due to a more ordered packing of the polymers after thermal rearrangement, as revealed by XRD and AFM study.

In order to improve the rigidity and planarity of the polythiophene polymers, fused thiophene monomers, such as thieno[3,2-*b*]thiophene, thieno[2,3-*b*]thiophene, dithieno[3,2-*b*:2′,3′-*d*]thiophene and thieno[2′,3′:4,5]thieno[3,2-*b*]thieno[2,3-*d*]thiophene, have been incorporated into the backbone in hope of further enhancement of the OFET performance. McCulloch et al. have synthesized a series of poly(2,5-bis(3-alkylthiophen-2-yl)thieno[3,2-*b*]thiophenes) (**P9** in Table 17.2) (McCulloch et al. 2006). All these polymers have an ionization potential close to 5.1 eV and show LC mesophases upon heating. The annealing post-treatment has also been proved essential in improving the long-range ordered packing of **P8** in the thin films. After annealing in the LC mesophase, much larger crystalline grain size and stronger X-ray diffraction peaks are observed for **P9b**. Mobility values of 0.2–0.7 $cm^2 V^{-1} s^{-1}$ have been obtained for this series of polymers. Another analogue with hexadecyl chains (**P9d**) is tested by Umeda et al. (Umeda et al. 2009). After treatment of the dielectric interface with self-assembled monolayer of various silane reagents, the mobility of **P9d** can go up to 1.0 $cm^2 V^{-1} s^{-1}$. Moreover, this series of polymers has demonstrated good ambient stability and hence is promising candidates for real applications. To further validate the successful design strategy for this type of polymers, McCulloch and co-workers also incorporated thieno[2,3-*b*]thiophene monomer to prepare **P10** (Heeney et al. 2005). **P10** shows an absorption maximum at 470 nm, which is bathochromically shifted for about 80 nm compared with that of **P9**. This is mainly due to the disruption of π-conjugation by the presence of thieno[2,3-*b*]thiophene monomers. **P10** shows two endotherms in DSC scans, one at room temperature and the other around 180 °C, corresponding to the side-chain melting and main chain melting respectively. Annealing at 100–120 °C improves the performance up to a factor of 2 and mobility values up to 0.15 $cm^2\ V^{-1}\ s^{-1}$ are obtained for this series of polymer. It is worthy to highlight that all these polymers are compatible with solution processing techniques such as roll-to-roll printing and ink jet printing. Baklar et al. (2010) have demonstrated that for OFET devices fabricated with **P9b** using ink-jet printing method, the mobility can be maintained in the range of 0.01–0.1 $cm^2\ V^{-1}\ s^{-1}$ for printed structures.

Kong and co-workers incorporated a 3,6-dimethylthieno[3,2-*b*]thiophene monomer into the polythiophene polymer (**P11** and **P12** in Table 17.2) (Kong et al. 2009). The extra two methyl groups on the thieno[3,2-*b*]thiophene monomer effectively disrupt the π-conjugation along the backbone due to the steric congestion at the linkage position. As a result, the absorption maximum of **P11** ($\lambda_{max}$ at 400 nm) has a bathochromic shift for about 150 nm compared with that of **P9** ($\lambda_{max}$ at 547 nm). **P11** exhibits two endotherms at 85 and 131 °C, both of which are lower than that of **P9**. Due to the distorted π-conjugation and the rather weak π-π stacking as reflected by featureless XRD diffraction patterns, the OFET device of **P11** only displays a low mobility of $8.1 \times 10^{-7}\ cm^2\ V^{-1}\ s^{-1}$ with a low $I_{on}/I_{off}$ ratio of $10^1$. By inserting two more thiophene spacers into the polymer backbone, **P12** can be constructed to relief the steric effect caused by the two methyl groups. The absorption maximum of **P12** goes to 600 nm which indicates a much better π-conjugation along the backbone. The pronounced π-conjugation also allows a more effective inter-chain π-π stacking which facilitates the charge transport in the thin film. For **P12**, the mobility is enhanced to 0.03 $cm^2\ V^{-1}\ s^{-1}$, about $10^4$ times of improvement from **P11**. The favourable structural order of the polymer chain which allows better π-π interaction is hence demonstrated to play an essential role in determining the mobility of the material.

Dithieno[3,2-*b*:2′,3′-*d*]thiophene moiety is inserted into the polythiophene polymer by Li et al. to explore its potential as OFET materials (**P13** in Table 17.2) (Li et al. 2008). The absorption onset of **P13** goes to 620 nm, showing a more conjugated backbone compared with **P7** and **P9**. However, due to the rigid backbone, no LC phase is observed for this polymer by DSC measurement. The thin film of **P13** shows strong X-ray diffraction after annealing at 160 °C and high mobility value up to 0.3 $cm^2\ V^{-1}\ s^{-1}$ with $I_{on}/I_{off}$ ratio $>10^7$ has been obtained. A more rigid fused thiophene monomer 3,7-dialkylthieno[2′,3′:4,5]thieno[3,2-*b*]thieno[2,3-*d*] thiophene is incorporated into polythiophenes as well (**P14** in Table 17.2) (Fong et al. 2008). As a result of the more rigid backbone, **P14** does not exhibit any LC mesophase upon heating to 350 °C. The *as-spun* films of **P14** reveals a good field-effect mobility of 0.33 $cm^2\ V^{-1}\ s^{-1}$, with $I_{on}/I_{off}$ ratio $>10^5$ and a $V_T$ of −10 V.

Here it is worth noting that the liquid crystalline property of a hairy-rod type polymer is closely related to the rigidity of the central backbone and the number and flexibility of the surrounding aliphatic chains. Liquid crystalline mesophase occurs at lower temperature with less rigidity of the central backbone and more flexibility of the side chains. However, long branched side-chains normally cause an amorphous nature to the polymer mainly due to the disruption of the ordering in the aliphatic region (Mondal et al. 2010). These empirical rules are very helpful to elucidate the LC property of conjugated polymers.

Other than thiophene based monomers, a variety of other type of electron rich donors and electron deficient acceptors can be incorporated into the polythiophene backbone to form copolymers. As thiophene-based donors are copolymerized with electron-rich moieties, the generated polymers are typically p-type under a normal OFET device configuration. As thiophenes are copolymerized with electron-deficient moieties, ambipolar materials and n-type materials can be generated. The nature of the charge carrier is essentially determined by a number of factors, such as the frontier orbital levels of the semiconductor material, the work function of the electrode, and the treatment on the dielectric interface. All these factors contribute to the final performance of the material in the transistor device. Nevertheless, by appropriate design of the polymers, for example, by judicious choice of different types of comonomers with different degree of electron richness, one is able to tune the charge carrier types in a polymer semiconductor. This will be illustrated by the following examples (**P15–P17**).

Chen and co-workers have prepared two polymers (**P15** in Table 17.3) copolymerized between 4,4′-dihexadecyl-2,2′-bithiophene and a dithieno

**Table 17.3** Summary of a series of thiophene copolymer as p-type, ambipolar and n-type OFET materials

| | Structure | Phase transition | HOMO (eV) | LUMO (eV) | Mobility ($cm^2\ V^{-1}\ s^{-1}$) |
|---|---|---|---|---|---|
| **P15** | 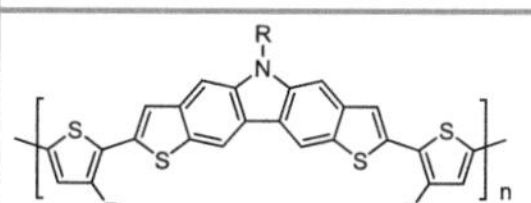 **P15a** R = $C_{12}H_{25}$ **P15b** R = $C_{16}H_{33}$ | Cr $\xrightarrow{147^{\circ}C}$ LC | **−5.0** | **−2.6** | **P15a**: $\mu_h$ 0.2 **P15b**: $\mu_h$ 0.39 |
| **P16** | 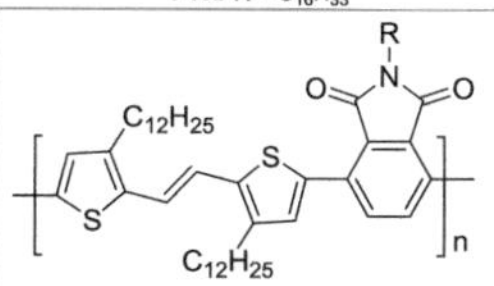 **P16a** R = 2-ethylhexyl **P16b** R = $C_{12}H_{25}$ | – | −**5.2** | −**3.3** | **P16a**: $\mu_h$ 0.75; μe 0.06 **P16b**: $\mu_h$ 0.068; $\mu_e$ 0.05 |
| **P17** | 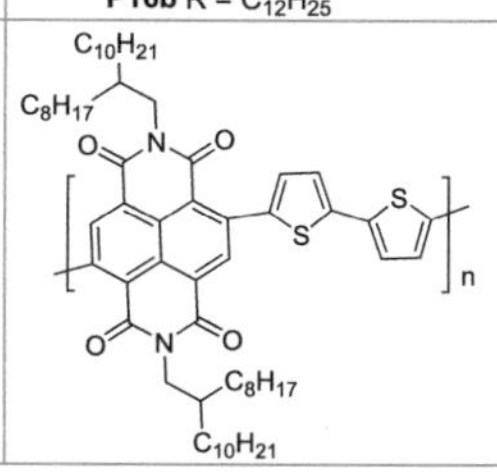 | – | −**5.45** | −**4.0** | $\mu_e > 0.1$ up to 0.85 |

[3,2-*b*:6,7-*b*]carbazole monomer (Chen et al. 2012). The dithienocarbazole monomer is electron rich in nature and hence the prepared polymer is intrinsically electron rich and behaves as a p-type semiconductor. The absorption maximum for **P15** is located at ca. 480 nm in the solution. The HOMO and LUMO energy levels of **P15** are estimated to be ~−5.0 eV and ~2.6 eV. For **P16b** with three hexadodecyl side chains a phase transition at 147 °C was observed corresponding to the crystalline to LC mesophase transitions. For **P16a**, as one hexadodecyl chain is replaced with a shorter dodecyl chain, no such phase transition is observed. Thermal annealing post-treatment is found essential to enhance the OFET performance of this series of polymers. For example, for **P16a**, the mobility is only ca. 0.01 $cm^2\ V^{-1}\ s^{-1}$ for the *as-spun* thin film on an octadecyltrichlorosilane (ODTS) treated dielectric surface. However, after the thin film is annealed at 150 °C, the mobility increases significantly to 0.2 $cm^2\ V^{-1}\ s^{-1}$. For both polymers, the optimal OFET characteristics are obtained on ODTS treated dielectric with annealing at 150 °C and the highest mobility values obtained for **P16a** and **P16b** are 0.2 and 0.39 $cm^2\ V^{-1}\ s^{-1}$,with $I_{on}/I_{off}$ ratio $>10^6$ and $V_T \sim -10$ V, respectively.

A series of copolymer between dithienylvinylene and phthalimide (**P17** in Table 17.3) is reported by Kim et al. (2013). For this series of polymers, the electron withdrawing phthalimide is inserted into the polythiophene backbone. Both polymers display a melt-like thermal transition at 260 °C and 293 °C, respectively and no LC mesophase is observed. The HOMO and LUMO energy levels of the polymers are determined to be ~−5.2 and ~−3.3 eV, respectively. Both polymers exhibit ambipolar charge transport behaviour in the OFET transistor device. For **P17a**, the best hole mobility ($\mu_h$) of 0.75 $cm^2\ V^{-1}\ s^{-1}$ is obtained after annealing at 200 °C and the best electron mobility ($\mu_e$) of 0.06 $cm^2\ V^{-1}\ s^{-1}$ is obtained after annealing at 250 °C. For **P17b**, the best hole mobility ($\mu_h$) of 0.07 $cm^2\ V^{-1}\ s^{-1}$ is obtained after annealing at 300 °C and the best electron mobility ($\mu_e$) of 0.05 $cm^2\ V^{-1}\ s^{-1}$ is obtained after annealing at 250 °C. The variation of the hole and electron mobility due to different annealing temperatures might be correlated to the structural reorganization during annealing process, which modifies the charge transport pathways differently for holes and electrons.

By incorporating more electron deficient moieties into the polythiophene backbone, the generated polymers would become more and more favourable for electron transport. This has been demonstrated by Yan et al. (2009). A naphthalene diimide and bithiophene copolymer (**P17**) has been prepared and implemented as n-type OFET semiconductors. **P17** possesses two long branched alkyl chains in each repeating unit to guarantee its solubility. No thermal transition below 300 °C is observed, mainly due to the rigid backbone and amorphous nature of the aliphatic chains. A low LUMO energy level at −4.0 eV is estimated, making **P17** very favourable for electron transport. A variety of dielectric substrates have been tested and generally **P17** exhibits excellent n-type behaviour with an electron mobility >0.1 $cm^2\ V^{-1}\ s^{-1}$ with the best data up to 0.85 $cm^2\ V^{-1}\ s^{-1}$, $I_{on}/I_{off} > 10^6$. It should be noted that all these promising data is obtained by solution based processing methods including spin coating, gravure, flexography and ink-jet printing, thus demonstrating great versatility and usefulness of this material.

**Table 17.4** Summary of thiazole containing polymers as OFET materials

| | Structure | Phase transition | Mobility ($cm^2\ V^{-1}\ s^{-1}$) |
|---|---|---|---|
| **P18** | 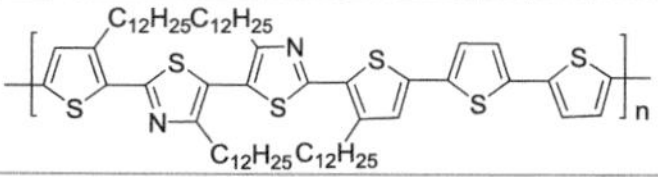 | Cr $\xrightarrow{150°C}$ LC $\xrightarrow{195°C}$ I | 0.33 |
| **P19** | 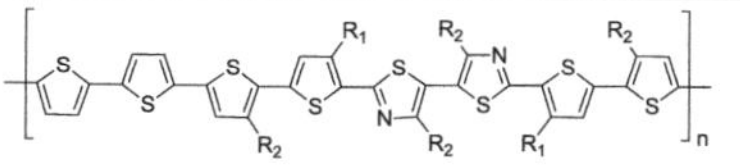 **P19a** $R_1 = R_2 = C_8H_{17}$ **P19b** $R_1 = R_2 = C_{12}H_{25}$ | **P19a** Cr $\xrightarrow{169°C}$ LC/I<br>**P19b** Cr $\xrightarrow{102°C}$ LC/I | **P19a**: 0.25<br>**P19b**: 0.1 |
| **P20** | 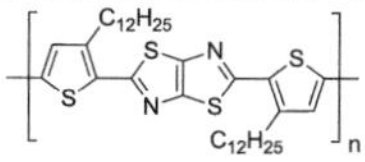 P20a Insoluble 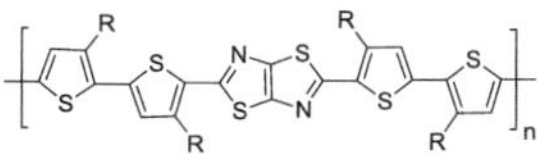 **P20b** R = $C_{12}H_{25}$ **P20c** R = $C_{14}H_{29}$ | **P20b** Cr $\xrightarrow{55°C}$ LC $\xrightarrow{303°C}$ I<br>**P20c** Cr $\xrightarrow{70°C}$ LC $\xrightarrow{288°C}$ I | **P20b**: 0.23<br>**P20c**: 0.3 |
| **P21** | 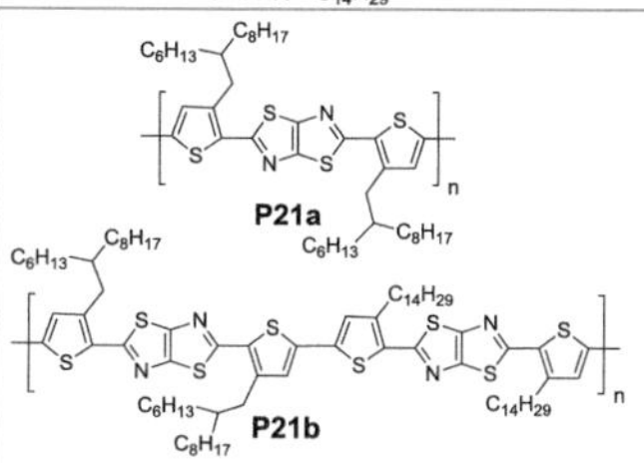 | Amorphous | **P21a**: $2.7 \times 10^{-4}$<br>**P21b**: 0.13 |

### 17.3.2.3 Thiazole Based LC Polymer as OFET Materials

Thiazole and related conjugated systems can be considered as the aza-analogue of thiophene and fused thiophenes. Replacement of carbon atoms with nitrogen atoms is a routinely used strategy to tune the electron affinity of a conjugated system and to prepare electron deficient systems. The HOMO energy level of the thiazole containing polymers hence is decreased compared with that of their thiophene analogues, thus further improving the ambient stability of the OFET device. Incorporation of thiazole and related derivatives has been accomplished to prepare a series of thiazole containing polymers as OFET materials (Table 17.4). Through appropriate design strategy, these thiazole containing polymers display a liquid crystalline mesophase. Annealing of these LC polymers has been found to be essential in improving the OFET performance of this series of polymers as well.

As the carbon atoms are replaced by nitrogen atoms, the polymers become more polar and the inter-chain interaction becomes stronger due to the presence of potential N–H hydrogen bonding. Therefore thiazole containing polymers typically exhibit reduced solubility compared with their thiophene analogues. For example, **P9** with two dodecyl chains in each repeating unit is found to be very soluble in common organic solvents. However, as the central thieno[3,2-*b*]thiophene moiety is replaced by thiazole[5,4-*d*]thiazole, **P20a** becomes insoluble in almost any organic solvents even under high temperature. Due to this solubility difference, thiazole containing polymers typically contains more flexible chains in each repeating unit to guarantee the solution processability of the final polymer.

A bithiadiazole containing donor-acceptor type polymer **P18** is prepared by Kim et al. (2009). In each repeating unit, there are four dodecyl side-chains to solubilize the polymer. The HOMO energy level of **P18** (−5.19 eV) is higher than that of regioregular P3HT (−4.9 to −5.0 eV), mainly due to the presence of thiadiazole units in the backbone. **P18** exhibits two endotherms at 150 °C and 195 °C, corresponding to the side chain melting and main chain melting, respectively. Grazing incidence angle X-ray diffraction (GIXRD) study reveals that **P18** packs on the gate dielectric in a side-on manner, that is, the conduction pathway is parallel to the dielectric surface, which is very beneficial for charge transport in the device. Annealing at 100 °C (crystal phase) to 180 °C (LC phase) has improved such favourable packing of the polymers. The OFET mobility is found closely related to the annealing temperature. Highest field-effect mobility of 0.33 $cm^2\ V^{-1}\ s^{-1}$ is obtained after annealing at 180 °C, which is about 1 order magnitude higher than that of the *as-spun* films. After inserting two more alkylthiophene units into the repeating unit of **P18**, **P19** can be prepared (Lee et al. 2013). The HOMO energy level of **P19** (−5.01 eV) does not vary compared with that of **P18. P19a** with six octyl side-chains in each repeating unit melts at 169 °C while **P19b** with six dodecyl side-chains melts at 102 and 144 °C. For this series of polymers, the highest mobility is 0.25 $cm^2\ V^{-1}\ s^{-1}$ for **P19a** after annealing at 160 °C and 0.1 $cm^2\ V^{-1}\ s^{-1}$ for **P19b** after annealing at 140 °C.

Another intriguing thiazole based building block is thiaozolo[5,4-*d*]thiazole, which resembles thieno[3,2-*b*]thiophene. This novel building block is incorporated into polythiophene system by McCulloch and co-workers (Osaka et al. 2007, 2009). Ideally **P20a** has the most concise and elegant chemical structure as OFET materials. However, this polymer turns out to be insoluble in organic solvents. To address this problem, two more alkyl-thiophene units are introduced to help dissolve the polymer. **P20b** and **P20c** are prepared to this end. The phase transition temperatures are at 55 °C and 303 °C for **P20b**, 70 °C and 288 °C for **P20c**, respectively. The crystalline to LC mesophase transition temperature for **P20c** is higher than **P20b**, presumably due to stronger inter-chain interactions and hence for **P20c** it requires higher energy for the longer chain to melt. Once the chains are completely melted, they behave as solvents surrounding the rigid backbone. The longer chains have better solubilizing ability and hence the LC mesophase to isotropic transition for **P20c** is lower than that of **P20b**. Bottom-gate, bottom-contact OFET devices for **P20b** and **P20c** are fabricated and annealed at 150 °C (LC mesophase) to improve their performance. Higher (>200 °C) annealing temperature degrades the device performance. Best performance is achieved as 0.11–0.23 $cm^2\ V^{-1}\ s^{-1}$, with $I_{on}/I_{off}$ ratio at $10^5$–$10^7$ and $V_T$ at −22 to −31 V for **P20b** and 0.20–0.30 $cm^2\ V^{-1}\ s^{-1}$, with $I_{on}/I_{off}$ ratio at $10^5$–$10^7$ and $V_T$ at −23 to −30 V for **P20c**.

To solve the solubility problem of **P20a**, McCulloch and co-workers utilize branched alkyl chains to solubilize the thiazolo[5,4-*d*]thiazole containing polymers and a series of such polymers (**P21**) has been prepared (Osaka et al. 2010). Both polymers have good solubility in chlorinated organic solvents. Due to the presence of long, branched alkyl chains at the periphery, both **P21a** and **P21b** are essentially amorphous. The HOMO energy level of **P21a** and **P21b** is at −5.2 to −5.3 eV, about 0.1 eV lower than that of **P9**. Bottom-gate bottom-contact OFET devices are fabricated and annealed at 150 °C for these two polymers. The best mobility data is $2.7 \times 10^{-4}$ and 0.13 $cm^2\ V^{-1}\ s^{-1}$ for **P21a** and **P21b**, respectively. The much lower mobility of **P21a** is due to the absence of strong π-π stacking, as revealed by grazing incidence wide-angle X-ray scattering study. This work highlights that the solubilizing chains not only help improve the solubility, but also play a vital role in the π-π stacking and morphology control of the polymer film. The detailed mechanism and methodology to control such influence, however, are still unrevealed.

## 17.4 Summary

In this chapter liquid crystalline polymers for OFET materials have been summarized and discussed. The application of LC materials as organic electronics is still in its infancy and there are still a lot to discover, especially about how the LC property can be beneficial for the performance of the device. For organic field-effect transistor materials, it has been clearly proven that thermal annealing of the LC polymers can help improve the morphology and π-π stacking of the polymers in the films, hence to enhance the overall device performance. A lot of LC conjugated polymers have been demonstrated as high performance OFET materials with good stability under ambient conditions. All these inspiring results have paved the way for the potential commercialization of such materials. There should be no wonder that more and more LC conjugated polymer materials will be synthesized and applied as organic electronics materials, especially as OFET materials, in the near future.

## References

Allard S, Forster M, Souharce B, Thiem H, Scherf U (2008) Organic semiconductors for solution-processable field-effect transistors (OFETs). Angew Chem Int Ed 47:4070–4098

Anthony JE, Facchetti A, Heeney M, Marder SR, Zhan X (2010) n-Type organic semiconductor in organic electronics. Adv Mater 22:3876–3892

Arias AC, MacKenzie JD, McCulloch I, Rivnay J, Salleo A (2010) Materials and applications for large area electronics: solution-based approaches. Chem Rev 110:3–24

Baklar M, Wöbkenberg PH, Sparrowe D, Goncalves M, McCulloch I, Heeney M, Anthopoulos T, Stingelin N (2010) Ink-jet printed p-type polymer electronics based on liquid-crystalline polymer semiconductors. J Mater Chem 20:1927–1931

Blom PWM, de Jong MJM, van Munster MG (1997) Electric-field and temperature dependence of the hole mobility in poly(p-phenylene vinylene). Phys Rev B 55:R656

Chandrasekhar S, Sadashiva BK, Suresh KA (1977) Liquid crystals of disc-like molecules. Pramana 9:471–480

Chen Y, Liu C, Tian H, Bao C, Zhang X, Yan D, Geng Y, Wang F (2012) Novel conjugated polymers based on dithieno[3,2-*b*:6,7-*b*]carbazole for solution processed thin-film transistors. Macromol Rapid Commun 33:1759–1764

Chung DS, Lee SJ, Park JW, Choi DB, Lee DH, Park JW, Shin SC, Lim YH, Kwon SK, Park CE (2008) High performance amorphous polymeric thin-film transistors based on poly[(1,2-bis (2′-thienyl)vineyl-5′,5″-diyl)-alt-(9,9-dioctylfluorene-2,7-diyl)] semiconductors. Chem Mater 20:3450–3456

Dong S, Tian H, Song D, Yang Z, Yan D, Geng Y, Wang F (2009) The first liquid crystalline phthalocyanine derivative capable of edge-on alignment for solution processed organic thin-film transistors. Chem Commun 21:3086–3088

Fong HH, Pozdin VA, Amassian A, Malliaras GG, Smilgies DM, He M, Gasper S, Zhang F, Sorensen M (2008) Tetrathienoacene copolymers as high mobility, soluble organic semiconductors. J Am Chem Soc 130:13202–13203

Fukuda M, Sawada K, Yoshino K (1993) Synthesis of fusible and soluble conducting polyfluorene derivatives and their characteristics. J Polym Sci Part A Polym Chem 31:2465–2471

Heeney M, Bailey C, Giles M, Shkunov M, Sparrowe D, Tierney S, Zhang W, McCulloch I (2004) Alkylidene fluorene liquid crystalline semiconducting polymers for organic field-effect transistor devices. Macromolecules 37:5250–5256

Heeney M, Bailey C, Genevicius K, Shkunov M, Sparrowe D, Tierney S, McCulloch I (2005) Stable polythiophene semiconductors incorporating thieno[2,3-*b*]thiophene. J Am Chem Soc 127:1078–1079

Kahng D, Atalla MM (1960) IRE solid-state devices research conference. Carnegie Institute of Technology, Pittsburgh

Kaszynski P, Dougherty DA (1993) Synthesis and properties of diethyl 5,10-dihetera-5, 10-dihydroindeno[2,1-a]indene-2,7-dicarboxylates. J Org Chem 58:5209–5220

Kim DH, Lee BL, Moon H, Kang HM, Jeong EJ, Park JI, Han KM, Lee S, Yoo BW, Koo BW, Kim JY, Lee WH, Cho K, Becerril HA, Bao Z (2009) Liquid-crystalline semiconducting copolymers with intramolecular donor-acceptor building blocks for high-stability polymer transistors. J Am Chem Soc 131:6124–6132

Kim J, Lim B, Baeg KJ, Noh YY, Khim D, Jeong HG, Yun JM, Kim DY (2011) Highly soluble poly(thienylenevinylene) derivatives with charge carrier mobility exceeding 1 $cm^2V^{-1}s^{-1}$. Chem Mater 23:4663–4665

Kim J, Baeg KJ, Khim D, James DT, Kim JS, Lim B, Yun JM, Jeong HG, Amegadze PSK, Noh YY, Kim DY (2013) Optimal ambipolar charge transport of thienylenevinylene-based polymer semiconductors by changes in conformation for high-performance organic thin film transistors and inverters. Chem Mater 25:1572–1583

Kinder L, Kanicki J, Petroff P (2004) Structural ordering and enhanced carrier mobility in organic polymer thin film transistors. Synth Met 146:181–185

Koezuka H, Tsumura T, Ando T (1987) Field-effect transistor with polythiophene thin film. Synth Met 18:699–704

Kong H, Jung YK, Cho NS, Kang IN, Park JH, Cho S, Shim HK (2009) New semiconducting polymers containing 3,6-dimethyl(thieno[3,2-*b*]-thiophene or selenopheneno[3,2-*b*] selenophene) for organic thin-film transistors. Chem Mater 21:2650–2660

Košata B, Kozmik V, Svoboda J (2003) Novel liquid crystals based on [1]benzothieno[3,2-b][1] benzothiophene. Liq Cryst 30:603–610

Laschat S, Baro A, Steinke N, Giesselmann F, Hägele C, Scalia G, Judele R, Kapatsina E, Sauer S, Schreivogel A, Tosoni M (2007) Discotic liquid crystals: from tailor-made synthesis to plastic electronics. Angew Chem Int Ed 46:4832–4887

Lee J, Chung JW, Jang J, Kim DH, Park JI, Lee E, Lee BL, Kim JY, Jung JY, Park JS, Koo B, Jin YW, Kim DH (2013) Influence of alkyl side chain on the crystallinity and trap density of states in thiophene and thiazole semiconducting copolymer based inkjet-printed field-effect transistors. Chem Mater 25:1927–1934

Li J, Qin F, Li CM, Bao Q, Chan-Park MB, Zhang W, Qin J, Ong BS (2008) High-performance thin-film transistors from solution-processed dithienothiophene polymer semiconductor nanoparticles. Chem Mater 20:2057–2059

Lilienfeld JE (1930) Method and apparatus for controlling electric currents. US Patent 1745175 A.

Lim E, Jung B, Lee J, Shim H, Lee J, Yang YS, Do L (2005) Thin-film morphologies and solution-processable field-effect transistor behavior of a fluorene-thieno[3,2-b]thiophene-based conjugated copolymer. Macromolecules 38:4531–4535

Lu K, Liu Y (2010) Polythiophenes: important conjugated semiconducting polymers for organic field-effect transistors. Curr Org Chem 14:2017–2033

McCulloch I, Heeney M, Bailey C, Genevicius K, MacDonald I, Shkunov M, Sparrowe D, Tierney S, Wagner R, Zhang W, Chabinyc ML, Kline RJ, McGehee MD, Toney MF (2006) Liquid-crystalline semiconducting polymers with high charge-carrier mobility. Nat Mater 5:328–333

McCulloch I, Heeney M, Chabinyc ML, DeLongchamp D, Kline RJ, Cölle M, Duffy W, Fischer D, Gundlach D, Hamadani B, Hamilton R, Richter L, Salleo A, Shkunov M, Sparrowe D, Tierney S, Zhang W (2009) Semiconducting thienothiophene copolymers: design, synthesis, morphology, and performance in thin-film organic transistors. Adv Mater 21:1091–1109

Minemawari H, Yamada T, Matsui H, Tsutsumi JY, Haas S, Chiba R, Kumai R, Hasegawa T (2011) Inkjet printing of single-crystal films. Nature 475:364–367

Mondal R, Becerril HA, Verploegen E, Kim D, Norton JE, Ko S, Miyaki N, Lee S, Toney MF, Brédas JL, McGehee MD, Bao Z (2010) Thiophene-rich fused-aromatic thienopyrazine acceptor for donor-acceptor low band-gap polymers for OTFT and polymer solar cell applications. J Mater Chem 20:5823–5834

Murphy AR, Fréchet JMJ (2007) Organic semiconducting oligomers for use in thin film transistors. Chem Rev 107:1066–1096

Ong BS, Wu Y, Liu P, Gardner S (2004) High-performance semiconducting polythiophenes for organic thin-film transistors. J Am Chem Soc 126:3378–3379

Osaka I, Sauvé G, Zhang R, Kowalewski T, McCullough RD (2007) Novel thiophene-thiazolothiazole copolymers for organic field-effect transistors. Adv Mater 19:4160–4165

Osaka I, Zhang R, Sauvé G, Smilgies DM, Kowalewski T, McCullough RD (2009) High-lamellar ordering and amorphous-like π-network in short-chain thiazolothiazole-thiophene copolymers lead to high mobilities. J Am Chem Soc 131:2521–2529

Osaka I, Zhang R, Liu J, Smilgies DM, Kowalewski T, McCullough RD (2010) Highly stable semiconducting polymers based on thiazolothiazole. Chem Mater 22:4191–4196

Redecker M, Bradley DDC, Inbasekaran M, Woo EP (1998) Nondispersive hole transport in an electroluminescent polyfluorene. Appl Phys Lett 73:1565–1567

Redecker M, Bradley DDC, Inbasekaran M, Woo EP (1999) Mobility enhancement through homogeneous nematic alignment of a liquid-crystalline polyfluorene. Appl Phys Lett 74:1400–1402

Shirakawa H, Louis EJ, MacDiarmid AG, Chiang CK, Heeger AJ (1977) Synthesis of electrically conducting organic polymers: halogen derivatives of polyacetylene, (CH)x. J Chem Soc Chem Commun 16:578–580

Sirringhaus H, Brown PJ, Friend RH, Nielsen MM, Bechgaard K, Langeveld-Voss BMW, Spiering AJH, Janssen RAJ, Meijer EW, Herwig P, de Leeuw DM (1999) Two-dimensional charge transport in self-organized, high-mobility conjugated polymers. Nature 401:685–688

Sirringhaus H, Kawase T, Friend RH, Shimoda T, Inbasekaran M, Wu W, Woo EP (2000a) High-resolution inkjet printing of all-polymer transistor circuits. Science 290:2123–2126

Sirringhaus H, Wilson RJ, Friend RH, Inbasekaran M, Wu W, Woo EP, Grell M, Bradley DDC (2000b) Mobility enhancement in conjugated polymer field-effect transistors through chain alignment in a liquid crystalline phase. Appl Phys Lett 77:406–408

Takimiya K, Kunugi Y, Otsubo T (2007) Development of new semiconducting materials for durable high-performance air-stable organic field-effect transistors. Chem Lett 36:578–583

Takimiya K, Osaka I, Mori T, Nakano M (2014) Organic semiconductors based on [1]benzothieno [3,2-b][1]benzothiophene substructure. Acc Chem Res 47(5):1493–1502. doi:10.1021/ar400282g

Umeda T, Kumaki D, Tokito S (2009) Surface-energy-dependent field-effect mobilities up to 1 $cm^2$/Vs for polymer thin-film transistor. J Appl Phys 105:024516

Veres J, Ogier S, Lloyd G (2004) Gate insulators in organic field-effect transistors. Chem Mater 16:4543–4555

Wang C, Dong H, Hu W, Liu Y, Zhu D (2012) Semiconducting π-conjugated systems in field-effect transistors: a material odyssey of organic electronics. Chem Rev 112:2208–2267

Warman JM, de Haas MP, Dicker G, Grozema FC, Piris J, Debije MG (2004) Pulse-radiolysis time-resolved microwave conductivity: π-bond-conjugated polymers versus π-π-stacked discotics. Chem Mater 16:4600–4609

Würthner F (2001) Plastic transistors reach maturity for mass applications in microelectronics. Angew Chem Int Ed 40:1037–1039

Xie L, Yin CR, Lai WY, Fan QL, Huang W (2012) Polyfluorene-based semiconductors combined with various periodic table elements for organic electronics. Prog Polym Sci 37:1192–1264

Yan H, Chen Z, Zheng Y, Newman C, Quinn JR, Dötz F, Kastler M, Facchetti A (2009) A high-mobility electron-transporting polymer for printed transistors. Nature 457:679–687

Zhao N, Botton GA, Zhu S, Duft A, Ong BS, Wu Y, Liu P (2004) Microscopic studies on liquid crystal poly(3,3‴-dialkylquaterthiophene) semiconductor. Macromolecules 37:8307–8312

# Chapter 18
# Azobenzene-Containing Liquid Single Crystal Elastomers for Photoresponsive Artificial Muscles

**Jaume Garcia-Amorós and Dolores Velasco**

## 18.1 What Are Liquid Crystals?

Substances in nature generally exist as solids, liquids, and gases. However, besides these three main states, matter may exhibit a large variety of other intermediate phases, including gels, superfluids and liquid crystals (LCs).

In solid materials, intermolecular forces hold the constituting molecules into fixed positions giving rise to well-defined crystalline lattices. On the other hand, in the liquid state, such interactions are overcome by the vigorous vibration of molecules which force them to move randomly. The liquid crystal state, also known as mesophase, is an intermediate phase of condensed matter which combines the fluidity typical of ordinary liquids and the characteristic molecular arrangement of crystalline solids (Gray 1998). In fact, within a mesophase, all molecules tend to point to a common direction, called director, ***n*** (Fig. 18.1). This is in contrast to the isotropic liquid state where molecules show no organisation.

Liquid-crystalline materials may display different phases depending on their intrinsic type of ordering. In the simplest one, the nematic phase, the constituting molecules, generally denoted as mesogens, are approximately oriented parallel to their longest axis because of their rod-like geometry. Indeed, in the nematic phase, mesogens exhibit long-range orientational order but lack of positional one. Such feature makes the nematic mesophase the less ordered and less viscous among all the liquid-crystalline phases. As a result, a much easier modification of the mesogens alignment by applying suitable external stimuli should be possible in the nematic mesophase in comparison to other liquid-crystalline phases where positional order might be present as well (*i.e.* smectic phase). Therefore, liquid

J. Garcia-Amorós • D. Velasco (✉)
Grup de Materials Orgànics, Institut de Nanociència i Nanotecnologia (IN$^2$UB), Departament de Química Orgànica, Universitat de Barcelona, Martí i Franquès 1, E-08028 Barcelona, Spain
e-mail: dvelasco@ub.edu



V.K. Thakur, M.R. Kessler (eds.), *Liquid Crystalline Polymers*,
DOI 10.1007/978-3-319-20270-9_18

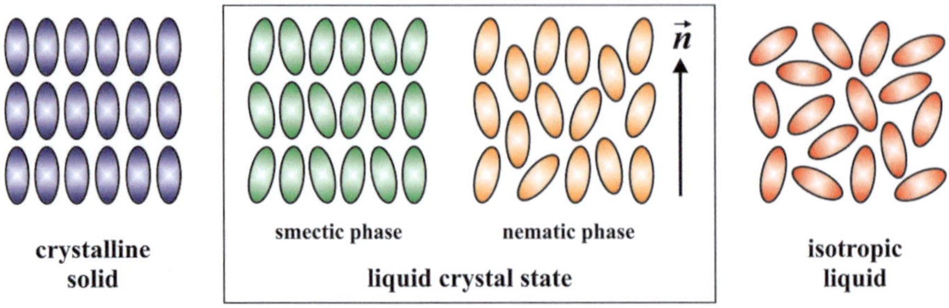


**Fig. 18.1** Schematic representation of the molecular distribution in crystalline solids, smectic and nematic liquid-crystalline mesophases and isotropic liquids, respectively

crystals showing nematic mesophases are by far the ones having a much greater impact in modern technology and, thus, the most commonly used for applications. Accordingly, this chapter will focus its discussion mainly on the actuating abilities of nematic liquid-crystalline materials.

## 18.2 Azobenzenes: Excellent Light-Sensitive Molecules for Photoactuation in Liquid-Crystalline Materials

Nematic LCs are materials that sense their surroundings. Indeed, nematic liquid-crystalline materials modify their characteristic features as a response to a great variety of environmental input energies, including heat, light, electric currents and magnetic fields. These perturbations induce variations in the organisation of the system at the molecular scale, which translates in the modulation of the macroscopic properties of the whole material (Fleischmann and Zentel 2013). Such feature is the basis of the successful and wide applicability of nematic LCs within actual technology.

Among all the different external stimuli mentioned above, light is the most attracting trigger for actuation because not only it is an unlimited, ubiquitous, free, and sustainable source of energy but also it permits the quick and remote activation of specific areas of the probe in a non-invasive fashion. In this way, light is an excellent fuel to swap the system between two different (meta)stable states with markedly different properties and, therefore, to enable photoactuation. In fact, both the convenience and practicality of optical stimulation has led, nowadays, photoresponsive nematic LCs to the top level among all activatable liquid-crystalline materials.

For photoactuation in nematic LCs to take place, it is essential that the constituting mesogenic molecules are intrinsically photoactive. If this is not the case, the material must be then provided with suitable photoactive molecules to endow it with light sensitivity. Among the wide palette of organic photochromes known currently, azobenzene and its substituted derivatives are doubtless the most used for

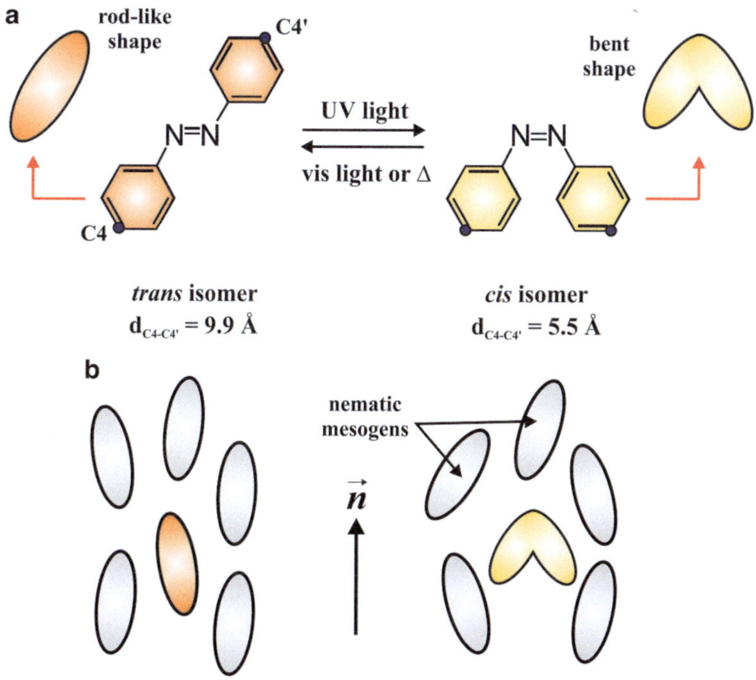


**Fig. 18.2** Reversible isomerisation of azobenzenes (**a**) and schematic representation of the disorganization produced in the host nematic mesophase as a result of the isomerisation of the photoactive azo dye (**b**)

photoactuation, mainly because of their attractive photochemical signature. Azobenzenes, which are *T*-type photochromes (thermally reversible), display a totally clean and reversible photoisomerisation process between their *trans* and *cis* isomers of different stability (Fig. 18.2a). Generally, UV-irradiation triggers the *trans*-to-*cis* isomerisation process whereas the *cis*-to-*trans* conversion can be carried out by illuminating the system with visible light. Moreover, and owing to the low stability of the *cis* isomer, it also reverts back to the thermodynamically stable *trans* form isothermally in the dark, spontaneously (Rau 1990).

The successful use of azobenzenes as light-sensitive molecules for photoactuation in nematic liquid-crystalline materials does not arise only from their spectroscopic features but also from the drastic geometrical change that they experience upon absorption of light. Before irradiation, the *trans* isomer of the azobenzene chromophore displays a rod-like geometry with a distance between the two carbon atoms in the positions 4 and 4′ of its aromatic rings of 9.9 Å. The linear shape of the *trans* isomer, which is similar to the one of the mesogens, allows its easy introduction in both low- and high-molecular mass nematic LCs without causing the destruction of the host mesophase. Upon illumination of the system with light of the appropriate wavelength, the bent *cis* isomer of the azo dye is generated in the sample, which shows a distance between the two distal carbon atoms of only 5.5 Å. As a result, the orientation of all the mesogenic molecules in the sample is modified dramatically owing to the much larger free volume requirement for the *cis* isomer in comparison to the one of its *trans* counterpart (domino effect). Hence, the *cis* azo moiety acts as an impurity decreasing the nematic order (Fig. 18.2b). Moreover, if the irradiation of the sample is carried out at a constant temperature, *T*, which is close enough to its nematic-to-isotropic transformation temperature, $T_{N-I}$, the system might reach the completely disorganized liquid phase isothermally as a result of a photoinduced nematic-to-isotropic phase transition (Prasad et al. 2004; Yu and Ikeda 2004; Garcia-Amorós et al. 2009). In both instances, once the irradiation is ceased, the system recovers its initial state due to the thermal *cis*-to-*trans* isomerisation of the azo photochrome in the dark. Remarkably, such effect has been widely exploited to induce large variations in the macroscopic dimensions of nematic liquid-crystalline materials exclusively under optical control by integrating multiple azobenzene chromophores in suitable LC matrixes, like polymers or elastomers (Broer et al. 2011; van Oosten et al. 2008). In fact, this strategy has led to achieve a great variety of light-triggered artificial muscle-like actuators displaying unique and outstanding abilities. In this context, it will be the main aim of the present chapter to provide the reader with a broad overview on the photo-mechanical effect taking place in materials exhibiting artificial muscle-like actuation where the azobenzene isomerisation is the key step for such behaviour.

## 18.3 Artificial Muscle-Like Actuators with Polysiloxane-Based Liquid Single Crystal Elastomers: Thermo- and Photo-Mechanical Effect

Nature's designs and capabilities have always served as models for humans to engineer a great variety of novel materials. Specifically, the multifunctionality of biological muscles, which seems to be the key to successful locomotion in nature, has inspired specialists in bionics and biomimetics to develop functional and smart materials that emulate their abilities, *i.e.* artificial muscle-like actuators.

Photoactive artificial muscle-like actuators are energy transducers capable to convert light into different mechanical quantities such as displacement, strain, velocity and stress (Koerner et al. 2008). Besides of light, artificial muscle-like actuators can be also triggered with other input energies like electric fields, diffusion of ions, pneumatic forces and heat. However, in all these instances, the photoactuating material should be soft and deform easily upon the application of the corresponding external stimulus. In particular, polymers have several valuable advantages for this aim in comparison to other materials; they are lightweight, inexpensive, pliable, fracture tolerant and easily implementable in devices.

Polymers may also display liquid-crystalline properties. Indeed, polymeric systems exhibiting mesophases can be either linear, denoted as linear liquid-crystalline polymers (LCPs), or cross-linked, named liquid-crystalline elastomers (LCEs). The main difference between LCPs and LCEs is the presence of a cross-linking agent in the latter, which links the polymer chains together and gives rise to the unique elastic properties of elastomeric materials (de Jeu 2012). Although both LCPs and LCEs display the characteristic molecular organization of liquid-crystalline materials, this ordering does not reach the macroscopic scale thereby extending up only to microscopic domains. In this way, photoactive LCPs and LCEs are generally multidomain systems where the director changes abruptly from one domain to another. As a result, when these photoactive systems are illuminated with non-polarised light, they undergo isotropic deformations and, therefore, the material does not expand or contract to an appreciable extent. In spite of this, outstanding and remarkable photoactuating abilities have been reported during the last decade by Ikeda et al. working with polydomain azobenzene-based liquid-crystalline elastomeric materials. In fact, they have successfully achieved a great variety of three-dimensional motions in metacrylate-based polydomain elastomeric materials by exposing them to polarised light of the appropriate wavelength (Yamada et al. 2008; Yu and Ikeda 2011; Ikeda and Ube 2011; Priimagi et al. 2012).

Nevertheless, the use of non-polarized light is generally preferable for technological applications. Under these conditions, it is strictly necessary the liquid crystal order to be extended along the whole material only in one dimension, that is, monodomain samples are required. Liquid single crystal elastomers (LSCEs) are weakly cross-linked polymer networks exhibiting a macroscopic orientation of the director. In fact, in LSCEs, the director points to a unique direction within the whole sample with the material being essentially disordered in the two perpendicular axes (Warner and Terentjev 2007). Several techniques can be used to convert LCEs into LSCEs, for instance, the use of boundaries and the application of electric and magnetic fields along a determined spatial direction. However, the most efficient method to endow LSCEs with their characteristic macroscopic orientation comes from the uniaxial stretching of a partially cross-linked elastomeric sample during its preparation.

A great variety of artificial muscle-like actuators based on functional liquid-crystalline elastomeric materials have been prepared successfully using different polymer backbones (Lee et al. 2012; Shankar et al. 2013; Pei et al. 2014). Amongst all of them, silicone is doubtlessly the most commonly used one, not only owing to its many advantageous features (*e.g.* constancy of properties over a wide range of temperature which leads to a large operating temperature range, hydrophobicity, excellent stability, suitable mechanical and anti-adhesive properties, and low toxicity) but also because of the low glass transition temperature of their derivate polymers, which enables photoactuation at room temperature.

Polysiloxane-based LSCEs are generally synthesized following the procedure in three steps developed for the first time by Küpfer and Finkelmann (1991). In the first step of the process, the different monomers, *i.e.* mesogen/s, cross-linker/s and comonomer/s, react through their terminal olefin with the Si-H groups of the main polysiloxane backbone *via* a Pt-catalysed hydrosilylation reaction which takes place at 5000 rpm at 70 °C for 2 h by means of the spin-casting technique. It should be noticed that this hydrosilylation reaction is stopped at its early stage in order to obtain a stable but partially cross-linked elastomeric system that can be macroscopically oriented further. In the second step of the process, the macroscopic orientation of the different nematic directors is performed by applying a uniaxial force to the elastomeric sample along its longest axis. After this stage, a monodomain LCE is obtained, *i.e.* an LSCE. It should be remarked that a proper macroscopic orientation of the directors in the LSCE is crucial since both the macroscopic dimensions and actuation abilities of the final photoactuator will be directly determined by the degree of order induced on the elastomer sample after this step. In the third and final step of the synthesis, the previously generated anisotropy is fixed by completing the cross-linking reaction at 70 °C (48 h) without removing the applied force. Afterwards, the resulting LSCE is purified by a swelling-deswelling process in order to remove both the catalyst and the non-reacted monomers from the network.

Owing to the presence of cross-linking points within their polymeric network, LSCEs are unique materials which combine the orientational order of liquid crystals and the elasticity of conventional rubbers. As a result of this coupling, their macroscopic dimensions can be easily modified by subtle variations in their intrinsic state of ordering through the application of different external stimuli leading, in this way, to artificial muscle-like actuation. The potential use of LSCEs as artificial muscles was first predicted theoretically by the French physicist de Gennes (1969). Years later, Wermter and Finkelmann (2001) demonstrated experimentally that polysiloxane-based nematic LSCEs suffer indeed a dramatic spontaneous contraction along the director direction when they are heated over their nematic-to-isotropic phase transition temperature due to the mesogens misalignment that occurs when driving the system from the ordered nematic phase to the completely disordered isotropic state. Once the temperature is lowered below $T_{N\text{-}I}$, the system recovers its original dimensions. Such effect is generally known as thermo-mechanical effect (Fig. 18.3a).

On the other hand, LSCEs that contain isomerisable azobenzenes as light-sensitive molecules can also undergo anisotropic macroscopic deformations upon exposure to non-polarised light of the appropriate wavelength (*i.e.* photo-mechanical effect, Fig. 18.3b). Photo-mechanical effect in nematic LSCEs was first observed by Finkelmann et al. (2001) and, since then, it has been deeply investigated from both the theoretical and the experimental point of view.

Photo-mechanical effect in LSCEs is generally studied in the laboratory by means of opto-mechanical experiments. In these experiments, a rectangular elastomeric sample is glued within two metallic rigid clamps and fixed inside the measurement cell, which is further thermostated at the corresponding temperature. When the photoactive LSCE is illuminated with light of the appropriate wavelength, the azo photochrome changes its geometry from linear to bent due to its *trans*-to-*cis* photoisomerisation. As a result, the mesogen molecules that are close to the ones of the azo dye become disorganised and induce a decrease in the local order parameter. Such feature should translate in a shortening of the LSCE in the director direction but, since the elastomeric sample is fixed by both ends, the

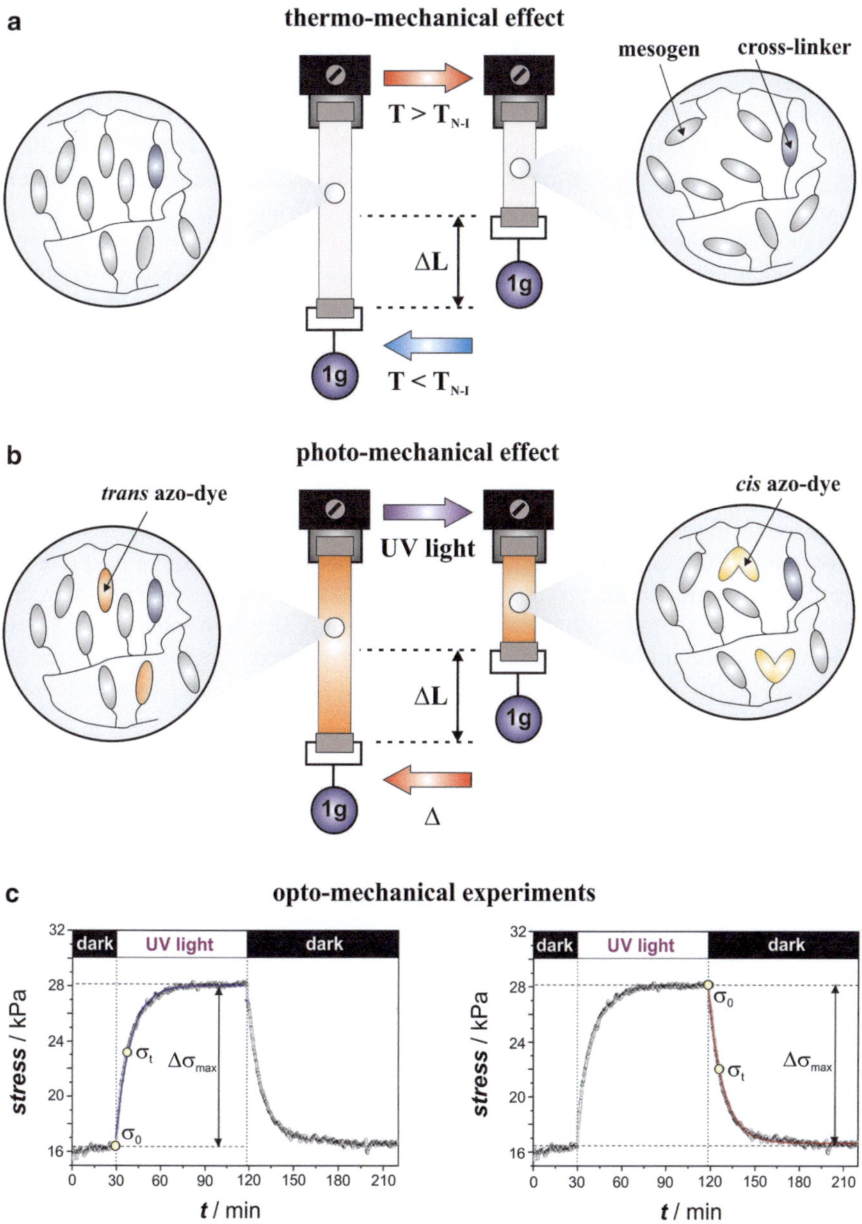


**Fig. 18.3** Thermo- (**a**) and photo-mechanical (**b**) effect in LSCEs. Representative opto-mechanical experiment (**c**) in nematic liquid single crystal elastomers: increase of the internal stress generated in the network upon irradiation with UV light (**c**, *left*) and decrease of the stress with the time in the dark at a constant temperature, $T$ (**c**, *right*)

appearance of a retractile force in the elastomer is observed instead. On turning off the light, the system recovers its initial state due to the thermal back isomerisation of the azo dye.

Importantly, opto-mechanical experiments afford a direct way to test the artificial muscle-like actuating abilities of photoactive elastomeric materials. Specifically, opto-mechanical experiments are based on the measurement of the evolution of the internal stress generated inside the LSCE with the time. A representative opto-mechanical experiment for a photoactive LSCE is shown in Fig. 18.3c. Upon illumination of the system, the internal stress created in the elastomer, $\sigma$, grows until the corresponding photostationary state is reached. The evolution of the stress in the elastomer during the irradiation process can be described mathematically by a growing monoexponential function (18.1):

$$\sigma_{\mathrm{t}} - \sigma_0 = \Delta\sigma_{\max} \cdot \left(1 - \exp\left[-t/\tau^{\mathrm{irrad}}\right]\right) \tag{18.1}$$

where $\Delta\sigma_{\max}$ and $\tau^{\mathrm{irrad}}$ correspond to the maximum opto-mechanical response produced by the artificial muscle-like actuator and the time required by the network to produce it, respectively. When the irradiation ceases, the thermal back *cis*-to-*trans* isomerisation of the azo dye occurs and the stress starts to diminish with the time until the pre-irradiation stress value is recovered. In this instance, the thermal relaxation of the artificial muscle can be fitted to a decreasing monoexponential function (18.2):

$$\sigma_{\mathrm{t}} - \sigma_0 = \Delta\sigma_{\max} \cdot \exp\left(-t/\tau^{\mathrm{th}}\right) \tag{18.2}$$

where $\tau^{\mathrm{th}}$ is the time required by the LSCE to recover the corresponding initial state isothermally in the dark (Cviklinski et al. 2002). It should be remarked that these three parameters, $\Delta\sigma_{\max}$, $\tau^{\mathrm{irrad}}$ and $\tau^{\mathrm{th}}$, are crucial to fully characterize the actuation abilities of LSCE-based photoactive artificial muscles. Indeed, such parameters inform about both the efficiency and response time of the final actuator, respectively, which are key parameters in the overall performance of materials with artificial muscle-like abilities. In this regard, this chapter will describe in the next two sections some of the investigations that have been performed during recent years not only to enhance the mechanical efficiency of materials with artificial muscle-like abilities but also to improve their response time.

## 18.4 Mechanical Efficiency of Polysiloxane Azobenzene-Based Artificial Muscle-Like Actuators

Liquid single crystal elastomers consist on high-molecular mass liquid crystals where both the mesogenic molecules and the photoactive azo moieties can be connected either head-to-tail, forming the polymeric main chain (main-chain

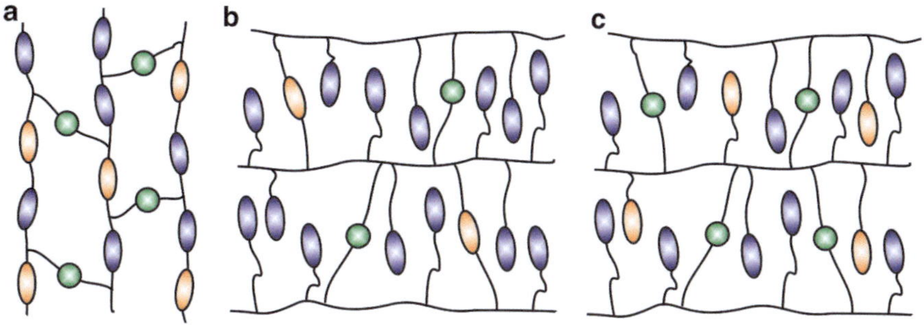


**Fig. 18.4** Different connections of the azo chromophore to the elastomeric network: forming part together with the mesogenic molecules of the main polymer backbone in main-chain LSCEs (**a**) and acting as a photoactive cross-linker (**b**) or pendant group (**c**) in side-chain LSCEs

LSCEs), or attached as side-chain groups to the main polymer backbone (side-chain LSCEs), generally *via* flexible spacers (Finkelmann 1987). In fact, the connectivity between the photoactive azo chromophore and the elastomeric network is one of the main factors controlling the opto-mechanical response of artificial muscle-like actuators (Sánchez-Ferrer 2011; Sánchez-Ferrer et al. 2011). Therefore, this section will be focused on the description of the actuating abilities and photo-mechanical efficiency of light-triggered artificial muscle-like materials where the azobenzene chromophore can be connected to the elastomeric network in three different manners (Fig. 18.4): through a head-to-tail connection with the mesogenic molecules thereby forming together the main polymer backbone in main-chain LSCEs (Fig. 18.4a), as a photoactive cross-linker in side-chain LSCEs (Fig. 18.4b) and as a photoactive pendant group in side-chain LSCEs (Fig. 18.4c).

### *18.4.1 Opto-Mechanical Efficiency of Azobenzene-Containing Main-Chain LSCEs*

Main chain LCPs were first synthesized in the early 1980s by Aguilera and Ringsdorf (1984). Years later, Donnio et al. (2000) obtained the first main-chain LSCE by cross-linking the liquid-crystalline polymer chains with a flexible siloxane-based cross-linker. Since then, a great variety of thermally-driven artificial muscle-like actuators with outstanding mechanical responses based on this type of elastomeric materials have been obtained. In addition, the introduction of azobenzene chromophores in small proportions (5 % mol) in main-chain nematic LSCEs has also led to the development of light-triggered artificial muscle-like actuators capable to produce opto-mechanical responses up to *ca.* 112 kPa at 308 K upon illumination with UV light (Sánchez-Ferrer and Finkelmann 2013) and being, therefore, more efficient than those actuators arising from their side-chain counterparts.

This mechanical differential behaviour can be easily explained by the fact that, in main-chain LSCEs, all the azobenzene molecules are part of the main polymer backbone. In this way, the generation of the bent *cis* isomer of the azo chromophore in the elastomeric network upon suitable illumination not only decreases the liquid-crystalline order at the molecular level but also affects dramatically the conformation of the main polymer backbone. On the other hand, in side-chain LSCEs, the different monomers are connected to the main polymer backbone through flexible spacers, which decreases significantly the coupling between the geometrical change suffered by the azo dye upon isomerisation and the main polymer matrix. Therefore, in side-chain LSCEs, the photo-mechanical effect is mainly due to the alteration of the ordering of the system. As a result, the photo-mechanical effect becomes less notable in side-chain LSCEs.

Main-chain LSCEs are invaluable materials for performing artificial-muscle like actuation. However, very slight variations in their synthetic preparation or chemical composition can affect dramatically not only the yield of the reaction and the growth of the polymeric main chain but also the thermal range of stability of the LC phase. For this reason, artificial muscle-like actuators based on nematic side-chain LSCEs have been by far more explored than those based on their main-chain counterparts.

### *18.4.2 Mechanical Efficiency of Side-Chain LSCEs Where the Azo Photochromes Act as Photoactive Cross-Linking Points*

As a main difference with main-chain LSCEs, in side-chain LSCEs, the photoactive azo chromophores, which are responsible for the observed photo-mechanical effect, are connected to the polymeric main chain *via* flexible spacers. Therefore, it should be expected that the length of such flexible spacer, in other words, the flexible alkyl chain that links the azobenzene core to the main polysiloxane backbone, is one of the main factors affecting the mechanical response produced by artificial muscle-like actuators based on this type of elastomeric materials (Garcia-Amorós et al. 2011a).

In order to illustrate this concept, five photoactive polysiloxane-based side-chain nematic LSCEs (**EAZO-n**, Fig. 18.5a) that contain 4,4-dialkoxyazobenzenes as photoactive cross-linkers, which hold spacers of different lengths, will be considered. Specifically, all of them are composed by the nematic mesogen 4-methoxyphenyl-4-(3-butenyloxy)benzoate (**M4OMe**, 90 % mol), the isotropic cross-linker 1,4-di-(10-undecenyloxy)benzene (**V1**, 5 % mol) and a 4,4-dialkoxyazobenzene as a light-sensitive cross-linker (**AZO-n**, 5 % mol).

As it has been anticipated above, the maximum opto-mechanical response produced by the different LSCEs upon irradiation with UV light ($\lambda_{irrad} = 380$ nm), $\Delta\sigma_{max}$, shows a clear dependence with the number of methylene units of the flexible

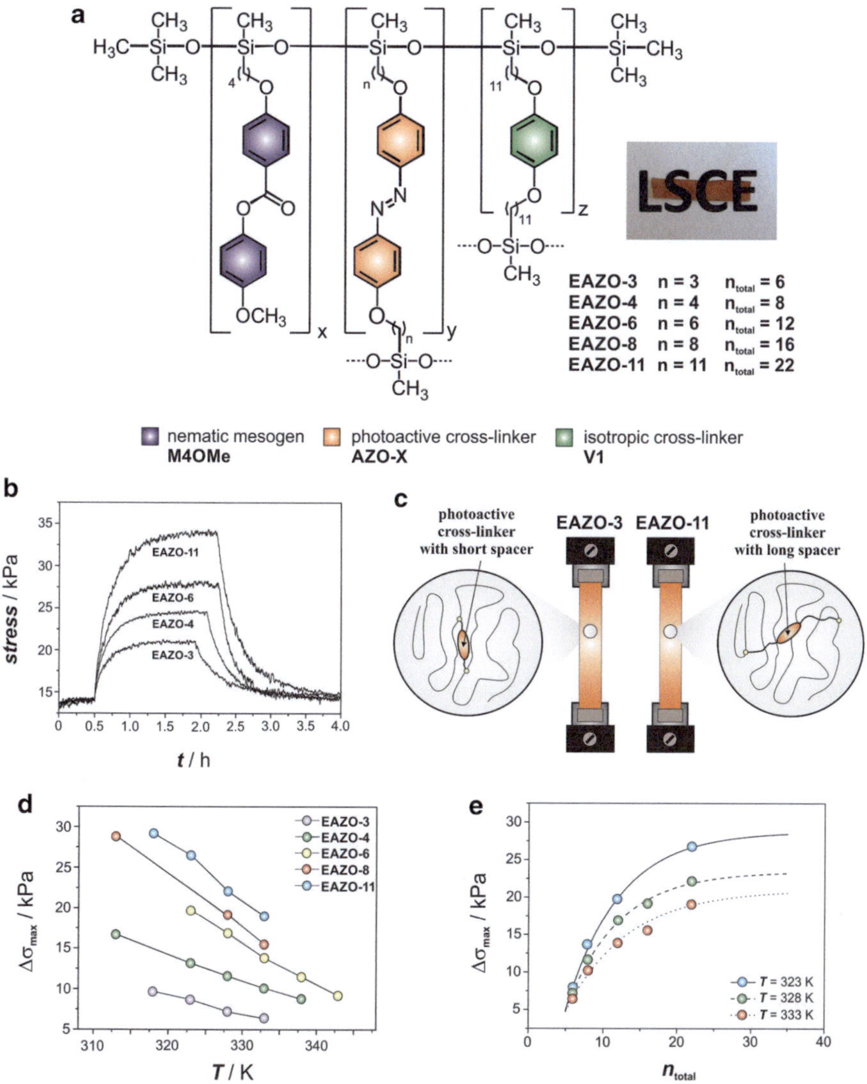


**Fig. 18.5** Chemical composition of the different nematic LSCEs **EAZO-n** (**a**), opto-mechanical experiments for the different elastomers under UV-irradiation ($\lambda_{irrad} = 380$ nm) at 333 K (**b**), schematic model of the photo-active cross-linker binding points within the elastomeric network (**c**), evolution of the maximum opto-mechanical response, $\Delta\sigma_{max}$, with the temperature (**d**), and evolution of the maximum opto-mechanical response with the total number of methylenic units of the flexible spacer, $n_{total}$, at different temperatures (**e**). Reproduced from Garcia-Amorós et al. (2011a) by permission of the Royal Society of Chemistry (RSC)

spacer (Fig. 18.5b). Specifically, more efficient light-triggered artificial muscle-like actuators are obtained when photoactive azo cross-linkers bearing longer flexible spacers are used. Indeed, LSCE **EAZO-11**, which holds the longest spacer, shows a maximum opto-mechanical response of 19.0 kPa at 333 K, while a threefold lower value of 6.4 kPa is obtained for the nematic elastomer **EAZO-3**, which contains the shortest one. This trend can be understood by the fact that the two terminal carbon atoms of the azo cross-linker that holds a long flexible spacer are connected to two more distant points of the main polysiloxane backbone (Fig. 18.5c), which should be brought closer once the *trans*-to-*cis* isomerisation of the azo cross-linker occurs. This effect becomes less intense in those systems containing azo cross-linkers with short flexible spacers. Accordingly, the largest constraint of the elastomer after photoisomerisation and, therefore, the highest opto-mechanical response will occur in that elastomer bearing the cross-linker with the longest spacer.

On the other hand, the maximum opto-mechanical response of artificial muscle-like actuators based on LSCEs containing 4,4-dialkoxyazobenzenes as light-sensitive cross-linkers exhibits also a clear dependence on the operating temperature. In this way, a gradual decrease of the maximum opto-mechanical response developed by the artificial muscle-like actuator is observed on raising up the temperature (Fig. 18.5d). It is well known that the magnitude of the opto-mechanical response produced by the network upon irradiation is directly related with the *cis* isomer population present in the corresponding photostationary state. It should be noticed that during the irradiation of the elastomeric sample, the *trans*-to-*cis* photoisomerisation reaction competes always with the thermal *cis*-to-*trans* back isomerisation, which proceeds faster at high temperatures. In this way, the photostationary state reached upon irradiation will be poorer in *cis* isomer at high temperatures and, consequently, the efficiency of the photo-mechanical effect will be lesser. Considering a linear relationship between $\Delta\sigma_{max}$ and $T$, it can be stated that LSCEs containing such azo chromophores as photoactive cross-linkers can afford responses up to 60 kPa at room temperature.

It is also remarkable that the maximum opto-mechanical response generated by the artificial muscle-like actuator at each operating temperature shows a monoexponential growing dependence on the total methylene units of the azo cross-linker flexible spacer, $n_{total}$ ($n_{total} = 2n$, where $n$ is the number of methylene units of the alkoxyl chain). Indeed, a different *plateau* value for the mechanical response of the system at each temperature is achieved. In fact, from the $\Delta\sigma_{max}$ *vs.* $n_{total}$ plots displayed in Fig. 18.5e, it can be nicely observed that there is a threshold length value of *ca.* $n_{total} \approx 30$ methylene units, that is, 15 carbon atoms in each alkoxyl chain of the flexible spacer, after which the opto-mechanical response of the artificial muscle-like actuator is not enhanced anymore. This effect can be easily understood taking into account that flexible spacers with such a great number of methylene units can exhibit a large number of possible conformational isomers diminishing the potential maximum mechanical efficiency of the photoactuator.

Briefly, the analysis of the characteristic times $\tau^{irrad}$ and $\tau^{th}$ for these elastomers yields that the rate of both processes is independent of the length of the azo cross-linker spacer. In this way, the relaxation time for the irradiation process, $\tau^{irrad}$, ranges from 15 to 21 min at 323 K and the one for the reverse process, $\tau^{th}$, falls between 29 and 35 min at the same temperature. Interestingly, both processes become slightly slower for elastomer **EAZO-3**, probably because of the low flexibility of its spacer.

### 18.4.3 Mechanical Efficiency in Side-Chain LSCEs Where the Azo Moieties Operate as Photoactive Pendant Groups

In side-chain LSCEs, the azo chromophores can be connected to the main polysiloxane backbone as photoactive cross-linkers, as it has been described in the previous subsection, or simply as photoactive pendant side groups. In order to describe the latter case, the opto-mechanical response of a nematic LSCE containing 4-cyano-4′-(5-hexenyloxy)azobenzene as a photoactive pendant group (**EAZO-CN**, Fig. 18.6a) will be discussed as a representative model. Such elastomer is composed by the nematic mesogen 4-methoxyphenyl-4-(3-butenyloxy)-benzoate (**M4OMe**, 85 % mol), the isotropic cross-linker 1,4-di-(10-undecenyloxy)benzene (**V1**, 10 % mol) and the light-sensitive co-monomer **AZO-CN** (5 % mol).

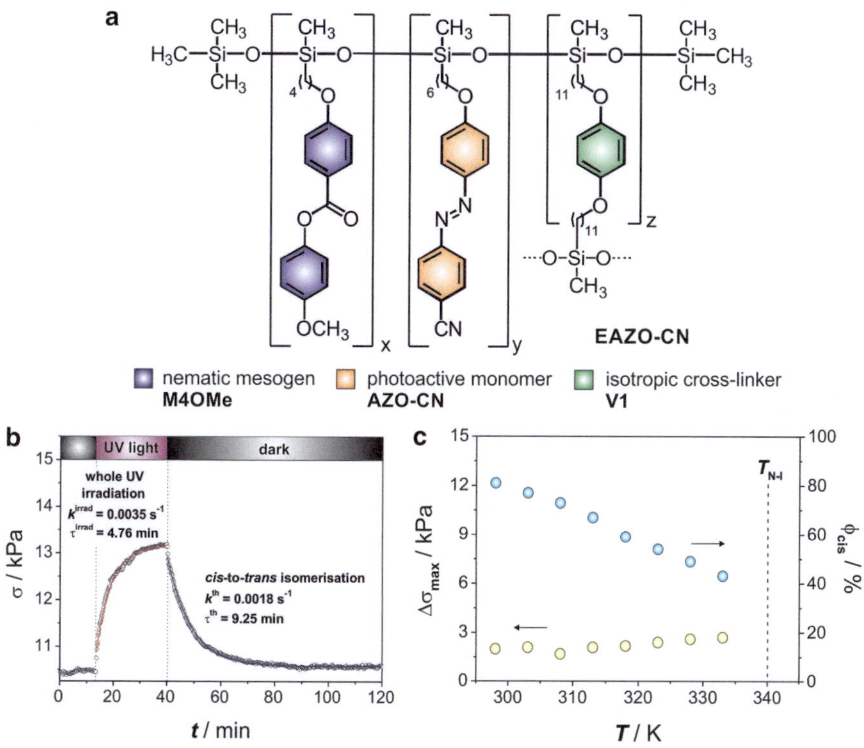


**Fig. 18.6** (**a**) Chemical composition of the nematic LSCE **EAZO-CN**. Opto-mechanical experiment at 328 K (**b**) and evolution of the maximum opto-mechanical response with both the calculated quantum yield and temperature (**c**) for the nematic LSCE **EAZO-CN**. Reproduced from Garcia-Amorós et al. (2014) by permission of the Royal Society of Chemistry (RSC)

The opto-mechanical experiment for **EAZO-CN** at 328 K is shown in Fig. 18.6b. Under these conditions, such system generates a maximum opto-mechanical of 2.5 kPa, which is roughly tenfold lower than that reported in the previous subsection for side-chain LSCEs where the azo chromophore acts as a cross-linking unit (*cf.* 7–22 kPa at 328 K, see Fig. 18.5d). As it has been abovementioned, the maximum opto-mechanical response generated by the artificial muscle-like actuator is generally related to the photoisomerisation quantum yield ($\Phi_{cis}$), that is, the *cis* isomer population present in the corresponding photostationary state. In this way, more efficient actuators should be expected when working at lower temperatures, as it has been described in the previous subsection for **EAZO-n** (see Fig. 18.5c). Surprisingly, when the azo chromophore is attached to main polysiloxane backbone as a pendant group, the maximum opto-mechanical response displayed by the photoactuator is independent of the operation temperature of the system. Specifically, for **EAZO-CN**, $\Delta\sigma_{max}$ is *ca.* 2.5 kPa in the 298–333 K range (Fig. 18.6c).

The differential behaviour observed between **EAZO-n** and **EAZO-CN** can be related to the fact that when azo cross-linkers are used, the decrease of the local order parameter is due both to the change in their molecular shape upon irradiation, and to the additional mechanical response created by bringing the polymer chains closer by a cooperative effect since they are connected to main polymer matrix through two different alkyl chains. The latter feature is not enabled when the azo dye is incorporated only as a side-chain comonomer. Besides, prior investigations have also shown that the elasticity modulus, which becomes higher at lower temperatures, plays a key role in the relative length of the liquid-crystalline network and, thus, in the mechanical behaviour. As an overall outcome, a constant opto-mechanical response is obtained for the nematic elastomer **EAZO-CN** over a wide temperature range (Garcia-Amorós et al. 2014).

Finally, it should be also pointed out that **EAZO-CN** exhibited a reversible kinetic behaviour in the time scale of hours, in other words, this artificial muscle-like actuator required 0.3 h for developing its maximum opto-mechanical response upon irradiation with UV light and 1.5 h to recover its initial state in the dark at 298 K.

## 18.5 Response Time of Polysiloxane Azobenzene-Based Artificial Muscle-Like Actuators

In the previous section of this chapter, we have shown several photoactive LSCEs displaying efficient opto-mechanical responses. However, in all instances, they need several hours not only to develop their maximum mechanical response but also to recover their initial dimensions in the dark at ambient temperature. While mechanical efficiency is controlled mainly by the connectivity between the photoactive azo chromophore and the elastomeric network, the time the network requires to recover its initial dimensions is mostly determined by the rate of the thermally-activated *cis*-to-*trans* isomerisation process of the azo dye used which, in turn, depends on its chemical functionalization. However, as it will be described later, the thermal isomerisation rate of the azo chromophore used plays also an important role in the mechanical response of the final photoactuator. In this line, and since the thermal relaxation time of artificial muscle-like actuators is also a key parameter to consider in the overall performance of such functional materials, in the last section of this chapter, we will focus our discussion on the two different strategies reported heretofore to develop photoactive artificial muscle-like actuators with low thermal relaxation times.

### *18.5.1 Photoactive Artificial Muscle-Like Actuators Based on Push-Pull Azoderivatives*

As it has been just aforementioned, for fast azobenzene-based artificial muscle-like actuators to be achieved, it is essential that the azo-chromophore used returns to its thermodynamically stable *trans* form in the dark as fast as possible. In this way, azobenzenes that undergo their thermal isomerisation through the rotational mechanism, of which push-pull azoderivatives are the most well-known example, will be valuable candidates for this aim since they are endowed with high thermal isomerisation rates at room temperature. In fact, the two only polysiloxane-based photoactive artificial muscle-like actuators exhibiting low relaxation times published so far are based on this strategy. Indeed, these systems contain the well-known push-pull azo-dyes 4-amino- (Camacho López et al. 2004) and 4-*N*,*N*-dimethylamino-4′-nitroazobenzene (Harvey and Terentjev 2007) as photo-active molecules, which are doped into host elastomeric networks but not covalently bonded to the polymeric structure decreasing thereby the stability of the final photo-actuator.

In order to overcome this drawback it has been reported recently that the incorporation of the push-pull azo dyes 4-(5-hexenyloxy)-4′-nitroazobenzene and 4-(5-hexenyloxy)-4′-methoxy-2′-nitroazobenzene as side-chain covalently-bonded co-monomers in nematic LSCEs (Fig. 18.7a) allows decreasing more than $10^3$ times the relaxation time that such azo chromophores show when they are just dissolved in common isotropic solvents (between 16 and 92 min). As a result, the corresponding nematic LSCEs **EAZO-2NO$_2$** and **EAZO-4NO$_2$** exhibit a reversible kinetic behaviour in the time scale of seconds at 298 K, that is, they need only 2–4 s to develop their maximum opto-mechanical response under irradiation with blue light ($\lambda_{irrad} = 450$ nm) and 3–4 s to recover their corresponding initial state in the dark (Fig. 18.7b).

It should be also pointed out that the introduction of **AZO-2NO$_2$** in a nematic linear polymer (**CPAZO-2NO$_2$**) afforded the same relaxation time than the one registered in the nematic LSCE **EAZO-2NO$_2$**. Therefore, it can be concluded that

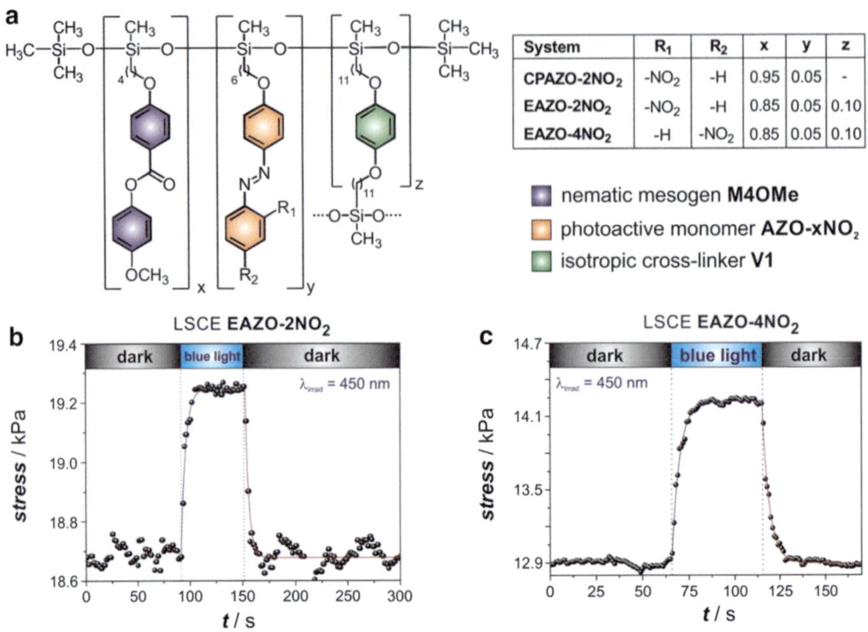


**Fig. 18.7** (**a**) Chemical composition of the nematic LCP (**CPAZO-2NO$_2$**) and LSCEs **EAZO-2NO$_2$** and **EAZO-4NO$_2$** and opto-mechanical experiments for both nematic elastomers **EAZO-2NO$_2$** (**b**) and **EAZO-4NO$_2$** (**c**) upon irradiation with blue light of $\lambda_{irrad} = 450$ nm at 298 K. Reproduced from with permission of Wiley-VCH. Copyright 2011 Wiley-VCH Verlag GmbH & Co. KGaA

the thermal isomerisation process for this type of azoderivatives is dramatically accelerated when they are chemically bonded to a nematic polymeric system (either LCPs or LSCEs) affording thereby actuators with low response times.

Finally, it should be pointed out that, although both artificial muscle-like actuators **EAZO-4NO$_2$** and **EAZO-2NO$_2$** exhibited fast relaxation times in the time scale of seconds, they displayed low maximum opto-mechanical responses of only 1.4 and 0.7 kPa, respectively. In fact, the fast thermal relaxation of these systems is the responsible of the low internal stress generated in both elastomers due to the low *cis* isomer content in the corresponding photostationary state. Accordingly, the slightly faster thermal isomerisation for **EAZO-2NO$_2$** accounts for the lower mechanical response produced by this actuator. Nevertheless, it should be taken into account that higher mechanical responses may be produced when working with more powerful light sources due to an enhancement in the *trans*-to-*cis* photoisomerisation quantum yield.

### 18.5.2 *Photoactive Artificial Muscle-Like Actuators Based on Azophenolic Dyes*

Besides push-pull azoderivatives, azophenolic dyes are also endowed with rapid thermal *cis*-to-*trans* isomerisation processes at room temperature, owing to their capability to establish the well-known azo-hydrazone tautomeric equilibrium, which allows, in turn, their thermal back reaction to proceed through the rotational isomerisation mechanism (Garcia-Amorós et al. 2010). In this way, azophenols are also valuable chromophores to be covalently-attached either in LCPs or LSCEs to obtain stable and fast responding photosensitive materials.

However, before moving on to LSCE materials, we will discuss briefly the isomerisation of azophenols in both isotropic and liquid-crystalline environments. Specifically, 4-(5-hexenyloxy)-4′-hydroxyazobenzene (**AZO-OH**) has been chosen as a representative example since it will be further used as a comonomer in LSCEs. The *cis* isomer of **AZO-OH** exhibits relaxation times of 311 and 405 ms in ethanol and acetonitrile at 298 K, respectively (Fig. 18.8b). Otherwise, its isomerisation rate is more than 1000-fold slower when it is dissolved in toluene at the same temperature ($\tau^{th} = 23$ min). In both ethanol and acetonitrile, the thermal back reaction of **AZO-OH** proceeds through a solvent-assisted tautomerisation to yield a hydrazone-like intermediate. Subsequently, a rotation around the N–N bond of the hydrazone-like intermediate can occur regenerating thereby the thermodynamically stable *trans* isomer. Such an intermediate cannot be promoted when the azo dye is dissolved in toluene.

Interestingly, when **AZO-OH** is part of a linear copolymer (**CPAZO-OH**, Fig. 18.8a), a toluene-isomerising solution of this system shows a relaxation time of the same order ($\tau^{th} = 254$ ms) than the ones registered for **AZO-OH** in ethanol and acetonitrile (*cf.* 311 and 405 ms, respectively) and very far away from the one registered for *cis*-**AZO-OH** in toluene (*cf.* 23 min). Remarkably, this very fast thermal isomerisation rate is observed even in the nematic state of the very same copolymer **CPAZO-OH** ($\tau^{th} = 333$ ms at 298 K), where no solvent is present. Therefore, these results evidence that when azophenol monomers are covalently bonded to the polysiloxane backbone, although they are present at a low concentration (10 % mol, see Fig. 18.8a), they still are able to interact efficiently with each other through intermolecular hydrogen bonding due to their forced spatial proximity. This assumption is corroborated by the fact that the linear homopolymer **PAZO-OH** (Fig. 18.8a) shows a relaxation time of only 17 ms in ethanol at 298 K (Fig. 18.8c), which is *ca.* 20-fold lower than the one registered for the corresponding monomer in the same solvent (*cf.* 311 ms). Indeed, such feature evidences that, besides the possibility of the azophenolic monomers to establish hydrogen bonding with the surrounding solvent molecules, a cooperative and efficient interaction between neighbouring azophenol moieties occurs. Moreover, it is highly remarkable that such dye–dye interaction can be even observed in the glassy state of the homopolymer **PAZO-OH**, registering thereby a relaxation time of 125 ms for such system. According to these results, it is possible to transfer the fast thermal isomerisation rate of the hydroxyl-substituted azo monomers to both

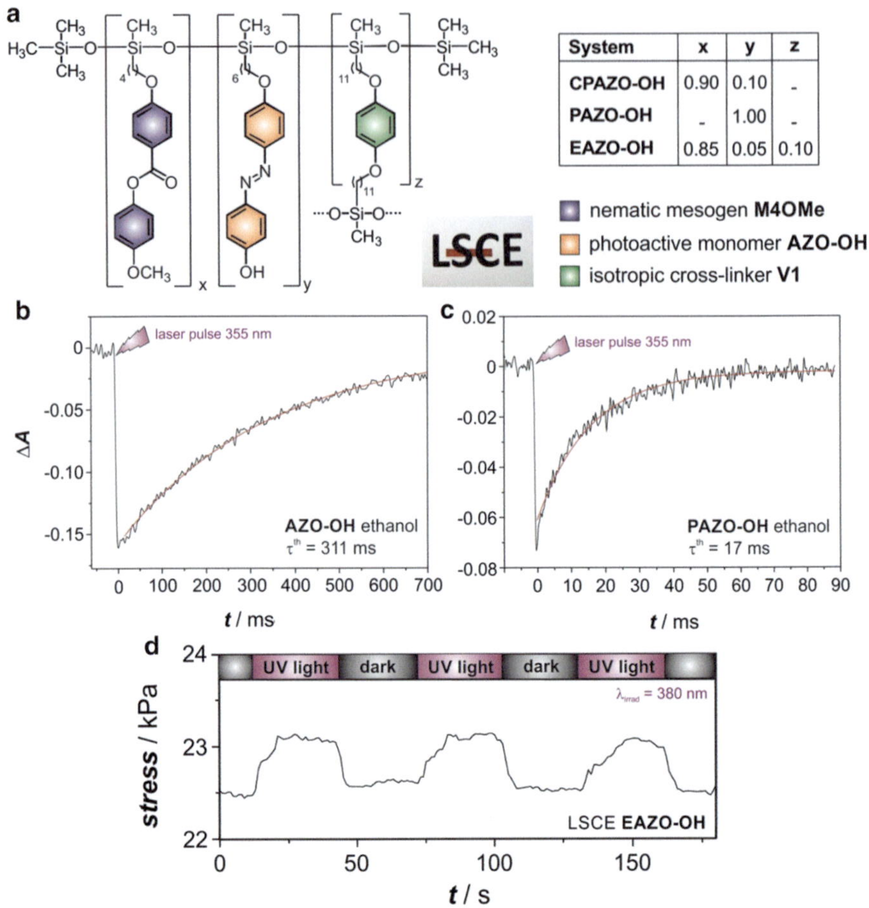


| System | x | y | z |
|---|---|---|---|
| CPAZO-OH | 0.90 | 0.10 | - |
| PAZO-OH | - | 1.00 | - |
| EAZO-OH | 0.85 | 0.05 | 0.10 |

**Fig. 18.8** Chemical composition of the nematic LCPs (**CPAZO-OH** and **PAZO-OH**) and LSCE **EAZO-OH** (**a**), transient absorption generated by laser pulsed irradiation with UV light ($\lambda_{irrad} = 355$ nm, $\lambda_{obs} = 370$ nm) for both **AZO-OH** (**b**) and **PAZO-OH** (**c**) in ethanol at 298 K (c = 20 μM) and opto-mechanical experiment (**d**) for **EAZO-OH** at 298 K ($\lambda_{irrad} = 380$ nm). Reproduced from Garcia-Amorós and Velasco (2014) and by permission from both the Royal Society of Chemistry (RSC) and Org. Lett.© 2011, American Chemical Society

nematic and glassy polymers generating thereby fast responding solvent-free photoactive systems (Garcia-Amorós and Velasco 2014).

In this line, one can take advantage of such intermolecular interactions to generate elastomeric materials, where no solvent is also present, with low response times. In order to describe properly the influence of the phenolic group in the opto-mechanical properties of the final photoactuator a co-elastomer, which contains the nematic monomer **M4OMe** (85 % mol), the isotropic cross-linker **V1** (10 % mol) and the photo-active hydroxyazoderivative **AZO-OH** (5 % mol), will be considered

(Fig. 18.8a). It should be pointed out that this liquid-crystalline co-elastomer contains only a small proportion of the azo moiety (5 % mol) in order to not overly disrupt the nematic order of the elastomer. However, the azo-dye content used is high enough for hydrogen bonding being established between the hydroxyazobenzene monomers. Indeed and, as a result of this interaction, the resulting artificial muscle-like actuator **EAZO-OH** exhibits a thermal relaxation time ($\tau^{th} = 1$ s, Fig. 18.8d) of the same order of magnitude as that of the azo-monomer dissolved in a protic isotropic solvent ($\tau^{th} = 311$ s). Nevertheless, although **EAZO-OH** exhibits a fast relaxation time, it shows a maximum opto-mechanical response of only 0.6 kPa at 298 K. Therefore, and as it has been described previously for both **EAZO-4NO$_2$** and **EAZO-2NO$_2$**, the fast thermal isomerisation of the monomer **AZO-OH** stands for the low mechanical efficiency of its nematic elastomer **EAZO-OH**.

## 18.6 Conclusions and Future Perspectives

Photo-active liquid single crystal elastomers (LSCEs) are ideal materials for artificial muscle-like actuation. This arises from the fact that their macroscopic dimensions can be easily modified by applying suitable input energies, such as light. For photoactuation to be achieved, light-sensitive molecules should be integrated in the elastomeric network. Among all known chromophores, azobenzenes are the most commonly used for photoactuation because of not only their reversible spectroscopic features but also due to their dramatic geometrical change upon isomerisation.

Two different key parameters should be considered for characterising properly the artificial muscle-like actuation of elastomeric materials: first, the maximum mechanical response that they are able to generate upon illumination and, second, the time required to produce it and that to recover the initial state further. The connectivity between the azo chromophore and the elastomeric network is one of main factors controlling the mechanical efficiency of the artificial muscle-like actuator. Specifically, main-chain LSCEs, where the azo dye forms, together with the mesogenic molecules, the main polymer network, are the most efficient materials. Considering their side-chain counterparts, those actuators bearing the azo chromophore as light-sensitive cross-linkers are more efficient than those that have it as a side-chain pendant group. Moreover, those elastomers holding photoactive azo cross-linkers which long flexible spacers yield more efficient artificial muscles. On the other hand, photoactive LSCEs containing azo chromophores as side-chain pendant groups show the very same opto-mechanical response independently of the operating temperature of the system. In addition, the use of non-push-pull azoderivatives, affords high mechanically-efficient but slow-responding artificial muscle-like actuators. Such systems can be used as bistable artificial muscles.

The response time of the artificial muscle-like actuator is mostly determined by the thermal isomerisation rate of the azo dye used. Specifically, the introduction of both alkoxynitro-substituted azobenzenes and azophenols yield fast-responding artificial muscle-like actuators with response times in the second time scale. However, such materials exhibit low mechanical efficiencies due to the thermal back reaction of the azo photochrome.

As a final remark, photofunctional smart materials, especially those for artificial muscle-like actuators, promise to make a significant impact in modern technology, since light is an endless, cheap and environmentally friendly energy source. Nowadays, researchers continue to develop new photo-active organic compounds in order to improve the properties of such materials and bring them closer to those created by Nature. However, in any instance, any future improvement will arise from a meticulous optimization of the system at the single-molecular level, that is, through the generation of appropriate molecular entities to endow the final material with the desired abilities. Likewise, this chapter has intended to present the reader a modest overview of the present efforts being put forth along these lines and we strongly hope that this purpose has been achieved.

**Acknowledgements** Financial support from the *Ministerio de Economía y Competitividad* (Spain) through grant CTQ2012-36074 is acknowledged. J. Garcia-Amorós is also grateful for a *Beatriu de Pinós* post-doctoral grant from the *Generalitat de Catalunya* (Spain, grant 2011 BP-A2-00016).

## References

Aguilera C, Ringsdorf H (1984) Thermotropic polyesters with mesogenic groups based on substituted hydroquinone units and highly flexible siloxan spacer in the main chain. Polym Bull 12:93–98

Broer D, Crawford GP, Zumer S (2011) Cross-linked liquid crystalline systems: from rigid polymer networks to elastomers (liquid crystals book series). CRC Press, Boca Raton

Camacho López M, Finkelmann H, Palffy-Muhoray P, Shelley M (2004) Fast liquid-crystal elastomer swims into the dark. Nat Mater 3:307–310

Cviklinski J, Tajbakhsh AR, Terentjev EM (2002) UV isomerisation in nematic elastomers as a route to photo-mechanical transducer. Eur Phys J E 9:427–434

de Gennes PG (1969) Possibilites offertes par la reticulation de polymeres en presence d'un cristal liquide. Phys Lett 28A:725–726

de Jeu WH (2012) Liquid crystal elastomers: materials and applications, advances in polymer science series. Springer, Berlin

Donnio B, Wermter H, Finkelmann H (2000) A simple and versatile synthetic route for the preparation of main-chain liquid-crystalline elastomers. Macromolecules 33:7724–7729

Finkelmann H (1987) Liquid crystal polymers, Chap 1. In: Gray GW (ed) Thermotropic liquid crystals. Wiley, Chichester, pp 1–27

Finkelmann H, Nishikawa E, Pereira GG, Warner M (2001) A new opto-mechanical effect in solids. Phys Rev Lett 87:015501-1–015501-4

Fleischmann EK, Zentel R (2013) Liquid-crystalline ordering as a concept in materials science: from semiconductors to stimuli-responsive devices. Angew Chem Int Ed 52:8810–8827

Garcia-Amorós J, Velasco D (2014) Understanding the fast thermal isomerisation of azophenols in glassy and liquid-crystalline polymers. Phys Chem Chem Phys 16:3108–3114

Garcia-Amorós J, Szymczyk A, Velasco D (2009) Nematic-to-isotropic photo-induced phase transition in azobenzene-doped low-molar liquid crystals. Phys Chem Chem Phys 11:4244–4250

Garcia-Amorós J, Sánchez-Ferrer A, Masaad WA, Nonell S, Velasco D (2010) Kinetic study of the fast thermal cis-to-trans isomerisation of para-, ortho- and polyhydroxyazobenzenes. Phys Chem Chem Phys 12:13238–13242
Garcia-Amorós J, Finkelmann H, Velasco D (2011a) Influence of the photo-active azo cross-linker spacer on the opto-mechanics of polysiloxane elastomer actuators. J Mater Chem 21:1094–1101
Garcia-Amorós J, Martínez M, Finkelmann H, Velasco D (2014) Photoactuation and thermal isomerisation mechanism of cyanoazobenzene-based liquid crystal elastomers. Phys Chem Chem Phys 16:8448–8454
Gray GW (1998) Handbook of liquid crystals. Wiley-VCH, Weinheim
Harvey CLM, Terentjev EM (2007) Role of polarization and alignment in photoactuation of nematic elastomers. Eur Phys J E Soft Matter 23:185–189
Ikeda T, Ube T (2011) Photomobile polymer materials: from nano to macro. Mater Today 14:480–487
Koerner H, White TJ, Tabiryan NV, Bunning TJ, Vaia RA (2008) Photogenerating work from polymers. Mater Today 11:34–42
Küpfer J, Finkelmann H (1991) Nematic liquid single crystal elastomers. Makromol Chem Rapid Commun 12:717–726
Lee KM, Wang DH, Koerner H, Vaia RA, Tan LS, White TJ (2012) Enhancement of photogenerated mechanical force in azobenzene-functionalized polyimides. Angew Chem Int Ed 51:4117–4121
Pei Z, Yang Y, Chen Q, Terentjev EM, Wei Y, Ji Y (2014) Mouldable liquid-crystalline elastomer actuators with exchangeable covalent bonds. Nat Mater 13:36–41
Prasad SK, Nair GG, Sandhya KL, Rao DSS (2004) Photoinduced phase transitions in liquid-crystalline systems. Curr Sci 86:815–823
Priimagi A, Shimamura A, Kondo M, Hiraoka T, Kubo S, Mamiya JI, Kinoshita M, Ikeda T, Shishido A (2012) Location of the azobenzene moieties within the cross-linked liquid-crystalline polymers can dictate the direction of photoinduced bending. ACS Macro Lett 1:96–99
Rau H (1990) Photochemistry and photophysics. CRC Press, Boca Raton
Sanchez-Ferrer A (2011) Light-induced disorder in liquid-crystalline elastomers for actuation. Proc SPIE 8107:810702-1–810702-8
Sánchez-Ferrer A, Finkelmann H (2013) Opto-mechanical effect in photoactive nematic main-chain liquid-crystalline elastomers. Soft Matter 9:4621–4627
Sánchez-Ferrer A, Merekalov A, Finkelmann H (2011) Opto-mechanical effect in photoactive nematic side-chain liquid-crystalline elastomers. macromol. Rapid Commun 32:671–678
Shankar MR, Smith ML, Tondiglia VP, Lee KM, McConney ME, Wang DH, Tan LS, White TJ (2013) Contactless, photoinitiated snap-through in azobenzene-functionalized polymers. Proc Natl Acad Sci 110:18792–18797
van Oosten CL, Bastiaansen CWM, Broer DJ (2008) Printed artificial cilia from liquid-crystal network actuators modularly driven by light. Nat Mater 8:677–682
Warner M, Terentjev EM (2007) Liquid crystal elastomers. Clarendon, Oxford
Wermter H, Finkelmann H (2001) Liquid crystalline elastomers as artificial muscles. e-Polymers 13:1–13
Yamada M, Kondo M, Mamiya JI, Yu Y, Kinoshita M, Barrett CJ, Ikeda T (2008) Photomobile polymer materials: towards light-driven plastic motors. Angew Chem Int Ed 47:4986–4988
Yu Y, Ikeda T (2004) Alignment modulation of azobenzene-containing liquid crystal systems by photochemical reactions. J Photochem Photobiol C Photochem Rev 5:247–265
Yu H, Ikeda T (2011) Photocontrollable liquid-crystalline actuators. Adv Mater 23:2149–2180

# Chapter 19
# Liquid Crystalline Epoxy Resin Based Nanocomposite

**Sheng-Hao Hsu, Min-Huey Chen, and Wei-Fang Su**

Composites made from epoxy resin (ER) and reinforcing filler have been used widely in many industries such as aerospace, building, marine, electricity, electronic, medical devices, etc. They require the material exhibiting high mechanical strength, dimensional stability, chemical resistance, temperature resistance, fire resistance, ease of fabrication, etc. The liquid crystalline epoxy resin (LCER) is known to be a self-reinforcing resin which exhibits exceptional mechanical and thermal properties due to the presence of highly ordered nano and micro domains and crosslinking capability in the resin (Douglas 2002; Mossety-Leszczak and Włodarska 2011). The ordered domains are built upon the liquid crystalline mesogen in the resin. The crosslinking capability makes the resin with high thermal and mechanical properties. By incorporating nano reinforcing filler into the LCERs to form nanocomposite, the properties of LCERs can be further enhanced. In this chapter we are going to discuss (1) chemical and physical properties of LCERs (2) processing of nanocomposite (3) applications and properties of nanocomposites and then end with (4) conclusions and future perspective.

---

S.-H. Hsu
Graduate Institute of Clinical Dentistry, School of Dentistry, National Taiwan University, Taipei, Taiwan

Department of Materials Science and Engineering, National Taiwan University, Taipei, Taiwan

M.-H. Chen
Graduate Institute of Clinical Dentistry, School of Dentistry, National Taiwan University, Taipei, Taiwan

W.-F. Su (✉)
Department of Materials Science and Engineering, National Taiwan University, Taipei, Taiwan
e-mail: suwf@ntu.edu.tw

 
V.K. Thakur, M.R. Kessler (eds.), *Liquid Crystalline Polymers*,
DOI 10.1007/978-3-319-20270-9_19

## 19.1 Chemical and Physical Properties of LCERs

The ER contains two oxirane rings at the end of molecule that can be crosslinked (cured) with curing agent to form three dimensional networks. If the main chain of the ER contains rigid rod mesogen, the liquid crystalline domain can be formed in the resin. The common mesogens are biphenyl (Su 1993; Ochi et al. 1997), benzaldehyde azine (Carfagna et al. 1994d), bisphenol-oxadiazol (Balamurugan and Kannan 2009), naphthalene (including binaphthalene) (Carfagna et al. 1994b; Giamberini et al. 1997; Su et al. 2002a; Duann et al. 2004), α-methylstilbene (Barclay et al. 1992; Lin et al. 1994; Amendola et al. 1995, 1996; Yamamoto et al. 2014), azobenzene (Castell et al. 2001; Wlodarska et al. 2009; Zhou et al. 2012) and bisphenol-S (Huo et al. 2008), azomethine (Mormann et al. 1997; Su et al. 2000; Castell et al. 2003a) or ester (benzoate) (Jahromi et al. 1994; Shiota and Ober 1996; Giamberini et al. 1997). The mesogen and two oxirane rings are linked by flexible spacers to form LCER. Table 19.1 summaries LCERs reported in the literatures. The chemical structure and length of spacer affect the phase transition temperature ($T_{LC}$) and mesophase range temperature ($T_{MR}$, the range of temperature between $T_{LC}$ and clear point ($T_{CP}$)) of LCERs (Jahromi et al. 1994; Castell et al. 2001, 2003a, b; Yamamoto et al. 2014). Generally, a longer the flexible spacer results in a lower the $T_{LC}$. Besides, the flexible spacer has been inserted between two mesogens to form dimeric LCERs (Shiota and Ober 1996; Choi et al. 2000).

The LCER can be cured by either conventional or liquid crystalline curing agent to form three dimensional networks. Table 19.2 summaries the chemical structures of curing agents those have been used to cure the LCERs. To obtain liquid crystalline network, one needs to determine an appropriate set of cure cycles by constructing isotropic-liquid crystalline time-temperature transformation (TTT) diagram of liquid crystalline epoxy resin system (LCER system, a formulated LCER and curing agent mixture) as shown in Fig. 19.1. The S-shaped curve is generally observed for the LCER system because the LC transformation is present. That transformation can be observed during curing process using polarized optical microscope equipped with a hot stage.

Figure 19.2a illustrates the formation of LCER system by mixing LCER and curing agent homogeneously upon heating. When the LCER system is cured above $T_{CP}$, the LCER molecule is in the isotropic state, the cured resin will not have any assembled molecules as shown in Fig. 19.2b which is similar to the cured conventional ER system. When the curing process of an LCER system is conducted in the temperature range of $T_{MR}$, the molecules are arranged to form polydomain structure of fixed LC phase in the cured resin. The average size of polydomain structure is affected by the temperature history of curing process. Generally, while an LCER system is cured at a lower temperature in $T_{MR}$, a larger polydomain size of LC phase will be obtained (Harada et al. 2009). The polydomain structure of LCER system forms spontaneously in $T_{MR}$, and enhances the properties of materials. However, the LC domains are disoriented and random dispersed throughout the cured LCER sample (Fig. 19.2c). That results in isotropic mechanical and thermal properties (Li and Kessler 2013). To obtain anisotropic properties, the polydomain

**Table 19.1** Chemical structures of conventional and liquid crystalline epoxy resins

| Conventional ER: DGEBA |
| --- |
| 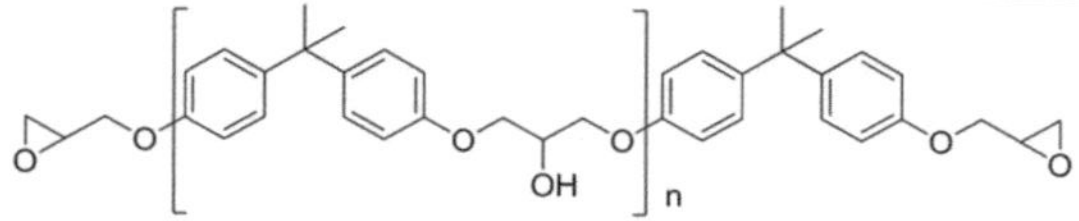 n≥0 |

| LCERs | Abbreviation | References |
| --- | --- | --- |
| 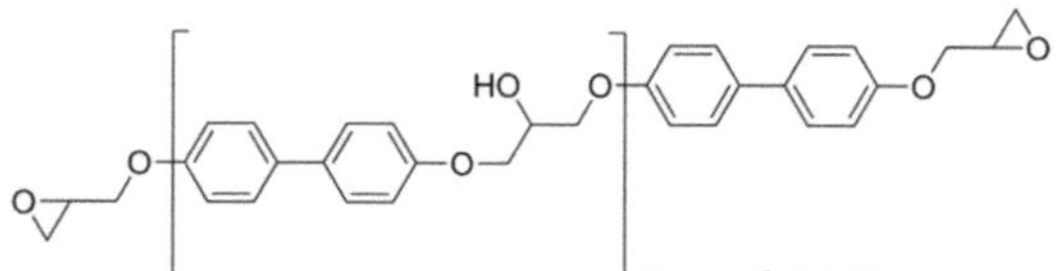 0 < n <1 | BP | Su (1993), Ochi et al. (1995) |
| 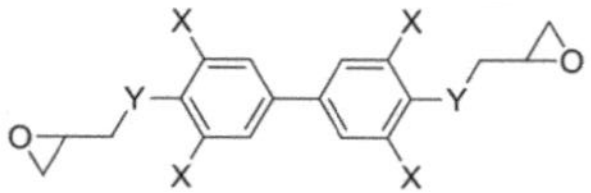 X= H or $CH_3$; Y=O, $CH_2O$, $OCH_2CH_2O$ or $OCH_2CH_2CH_2O$ | | Grebowicz (1996), Lee et al. (1998) |
| 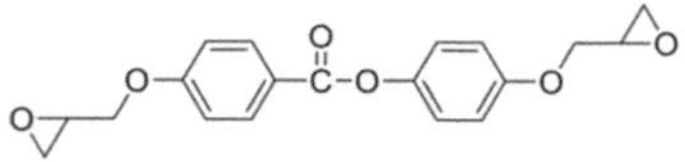 | | Giamberini et al. (1997), Carfagna et al. (2000a) |
| 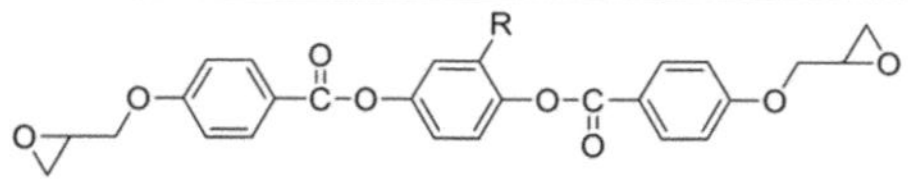 R= H, $CH_3$ or Cl | | Lee and Jang (1998), Lee et al. (1999) |
| 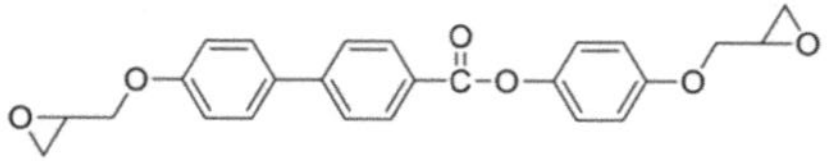 | | Giamberini et al. (1997), Wang et al. (2011) |

Table 19.1 (continued)

| LCERs | Abbreviation | References |
|---|---|---|
| 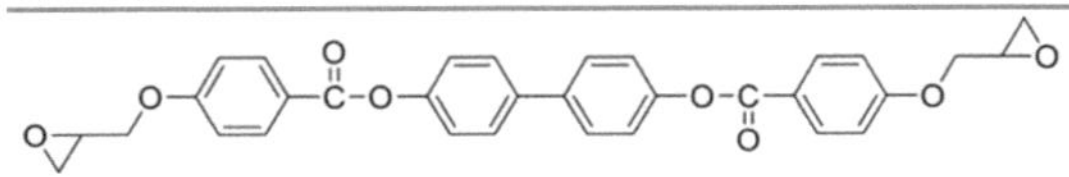 | | Lee et al. (1998) |
| 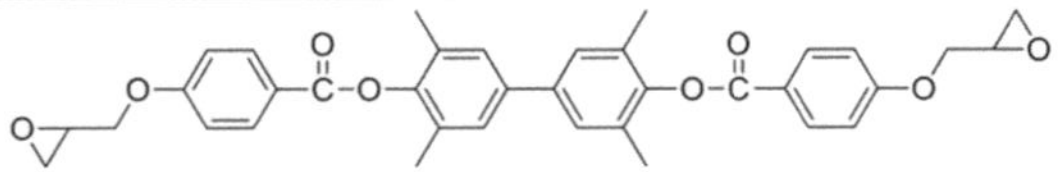 | | Cai et al. (2007), Liu et al. (2011a) |
| 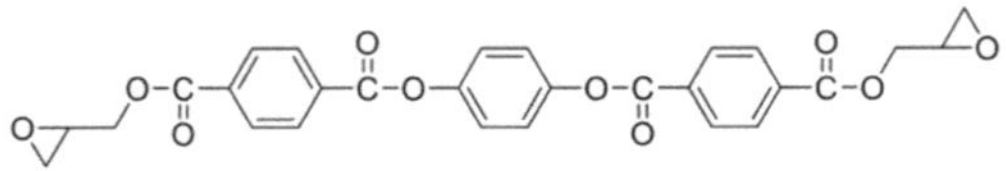 | | Lee et al. (1999), Lee and Jang (2006) |
| 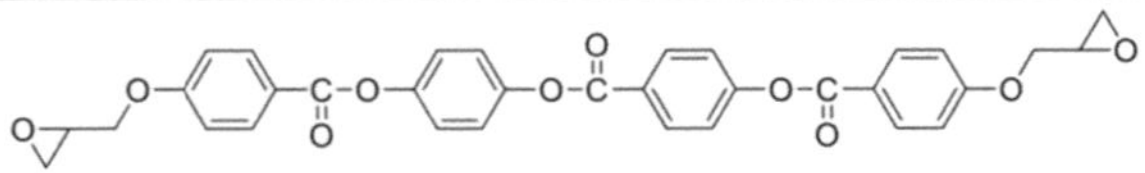 | | Lee and Jang (2006) |
| 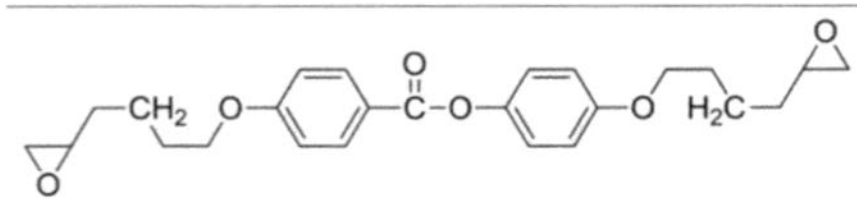 | | Jahromi et al. (1994) |
| 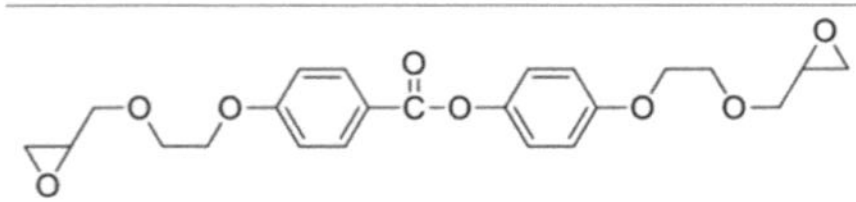 | | Jahromi et al. (1994) |
| 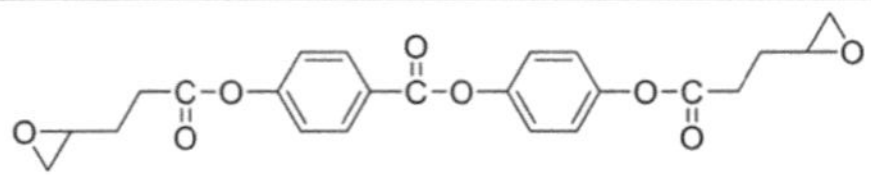 | | Wlodarska et al. (2009) |
| 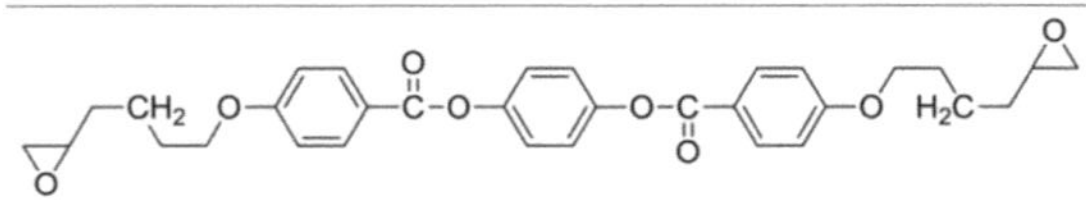 | | Jahromi et al. (1994), Shiota and Ober (1996) |

| | | |
|---|---|---|
| 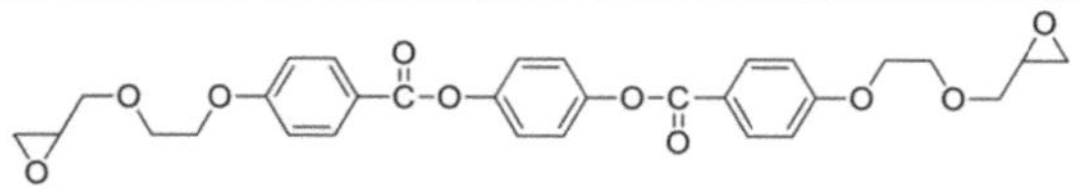 | | Jahromi et al. (1994) |
| 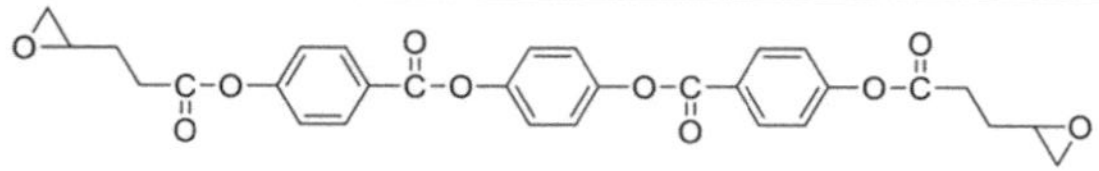 | | Wlodarska et al. (2009) |
| 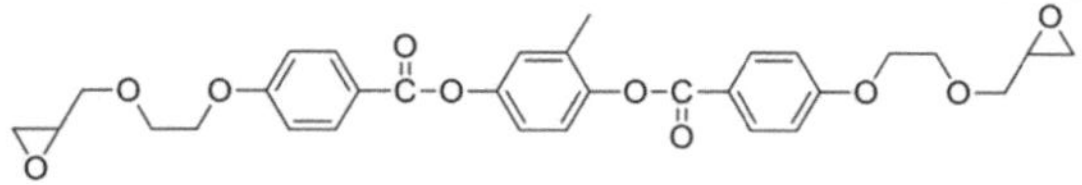 | | Liu et al. (2008) |
| 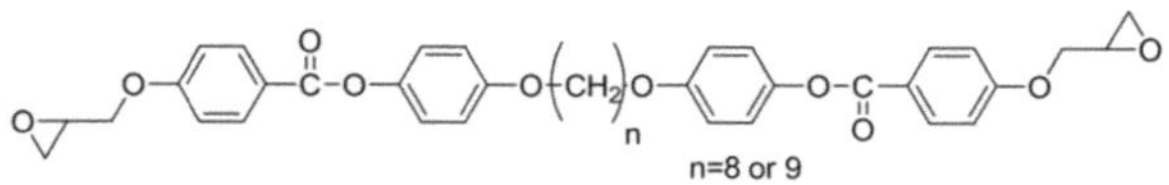 | | Shiota and Ober (1996, 1997) |
| 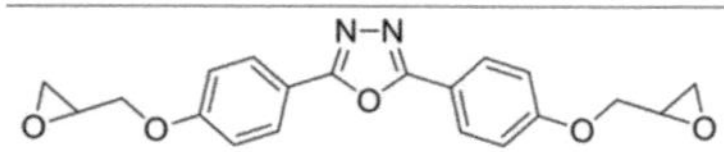 | | Balamurugan and Kannan (2009) |
| 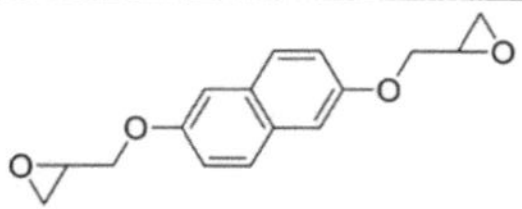 | BEPN | Giamberini et al. (1997), Su et al. (2002a) |
| 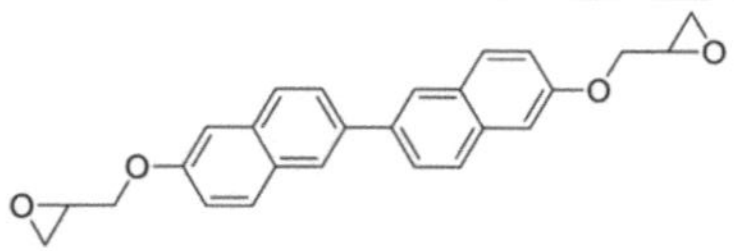 | | Carfagna et al. (1994b) |
| 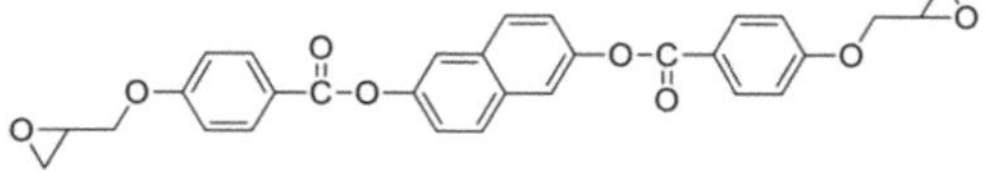 | | Lee and Jang (1999) |

**Table 19.1** (continued)

| LCERs | Abbreviation | References |
|---|---|---|
| 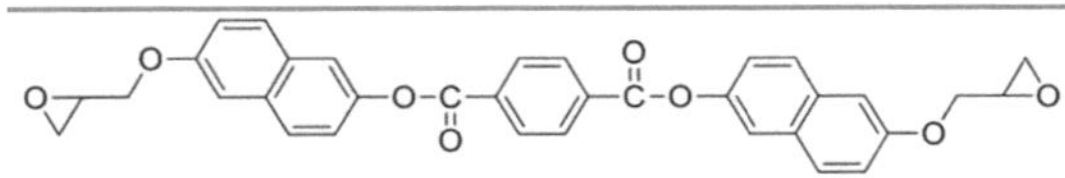 | | Le Hoang et al. (2013) |
| 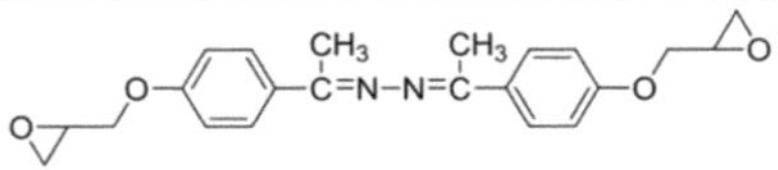 | | Carfagna et al. (1994d) |
| 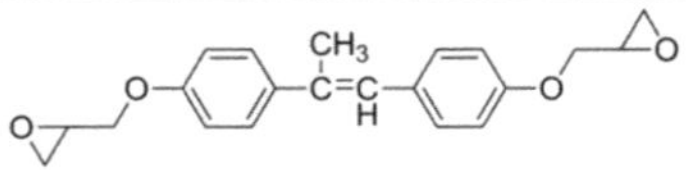 | | Barclay et al. (1992), Lin et al. (1994), Ortiz et al. (1998) |
| 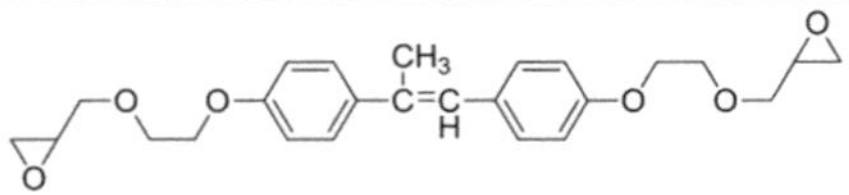 | | Yamamoto et al. (2014) |
| 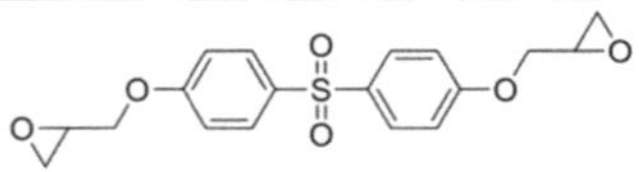 | | Huo et al. (2008) |
| 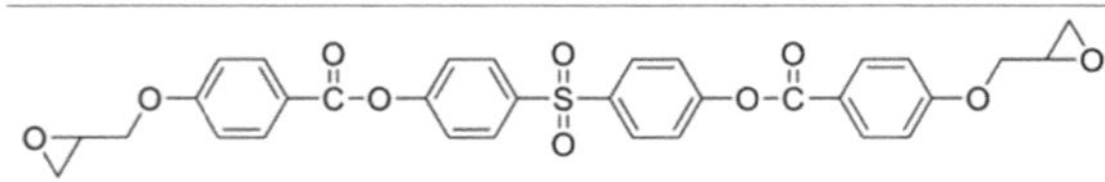 | | Huo et al. (2008) |
| 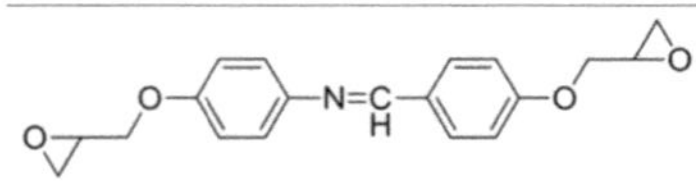 | | Su et al. (2000, 2002b) |
| 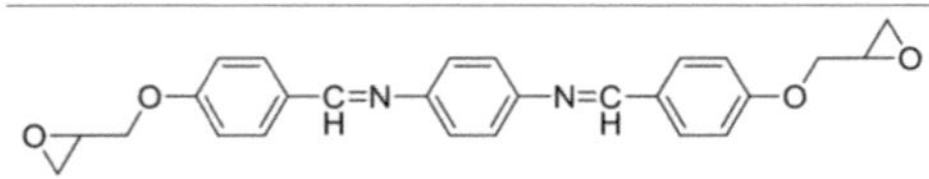 | | Castell et al. (2003b), Vengatesan et al. (2011) |

| | | |
|---|---|---|
| 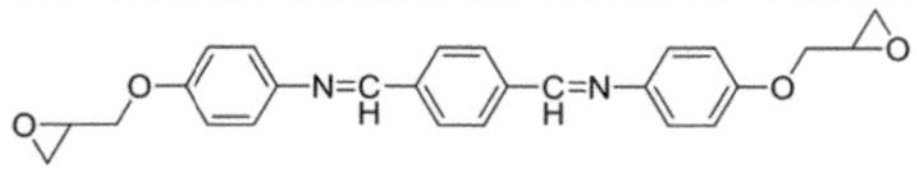 | | Ochi et al. (2004), Mija and Cascaval (2009) |
| 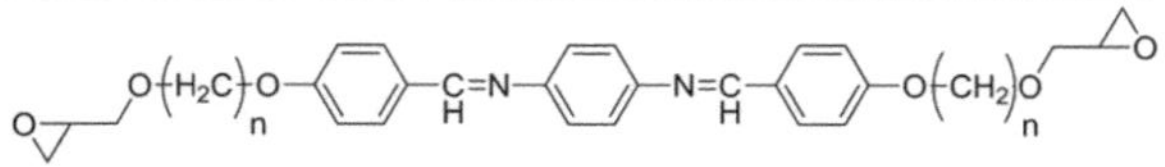 n=2, 4, 6, 8, 10 | | Castell et al. (2003b) |
| 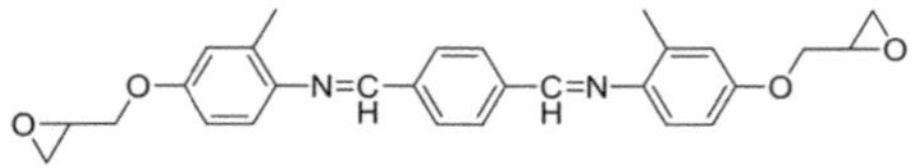 | DGETAM | Harada et al. (2004, 2013) |
| 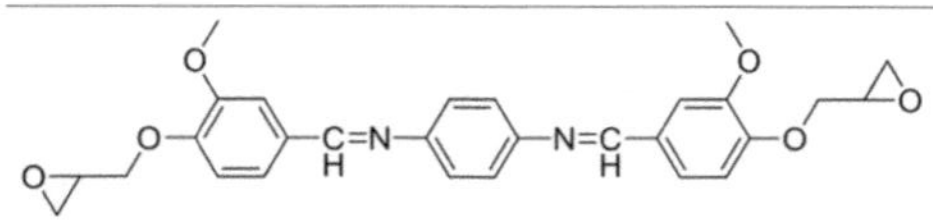 | | Vengatesan et al. (2011) |
| 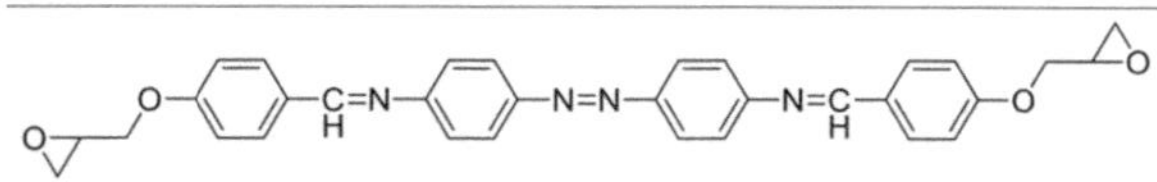 | | Castell et al. (2001) |
| 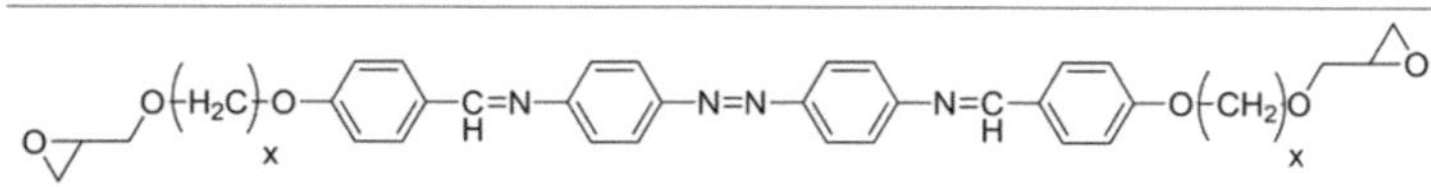 x=2, 4, 6, 8, 10 | | Castell et al. (2001) |
| 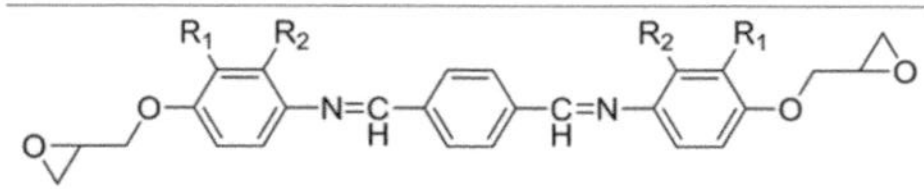 R1 and R2 are H or $CH_3$ | | Mormann et al. (1997) |
| 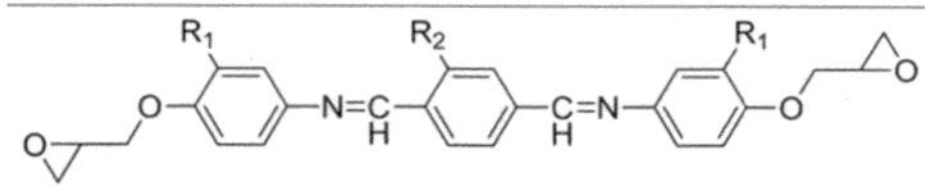 R1 and R2 are H, $CH_3$, $OCH_3$ or Cl | | Mormann et al. (1997) |

**Table 19.1** (continued)

| LCERs | Abbreviation | References |
|---|---|---|
| 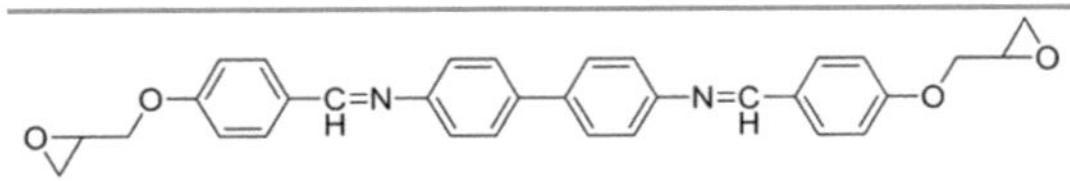 | | Castell et al. (2004) |
| 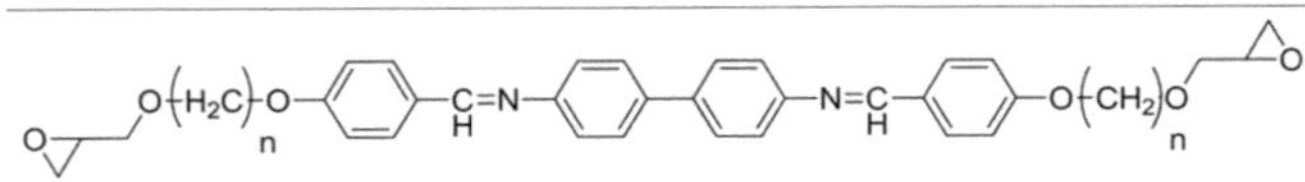 n=2 4, 6, 8, 10 | | Castell et al. (2004) |
| 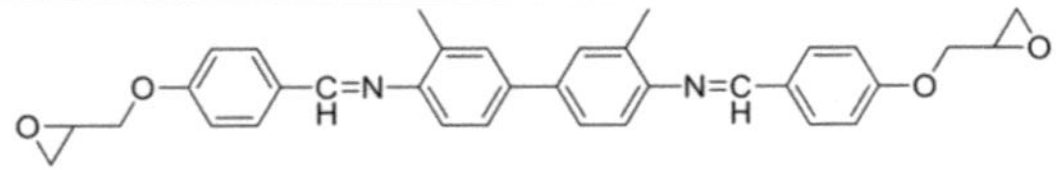 | | Gao et al. (2007) |
| 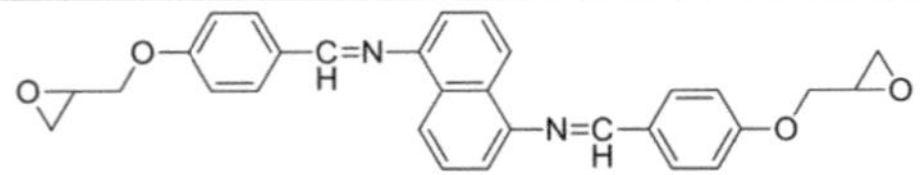 | | Castell et al. (2003a) |
| 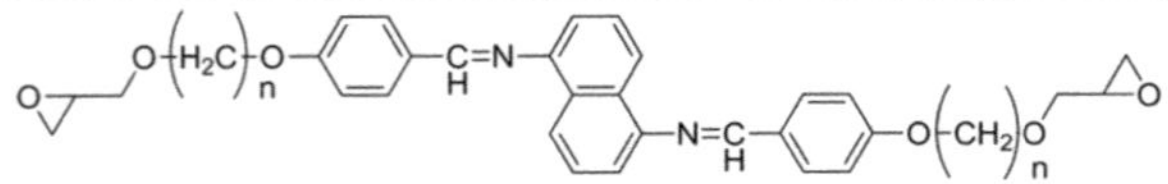 n=2, 4, 6, 8, 10 | | Castell et al. (2003a) |
| 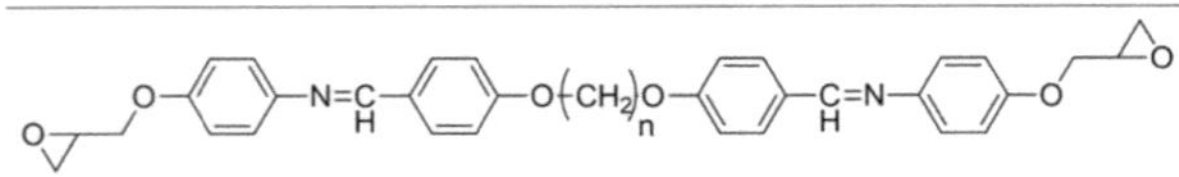 n= 6, 7, 8, 9 | | Choi et al. (2000) |
| 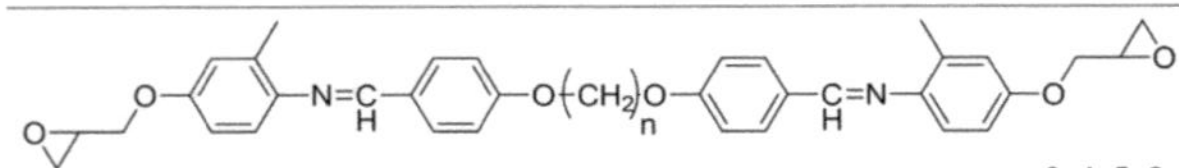 n=3, 4, 5, 6 | | Ochi et al. (2004) |

| | | |
|---|---|---|
| 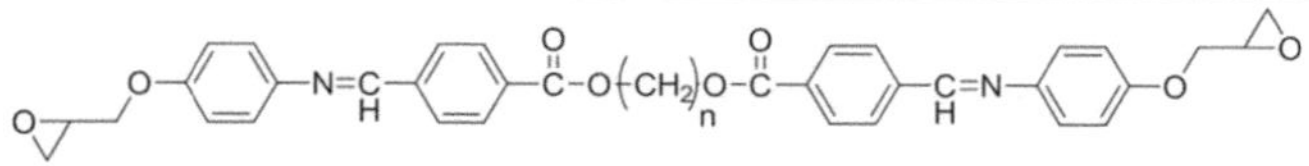 n=3, 4, 5, 6,7 ,8, 10, 12 | | Ribera et al. (2001, 2002) |
| 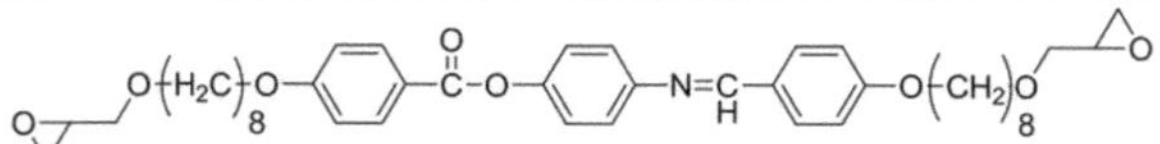 | | Xiao et al. (2012) |
| 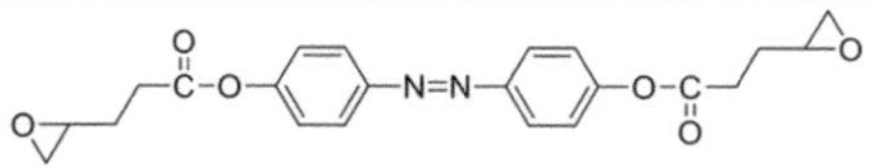 | | Wlodarska et al. (2009) |
| 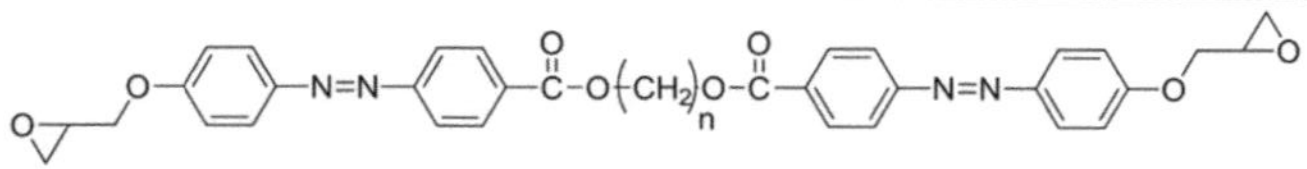 n=3, 4, 5, 6 | | Zhou et al. (2012) |

**Table 19.2** Chemical structures of curing agents for LCERs

| Curing agent | Abbreviation | References |
|---|---|---|
| 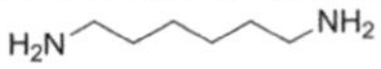 | | Shiota and Ober (1996) |
| 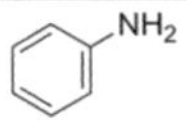 | | Liu et al. (1997) |
| 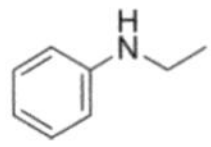 | | Liu et al. (1997) |
| 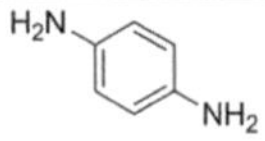 | p-PDA | Shiota and Ober (1996), Ochi et al. (1997) |
| 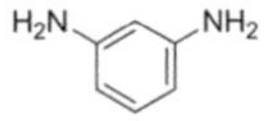 | m-PDA | Harada et al. (2009, 2013) |
| 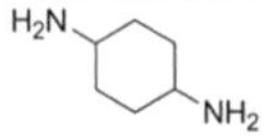 | | Carfagna et al. (1994a) |
| 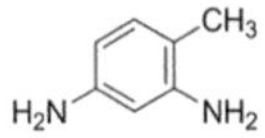 | | Carfagna et al. (1994c), Amendola et al. (1996), Ribera et al. (2001) |
| 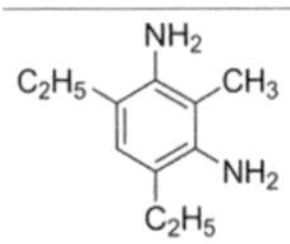 | | Liu et al. (2008) |
| 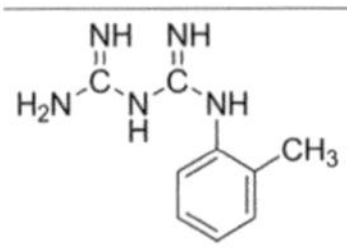 | | Szczepaniak et al. (1997) |
| 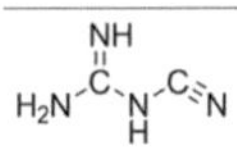 | | Su et al. (2002b) |
| 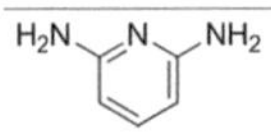 | | Grebowicz (1996), Vyazovkin et al. (2003) |
| 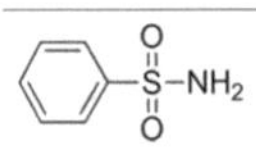 | | Liu et al. (1997) |
| 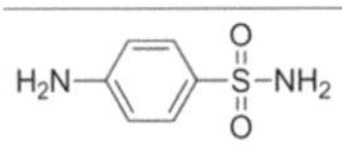 | SAA | Lin et al. (1994), Shiota and Ober (1996), Li and Kessler (2014) |
| 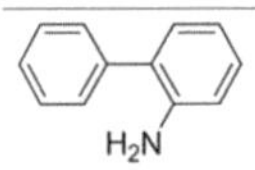 | | Mija and Cascaval (2009) |
| 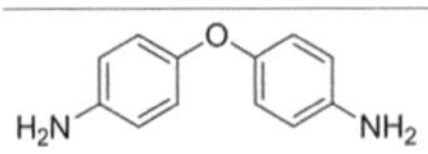 | | Huo et al. (2008), Balamurugan and Kannan (2010) |
| 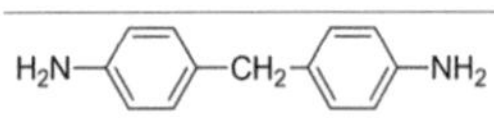 | | Barclay et al. (1992), Ochi et al. (1995) |
| 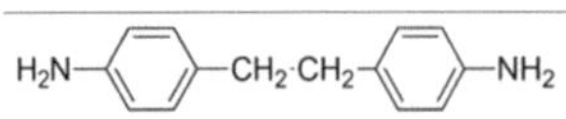 | DDE | Ochi et al. (1997), Harada et al. (2013), Yamamoto et al. (2014) |

(continued)

**Table 19.2** (continued)

| Curing agent | Abbreviation | References |
|---|---|---|
| 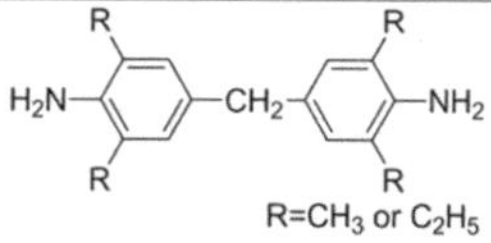 | | Ochi et al. (1997) |
| 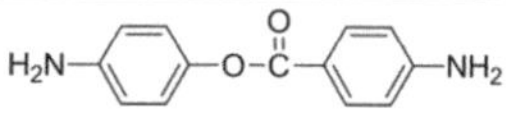 | | Lee and Jang (1998, 1999), Lee et al. (1998) |
| 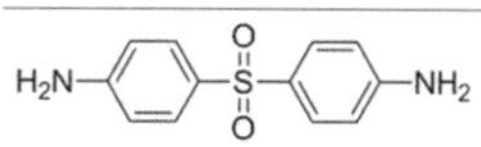 | DDS | Su (1993), Lee and Jang (1999), Liu et al. (2011b), Le Hoang et al. (2013) |
| 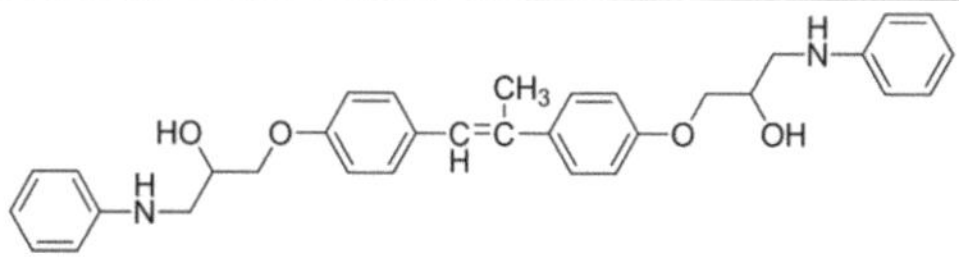 | | Lin et al. (1994, 1997) |
| 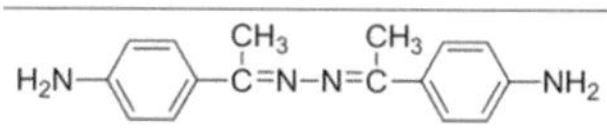 | | Carfagna et al. (1994d), Grebowicz (1996) |
| 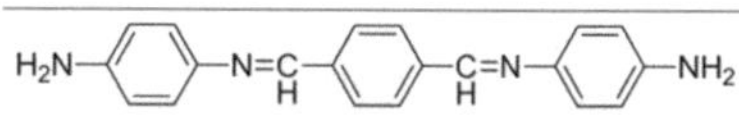 | | Harada et al. (2006) |
| 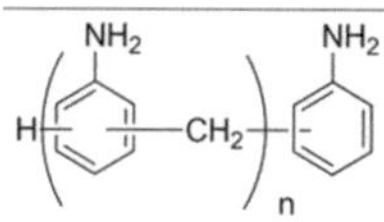 | | Ochi et al. (1995) |
| 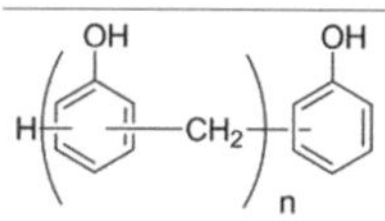 | | Ochi et al. (1995, 1997) |
| 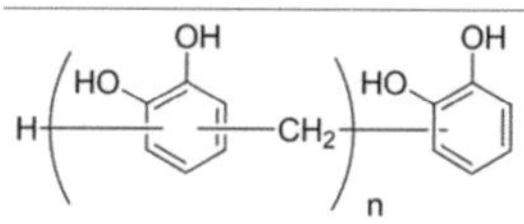 | | Ochi et al. (1995) |
| 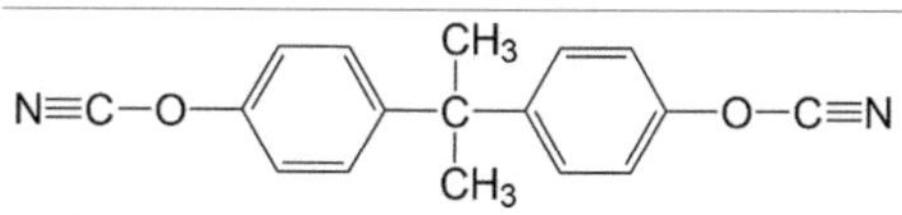 | B-10 | Su and Chuang (2002) |
| 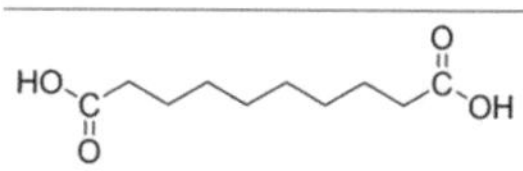 | | Giamberini et al. (1997, 1995) |
| 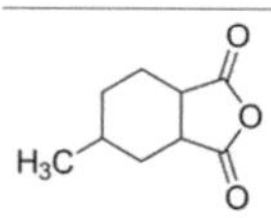 | | Xiao et al. (2012) |
| 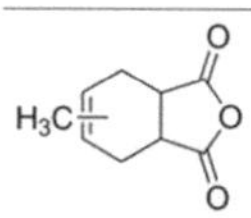 | MCHDA | Su et al. (2000), Punchaipetch et al. (2001) |

(continued)

**Table 19.2** (continued)

| Curing agent | Abbreviation | References |
|---|---|---|
| 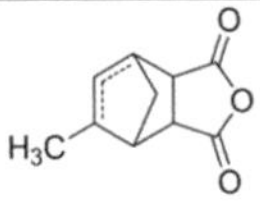 | | Su et al. (2002a) |
| 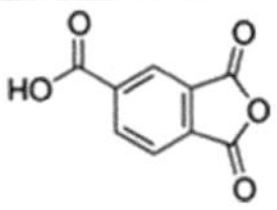 | BTA | Smith et al. (1999) |
| 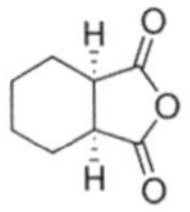 | HPA | Smith et al. (1999) |
| 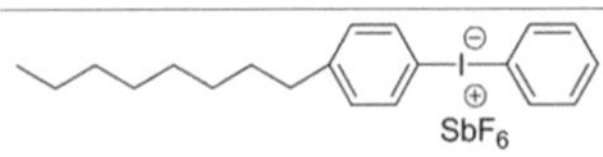 | OPIA | Hsu et al. (2012a) |

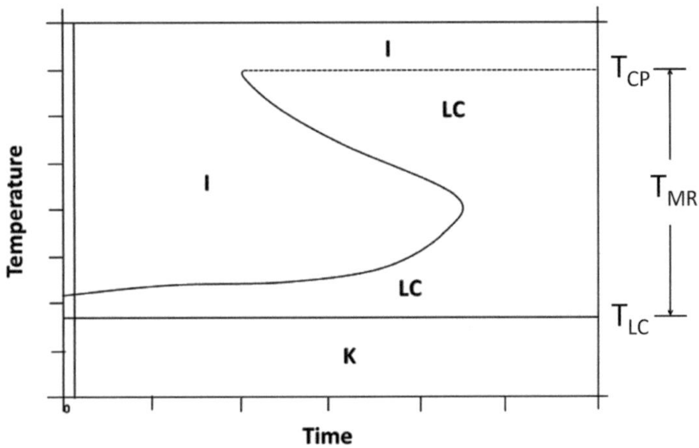


**Fig. 19.1** Isotropic-liquid crystalline time-temperature transformation (TTT) diagram of liquid crystalline resin system. *I* isotropic phase, *LC* liquid crystalline phase, and *K* crystalline phase of the LCER-curing agent mixture

structure has to be oriented by external force such as electric or magnetic field during cure in the $T_{MR}$ of LCER system (Fig. 19.2d).

The development LC phase in the cured resin depends on the type of curing agent and curing condition (Ortiz et al. 1998). The presence of LC phase in the resin system has dramatic effect on the curing rates. The activation energy can be lower in the liquid crystalline phase due to a decreased viscosity and thus a higher rate constant (Schnurpfeil et al. 2001; Liu et al. 1997, 1998). The rate-enhancement is associated with an increased concentration of reactive groups at the boundaries in the liquid crystalline phase. The structures of the resulting 3D networks are very complicated which depend on the interplay among the ER structure, curing agent structure

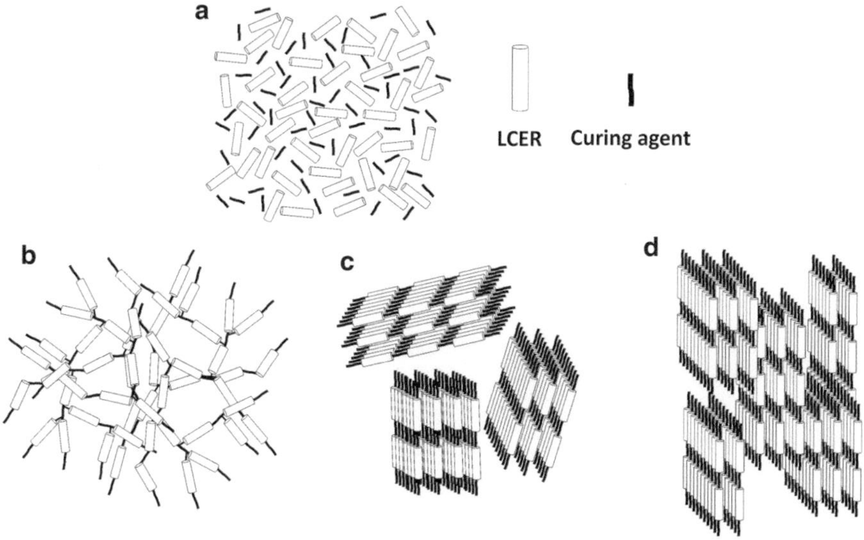


**Fig. 19.2** Illustration of molecular arrangements of LCER system under different condition (**a**) heating and melting, (**b**) curing in isotropic phase, (**c**) curing in $T_{MR}$ with the formation of polydomain structure (**d**) curing in $T_{MR}$ with applied electrical or magnetic field with the formation of ordered-polydomain structure

and the effect of crosslinking. It is interesting to note that the order of the network is mostly greater than that of the pristine ER because the presence of curing agent can improve the flow and alignment of the LCER during the curing process. The specific molecular interaction between either the mesogen of LCER or curing agent will result different LC phases in the cured resin such as nematic phase forms initially and then the LCER cured with a smectic network. This phenomenon has been observed in para phenylene diamine cured ER containing 1,4-bibenzoyl oxybenzene mesogen (Mallon and Adams 1993).

The physical properties of cured LCER depend on the chemical structure, composition of the LCER and the curing agent. They also depend on the curing schedule. The formal names of LCERs and curing agents are rather long, for the ease of identifying, abbreviations of LCERs and the curing agents are listed in the tables and they are used in the following discussions. All the properties present below for cured LCER and conventional ER are based on the stoichiometric ratio of resin and curing agent.

### *19.1.1 Effect of Substituent and Spacer on $T_{LC}$ of LCER*

The $T_{LC}$ of LCER resin will be affected by placing substitute or spacer on its main chain. Table 19.3 shows some examples. The resin 1 has the shortest aspect ratio of mesogen among five resins. The LCER exhibits monotropic liquid crystalline due

**Table 19.3** Chemical structure effect on the $T_{LC}$ of LCERs

| No | Resin | $T_m$ | $T_{LC}$ | $T_{CP}$ (°C) | References |
|---|---|---|---|---|---|
| 1 | 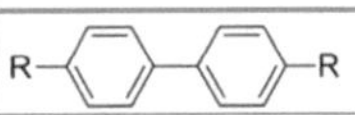 | 115 | 135 | 152 | Su (1993) |
| 2 | 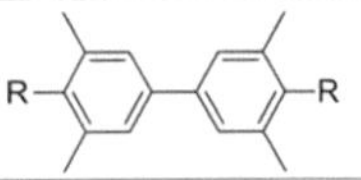 | 104–110 | – | – | Su et al. (1998) |
| 3 | 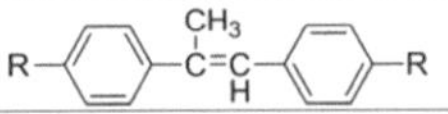 | 53 | 107 | 128 | Yamamoto et al. (2014) |
| 4 | 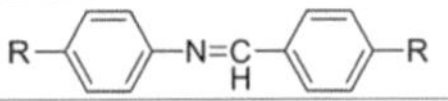 | 120 | 126 | 127 | Su et al. (2000) |
| 5 | 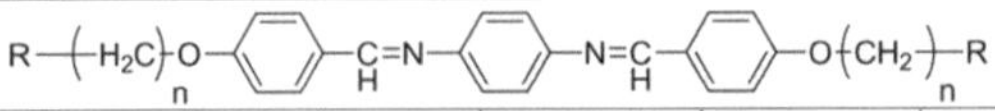 | | | | Castell et al. (2003b) |
| | x = 0 | 197 | 257 | 355 | |
| | x = 2 | 170 | 240 | 349 | |
| | x = 4 | 93 | 139–221 | 341 | |
| | x = 6 | 98 | 140–205 | 352 | |
| | x = 8 | 107 | 139–191 | 312 | |
| | x = 10 | 111 | 147–179 | 291 | |

to its mesogen is short (aspect ratio <5) such as the resin 1, 3 and 4. As compared with the resin 1, the resin 2 contains four methyl groups on the biphenyl mesogen. The methyl substitutes restrict the mesogen packing, so this resin does not exhibit LC phase upon heating and its melting point is decreased. The resin 3 contains additional methyl ethylene double bond in the biphenyl mesogen, so lowered melting temperature ($T_m$), $T_{LC}$ and $T_{CP}$ are observed as compared with the resin 1. The result is due to the increase of free volume in the resin. The resin 4 also contains a double bond in between the biphenyl group, but this mesogen is not as bulky as that of the resin 3, the $T_m$ and $T_{LC}$ are higher. When the length of mesogen of the resin 4 is extended with one more rigid rod unit to become the resin 5, its three phase transition temperatures are increased to quite high near 200 °C for $T_m$ and above 300 °C for $T_{CP}$. These temperatures are too high for ease of processing. The phase transition temperature can be reduced by inserting the flexible spacer $–CH_2–$ between mesogen and epoxy ring. The decreasing in phase transition temperature follows the size increasing in flexible spacer. However, the $T_m$ reduces to the lowest at 93 °C when four units of spacer are used, then it increases somewhat with increasing spacer size to 10 units. The result is probably due to the chain entanglement and chain folding of long length spacer. The range of LC phase is increased to close to 100 °C by the addition of flexible spacing which will help the processing window at the liquid crystalline state. Both nematic and smectic phases are present in the LC phase with spacer length equal or larger than four units.

### 19.1.2 Mechanical and Thermal Properties of Cured LCERs

The modulus of glassy state of polymer is governed by bond deformations and rotations which will be relatively unaffected by the liquid crystalline state. However, the liquid crystalline molecules tend to be aligned to have high density that will increase the modulus of cured LCER, if the resin system is cured at the LC state. The modulus of cured LCER can be further increased under the external field such as electrical field or magnetic field. For the resin system of BP-MCHDA, the storage modulus is increased by 6 % as compared with the conventional bisphenol A ER system DGEBA-MCHDA (Su et al. 2000). The cure schedule has a big effect on the storage modulus of LCER. For instance, the resin system of BP-SAA has a wide range of LC state from 120 to 190 °C. At 125 °C/16 h curing, the storage modulus of LCER is slight lower than that of conventional DGEBA-MCHDA system (Su et al. 2000). The result is possible from not fully cured LCER system. The storage modulus of this LCER system can be increased from 3975 to 4159 MPa when the curing temperature is increased from 170 to 190 °C for 12 h. In contrast, if the resin is cured in isotropic state at 200 °C for 12 h, the modulus is reduced to 3422 MPa (Li et al. 2013). When the resin is cured under strong magnetic field of 12 T, the storage modulus of oriented LCER is increased from 2500 to 4800 MPa as compared with unoriented LCER (Li and Kessler 2013). The dynamic modulus in the rubbery state of the cured LCER system above what is predicted from rubber elasticity theory has been observed in many LCER systems (Su et al. 2000; Ortiz et al. 1998). This is the characteristics of self-reinforcing feature from aligned liquid crystalline molecule in the resin.

The fracture toughness of cured LCER system is higher than that of conventional ER system, because the crack appears to grow preferentially at the boundaries of the liquid crystalline domain, resulting in crack deflection and segmenting. The failure occurs by formation of individual microcracks, which grow and coalesce, eventually resulting in failure. On the other hand, the isotropic ER fails by formation and growth of a single crack (Ortiz et al. 1998, 2000; Sue et al. 1997; Robinson et al. 2002; Carfagna et al. 2000b). Again, the curing condition has big effect on the fracture toughness of cured LCER system. If the LCER system is cured at the LC state, the fracture toughness is higher than that cured at the isotropic state. The fracture toughness of DGETAM/DDE can be increased by 100 % while the direction of the crack growth is vertical to the alignment of LC phase (Harada et al. 2004). Furthermore, if the curing state favors large crystalline domain, the fracture toughness is further increased as shown in Fig. 19.3 (Harada et al. 2009). On the contrary, the flexural strength of LCER system is lower than that of conventional ER system, because the rigid and extended nature of the network segments prevents plastic deformation (Earls et al. 1993).

The characteristics of aligned liquid crystalline molecule in the LCER results in higher density and less free volume in the resin which provide the resin with higher $T_g$ and higher thermal stability as compared with conventional ER after reacting with curing agent. The extent of increase depends on the chemical structure of resin and

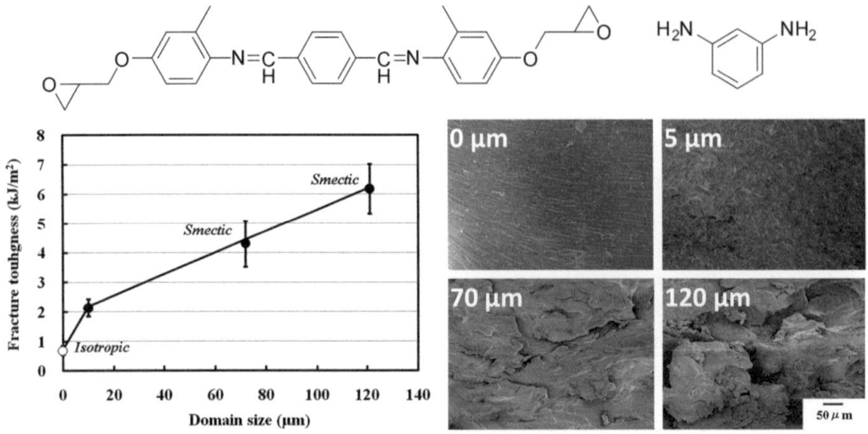


**Fig. 19.3** Fracture toughness of the DGETAM/m-PDA system is increased with domain size (*left*), corresponding to the increasing LC domain size, the roughness of their fractured surfaces drastically increased (*right*) (Harada et al. 2009)

**Table 19.4** Comparison of thermal properties of LCERs and DGEBA cured by different curing agents

| Sample | $T_g$ (°C)[a] | CTE (ppm)[b] | $T_d$ (°C)[c] |
|---|---|---|---|
| DGEBA/MCHDA | 74 | 756 | 267 |
| BP/MCHDA | 103 | 224 | 311 |
| AM/MCHDA | 151 | 226 | 298 |
| DGEBA/SAA | 152 | 137 | 323 |
| BP/SAA | 217 | 21 | 336 |
| AM/SAA | 219 | 72 | 332 |
| DGEBA/B-10 | 237 | 86 | 339 |
| BP/B-10 | 253 | 74 | 335 |
| AM/B-10 | 289 | 68 | 321 |

[a]$T_g$ is measured by dynamic mechanical analysis
[b]CTE: coefficient of thermal expansion, measured by thermomechanical analyzer
[c]$T_d$ is determined by the temperature at 5 % weight loss using thermogravimetric analysis

type of curing agent (Su et al. 2002a, b, c; Su and Chuang 2002). In general, the LCER cured by aromatic type curing agent such as SAA and cyanate ester B10 provides higher $T_g$, lower coefficient of thermal expansion (CTE) and higher thermal decomposition temperature $T_d$ than those cured by aliphatic type curing agent such as MCHDA. Thermal properties of BP and AM LCER system are summarized in Table 19.4. They are cured by MCHDA, SAA and B-10 respectively and the results are compared with the conventional ER DGEBA. As expected, both BP and AM have shown better thermal properties than DGEBA regardless the curing agent type. The B10 cured epoxy exhibits the best thermal properties due to the formation of cyanurate ring structure (Su and Chuang 2002). Furthermore, the thermal conductivity and the thermal expansion of the LCER system containing oriented-polydomain structure exhibit dramatic anisotropic improvement at the direction parallel to that of LC phase alignment (Harada et al. 2003; Li and Kessler 2013).

The fire resistance of LCERs has been found to be superior to that of conventional ER due to ordered and compact structure. The char yield of azomethine based pristine LCER is higher than 50 % at 600 °C. (Castell et al. 2001, 2003a). For SAA cured ERs, the residual weight % for DGEBA, BP and AM is 29.05, 46.01 and 61.10, respectively at 450 °C (Su et al. 2000). The limiting oxygen index (LOI) of, DGEBA, AM, BP and BEPN is 24, 26, 30, 30, respectively (Duann et al. 2004; Chiang and Ma 2002). The azomethine based LCERs apparently exhibit highest char yield but lowest LOI among the LCERs discussed. The additional –CH=N– linkage in the phenyl mesogen group is easy to form char but less fire resistant than the pure phenyl based mesogen. Nevertheless, its LOI is still higher than the conventional epoxy resin.

### *19.1.3 Permeability of Cured LCER*

The cured LCER usually exhibits much lower permeability, diffusion coefficient and sorption coefficient toward permeants such as moisture, solvents as compared with cured ER (Vittoria et al. 1991; Weinkauf and Paul 1992). The results can be understood from a two-phase model, the permeant penetrates through the crystalline domains extremely slow as compared with the amorphous phase, so the permeant only diffuse predominately through the domain boundaries. The LCER is known to have aligned and densely packed crystalline domains in the cured LCER system which will result in low permeability. The Sumitomo Chemical Company has taken the advantage of this property and developed biphenyl LCER resin into the IC packaging and encapsulation materials (Hirano et al. 1999).

## 19.2 Processing of Nanocomposite

The emerging of nanotechnology at the end of twentieth century, many new nanomaterials have been developed such as nanoparticles, nanorods, nanowires, nanotubes, and nanoplates. By using nanomaterial as reinforcing filler and incorporating it into the polymer matrix, one can obtain polymer nanocomposite. There are many review articles and textbooks devoted to this subject (Twardowski 2007; Mittal 2010, 2013). However, the researches and discussions on nanocomposites made from LCER and nanofiller are relatively few.

### 19.2.1 Nanofiller

The most common nanofillers are inorganic oxides and carbon based nanofillers with different geometries. We will concentrate on the $SiO_2$ nanoparticle and multiwall carbon nanotube (CNT) here. Both types of fillers exhibit rather poor compatibility with the hydrophobic LCER, so they have to be chemically modified with appropriate chemical functionality for LCER. The chemical functionality on the nanofiller can bond with the resin either by van der Waal forces or chemical bonds. The former is called heterogeneous nanocomposite and the latter is called homogeneous nanocomposite.

For the homogeneous CNT nanocomposite, the CNTs are first treated with strong nitric acid to have carboxylic acid functionality. The carboxylic acid group of CNT can react further with the epoxy functional group of LCER to become epoxy modified CNT which can be easily dispersed in the resin matrix (Chen et al. 2011; Hsu et al. 2012b). For the homogeneous $SiO_2$ nanocomposite, the $SiO_2$ nanoparticles are first reacted with silane containing epoxy functionality. Then the epoxide-functionalized $SiO_2$ can be blended easily with the resin/curing agent to form nanocomposite resin system to be cured by heat or radiation (Hsu et al. 2012a).

### 19.2.2 Processing

The chemical modified nanofillers are mixed with LCERs using high shear force with or without using solvent, which is similar to the mixing with conventional ER system. The solvent reduces the viscosity of resin and ease the dispersion of nanofiller, but it will impose the pollution problem and consume energy for handling and removal of the solvent. Both solventless and solvent based resin systems are reported in the literatures. The rigid rod characteristic of LC moiety in the LCER makes the viscosity of melting LCER low due to the absence of chain entanglement. However, high viscosity is usually observed in conventional ER due to the entangled flexible moiety.

Figure 19.4 shows the morphology change of epoxide-functionalized CNT (ef-CNT)/BP/DDS nanocomposite resin system upon curing observed by polarized optical microscopy (POM) and scanning electron microscopy (SEM). The LC domain is clearly observed in the pristine BP resin (Fig. 19.4a). Figure 19.4b shows the crystalline morphology of cured BP/DDS resin system. The crystalline size is smaller than the LC domain of pristine BP resin due to the presence of DDS. The ef-CNT filler induces the growth of LC domain of BP resin along its long axis (Fig. 19.4c). As compared with the radially aligned crystalline morphology of BP/DDS resin, the elongated large crystalline domain can be obtained in the confined z-axis environment due to the incorporation of 2 wt% ef-CNT (Fig. 19.4d). Without any z-axis confinement to, a part of crystalline domain will grow into spherical crystalline (Fig. 19.4e–g) and interconnect with 2 wt% ef-CNT (Fig. 19.4h) in the cured BP/DDS resin system.

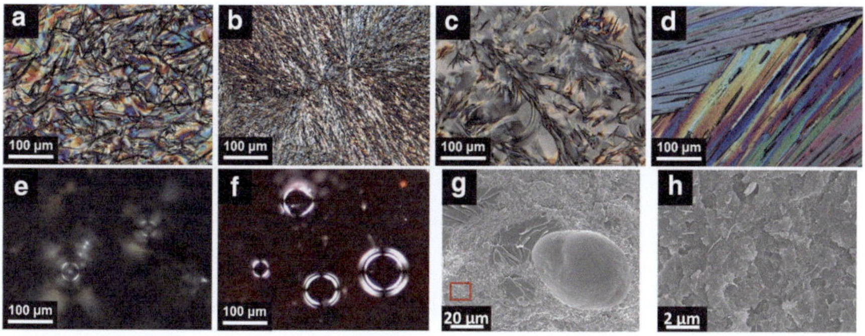


**Fig. 19.4** POM images of (**a**) LC phase of pristine BP resin, (**b**) cured BP/DDS resin, (**c**) LC phase of ef-CNT/BP (**d**) cured ef-CNT/BP/DDS with z-axis confinement during curing, (**e**) cured ef-CNT/BP/DDS bulk and (**f**) fracture surface of cured ef-CNT/BP/DDS bulk. SEM images of (**g**) fracture surface of cured ef-CNT/BP/DDS nanocomposite and (**h**) the magnified *red square* outline area of (**g**)

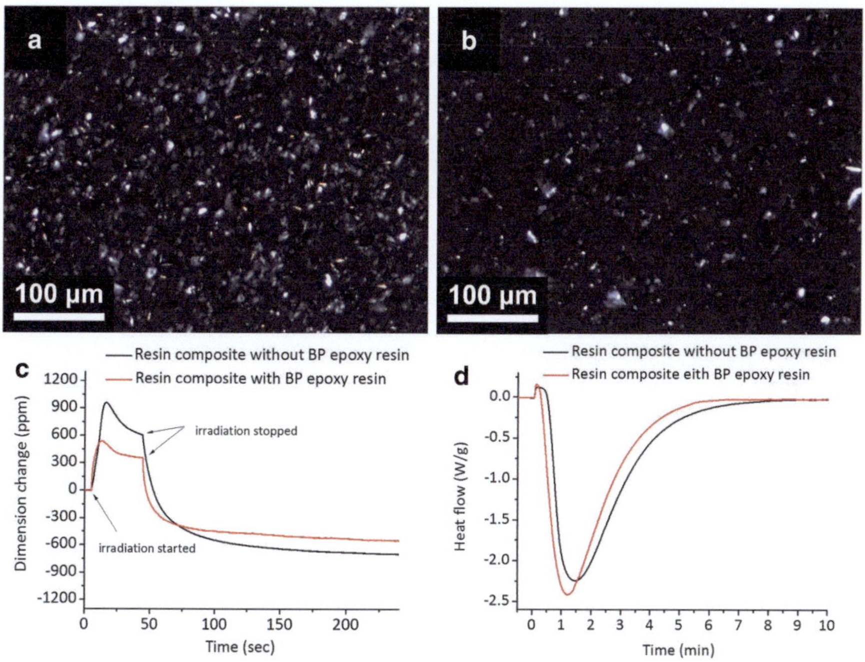


**Fig. 19.5** POM images of (**a**) LCER/Lewis acid photocatalyst (OPIA), (**b**) radiation cured LCER. Comparison of (**c**) dimension change and (**d**) curing rate between LCER and ER

Figure 19.5 shows the properties of homogeneous nanocomposite made from LCER/$SiO_2$ system. Figures 19.5a, b are the POM photos of uncured and cured nanocomposite respectively. The crystallines of LCER are preserved after the radiation curing (Hsu et al. 2012b). The phase transformation is occurred during the curing of the LCER system, the melted phase decreases the extent of polymerization shrinkage (Fig. 19.5c). Thus the dimensional change of LCER system is less than that of conventional ER system which is a very useful feature in the actual applications such as encapsulation and restoration. The curing of LCER system is an exothermic reaction (Fig. 19.5d). It is also interested to observe and confirm the faster reaction rate in the LCER curing as discussed previously.

## 19.3 Applications and Properties of Nanocomposites

The cured nanocomposites from LCER have exhibited exceptional mechanical and thermal properties with low permeation property. They have wide applications in various industries and medical technology.

### *19.3.1 Encapsulation Materials for Electrical and Electronic Applications*

The capability of phase transformation of LCER at elevated temperature is used by Hitachi Company as heat sink or thermal conducting encapsulants in the electronic industry (Tanaka et al. 2007). The high $T_g$ and high thermal stability of LCER have been found in the application of printed wiring board and IC-encapsulants (Hirano et al. 1999). The inorganic nanoparticles such as $SiO_2$, BN are usually added into the LCER system at concentration higher than 60 wt% to dissipate the heat and increase the mechanical strength of the resin.

The highly compact layer structure can be formed in the cured LCER which will provide high resistance in the high voltage environment. Siemens Company has taken the advantage of these properties and applied in the insulations of coils and transformers to withstand the environment of high temperature and high electrical stress (Smith et al. 1999). Table 19.5 shows the comparison between cured LCER and ER in terms of their thermal, electrical and mechanical properties. The reinforcing filler such as mica is frequently used in the electrical application to endure high electrical stress (Lee and Neville 1967). With the presence of mica in resin, the high voltage electrical wave goes through the boundary between plate filler and resin which retards the electrical breakdown as compared with the cured resin without reinforcing filler.

The aligned mesogen molecule in the LCER functions as mica to show exceptional dielectric strength and low dissipation factor.

**Table 19.5** Comparison of thermal, electrical and mechanical properties of cured LCER and ER for electrical applications

| Sample | $T_g$ (°C) | CTE (ppm) | Dielectric strength (KV/10 mil) | Dissipation factor @1 MHz | Tensile strength (Mpa) | Elong. at Tear | E-modulus (MPa) |
|---|---|---|---|---|---|---|---|
| BP/DDS | 231 | 56.9 | – | – | – | – | – |
| DGEBA/DDS | 200 | 67.7 | – | – | – | – | – |
| BP/BTA | 183 | – | 12.38 | 0.0319 | – | – | – |
| DGEBA/BTA | 127 | – | 10.89 | 0.0372 | – | – | – |
| BP/HPA | – | – | – | – | 65 | 3.2 | 3070 |
| DGEBA/HPA | – | – | – | – | 76 | 6.4 | 2346 |

### *19.3.2 Light Weight Structure Material*

The excellent properties synergistically derived from the interactions among the resin and carbon nanotubes (CNT) are ideal for the application in light weight structure such as airplane body. In addition, CNTs have remarkable advantages over conventional carbon fibers in composite processing due to their high flexibility. The incorporation of CNTs into ERs not only improves the electrical, thermal and mechanical properties of ERs but also modifies the processing behaviors. Since LCER can be further reinforced by a second phase, and the alignment of the mesogen phase of LCER is expected to be enlarged and elongated due to the presence of high aspect ratio of second phase, such as CNTs (Fig. 19.4). Therefore, it is possible to fabricate a composite possessing high mechanical properties, good thermal stability and electrical conductivity for aeronautic and astronautic engineering by utilizing both LCER and CNTs.

The ef-CNT/BP/DDS system has developed for this application (Chen et al. 2011; Hsu et al. 2012b). The incorporation of ef-CNT in the LCER system induces the formation of crystalline phase in the composite, thus greatly enhances the thermal and mechanical properties of the composite as compared with the pristine LCER system (Table 19.6). Besides, the incorporation of thermal conductive CNT has accelerated the curing reaction of ER. However, the excessive amount of ef-CNT in matrix results in reaching vitrification stage quickly and lowering the degree of conversion. The mechanical properties are decreased when the amount of CNT is >2.00 wt%.

### *19.3.3 LCER as Biomaterials for Medical Application*

The high mechanical strength and low permeation characteristics of LCER nanocomposite are suitable for medical applications. The application in dentistry

**Table 19.6** Summary of thermal and mechanical properties of ef-CNT/BP/DDS composites (Hsu et al. 2012b)

| wt% of ef-CNT | Properties | | | | | | |
|---|---|---|---|---|---|---|---|
| | $_{TMA}T_{g1}$ (°C) | $_{TMA}T_{g2}$ (°C) | $_{DMA}T_{g}$ (°C) | Storage modulus at 50 °C (GPa) | $T_d$ (°C) | Residual at 600 °C (wt%) | Hardness [Hv(±SD)] |
| 0.00 | 191.30 | 204.59 | 202.91 | 2.27 | 347.94 | 29.28 | 16.62(0.48) |
| 0.50 | 196.79 | 230.70 | 254.86 | 2.96 | 358.44 | 26.60 | 20.99(0.50) |
| 1.00 | 204.81 | 229.44 | 271.94 | 3.04 | 354.34 | 28.15 | 24.52(0.80) |
| 2.00 | 224.71 | 238.65 | 272.87 | 3.20 | 361.71 | 29.09 | 27.14(1.22) |
| 4.00 | 201.33 | 245.67 | 258.61 | 2.89 | 357.89 | 32.19 | 27.86(2.20) |
| 10.00 | 184.01 | 241.63 | 205.91 | 2.41 | 357.94 | 37.20 | 28.02(1.80) |

has been studied including direct or indirect dental restoration, dental core-post system, and dental brace etc. The LCER nanocomposite can be polymerized with various curing agents to form the polymer with low shrinkage.

The $SiO_2$ nanoparticle reinforced LCER has been developed into dental restorative material and orthodontic device (Hsu et al. 2012a; Tai et al. 2014; Su et al. 2012). The nanoparticles of 70~100 nm $SiO_2$ were treated with γ-glycidoxypropyl trimethoxysilane to have epoxy functionality, then they are blended with BP LCER and cyclic aliphatic ER, photoinitiator to be cured with visible light. Figure 19.6 shows the SEM photos of fracture surface and cross sectional surface of cured resin with and without LCER. The morphology changes are drastic from featureless surface (Fig. 19.6a) to intercalating layer structure with nanoparticles (Fig. 19.6b) when the resin composition contains only 6.6 wt% of BP. The nanoparticles are well dispersed in the resin matrix (Fig. 19.6c) but elongated LC domains are present in the resin containing BP (Fig. 19.6d).

Figure 19.7 shows the comparison of properties of cured nanocomposite resins without LCER (E-BP00ZL series) and with LCER (E-BP07ZL series). As expected, the hardness, flexural modulus, flexural strength are increased with increasing the amount of $SiO_2$ hard filler. The nanocomposites containing BP all reveal higher performance than that of nanocomposites without BP except the flexural strength (Fig. 19.7c) when the amount of silica is at or higher than 41.85 vol% (~60 wt%). The high amount of filler causes the phase separation to occur, the rigid and extended nature of the network segments in LC domain prevents plastic deformation as discussed earlier. The water absorption of BP containing nanocomposites is much lower than the ones without BP because of the dense characteristic of LC domain with reduced water diffusion rate toward the composite. The performance of LC nanocomposites is either comparable or better than commercial dental restorative materials (Fig. 19.7d).

The toxicity of LCER nanocomposite is lower than that of nanocomposite contains no LCER and commercial dental restorative materials by both MTT and LDH tests (Fig. 19.8). Therefore, the LCER nanocomposite is a good biomaterial for medical applications.

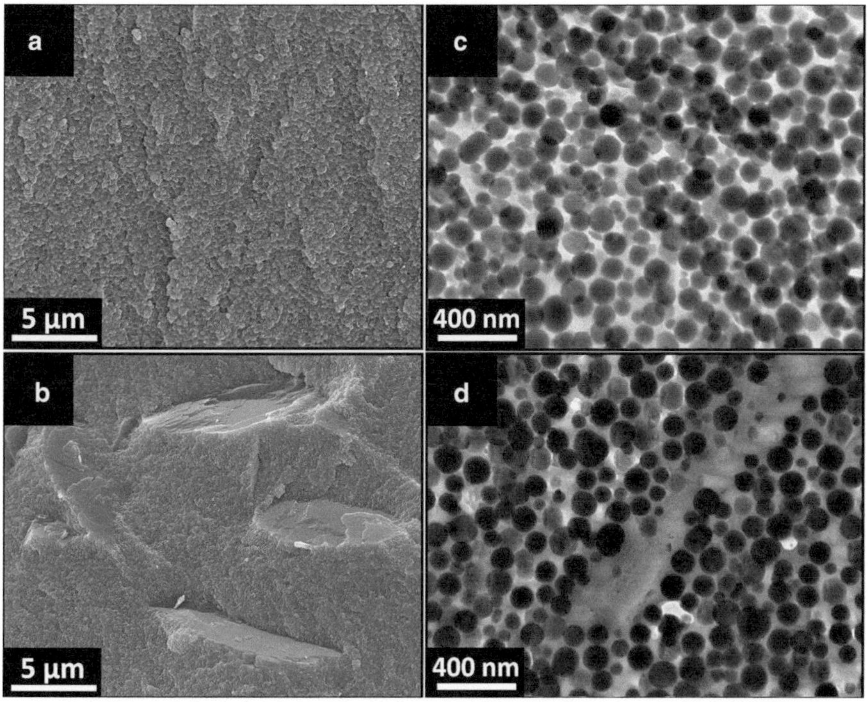


**Fig. 19.6** SEM images of fracture surface of cured nanocomposite resins containing (**a**) no liquid crystalline resin, (**b**) and 6.6 wt% BP resin in the composition of resin matrix. (**c**) and (**d**) are the TEM images of the cured composite resins corresponding to (**a**) and (**b**), respectively

## 19.4 Conclusions and Future Perspective

The development and progress of LCER technology and its applications are reviewed. The chemical structure effects on the liquid crystalline state of the LCER are discussed. The performance of the cured LCER system depends on the chemical structures of the resin and curing agent, and curing schedule. In general, the materials made from LCER exhibits superior chemical, mechanical, electrical, thermal properties than those made from conventional ER due to the presence of highly ordered arrangement of molecules at LC state. The resin also shows extremely low permeation and low toxicity behaviors. By incorporating nanofiller into the LCER to make into nanocomposite, the performance of the material is further improved. The nanocomposite has potential applications in light weight structure, electrical insulation, electronic encapsulation, medical device, etc. To realize the full capability of LCERs into actual applications, further development should pursue to have low melting temperature LCERs for ease of molecular alignment, new LC curing agents with comparable $T_{LC}$ as that of LCER, new surface modification technique for reinforcing filler and new types of nanocomposites. Although the reports on the optical properties of LCER systems

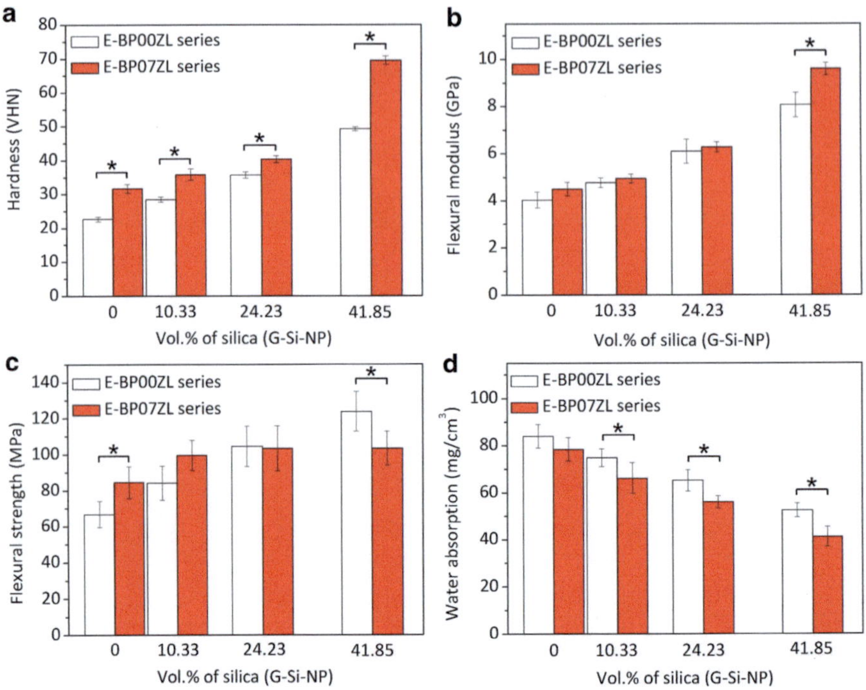


**Fig. 19.7** Comparison of properties of cured nanocomposite resins without liquid crystalline resin (E-BP00ZL series) and with liquid crystalline resin (E-BP07ZL series): (**a**) hardness, (**b**) flexural modulus, (**c**) flexural strength and (**d**) water absorption. *Asterisk* denotes significant difference ($p < 0.05$) between two series is achieved (Hsu et al. 2012a)

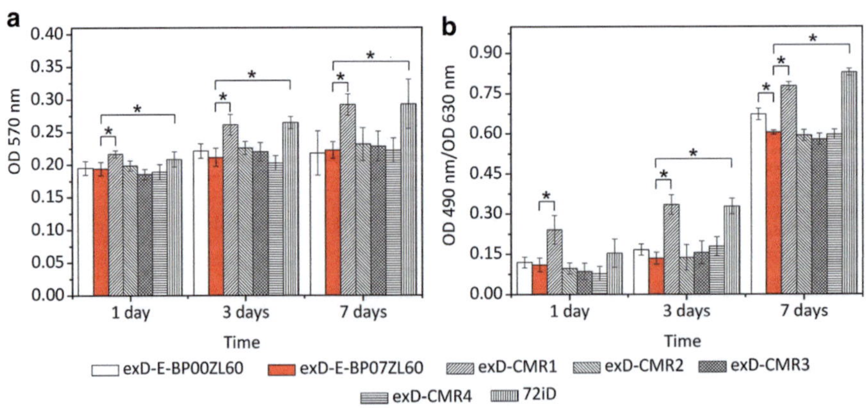


**Fig. 19.8** Results of MTT (**a**) and LDH (**b**) assays of MG-63 cells on the extracts of cured dental restorative composite resins for 7 days. *Asterisk* denotes a significant difference ($p < 0.05$) is achieved as compared to the extracts of cured nanocomposite resin containing no liquid crystalline resin (E-BP00ZL60) (Hsu et al. 2012a)

are limited, they are fruitful areas and worth to explore more extensively for actual applications such as high refractive index material, waveguide, optical switch, nonlinear optic, etc.

**Acknowledgement** The authors would like to thank Miss Shiow-Wei Wang, Miss Yin-Hsi Lin and Mr. Chun-Fu Lu for their help in the preparation of this manuscript.

## References

Amendola E, Carfagna C, Giamberini M, Pisaniello G (1995) Curing reactions of a liquid-crystalline epoxy-resin based on the diglycidyl ether of 4,4′-dihydroxy-alpha-methylstilben. Macromol Chem Phys 196:1577–1591

Amendola E, Carfagna C, Giamberini M, Komitov L (1996) Anisotropic liquid crystalline epoxy thermoset. Liq Cryst 21:317–325

Balamurugan R, Kannan P (2009) Synthesis and properties of a liquid crystalline thermoset epoxy resin containing 1,3,4-oxadiazole groups. High Perform Polym 21:251–264

Balamurugan R, Kannan P (2010) 1,3,4-Oxadiazole epoxy resin-based liquid crystalline thermosets and their cure kinetics. J Mater Sci 45:1321–1327

Barclay GG, Ober CK, Papathomas KI, Wang DW (1992) Liquid-crystalline epoxy thermosets based on dihydroxymethylstilbene—synthesis and characterization. J Polym Sci Polym Chem 30:1831–1843

Cai ZQ, Sun JZ, Zhou QY, Xu JL (2007) Synthesis and characterization of a novel liquid-crystalline epoxy resin combining biphenyl and aromatic ester-type mesogenic units. J Polym Sci Polym Chem 45:727–735

Carfagna C, Amendola E, Giamberini M (1994a) Liquid-crystalline epoxy-resins—curing of blends of mesogenic and nonmesogenic epoxy monomers. J Mater Sci Lett 13:126–128

Carfagna C, Amendola E, Giamberini M (1994b) Liquid-crystalline epoxy-resins containing binaphthyl group as rigid block with enhanced thermal-stability. Macromol Chem Phys 195:2307–2315

Carfagna C, Amendola E, Giamberini M (1994c) Rigid-rod networks—liquid-crystalline epoxy-resins. Compos Struct 27:37–43

Carfagna C, Amendola E, Giamberini M, Filippov AG (1994d) Liquid-crystalline epoxy-resins—a glycidyl-terminated benzaldehyde azine cured in the nematic phase. Macromol Chem Phys 195:279–287

Carfagna C, Acierno D, Di Palma V, Amendola E, Giamberini M (2000a) Composites based on carbon fibers and liquid crystalline epoxy resins, 1—Monomer synthesis and matrix curing. Macromol Chem Phys 201:2631–2638

Carfagna C, Meo G, Nicolais L, Giamberini M, Priola A, Malucelli G (2000b) Composites based on carbon fibers and liquid crystalline epoxy resins, 2—dynamic-mechanical analysis and fracture toughness behavior. Macromol Chem Phys 201:2639–2645

Castell P, Galia M, Serra A (2001) Synthesis of new epoxy liquid-crystalline monomers with azo groups in the central mesogenic core. Crosslinking with amines. Macromol Chem Phys 202:1649–1657

Castell P, Serra A, Galia M (2003a) New liquid-crystal line thermosets from liquid-crystalline bisazomethynic epoxy resins with naphthylene disruptors in the central core. J Polym Sci Polym Chem 41:1536–1544

Castell P, Serra A, Galia M, Giamberini M, Carfagna C (2003b) Anisotropic thermosets from liquid-crystalline azomethynic epoxy resins and primary aromatic diamines. J Polym Sci Polym Chem 41:1–12

Castell P, Serra A, Galia M (2004) Liquid-crystalline thermosets from liquid-crystalline epoxy resins containing bisazomethinebiphenylene the central core: copolymerization with a nonmesomorphic epoxy resin. J Polym Sci Polym Chem 42:3631–3643

Chen S, Hsu SH, Wu MC, Su WF (2011) Kinetics studies on the accelerated curing of liquid crystalline epoxy resin/multiwalled carbon nanotube nanocomposites. J Polym Sci Polym Phys 49:301–309

Chiang CL, Ma CCM (2002) Synthesis, characterization and thermal properties of novel epoxy containing silicon and phosphorus nanocomposites by sol-gel method. Eur Polym J 38:2219–2224

Choi EJ, Ahn HK, Lee JK, Jin JI (2000) Liquid crystalline twin epoxy monomers based on azomethine mesogen: synthesis and curing with aromatic diamines. Polymer 41:7617–7625

Douglas EP (2002) Liquid crystalline thermosets. Encyclopedia of polymer science and technology, vol 3. Wiley, New York. doi:10.1002/0471440264

Duann YF, Liu TM, Cheng KC, Su WF (2004) Thermal stability of some naphthalene- and phenyl-based epoxy resins. Polym Degrad Stabil 84:305–310

Earls JD, Hefner RE, Puckett PM (1993) Mesogenic epoxy compounds. U.S. Patent 5,270,405

Gao Z, Yu Y, Xu Y, Li S (2007) Synthesis and characterization of a liquid crystalline epoxy containing azomethine mesogen for modification of epoxy resin. J Appl Polym Sci 105:1861–1868

Giamberini M, Amendola E, Carfagna C (1995) Curing of a rigid-rod epoxy resin with an aliphatic diacid—an example of a lightly cross-linked liquid-crystalline thermoset. Macromol Rapid Commun 16:97–105

Giamberini M, Amendola E, Carfagna C (1997) Lightly crosslinked liquid crystalline epoxy resins: the effect of rigid-rod length and applied stress on the state of order of the cured thermoset. Macromol Chem Phys 198:3185–3196

Grebowicz JS (1996) On the formation of liquid crystalline texture in epoxy resins. Macromol Symp 104:191–221

Harada M, Ochi M, Tobita M, Kimura T, Ishigaki T, Shimoyama N, Aoki H (2003) Thermal-conductivity properties of liquid-crystalline epoxy resin cured under a magnetic field. J Polym Sci Polym Phys 41:1739–1743

Harada M, Ochi M, Tobita M, Kimura T, Ishigaki T, Shimoyama N, Aoki H (2004) Thermomechanical properties of liquid-crystalline epoxy networks arranged by a magnetic field. J Polym Sci Polym Phys 42:758–765

Harada M, Watanabe Y, Tanaka Y, Ochi M (2006) Thermal properties and fracture toughness of a liquid-crystalline epoxy resin cured with an aromatic diamine crosslinker having a mesogenic group. J Polym Sci Polym Phys 44:2486–2494

Harada M, Sumitomo K, Nishimoto Y, Ochi M (2009) Relationship between fracture toughness and domain size of liquid-crystalline epoxy resins having polydomain structure. J Polym Sci Polym Phys 47:156–165

Harada M, Ando J, Hattori S, Sakurai S, Sakamoto N, Yamasaki T, Masunaga H, Ochi M (2013) In-situ analysis of the structural formation process of liquid-crystalline epoxy thermosets by simultaneous SAXS/WAXS measurements using synchrotron radiation. Polym J 45:43–49

Hirano Y, Akiba M, Yokota A, Nakamura H, Naitoh S (1999) Epoxy resin, resin composition, and resin-encapsulated semiconductor device. U.S. Patent 5,939,473

Hsu SH, Chen RS, Chang YL, Chen MH, Cheng KC, Su WF (2012a) Biphenyl liquid crystalline epoxy resin as a low-shrinkage resin-based dental restorative nanocomposite. Acta Biomater 8:4151–4161

Hsu SH, Wu MC, Chen S, Chuang CM, Lin SH, Su WF (2012b) Synthesis, morphology and physical properties of multi-walled carbon nanotube/biphenyl liquid crystalline epoxy composites. Carbon 50:896–905

Huo L, Gao J, Du Y, Chai Z (2008) Synthesis and characterization of the two novel liquid crystalline epoxy resins based on bisphenol-S mesogen. J Appl Polym Sci 110:3671–3677

Jahromi S, Lub J, Mol GN (1994) Synthesis and photoinitiated polymerization of liquid-crystalline diepoxides. Polymer 35:622–629

Le Hoang S, Shin S, Lim D-G, Nguyen Ngoc T, Bui Thanh S, Bae J-Y (2013) A novel wholly aromatic liquid crystalline epoxy monomer based on 2,7-naphthalene mesogen: Synthesis and curing with aromatic diamine. Macromol Res 21:1105–1109

Lee JY, Jang JS (1998) Effect of substituents on the curing of liquid crystalline epoxy resin. J Polym Sci Polym Chem 36:911–917

Lee JY, Jang JS (1999) Synthesis and curing of liquid crystalline epoxy resin based on naphthalene mesogen. J Polym Sci Polym Chem 37:419–425

Lee JY, Jang JS (2006) The effect of mesogenic length on the curing behavior and properties of liquid crystalline epoxy resins. Polymer 47:3036–3042

Lee H, Neville K (1967) Handbook of epoxy resins. McGraw-Hill, New York

Lee JY, Jang JS, Hwang SS, Hong SM, Kim KU (1998) Synthesis and curing of liquid crystalline epoxy resins based on 4,4′-biphenol. Polymer 39:6121–6126

Lee JY, Jang JS, Hong SM, Hwang SS, Kim KU (1999) Relationship between the structure of the bridging group and curing of liquid crystalline epoxy resins. Polymer 40:3197–3202

Li YZ, Kessler MR (2013) Liquid crystalline epoxy resin based on biphenyl mesogen: effect of magnetic field orientation during cure. Polymer 54:5741–5746

Li YZ, Kessler MR (2014) Creep-resistant behavior of self-reinforcing liquid crystalline epoxy resins. Polymer 55:2021–2027

Li YZ, Badrinarayanan P, Kessler MR (2013) Liquid crystalline epoxy resin based on biphenyl mesogen: thermal characterization. Polymer 54:3017–3025

Lin QH, Yee AF, Earls JD, Hefner RE, Sue HJ (1994) Phase-transformations of a liquid crystalline epoxy during curing. Polymer 35:2679–2682

Lin QH, Yee AF, Sue HJ, Earls JD, Hefner RE (1997) Evolution of structure and properties of a liquid crystalline epoxy during curing. J Polym Sci Polym Phys 35:2363–2378

Liu JP, Wang CC, Campbell GA, Earls JD, Priester RD (1997) Effects of liquid crystalline structure formation on the curing kinetics of an epoxy resin. J Polym Sci Polym Chem 35:1105–1124

Liu JP, Wang CC, Campbell GA, Earls JD, Priester RD (1998) The use of tetraalkylphosphonium-tetrafluoroborate-tetrafluoroboric acid in the curing of a liquid crystalline epoxy resin. J Polym Sci Polym Chem 36:1457–1471

Liu GD, Zhou B, Zhao DM, Li Q, Gao JG (2008) Novel triaromatic ester mesogenic liquid crystalline epoxy resin containing both methyl substituent and ethoxy flexible spacer: synthesis and curing. Macromol Chem Phys 209:1160–1169

Liu YL, Cai ZQ, Wang WC, Wen XF, Pi PH, Zheng DF, Cheng J, Yang ZR (2011a) Mechanical properties and morphology studies of thermosets from a liquid-crystalline epoxy resin with biphenol and aromatic ester groups. Macromol Mater Eng 296:83–91

Liu YL, Cai ZQ, Wen XF, Pi PH, Zheng DF, Cheng J, Yang ZR (2011b) Thermal properties and cure kinetics of a liquid crystalline epoxy resin with biphenyl-aromatic ester mesogen. Thermochim Acta 513:88–93

Mallon JJ, Adams PM (1993) Synthesis and characterization of novel epoxy monomers and liquid crystal thermosets. J Polym Sci Polym Chem 31:2249–2260

Mija A, Cascaval CN (2009) Liquid crystalline epoxy thermoset with azomethine mesogen. Polimery 54:786–789

Mittal V (2010) Optimization of polymer nanocomposite properties. Wiley-VCH, Weinheim

Mittal V (2013) Thermoset nanocomposites. Wiley-VCH, Weinheim

Mormann W, Broche M, Schwarz P (1997) Mesogenic azomethine based diepoxides—monomers for the synthesis of "liquid crystal" thermoset networks. Macromol Chem Phys 198:3615–3626

Mossety-Leszczak B, Włodarska M (2011) Liquid crystallinity in polymers—liquid crystalline epoxy resins. In: Iwan A, Schab-Balcerzak E (eds) Liquid crystalline organic compounds and polymers as materials of the XXI century: from synthesis to applications, Transworld Research Network, Kerala

Ochi M, Tsuyuno N, Sakaga K, Nakanishi Y, Murata Y (1995) Effect of network structure on thermal and mechanical-properties of biphenyl-type epoxy-resins cured with phenols. J Appl Polym Sci 56:1161–1167
Ochi M, Shimizu Y, Nakanishi Y, Murata Y (1997) Effect of the network structure on thermal and mechanical properties of mesogenic epoxy resin cured with aromatic amine. J Polym Sci Polym Phys 35:397–405
Ochi M, Hori D, Watanabe Y, Takashima H, Harada M (2004) Bonding properties of epoxy resins containing two mesogenic groups. J Appl Polym Sci 92:3721–3729
Ortiz C, Kim R, Rodighiero E, Ober CK, Kramer EJ (1998) Deformation of a polydomain, liquid crystalline epoxy-based thermoset. Macromolecules 31:4074–4088
Ortiz C, Belenky L, Ober CK, Kramer EJ (2000) Microdeformation of a polydomain, smectic liquid crystalline thermoset. J Mater Sci 35:2079–2086
Punchaipetch P, Ambrogi V, Giamberini M, Brostow W, Carfagna C, D'Souza NA (2001) Epoxy plus liquid crystalline epoxy coreacted networks: I. Synthesis and curing kinetics. Polymer 42:2067–2075
Ribera D, Mantecon A, Serra A (2001) Synthesis and crosslinking of a series of dimeric liquid crystalline epoxy resins containing imine mesogens. Macromol Chem Phys 202:1658–1671
Ribera D, Mantecon A, Serra A (2002) Liquid-crystalline thermosets from mesogenic dimeric epoxy resins by tertiary amine catalysis. J Polym Sci Polym Chem 40:3916–3926
Robinson EJ, Douglas EP, Mecholsky JJ (2002) The effect of stoichiometry on the fracture toughness of a liquid crystalline epoxy. Polym Eng Sci 42:269–279
Schnurpfeil G, Harder A, Schroder H, Wohrle D, Hartwig A, Hennemann OD (2001) Synthesis and photoinitiated polymerization of nematic liquid-crystalline diepoxides. Macromol Chem Phys 202:180–187
Shiota A, Ober CK (1996) Synthesis and curing of novel LC twin epoxy monomers for liquid crystal thermosets. J Polym Sci Polym Chem 34:1291–1303
Shiota A, Ober CK (1997) Analysis of smectic structure formation in liquid crystalline thermosets. Polymer 38:5857–5867
Smith JDB, Schoch J, K. F., A. SWF (1999) Electrical insulation using liquid crystal thermoset epoxy resins. U.S. Patent 5,904,984
Su WFA (1993) Thermoplastic and thermoset main-chain liquid-crystal polymers prepared from biphenyl mesogen. J Polym Sci Polym Chem 31:3251–3256
Su WFA, Chuang CM (2002) Effects of chemical structure changes on curing reactions and thermal properties of cyanate ester-cured rigid-rod epoxy resins. J Appl Polym Sci 85:2419–2422
Su WFA, Schoch KF, Smith JDB (1998) Comparison of cure conditions for rigid rod epoxy and bisphenol a epoxy using thermomechanical analysis. J Appl Polym Sci 70:2163–2167
Su WFA, Chen KC, Tseng SY (2000) Effects of chemical structure changes on thermal, mechanical, and crystalline properties of rigid rod epoxy resins. J Appl Polym Sci 78:446–451
Su WF, Fu YC, Pan WP (2002a) Thermal properties of high refractive index epoxy resin system. Thermochim Acta 392:385–389
Su WF, Huang HW, Pan WP (2002b) Thermal properties of rigid rod epoxies cured with diaminodiphenylsulfone and dicyandiamide. Thermochim Acta 392:391–394
Su WF, Lee YC, Pan WP (2002c) Thermal properties of phthalic anhydride- and phenolic resin-cured rigid rod epoxy resins. Thermochim Acta 392:395–398
Su WF, Hsu SH, Tai YY, Chen MH (2012) Liquid crystalline epoxy nanocomposite material and application thereof U.S. Patent 8,309,632
Sue HJ, Earls JD, Hefner RE (1997) Fracture behaviour of liquid crystal epoxy resin systems based on the diglycidyl ether of 4,4′-dihydroxy-alpha-methylstilbene and sulphanilamide. 1. Effects of curing variations. J Mater Sci 32:4031–4037
Szczepaniak B, Frisch KC, Penczek P, Rejdych J, Winiarska A (1997) Curing of liquid crystalline epoxy resins with a biguanide. J Polym Sci Polym Chem 35:2739–2745

Tai YY, Hsu SH, Chen RS, Su WF, Chen MH (2014) Liquid crystalline epoxy nanocomposite material for dental application. J Formos Med Assoc 114:46–51

Tanaka S, Takezawa Y, Takahashi H (2007) Epoxy compound and cured epoxy resin product. U.S. Patent 7,279,574

Twardowski TE (2007) Introduction to nanocomposite materials: properties, processing. Characterization. DEStech Publications, Inc., Lancaster

Vengatesan MR, Devaraju S, Alagar M (2011) Studies on thermal, mechanical and morphological properties of organoclay filled azomethine modified epoxy nanocomposites. High Perform Polym 23:3–10

Vittoria V, Russo R, Decandia F, Magagnini PL, Bresci B (1991) Transport-properties of a side-chain liquid-crystalline polymer—poly(4-acryloyloxy biphenyl). J Polym Sci Polym Phys 29:1163–1168

Vyazovkin S, Mititelu A, Sbirrazzuoli N (2003) Kinetics of epoxy-amine curing accompanied by the formation of liquid crystalline structure. Macromol Rapid Commun 24:1060–1065

Wang HM, Zhang YC, Zhu LR, Zhang BL, Zhang YY (2011) Synthesis and curing behavior of a novel liquid crystalline epoxy resin. J Therm Anal Calorim 103:1031–1037

Weinkauf DH, Paul DR (1992) Gas-transport properties of thermotropic liquid-crystalline copolyesters. 1. The effects of orientation and annealing. J Polym Sci Polym Phys 30:817–835

Wlodarska M, Bak G, Mossety-Leszczak B, Galina H (2009) Influence of liquid crystalline ordering on the properties of selected cured nematic epoxy materials. J Mater Process Tech 209:1662–1671

Xiao Y, Cao L, Feng JH, Yuan LJ (2012) Synthesis and curing of liquid crystalline epoxy resin based on asymmetric mesogen. J Appl Polym Sci 126:527–535

Yamamoto H, Fujita A, Harada M, Ochi M (2014) Synthesis and characterization of novel liquid crystalline epoxy resin with low melting point. Mol Cryst Liq Cryst 588:41–50

Zhou DW, Lu MG, Liang LY, Shen TF, Xiao WP (2012) Curing behavior of Azo-containing twin liquid crystalline epoxy resins with anhydride. Polym Eng Sci 52:1375–1382

# Chapter 20
# Synthesis of Functional Liquid Crystalline Polymers for Exfoliated Clay Nanocomposites

**Wenyi Huang**

## 20.1 Introduction

Polymer/clay (i.e., layered silicate) nanocomposites have attracted enormous attention from both academia and industry over the past three decades, due to their capability of improving the mechanical and/or physical properties (e.g. gas barrier, heat resistance, and low flammability) of the polymer matrix (Sinha Ray and Okamoto 2003; Okada and Usuki 2006; Han 2010). These important features are not shared by conventional polymer/particle composites, which require high loading of fillers. Instead, the layered silicate platelets have a thickness of approximately 1 nm and a very large aspect ratio (e.g. 10–1000) with a large surface area (Sinha Ray and Okamoto 2003). As a result, a very low loading of clay is needed for polymer/clay nanocomposites, supposed that all the layered silicates can be well dispersed in the polymer matrix. The degree of dispersion of nanoclays in the polymer matrix determines the morphology and properties of nanocomposites.

Basically, there are three types of dispersion morphology for polymer/clay nanocomposites, including intercalation, flocculation, and exfoliation. Among these, exfoliation is highly preferred for most industrial applications. However, it is not an easy task to attain such an ideal dispersion characteristic of exfoliation of nanoclays in the polymer. Thus far, a variety of clay nanocomposites have been reported using thermoplastic polymers, to name a few, polypropylene (Kawasumi et al. 1997; Nam et al. 2001), polystyrene (Hoffmann et al. 2000), polycarbonate (Lee and Han 2003a), poly(ε-caprolactone) (Di et al. 2003), nylon 6 (Reynaud et al. 2001), poly(vinylpyridine) (Percy et al. 2000), poly(ethylene oxide) (Zhang and Archer 2002), poly(ethylene-*ran*-vinyl acetate-*ran*-vinyl alcohol) (Lee and Han 2003b), and olefinic block copolymers (Silva et al. 2001). However, most of these

W. Huang (✉)
Materials Science & Engineering Core R&D, The Dow Chemical Company, Midland, MI 48667, USA
e-mail: polymerhuang@yahoo.com



V.K. Thakur, M.R. Kessler (eds.), *Liquid Crystalline Polymers*,
DOI 10.1007/978-3-319-20270-9_20

systems did not exhibit exfoliated dispersion of nanoclays in the polymer presumably owing to the lack of desirable compatibility between the polymer matrix and organoclay employed. Thus, it is of fundamental significance to gain in-depth insight into the exfoliation mechanism of nanoclays in the polymer.

Liquid crystalline polymers (LCPs) are referred to as a unique family of polymers that is built up by rigid rod-like mesogenic units, leading to the formation of lyotropic or thermotropic mesophases (Donald and Windle 1992). The intriguing feature of mesophase lies in that it is described by a positional order parameter arising from its geometrical anisotropicity in space (Collings and Hird 1997). LCPs encompass excellent mechanical properties, dimensional stability, thermal stability coupled with the absence of creep and shrinkage. As a consequence, they are ideal candidates for high-performance applications (Donald and Windle 1992).

It is of great interest to incorporate nanoparticles like nanoclays into LCPs, due to their potentials for enhancing many intrinsic properties of LCPs and thus expanding their application areas. In order to reach this ideal situation, nanoclays must have a high degree of dispersion in LCPs. However, it is rather difficult for rigid LCP molecules to enter the gallery of layered silicates, and thus poor dispersion, intercalation at best, are often observed. To date, only a few papers (Vaia and Giannelis 2001; Chang et al. 2002; Zhang et al. 2003; Shen et al. 2005; Bandyopadhyay et al. 2009, 2010) reported on the preparation of organoclay nanocomposites based on LCPs, but more often than not, only intercalation of organoclay aggregates was observed. The root cause may be due to the absence of strong attractive interactions between nanoclays and LCPs, which are determined by the chemical structure of a surfactant residing at the clay surface, clay gallery distance, as well as the chemical structure of LCPs.

To achieve high degree of dispersion of nanoclays in LCPs, the foremost thing is to match the chemical affinity between nanoclays and LCPs and thus improve their compatibility. Since most of commercially available LCPs are hydrophobic, it is important to transform the hydrophilic characteristic of pristine clays into hydrophobic one by replacing the hydrated metal cations from the interlayers with organic cations such as alkylammonium (Usuki et al. 1993). Nevertheless, there does not exist such a general-purpose surfactant that can modify the surface properties of natural clay for all the LCPs. On the other hand, functional groups are often introduced into LCPs for matching the surface chemistry of pristine clays or organoclays. If properly designed for functional LCPs, it may be feasible to prepare exfoliated clay nanocomposite based on pristine clays.

Although the mechanical mixing is important for achieving exfoliation of nanoclays in the polymer, it is more desirable to provide direct driving force between the polymer and nanoclays for the process of exfoliation. The objective of this chapter focuses on the understanding of fundamental mechanism of exfoliation induced by specific interactions. Here, the specific interactions are referred to as the non-covalent bonds such as hydrogen bonding, ionic interactions, electrostatic interactions, dipole-dipole interactions, $\pi-\pi$ interactions, and metal-ion coordination, and they have been recognized as being of crucial importance in inducing supramolecular structures in nature (Lehn 1988; Balzani and De Cola 1992).

Among these, hydrogen bonding represents the most important intermolecular interactions between macromolecular chains, for instance, the cooperative hydrogen bonding within polyamides results in a variety of crystalline structures (Wunderlich 1973). Notice that hydrogen bonding or ionic interactions between dissimilar polymer pairs can also induce the formation of miscible polymer blends, complexes and networks (Coleman et al. 1991). It should be pointed out that metal-ion coordination is another unique strong bonding that can give rise to very stable supramolecular structures (Manners 2004). These fundamental principles of supramolecular chemistry extracted from the nature have been leveraged to supramolecular polymers and even more complicated structures like LCP/clay nanocomposites presented in this chapter.

The goal of this chapter is to demonstrate the importance of specific interactions in inducing exfoliation of nanoclays in liquid crystalline polymers. For this, the rationale of molecular design for functional liquid crystalline polymers with heterocyclic groups such as pyridyl, azopyridine, and terpyridine groups was elaborated on. Two main mechanisms, hydrogen bonding induced exfoliation and Coulombic interactions induced exfoliation, were detailed and corroborated with robust evidences. The rheological properties of exfoliated polymer/clay nanocomposites are also discussed.

## 20.2 Fundamental Mechanism for the Exfoliation of Nanoclays in Functional Liquid Crystalline Polymers

### *20.2.1 Rationale of Molecular Design for Functional Liquid Crystalline Polymers*

In nature, the life is encoded by deoxyribonucleic acid (DNA), which encompasses all the information for self-replication, development, and survival of the organism. Indeed, DNA is a polymer having many repetitive units comprised of a sugar molecule, a heterocyclic base, and a phosphate group that links the units together. Noticeably, the heterocyclic bases found in DNA include adenine (A), guanine (G), cytosine (C), and thymine (T), which can pair highly selectively to each other via specific interactions with A::G and C::T (Champe and Harvey 1994).

Among all the heterocyclic bases, pyridine is an important and intriguing nitrogenous base, and it possesses six delocalized electrons in its $\pi$ system. Most interestingly, the two non-bonding electrons on nitrogen are in a $sp^2$ orbital perpendicular to the plane of the ring, and they do not interact with the $\pi$ electrons of the aromatic system. As a result, the lone pair of electrons in $sp^2$ orbital accounts for both basicity and strong permanent dipole of pyridine, leading to its great propensity to abstract a proton. In this aspect, pyridine can be regarded as a strong "proton acceptor", and thus it is ready to form strong hydrogen bonds (H-bonds) with "proton donor" molecules (Joule and Mills 2000). The energy of an H-bond in the gas phase is typically in the range of 2–20 kcal/mol, which is much weaker than that of covalent bonds, but significantly larger than dipolar or London dispersion force energy (<2 kcal/mol). It should be mentioned that the electronic attraction will be amplified leading to stronger H-bonds (10–45 kcal/mol), if either the donor or acceptor is charged (Scheiner 1997). This is especially true when pyridine is quaternarized into pyridium, where ionic interactions prevail. Moreover, the lone pair of electrons in pyridine (Lewis base) paves a way to form complexes with transition metal ions (Lewis acid) via coordination bonds (Tomasik 1985). On the other hand, 2,2′:6′,2″-terpyridine has recently been recognized as a very powerful chelating ligand for a variety of transition-metal ions in low oxidation state, forming very stable octahedral complexes to construct a broad spectrum of metallo-supramolecular polymers (Schubert and Eschbaumer 2002). Because of the extreme flexibility of heterocyclic molecules, they have been extensively employed in the new fields ranging from supramolecular chemistry (Steed and Atwood 2000), self-assembled liquid crystals (Kato and Mizoshita 2002), to supramolecular polymers (Ciferri 2000), and their importance has been widely recognized from experiment to theory (Ciferri 1999).

In view of the flexibility and versatility of nitrogen-containing heterocyclic derivatives, Huang and Han (2006a, b, c) synthesized three segmented main-chain thermotropic liquid crystalline polymers functionalized with heterocyclic groups, namely PyHQ12 having pendent pyridyl group, PABP having side-chain azopyridyl group with flexible spacer, and PTBP having side-chain terpyridine group, which was further reacted to become ruthenium complex. Two mechanisms will be elaborated on below to demonstrate how specific interactions induce the exfoliation of nanoclays in these functional LCPs.

### *20.2.2 Hydrogen Bonding Induced Exfoliation of Nanoclays in Functional Liquid Crystalline Polymers*

The first fundamental question arises as to whether the functionality is necessary for achieving the exfoliation of nanoclays in LCPs. For this purpose, two organoclays used contain surfactants with and without hydroxyl groups, while two LCPs with a similar chemical structure were synthesized, where one LCP had phenyl group and the other has pyridyl group. This experimental design aimed to demonstrate that compatibility, via specific interactions, between a LCP and an organoclay is essential to achieve a very high degree of dispersion of organoclay aggregates. Specifically, these two organoclays are Southern Clay products: (i) Cloisite 30B treated with a surfactant (MT2EtOH) having the chemical structure methyl, tallow, bis-2-hydroxyethyl, quaternary ammonium chloride, and (ii) Cloisite 15A treated with a surfactant (2M2HT) having the chemical structure, methyl, hydrogenated tallow, quaternary ammonium chloride. The chemical structures of the surfactants, MT2EtOH and 2M2HT, are given below (Huang and Han 2006a).

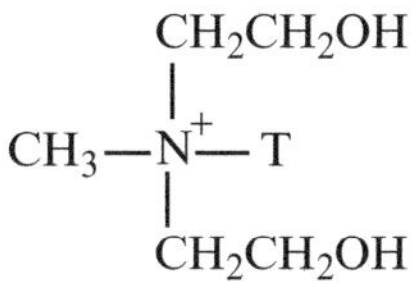

MT2EtOH

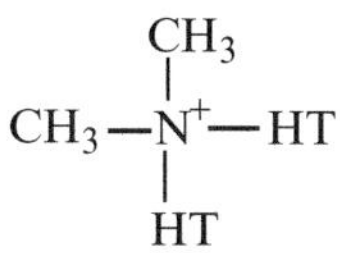

2M2HT

In the chemical structure of MT2EtOH, $N^+$ denotes quaternary ammonium chloride and T denotes tallow consisting of ca. 65 % C18, ca. 30 % C16, and ca. 5 % C14, and in the chemical structure of 2M2HT, $N^+$ denotes quaternary ammonium chloride and HT denotes hydrogenated tallow consisting of ca. 65 % C18, ca. 30 % C16, and ca. 5 % C14. It should be mentioned that 100 % of $Na^+$ ions in natural clay (montmorillonite) have been exchanged. Note that Cloisite 30B has hydroxyl groups while Cloisite 15A does not, as confirmed by Fourier transform infrared (FTIR) spectroscopy (Huang and Han 2006a).

The literature reported that only intercalation of organoclay aggregates happened in nanocomposites based on main-chain LCPs (Vaia and Giannelis 2001; Chang et al. 2002; Zhang et al. 2003; Shen et al. 2005; Bandyopadhyay et al. 2009, 2010). Such experimental observations are not surprising in that the LCPs employed in both studies did *not* have any functionality. To demonstrate whether functionality is necessary to obtain highly dispersed nanocomposites based on LCPs, Huang and Han (2006a) synthesized two main-chain LCPs. The first LCP, which hereafter is referred to as PyHQ12, has pendent pyridyl group with the chemical structure,

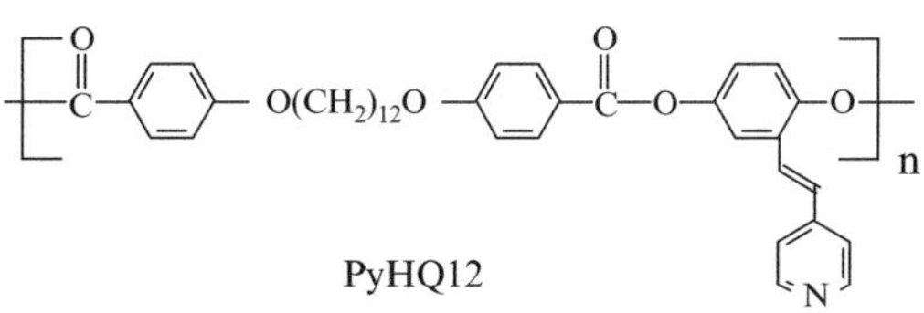

PyHQ12

The detailed synthetic procedures and characterization of PyHQ12 are given in the literature (Huang and Han 2006d). The rationale behind the synthesis of

PyHQ12 lies in the expectation that the pendent pyridyl group in PyHQ12 may form hydrogen bonds with the polar group in a surfactant residing at the surface of an organoclay, giving rise to highly dispersed nanocomposites. For comparison, the other LCP, which is hereafter referred to as PSHQ12, has the chemical structure (Huang and Han 2006a),

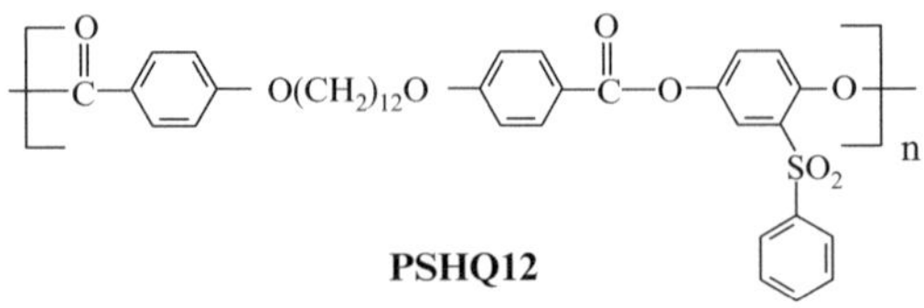

**PSHQ12**

It is worth noting that PSHQ12 has exactly the same backbone as PyHQ12. Consequently, PyHQ12 and PSHQ12 were used to prepare nanocomposites with two different organoclays, one with hydroxyl groups (Cloisite 30B) and the other without polar groups (Cloisite 15A).

X-ray diffraction (XRD) patterns in Fig. 20.1 clearly indicate that PyHQ12/Cloisite 30B nanocomposite has a very high degree of dispersion of organoclay aggregates, as evidenced by featureless XRD pattern (Huang and Han 2006a). In contrast, PyHQ12/Cloisite 15A nanocomposite has a $d_{001}$ spacing of 3.4 nm, an increase of 0.3 nm (9.7 % increase) over the $d_{001}$ spacing (3.1 nm) of Cloisite 15A, suggesting that the nanocomposite might have intercalated Cloisite 15A aggregates.

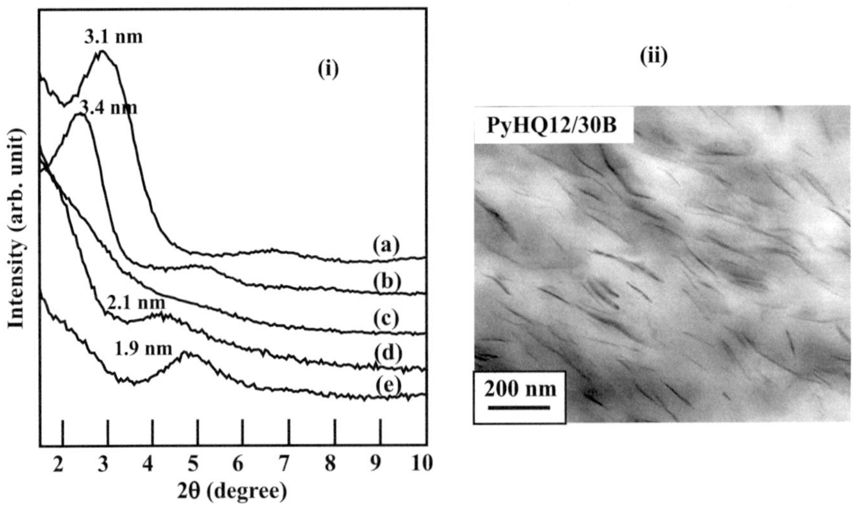


**Fig. 20.1** Dispersion characteristics of nanocomposites: (**i**) XRD patterns for (*a*) Cloisite 15A, (*b*) PyHQ12/Cloisite 15A nanocomposite, (*c*) PyHQ12/Cloisite 30B nanocomposite, (*d*) PSHQ12/Cloisite 30B nanocomposite, and (*e*) Cloisite 30B. (**ii**) TEM image of PyHQ12/Cloisite 30B nanocomposite (Reprinted with permission from Huang and Han (2006a). Copyright 2006 American Chemical Society)

Similarly, PSHQ12/Cloisite 30B nanocomposite has a $d_{001}$ spacing of 2.1 nm, an increase of 0.2 nm (10.5 % increase) over the $d_{001}$ spacing (1.9 nm) of Cloisite 30B, suggesting that the nanocomposites might have intercalated Cloisite 30B aggregates. It is very clear from the above observations that PyHQ12/Cloisite 30B nanocomposite has XRD patterns, which are distinctly different from the XRD patterns of the two other organoclay nanocomposites. Furthermore, the dispersion characteristics of LCP/clay nanocomposites were confirmed by transmission electron microscopy (TEM). Figure 20.1 displays that PyHQ12/Cloisite 30B nanocomposite has a very high degree of dispersion of Cloisite 30B aggregates in the matrix (PyHQ12), whereas PyHQ12/Cloisite 15A and PSHQ12/Cloisite 30B nanocomposites have large aggregates of organoclay, indicating poor dispersion of organoclay aggregates (Huang and Han 2006a). Therefore, there is a good correlation between the XRD patterns and the TEM images.

The origin of the differences in the dispersion characteristics between PyHQ12/Cloisite 30B and PSHQ12/Cloisite 30B nanocomposites, and between PyHQ12/Cloisite 30B and PyHQ12/Cloisite 15A nanocomposites is explained in terms of the attractive interactions, as determined by FTIR spectroscopy, between the pendent pyridyl group in PyHQ12 and the hydroxyl groups in the surfactant MT2EtOH residing at the surface of organoclay Cloisite 30B (Huang and Han 2006a). What is remarkable in the FTIR spectra for the PyHQ12/Cloisite 30B nanocomposite is that the rather strong absorption peak at a wavenumber of 2570 $cm^{-1}$ persists even at a temperature as high as 180 °C. Nevertheless, there were little or no attractive interactions between PSHQ12 and Cloisite 30B, and between PyHQ12 and Cloisite 15A. These observations explain why the previous investigators only observed intercalation of organoclay nanocomposites based on a LCP without functionality.

Unfortunately, a considerable loss in the degree of liquid crystallinity of PyHQ12 was observed in the PyHQ12/Cloisite 30B nanocomposite by differential scanning calorimetry and polarized optical microscopy, while the nanocomposite has a very high degree of dispersion of organoclay aggregates (Huang and Han 2006a). This observation can be interpreted by the argument that attractive interactions, via hydrogen bonding, between the pendent pyridyl group in PyHQ12 and the hydroxyl groups in the surfactant MT2EtOH residing at the surface of Cloisite 30B might have restricted the mobility of PyHQ12 chains in the nanocomposite, and consequently the orientation of the nematic mesophase might have been severely restricted. Note that the formation of hydrogen bonds between PyHQ12 and Cloisite 30B in the PyHQ12/Cloisite 30B nanocomposite is random in space (see Fig. 20.2), making the orientation of PyHQ12 chains very difficult (Huang and Han 2006a). Then, it was realized that this was due to the proximity of the pendent pyridyl group to the mesogenic main-chain backbone. In other words, the pendent pyridyl group was located so close to the mesogenic main-chain backbone of PyHQ12 that hydrogen bonding between the pendent pyridyl group in PyHQ12 and the hydroxyl group in the surfactant residing at the surface of organoclay

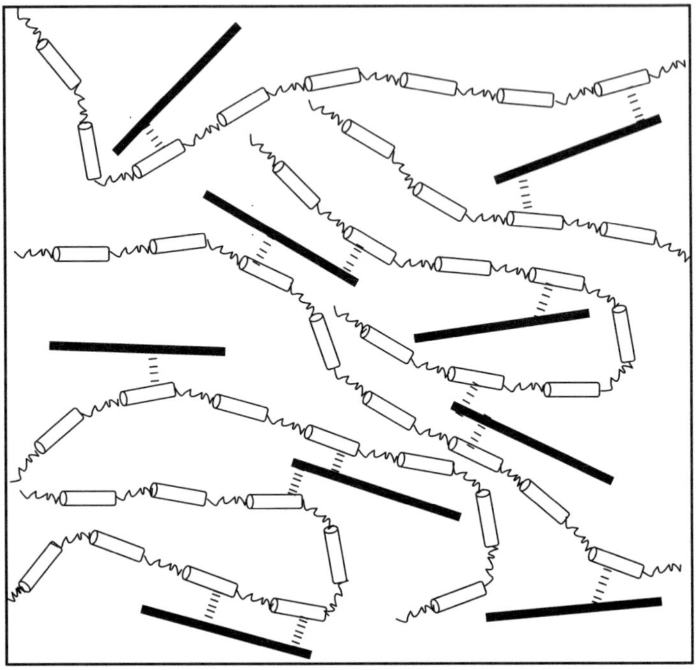

**Fig. 20.2** Scheme describing the distributions of PyHQ12 chains and organoclay Cloisite 30B in the highly dispersed PyHQ12/Cloisite 30B nanocomposite, where the *dark sticks* represent Cloisite 30B platelets, the *thin cylinders* (which denote mesogens) plus *waved lines* (which denote flexible spacers) represent PyHQ12 chains, and the *short dashed lines* represent hydrogen bonds between the pendent pyridyl group in PyHQ12 and the hydroxyl groups in the surfactant MT2EtOH residing at the surface of Cloisite 30B (Reprinted with permission from Huang and Han (2006a). Copyright 2006 American Chemical Society)

greatly restricted the mobility (thus orientation) of the mesogenic main chain of PyHQ12. Thus the functionality in a LCP is necessary to obtain highly dispersed nanocomposites, but at the same time there is a possibility to lose some degree of liquid crystallinity in the LCP. Another challenge is to design and synthesize a functional LCP that would give rise to a very high degree of dispersion of organoclay aggregates without a loss in the degree of liquid crystallinity of the LCP in organoclay nanocomposites prepared.

To overcome this undesirable feature aforementioned, another functional LCP, which is hereafter referred to as PABP and has side-chain azopyridine group with flexible spacer, was synthesized in such a way that the pyridyl group in the side-chain azopyridine of PABP was located sufficiently far away from the mesogenic main chain. The chemical structure of PABP is given as follows (Huang and Han 2006b),

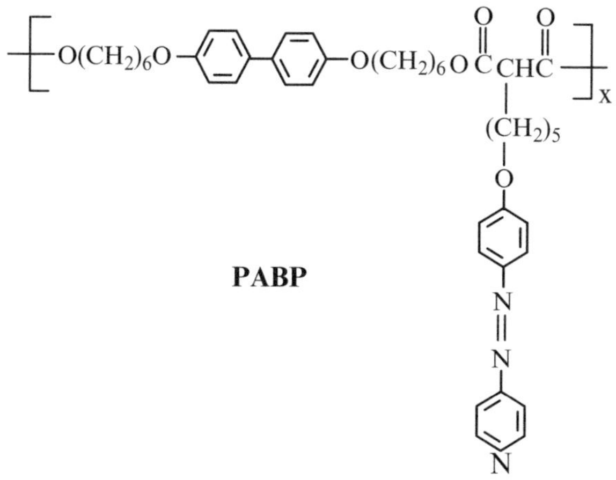

**PABP**

The detailed synthetic procedures and characterizations of PABP are given in the literature (Huang and Han 2006d). In the preparation of nanocomposites, an organoclay (Cloisite 15A) treated with a surfactant without polar group and the other organoclay (Cloisite 30B) treated with a surfactant having hydroxyl group were used to blend with PABP.

The XRD patterns and TEM image in Fig. 20.3 have demonstrated unambiguously the presence of specific interactions, via hydrogen bonding, between PABP and Cloisite 30B, which then gave rise to highly dispersed aggregates of Cloisite 30B (Huang and Han 2006b). In contrast, PABP/Cloisite 15A nanocomposite only shows intercalation of Cloisite 15A aggregates (Huang and Han 2006b). It should

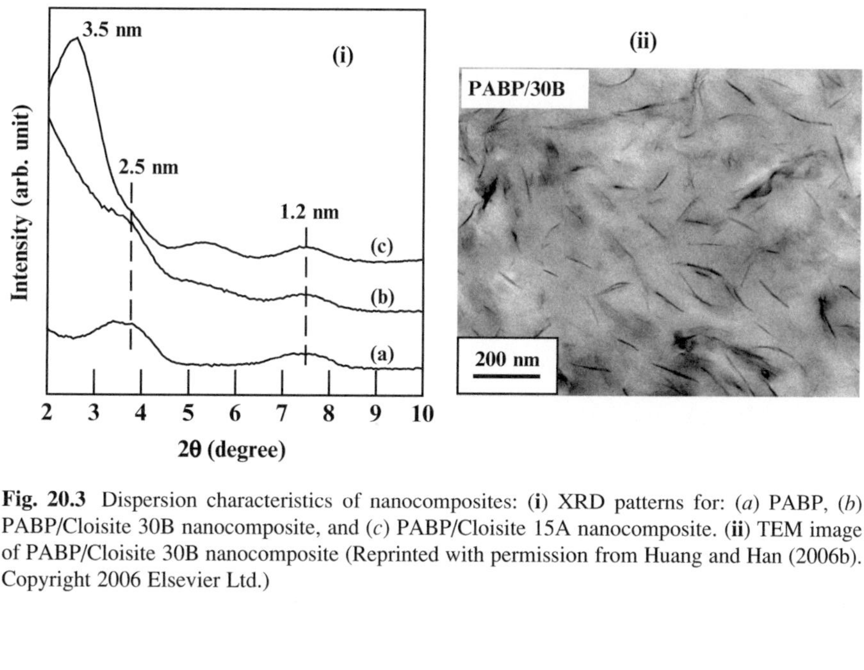


**Fig. 20.3** Dispersion characteristics of nanocomposites: (**i**) XRD patterns for: (*a*) PABP, (*b*) PABP/Cloisite 30B nanocomposite, and (*c*) PABP/Cloisite 15A nanocomposite. (**ii**) TEM image of PABP/Cloisite 30B nanocomposite (Reprinted with permission from Huang and Han (2006b). Copyright 2006 Elsevier Ltd.)

be mentioned that the two XRD peaks at 2θ of ca. 3.75° and 7.40° for PABP in Fig. 20.3 are attributable to smectic layered mesophase structure of PABP. Notice in Fig. 20.3 that the *d* spacing of Cloisite 15A in PABP/Cloisite 15A nanocomposite has increased only slightly from 3.1 to 3.5 nm, but interestingly the PABP/Cloisite 30B nanocomposite shows virtually featureless XRD patterns, indicative of high degree of dispersion.

*In situ* FTIR spectroscopy provides strong evidence that hydrogen bonds were formed between the pyridyl group in the side-chain azopyridine of PABP and the hydroxyl groups in the surfactant MT2EtOH residing at the surface of Cloisite 30B in the PABP/Cloisite 30B nanocomposite, while very weak interaction existed in the PABP/Cloisite 15A nanocomposite (Huang and Han 2006b). This study has further confirmed that successful preparation of LCP/organoclay nanocomposites having a very high degree of dispersion of organoclay aggregates depends very much on the extent of compatibility between an organoclay and a LCP. That is, a mismatch or lack of compatibility between an organoclay and a LCP would not produce nanocomposites having a very high degree of dispersion of organoclay aggregates. This has been illustrated in the PABP/Cloisite 15A nanocomposite in that PABP and the surfactant 2M2HT residing at the surface of Cloisite 15A do not have sufficient attractive interactions and thus the nanocomposite gives rise to only intercalation of Cloisite 15A aggregates.

Most importantly, the mesophase structure (liquid crystallinity) of PABP in the PABP/Cloisite 30B nanocomposite is little affected by the hydrogen bonding between PABP and Cloisite 30B, as evidenced by differential scanning calorimetry and polarized optical microscopy (Huang and Han 2006b). This is attributed to the fact that the pyridyl group in the side-chain azopyridine of PABP is located sufficiently far away from the mesogenic main chain, and thus the hydrogen bonds formed between the pyridyl group at the end of the side-chain azopyridine with five methylene units, which lie in the middle of flexible spacers having twelve methylene units, and the hydroxyl groups in the surfactant MT2EtOH residing at the surface of Cloisite 30B apparently have not interfered with the mobility (thus the orientation) of the mesogenic main chain of PABP. This observation is depicted schematically in Fig. 20.4, in which the dark ellipsoidal areas represent the mesogens of the main chain interconnected by the flexible spacers (shown by thin waved lines), and the dark sticks represent Cloisite 30B platelets that form hydrogen bonds with the pyridyl group located at the end of side-chain azopyridyl group (Huang and Han 2006b). Notice in Fig. 20.4 that the side-chain azopyridine is connected to the main chain through five methylene units (shown by thin waved lines). The situation depicted in Fig. 20.4 is quite different from that depicted in Fig. 20.2, in which the pendent pyridyl group was connected to the mesogenic main chain of PyHQ12 only through a vinylene unit and thus the hydrogen bonds formed between the pendent pyridyl group and the hydroxyl groups in the surfactant MT2EtOH residing at the surface of Cloisite 30B might have interfered with the orientation of the mesogenic main chain, consequently disrupting the mesogenic

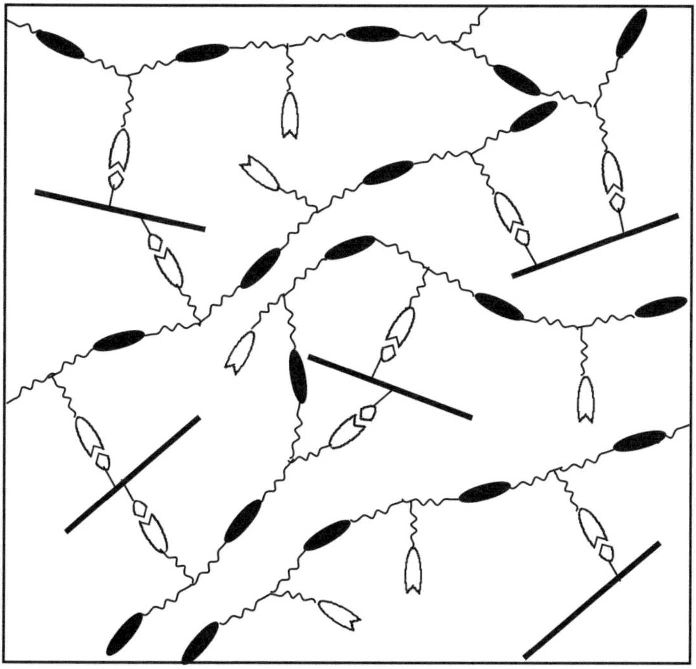

**Fig. 20.4** Scheme describing the distributions of PABP and Cloisite 30B in the highly dispersed PABP/Cloisite 30B nanocomposite, where the *dark sticks* represent Cloisite 30B platelets, the *dark ellipsoids* represent the main-chain mesogens, *waved lines* represent flexible spacers, the empty notched ellipsoids represent the azopyridine in the side chain of PABP, and the empty notched ellipsoids with caps represent hydrogen bonds between the pyridyl group in the side-chain azopyridine of PABP and the hydroxyl groups in the surfactant MT2EtOH residing at the surface of Cloisite 30B (Reprinted with permission from Huang and Han (2006b). Copyright 2006 Elsevier Ltd.)

structure of PyHQ12. To this end, the ultimate goal to obtain a highly dispersed organoclay nanocomposite based on a segmented main-chain LCP without sacrificing the inherent characteristics (liquid crystallinity) of LCP has been accomplished.

### *20.2.3 Coulombic Interactions Induced Exfoliation of Nanoclays in Liquid Crystalline Polymers*

As mentioned above, compatibility represents one of the most important issues when one designates to achieve a very high degree of dispersion of clay aggregates in the polymer matrix. The attractive interactions (or compatibility) induced by hydrogen bonding between the polymer and organoclay selected has demonstrated to be very effective in achieving exfoliation of the aggregates of the organoclay in LCPs.

It should be mentioned that the organoclays used in the aforementioned studies were prepared by treating natural clays with surfactants.

Montmorillonite (MMT) is a type of smectic natural clay, which tends to swell when exposed to water because of its hydrophilicity (Villemore 1990). Thus, in its pristine state MMT is only miscible with hydrophilic polymers, such as poly (ethylene oxide) (PEO) (Aranda and Ruiz-Hitzky 1992) or poly(vinyl alcohol) (PVA) (Greenland 1963). To render the compatibility between MMT and hydrophobic polymers, in most cases MMT was treated with a surfactant through ion-exchange reactions to convert the hydrophilic silicate surface into an organophilic one (Fukushima et al. 1988; Usuki et al. 1993). Unfortunately, very high degree of dispersion seldom happens in conventional polymer/organoclay nanocomposites, primarily due to the lack of attractive interactions between the surfactant and polymers. Therefore, one must design and synthesize a polymer that can have strong specific interactions with the selected organoclay, such that the chemically modified clay can have strong attractive interactions with the selected polymer (Huang and Han 2006a, b). The next question arises to as whether it is possible to exfoliate natural clay (e.g. Montmorillonite) in a functional polymer without the necessity to treat the clay with a surfactant.

Huang and Han (2006c) have made breakthrough in harnessing the specific interactions between positively charged metal-ion center in a LCP and negatively charged clay surface to induce the exfoliation of MMT in a LCP. For this, a segmented main-chain liquid-crystalline polymer having side-chain terpyridine group (PTBP) was first synthesized. PTBP was then reacted with a monocomplex (6TPy-$RuCl_3$), which was formed between a terpyridine (6-(2,2′:6′,2″-terpyridyl-4′-oxy)hexane (6TPy) and ruthenium chloride ($RuCl_3 \cdot 3H_2O$), to afford a ruthenium metal complex $[Ru^{II}$(PTBP)(6TPy)]$(PF_6)_2$, which is referred to as PTBP-Ru-6TPy thereafter. The reaction scheme is given in Fig. 20.5 and the detailed synthetic procedures can be found in the original paper (Huang and Han 2006c). (PTBPRu-6TPy)/MMT nanocomposite was prepared by mixing a predetermined amount of PTBP-Ru-6TPy dissolved in a co-solvent of *N,N*-dimethylformamide (DMF) and $H_2O$ (10:1, v/v) and MMT aggregates suspended in a mixture of DMF and $H_2O$ (10:1, v/v) under vigorous stirring. The solvent in the mixture was evaporated slowly under constant stirring for 2 days at ambient temperature. For comparison, two organoclays (Cloisite 30B and Cloisite 15A) were also used to prepare nanocomposites with PTBP-Ru-6TPy through the similar experimental protocol.

Remarkably, the MMT aggregates in the nanocomposite were very well dispersed in the matrix of PTBP-Ru-6TPy, as evidenced by the featureless XRD patterns TEM image in Fig. 20.6 (Huang and Han 2006c). In the absence of ruthenium complex, the *d*-spacing of MMT in PTBP/MMT nanocomposite has increased only slightly from 1.1 to 1.3 nm, indicating that there would be little chance to obtain a highly dispersed MMT in PTBP. On the other hand, the *d*-spacing of organoclay aggregates in (PTBP-Ru-6TPy)/(Cloisite 30B) nanocomposite has only increased from 1.9 to 4.4 nm, and that of (PTBP-Ru-6TPy)/(Cloisite 15A) nanocomposite has slightly increased from 3.1 to 3.5 nm. The poor dispersion of Cloisite 15A or 30B in PTBP-Ru-6TPy means that the chance of

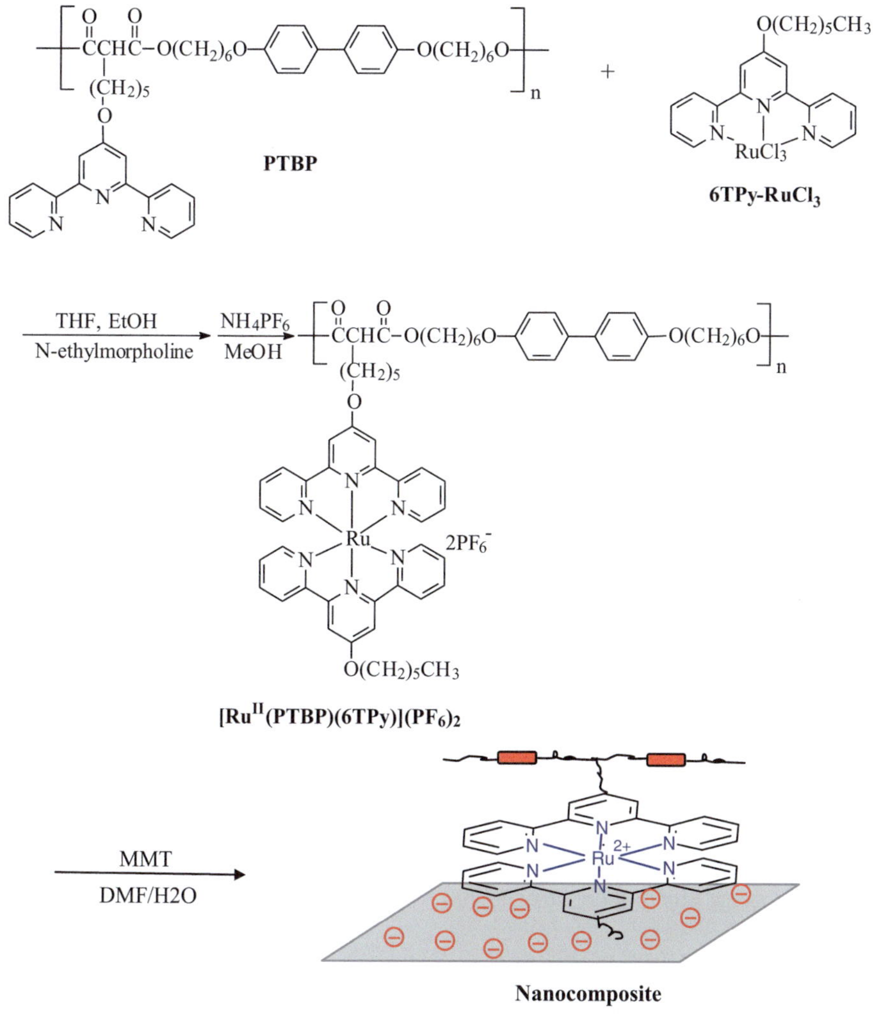


**Fig. 20.5** Reaction scheme for synthesizing [Ru$^{II}$(PTBP)(6TPy)]$(PF_6)_2$ complex, and its nanocomposite with Montmorillonite (Reprinted with permission from Huang and Han (2006c). Copyright 2006 American Chemical Society)

direct interactions between the positively charged ruthenium center in PTBP-Ru-6TPy and the negatively charged clay surface has been largely diminished, because of the shielding effect of positively charged surfactants.

The UV-vis spectra indicate that both red-shifted amount of metal-to-ligand charge-transfer band and the reduced amount of $\pi-\pi^*$ band of (PTBP-Ru-6TPy)/MMT nanocomposite have been largely increased, which can be attributed to the Coulombic interactions between the positively charged ruthenium center in PTBP-Ru-6TPy and the negatively charged MMT surfaces, as schematically represented in Fig. 20.5 (Huang and Han 2006c). If this kind of Coulombic interactions were

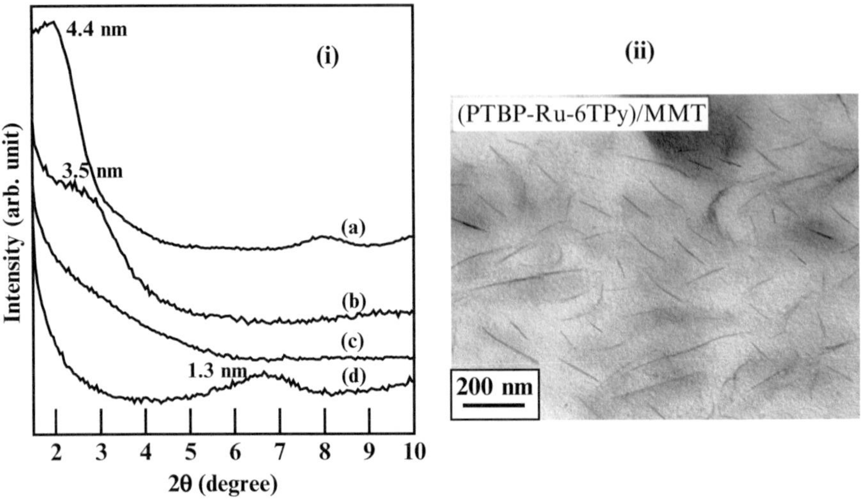


**Fig. 20.6** Dispersion characteristics of nanocomposites: (**i**) XRD patterns of (*a*) (PTBP-Ru-6TPy)/(Cloisite 30B) nanocomposite, (*b*) (PTBP-Ru-6TPy)/(Cloisite 15A) nanocomposite, (*c*) (PTBP-Ru-6TPy)/MMT nanocomposite, and (*d*) PTBP/MMT nanocomposite. (**ii**) TEM image of (PTBP-Ru-6TPy)/MMT nanocomposite (Reprinted with permission from Huang and Han (2006c). Copyright 2006 American Chemical Society)

absent or very small due to the shielding effect of a surfactant (MT2EtOH or 2M2HT) residing at the surface of Cloisite 30B or Cloisite 15A, a high degree of exfoliation of organoclay aggregates in (PTBP-Ru-6TPy)/(Cloisite 30B) and (PTBP-Ru-6TPy)/(Cloisite 15A) nanocomposites would not be possible. This is the reason why exfoliation does not occur in both (PTBP-Ru-6TPy)/(Cloisite 30B) and (PTBP-Ru-6TPy)/(Cloisite 15A) nanocomposites. The conclusion made by UV-vis spectra was further substantiated by FTIR spectroscopy.

## 20.3 Rheological Behaviors of Exfoliated Functional Liquid Crystalline Polymer/Clay Nanocomposites

Liquid crystalline polymers themselves are complex fluids due to the formation of mesophase domains arising from their rigid molecular structure, and the incorporation of nanoclays into LCPs makes the rheological behaviors of these systems even more complicated. Rheological properties of LCP/clay nanocomposites depend not only on the chemical structure and mesophase of LCPs but also on the size, shape, aspect ratio, surface modification, and dispersion characteristics of nanoclays. Although XRD patterns and polarized optical microscopy are powerful tools to quantify the dispersion characteristics of nanoclays in LCPs, the obtained information may not represent the bulk properties of nanocomposites at length (Sinha Ray 2006). In this sense, the rheological investigation of LCP/clay nanocomposites may generate useful information associated with structural variation and the degree of dispersion of nanoclays in LCPs, especially when there have strong attractive interactions between LCPs and exfoliated nanoclays. In other words, the rheological properties of these systems may change significantly with favorable interactions between LCPs and nanoclays, as compared to those systems without strong attractive interactions or having strong particle-particle attractions.

Dynamical rheological measurements were conducted on Advanced Rheometric Expansion System (ARES, TA Instrument) to understand the rheological properties of functional LCP/clay nanocomposites (Huang and Han 2006a). As given in Fig. 20.7, the values of $|\eta^*|$ for both PyHQ12 and PSHQ12 initially decrease with increasing temperature, going through a minimum, followed by an increase going through a maximum, and then decrease again with a further increase in temperature. This is attributed to the morphological change of LCPs with the temperature. However, the value of $|\eta^*|$ for PyHQ12/Cloisite 30B nanocomposite decreases steadily with increasing temperature, because the formation of hydrogen bonds in the PyHQ12/Cloisite 30B nanocomposite has disrupted the orientation characteristics of PyHQ12, while giving rise to a very high degree of dispersion of Cloisite 30B aggregates in the nanocomposite. On the other hand, because of the intercalated morphology, the $|\eta^*|$ for PyHQ12/Cloisite 15A nanocomposite initially decreases rapidly with increasing temperature to ca. 145 °C, remains more or less constant as the temperature is increased to ca. 160 °C, and then decreases again with a further increase of temperature. As expected, the temperature dependence of $|\eta^*|$ for PSHQ12/Cloisite 30B nanocomposite is virtually identical to that for neat PSHQ12, owing to the absence of attractive interactions and poor dispersion characteristics of nanoclays. These observations have a good correlation with the

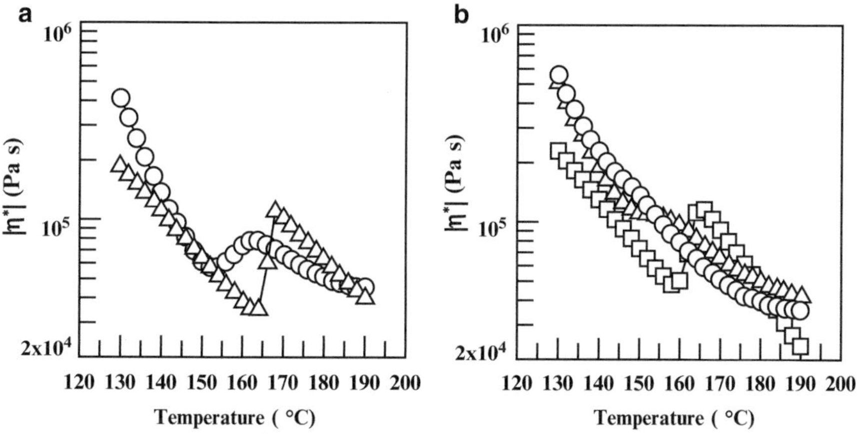


**Fig. 20.7** Variations of $|\eta^*|$ with temperature during isochronal dynamic temperature sweep experiments at $\omega = 0.1$ rad/s for (**a**): (*open circle*) PyHQ12 and (*open triangle*) PSHQ12, and for (**b**): (*open circle*) PyHQ12/Cloisite 30B nanocomposite, (*open triangle*) PyHQ12/Cloisite 15A nanocomposite, and (*open square*) PSHQ12/Cloisite 30B nanocomposite (Reprinted with permission from Huang and Han (2006a). Copyright 2006 American Chemical Society)

conclusions made by polarized optical microscopy, differential scanning calorimetry, and X-ray diffraction technique (Huang and Han 2006a).

The linear dynamical viscoelasticity given in Fig. 20.8 demonstrates that there is a distinct difference in temperature dependence of log G′ versus log G″ plots between PyHQ12/Cloisite 30B and PSHQ12/Cloisite 30B nanocomposites, while PSHQ12/Cloisite 30B and PyHQ12/Cloisite 15A nanocomposites have similar temperature dependence of log G′ versus log G″ plots. For homogenous polymers in the isotropic state, the slope of log G′ versus log G″ plot is 2 in the terminal region (Han 2010). However, the slope of log G′ versus log G″ plots for PyHQ12/Cloisite 30B nanocomposite in the isotropic region shows a strong deviation from 2, which is attributed to the strong attractive interactions between PyHQ12 and Cloisite 30B as well as the high degree dispersion of nanoclays in PyHQ12. In contrast, the slope of log G′ versus log G″ plots for PSHQ12/Cloisite 30B

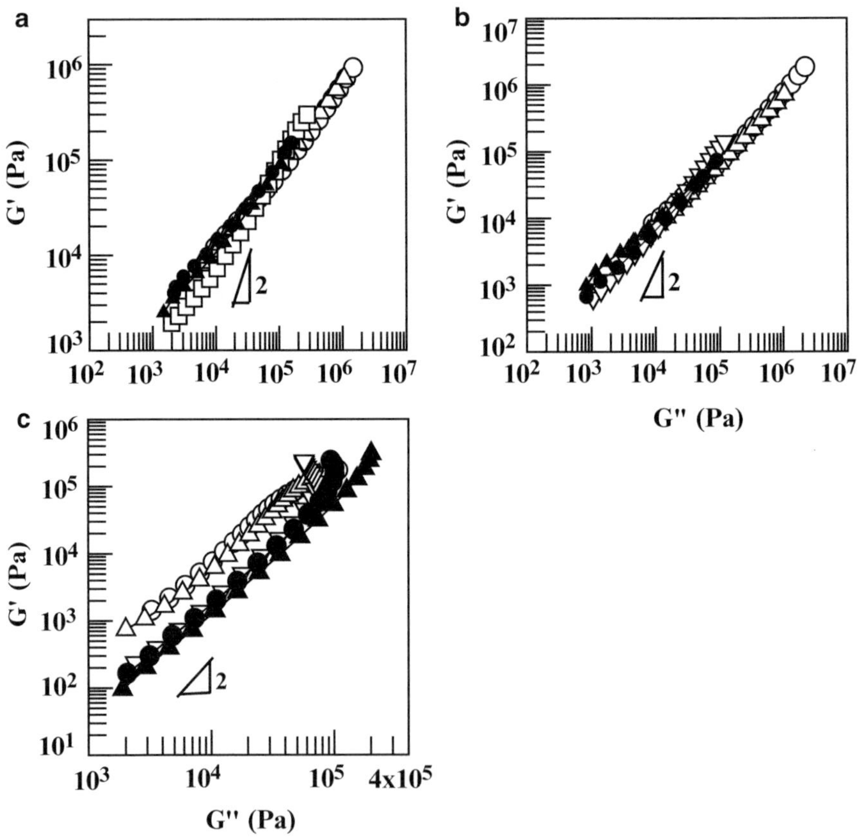


**Fig. 20.8** Log G′ versus log G″ plots for (**a**) PyHQ12/Cloisite 30B nanocomposites, (**b**) PyHQ12/Cloisite 15A nanocomposites, and (**c**) PSHQ12/Cloisite 30B nanocomposites at various temperatures (°C): (*open circle*) 130, (*open triangle*) 140, (*open square*) 155, (*open inverted triangle*) 160, (*filled circle*) 180, and (*filled triangle*) 190 (Reprinted with permission from Huang and Han (2006a). Copyright 2006 American Chemical Society)

nanocomposite is very close to 2, which is attributed to the poor dispersion of Cloisite 30B in PSHQ12. To conclude, the high-degree dispersion of nanoclays in LCPs is a result of attractive interactions (e.g. hydrogen bonding) between a functional LCP and an organoclay (Huang and Han 2006a). At the same loading of nanoclays, nanocomposites with high-degree dispersion of nanoclays exhibit different rheological behaviors compared to their agglomerated counterparts. The theory for homogeneous systems is invalid for exfoliated LCP/clay nanocomposites, even in their isotropic state. This implies that complex constitutive equation over the heterogeneous phase should be adopted to account for attractive interactions between the polymer and clay nanoplatelets.

For PABP/organoclay nanocomposites, the temperature dependence of G′ during isochronal temperature sweep experiments at an angular frequency of 0.1 rad/s is illustrated in Fig. 20.9 (Huang and Han 2006b). Noticeably, the values of G′ for the PABP/Cloisite 30B nanocomposite at temperatures above the clearing temperature of PABP are exceedingly large (more than two orders of magnitude) compared to those for PABP, and about one order of magnitude larger than those for the PABP/Cloisite 15A nanocomposite. These observations can be explained from the point of view that strong attractive interactions exist, via hydrogen bonding, between the pyridyl group in the side-chain azopyridine and the hydroxyl group in the surfactant MT2EtOH residing at the surface of organoclay Cloisite 30B. The extent of hydrogen bonding in the PABP/Cloisite 30B nanocomposite is still strong even at temperatures above the clearing temperature (ca. 146 °C) of PABP. It is fair to state that

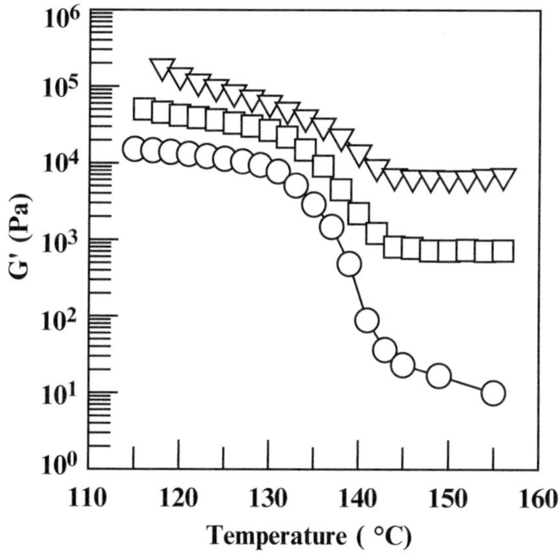


**Fig. 20.9** Variations of G′ with temperature during isochronal dynamic temperature sweep experiments at $\omega = 0.1$ rad/s for (*open circle*) PABP, (*open square*) PABP/Cloisite 15A nanocomposite, and (*open inverted triangle*) PABP/Cloisite 30B nanocomposite (Reprinted with permission from Huang and Han (2006b). Copyright 2006 Elsevier Ltd.)

the greater the strength of hydrogen bonding, the better the compatibility between the organoclay Cloisite 30B and the matrix PABP in the PABP/Cloisite 30B nanocomposite, giving rise to an improved dispersion of organoclay aggregates in the nanocomposite owing to the larger surface areas available (Huang and Han 2006b). In turn, the larger the surface areas available, the greater will be the values of $G'$ in the nanocomposites. Therefore, it can be concluded that an investigation of linear dynamic viscoelasticity of organoclay nanocomposites based on functional LCPs is very useful to determine whether the functional LCPs employed are effective to achieve a high degree of dispersion of organoclay aggregates.

## 20.4 Conclusions and Perspectives

This chapter elaborates on the fundamental mechanism of specific interactions-induced exfoliation of nanoclays, either pristine or chemically treated, in functional liquid crystalline polymers. The first type of driving force for exfoliating nanoclays in LCPs is hydrogen bonding, which is formed between pyridyl functional groups in LCPs and the hydroxyl groups of surfactants residing at the surface of an organoclay. The second type of driving force for exfoliating nanoclays in LCPs is Coulombic interactions, which are induced between the positively charged ruthenium complex center in a LCP and the negatively charged clay surface. It has been demonstrated that the functionality in LCPs is necessary for achieving high degree of dispersion of nanoclays in LCPs. Because of the proximity of the pendent pyridyl group to the mesogenic main-chain backbone, the liquid crystallinity of LCPs may be lost if the polymer has strong attractive interactions with nanoclays. This issue has been addressed by moving the pyridyl functional groups away from the main-chain mesogenic units and introducing flexible spacer in the side chain of the polymer.

In addition to X-ray diffraction pattern and transmission electron microscopy, rheological measurements, when properly interpreted, can be used as a powerful tool for a better understanding of the dispersion characteristics of LCP/clay nanocomposites. Specifically, dynamical rheological measurements have been investigated to show how the hydrogen bonds formed between the pyridyl group in a functional LCP and the hydroxyl groups of a surfactant residing at the surface of Cloisite 30B affected the dynamic viscoelastic properties of LCP/clay nanocomposites. However, the existing rheological model based on continuum may not be sufficient enough to accurately predict the rheological behaviors of exfoliated LCP/clay nanocomposites. Therefore, it is highly imperative to develop a new mathematical model, which accounts for the specific interactions at molecular level between the polymer and clay as well as the morphological variations in the nanocomposites.

The key learnings obtained from the fundamental understanding of exfoliation mechanism of nanoclays in functional liquid crystalline polymers may be leveraged to other polymer systems. It is important to match the functionality in the polymer with the functionality of clays in order to achieve the maximized attractive interactions. Besides the hydrogen bonding and Coulombic interactions, other specific interactions such as ionic interactions, electrostatic interactions, dipole-dipole interactions, $\pi-\pi$ interactions, and metal-ion coordination may be also utilized to induce the exfoliation of nanoclays in the polymers. When the functional polymers are properly designed, it is possible to prepare exfoliated nanocomposites from natural clay without the necessity to modify clay surface, for example, the functional LCP with ruthenium metal complex present in this chapter. The same idea can be generated to polymeric ionomers, which also contain positively charged ions and thus produce strong ionic interactions between negatively charged surfaces of natural clays. Because most of commercially available thermoplastic polymers do not possess functional groups, it is wise to introduce the third component (e.g. compatibilizer) to promote the attractive interactions between thermoplastic polymers and clays.

Furthermore, the strategies developed for dispersing nanoclays in LCPs are also useful for other nanoparticle systems, especially for other 2-D nanoplatelets like graphenes. Always, the compatibility between the polymer and nanoparticle is of utmost importance to attain high degree of dispersion of nanoparticles in the polymer. In most cases, it is necessary to modify either polymer or nanoparticle, or both of them with functional groups in order to create strong attractive interactions.

## References

Aranda P, Ruiz-Hitzky E (1992) Poly(ethylene oxide)-silicate intercalation materials. Chem Mater 4:1395

Balzani V, De Cola L (1992) Supramolecular chemistry. Kluwer, Boston

Bandyopadhyay J, Ray SS, Bousmina M (2009) Viscoelastic properties of clay-containing nanocomposites of thermotropic liquid-crystal polymer. Macromol Chem Phys 210 (2):161–171

Bandyopadhyay J, Ray SS, Bousmina M (2010) Structural analysis of liquid crystal polymer based nanocomposites by x-ray scattering. Macromol Chem Phys 211(15):1632–1639

Champe PC, Harvey RA (1994) Lippincott's illustrated reviews: biochemistry. J. B. Lippincott Co., Philadelphia

Chang J-H, Seo B-S, Hwang D-H (2002) An exfoliation of organoclay in thermotropic liquid crystalline polyester nanocomposites. Polymer 43:2969

Ciferri A (1999) Supramolecular liquid crystallinity as a mechanism of supramolecular polymerization. Liq Cryst 26:489

Ciferri A (2000) Supramolecular polymers. Marcel Dekker, New York

Coleman MM, Graff JF, Painter PC (1991) Specific interactions and the miscibility of polymer blends: practical guides for predicting and designing miscible polymer mixtures. Technomic, Lancaster

Collings PJ, Hird M (1997) Introduction to liquid crystals chemistry and physics. Taylor and Francis, London

Di Y, Iannace S, Di Maio E, Nicolais L (2003) Nanocomposites by melt intercalation based on polycaprolactone and organoclay. J Polym Sci Polym Phys Ed 41:670
Donald AM, Windle AH (1992) Liquid crystalline polymers. University Press, Cambridge
Fukushima Y, Okada A, Kawasumi M, Kurauchi T, Kamigaito O (1988) Swelling behaviour of montmorillonite by poly-6-amide. Clay Miner 23:27
Greenland DJ (1963) Adsorption of poly(vinyl alcohols) by montmorillonite. J Colloid Sci 18:647
Han CD (2010) On the mechanism leading to exfoliated nanocomposites prepared by mixing. Adv Polym Sci 231:1–75
Hoffmann B, Dietrich C, Thomann R, Friedrich C, Mulhaupt R (2000) Morphology and rheology of polystyrene nanocomposites based upon organoclay. Macromol Rapid Commun 21:57
Huang W, Han CD (2006a) Dispersion characteristics and rheology of organoclay nanocomposites based on a segmented main-chain liquid-crystalline polymer having pendent pyridyl group. Macromolecules 39(1):257–267
Huang W, Han CD (2006b) Dispersion characteristics and rheology of organoclay nanocomposites based on a segmented main-chain liquid-crystalline polymer having side-chain azopyridine with flexible spacer. Polymer 47(12):4400–4410
Huang W, Han CD (2006c) Ruthenium(ii) complex-induced dispersion of montmorillonite in a segmented main-chain liquid-crystalline polymer having side-chain terpyridine group. Macromolecules 39:8207–8209
Huang W, Han CD (2006d) Synthesis of combined main-chain/side-chain liquid-crystalline polymers via self-assembly. Macromolecules 39(14):4735–4745
Joule JA, Mills K (2000) Heterocyclic chemistry. Blackwell Science, Malden
Kato T, Mizoshita N (2002) Self-assembly and phase segregation in functional liquid crystals. Curr Opin Solid State Mater Sci 6:579
Kawasumi M, Hasegawa N, Kato M, Usuki A, Okada A (1997) Preparation and mechanical properties of polypropylene-clay hybrids. Macromolecules 30:6333
Lee KM, Han CD (2003a) Effect of hydrogen bonding on the rheology of polycarbonate/organoclay nanocomposites. Polymer 44:4573
Lee KM, Han CD (2003b) Rheology of organoclay nanocomposites: effects of polymer matrix/organoclay compatibility and the gallery distance of organoclay. Macromolecules 36:7165
Lehn JM (1988) Supramolecular chemistry—scope and perspectives molecules, supermolecules, and molecular devices. Angew Chem Int Ed Engl 27:89
Manners I (2004) Synthetic metal-containing polymers. Wiley-VCH, Weinheim
Nam PH, Maiti P, Okamoto M, Kotaka T, Hasegawa N, Usuki A (2001) A hierarchical structure and properties of intercalated polypropylene/clay nanocomposites. Polymer 42:9633
Okada A, Usuki A (2006) Twenty years of polymer-clay nanocomposites. Macromol Mater Eng 291:1449–1476
Percy MJ, Barthet C, Lobb JC, Khan MA, Lascelles SF, Vamvakaki M, Armes SP (2000) Synthesis and characterization of vinyl polymer-silica colloidal nanocomposites. Langmuir 16:6913
Reynaud E, Jouen T, Gauthier C, Vigier G, Varlet J (2001) Nanofillers in polymeric matrix: a study on silica reinforced pa6. Polymer 42:8759
Scheiner S (1997) Hydrogen bonding: a theoretical perspective. Oxford University Press, New York
Schubert US, Eschbaumer C (2002) Macromolecules containing bipyridine and terpyridine metal complexes: towards metallosupramolecular polymers. Angew Chem Int Ed Engl 41:2892
Shen M-M, Lu M-G, Chen Y-L, Ha C-Y (2005) Nanocomposites based on liquid-crystalline epoxy-clay: synthesis and morphology. Polym Int 54(8):1163–1168
Silva AS, Mitchell CA, Tse MF, Wang H-C, Krishnamoorti R (2001) Templating of cylindrical and spherical block copolymer microdomains by layered silicates. J Chem Phys 115:7166
Sinha Ray S (2006) Rheology of polymer/layered silicate. J Ind Eng Chem 12:811–842
Sinha Ray S, Okamoto M (2003) Polymer/layered silicate nanocomposites: a review from preparation to processing. Prog Polym Sci 28:1539–1641

Steed JW, Atwood JL (2000) Supramolecular chemistry. Wiley, New York

Tomasik P (1985) Pyridine-metal complexes (Newkome GR, Strekowski L). Wiley, New York

Usuki A, Kojima Y, Kawasumi M, Okada A, Fukushima Y, Kurauchi T, Kamigaito O (1993) Synthesis of nylon 6-clay hybrid. J Mater Res 8:1179

Vaia RA, Giannelis EP (2001) Liquid crystal polymer nanocomposites: direct intercalation of thermotropic liquid crystalline polymers into layered silicates. Polymer 42:1281

Villemore G (1990) Effect of negative surface-charge densities of smectite clays on the adsorption isotherms of racemic and enantiomeric tris (2,2′-bipyridyl) ruthenium(ii) chloride. Clays Clay Miner 38:622

Wunderlich B (1973) Macromolecular physics. Academic, New York

Zhang Q, Archer LA (2002) Poly(ethylene oxide)/silica nanocomposites: structure and rheology. Langmuir 18:10435

Zhang G, Jiang C, Su C, Zhang H (2003) Liquid-crystalline copolyester/clay nanocomposites. J Appl Polym Sci 89(12):3155–3159

# Chapter 21
# Liquid Crystalline Polymers as Tools for the Formation of Nanohybrids

**B. Lonetti, M. Mauzac, C. Mingotaud, M.L. Kahn, A.-F. Mingotaud, K. Soulantica, H.H. Nguyen, N. Lauth-de Viguerie, and J.-D. Marty**

## 21.1 Introduction

Liquid crystals (LCs) have been extensively exploited in display technologies throughout the past decades. However, they also play an important role in numerous sectors including organic transistors, sensors, light modulators, bio-medical applications and more. Unsurprisingly, due to their organization at the molecular level, LCs have also found an important place in nanoscience. In recent years, intimate mixtures of LCs and nanostructures have arisen great interest in the scientific community in order to obtain materials that manifest simultaneous nanoparticle-related properties and liquid crystalline behavior (Saliba et al. 2013; Hegman et al. 2007).

First studies have involved the doping of LC materials with preformed nanoparticles. As trivial as it may sound, a homogeneous material involving nanoparticles (NPs) and LCs is not straightforward to obtain. Indeed, the insertion of NPs can disrupt the LC order and can generate various types of defects e.g., hyperbolic hedgehogs, Saturn rings or boojums.... Those defects were taken into account to generate anisotropic colloidal structures. The incorporation in a homogeneous way of NPs within a LC medium often requires surface modification of

B. Lonetti • M. Mauzac • C. Mingotaud • A.-F. Mingotaud • H.H. Nguyen
N. Lauth-de Viguerie • J.-D. Marty (✉)
Laboratoire des Interactions Moléculaires et Réactivités Chimiques et Photochimiques, Université de Toulouse, CNRS UMR 5623, Toulouse Cedex 09 31062, France
e-mail: marty@chimie.ups-tlse.fr

M.L. Kahn
Laboratoire de Chimie de Coordination, CNRS UPR 8241, Université de Toulouse, 205, Route de Narbonne, Toulouse 31062, France

K. Soulantica
Laboratoire de Physique et Chimie de Nano-Objets, Université de Toulouse, INSA, UPS, LPCNO, CNRS UMR 5215, 135 Avenue de Rangueil, Toulouse Cedex 9 31077, France



V.K. Thakur, M.R. Kessler (eds.), *Liquid Crystalline Polymers*,
DOI 10.1007/978-3-319-20270-9_21

the former. For this reason, many studies have proposed coating of NPs with LC ligands, by using for instance ligand exchange processes. Doping of LCs with NPs was found to improve the physical properties of LCs and can give rise to applications in various technologies such as highly-sensitive LC-based sensors. In addition, LCs exhibit a certain degree of order by nature, which can be exploited either in order to organize NPs in two- or three-dimensions or as templating agents.

In addition to this *ex situ* strategies to obtain NPs embedded in LCs media, direct growth of NPs within a LC phase was also reported (Saliba et al. 2013; Hegman et al. 2007). This pathway requires doping of a LC by an adequate precursor that will be subsequently modified *in situ* (reduction, oxidation, hydrolysis...) to obtain NPs. This strategy also presents several issues concerning the compatibility of a chosen precursor with a LC phase and the way the precursor can be directly modified *in situ* without disrupting LC order. These issues have been circumvented by the use of lyotropic LC mesophases which have been used to obtain nanostructured silica materials useful in a wide range of applications. In contrast, thermotropic materials have scarcely been reported as media for the *in situ* growth of NPs.

This chapter will focus on the use of liquid crystalline polymers (LCPs) and elastomers (LCEs) in order to obtain intimate structures of NPs and LCs. We will first describe the doping of NPs in LC media and in a second section, the direct growth within a LC matrix. This overview will therefore highlight the various strategies to incorporate different types of NPs, e.g. metallic or metal oxides, in different types of LCs.

## 21.2 Stabilization of Preformed Nanoparticles

The *ex situ* procedure to get hybrid LCPs-NPs composites is usually preferred when nanoparticles size and shape control are necessary for precise practical applications, i.e. solar cells, optical sensing devices or actuation, which can take advantage from the collective behaviour of ordered NPs.

One of the most crucial points of this approach is the homogeneity of the NPs dispersion in the LC matrix. Especially, when high NPs weight fractions are necessary, nanoparticle-nanoparticle interactions should be at least partially counter-balanced by nanoparticle-polymer interactions. For this reason, in order to obtain homogeneous mixtures, the most popular strategy is to render inorganic nanoparticle as compatible to the liquid crystalline polymer matrix as possible. This can be achieved by using mesogens on the polymer chain with functional groups able to interact with NPs or surrounding the NPs with liquid crystalline ligands. This is true whichever the NPs used. The NP choice usually depends on the desired application, the LCP is then accordingly tailored.

### *21.2.1 Hybrid Liquid Crystalline Polymers*

Shandryuk et al. reported about the dispersion, inside a side-chain liquid crystal polymer, of CdSe spherical quantum dots (QDs) (Shandryuk et al. 2008) and CdS nanorods (Ezhov et al. 2011), which share the anisotropic character with LCs. In both cases, they chose a poly[4-(n-acryloyloxyalkoxy)]benzoic acid with a smectic C-type phase, since the presence of alkoxybenzoic acid functionalities should ensure the interaction with the inorganic fillers. In the case of CdSe, the interaction between the QDs and the polymer was clearly shown by infrared analysis: upon increasing the concentration of CdSe, the intensities of bands related to the carbonyl group $\nu$(CO) in the H-bonded dimers decreased while new bands relative to carboxylate anions appeared. X-ray scattering spectra showed a change of the interlayer spacing due to the presence of CdSe NPs, so the authors suggested a composite nanostructure where polymer layers alternate with QDs highly charged layers (See Fig. 21.1a, b). The LCP-NP interactions were also confirmed by the change in the thermal behaviour of the LC in the composite. Up to 20 wt% QDs

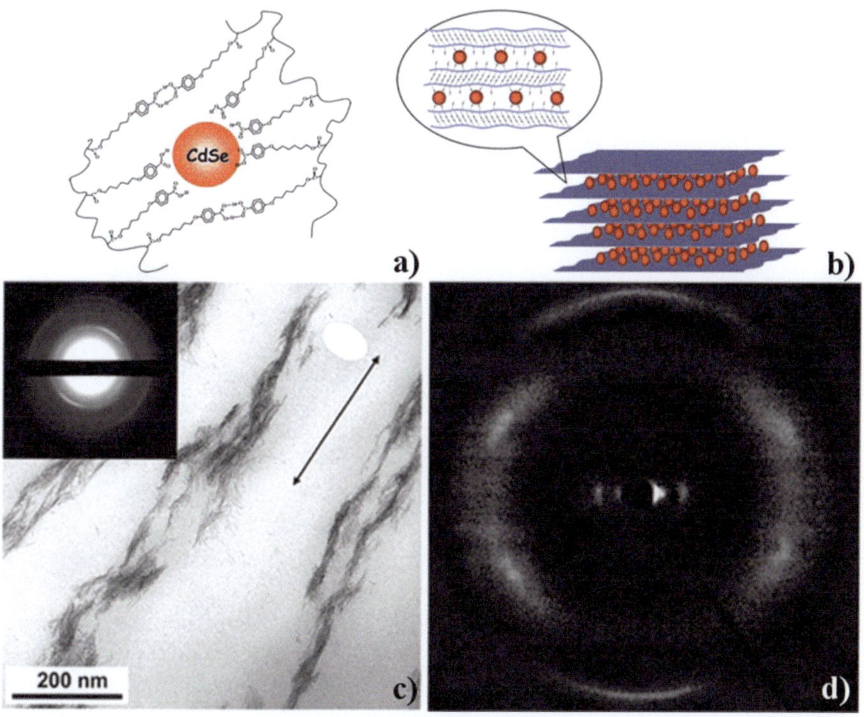


**Fig. 21.1** Schematic representation of (**a**) a side-chain LC polymer stabilizing CdSe NPs via alkoxy benzoic functionalities and (**b**) the localization of CdSe between polymer nanolayers (adapted with permission from Shandryuk et al. 2008). (**c**) TEM image and (**d**) X-ray pattern (*right*) of uniaxially oriented films of the liquid crystal polymer with CdS nanorods (adapted with permission from Ezhov et al. 2011)

content, the LC-Isotropic transition temperature dropped by some degrees and the associated transition enthalpy decreased. At 40 wt% QD content, the LC transition disappeared. The NPs content in a LC matrix is a limiting factor, so it must be carefully chosen in order to preserve the liquid crystalline properties of the final composite materials. The optical properties of these materials were studied (Vasilets et al. 2011) and the good insertion of the NPs inside the LC matrix was confirmed.

The case of the CdS nanorods described by the same authors is less straightforward (Ezhov et al. 2011). The IR spectra of the composite showed no sign of the presence of the ligands used for the nanorods synthesis, suggesting that they have been replaced by the functional groups of the polymer. As for the case of spherical NPs, when nanorods were incorporated in the polymer matrix, the number of H-bonded dimers in the composite decreased. The X-ray analysis showed only a change of 0.5 nm in the interlayer spacing, which is quite small. Moreover, the LC thermal behavior was only slightly influenced by the presence of the nanorods. The authors proposed weak interactions between the nanorods and the LC matrix. This was first confirmed by TEM images showing 100–200 nm low density domains of nanorods evenly distributed in the matrix. Interestingly, the uniaxial stretching of the composite films caused the formation of wire-like structures oriented along the deformation (Fig. 21.1), thus confirming an interaction between the nanorods and the mesogens. They furthermore compared the LC matrix with an amorphous one made of norbornene-methyl methacrylate copolymer without functional groups. They found a significant phase separation with larger and much denser domains in the absence of interacting groups. The good dispersion of the nanorods in the LC matrix was also proved by the optical properties of the composites: the CdS nanorods emission peak in the LC matrix and the peak of a homogeneous solution were nearly the same, whereas it was red-shifted in the amorphous matrix.

In the examples reviewed above, the LC character helps to organize the inorganic filler: the NP location and organization was a key factor for sensing applications. Therefore the authors are concerned about NP good dispersion which was obtained thanks to specific NP-LC interactions. The necessity to adapt the NP content is also discussed. Similarly, when doped liquid crystalline block copolymers are assessed for hybrid photovoltaic devices (Li et al. 2012; Shi et al. 2013), the NP concentration, location and order must be carefully chosen, since they directly influence the performances of the device. More precisely, in this kind of application, the main issue is to achieve large interfacial areas and highly ordered nanostructures in which the semiconductor NPs are ideally located. Liquid crystalline block-copolymers are well suited to favour oriented ordered domain formation, and the LC matrix-nanoparticle interface is crucial. Besides, a high NP content is necessary, in order to guarantee a sufficiently high current density, albeit up to a certain limit in order to avoid the disruption of the liquid crystalline order. For these examples a good compatibility is even more essential.

Thus, ZnO QDs were dispersed in a liquid crystal donor-acceptor copolymer poly[3-(6-(cyanobiphenyloxy)thiophene)-*alt*-4,7-(benzothiadiazole)], P3HbpT-BTD (Li et al. 2012) (Fig. 21.2b). The composite films were prepared with a high

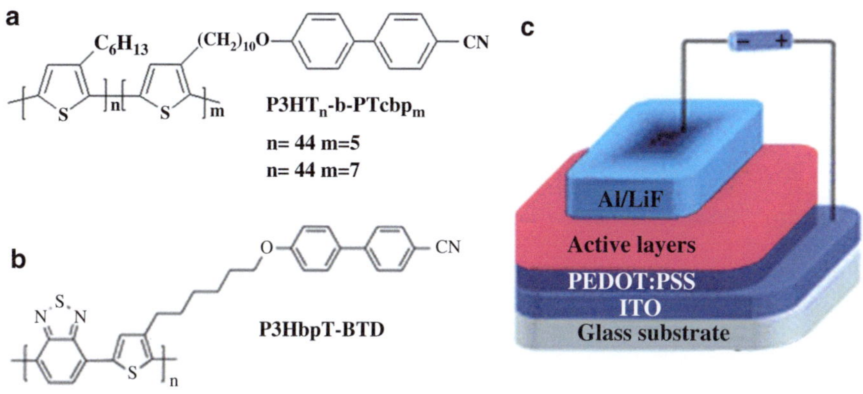


**Fig. 21.2** Chemical formula of the (**a**) polythiophene-*b*-poly{3-[10-(4′-cyanobiphenyloxy)-decyl]thiophene} (P3HT-b-PTcbp) (adapted with permission from Shi et al. 2013) and (**b**) poly [3-(6-(cyanobiphenyloxy)thiophene)-*alt*-4,7-(benzothiadiazole)], P3HbpT-BTD (adapted with permission from Li et al. 2012) used in the active layer of a hybrid solar cell device schematized in (**c**) (adapted with permission from Li et al. 2012)

dopant content, (1:3 wt% P3HbpT-BTD: QDs), by spin-coating and annealing of highly concentrated dichlorobenzene solutions. The authors observed a relation between the annealing conditions, the composite film microstructure and the device performance. In particular, films annealed while being in the mesomorphous state achieved power conversion efficiencies comparable to the best values in the literature (ca. 2 %). X-ray diffraction and TEM images showed highly ordered ZnO domains and well developed interpenetrating networks if annealing was conducted in the liquid crystal state; on the contrary, films annealed while in the crystalline or isotropic state were characterized by extensive ZnO aggregation. Therefore, the enhanced charge separation and transport efficiency could be attributed to the well-ordered percolated network.

The performances of such device were improved by the same authors by allowing a more intimate mixture of the liquid crystalline block-copolymer and inorganic NPs (Shi et al. 2013). As already mentioned, in order to have good charge separation, the NPs must ideally be located at the heterojunction. This situation is more likely to be achieved if the polymer possesses groups allowing cooperative interactions with the QDs. For this purpose, the authors compared the power conversion efficiencies of hybrid solar cells formed by conjugated rod–rod diblock copolymers containing cyanobiphenyl mesogenic pendant groups, polythiophene-*b*-poly{3-[10-(4′-cyanobiphenyloxy)-decyl]thiophene} (P3HT-*b*-PTcbp) (Fig. 21.2a), as electron donors, and by ZnO or CdS QDs as electron acceptors. They fabricated devices where the QDs surfaces were modified or not by liquid crystalline ligands, 4′-hydroxy-[1,1′-biphenyl]-4-carbonitrile (cbp). The role of the liquid crystalline ligands was to guarantee a better miscibility with the polymer owing to the intermolecular interactions arising from their similar structure. All the devices containing QDs with liquid crystalline ligands possessed better power energy conversion. The authors attributed the enhanced performances to the improved

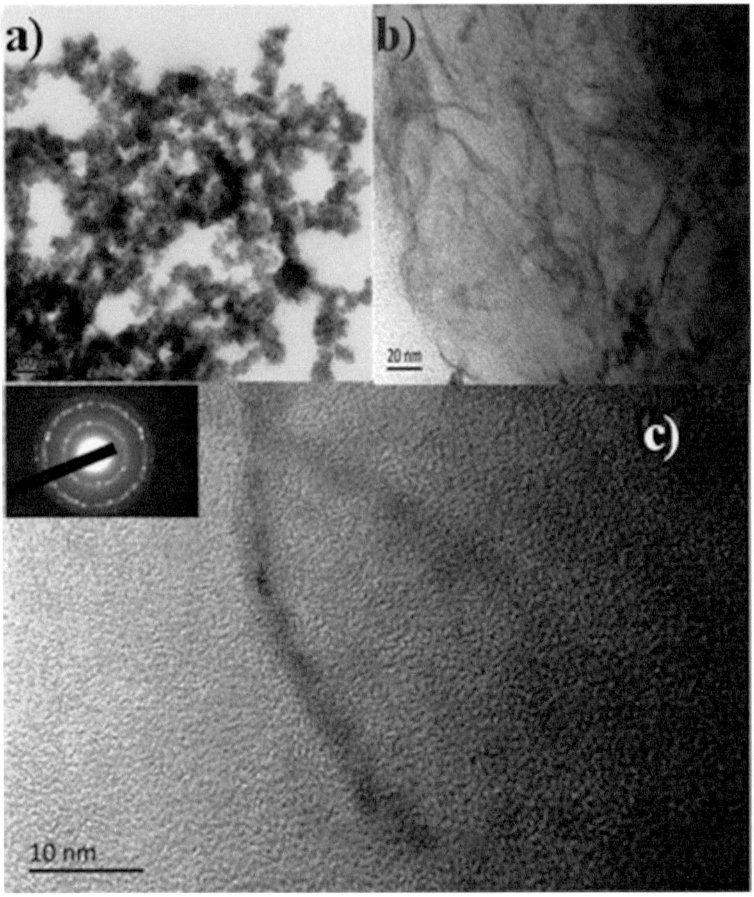


**Fig. 21.3** TEM images of (**a**) P3HT-b-PTcbp/ZnO, and (**b**) P3HT-b-PTcbp/cbp@ZnO. (**c**) HRTEM images of the self-assembly of cbp@ZnO in P3HT-b-PTcbp matrix. The concentration of polymer was 10 mg/mL, the polymer:ZnO ratio 1:2 (adapted with permission from Shi et al. 2013)

order of the active layer due to the presence of liquid crystalline moieties on the conducting polymer and to the better dispersion of nanoparticle, evidenced by TEM images, when liquid crystalline cbp are present on the QDs surface (Fig. 21.3).

### *21.2.2 Hybrid Liquid Crystalline Polymers and Elastomers for Soft Actuation*

Another interesting approach is to exploit the interactions between LCs and NPs in order to elaborate a responsive material. Liquid crystal groups exhibit an anisotropic response to a magnetic field: depending on their magnetic susceptibility, the mesogens assume a parallel or perpendicular orientation with respect to it. Since the 1970s when F. Brochard and P.J. de Gennes (Brochard and de Gennes 1970) suggested to use magnetic particles in order to enhance this magneto-orientational response of LCs, investigations have been dedicated to the better understanding of the phenomenon, with the nanoparticle size and concentration effects being the main inquiry. The interest in this subject is still alive (Mertelj et al. 2013). Starting from these considerations, liquid crystalline silicones were doped with Co nanorods (Zadoina et al. 2009; Riou et al. 2014) aiming at (i) enhancing the anchoring efficiency of LCs on the surface of NPs by the use of rod shaped nanoparticles and/or by the insertion of interacting groups (ii) guaranteeing higher torque with smaller NPs, by using a strong magnetic material such as metallic Co instead of weak magnetic iron oxide NPs. Their composite material possessed both liquid crystalline and magnetic properties, the samples being ferromagnetic in the smectic state. The almost square hysteresis curves of those materials proved that cobalt nanorods could be efficiently aligned in a magnetic field (3 T) during cooling down from the isotropic state to the smectic phase. An effect due to the anisotropic matrix, which was also sensitive to a magnetic field, was observed: a quite low magnetic field (50 mT) was sufficient to orientate the nanorods when they were dispersed in the LCP compared to an amorphous matrix. This behavior was interpreted as a possible signature of cooperative effects between the LCP and the magnetic nanorods (Zadoina et al. 2009). The microscopic structure of these composites was also studied by X-ray scattering. Indeed, even if they are homogeneous at the macroscopic level, some inhomogeneities existed at microscopic level and the nanorods were organized in bundles possessing a hexagonal order in the planes perpendicular to the bundle axis (Fig. 21.4).

In this particular case, the strong ferromagnetic interactions between nanorods are quite difficult to be completely overcome even when liquid crystalline lateral groups are carboxylic acids interacting with nanorods. Nevertheless, the lateral distances between the nanorods were larger than their diameter indicating that polymer chains surrounded them (Riou et al. 2014). The authors then examined the orientational coupling between liquid crystalline groups and nanorods. For this purpose, they compared different composites, with and without groups interacting with the nanorods and they measured the nematic order parameter of the mesomorphous groups under a magnetic field: the polymer containing carboxylic groups was more easily oriented in the presence of magnetic nanorods at low magnetic field, while the effect was low, almost inexistent, when no ligand was present on the liquid crystalline matrix (Riou et al. 2014).

Aiming always at obtaining a magneto-responsive material, liquid crystalline elastomers (LCEs) were doped with magnetic NPs. In this kind of materials, the inorganic and sensitive to magnetic field component should act on the elastic, deformable matrix producing phenomena such as shape change or tuneable and anisotropic elastic modulus. Vice versa, the elastic force acting on the NPs could orientate them and change the magnetic properties of the composite material with respect to the free NPs. These two approaches were followed by different teams (Song et al. 2007; Riou et al. 2012; Garcia-Marquez et al. 2011;

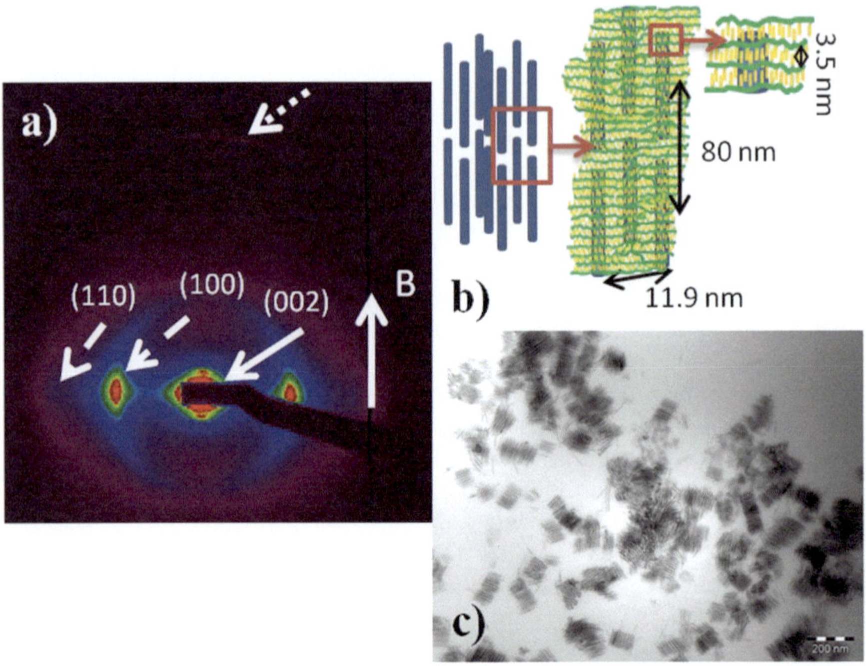


**Fig. 21.4** (**a**) SAXS pattern of the polymer composite containing carboxylic interacting groups doped with 0.8 %wt Co nanorods. The *solid white arrow* points to the (002) reflection due to the organization of the Co nanorods in rows within the bundles whereas the *dashed arrows* point to the (100) and (110) reflections of the hexagonal 2-dimensional lattice. The *dotted white arrow* points to the smectic reflection from the mesogens; (**b**) schematic representation of the nanorod bundles in the composite; (**c**) typical TEM picture of a magnetic liquid crystal composite (adapted with permission from Riou et al. 2012)

Haberl et al. 2013a, b). The composite material homogeneity and the shape change amplitude are essential characteristics to be taken into account in the design of these materials. The presence of chemical interactions between the magnetic component and the matrix is crucial when a good compatibility between the inorganic particles and the organic medium and an efficient response to an external stimulus are pursued. The components of the LCE composites used in the following examples are summarized in Fig. 21.5.

In the first example of a monodomain magnetic elastomer, dopamine anchored ferrite NPs with an average diameter of 6 nm were covalently bonded to a liquid crystalline polysiloxane elastomer. Transparent LC films were obtained by spin-coating. When the film was additionally annealed in a 0.35 T magnetic field, more transparent LCEs with a 0.42 nematic order parameter were obtained. The LC films showed super-paramagnetic behavior at 300 K. This example proves that the magnetic NPs do not hamper the elastomer from retaining its liquid crystalline properties and the mesogens orientation inside the matrix. The composites

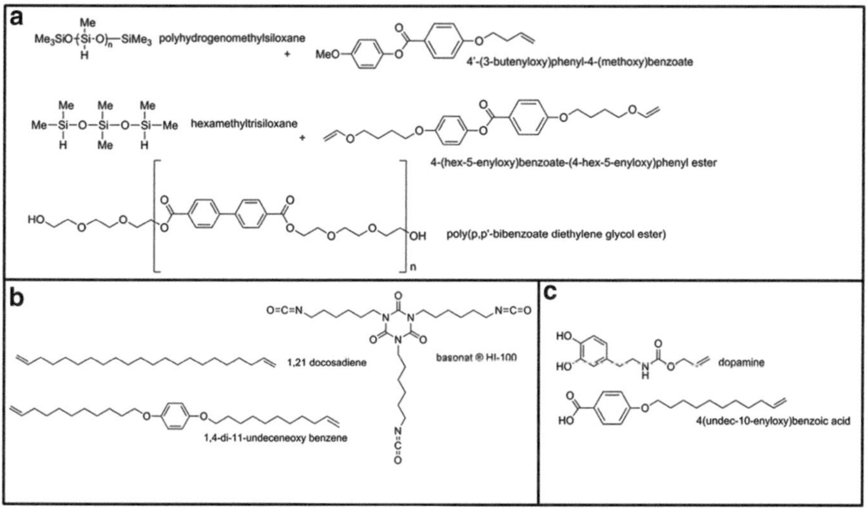


**Fig. 21.5** (**a**) Components of the liquid crystal polymer chains; (**b**) cross-linking agents; (**c**) NPs ligands allowing them to work as crosslinkers

contained 7 %wt of ferrite NPs, but no information was given about the effective content of iron oxide.

Based on the same idea where the magnetic NPs are directly linked to the elastomer matrix and play the role of cross-linking agents, a main chain liquid crystalline elastomer (MC-LCE) doped with iron oxide NPs was synthesized by Donnio's group (Garcia-Marquez et al. 2011). MC-LCEs were chosen as they can undergo more important shape changes than their comb-like homologues. The MC-LCE was synthesized *via* a one step reaction, by mixing the di-olefinic mesogen with the hexamethyl trisiloxane spacer and 3.3 nm ferrite NPs covered with 4-(undec-10-enyloxy) benzoic acid as cross-linkers. The hydrosilylation reaction was activated by the Karstedt's catalyst at 40 °C. The authors proved the feasibility of the synthesis, and the mesomorphous and magnetic behavior of these composites. Their next objective being the synthesis of monodomain samples in order to achieve a magneto-responsive behaviour.

As in the case of the LC polymer mentioned above, Riou et al. used anisotropic ferromagnetic NPs (Riou et al. 2012) in order to achieve a strong orientational coupling between magnetic nanoparticles and the liquid crystal medium and maximize the torque induced by the applied magnetic field on the medium. They described the synthesis of a LCE and showed the potential for this new kind of material. In order to obtain the desired magneto-responsive character, the synthesis of monodomain samples, where Co nanorods were oriented parallel to the nematic vector, must be further developed.

The research performed in the examples described above aimed at achieving a shape change by the wireless action of a magnetic field. On the contrary, a LCE possessing a reversible magnetic memory was synthetized by Harberl et al. by using elongated iron oxide NPs (7:3 maghemite:hematite) (Harberl et al. 2013). The NPs' surface was functionalized with amino groups so that the tri-functional isocyanate cross-linker could react simultaneously with the LCP and the NPs in order to create a tridimensional network. Different composites with a NPs content ranging from 0.5 to 10 %wt were synthesized and their elastic properties were characterized. The samples were stretched in the isotropic state and then cooled down to room temperature in order to keep the deformation (state "ON"). In this state, the samples exhibited an anisotropic susceptibility due to the orientation distribution of the magnetic NPs induced by strain. This anisotropy could be erased by heating up the sample in the isotropic state (state "OFF") (Fig. 21.6). These two states could then

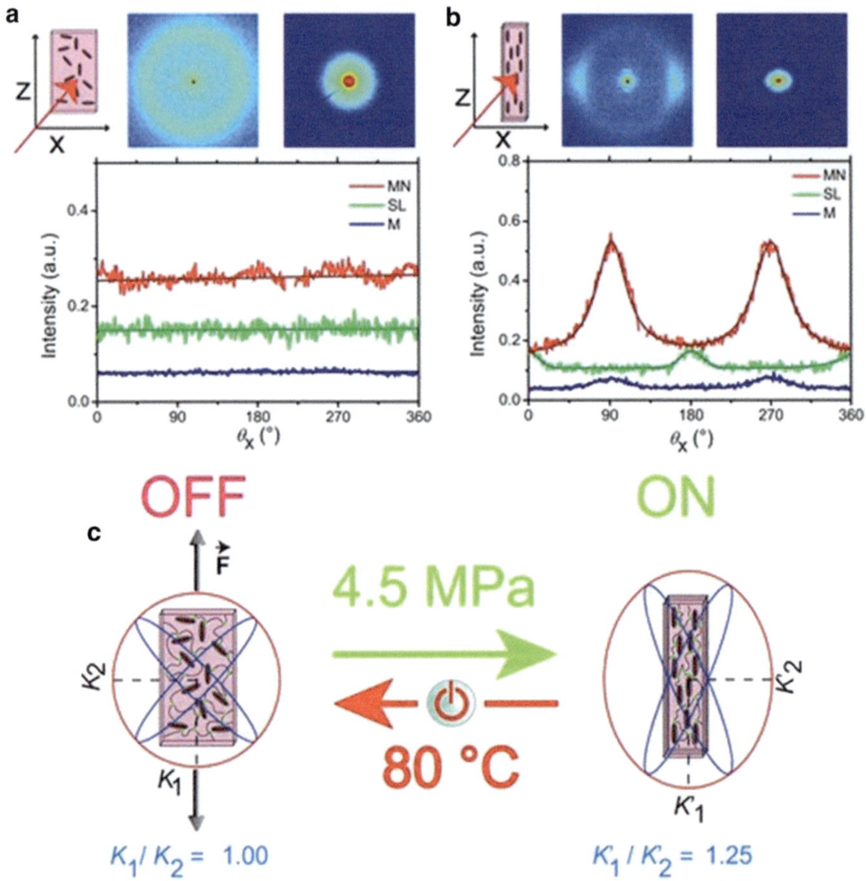


**Fig. 21.6** Wide angle (*light blue*) and small angle (*dark blue*) X-ray scattering patterns and the azimuthal intensities at the correlation lengths of the magnetic NPs (*red curve*), smectic layers (*green curve*) and mesogens (*blue curve*) of the LCE in (**a**) the OFF state and (**b**) the ON state. The *red arrow* is the X-ray beam direction. (**c**) Schematic representation of the magnetic susceptibility components in the OFF and ON states (adapted with permission from Haberl et al. 2013a)

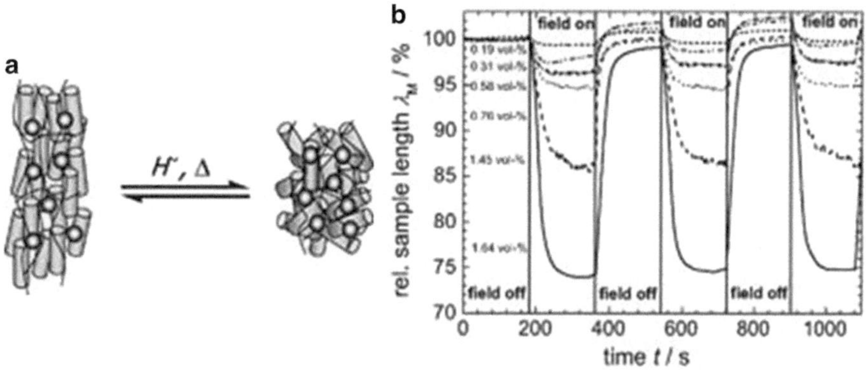


**Fig. 21.7** (**a**) Scheme of the LCE contraction induced by magnetic heating and (**b**) relative lengths of the LCE containing various NPs content in an alternative magnetic field of 300 kHz, 43 kA $m^{-1}$ (adapted with permission from Kaiser et al. 2009)

be magnetically read out; this kind of materials pave the way towards applications such as strain sensors or magnetic switches.

Another approach for inducing wireless shape change is to exploit the magnetic heating by iron oxide ($Fe_2O_3$) in order to induce a nematic/isotropic transition in a monodomain LCE (Kaiser et al. 2009). In this example, hydrophobic *N*-oleoylsarcosine stabilized NPs, compatible with the organic medium, were used, and no special care to guarantee interactions with the liquid crystalline groups was necessary. Toluene NP solutions were mixed to the polyhydrogenomethylsiloxane, PHMS, the mesogen 4′-(3-butenyloxy)phenyl-4-(methoxy)benzoate and the cross-linker 1,4-di-11-undecenyloxy benzene. The hydrosilylation reaction was conducted at 70 °C in the presence of a Pt based catalyst. The monodomain elastomer (order parameter of 0.63 as deduced from X-ray experiments) was obtained through a two step cross-linking procedure. The sample reversible contraction after application of an alternating magnetic field increased with increasing nanoparticle concentration up to a maximum of 74 % in the presence of 1.64 % NPs (volume fraction). Furthermore, the actuation depended on the magnetic field intensity, which can then be modified in order to adjust the final dimensions (Fig. 21.7).

Finkelmann group validated the concept of shape change in hybrid, uniformly oriented LCEs, through magnetic heating (Chambers et al. 2006). Even if this work is beyond the scope of this review, it is noteworthy that wireless heating has also been successful using IR light as demonstrated by Marshall et al. (2012). In order to have any practical application, the response time and the efficiency should be improved in such devices.

Remarkable work has been directed to push the performance limits of traditional organic materials by doping monodomain LCE with other inorganic NPs. For instance, Montazami et al. (2012) inserted 3 nm gold NPs in a monodomain LCE in concentrations up to 0.1 % mol. Doping stiffened the elastomer, but the actuator strain rate was significantly improved at higher nanoparticle concentrations, due to the enhanced thermal conductivity and by consequence to the more efficient heat transfer to the matrix.

Domenici et al. successfully created a conductive layer of $PbTiO_3$ (Domenici et al. 2010) or molybdenum based nanowires (Domenici et al. 2011) in monodomain LCEs. Those elastomers retained the typical shape memory property of monodomain LCEs, even though their maximum elongation was slightly lower than in the absence of nanoparticles. Despite the fact that the preparation method was the same, spherical NPs and nanowires were differently aligned inside the elastomer. The possibility of electromechanical actuation of these materials was suggested and in the case of ferroelectric nanoparticles another mechanism for actuation, i.e. by applying external voltage perpendicularly to the thin layer of nanoparticles, was proposed.

## 21.3 *In Situ* Synthesis of Nanoparticles

As previously mentioned, successful hybridization of NPs within a liquid crystalline medium depends on several factors including NP size/concentration, chemical compatibility, topological defects and so on. For this reason, research groups in this field started focusing on the possibility of growing NPs directly within the LC instead of using preformed particles. However, the direct synthesis of NPs in LCs generally presents several difficulties:

1. The first one concerns the retention of the liquid crystalline phase upon addition of the NP precursor that could disrupt the LC phase.
2. The second one concerns the development of an *in situ* procedure to generate NPs within a LC medium. For this, in order to avoid the diffusion of additional reactants within the LC medium, most studies involve the *in situ* reduction of metal precursors *via* oxidation of the LC medium in order to obtain the desired NPs. In other cases, sputtering or electrodeposition techniques were used. Alternatively the diffusion of small molecules in a gas phase (water, dihydrogen) have been also reported.
3. Last, if the interactions between the formed NPs and the LC phase is poor, segregation of the formed NPs may occur.

Another advantage of this *in situ* synthesis is that the LC anisotropic medium may induce a morphology-controlled growth of NPs, *i.e.* control over size and shape. Whereas this last approach has been often reported in the case of lyotropic liquid crystals, (Saliba et al. 2013; Hegman et al. 2007) the use of thermotropic liquid crystals is seldomly found in literature. For this specific purpose, the use of LC polymers has demonstrated its better efficiency than the use of small LC molecules. That will be discussed in the following paragraphs.

### 21.3.1 Solvent Mediated In Situ Formation of Nanoparticles/Liquid Crystal Hybrids

In order to obtain homogeneous dispersions of inorganic NPs in liquid crystalline polymer matrices, another possibility is the *in situ* synthesis of the NPs. Zadoina et al. adapted the synthesis of Co NPs starting from bis(di(trimethylsilyl)amido) cobalt(II), $Co[N(SiMe_3)_2]_2$, in the presence of saturated fatty acid and amine ligands (Zadoina et al. 2011). They successfully obtained spherical and/or rod-like Co NPs inside a silicone based LCP. The precursor decomposition was carried under 3 bar $H_2$ in toluene solution varying the ratio between the Co precursor, the carboxylic acid side-chain groups on the polymer and the hexadecyl amine. For a low content of the acid ligand, anisotropic NPs started to form, and a majority of nanorods formed when hexadecyl amine ligand was added in the solution at Co:acid ligand:amine ligand molar ratio 1:0.16:1. All NPs were well dispersed inside the matrix which retained its mesomorphous character. The ferromagnetic properties of Co NPs were also studied. By using a similar strategy, Domracheva et al. (2011) prepared superparamagnetic $\gamma$-$Fe_2O_3$ nanoparticles by oxidation of a second generation polypropylene imine LC dendrimer–iron (II) complex in THF solution. NPs with an average diameter of about 2.5 nm encapsulated within the LC dendrimer were obtained. After removal of the solvent, those particles presented a liquid crystalline columnar phase, identified by X-ray diffractometry measurements. The electron magnetic resonance spectra of those NPs within the LC phase were then compared to the one of bulk $\gamma$-$Fe_2O_3$ demonstrating that, in contrast to bulk $\gamma$-$Fe_2O_3$, LC NPs possessed enhanced effective magnetic and uniaxial anisotropy induced by the particle surface and shape effect.

### 21.3.2 Solvent-Free In Situ Formation of Nanoparticles/Liquid Crystal Hybrids

Lee and Jin (2003) were the first to report the growth of metallic NPs inside a thermotropic LC based on polymeric materials. A polyester matrix polymer consisting of hydrophilic PEG branches attached to a hydrophobic aromatic polyester backbone was used for his purpose (see Table 21.1 and Fig. 21.8a). This structure tended to be organized in smectic-like domains with the PEG branches occupying the inter-layer domains. Gold nanoparticles were obtained in a two-step procedure. First an aqueous solution of $HAuCl_4$ was allowed to diffuse into the hydrophilic domains. Then, these gold ions were reduced to Au NPs upon addition of a hydrazine solution as shown on TEM micrographs in Fig. 21.8b. The mesomorphous properties of the LC/NP hybrid were not discussed.

The procedure used to form the NPs necessitates the addition of a limited quantity of water to the liquid crystal materials in order to introduce the NP precursor and the reducing agent. Therefore a partial disruption of the mesomorphic

**Table 21.1** Highlighted examples of thermotropic LCs used for direct synthesis of NPs

| LC type | NP type/Reference |
|---|---|
| 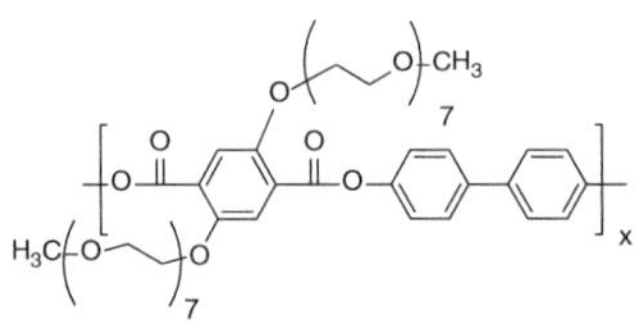 | Au NPs<br>(7–14 nm)<br>Lee and Jin (2003) |
| 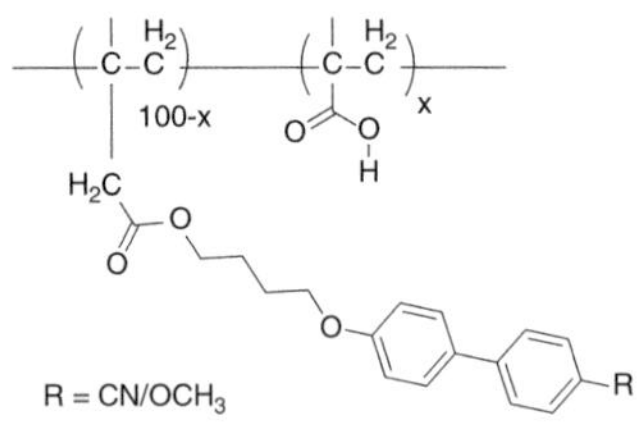 | Ag NPs<br>(5–30 nm)<br>Barmatov et al. (2004) |
| 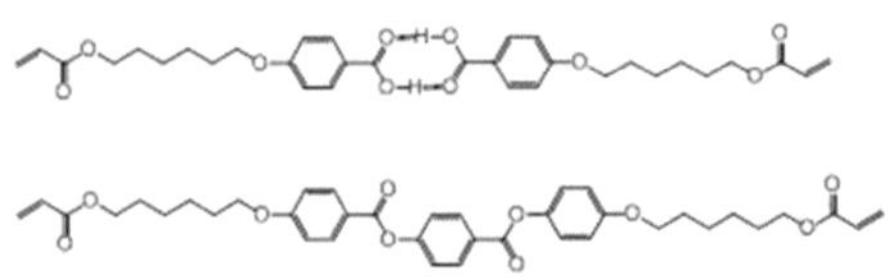 | Ag Nps<br>(5 nm)<br>Dasgupta et al. (2013) |
| 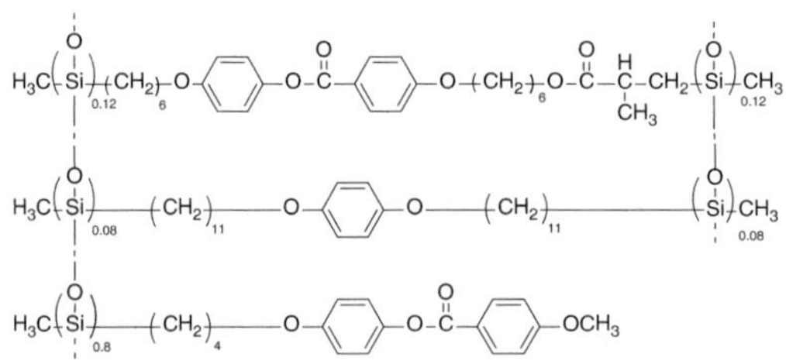 | Au NPs<br>(3–50 nm)<br>Gascon et al. (2005) |
| 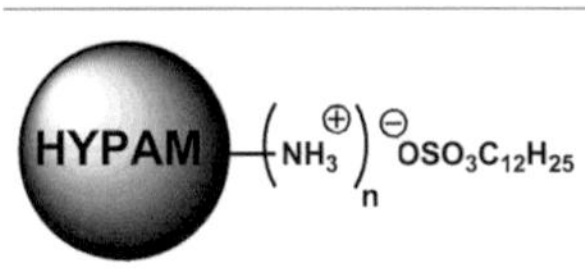 | AuNPs<br>(2–5 nm)<br>Nguyen et al. (2014) |
| 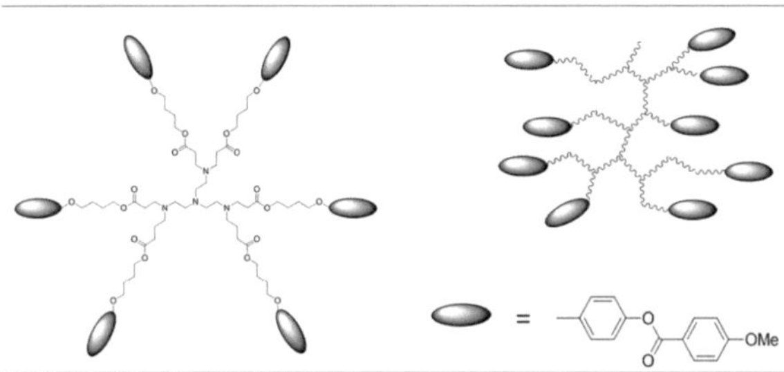 | ZnO<br>Nanoworms<br>Nanowires<br>Saliba et al. (2011, 2012) |

order at localized sites during the nanocomposite preparation could occur. To avoid this, Barmatov et al. (2004) used a different strategy that involved the direct reduction of the metal precursor within the LC matrix. LC polymer with cyano- or methoxy-biphenylmesogenic side-fragments were used in this study (Fig. 21.8c

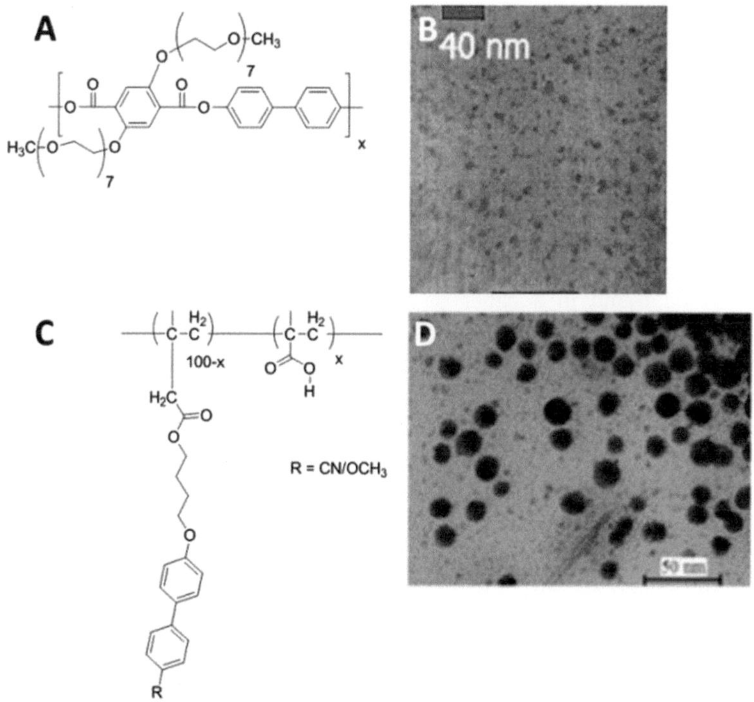


**Fig. 21.8** Structure of DiPEG-BP LC polymer (**a**) and TEM micrograph of Ag NPs within (**b**) (Adapted with permission from Lee and Jin (2003)). Molecular structure of side chain LC polymer used by Barmatov et al. (2004) incorporating cyanobiphenyl or methoxybiphenyl mesogens (**c**) and TEM micrograph of Ag NPs synthesized within (**d**)

and Table 21.1). A silver-olefin complex was first mixed with a solution of polymer in THF and the solvent was then evaporated. After heating, $Ag^+$ ions were reduced by olefin ligands, inducing the formation of isotropic Ag NPs with sizes in the 5–30 nm range (see Fig. 21.8d). Since NPs presented large size distribution, the control over the NP morphology was somehow limited. Moreover, the presence of NPs had a strong and negative influence on the mesophase stability: the composites based on cyanobiphenyl mesogens lost the mesomorphic properties above 2 wt% silver content whereas the ones based on methoxybiphenyl entities remained relatively stable up to 15 wt% of silver NPs. The drastic decrease of mesophase stability in the first case was thought to be related to the strong interactions between terminal cyano groups and silver surfaces, significantly disturbing the packing of the mesogenic groups and resulting in the rapid loss of LC behaviour.

### *21.3.3 Effect of LC Organization on Nanoparticles Growth*

Until now, the discussed examples generally focused on the incorporation of NPs in LCs to form LC nanocomposites. However, not much attention was given to the control of NP morphology using the LC as a template. The first attempts towards this control have involved the use of mixture of ionic liquids and NPs precursors presenting mesomorphic properties leading, after reduction, to gold, copper or silver nanoparticles (Taubert 2004; Dobbs et al. 2006). Particle size and thickness can be tuned by varying the reaction temperature. While nanospheres were obtained under isotropic conditions, anisotropic structures (copper platelets, leaf-like Au nanostructures) were obtained in a LC state. However, none of those examples have shown a direct relation between the LC structure and the morphology of the synthesized NPs. A clear correlation between the LC phase and the NP size and shape can be found in a paper of Mallia et al. (2007). The system involved amphiphilic low-molecular-weight mesogens, which were capable of reducing the metallic salt without the need of an external reducing agent. The gold ions interacted with the ammonium groups in the liquid crystal molecule. The size and shape control was achieved by either varying the concentration of the gold chloride precursor or by changing the organization of LC groups.

In this context, liquid crystalline polymers/elastomers were demonstrated to be the materials of choice in order to gain control over the morphology of the *in situ* generated NPs. Nguyen et al. (2014) have used branched thermotropic liquid crystals obtained from the ionic interactions between hyperbranched polyamidoamine of different molar masses and a surfactant (sodium dodecyl sulfate). These ionic complexes presented columnar and lamellar phases and these phases were used to generate gold NPs *in situ*. The ionic nature of these mesophases allowed enhancing the interaction of the salt precursor with the liquid crystalline phases to obtain homogeneous materials. In this work, $HAuCl_4$ was first mixed with crude polymer and let to reach a chosen temperature. A flux of dihydrogen was then introduced to ensure the reduction of the gold salt, leading to the formation of NPs. The templating effect of the mesophase resulted in the formation of NPs with a spherical shape. As illustrated in Fig. 21.9, their size was dictated by the local organization of the mesophase (as already observed by Mallia et al. (2007)), but also by the average molar mass of the hyperbranched complex.

Gascon et al. (2005) demonstrated that not only the liquid crystalline nature can have a strong influence on the nanoparticle size but also the local organization of a given mesophase. For this, liquid crystalline elastomers, based on side chain polysiloxane (see Table 21.1) with the same chemical structure but different macroscopic organizations (polydomain and monodomain), were synthesized. Whereas, in the smectic phase of a monodomain polymer, Au NPs of 3 nm were obtained, larger Au NPs of 50 nm were observed when the same experiment was performed in the isotropic state as shown in Fig. 21.10. The differences observed may be related to the formation of NPs within the structure defects of the LC polydomain samples presenting, due to their intrinsic nature, more defects than monodomain materials.

More recently Dasgupta et al. (2013) synthesized nanoporous smectic liquid crystalline polymer networks and used them as a template for the photochemical synthesis of silver nanoparticles. They showed that the diameter of the NPs was controlled by the length of the cross-linking agent used during synthesis. Indeed this length allowed controlling the periodicity of the layered morphology and therefore

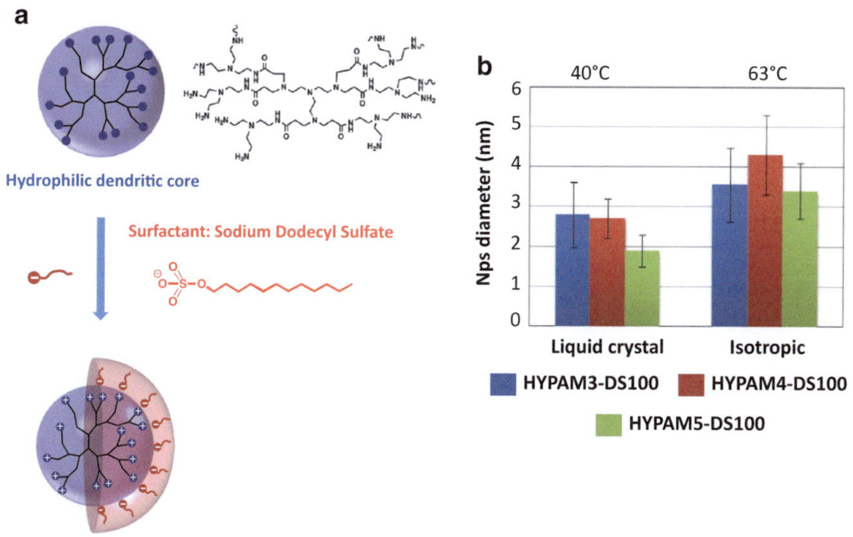


**Fig. 21.9** (**a**) Synthesis of hyperbranched polyamidoamine-dodecyl sulfate (HYPAM-DS) complexes. (**b**) Average diameter of gold nanoparticles synthesized within the HYPAM-DS polymers as a function of the polymer organization (lamellar LC at 40°C or isotropic at 63°C) and of the molar mass of the hyperbranched complex: three hyperbranched polymers with increasing molar masses (HYPAM 3, HYPAM4 and HYPAM 5) and the same level of functionalization (100 %) with dodecyl sulfate were studied namely HYPAM3-DS100, HYPAM4-DS100 and HYPAM5-DS100. Adapted from Nguyen et al. (2014)

of the *in situ* generated silver nanoparticles. Those materials have been used to fabricate patternable organic/inorganic hybrid materials.

A new strategy to improve the control of NP growth within the LC mesophase has been published by S. Saliba et al. (2011, 2012). This approach involved the *in situ* hydrolysis of dicyclohexylzinc, [Zn($Cy_2$)] leading to ZnO NPs with volatile cyclohexane as the unique side product. The use of LC compounds having an oligomeric or a polymeric structure that contains various amine and amide chemical functions, favoured interactions between the LC and [Zn($Cy_2$)]/ZnO NPs. Those interactions were evidenced by experiments in solution and by solid state $^{13}$C-NMR studies in LC state and presumably limit the disruption of the LC organization and increase the templating effect of the LC phase. Two different types of LCs which differed by their backbone type were investigated (see Fig. 21.11). Both targeted LCs exhibited a mesophase (*i.e.* nematic phase) close to- or at room temperatures. Under isotropic conditions, hydrolysis of the [Zn $(Cy)_2$] precursor led to isotropic NPs. At these given temperatures, growth control could be achieved, the average size of the NPs being around $5.4 \pm 0.7$ and $2.7 \pm 0.3$ nm, respectively.

When the experiments were performed in the nematic phase state of the LC compound (*i.e.* at lower temperatures), anisotropic ZnO structures, were obtained. As shown in Fig. 21.11, either nano-worm-like or nano-wire structures were grown

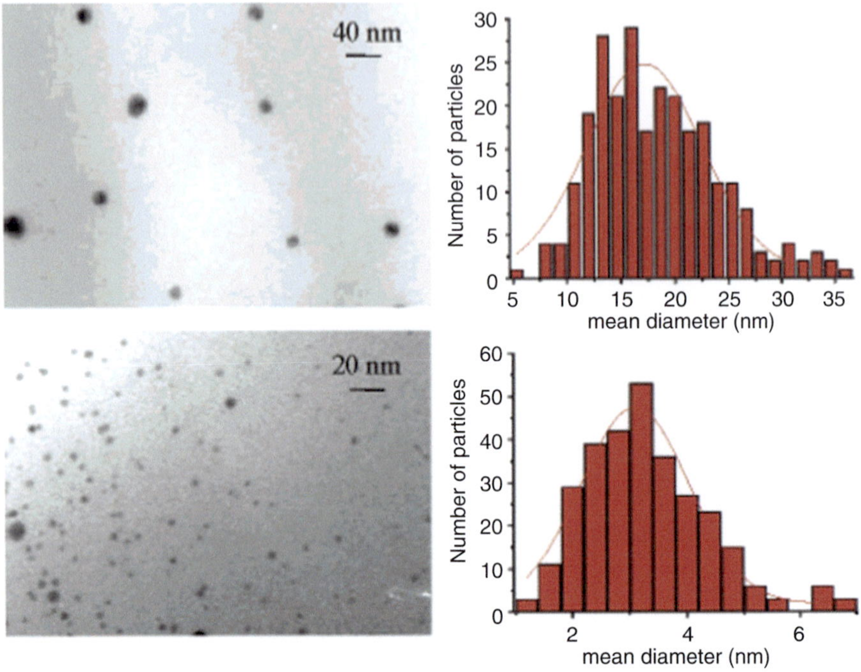


**Fig. 21.10** TEM micrographs and corresponding histograms illustrating the particle size distribution of gold NPs embedded in composites formed in a polydomain (*up*) or monodomain (*bottom*) smectic phase. Adapted with permission from Gascon et al. (2005)

in branched LC and hyperbranched LC. They exhibited an average width of $2.5 \pm 0.2$ and $2.7 \pm 0.4$ nm, respectively. The lengths vary from a few nanometers up to around 100 nm when the branched LC was used, while lengths from ca. 10 to 200 nm were obtained when the hyperbranched LC was used. Thus, a direct correlation between the structural characteristics of the LC and the morphology of the nanostructures, was demonstrated. The new LC/NPs composites presented, at the same time, LC properties and optical properties originating from ZnO. This new *in situ* strategy paves the way for new LC/NP composites of controllable and stable properties.

From the above, in order to tailor the NP morphology using a LC phase, we suggest that different criteria should be fulfilled. Primarily, the chemical reaction leading to the NPs should not disrupt the LC organization. Thus, the LC molecules should not play the role of reactants and side products should be avoided as much as possible throughout the NP formation. Next, interactions between the LC molecules, the NP precursor and possibly the synthesized NPs, should favour the templating effect of the LC phase. Finally, the use of relatively high viscosity LCs should prevent a fast disruption of the organization during the NP formation.

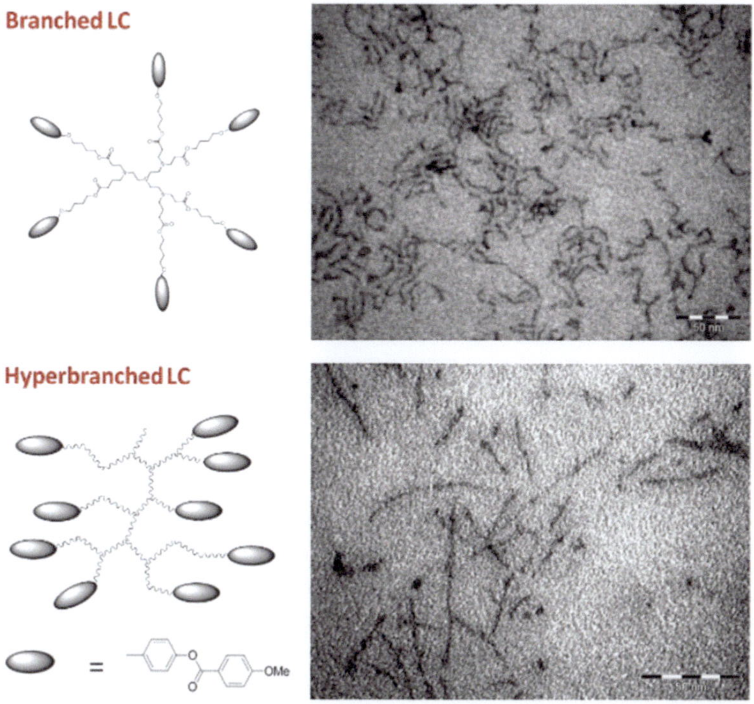


**Fig. 21.11** TEM micrographs of ZnO NPs synthesized under various conditions: (**a**) in the presence of branched liquid crystalline molecules at 5 °C (scale: 50 nm) (**b**) in hyperbranched LC at 45 °C (scale: 50 nm) or at 5 °C (scale: 50 nm). Adapted from Saliba et al. (2011, 2012)

## 21.4 Conclusions and Future Perspectives

NPs/LC hybrids have been the subject of many reports in the last few years. Such materials can be obtained either by *ex situ* or *in situ* strategies. The former is often the most appropriate strategy, when monodisperse NPs are needed for a given application. Nevertheless this may imply tedious synthetic steps to render NPs compatible with a LC medium. In this context, LC polymers or elastomers can incorporate in their structure specific moieties that can make them suitable to interact with the preformed NPs and allow increasing colloidal stability within a LC medium. The mesophases obtained from such polymers can be considered as good candidates for the organization of the NPs within the LC structure in a precise and controlled fashion, which is required for instance when dealing with optical properties of those assemblies. Alternatively, incorporation of NPs (for instance with magnetic properties) within the LC structure has allowed to modify the LC behavior of those materials obtaining soft actuators. *In situ* strategies have also been used with such polymers. In these cases the main issues concern the difficulties to perform chemistry in a LC medium without disrupting the LC order. These difficulties can be circumvented by using physical methods, by using LC as a reactant or even by diffusion of reactants in a gas phase within the LC structure. By this strategy, LC polymers/elastomers have demonstrated their specific and critical role (comparatively to small LC molecules) to achieve a control over NPs growth and over their final morphology. Until now, most studies have focused on the use of lamellar and nematic mesophase obtained from such LC polymers/elastomers. Therefore LC polymers exhibiting different mesophases (cubic, columnar, ...) could be the materials of choice for inducing specific growth mechanisms and specific arrangements of the NPs within those hybrids.

## References

Barmatov EB, Pebalk DA, Barmatova MV (2004) Influence of silver nanoparticles on the phase behavior of side-chain liquid crystalline polymers. Langmuir 20:10868–10871

Brochard F, de Gennes PG (1970) Theory of magnetic suspensions in liquid crystals. J Phys 31:691–708

Chambers M, Zalar B, Remskar M, Zumer S, Finkelmann H (2006) Actuation of liquid crystal elastomers reprocessed with carbon nanoparticles. Appl Phys Lett 89:243116.1–243116.3

Dasgupta D, Shishmanova IK, Ruiz-Carretero A, Lu K, Verhoeven M, van Kurigen HPC, Portale G, Leclere P, Basttiaansen CWM, Broer DJ, Schenning PHJ (2013) Patterned silver nanoparticles embedded in a nanoporous smectic liquid crystalline polymer network. J Am Chem Soc 135:10922–10925

Dobbs W, Suisse JM, Douce L, Welter R (2006) Electrodeposition of silver particles and gold nanoparticles from ionic liquid-crystal precursors. Angew Chem Int Ed 45:4179–4182

Domenici V, Zupancic B, Laguta VV, Belous AG, V'Yunov OI, Remskar M, Zalar B (2010) PbTiO3 nanoparticles embedded in a liquid crystalline elastomer matrix: structural and ordering properties. J Phys Chem C 114:10782–10789

Domenici V, Conradi M, Remskar M, Virsek M, Zupancic B, Mrzel A, Chambers M, Zalar B (2011) New composite films based on MoO3-x nanowires aligned in a liquid single crystal elastomer matrix. J Mater Sci 46:3639–3645

Domracheva NE, Pyataev AV, Manapov RA, Gruzdev MS (2011) Magnetic resonance and Mössbauer studies of superparamagnetic $\gamma$-$Fe_2O_3$ nanoparticles encapsulated into liquid-crystalline poly(propylene imine) dendrimers. ChemPhysChem 12:3009–3019

Ezhov AA, Shandryuk GA, Bondarenko GN, Merekalov AS, Abramchuk SS, Shatalova AM, Manna P, Zubarev E, Talroze RV (2011) Liquid-crystalline polymer composites with CdS nanorods: structure and optical properties. Langmuir 27:13353–13360

Garcia-Marquez A, Demortiere A, Heinrich B, Guillon D, Begin-Colin S, Donnio B (2011) Iron oxide nanoparticle-containing main-chain liquid crystalline elastomer: towards soft magnetoactive networks. J Mater Chem 21:8994–8996

Gascon I, Marty JD, Gharsa T, Mingotaud C (2005) Formation of gold nanoparticles in a side-chain liquid crystalline network: influence of the structure and macroscopic order of the material. Chem Mater 17:5228–5230

Haberl JM, Sanchez-Ferrer A, Mihut AM, Dietsch H, Hirt AM, Mezzenga R (2013a) Liquid-crystalline elastomer-nanoparticle hybrids with reversible switch of magnetic memory. Adv Mater 25:1787–1791

Haberl JM, Sanchez-Ferrer A, Mihut AM, Dietsch H, Hirt AM, Mezzenga R (2013b) Strain-induced macroscopic magnetic anisotropy from smectic liquid-crystalline elastomer-maghemite nanoparticle hybrid nanocomposites. Nanoscale 5:5539–5548

Hegman T, Qi H, Marx VM (2007) Nanoparticles and liquid crystals: synthesis, self-assembly, defect formation and potential applications. J Inorg Organomet Polym Mater 17:483–508

Kaiser A, Winkler M, Krause S, Finkelmann H, Schmidt AM (2009) Magnetoactive liquid crystal elastomer nanocomposites. J Mater Chem 19:538–543

Lee JW, Jin J (2003) Formation of gold nanoparticles within a liquid crystalline polymeric matrix. J Nanosci Nanotechnol 3:219–221

Li F, Chen W, Chen YW (2012) Mesogen induced self-assembly for hybrid bulk heterojunction solar cells based on a liquid crystal D-A copolymer and ZnO nanocrystals. J Mater Chem 22:6259–6266

Mallia VA, Vemula PK, John G, Kumar A, Ajayan PM (2007) In situ synthesis and assembly of gold nanoparticles embedded in glass-forming liquid crystals. Angew Chem Int Ed 46:3269–3274

Marshall JE, Ji Y, Torras N, Zinoviev K, Terentjev EM (2012) Carbon-nanotube sensitized nematic elastomer composites for IR-visible photo-actuation. Soft Matter 8:1570–1574

Mertelj A, Lisjak D, Drofenik M, Copic M (2013) Ferromagnetism in suspensions of magnetic platelets in liquid crystal. Nature 504:237–241

Montazami R, Spillmann CM, Naciri J, Ratna BR (2012) Enhanced thermomechanical properties of a nematic liquid crystal elastomer doped with gold nanoparticles. Sens Actuators A Phys 178:175–178

Nguyen HH, Valverde Serrano C, Lavedan P, Goudounèche D, Mingotaud AF, Lauth-de Viguerie N, Marty JD (2014) Mesomorphic ionic hyperbranched polymers: effect of structural parameters on liquid-crystalline properties and on the formation of gold nanohybrids. Nanoscale 6:3599–3610

Riou O, Zadoina L, Lonetti B, Soulantica K, Mingotaud AF, Respaud M, Chaudret B, Mauzac M (2012) In situ and ex situ syntheses of magnetic liquid crystalline materials: a comparison. Polymers 4:448–462

Riou O, Lonetti B, Davidson P, Tan RP, Cormary B, Mingotaud AF, Di Cola E, Respaud M, Chaudret B, Soulantica K, Mauzac M (2014) Liquid crystalline polymer-co nanorod hybrids: structural analysis and response to a magnetic field. J Phys Chem B 118:3218–3225

Saliba S, Coppel Y, Achard MF, Mingotaud C, Marty JD, Kahn ML (2011) Thermotropic liquid crystals as templates for anisotropic growth of nanoparticles. Angew Chem Int Ed 50:12032–12035

Saliba S, Coppel Y, Mingotaud C, Marty JD, Kahn ML (2012) ZnO/liquid crystalline nanohybrids: from solution properties to the control of the anisotropic growth of nanoparticles. Chem Eur J 18:8084–8091

Saliba S, Mingotaud C, Kahn ML, Marty JD (2013) Liquid crystalline thermotropic and lyotropic nanohybrids. Nanoscale 5:6641–6661

Shandryuk GA, Matukhina EV, Vasil'ev RB, Rebrov A, Bondarenko GN, Merekalov AS, Gas'kov AM, Talroze RV (2008) Effect of H-bonded liquid crystal polymers on CdSe quantum dot alignment within nanocomposite. Macromolecules 41:2178–2185

Shi YQ, Li F, Tan LC, Chen YW (2013) Hybrid bulk heterojunction solar cells based on the cooperative interaction of liquid crystals within quantum dots and diblock copolymers. ACS Appl Mater Interfaces 5:11692–11702

Song HM, Kim JC, Hong JH, Lee YB, Choi J, Lee JI, Kim WS, Kim JH, Hur NH (2007) Magnetic and transparent composites by linking liquid crystals to ferrite nanoparticles through covalent networks. Adv Funct Mater 17:2070–2076

Taubert A (2004) CuCl nanoplatelets from an ionic liquid-crystal precursor. Angew Chem Int Ed 43:5380–5382

Vasilets VN, Savenkov GN, Merekalov AS, Shandryuk GA, Shatalova AM, Tal'roze RV (2011) Immobilization of quantum dots of cadmium selenide on the matrix of a graft liquid-crystalline polymer. Polym Sci Ser A 53:521–526

Zadoina L, Lonetti B, Soulantica K, Mingotaud AF, Respaud M, Chaudret B, Mauzac M (2009) Liquid crystalline magnetic materials. J Mater Chem 19:8075–8078

Zadoina L, Soulantica K, Ferrere S, Lonetti B, Respau M, Mingotaud AF, Falqui A, Genovese A, Chaudret B, Mauzac M (2011) In situ synthesis of cobalt nanoparticles in functionalized liquid crystalline polymers. J Mater Chem 21:6988–6994

**ERRATUM**

# New Liquid Crystalline Poly(azomethine esters) Derived from PET Waste Bottles

**Ahmed Mohammed Issam and A.K. Nurul Khizrien**


V.K. Thakur, M.R. Kessler (eds.), *Liquid Crystalline Polymers: Processing and Applications*, DOI 10.1007/978-3-319-20270-9_12

DOI 10.1007/978-3-319-20270-9_12

On the Title page of Chapter 12, the first author name was incorrectly set as "Ahmed Mohammed Issam" but it should read "**Issam Ahmed Mohammed**".

The online version of the original chapter can be found at http://dx.doi.org/10.1007/978-3-319-20270-9_12


V.K. Thakur, M.R. Kessler (eds.), *Liquid Crystalline Polymers*, DOI 10.1007/978-3-319-20270-9_22

# Index




V.K. Thakur, M.R. Kessler (eds.), *Liquid Crystalline Polymers*,
DOI 10.1007/978-3-319-20270-9

MIX
Papier aus verantwortungsvollen Quellen
Paper from responsible sources
FSC® C105338

If you have any concerns about our products,
you can contact us on
**ProductSafety@springernature.com**

In case Publisher is established outside the EU,
the EU authorized representative is:
**Springer Nature Customer Service Center GmbH**
**Europaplatz 3, 69115 Heidelberg, Germany**

Printed by Libri Plureos GmbH
in Hamburg, Germany